D1790850

WITHDRAWN
BY
JEFFERSON COUNTY PUBLIC LIBRARY
Lakewood, CO

R 503 MCGRAW HILL YEARBOOK OF S
       MCGRAW HILL YEARBOOK OF SCIE
       NCE AND TECHNOLOGY
       VOL 1996   C1212123335

CL

DEC 1995

REFERENCE ONLY

McGRAW-HILL YEARBOOK OF
Science & Technology

1996

McGRAW-HILL YEARBOOK OF
Science & Technology

1996

Comprehensive coverage of recent events and research as compiled by the staff of the McGraw-Hill Encyclopedia of Science & Technology

McGraw-Hill, Inc.
New York San Francisco Washington, D.C. Auckland Bogotá Caracas Lisbon London Madrid
Mexico City Milan Montreal New Delhi San Juan Singapore Sydney Tokyo Toronto

McGRAW-HILL YEARBOOK OF SCIENCE & TECHNOLOGY
Copyright © 1995 by McGraw-Hill, Inc.
All rights reserved. Printed in the United States of America.
Except as permitted under the United States Copyright Act of 1976,
no part of this publication may be reproduced or distributed in any
form or by any means, or stored in a database or retrieval system,
without prior written permission of the publisher.

1 2 3 4 5 6 7 8 9 0   DOW/DOW   9 0 0 9 8 7 6 5

Library of Congress Cataloging in Publication data

McGraw-Hill yearbook of science and technology.
1962-   . New York, McGraw-Hill Book Co.

   v. illus. 26 cm.
   Vols. for 1962-   compiled by the staff of the
McGraw-Hill encyclopedia of science and technology.
   1. Science—Yearbooks. 2. Technology—
Yearbooks. 1. McGraw-Hill encyclopedia of
science and technology.
Q1.M13   505.8   62-12028

Printed on acid-free paper.

ISBN 0-07-051772-X
ISSN 0076-2016

International Editorial Advisory Board

Dr. Neil Bartlett
Professor of Chemistry
University of California, Berkeley

Dr. Richard H. Dalitz
Department of Theoretical Physics
Oxford University, England

Dr. Freeman J. Dyson
Institute for Advanced Study
Princeton, New Jersey

Dr. Leon Knopoff
Institute of Geophysics and Planetary Physics
University of California, Los Angeles

Dr. H. C. Longuet-Higgins
Royal Society Research Professor, Experimental Psychology
University of Sussex, Brighton, England

Dr. Alfred E. Ringwood
Professor of Geochemistry
Australian National University, Canberra

Dr. Arthur L. Schawlow
Professor of Physics
Stanford University

Dr. Koichi Shimoda
Department of Physics
Keio University, Tokyo

Dr. A. E. Siegman
Director, Edward L. Ginzton Laboratory
Professor of Electrical Engineering
Stanford University

Prof. N. S. Sutherland
Professor of Experimental Psychology
University of Sussex, Brighton, England

Dr. Hugo Theorell
Nobel Institute
Stockholm, Sweden

Lord Todd of Trumpington
Professor of Organic Chemistry
Cambridge University, England

Dr. George W. Wetherill
Director, Department of Terrestrial Magnetism
Carnegie Institution of Washington

Dr. E. O. Wilson
Professor of Zoology
Harvard University

Dr. Arnold M. Zwicky
Professor of Linguistics
Ohio State University

Editorial Staff

Sybil P. Parker, Editor in Chief

Katherine Moreau, Senior Editor
Jonathan Weil, Editor
Betty Richman, Editor
Glenon C. Butler, Editor
Patricia W. Albers, Editorial Administrator
Frances P. Licata, Editorial Assistant

Ron Lane, Art Director
Vincent Piazza, Assistant Art Director
Angelika Fuellemann, Art Production Assistant

Joe Faulk, Editing Manager
Ruth W. Mannino, Senior Editing Supervisor

Thomas G. Kowalczyk, Production Manager
Suzanne W. B. Rapcavage, Senior Production Supervisor

Suppliers: North Market Street Graphics, Lancaster, Pennsylvania, generated the line art, and composed the pages in Times Roman, Helvetica Condensed Black, and Helvetica Condensed Bold.

The book was printed and bound by R. R. Donnelley & Sons Company, The Lakeside Press at Willard, Ohio.

Consulting Editors

Prof. R. McNeill Alexander. *Deputy Head. Department of Pure and Applied Biology, University of Leeds, England.* COMPARATIVE VERTEBRATE ANATOMY AND PHYSIOLOGY.

Prof. Eugene A. Avallone. *Consulting Engineer: Professor Emeritus of Mechanical Engineering: City College of the City University of New York.* MECHANICAL ENGINEERING.

A. E. Bailey. *Formerly, Superintendent of Electrical Science, National Physical Laboratory, London, England.* ELECTRICITY AND ELECTROMAGNETISM.

Prof. William P. Banks. *Chairman, Department of Psychology, Pomona College, Claremont, California.* PHYSIOLOGICAL AND EXPERIMENTAL PSYCHOLOGY.

Dr. Allen J. Bard. *Department of Chemistry, University of Texas, Austin.* PHYSICAL CHEMISTRY.

Dr. Alexander Baumgarten. *Director, Clinical Immunology Laboratory, Yale-New Haven Hospital, New Haven, Connecticut.* IMMUNOLOGY AND VIROLOGY.

Prof. Richard D. Berger. *Plant Pathology Department, University of Florida, Gainesville.* PLANT PATHOLOGY.

Dr. Robert T. Beyer. *Hazard Professor of Physics, Emeritus, Brown University, Providence, Rhode Island.* ACOUSTICS.

Prof. S. H. Black. *Department of Medical Microbiology and Immunology, Texas A&M University, College Station.* MEDICAL MICROBIOLOGY.

Prof. Anjan Bose. *Director, School of Electrical Engineering and Computer Science, Washington State University, Pullman.* ELECTRIC POWER ENGINEERING.

Ronald Braff. *Principal Engineer. MITRE Corporation/Center for Advanced Aviation System Development, McClean, Virginia.* NAVIGATION.

Dr. Chaim Braun. *Bechtel Corporation, Gaithersburg, Maryland.* NUCLEAR ENGINEERING.

Robert D. Briskman. *President, CD Radio, Inc., Washington, D.C.* TELECOMMUNICATIONS.

Michael H. Bruno. *Graphic Arts Consultant, Sarasota, Florida.* GRAPHIC ARTS.

Dr. John F. Clark. *Academic Program Chairman and Professor, Space Technology, Florida Institute of Technology, Titusville.* SPACE TECHNOLOGY.

Prof. David L. Cowan. *Chairman, Department of Physics and Astronomy, University of Missouri, Columbia.* CLASSICAL MECHANICS AND HEAT.

Dr. C. Chapin Cutler. *Ginzton Laboratory, Stanford University, California.* RADIO COMMUNICATIONS.

Dr. Gene Dresselhaus. *Frances Bitter National Magnetic Laboratory, Massachusetts Institute of Technology, Cambridge.* SOLID-STATE PHYSICS.

Dr. Jay S. Fein. *Division of Atmospheric Sciences, National Science Foundation, Arlington, Virginia.* METEOROLOGY AND CLIMATOLOGY.

Dr. William K. Ferrell. *Professor Emeritus, College of Forestry, Oregon State University, Corvallis.* FORESTRY.

Dr. Barbara Gastel. *Department of Journalism, Texas A&M University, College Station.* MEDICINE AND PATHOLOGY.

Prof. Lawrence Grossman. *Department of Geophysical Science, University of Chicago, Illinois.* GEOCHEMISTRY.

Prof. Lawrence I. Grossman. *Associate Director, Center for Molecular Medicine and Genetics, Wayne State University, Detroit, Michigan.* GENETICS AND EVOLUTION.

Dr. Ralph E. Hoffman. *Associate Professor, Yale Psychiatric Institute, Yale University School of Medicine, New Haven, Connecticut.* PSYCHIATRY.

Prof. Stephen F. Jacobs. *Professor of Optical Sciences, University of Arizona, Tucson.* ELECTROMAGNETIC RADIATION AND OPTICS.

Dr. S. C. Jong. *Senior Staff Scientist and Program Director, Mycology and Protistology Program, American Type Culture Collection, Rockville, Maryland.* MYCOLOGY.

Prof. Cornelis Klein. *Department of Earth and Planetary Sciences, University of New Mexico, Albuquerque.* GEOLOGY (MINERALOGY AND PETROLOGY).

Prof. Karl E. Lonngren. *Department of Electrical and Computer Engineering, University of Iowa, Iowa City.* PHYSICAL ELECTRONICS.

Dr. Philip V. Lopresti. *Engineering Research Center, AT&T, Princeton, New Jersey.* ELECTRONIC CIRCUITS.

Dr. Michael L. McKinney. *Department of Geological Sciences, University of Tennessee, Knoxville.* INVERTEBRATE PALEONTOLOGY.

Prof. Marcia McNutt. *Department of Earth, Atmospheric, and Planetary Sciences, Massachusetts Institute of Technology, Cambridge.* GEOPHYSICS.

Dr. George L. Marchin. *Associate Professor of Microbiology and Immunology, Division of Biology, Kansas State University, Manhattan.* BACTERIOLOGY.

Prof. Melvin Marcus. *Department of Geography, Arizona State University, Tempe.* PHYSICAL GEOGRAPHY.

Consulting Editors (continued)

Dr. Henry F. Mayland. *Research Soil Scientist, Snake River Conservation Research Center, USDA-ARS, Kimberly, Idaho.* SOILS.

Prof. Randy Moore. *Dean, Buchtel College of Arts and Sciences, University of Akron, Ohio.* PLANT ANATOMY AND PLANT PHYSIOLOGY.

Prof. Conrad F. Newberry. *Department of Aerospace and Astronautics, Naval Postgraduate School, Monterey, California.* AERONAUTICAL ENGINEERING AND PROPULSION.

Dr. Gerald Palevsky. *Consulting Professional Engineer, Hastings-on-Hudson, New York.* CIVIL ENGINEERING.

Prof. Jay M. Pasachoff. *Director, Hopkins Observatory, Williams College, Williamstown, Massachusetts.* ASTRONOMY.

Prof. David J. Pegg. *Department of Physics and Astronomy, University of Tennessee, Knoxville.* ATOMIC, MOLECULAR, AND NUCLEAR PHYSICS.

Dr. William C. Peters. *Professor Emeritus, Mining and Geological Engineering, University of Arizona, Tucson.* MINING ENGINEERING.

Prof. W. D. Russell-Hunter. *Professor of Zoology, Department of Biology, Syracuse University, New York.* INVERTEBRATE ZOOLOGY.

Dr. Andrew P. Sage. *First American Bank Professor and Dean, School of Information Technology and Engineering, George Mason University, Fairfax, Virginia.* CONTROL AND INFORMATION SYSTEMS.

Prof. Susan R. Singer. *Department of Biology, Carleton College, Northfield, Minnesota.* DEVELOPMENTAL BIOLOGY.

Mel Schwartz. *Materials Consultant, United Technologies Corporation, Stratford, Connecticut.* MATERIALS SCIENCE AND ENGINEERING.

Prof. Marlin U. Thomas. *Head, School of Industrial Engineering, Purdue University, West Lafayette, Indiana.* INDUSTRIAL AND PRODUCTION ENGINEERING.

Prof. John F. Timoney. *Department of Veterinary Science, University of Kentucky, Lexington.* VETERINARY MEDICINE.

Prof. Romeo T. Toledo. *Department of Food Science and Technology, University of Georgia, Athens.* FOOD ENGINEERING.

Prof. Joan S. Valentine. *Department of Chemistry and Biochemistry, University of California, Los Angeles.* INORGANIC CHEMISTRY.

Dr. Blaire Van Valkenburgh. *Department of Biology, University of California, Los Angeles.* VERTEBRATE PALEONTOLOGY.

Prof. Frank M. White. *Department of Mechanical Engineering, University of Rhode Island, Kingston.* FLUID MECHANICS.

Prof. Richard G. Wiegert. *Institute of Ecology, University of Georgia, Athens.* ECOLOGY AND CONSERVATION.

Prof. Frank Wilczek. *Institute for Advanced Study, Princeton, New Jersey.* THEORETICAL PHYSICS.

Prof. W. A. Williams. *Department of Agronomy and Range Science, University of California, Davis.* AGRICULTURE.

Contributors

A list of contributors, their affiliations, and the titles of the articles they wrote appears in the back of this volume.

Preface

The *1996 McGraw-Hill Yearbook of Science & Technology* continues a long tradition of presenting outstanding recent achievements in science and engineering. Thus it serves both as an annual review of what has occurred and as a supplement to the *McGraw-Hill Encyclopedia of Science & Technology,* updating the basic information in the seventh edition (1992) of the Encyclopedia. It also provides a preview of advances that are in the process of unfolding.

The Yearbook reports on topics that were judged by the consulting editors and the editorial staff as being among the most significant recent developments. Each article is written by one or more authors who are specialists on the subject being discussed.

The *McGraw-Hill Yearbook of Science & Technology* continues to provide librarians, students, teachers, the scientific community, and the general public with information needed to keep pace with scientific and technological progress throughout our rapidly changing world.

Sybil P. Parker
Editor in Chief

McGRAW-HILL YEARBOOK OF
Science & Technology

1996

A–Z

Acoustic thermometry

The Acoustic Thermometry of Ocean Climate (ATOC) program is designed to measure global trends in ocean temperature. The oceans play a central role in global climate due to ocean storage of both greenhouse gases and heat. Atmospheric warming on an Earth without oceans would be at two or three times the rate of an Earth with oceans, other factors remaining the same. It is important therefore to measure directly the ocean's role at ocean-basin and transglobal scales rather than to infer it from the very uncertain calculations of planetary heat budgets.

Long-range measurement. Local measurements of ocean temperature trends by traditional soundings are difficult because of the large local variability of temperature. At a depth of 1 km (0.6 mi) the greenhouse-induced warming is expected to be of the order of 0.004°C (0.007°F) per year in the 1990s, compared with a root-mean-square variability of 1°C (1.8°F), arising mostly from mesoscale eddies. Detection of such a small, long-term trend in the presence of this mesoscale noise would, in principle, take centuries. However, if an average temperature can be measured over a range scale of 10,000 km (6000 mi), the mesoscale noise would be statistically averaged and the situation much more favorable.

ATOC has proposed a method of measuring acoustic travel times over long ranges. The technique is based on two considerations: Sound speed is a function of temperature (that is, sound travels faster in warmer water); and the ocean sound (sofar) channel, typically at 1000 m (3300 ft) depth, provides a very efficient waveguide. The 1991 Heard Island Feasibility Test (HIFT) demonstrated that coded low-frequency acoustic signals could be recorded at adequate signal levels at distances up to 16,000 km (10,000 mi), thus enabling acoustic transmissions to be used as an integrating, or averaging, thermometer. Expected changes in travel time from greenhouse warming of 0.004°C (0.007°F) per year over such long paths are of the order of 2 s in 10 years. If coherent paths can indeed be tracked, the expected precision of individual measurements is entirely adequate, between 0.01 and 0.1 s.

Although the variability of the mesoscale noise can be adequately suppressed in the long-range averages (to something like 0.1 s root-mean-square), there is still the problem of ocean-basin-scale ambient variability, which also appears as noiselike variability in the long-term climate trend. A significant part of the ATOC program is to design, specify, and verify the performance of a global network of acoustic sources and receivers from which climate change can be detected in the presence of this large-scale ambient variability, using coupled atmosphere-ocean models for guidance. Acoustic thermometry data can also be assimilated back into those models to validate and improve their performance as tools for climate change prediction.

Ten-year program. A time series of acoustic travel with a large sample set is required to detect and resolve gyre- or larger-scale temperature changes. The minimum length of that series will vary from basin to basin. If, as is now believed, about 10 years are required, it would be appropriate to make the earliest possible start at achieving global coverage, even if it is sparse to start with, since great uncertainty currently exists regarding the temporal and spatial scales of response of global warming to the production of greenhouse gases. The empirical data derived in the ATOC program could be significant in resolving extant issues of planetary-scale temporal response times. The plan is to build a network of sources and receivers that by the middle of the 10-year program would be acquiring sufficient data to begin improving the predictive capability of global climate models.

The proposed network could also serve a broader community of ocean scientists. For example, in addition to routinely collecting temperature information, reciprocal acoustic paths between source-receiver pairs can yield information about mass transport. It is hoped that the data set to be produced in this program will be especially synergistic to satellite altimetry data sets to be obtained in the TOPEX and EOS (Earth Observing System) programs over the next few years.

Two-year demonstration phase. HIFT lasted only 5 days, but even so was able to show that electrically driven sources can be used to send signals worldwide, that the signals can be received with adequate signal-to-noise ratios and possess sufficient phase stability to allow the signal processing required to measure accurate travel times along discrete ocean paths, and that the signals themselves do not significantly disturb marine mammals. (They did not suffer physical harm, nor were they deprived of their habitat.) The current phase of ATOC is designed to progress from a single moving source, transmitting for 5 days, to two fixed sources, a large number of receivers, and at least 1 year of transmissions. The 2-year demonstration phase (see **illus.**) takes the first steps required to design, develop, and deploy an efficient and cost-effective global monitoring system of the type required for a 10-year monitoring program. This phase will address the scientific and technical issues that remain before a complete worldwide system can sensibly be established.

Three acoustic sources and four vertical line array receivers have been built and tested, and seven NAVFAC (U.S. Navy facility) arrays instrumented to receive and record ATOC's signals around the northern rim of the Pacific. The heart of the current effort consists of a field program to deploy two sources in the Pacific Ocean (the third

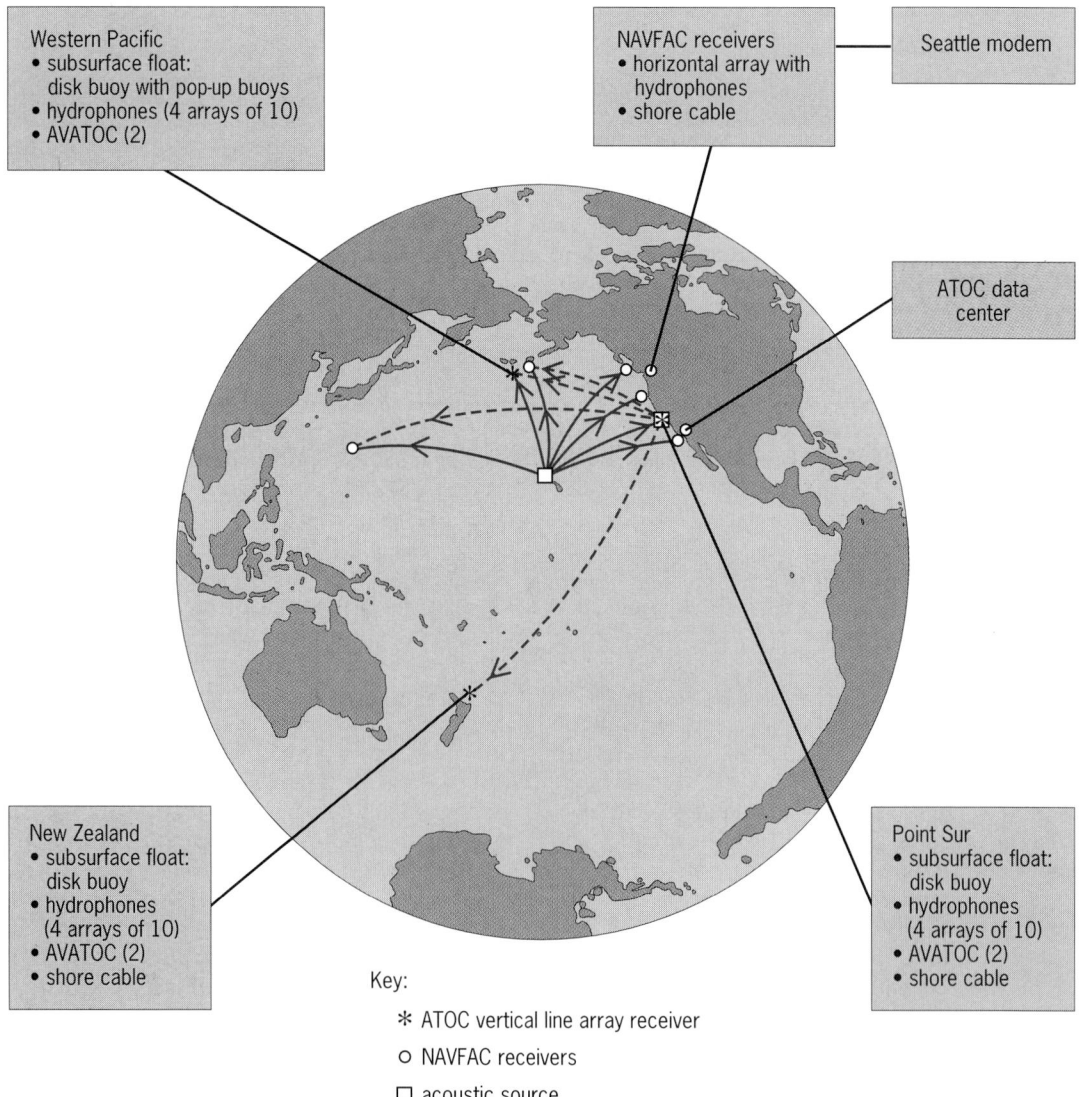

ATOC demonstration network in the Pacific Ocean. Acoustic paths are shown from sources off Kauai, Hawaii, and Point Sur, California, to various receivers. The pop-up buoys, released periodically on command, carry acoustic reception data to the surface recovery vessel. To avoid crowding, not all NAVFAC receivers are shown; locations of those receivers shown are approximate. AVATOC = advanced vertical array for thermometry of ocean climate.

source is reserved, as a spare); to make oceanographic measurements over a 10,000-km (6000-mi) path to New Zealand; and to collect data with existing Navy acoustic receiver systems, a fixed vertical array near Hawaii and one near Raratonga, to resolve mode arrival issues. (The third vertical array may be deployed near northeastern New Zealand or near Adak, Alaska, in the Aleutian Islands, and the fourth is reserved as a spare.) Together these receivers will allow acquisition of data at a variety of ranges from 5000 to 10,000 km (3000 to 6000 mi) on receiving arrays having both horizontal and vertical directivity.

Technical approach. The coherence and signal-to-noise ratios observed in HIFT permit transmission bandwidths of 35 Hz at 75-Hz center frequency, and a source level of 195 dB with respect to a sound pressure of 1 micropascal at a distance of 1 m (3 ft) from the source (about 260 W acoustic output, equivalent to 133 dB in air). This is a reduction by a factor of 100 from the 215 dB (26,000 W) used in HIFT. The signals will be 42 phase-modulated, binary-coded sequences, each 27 s long, resulting in a 20-min transmission.

Receiver design and development. The receivers for the demonstration phase have two objectives: to establish the limits of how well paths can be resolved and tracked for signals which have propagated over 5000–10,000 km (3000–6000 mi), from the north Pacific to New Zealand; and to specify the simplest receiver design consistent with the goals of a future worldwide ATOC network. ATOC will accomplish these goals by combining the use of existing NAVFAC arrays with the deployment of a small number of vertical line arrays.

A critical examination of the data from basin-scale acoustic tomography experiments with ranges in excess of 4000 km (2500 mi) also suggests that path identification is ambiguous. One purpose of the receiver arrays for the demonstration phase is to provide the spatial resolution required for understanding these path-identification issues.

A second reason for using arrays during the demonstration phase is related to the 100-fold reduction in source level from that of HIFT, mentioned above. In this manner, it is possible to reduce the potential impact of the sound on marine mammals and to reduce the high power required to drive the HIFT sources. The array gain by the ATOC 40-hydrophone arrays is a means of recovering some of the loss in signal-to-noise ratio resulting from the reduction in source level.

Source and receiver array locations. A 75-Hz acoustic source, cabled to shore, will be deployed on the ocean bottom, at a depth near that of the sound-channel axis (800 m or 2600 ft), 13 km (8 mi) north of Kauai, Hawaii. This location was selected because of its excellent view toward the north Pacific and because axial-depth water is available near shore, thereby minimizing cable runs and providing a steep enough slope to reduce acoustic interaction with the bottom. Transmissions from this source will be received by a variety of Navy receivers (see illus.), thus providing a rich set of acoustic paths throughout the north Pacific. Instrumentation has been designed, developed, and installed at the appropriate NAVFACs to acquire these data, beamform and correlation-process them for time-of-arrival statistics, store them, and transmit portions of them automatically over telephone modems to a receiving facility in Seattle, Washington.

In addition to reception by the (horizontal) Navy arrays, vertical receiving arrays have been developed and will be deployed near Hawaii, Raratonga, northeastern New Zealand, and possibly Adak.

A second 75-Hz source will be installed at 980-m (3300-ft) depth, 88 km (48 mi) offshore from Pillar Point. This source will allow reception of signals at trans-Pacific ranges of approximately 10,000 km (6000 mi), almost twice that obtainable with the Hawaii source and NAVFAC receivers alone. Ranges of this magnitude are certain to be part of the global system. Further, this source will allow ensonification of the southern Pacific Ocean, and reception by New Zealand and Australian international partners.

An important part of the field program will consist of oceanographic measurements to validate acoustic results. These sections will be carried out with instruments that will take vertical profiles of conductivity (from which salinity is calculated), temperature, and depth (CTDs), as well as expendable bathythermographs (XBTs). Several CTD/XBT sections are to be conducted along selected transmission paths. The program will consist of making slices across the ocean along which research vessels will stop periodically to make CTD measurements, and will deploy XBTs while the ship is under way. The Point Sur to New Zealand path is particularly interesting for this purpose.

Marine mammal research program. A companion program is under way in Hawaii and California to assess the effects, if any, of ATOC's low-frequency sounds on marine mammals. Although most marine mammals do not hear well at low frequency, some large whales and seals may be acoustically sensitive enough, and capable of diving deeply enough, to be affected by ATOC's signals. The marine mammal research program (MMRP) is designed to observe marine mammals, both visually and acoustically, and to detect even subtle effects of the sounds on their behavior. The program is conducting baseline studies at the two source sites, well before the acoustic sources are installed, to establish the animals' natural behavior and their reactions to other sources of artificial noise in the ocean, such as ships, recreational craft, and helicopters.

For background information SEE CLIMATE MODELING; GREENHOUSE EFFECT; HEAT BALANCE, TERRESTRIAL ATMOSPHERIC; OCEAN CIRCULATION; SEAWATER; UNDERWATER SOUND in the McGraw-Hill Encyclopedia of Science & Technology.

Andrew M. G. Forbes

Bibliography. A. Baggeroer and W. Munk, The Heard Island feasibility test, *Phys. Today*, 45(9):22–30, September 1992.

Agricultural soil and crop practices

Tillage is performed in arable cropping systems for many reasons, including burial of crop residues and weeds for disease and insect control; incorporation of fertilizers and chemicals; creation of aggregates and a condition of macroporosity for improved aeration, water infiltration, and root growth; promotion of soil drying and warming; and reduction of weed competition at planting and lay-by (final cultivation and spraying). Tillage depth can range from shallow operations that barely scrape the surface 1–2 cm (0.4–0.8 in.) of soil to deep operations that disturb or even invert soil to depths of 0.5 m (1.6 ft). Subsoiling (sometimes called ripping or deep chiseling) is deep tillage using implements that produce little or no inversion of the soil profile (see **illus.**).

Purpose of subsoiling. Subsoiling is noninversive tillage aimed solely at mitigating physical and chemical problems occurring deep in the soil. It generally has very little impact on properties of the surface soil layer, commonly called the plow layer (typically 0–20 cm or 0–8 in.). Noninversive tillage (such as subsoiling or chiseling) breaks up soil without moving appreciable amounts vertically from one layer to another (unlike plowing, which inverts the soil, transferring the soil on top to the bottom of the depth plowed, and vice versa). Subsoiling is performed where tillage-amendable soil constraints exist below the depth of primary surface tillage (typically 20–50 cm or 8–20 in.).

Subsoiling reduces rooting-restrictive or drainage-restrictive subsoil layering or compaction. Subsoil layers result from stratified deposition or in-place development of soils through weathering. Subsoil compaction can result from natural consolidation, or it can be caused by traffic and tillage transference of compressive forces to depths below the reach of primary surface tillage. Tillage-induced or traffic-induced compaction commonly results from field operations in the spring or fall when soils are wet. Although most farmers know that traffic and tillage on wet soils increase the risk of compaction, they are often forced by weather or other logistical or economic pressures to proceed within rigid time constraints. Thus, entry is forced onto wet soils, which have a greater potential for compaction.

Efficacy assessment. The effectiveness of subsoiling implements is assessed by measuring the changes in soil properties and the effects on crop performance. Subsoiling decreases profile bulk density (dry weight per unit volume) and soil strength, as measured by cone index or penetration resistance—the force required (megapascals) for penetration of a 13-mm-diameter (0.5-in.) 30° stainless steel cone. Subsoiling increases soil porosity and the rate and capacity of water infiltration, which may or may not result in improved crop performance, depending on the severity of subsoil limitations and the amount of crop stress during the growing season. One of the greatest difficulties in recommending subsoiling operations is the inadequacy of soil diagnostic criteria to predict subsoiling efficacy for a given crop, climate, and management system. Subsoiling is usually an annual requirement, because the environments and cropping systems prone to subsoil compaction tend to promote subsoil reconsolidation.

Subsoiling methods. Subsoiling can be done as a broadcast operation, that is, the entire subsoil is disrupted to a given depth. However, it often is restricted to the soil zone immediately beneath planted crop rows. This practice, known as in-row subsoiling, reduces subsoiling's costly horsepower and energy requirements.

In-row subsoiling. Horsepower/energy requirements and subsoiling effectiveness vary with soil properties and subsoiler design. The baseline power requirement increases with the depth, number, contact surface area, and perpendicularity to the direction of travel of the shanks, and with the bulk density, clay content, and dryness of the soil. For subsoilers penetrating 0.30–0.45 m (0.98–1.5 ft) into the soil, a range of approximately 2.2–3.0 W (30–40 horsepower) is typically required per subsoil shank.

In-row subsoiling is also used for deep injection of soil amendments or fertilizers. Slurried lime, for example, has been injected to improve the pH of acid subsoils. Nitrogen or phosphorous fertilizers injected into zones directly below the planted row can improve early nutrient interception by the rapidly expanding seedling root systems. Deep-injected organic sludges provide placement of broad-spectrum low-analysis (dilute) fertilizer that also helps preserve the improved tilth in the subsoil zone that has been shattered by the subsoiling operation.

Subsoiling can be an independent tillage operation, or it can be combined with other practices to

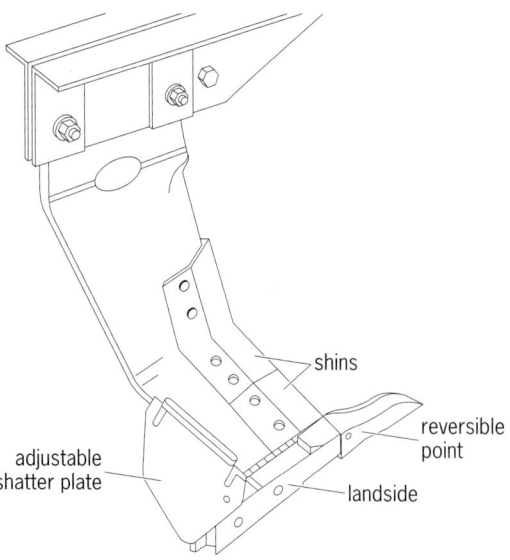

Subsoiling implement with winglike broad-angle lifting surfaces for offset loosening of zones to one side of the shaft. (*The Tye Company*)

Table 1. Effect of zone subsoiling on furrow-irrigated Russet Burbank potato tuber yield and grade in Kimberly, Idaho

	Yield, metric ton/ha (ton/acre)		Grade no. 1, %		Grade no. 1 > 284g* (10 oz), %		Grade no. 1, 114–284 g† (4–10 oz),%	
	1989	1990	1989	1990	1989	1990	1989	1990
Zone-subsoiled	39.4 (17.5)	41.9 (18.7)	62.6	64.2	29.1	14.9	33.5	49.3
Nonsubsoiled	36.3 (16.2)	37.7 (16.8)	57.2	56.5	27.0	12.9	30.2	43.6
Probability, %	NS‡	0.08	3.58	5.94	NS	NS	2.29	6.03

* USDA standard market quality grade no. 1.
† Weight limits used by most packers and processors as cutoffs for premium pay categories when buying a farmer's potato crop.
‡ NS = not significant.
SOURCE: R. E. Sojka et al., Zone-subsoiling effects on infiltration, runoff, erosion, and yields of furrow-irrigated potatoes, *Soil Tillage Res.*, 25:351–368, 1993.

reduce the number (and hence cost) of equipment passes over the field. Since timing and spatial placement of subsoiling greatly affect its efficacy, it is often combined with row-crop planting. Although this practice requires specially modified planting equipment, it prevents disrupted subsoil from being recompacted by intervening preplant field operations (such as fertilizer spreading, chemical application, and secondary surface tillage). It also guarantees precision placement of the shattered zone directly below the planted row and maximizes subsoil disruption during seed germination and vigorous early root exploration of the soil profile.

Zone subsoiling. A technique known as zone subsoiling is a sophisticated variation on the theme of in-row subsoiling. This term was coined to describe the pattern of noninversive in-row profile disruption accomplished with a unique subsoiler that has a winglike configuration. Unlike the straight or curved shanks of more standard subsoilers, this subsoiler resembles a subterranean wing or lifting surface (see illus.). Its lower half is curved laterally from the line of travel, allowing the subsoiler to reach under planted rows from the side. This unique feature allows delay of the subsoiling operation until several days, or even 1–2 weeks, after planting of certain slowly germinating crops, thus extending the period of maximum soil disruption and allowing greater flexibility in accessing optimal soil conditions for subsoiling. Both in-row subsoiling and zone subsoiling permit subsoil disruption where it is needed, under the crop row, while leaving interrow spaces undisturbed to provide support and traction for tractors and other field equipment whose tires run between the rows.

Slit tillage. Where the subsoil restrictive layer is relatively shallow, and where it overlies more friable subsoil, the horsepower and energy requirements of standard subsoiling operations are sometimes avoided by using a unique concept known as slit tillage. This type of subsoiling does not disrupt a large volume of the soil profile; instead it creates a very narrow slit, penetrating to below the depth of a restrictive layer. Roots follow the narrow slit through the restrictive layer and then branch out extensively when they reach the more favorable environment below. The slits can be stabilized in a few years by the decomposing roots of preceding crops. If combined with traffic-pattern control, slit tillage gradually improves crop performance, eventually matching the results of more disruptive subsoiling, without the larger power or energy requirements.

Advantages and disadvantages. Subsoiling of restrictive soils increases the extent of root exploration and improves water infiltration to the lower soil profile. These combined effects reduce plant water and nutrient stresses. More vigorous crop growth results, which generally increases yield and improves market quality at harvest (**Table 1**). Where subsoil restrictions to rooting and infiltration are particularly severe, the effects on yield can be directly related to the mean soil strength of the potential rooting volume.

Because subsoiling greatly increases infiltration, it can substantially reduce runoff from both rainfed and irrigated cropping systems. The result is an increase in the efficiency of water intake and a reduction in runoff and in the potential for soil erosion (**Table 2**).

Special precautions must be taken when subsoiling is performed on sloping ground in high-rainfall environments. Water can channel downslope through the subsoiler's openings. Thus, the surface soil behind the subsoil shank must be firmed sufficiently to prevent soil from washing away and seeds from washing away or subsiding deep into the soil when driving rain occurs before complete crop establishment.

Table 2. Effect of zone subsoiling on cumulative seasonal infiltration and soil loss for furrow-irrigated Russet Burbank potatoes grown in Kimberly, Idaho

	Infiltration, mm (in.)		Soil loss, kg/ha (lb/acre)	
	1989	1990	1989	1990
Zone-subsoiled	306 (11.9)	321 (12.6)	871 (976)	2604 (2918)
Nonsubsoiled	281 (11.1)	254 (10.0)	1154 (1293)	8450 (9469)
Probability, %	NS*	0.01	NS	0.10

* NS = not significant.
SOURCE: R. E. Sojka et al., Zone-subsoiling effects on infiltration, runoff, erosion, and yields of furrow-irrigated potatoes, *Soil Tillage Res.*, 25:351–368, 1993.

Subsoiling of soils with poor internal drainage (natural or artificial) in wet climates can increase infiltration enough to waterlog the soil profile, and thus can ultimately be more damaging to some crops than failure to disrupt root-restrictive layers. Damage may result either from direct effects of waterlogging on crop growth (restricted root aeration and increased disease susceptibility) or from indirect effects such as denitrification or leaching loss of applied fertilizers and chemicals or delayed warming of the wet soil.

Special considerations. Care must be exercised when selecting and configuring subsoiling equipment. The subsoiler must be compatible with other existing system components. In conservation tillage or no-till systems, the subsoiler must be designed to perform well in elevated amounts of plant residues, providing subsoil disruption with minimal surface disturbance. The subsoiler power requirements must be compatible with available equipment. If the cropping system includes rotation to crops at different row spacings, the subsoiler must be adjustable. If the land to be subsoiled contains buried tree stumps or rocks, the subsoiler will require sheer pins or tripping devices that allow subsoil shanks to ride over obstacles in order to avoid damage to planters, tool bars, tractor hitches, drive systems, and so forth. Subsoiler spacing must be close enough to disrupt soil sufficiently for the desired crop or soil response but far enough apart for soil to flow easily between the shanks. Finally, subsoiling should be restricted to the minimum depth needed to allow rooting and infiltration into unrestrictive subsoil. Excessively deep subsoiling needlessly increases tractor power (size) and fuel requirements, increasing operational costs, causing wheel-track surface compaction, and ultimately degrading surface soil aggregates and structure.

For background information SEE AGRICULTURAL SOIL AND CROP PRACTICES; SOIL in the McGraw-Hill Encyclopedia of Science & Technology.

Robert E. Sojka

Bibliography. M. R. Carter (ed.), *Conservation Tillage in Temperate Agroecosystems*, 1994; B. D. Soane and C. van Ouwerkerk (eds.), *Soil Compaction in Crop Production*, 1994; R. E. Sojka et al., Zone-subsoiling effects on infiltration, runoff, erosion, and yields of furrow-irrigated potatoes, *Soil Tillage Res.*, 25:351–368, 1993; R. E. Sojka, D. L. Karlen, and W. J. Busscher, A conservation tillage research update from the Coastal Plain Soil and Water Conservation Research Center of South Carolina: A review of previous research, *Soil Tillage Res.*, 21:361–376, 1991.

Aircraft noise

Several technologies, chiefly in the design of aircraft engines, have advanced so as to reduce aircraft noise to new levels of quiet, both in the environment surrounding the aircraft and within the passenger cabin. By permitting aircraft operations from noise-sensitive airports, quieter airliners can give travelers and cargo shippers freedom in departure times while relieving the surrounding community of noise.

Noise levels. An example is the McDonnell Douglas MD-90 midrange twin-engine airliner, which went into service in early 1995. The noise it produces is 22 dB below the Federal Aviation Administration's current noise requirements. The number refers to the cumulative total of the differences in sound level below a requirement for three measurements: approach, flyover, and sideline noise. The requirements themselves vary depending on the weight of the aircraft.

Testing to ascertain an aircraft's noise is done by flying carefully controlled flight paths over a calibrated array of microphones on the ground. These flights are conducted under stringent limitations on weather and background noise.

Engine design. Key to the low sound characteristics of a modern airliner such as the MD-90 are the engines and their installation. One factor in noise production is the length of the inlet, which is 38 cm (15 in.) longer on the MD-90 than needed, producing some weight and skin-friction drag penalties. The length increment cuts the forward projected noise by about 1–2 dB and provides straighter airflow into the engine. The inlet duct is treated to reduce sound emissions; its surface is perforated by small holes that lead to subsurface chambers. These cells absorb sound energy, functioning on the Helmholtz principle like some home audio enclosures that resonate at select frequencies, depending on hole and cavity geometry.

Acoustic treatment. Noise reduction features on the V2500-D5 turbofan engine for the MD-90 include the novel use of acoustic lining on surfaces surrounding the hot engine-core gas stream as well as the inlet. This lining is used on the surfaces of the central closing cone at the aft end of the engine core and the surrounding nozzle facing the cone. Low levels of particulates in the exhaust stream, resulting from more efficient combustion, allow the lining to remain clean and effective, thus reducing rearward noise on the order of 1–2 dB.

Nacelle. The engine nacelle also forms a continuous fan-airflow duct surrounding the core for its entire length and ending in a circular confluent nozzle. The nozzle muffles the high-speed core airflow by allowing it to mix with the slower, surrounding fan air for a lower overall velocity at the exit. The turbulence associated with this velocity relative to the outside air causes so-called jet noise. Earlier noise-reducing nozzle designs were less efficient, blocking the exhaust with assemblies that resembled cookie cutters, cambered surfaces that forced the flows to mix.

Turbomachinery. In designing the rotating turbomachinery for an inherently quiet engine, key parameters include the bypass ratio of fan-to-core airflow. The bypass ratio is determined mainly by aircraft thrust requirements for a given payload weight and range. The number of rotating engine blades and their width (chord) and angular spacing directly govern sound frequency. For the V2500, for example, blade number, size, and spacing were chosen to optimize thrust performance and minimize noise while staggering noise frequencies in order to avoid the predominance of one tone (siren effects) or the production of beats. One specific example was elimination of harmonic (round-multiple) relationships between the number of rotating blades and stationary vanes: The V2500 fan has 22 blades and its exit guide vanes number 60, for a ratio of 2.73; for its low-pressure compressor this ratio is between 1 and 2.

Electronics. Modern electronics also contribute to low noise by allowing accurate control of the engines. An example is the automatic cut-back feature that precisely regulates thrust as altitude changes during takeoffs over sensitive communities where noise abatement is required.

Engine location. The location of engines on an airplane also influences the noise they project. The aft-fuselage-mounted twin engines of the MD-90 are intrinsically quieter than similar engines on other aircraft which mount them below each wing. With aft-mounted engines, the fuselage shields communities on the right side of the aircraft from sideline noise of the left engine and vice versa. The whole community directly hears both engines with an underwing installation. In addition, the location of the engines above and behind the wing lessens forward noise because the engines are shielded by both the wing and the wake airflow streaming from the wing below the engines.

Wing-mounting advocates may argue that these effects may be offset somewhat by the higher airframe structural weight needed to support aft-mounted engines. With more weight, greater thrust performance, and thus greater potential noise, is required on takeoff.

With low-noise engines, the airframe is becoming a significant noise source. Another advantage of fuselage-mounted engines is that they allow the wing trailing-edge flaps and leading-edge slats to form a continuous-span high-lift system. These control surfaces are deployed to increase lift at low speeds during takeoff and landing. With fuselage-mounted engines, there is neither noise-generating disruption in airflow nor a complex wake pattern from lift discontinuities that would result from a wing-mounted engine. One example of such a discontinuity would be a cutout in a trailing-edge flap behind a wing-mounted engine to prevent the exhaust stream from impinging on the flap. Seals between the flaps or slats and the wing and adjacent hardware may improve lift as well as reduce airframe noise.

Cabin noise. Inside airliner cabins, the basic physical principle of destructive interference is being applied to quell the engine-related sounds from turboprop engines and fuselage-mounted turbofans. Such a global reduction in noise throughout a cabin was not possible until recent advances in digital microcomputing power produced control units small and economical enough for commercial aircraft use. The first active noise cancellation systems were operational on turboprop aircraft in late 1994, and on airliners with fuselage-mounted turbofans in late 1995.

The basic precept of active silencing involves generating a second sound source equal in frequency to an objectionable sound but shifted 180° out of phase. If the sound amplitudes are equal, the high-pressure region of one wave cancels the low-pressure area of the other. The resulting neutral pressure is perceived as silence. Microphone and other frequency data such as engine tachometer readings are fed to a central microprocessor which generates the antinoise signal. Feedback via microphones fine-tunes the sound cancellation.

The waves must not only cancel as a function of time (temporal matching) but do so throughout a desired volume of space, and therefore the position and shape of the wavefronts must also match. The importance of this condition is illustrated by the problem of quieting the community noise hum from an electrical power transformer in a distribution yard. Here noise is generated from the transformer core expanding and contracting with the alternating current. The noise frequency and its harmonics are synchronized to the line frequency, making the temporal match simple. The core motions are transmitted to the transformer case whose structure vibrates at various magnitudes, producing a complex wavefront radiating from it rather than the simple, spherical wavefront generated from a single speaker or point source.

Many speakers may thus be required in a noise cancellation system to produce a spatial match of the wavefront. The challenge is to produce an acceptable reduction with the fewest speakers and least complex controls, while keeping costs reasonable.

Previously, passive means were employed in attempts to alleviate aircraft cabin noise. Such measures consisted of heavy, lead-impregnated vinyl sheets in cabin walls or weighty tuned mass dampers, neither alternative being overly successful and both costing useful payload.

Prop-powered airliners not only have engine tones but the particular problem of noise from the air vortices shed by the passing blades repetitively rapping on the fuselage like a drum. The frequencies and harmonics generated by blade passage are typically 50–300 Hz. These low frequencies, with their correspondingly long wavelengths, are the most difficult to attenuate passively since the energy of a sound wave is proportional to its wavelength, as is illustrated by the way bass tones propagate from closed rooms.

However, active control is the easiest for these longer wavelengths, since any delay in processing a noise signal for temporal matching is a smaller fraction of the period between waves. Such small mismatches coupled with frequency data from engine tachometers make microprocessor-based cabin noise control possible. Noise reductions of better than 50% have been realized on turboprop airliners. Future improvements in processing power will make higher frequencies and complex random sounds, such as jet and boundary-layer noise, amenable to control with reasonably sized and priced avionics.

For rear-fuselage turbofan applications, the current technology is most effective in quelling aft cabin noise. Here, harmonics from rotating engine machinery that is not in perfect balance can produce noise and annoying beats. In tests on an MD-80, a reduction of 10 dB (75%) of tones (harmonics) was realized.

Active noise control will likely see future use in reducing fan noise in air-conditioning ducts and appliances such as range hoods as well as for road-vehicle interiors and mufflers.

For background information SEE ACOUSTIC NOISE; ACOUSTIC RESONATOR; ADAPTIVE SOUND CONTROL; AIRCRAFT NOISE; TURBOFAN in the McGraw-Hill Encyclopedia of Science & Technology.

Richard DeMeis

Bibliography. R. DeMeis, Quieting cabin noise, *Aerosp. Amer.*, pp. 20–21, February 1995; R. DeMeis, The greening of aircraft technology, *Aerosp. Amer.*, pp. 18–19, August 1994.

Anomalocaris

Anomalocaris is the largest known Cambrian predator. For many years, only the limbs and the unique circular jaw were known, because the strengthened cuticle composing them is more resistant to degradation than the rest of the body. Recent discoveries of complete specimens have shown that *Anomalocaris* belongs to a diverse and widely distributed group of metazoans of similar status to the arthropods.

Morphology. *Anomalocaris* was first discovered at the end of the last century, when large numbers of scattered segmented appendages were found in Middle Cambrian rocks on Mount Stephen in British Columbia. These fossils were misinterpreted as trunks of a shrimplike arthropod, and it was not until 1979 that they were recognized as limbs. The rest of *Anomalocaris* came to light in 1985, when new research on specimens from the famous Burgess Shale of British Columbia revealed examples with the limbs in place on the body.

The large limbs of *Anomalocaris* that commonly occur separated from the rest of the body project from the front of the head. Behind the limbs is the jaw, which is unlike that of any other known living or fossil animal, consisting of 32 radially arranged, overlapping plates with teeth on their inner margins. The teeth do not meet in the middle, but probably functioned in gripping and breaking prey. Like the segmented limbs, the jaw was composed of sclerotized decay-resistant cuticle, and it too sometimes became separated from the rest of the body. Isolated jaws were originally misinterpreted as jellyfish, and this mistake was corrected only with the discovery of specimens showing the jaw in place on the body (even though a number of examples of these so-called jellyfish clearly show a battery of inwardly facing teeth). The trunk of *Anomalocaris* bears a series of wide overlapping flaps down each side. Specimens recently discovered in British Columbia and China show that at least some of these animals also have three pairs of tail flaps at the rear with, in some cases, a pair of long slender appendages that trailed behind.

Distribution. *Anomalocaris* is not a rare and unusual Cambrian animal confined to a few localities in North America. Examples have been found in western Canada, Utah, California, and Pennsylvania, as well as in Poland, Chengjiang in Yunnan Province in south China, and Kangaroo Island in Australia. When these localities are plotted on a map showing the configuration of the continents during Cambrian times, it is clear that *Anomalocaris* lived in the tropics. Not all the examples of *Anomalocaris* now known can be accommodated within a single genus, but they are all assigned to one family, the Anomalocarididae. Anomalocarididae includes the genera *Peytoia* and *Cassubia*, and others that have yet to be formally named. Relationships within the family remain to be completely unraveled, so it is convenient to refer to these creatures collectively as Anomalocarididae. They range in age from the Polish and Chinese specimens which are early Cambrian (Atdabanian, just over 525 million years old according to recent estimates), to examples from Utah which are late Middle Cambrian, perhaps 10 m.y. younger.

The anomalocaridids are a product of the Cambrian radiation. It is not clear whether their disappearance near the end of the Middle Cambrian reflects the timing of their extinction, or whether examples were simply not preserved in younger rocks. These animals lacked biomineralized skeletons; thus their occurrence in the fossil record relies on conditions that inhibit the normal processes of degradation and decay. Exceptionally preserved fossil biotas of the type that might reveal the presence of anomalocaridids are rare in the Upper Cambrian and Ordovician.

New Chinese examples. The most completely known examples of *Anomalocaris* come from the Burgess Shale. The newly discovered Chengjiang specimens, however, rival them in importance. The site was discovered in 1984, and several localities are still being systematically excavated. The diversity of the fauna now approaches 100 species. Limbs of *Anomalocaris* were reported in 1991. Material described in 1994 revealed three different types of anomalocaridid. Two, a new species similar to *A. canadensis* from the Burgess Shale (**Fig. 1**) and a new unnamed genus, preserve details of the body. The third is represented only by the circular jaw. Specimens of this last type show an outer circle

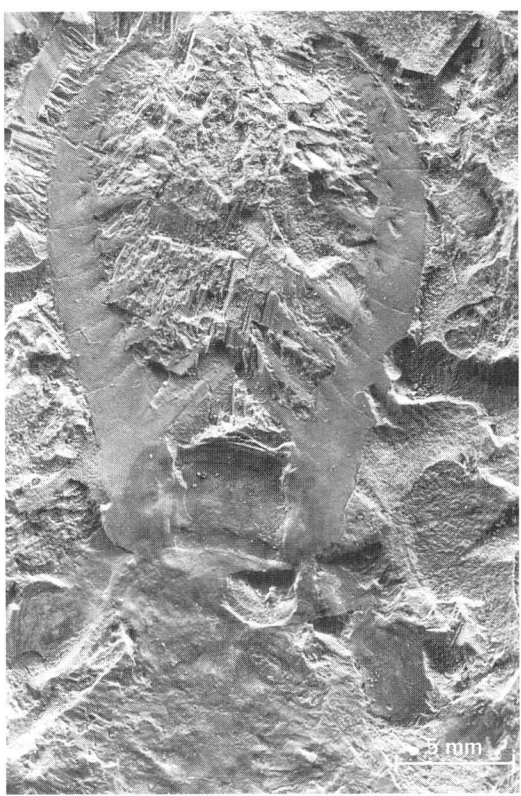

Fig. 1. Head and raptorial appendages of a new species of *Anomalocaris* from the Lower Cambrian Chengjiang fauna of China. (*From J.-Y. Chen, L. Ramsköld, and G.-Q. Zhou, Evidence for monophyly and arthropod affinity of Cambrian giant predators, Science, 264:1304–1308, 1994*)

of toothed plates surrounding numerous rows of teeth that lined the mouth (a similar arrangement is evident in some Burgess Shale specimens). These isolated jaws reach diameters of up to 9.8 in. (25 cm), which, extrapolating on the basis of complete anomalocaridids, indicates that this animal may have reached lengths of up to 6.6 ft (2 m).

Mode of life. There is no doubt that anomalocaridids were formidable predators. Their large size, spiny grasping appendages, powerful toothed jaw, and likely maneuverability in water all point to this conclusion. In addition, a census of the Burgess Shale fossils indicates that anomalocaridids represent a very small percentage of the total individual animals present (less than 0.1%), and this small representation is consistent with a position at the top of the food chain. Evidence for the diet of anomalocaridids is somewhat circumstantial. Trilobites with bites removed from the rear of the exoskeleton are known from several Cambrian localities, a number of which also yield anomalocaridids. In some cases, these bites are clearly shaped in the form of an asymmetrical "W," exactly the shape made by the radial jaw of *Anomalocaris*. However, it is unlikely that trilobites were a major part of the prey. They had a heavily calcified exoskeleton, and there is no sign of fragments of this cuticle in the gut traces of the complete anomalocaridids so far discovered. In any case, these wounded trilobites recovered, as evidenced by the healed margins of the fractured cuticles. It is more likely that anomalocaridids preyed on the much more diverse and abundant arthropods without a biomineralized cuticle, or on other soft-bodied animals such as worms.

Capture of prey. The new discoveries of anomalocaridids have revealed evidence of a diversity of strategies for prey capture. The raptorial appendages in *Anomalocaris* bear a series of paired spines alternating in length along the limb. In *Peytoia*, the appendages bear a row of elongate bladelike projections that are graduated in length, each with evenly spaced forward-facing spines. This tool would have been ideal for raking through fine sediment in search of prey. A new anomalocaridid from the Emu Bay Shale at Big Gully on Kangaroo Island in Australia has appendages with spiny projections (**Fig. 2**), each with large numbers of long overlapping fine spines that could have functioned in filtering small animals out of the sediment or the water. Thus, it appears that the anomalocaridids were ecologically diverse even by the Early Cambrian. They occurred in both shallow and deep water settings, suggesting a range of habitats. Also, they were large enough to have preyed on all the elements in the Cambrian biota; thus, the ecosystem was sufficiently complex to include predators at a secondary level.

Mobility. Anomalocaridids were probably fast, maneuverable swimmers. Functional analysis of the Burgess Shale *Anomalocaris* suggests that it swam by moving the paired flaps on either side of the trunk up and down like a series of hydrofoils or underwater wings to form a propulsive wave running backward along the body. A reversal of the wave would have caused *Anomalocaris* to stop and swim rather less efficiently forward. Different movements of the flaps on opposite sides of the body would have allowed the animal to turn. The hypothesis that *Anomalocaris* swam in this way was tested recently by using a working life-size model 31 in. (79 cm) long (**Fig. 3**). Experiments conducted in a swimming pool showed that the model could swim and maneuver by using the trunk flaps, as deduced on the basis of the Burgess Shale specimens. A propulsive wave running along the length of the body is a feature of swimming in living cuttlefish and other fishes. In these cases, however, a

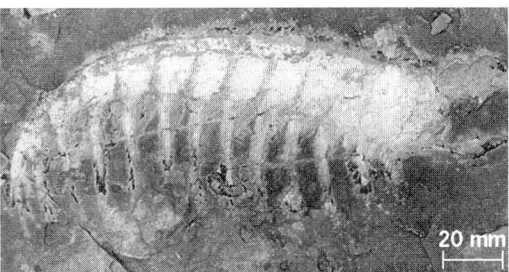

Fig. 2. Isolated head appendage of a new species of *Anomalocaris* from the Lower Cambrian Emu Bay Shale at Big Gully, Kangaroo Island, south Australia.

Fig. 3. Life-size (28 in. or 70 cm) swimming model of *Anomalocaris canadensis*. (*Courtesy of Nippon Hoso Kyokai*)

continuous fin is used rather than an overlapping series of flaps as in *Anomalocaris*. The morphological constraints imposed by a segmented exoskeleton would have eliminated the possibility of such an adaptation evolving in the anomalocaridids.

The recent discovery of anomalocaridids with tail flaps in British Columbia and China has prompted the suggestion that these creatures may have swum by flexing the trunk up and down in the carangiform mode employed by many fish. Such flexibility, however, would have been limited by the relatively wide spacing of the trunk articulations and the possible presence of dorsal plates that cover the trunk (tergites). In addition, if body undulations were used in locomotion, the flaps should be wider on the tail, where the greatest thrust would be generated, than near the anterior of the trunk as they are in *Anomalocaris*. Thus, although body movements may have occasionally been used in swimming, the main mode of propulsion was with the flaps on the trunk. The tail flaps functioned primarily as a rudder in steering and stabilizing and for generating some lift at the rear of the body.

Relationships. The anomalocaridids are a diverse group of giant Cambrian predators. When current research on newly discovered specimens from British Columbia, China, and Australia is complete, a much clearer idea of the extent of morphological variety among them will become available. Cladistic analyses show that the Burgess Shale creature *Opabinia* (which has five eyes and a long anterior proboscis armed with claws at the tip) belongs to the same clade. The segmented raptorial head appendages of anomalocaridids suggest an arthropod relationship, but their circular jaw and flaplike trunk appendages are unique. Together the anomalocaridids and *Opabinia* form a sister group to the arthropods as a whole. Thus, the significance of *Anomalocaris* is not just ecological, as a giant predator, but also phylogenetic, as representative of a high-ranking group of extinct Cambrian metazoans.

For background information *SEE* BURGESS SHALE; CAMBRIAN; METAZOA; TRILOBITA in the McGraw-Hill Encyclopedia of Science & Technology.

Derek E. G. Briggs

Bibliography. D. E. G. Briggs, Giant predators from the Cambrian of China, *Science*, 264:1283–1284, 1994; D. E. G. Briggs, D. H. Erwin, and F. J. Collier, *The Fossils of the Burgess Shale*, 1994; J.-Y. Chen, L. Ramsköld, and G.-Q. Zhou, Evidence for monophyly and arthropod affinity of Cambrian giant predators, *Science*, 264:1304–1308, 1994; H. B. Whittington and D. E. G. Briggs, The largest Cambrian animal, *Anomalocaris*, Burgess Shale, British Columbia, *Phil. Trans. Roy. Soc. London B*, 309:569–609, 1985.

Antigen

Foreign substances that can elicit an immune response in the host are called antigens. B lymphocytes (B cells) and T lymphocytes (T cells), the two antigen-specific cell types of the immune system, differ fundamentally in the forms of antigen to which they respond. B cells respond directly and specifically to intact protein antigens in solution, using cell surface antibodies as antigen receptors. T cells, in contrast, are unable to respond to intact antigens in solution. Instead, they recognize short peptides bound to major histocompatibility complex (MHC) class I and II molecules that are expressed on the surface of antigen presenting cells (APCs). Thus, for a T cell to respond to an intact protein antigen in solution, the antigen must first be taken up by antigen presenting cells and then degraded intracellularly into peptides, some of which are of a size suitable to bind to MHC molecules. However, if a peptide of suitable size possesses amino acid anchor residues that are complementary to one of the MHC molecules expressed by the antigen presenting cells, it can bind intracellularly to that molecule. The resulting complex of MHC molecule and peptide is then expressed on the cell surface of the antigen presenting cells, where it is available to present the antigen to T cells, that is, to stimulate those T cells which have a receptor specific for that particular peptide complementary to that particular MHC molecule. *SEE* CELLULAR IMMUNOLOGY.

MHC genes and gene products. The MHC is a short genetic region, about one-thousandth the length of the genome of higher animals, which encodes many molecules that are involved in antigen presentation. The MHC in humans is known as HLA; in the mouse, H-2. Two types of MHC molecules with central roles in antigen presentation, MHC classes I and II, are structurally similar but differ in their tissue distribution and in the sites of origin of the peptides they bind. Class I molecules are expressed on the surfaces of most cells, but class II molecules are expressed primarily on the surfaces of macrophages and B lymphocytes. Class I molecules present peptides processed from foreign, primarily viral, proteins expressed in the cytoplasm of the cell, whereas class II molecules present those peptides processed from proteins taken in from outside the cell.

In humans, there are three molecular forms (isoforms) of class I molecules and three of class II. Each isoform is highly polymorphic, that is, each has many genetically distinct forms in the population. Most individuals are heterozygous (that is, have two different allelic forms of each class I and II isoform). Although each iso- and allelic form of an MHC molecule can bind to only the limited subset of peptides with the appropriate anchor residues, the abundance of MHC isoforms and the high probability of heterozygosity for each isoform yields a sufficiently diverse repertoire of MHC class I and II molecules that peptides from many different foreign antigens can be bound and presented.

A major advance in the understanding of the structural basis of antigen processing and presentation was made possible by the elucidation of the structures of MHC class I and II molecules and their bound peptides by x-ray crystallography. Several features of these structures are critical to the understanding of antigen processing and presentation, such as the presence of a peptide-binding groove in the outermost extracellular domain of MHC molecules. This arrangement of the peptide-binding groove allows the T-cell receptor to contact both the bound peptide and the amino acids on the top surface of the walls of the peptide-binding groove of the MHC molecule. The peptide-binding groove of class I molecules accommodates only peptides of 8–10 amino acids; class II molecules accommodate peptides of 10–30 or more amino acids. The crystal structures also revealed that most of the amino acid positions that are polymorphic in MHC molecules are located around the binding groove, and are therefore the major determinants governing which peptides can be bound. In addition, it was also determined that the binding grooves of MHC molecules are mostly occupied by peptides degraded from intracellular self proteins. The peptides derived from foreign proteins, although critical for host defense, constitute only a small fraction of the peptides bound to MHC molecules. The MHC molecules do not appear to distinguish in binding between peptides derived from foreign and self proteins. Rather, the ability of T cells to respond to foreign antigens and not to self antigens is a function of the T cells themselves.

MHC peptide complexes. At least three MHC-encoded class I and II molecules assist in the formation of MHC class I and II peptide complexes.

TAP genes. The first indication that products of genes other than class I and II MHC genes are required for antigen presentation came from mutants of cultured cells with markedly reduced MHC class I cell surface expression and global defects as targets for class I restricted cytotoxic (so-called killer) lymphocytes. The genes affected in such mutants, *TAP 1* and *2*, map to the MHC class II region and encode the subunits of a peptide transporter (hence the name transporter associated with antigen processing). The *TAP* transporter is localized on the outer surface of the endoplasmic reticulum, a membrane-bound vesicle into which membrane proteins are inserted as they are synthesized. The *TAP* transporter serves to transport peptides into the endoplasmic reticulum generated from processing of cytoplasmic proteins. In the endoplasmic reticulum, those peptide products of processing that have the appropriate anchor residues and are of appropriate size become bound to newly synthesized class I heavy and light chains; the resulting trimeric complex is then transported to the cell surface. However, if peptides are not delivered to the endoplasmic reticulum, as in *TAP* mutants, heavy and light chain class I dimers form but are unstable and are inefficiently transported to the cell surface. The *TAP* transport mechanism, like class I and II molecules, does not distinguish between peptides degraded from self and foreign proteins.

LMP genes. Closely linked to the *TAP* genes in the MHC class II region are the genes *LMP 2* and *7*, which encode two subunits of the proteosome, a multisubunit protein-degrading enzyme that plays a major role in the degradation of cytoplasmic proteins. The other proteosome subunits are encoded elsewhere in the genome. The expression of *LMP 2* and *7* is generally limited to immunocompetent cells; in these cells, the inflammatory lymphokine γ-interferon increases their abundance, and increases that of MHC class I and II and *TAP* genes as well. Proteosomes lacking *LMP 2* and *7* are proteolytically active, and until recently, it was difficult to discern a role for *LMP 2* and *7* in antigen processing. However, in mice in which the *LMP 7* gene has been "knocked out" in the germ line by a method of genetic manipulation called homologous recombination, reduced cell surface class I expression and defective presentation of a minor histocompatibility antigen suggest that the *LMP 2* and *7* genes have roles in the generation of peptides for antigen presentation.

DM genes. Two genes whose role in antigen processing has most recently been identified are *DMA* and *DMB*. Studies with immunoselected mutants defective in class II restricted antigen presentation have established that the products of these genes are essential for normal class II restricted antigen presentation. The *DMA* and *DMB* genes map to the MHC class II region, close to the *TAP* and *LMP* genes, and they encode the subunits of an integral membrane protein. Although *DM* genes are distantly related in their sequences to both MHC class I and II genes, their pattern of expression in different cells is like that of class II molecules, and they also resemble class II genes in being upregulated by γ-interferon.

Modeling of *DM* structure based on amino acid sequences suggests that DM molecules, like class I and II molecules, have a peptide-binding groove, but there is no direct evidence on this point or on the possible nature of any peptides which may be bound to DM molecules. DM molecules differ from class I and II molecules in that they are apparently not expressed on the cell surface, and thus presumably are unable to present antigens to T cells.

To understand the function of DM molecules in antigen processing, it is first necessary to consider the maturation and intracellular trafficking of class II molecules. Shortly after their synthesis, the two polypeptide subunits of class II molecules become associated with a third polypeptide [the invariant (Ii) chain] in the endoplasmic reticulum. The invariant chain serves two functions: it prevents peptides from binding to class II molecules in the endoplasmic reticulum, and it directs the class II invariant chain complex to an acidic vescicular compartment, MIIC (MCH class II compartment), which shares characteristics with organelles known as lysosomes and endosomes, in which macromolecules are degraded. In this compartment, the invariant chain is removed from class II molecules, at least in part by a sequential series of proteolytic clips, finally yielding empty class II molecules free of invariant chains. The empty class II molecules can then bind peptides with the appropriate anchor residues that have been processed from exogenous proteins taken up by the antigen presenting cell. The resulting peptide–class II complex, expressed at the cell surface, is the ligand recognized by the T cell receptor.

The role of the *DM* gene in antigen processing and presentation has been primarily inferred from the altered characteristics, or phenotype of *DM* mutants. The pattern of peptides bound to class II molecules in these mutants differs drastically from that of nonmutant cells. It consists largely of a few peptides from one region of the invariant chain, in place of the complex mixture of peptides from endogenous and exogenous proteins found in nonmutant cells. It seems likely that the DM molecule has a required function in the removal of the last remnant of the invariant chain from class II molecules, and that removal of this remnant is required to permit class II molecules to bind peptides processed from exogenous proteins. A less likely possibility is that the prevalence of class II molecules with the bound invariant chain remnant in the mutants represents a secondary consequence of a deficiency of the peptides that normally replace invariant chain peptides bound to class II molecules, and hence, that the *DM* gene has a role in generating the pool of these peptides or in their transport.

Linkage and function. The clustering of genes with related functions in MHC is unusual in higher organisms. Whether it occurs elsewhere in the genome, and if so, how commonly, is for the most part unknown, because the number of genes with known functions that have been mapped in higher animals is only a fraction of the total number of genes. However, the genome mapping data available suggest that such clustering may be rare. Whether the clustering of functionally related genes in MHC has occurred by chance or results from evolutionary selective forces is an unanswered question.

For background information *SEE* ANTIBODY; ANTIGEN; GENE ACTION; HISTOCOMPATIBILITY; IMMUNOLOGY; PEPTIDE in the McGraw-Hill Encyclopedia of Science & Technology.

Donald Pious

Bibliography. H. J. Fehling et al., MHC expression in mice lacking the proteosome subunit, *Science,* 265:1234–1237, 1994; S. P. Fling, B. Arp, and D. Pious, HLA-DMA and -DMB genes are both required for MHC class II/peptide complex formation in antigen presenting cells, *Nature,* 368:554–558, 1994; D. M. Madden et al., The three-dimensional structure of HLA-B27 at 2.1 A resolution suggests a general mechanism for tight binding to MHC, *Cell,* 70:1035–1048, 1992; R. Srivastava, G. P. Ram, and R. Tyle (eds.), *Immunogenetics of the Major Histocompatibility Complex,* 1991; L. Van Kehr et al., TAP1 mutant mice are deficient in antigen presentation, surface class II molecules, and CD4 $\bar{8}^+$ T cells, *Cell,* 71:1205–1214, 1994.

Anyons

Anyons are particles obeying unconventional forms of quantum statistics. For many years it was believed that there were only two possible forms of quantum statistics, Bose-Einstein and Fermi-Dirac. Recent work has revealed that these are only two among a continuum of possibilities. Elementary excitations (quasiparticles) in the fractional quantum Hall effect are important examples of anyons.

Bosons and fermions. In quantum mechanics, there are important dynamical effects in the behavior of identical particles that have no classical analog. Such particles are a profound consequence of the fundamental rule of quantum theory: in order to calculate the probability that a given state will occur, it is necessary to add the amplitudes for all processes leading to that state, and then to square this total amplitude. Thus, in the case of two indistinguishable particles A and B, the amplitude for the process that leads to A arriving at point x while B arrives at point y must be added to the amplitude for the process that leads to A arriving at y while B arrives at x, the so-called exchange process, because the final states cannot be distinguished.

Actually, the simple recipe of adding the amplitude for the exchange process is appropriate only for particles obeying Bose-Einstein statistics, that is, bosons. The recipe for particles obeying Fermi-Dirac statistics, that is, fermions, is that the amplitude for the exchange process must be subtracted.

Properties of matter. The quantum statistics of a given type of particle drastically affects the properties of matter, in which many copies of that particle are present. From the definition of bosons it might be anticipated that it is particularly favorable to have—that there is a large probability for—configurations in which all the bosons are in the same state. This anticipation is borne out by more sophisticated analysis. Photons, the quanta of light, are prime examples of bosons. The possibility of laser action is a direct consequence of their boson character: In a laser beam, there are many photons all in the same state; they all represent light of a single frequency (color), and all move in the same direction. From the definition of fermions, on the contrary, it might be anticipated that it is very unfavorable to have two fermions in the same state. Indeed the first manifestation of fermion behavior to be discovered was the Pauli exclusion principle, which (in modern language) states that two electrons cannot occupy the same quantum state. The solidity of ordinary matter is mainly due to this effect: matter resists compression, because the electrons in it cannot be forced into the same state.

Anyon statistics. The primary definition of anyons posits other possible recipes for adding exchange processes. Actually, in defining anyons the analysis of exchange must be slightly refined to take account of the direction in which the exchange takes place. Thus, anyons of type θ are defined by the rule that the amplitudes for processes where A winds around B n times in a counterclockwise direction are added with a factor $e^{2in\theta}$, where $i = \sqrt{-1}$. In this definition, half-integer values of n are allowed; half a winding is an exchange. The values $\theta = 0$ and $\theta = \pi$ correspond to the previous definitions of bosons and fermions, respectively.

The more general possibilities for quantum statistics, corresponding to other possible values of the angle θ, are defined in a mathematically consistent way only for particles whose motion is restricted to two space dimensions. This requirement is not as crippling as it might at first appear, because many of the most interesting materials in modern condensed-matter physics are effectively two-dimensional, including the electronic components of computer circuitry and the copper-oxide layers fundamental to high-temperature superconductivity. Thus, the question arises as to whether the elementary excitations in these systems, the quasiparticles, are anyons. This question can only be settled by a deep investigation of specific cases, and is the subject of much research. It is already known with certainty that the quasiparticles in one fascinating class of states of matter, the fractional quantized Hall states, are anyons.

Generalized exclusion principle. A related but distinct generalization of quantum statistics takes its point of departure directly from the Pauli exclusion principle rather than from the recipe for combining exchange amplitudes: whereas adding one boson to a system does not limit the states available to the next boson, adding one fermion removes one state otherwise available to the next fermion. A more general possibility may be considered, that adding one g-on removes g states for the next, where $g = 1, 2, 3, \ldots$. This generalization is known to be realized in some exotic states of matter, and it too is the subject of much research.

For background information SEE BOSE-EINSTEIN STATISTICS; EXCLUSION PRINCIPLE; FERMI-DIRAC STATISTICS; HALL EFFECT; NONRELATIVISTIC QUANTUM THEORY; QUANTUM MECHANICS; QUANTUM STATISTICS in the McGraw-Hill Encyclopedia of Science & Technology.

Frank Wilczek

Bibliography. F. D. M. Haldane, "Fractional statistics" in arbitrary dimensions: A generalization of the Pauli principle, *Phys. Rev. Lett.*, 67:937–940, 1991; F. Wilczek, Anyons, *Sci. Amer.*, 264(5):58–64, May 1991; F. Wilczek (ed.), *Fractional Statistics and Anyon Superconductivity*, 1990.

Applications satellites

Soon after the launch of the space age it was recognized that the predictable orbits of satellites provide a valuable reference frame for global navigation systems. In the mid-1970s the first Global Positioning System (GPS) tracking network was established by the United States military to link the satellite reference frame to Earth. The military network includes five globally distributed tracking stations that provide 10-m (33-ft) orbit determinations. More recently, organizations from various nations established the International GPS Service (IGS) network, including more than 50 globally distributed tracking stations and providing 10-cm (4-in.) orbit determination in support of geodetic and geophysical research activities.

Atmospheric water vapor. The present Global Positioning System consists of a constellation of 24 satellites that transmit L-band radio signals to large numbers of users engaged in navigation, time transfer, and relative positioning. These signals are delayed and refracted by the gases constituting the atmosphere, including water vapor, as they propagate from GPS satellites to ground-based GPS receivers (**Fig. 1***a*). The delay introduced by water vapor, by virtue of its polar nature, is nearly proportional to the quantity of vapor integrated along the signal path. In applications where the receiver position is unchanging and accurately known, estimations of the delay introduced by water vapor can be transformed into an estimate of the vertically integrated water vapor overlying the receiver by using only surface pressure readings and GPS observations.

Water, again owing to its polar aspect, plays a pivotal role in atmospheric processes that act over a wide range of spatial and temporal scales. The unusually large latent energy associated with water's phase changes significantly affects the vertical stability of the atmosphere, the structure and evolution of storm systems, and the meridional energy balance of the atmosphere. Water vapor, the most variable and inhomogeneous of the major constituents of the atmosphere, also contributes more than any other component to the greenhouse effect. Conversely, the high albedo (reflecting power) associated with clouds limits the solar radiation available to heat the Earth and its atmosphere. Thus,

14 Applications satellites

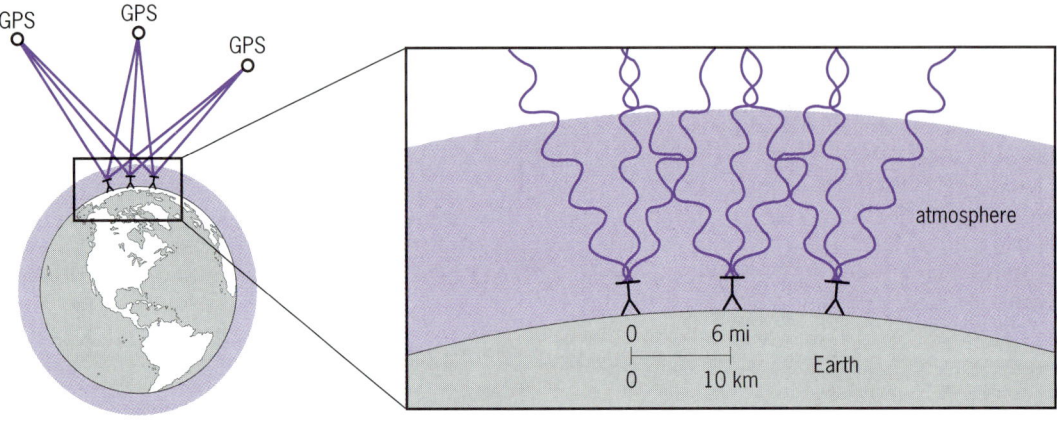

Fig. 1. Ground-based GPS meteorology. (*a*) GPS signals passing from GPS satellites through the intervening Earth's atmosphere to ground-based receivers. (*b*) GPS field experiment region (box) showing the locations of the GPS receivers (open circles), and remote GPS sites (closed circles) used in absolute water vapor estimation.

water vapor plays a crucial role in global climate change.

Given the present-day operational weather data system, inadequate resolution of the temporal and spatial variability in water vapor has been cited as the single greatest obstacle to improved short-range precipitation forecasts. Mesoscale numerical model simulations have shown that when model-predicted precipitable water vapor is relaxed toward an observed value, the model recovers the vertical structure of water vapor with an accuracy much greater than that from statistical retrieval based on climatology, leading to significantly improved short-range precipitation forecasts. By extrapolation it can be inferred that GPS water vapor data will be valuable to longer-range numerical forecasts.

Meteorological signals. A primary task of geodetic GPS algorithms is to correct the ranges between satellites and receivers so as to remove the delay (equivalent excess path length) introduced by the refractivity of Earth's atmosphere. The ionosphere introduces a delay that can be determined and removed by recording both of the frequencies transmitted by GPS satellites and by exploiting known dispersion relations for the ionosphere. The remaining delay due to the neutral atmosphere can be divided into two parts, known as hydrostatic delay and wet delay. Both delays are smallest for paths oriented along the zenith direction and increase approximately inversely with the sine of the elevation angle. Most expressions for the delay along a path of arbitrary elevation consist of the zenith delay multiplied by a value known as the mapping function, which describes the dependence on elevation angle.

The hydrostatic delay is a function of surface pressure, and typically reaches about 2.3 m (7.6 ft) in the zenith direction. It is possible to calculate the zenith hydrostatic delay to better than 1 mm (0.04 in.), given surface pressure measurements accurate to 0.3 millibar (30 pascals) or better. The basis for estimating the precipitable water in the atmosphere arises from the fact that the wet delay is closely related to the quantity of water vapor overlying the receiver. The zenith wet delay can be less than 20 mm (0.8 in.) in arid regions, and as large as 350 mm (14 in.) in humid regions. Significantly, the daily variability of the wet delay usually exceeds that of the hydrostatic delay by more than an order of magnitude, especially in temperate areas. Zenith wet delay can be estimated from GPS data as part of the overall least squares inversion for the coordinates of the GPS receivers, the orbital parameters of the GPS satellites, and other geodetic parameters of interest.

GPS field experiment. During May 1993, a field experiment was conducted to evaluate GPS measurement of atmospheric water vapor in a region noted for severe weather. During the experiment, a team of geodesists and meteorologists installed GPS receivers and microwave radiometers at six sites in Oklahoma, Kansas, and Colorado (Fig. 1b). The sites were chosen to take advantage of existing operational infrastructure at the wind profiler sites of the National Oceanic and Atmospheric Administration (NOAA).

To estimate the zenith wet delay, and hence precipitable water vapor, a knowledge of the station position and the satellite orbits is required. Precise orbits were obtained from the International GPS Service. During the field experiment, five or more satellites were typically visible at any time at each receiver site. For GPS networks with station separation smaller than ~500 km (~300 mi) the techniques used to estimate zenith delay are more sensitive to relative rather than absolute delays. This situation comes about because a GPS satellite observed by two or more receivers is viewed at almost identical elevation angles, causing the delay estimates to be highly correlated. Therefore, the estimates of zenith wet delay derived from a small network are subject to an unknown bias at each epoch. The value of the bias is constant across the whole network (that is, the bias varies in time but not in space). There are several possible approaches to estimating this bias, a task known as levering (the group of biased estimates of precipitable water vapor are levered until they all have the correct absolute value). One approach is to measure precipitable water vapor with a special ground-based water vapor radiometer at a single reference site, and use the GPS data to estimate precipitable water vapor relative to this reference site at any number of so-called secondary sites in the monitoring network.

An attractive alternate approach that eliminates the need for an independent measurement of precipitable water vapor at a reference site is to incorporate a few GPS stations that introduce baselines much longer than 500 km (300 mi). Provided GPS satellite orbit information of sufficient accuracy is available, it is possible to estimate a so-called absolute zenith water delay and, therefore, absolute precipitable water vapor from the GPS observations alone.

In comparing the data for precipitable water vapor derived from GPS and radiometers taken during the field experiment, an average agreement of 1–1.5 mm (0.04–0.06 in.) is found. It has been found that GPS measurements outperform those of the radiometers during periods of rain. Under these conditions the radiometer data are significantly degraded by the presence of water droplets on the windows of the microwave radiometers deployed in the field experiment. This degradation of the radiometer data represents a drawback in using a radiometer as a reference with which to lever a network of GPS receivers. It has also been demonstrated that the GPS estimation technique is not affected by precipitation.

From a logistical perspective, the weather experienced during the GPS field experiment was perhaps too active. While seeking dry lines and fronts, the participants and their equipment were not prepared for the onslaught of severe thunderstorms, flooding, and lightning. More than 100 tornadoes were reported in the general area of the GPS field experiment in association with a slow-moving dry line from May 6 to May 9, 1993. Dry lines are boundaries in the troposphere between dry air descending over the Rockies from the west and moist air approaching from over the Gulf of Mexico to the south. The data from the experiment show rapid changes in precipitable water vapor associated with the passage of dry lines and cold fronts across the network. The enhanced spatial resolving power represented by the GPS data is demonstrated when a comparison is made between objective analyses of precipitable water vapor with and without the GPS data. An analysis of data of precipitable water vapor obtained by using radiosondes shows air that is too dry over western Oklahoma, in conflict with the analyzed position of the dry line and with the GPS data. It was found that the best agreement between the position of the dry line and the analysis of precipitable water vapor occurs in the heart of the GPS network, where the greatest concentration of supplemental data resides. A high value of 32 mm (1.3 in.) for precipitable water vapor at Vici, Oklahoma, that was determined is consistent with thunderstorm activity observed in that area, and it illustrates the degree of local variability in the atmospheric water vapor distribution.

In addition to improved spatial resolution of atmospheric water vapor, it can be seen that the GPS data provide significantly improved temporal resolution of water vapor over that available from the operational radiosonde network. Changes in water vapor content that occur above the surface (for example, cold fronts in the middle troposphere) are not detectable by routine surface observations, and can have significant impact on the convective instability of the atmosphere. However, these changes are detectable in the GPS water vapor data. For example, a large decrease in precipitable water vapor precedes the passage of the surface cold front.

Prospects. The prospect of using GPS to measure atmospheric water vapor for research and operational weather forecasting is made especially promising by the fact that extensive networks of continuously operating GPS receivers are now being constructed around the world by geophysicists, geodesists, surveyors, and others for a variety of scientific, engineering, and civilian navigation and positioning applications. Current estimates indicate that as many as 1500 continuously operating GPS receivers will have been installed by 1997. Precipitable water vapor data from these receivers could be made available to the meteorological community at little incremental cost.

A spectrum of federal, state, and private organizations and their counterparts in other countries are establishing or planning to use GPS tracking sites and networks. For example, in the United States the Federal Aviation Administration (FAA) is constructing a considerable network to provide differential position data for aircraft runway approaches, and the U.S. and Canadian Coast Guards are installing networks for improved coastal navigation (**Fig. 2**). Other GPS networks are planned for dam and dike monitoring by the U.S. Army Corps of Engineers, and for surveying by state governments. NOAA has deployed GPS receivers at wind profiler sites and is evaluating deployment of more, pending development of real-time data-processing capabilities. Regional networks of continuous GPS receivers to monitor crustal strain for earthquake research and forecasting and for satellite tracking purposes are being established in the United States, Japan, Europe, and elsewhere.

Operational numerical weather prediction models routinely predict synoptic-scale (>500 km) circulations of the atmosphere with considerable success, but fail to predict precipitation patterns and severe weather with similar levels of reliability. This disparity in prognostic skill is a consequence of the formation of precipitation at scales much smaller than those resolved by present-day global models and of the lack of mesoscale data (currently limited in resolution to the spacing of the radiosonde network)

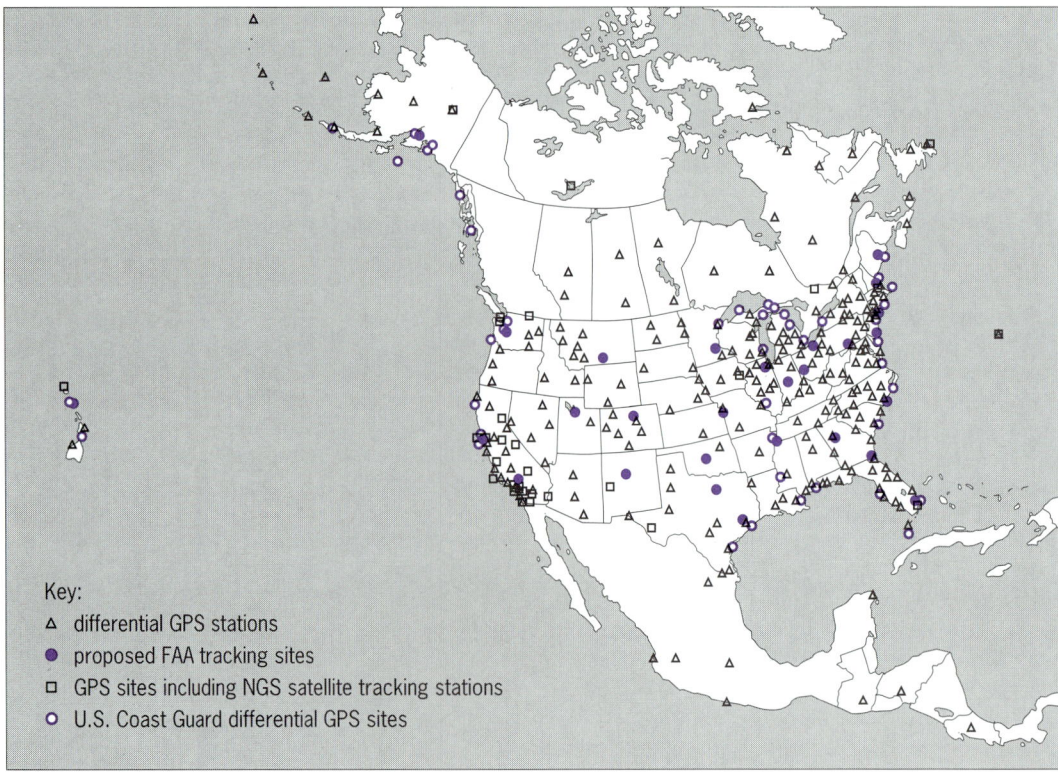

Fig. 2. Location of FAA proposed Aircraft Addressing and Reporting System (ACARS) differential GPS stations, proposed FAA aircraft tracking sites, selected continuously operating GPS sites (including NGS satellite tracking stations), and U.S. Coast Guard differential GPS sites.

with which to initialize high-resolution weather prediction models. Recent theoretical modeling studies indicate that assimilating precipitable water vapor observations into high-resolution models significantly improves precipitation forecasts.

Improved knowledge of the precipitable water vapor distribution helps to document the presence of sometimes elusive and potentially hazardous atmospheric structures such as surface dry lines and cold fronts in the middle troposphere. In conjunction with operationally available data sets (including surface, radar, radiosonde, and satellite data), GPS-derived precipitable water vapor data represent an important new resource for operational numerical weather prediction and basic research into small- and large-scale storm systems. Long-term, global measurement of precipitable water vapor from GPS networks could provide key insights into the hydrologic cycle and global climate change. An opportunity now exists to optimize the total value of the emerging GPS infrastructure. To accomplish this goal, the organizations that are establishing the infrastructure must adequately communicate and coordinate plans and objectives.

For background information SEE APPLICATIONS SATELLITES; ATMOSPHERE; GREENHOUSE EFFECT; SATELLITE NAVIGATION SYSTEMS; WEATHER FORECASTING AND PREDICTION in the McGraw-Hill Encyclopedia of Science & Technology.

Steven Businger

Bibliography. M. Bevis et al., GPS meteorology: Mapping zenith wet delays onto precipitable water, *J. Appl. Meteor.*, 33:379–386, 1994; M. Bevis et al., GPS meteorology: Remote sensing of atmospheric water vapor using the Global Positioning System, *JGR-Atm.*, 97:15,787–15,801, 1992; S. Businger, W. H. Bauman III, and G. F. Watson, Coastal frontogenesis and associated severe weather on 13 March 1986, *Month. Weath. Rev.*, 119:2224–2251, 1991; Y.-H. Kuo, Y.-R. Guo, and E. R. Westwater, Assimilation of precipitable water into mesoscale numerical model, *Month. Weath. Rev.*, 4:1215–1238, 1993; C. Rocken et al., Sensing atmospheric water vapor with the Global Positioning System, *Geophys. Res. Lett.*, 20:2631–2634, 1993.

Atmospheric aerosols

Atmospheric aerosols play an important role in climate and climate change, either directly by scattering and absorbing solar and infrared radiation or indirectly by influencing the microphysical properties of clouds and thereby modifying their radiative properties and lifetimes. The direct effects of aerosols on Earth's radiation balance depend on their single scattering albedo (reflecting power) and therefore on their composition. For example, the globally averaged direct aerosol climate forcing (that is, the change in reflected solar radiation) due to backscattering by anthropogenic sulfate aerosols (mainly from fossil fuel combustion) may range from -0.3 to -1.1 W/m^2. This forcing varies geographically, and it acts locally to mask greenhouse warming by infrared absorbing gases of anthropogenic origin. However, absorption of solar radiation by aerosols such as black (elemental) carbon can offset the cooling due to aerosol backscattering.

Link to clouds. Aerosols may also indirectly affect the radiation budget by changing the reflectivity of clouds when the aerosols act as cloud condensation nuclei. A critical link between aerosols and clouds is the nucleation process, in which supersaturation activates some fraction of the aerosols to form cloud drops. Changes in the concentrations of cloud condensation nuclei can change the number of cloud drops, thereby affecting the optical properties of the clouds. Ship-trail experiments have demonstrated that an increase in drop concentration can locally increase the reflectivity of shallow marine stratocumulus clouds. Studies from radiative-convective models have suggested that a global doubling of the concentration of cloud condensation nuclei would reduce the surface temperature by about 1 K (1.8°F). However, unlike most greenhouse gases, aerosols have an average lifetime of only days or weeks in the troposphere, and they appear in highly variable concentrations. The large spatial and temporal variabilities have made it difficult to quantify their effects on cloud radiative properties and the radiation budget. There are also complications as a result of the nonlinearity of cloud microphysics, where the cloud droplet number is a function of updraft velocity and aerosol characteristics. SEE APPLICATIONS SATELLITES.

Composition. Atmospheric aerosols are composed of varying amounts of numerous chemical species. To investigate the aerosol-cloud-climate interactions the aerosol sources, transformation, and removal processes must first be characterized on a global basis. Aerosols are derived in the atmosphere from a variety of sources. Over continents, the natural sources of aerosols are combustion (for example, forest fires), volcanoes, chemical reactions of natural trace gases that produce aerosols, soil and dust particles raised from the Earth's surface, and pollen and volatile hydrocarbons emitted by plants. Anthropogenic sources of aerosols include the combustion of fuels for transportation; the industrial activities and the biomass burning that produce emissions of the water-soluble inorganic species (for example, nitrate, sulfate, and ammonium), organic condensed species, and elemental carbon; and mineral dust from enhanced soil erosion. Water-soluble inorganic aerosols and some types of organic aerosols are formed through gas-to-particle conversion processes. Thus, nitrogen oxide (NO$_x$) and sulfur dioxide (SO$_2$) are emitted in the form of a gas. Sulfur dioxide is converted to aerosol sulfate ion (SO$_4^{2-}$) by both homogeneous and heterogeneous processes. Nitrogen oxide is converted to nitric acid (HNO$_3$) and aerosol nitrate ion (NO$_3^-$) by gas-phase reactions. Other aerosol types are directly emitted as particles. Over ocean areas, aerosol sources include sea salt, as well as oceanic emission of dimethyl sulfide. Dimethyl sulfide, which dominates the production of cloud condensation nuclei over the oceans, is present as a gas in the atmosphere and is converted to SO_2 and methanesulfonic acid via homogeneous gas-phase reaction. These emitted aerosols and aerosol precursors are distributed in the atmosphere through diffusion, turbulence, and transport of air masses. They are removed from the atmosphere by precipitation, coagulation, and sedimentation.

Gaseous sulfur emissions. Many of the aerosols that influence Earth's radiation balance are derived from gaseous sulfur emissions. Sulfate aerosols are formed through either gas-phase photochemical reactions of the emitted sulfur compounds or aqueous processes involving in-cloud oxidation of SO_2. Since anthropogenic sulfur emissions dominate natural emissions by a factor of 2 on a global average (with 90% of the anthropogenic sources in the Northern Hemisphere) and they have been significantly increased during the past several decades, a large increase in the concentration of sulfate aerosols has occurred, especially over and around industrialized regions. In 1991 it was estimated that the global radiative forcing due to scattering by anthropogenic sulfate aerosols is around -0.6 W/m^2. This calculation was based on the burden of sulfate simulated by a three-dimensional chemical-transport model along with estimates that had been made of the scattering and backscattering coefficients per unit mass concentration. In 1993 a research group used cloud fields from the community climate model of the National Center for Atmospheric Research (NCAR CCM2) with monthly mean sulfate distributions from a three-dimensional chemical-transport model, and they estimated the direct sulfate forcing at only -0.3 W/m^2. This group reported that their sensitivity study showed a 10% variation in the direct sulfate forcing as a result of changes to the chemical and physical nature of sulfate aerosols (for example, size distribution and chemical composition). They also indicated that the difference in the magnitude of the direct sulfate forcing between their results and those reported earlier by the other research group in 1991 was due to the difference in modeling of the optical properties of the sulfate aerosols. A recent study with a coupled three-dimensional chemistry-climate-mixed layer ocean model of 20-year duration shows that the temperature change from current-day direct forcing by anthropogenic sulfate aerosols and by carbon dioxide (CO_2) exhibit distinctively different regional patterns and estimates than the direct sulfate forcing between -0.6 and -0.9 W/m^2.

The indirect forcing by sulfate aerosols through cloud processes is more difficult to estimate, in part because of the large uncertainty in relating aerosol number distribution to sulfate mass concentration. The sulfate aerosol number distribution depends on the processes that form particulate sulfate. Sulfuric acid formed in the gas phase may either condense onto existing particles, causing them to grow, or form new particles through homogeneous nucleation. The resulting aerosol distribution is of importance in evaluating its impact on cloud optical properties. Recently, several research groups have developed parameterizations relating cloud drop concentration to sulfate mass or to aerosol concentration; they have used these parameterizations to develop estimates of the annual globally averaged indirect anthropogenic sulfate forcing in the ranges of -0.4 to -0.9 W/m^2 and -0.7 to -1.4 W/m^2. These parameterizations made use of measured relationships in continental and maritime clouds. However, these relationships are inherently noisy, yielding more than a factor of 2 variation in cloud drop concentration for a given aerosol number (or for a given sulfate mass) concentration. Therefore, the indirect sulfate forcing cannot be quantified until a more precise relationship between sulfur sources, aerosol distribution, and cloud droplet concentration is available.

Other types of emissions. In addition to aerosols formed from anthropogenic sulfur emissions, carbonaceous aerosols, either directly emitted by combustion or produced from condensable gaseous precursors released from biomass burning, also impact radiative balance. The total current forcing by carbonaceous aerosols produced during biomass burning has been estimated as -0.8 W/m^2. Biomass burning takes place mainly in tropical and subtropical South America, Asia, Africa, and Australia. Long-range transport of the absorptive black carbon aerosols to the Arctic is very important in altering the local temperature profile over its high albedo surface. The result is a net positive climate forcing from March to May.

Another key anthropogenic aerosol type is the mineral dust aerosols; these result from desertification, mainly in Africa. The mineral dust aerosols not only scatter and absorb solar radiation but also absorb infrared radiation. Therefore, calculation of radiative forcing by mineral dust aerosols is dependent on the size distribution of chemical composition and subject to large uncertainty.

Injection of aerosols from major volcanic eruptions into the stratosphere may also have a potential climatic influence because of the long residence time of stratospheric aerosols, typically months to years. Volatility measurements indicate that the long-lived volcanic aerosols in the stratosphere are composed primarily of about 75% sulfuric acid and 25% water. Volcanic aerosols tend to cool the Earth atmosphere system by reflecting more solar radiation back to space. However, the presence of volcanic aerosols results in local heating in the vicinity of the aerosol layer. Typically, a stratospheric warming by several degrees is observed, with most of the heating occurring through the absorption of infrared radiation. The height of the volcanic aerosol injection is also an important factor affecting the stratospheric temperature.

Assessment of effects. There are considerable uncertainties in the assessment of the effects of aerosols on climate, mainly due to incomplete knowledge of sources and sinks of aerosols and their consequential direct and indirect radiative effects. An understanding of aerosol-cloud-climate interactions requires expertise in ocean-to-air transport mechanisms, gas-phase and aqueous-phase chemistry, aerosol physics, cloud physics and chemistry, radiative transfer theory, and global climate modeling. Therefore, establishment of an International Global Aerosol Program (IGAP) should help in understanding and quantifying the perturbing effects of anthropogenic aerosols on climate and climate change.

For background information SEE AEROSOL; ALBEDO; ATMOSPHERE; CLOUD PHYSICS; GREENHOUSE EFFECT; NUCLEATION; TERRESTRIAL RADIATION in the McGraw-Hill Encyclopedia of Science & Technology.

Catherine C. Chuang

Bibliography. R. J. Charlson et al., Perturbation of the Northern Hemisphere radiative balance by backscattering from anthropogenic sulfate aerosols, *Tellus*, 43AB:152–163, 1991; P. Hobbs (ed.), *A Plan for an International Global Aerosol Program (IGAP)*, 1994; A. Jones, D. L. Roberts, and A. Slingo, A climate model study of indirect radiative forcing by anthropogenic sulphate aerosols, *Nature*, 370:450–453, 1994; J. E. Penner, R. Dickinson, and C. O'Neill, Effects of aerosol from biomass burning on the global radiation budget, *Science*, 256:1432–1434, 1992; K. E. Taylor and J. E. Penner, Response of the climate system to atmospheric aerosols and greenhouse gases, *Nature*, 369:734–737, 1994.

Atomic crystal

Techniques for laser cooling and trapping of atoms have recently been used to assemble atoms into crystallike structures. In contrast to solids, which are bound into crystals by strong interatomic forces, these atomic crystals are assembled by the tenuous light forces arising in an interference pattern formed by laser beams. The laser field is responsible for both cooling and capturing atoms in microscopic traps arranged in a regular pattern, a so-called optical lattice. Atomic crystals have densities approximately 10^{11} times lower than solids, and combine properties of single, isolated atoms with those of condensed matter. Possible applications include the pursuit of ever colder and denser atomic samples for atomic spectroscopy and atom interferometry, light-controlled deposition of atoms for nanofabrication, and the study of current issues in solid-state physics relating to quantum motion and many-body effects in a periodic potential.

Optical lattices. The light-induced force creating an atomic crystal can be separated into a friction force responsible for laser cooling and a conservative force forming the potential that holds the crystal together. The potential arises from the interaction of the light electric field with the induced atomic electric dipole, which changes the internal atomic energy by an amount proportional to the light intensity. Under conditions optimum for laser cooling, an atom rarely absorbs photons, and its motion is primarily affected by the potential that acts on the atom in its ground state. For a light frequency below atomic resonance, the ground-state energy is decreased and the light-induced potential has its minimum, where the light intensity is maximum. A one-dimensional array of microscopic optical traps can therefore be formed in a standing laser wave (**Fig. 1**). In an extension of this idea to three dimensions, three orthogonal standing waves can be made to overlap to form a three-dimensional array, or optical lattice, of atomic microtraps.

To capture atoms in a three-dimensional optical lattice it is necessary to cool them until their energy is insufficient to escape individual microtraps; this state typically corresponds to temperatures below a few tens of microkelvin (less than 50×10^{-6}°F above absolute zero). Such low temperatures are easily achieved by laser cooling in the laser field that forms the optical lattice, provided that the atoms have several Zeeman levels and the laser field has spatially varying polarization. A complication arises because an atom in a given Zeeman level may escape from its microtrap by transfer to another Zeeman level. This effect can be suppressed by a careful choice of atomic transition and laser polarization. Optical lattices have now been formed by using a variety of laser-beam geometries. However, of the many elements with suitable atomic structure only rubidium and cesium atoms have so far been bound into atomic crystals.

Spectroscopy. The formation of an atomic crystal is most clearly observed by detecting transitions between quantized states of the atomic center-of-mass motion. An individual atom in a microtrap is well approximated by a quantum-mechanical harmonic oscillator, and the quantized states of motion

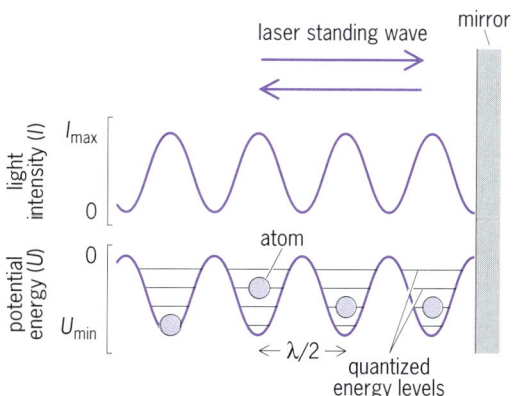

Fig. 1. Light-induced potential formed by a laser standing wave. A series of microtraps are separated by $\lambda/2$, one-half the laser wavelength. Atoms caught in traps occupy quantized vibrational levels.

form a series of vibrational levels. Each vibrational state is labeled by a quantum number n and separated from neighboring states by an energy $h\nu_{osc}$, where h is Planck's constant and ν_{osc}, the atomic oscillation frequency. The distribution of atoms among vibrational levels is roughly thermal, and the number of atoms in level n always exceeds the number of atoms in level $n + 1$.

Fluorescence spectroscopy. A typical fluorescence spectrum radiated from an atomic crystal (**Fig. 2***a*) consists of a central line at the frequency ν_L of the optical lattice and sidebands at frequencies $\nu_L \pm \nu_{osc}$. The interpretation of this spectrum is based on the harmonic-oscillator model (Fig. 2*b*). An atom initially in level n absorbs and re-emits a photon, and either returns to level n or makes a transition to one of the levels $n \pm 1$. If the final level is n, the emitted photon has frequency ν_L, and the process contributes to the central line. The effect is a close analogy to the Mössbauer effect, but occurs here at an optical frequency. If the final level is $n + 1$ or $n - 1$, the emitted photon has frequency $\nu_L - \nu_{osc}$ or $\nu_L + \nu_{osc}$, and the process contributes to the sidebands. The difference in sideband amplitude shows that transitions $n \to n + 1$ are more frequent than transitions $n + 1 \to n$, since more atoms occupy level n than level $n + 1$. The ratio of amplitudes determines the atomic temperature, which is typically in the range 1–30 μK (2–50 × 10^{-6}°F above absolute zero).

Stimulated Raman spectroscopy. This technique sends a probe beam through the atomic sample, and measures gain or absorption as the probe frequency ν_P is changed. A typical spectrum of Raman gain-absorption (**Fig. 3***a*) consists of a dispersively shaped line at frequency ν_L, and a pair of sidebands at frequencies $\nu_L \pm \nu_{osc}$.

The positions and signs of the sidebands are explained as follows (Fig. 3*b*): If the probe-beam frequency is $\nu_P = \nu_L - \nu_{osc}$, then the transition $n \to n + 1$ is induced by absorption of a lattice photon and stimulated emission of a probe photon. Conversely, the transition $n + 1 \to n$ is induced by absorption of a probe photon and stimulated emission of a lattice photon. Because more atoms are found in level n than in level $n + 1$, the first process dominates, and the probe beam is amplified. At a probe frequency $\nu_P = \nu_L + \nu_{osc}$ the processes are reversed, and the probe beam is absorbed.

The central feature is created by interference of light scattered at successive layers of the atomic crystal, an effect analogous to Bragg scattering and strongly indicative of long-range atomic order. From its width it can be estimated that atoms remain trapped in individual microtraps for several milliseconds, possibly longer. Stimulated Raman transitions and effects of long-range order are seen also in reflection, phase conjugation, and four-wave mixing.

Quantum motion. Fluorescence and stimulated Raman spectra are obtained in a regime where the physics of individual atoms in an optical lattice is well approximated by isolated quantum-mechanical harmonic oscillators. In some situations a large fraction of the atoms occupy the vibrational ground state, and therefore approach minimum-uncertainty wave packets. Thus, it may be possible to squeeze the atomic motion, in a manner formally analogous to the squeezing of light. It might then be possible to obtain atoms that have momentum or position uncertainty below that of the vibrational ground state. In the first case (momentum squeezing), atoms released from the optical lattice will appear much colder than unsqueezed atoms. In the second case (position squeezing), controlled deposition of

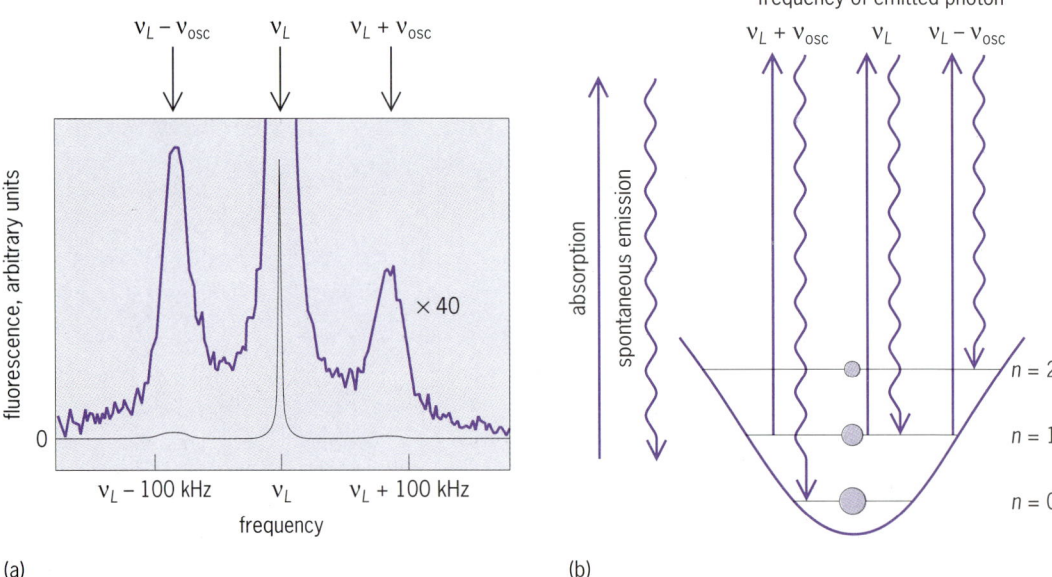

Fig. 2. Fluorescence spectroscopy of an atomic crystal. (*a*) Spectrum of fluorescence radiated from the crystal. (*b*) Interpretation based on harmonic-oscillator model. Arrows indicate atomic transitions during absorption and spontaneous emission. Sizes of circles indicate numbers of atoms in vibrational energy levels.

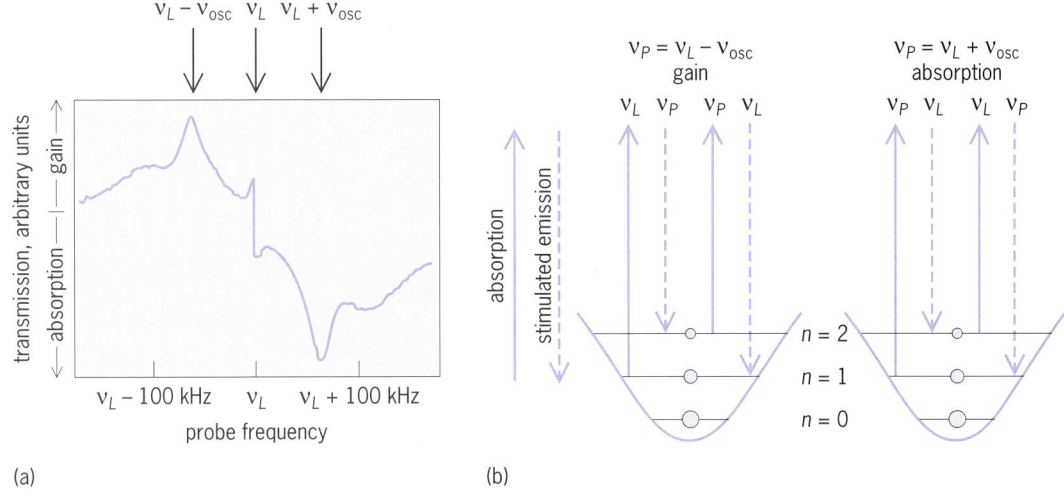

Fig. 3. Stimulated Raman spectroscopy of an atomic crystal. (*a*) Spectrum of probe transmission through the crystal (*after G. Grynberg et al., Quantized motion of cold cesium atoms in two- and three-dimensional optical potentials, Phys. Rev. Lett., 70:2249–2252, 1993*). (*b*) Interpretation based on harmonic-oscillator model. Arrows indicate atomic transitions during absorption and stimulated emission. Sizes of circles indicate numbers of atoms in vibrational energy levels.

the atoms onto a surface might allow the fabrication of nanoscale structures. Adiabatic expansion of atoms in their microtraps, achieved by slow attenuation of the optical lattice potential, provides an alternative method for obtaining very cold atoms. It is possible that squeezing or adiabatic expansion can produce atoms with a mean kinetic energy below that gained from a single photon scattering.

A different regime of atomic quantum motion may perhaps be studied in optical lattices that are so shallow that their microtraps support at most a single vibrational level. In such a system quantum-mechanical tunneling becomes likely, and atoms propagate through the optical lattice in close analogy to electrons in solids. Before this type of quantum motion can be observed it is necessary to suppress photon scattering from the light field creating the optical lattice. This result can in theory be achieved by forming the optical lattice with laser light of a frequency that is tuned below an atomic resonance by many thousand times the natural line width.

Atom-atom interactions. Experiments so far have produced atomic crystals containing a few times 10^{11} atoms per cubic centimeter (10^{12} atoms per cubic in.). This density is nearly 100 times below that of microtraps, or sites, in the underlying optical lattice. In contrast to a solid, an atomic crystal therefore consists almost entirely of vacancies, with a few atoms scattered randomly over lattice sites. Attempts to further increase the density of atoms have proved unsuccessful because of interactions that arise when atoms approach each other to within an optical wavelength. These interactions are in principle suppressed if the optical lattice is formed by light of a frequency far from atomic resonance. If atom-atom interactions can be controlled by this or other means, it may be possible to reach a regime where the occupation probability for a given lattice site is close to unity, and where many-body effects have significant influence on the behavior of the system. Based on analogies with solids, it might then be possible to observe collective behavior, for example, the formation of superlattices and clusters or phenomena such as excitons, phonons, and spin-density waves.

For background information SEE CRYSTAL STRUCTURE; FLUORESCENCE; HARMONIC OSCILLATOR; LASER COOLING; NONRELATIVISTIC QUANTUM THEORY; PARTICLE TRAP; QUANTUM MECHANICS; RAMAN EFFECT; SQUEEZED QUANTUM STATES in the McGraw-Hill Encyclopedia of Science & Technology.

Poul Jessen

Bibliography. G. P. Collins, Three-dimensional optical molasses binds a new type of crystal, *Phys. Today*, 46(6):17–19, June 1993; G. Grynberg et al., Quantized motion of cold cesium atoms in two- and three-dimensional optical potentials, *Phys. Rev. Lett.*, 70:2249–2252, 1993; A. Hemmerich and T. Hänsch, Two-dimensional atomic crystal bound by light, *Phys. Rev. Lett.*, 70:410–413 (1993); P. S. Jessen et al., Observation of quantized motion of Rb atoms in an optical field, *Phys. Rev. Lett.*, 69:49–52, 1992.

Atomic nucleus

Recent work that builds on ideas developed since the mid-1970s indicates that fermion dynamical symmetries may provide a method to apply the shell model systematically to all nuclei. These symmetries indicate that sets of states in nuclei differing considerably in proton and neutron number may in a certain sense be the same state, and thus are expected to have very similar properties.

Nuclear shell model. In the nuclear shell model the neutrons and protons (the nucleons) are assumed to move in a spherically symmetric potential resulting from the average interactions with other nucleons. This assumption is a good first approximation, but there are additional residual interactions among the nucleons that cannot be subsumed in such an average potential. A realistic calculation including these residual interactions is possible only for the lightest nuclei; the traditional shell model can be solved for the more general case only if the complexity is reduced by truncation, which requires a way to throw away enough of the problem to allow the remainder to be solved. (A new method under development, termed the Monte Carlo shell model, may also prove useful for general solutions of the shell model.) Truncation approaches have had limited success because the truncation must be implemented in such a way that methods are known to correct for the influence of the part of the problem that was discarded.

Since the mid-1970s a new approach to nuclear structure that exploits powerful symmetry principles has been developed and refined. This approach provides a new prescription for truncation of the nuclear shell model that reduces the problem to manageable dimensions, while allowing the influence of the excluded part of the calculation to be incorporated consistently.

Global and dynamical symmetries. Symmetries imply the existence of certain operations on an object that leave it unchanged. For example, the square in **Fig. 1** is symmetric under rotations in the plane by any integer multiple of 90°, and under reflections through vertical, horizontal, and diagonal planes perpendicular to the paper and passing through the center of the square. (It is also symmetric under additional operations that will not be discussed.) Such symmetries play important roles in many areas of science. For example, symmetries under various spatial operations are important in chemistry and in crystallography.

For physical systems of atomic dimensions or smaller, the appropriate description is in terms of the theory of quantum mechanics, where the dynamics of the system is dictated by a function called the hamiltonian. Physical systems may exhibit certain symmetries associated with transformations on the hamiltonian. Because the hamiltonian governs the dynamics, it is not surprising that symmetry operations associated with the hamiltonian lead not to spatial relations but to certain regularities in the spectrum and transition rates of the system.

It is useful to divide hamiltonian symmetries into two broad categories. The first category consists of global symmetries that impose general conservation laws. One such conservation law is associated with angular momentum, which is rigorously conserved in any isolated system. An acceptable hamiltonian must lead to conservation of angular momentum; otherwise the dynamics could create or destroy net angular momentum, in contradiction to all experimental observations.

As powerful as global symmetries are, a second category of hamiltonian symmetries has even more remarkable properties. These symmetries can specify detailed properties of the physical system. They may be termed dynamical symmetries. The identification of such symmetries can yield a remarkable return on a minimal investment: global symmetries specify only what is permitted; dynamical symmetries allow a prediction of what actually happens.

Interacting boson model. Until the mid-1970s, global symmetries were staple items in nuclear physics but dynamical symmetries were not widely appreciated. This view began to change with the introduction of the interacting boson model (IBM), and the realization that this theory was associated with a rich set of dynamical symmetries. The interacting boson model was successful in describing broad ranges of low-energy nuclear structure data, but it had some difficulties. The model assumed that the particles making up a complex nucleus belonged to the general class of particles called bosons, but neutrons and protons (if their internal structure is not considered) belong instead to a second class of particles called fermions, which have properties differing fundamentally from those of bosons. The qualitative justification for using bosons rather than fermions to describe nuclear structure is that a pair of nucleons may exhibit many of the properties of a boson, and nuclei often behave as if they are composed of pairs of correlated nucleons.

Fermion dynamical symmetries. The empirical success of the interacting boson model, coupled with reservations about bosons in nuclei, led to the natural question of whether a fermion hamiltonian could exhibit similar dynamical symmetries. This question is not trivial; a fermion theory is more realistic but much more complicated than a boson theory. This question was answered in two stages. The first was the introduction of a mathematical

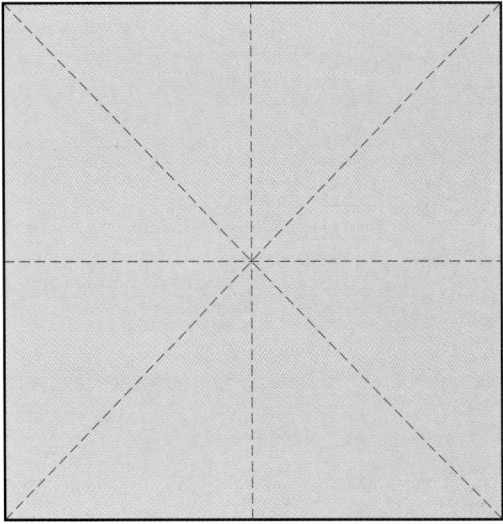

Fig. 1. Object with spatial symmetry. The square is invariant under rotations by integer multiples of π/2 radians (90°). The intersections of planes of reflection symmetry with the plane containing the square are indicated by broken lines.

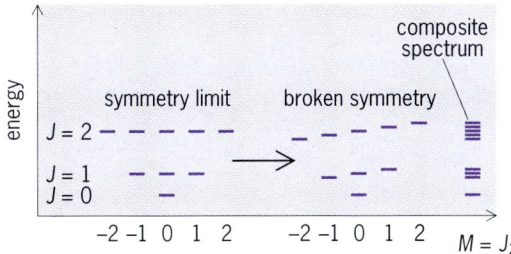

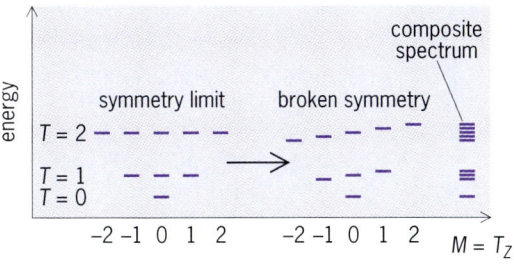

Fig. 2. Comparison of angular momentum and isotopic spin. (a) Each state of angular momentum J has $2J+1$ components labeled by the projection J_z. (b) Each state of isospin T has $2T+1$ components labeled by the projection T_z.

device termed the Ginocchio model, which was schematic but had two attributes of enduring interest: a hamiltonian consisting entirely of fermion operators, and dynamical symmetries very similar to those of the interacting boson model.

Thus, the Ginocchio model provided a proof of principle that fermion hamiltonians existed having many of the dynamical symmetries that are observed in nuclei. This proof of principle was expanded into a viable theory of nuclear structure in the fermion dynamical symmetry model (FDSM), which generalized and reinterpreted the Ginocchio model and provided the crucial connection between the fermion dynamical symmetries and the nuclear shell model. Extensive investigations with the fermion dynamical symmetry model have provided considerable insight into why a system of fermions can often behave as if it were a much simpler system of bosons with clearly realized dynamical symmetries. Furthermore, because this model starts from fermions, its dynamical symmetries provide a new truncation principle for the shell model that holds promise as a methodology for systematic shell-model calculations in all nuclei.

Unified picture of nuclear structure. An attractive feature of these dynamical symmetries is that they suggest relations among states lying in different nuclei. The basic idea is that a set of states or particles that are normally considered to be independent may actually correspond to a single state or particle, but with each viewed from different perspectives. Such ideas are attractive for their beauty and for their power. They are beautiful because they suggest that seemingly complicated systems may be simple if an appropriate mathematical vantage point is found; they are powerful because they imply that the solution of a single problem may contain the solution of many problems.

There is a prototype for such an approach. In the 1930s, various experimental observations indicated that the neutron and the proton behaved in many respects as if they were the same particle. To explain this phenomenon, an abstract quantity termed the isotopic spin, or isospin, was introduced, and it was proposed that the isospin was approximately conserved in the interactions of neutrons and protons. Thus a symmetry is implied that mathematically, but not physically, is exactly the same as the symmetry associated with physical spin for a particle (the angular momentum symmetry discussed above). The mathematical relationship between isospin and real angular momentum is illustrated in **Fig. 2**. Just as different magnetic substates of an angular momentum J may be viewed as the same intrinsic state but having a different projections on an axis of the physical space, the neutron and proton may be viewed as the same intrinsic particle, but with different projections on an axis in some abstract space that may be termed the isospace. Thus, there is a one-to-one mathematical analogy between spin and isotopic spin, even though physically the two concepts have nothing in common. Different values of the projection of isotopic spin, T_z, correspond to different proton numbers; thus, the isotopic spin states for a given isospin T and various values of T_z shown in Fig. 2 correspond to states that lie in different nuclei.

Isotopic spin is a useful symmetry for limited sets of states in light nuclei, but its utility is limited for heavier nuclei. However, the new hamiltonian symmetries discussed above can unify many states in broad ranges of nuclei, even for the heavy elements. In **Fig. 3**, the energy pattern for the states in

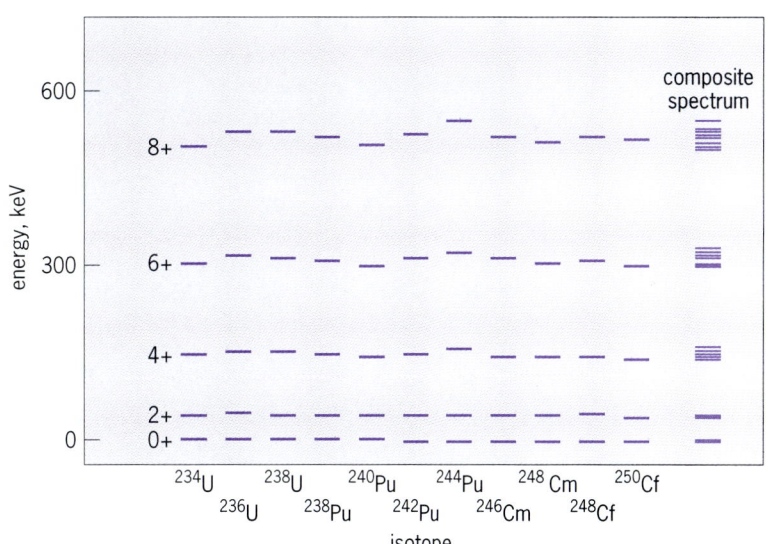

Fig. 3. Set of nearly identical rotational bands, with states labeled by the angular momentum quantum number and the parity (±). Experimental energies for isotopes of uranium (U), plutonium (Pu), curium (Cm), and californium (Cf) are shown. (*After M. W. Guidry et al., Some general constraints on identical band symmetries, Phys. Rev. C, 48:1739–1744, 1993*)

each of the nuclei shown is remarkably similar, a fact brought into sharp relief by collecting all states into a single spectrum on the right side. A comparison of Fig. 3 with that of an angular momentum in the presence of weak symmetry breaking (Fig. 2) suggests that a weakly broken symmetry that links many states in different nuclei may be responsible for this regularity.

The above is a simple example of a phenomenon termed the identical band problem. It has been demonstrated that fermion dynamical symmetries such as those found in the fermion dynamical symmetry model lead naturally to the kinds of similarities exhibited in Fig. 3. Whether such symmetries can account in general for identical bands remains to be seen, but the synthesis implied by a symmetry explanation for similarities such as those displayed in Fig. 3 provides an attractive basis for such an explanation.

For background information SEE ANGULAR MOMENTUM; ISOBARIC SPIN; NONRELATIVISTIC QUANTUM THEORY; NUCLEAR STRUCTURE; QUANTUM STATISTICS; SYMMETRY LAWS (PHYSICS) in the McGraw-Hill Encyclopedia of Science & Technology.

Michael W. Guidry

Bibliography. J. N. Ginocchio, A schematic model for monopole and quadrupole pairing in nuclei, *Ann. Phys.*, 126:234–276, 1980; M. W. Guidry et al., Some general constraints on identical band symmetries, *Phys. Rev. C*, 48:1739–1744, 1993; F. Iachello and A. Arima, *The Interacting Boson Model*, 1987; C.-L. Wu, D. H. Feng, and M. W. Guidry, The fermion dynamical symmetry model, *Adv. Nucl. Phys.*, 21:227–443, 1994.

Autoimmunity

Recent studies in autoimmunity have involved experimental, animal models, focusing on the role of infections in autoimmune diseases and the genetic basis of autoimmunity in mice.

Infection and autoimmunity. The elimination of pathogens is one of the fundamental tasks of the immune system. An array of nonspecific (including phagocytes) and specific (T cells, B cells) cells carry out this function. Under normal conditions, the immune system is capable of distinguishing between self and nonself, a prerequisite for eliminating pathogens without attacking the body's own components. Although autoreactive lymphocytes exist in healthy individuals, potent regulatory mechanisms control T- and B-cell unresponsiveness or tolerance to self antigens. Although genetic factors influence suscepibility to autoimmunity in humans and experimental animals, it has become evident that environmental factors contribute to disease penetrance. Infectious pathogens interact with the immune system of their host in many ways, adopting immunosuppressive tactics to escape quick elimination. These strategies often lead to a dynamic equilibrium between an invader and the immune response of the host, resulting in chronic infections. Epidemiological and clinical studies suggest that acute and chronic infections contribute to the pathogenesis of many immunologic diseases. Direct and indirect effects of pathogens on T and B cells can lead to the breakdown of tolerance and possibly result in autoimmune conditions.

Infections often precede the onset of clinical signs of autoimmune diseases. Many microorganisms have been implicated in the pathogenesis of autoimmune conditions such as multiple sclerosis, reactive arthritis, and rheumatoid arthritis. For example, demyelination of the nervous system is associated with a variety of viruses and bacteria. Viruses that have been associated with demyelination include *Lentivirus: Visna virus; Coronavirus:* murine hepatitis virus, *Togavirus; Paramyxovirus:* canine distemper virus; and *Picornavirus:* Theiler's murine encephalitis virus. Implicated nonviral pathogens include *Acanthamoeba, Borrelia, Brucella, Campylobacter, Hartmanella,* and *Trypanosoma*. In patients with multiple sclerosis, antibody titers are elevated against several viruses, including measles, influenza, herpes simplex, rubella, and Epstein-Barr. Recent advances in genetic engineering provide powerful tools to study the effects of currently unknown pathogens on the penetrance of autoimmune disease. For example, a transgenic mouse (having genetic material experimentally transferred to it) that expresses a T-cell receptor specific for myelin basic protein spontaneously develops experimental autoimmune encephalomyelitis (an animal model for multiple sclerosis) in a higher degree under pathogenic conditions than in a germ-free environment. Similarly, mice that have had interleukin 2 inactivated by the knockout technique and have been raised under germ-free conditions fail to produce an inflammatory bowel disease commonly seen in mice under less clean conditions. Taken together, these observations point to a crucial role of infectious microorganisms in the pathogenesis of autoimmune diseases.

Possible mechanisms. Several mechanisms have been proposed to explain the mechanism by which infectious pathogens influence the immune system to turn against the self structures of the body, thus initiating autoimmune disease.

Molecular mimicry. Immunological cross-reactivity, termed molecular mimicry, between antigens produced by pathogens and self structures of their host might lead to autoimmunity. Indeed, similarities between infectious organisms and self antigens have been detected. For example, there is molecular homology between proteins of measles virus or adenovirus and stretches of myelin basic protein, a structural element of the central nervous system assumed to be the target of autoreactive cells in multiple sclerosis. Moreover, the specificity of antigen recognition by T cells is not always absolute: it is sufficient for an antigen to share no more than 50% of the amino acids of myelin basic protein epitopes in order to induce autoreactivity. It is possible to induce experimental autoimmune encephalomyletis in mice with peptide antigens derived from infectious pathogens. Other antigenic cross-reactivities include neural structures and the bacillus *Campylobacter jejeuni* in the peripheral demyelinating disease Guillain-Barré syndrome and those between structural components of *Mycobacterium tuberculosis* and cartilage that occur in adjuvant-induced arthritis. SEE ANTIGEN.

Superantigens. Often, autoimmune diseases take a chronic course with frequent relapses and remissions of disease activity and clinical signs. Exacerbations of multiple sclerosis can be triggered by nonspecific viral infections of the upper respiratory tracts. Enterotoxins produced by the bacterium *Staphylococcus aureus* can induce relapses and exacerbations in mice that have recovered from paralysis induced by experimental autoimmune encephalomyletis. Similarly, a mycoplasma-derived antigen contributes to the pathogenesis of arthritis. Staphylococcal enterotoxins belong to a class of infectious pathogen-derived substances known as superantigens, which activate a wide range of T cells irrespective of their antigen specificity. This property enables superantigens to react with T cells under various inflammatory conditions, thus intensifying inflammation and facilitating relapses of clinical disease. Since autoimmune conditions such as arthritis, diabetes, and multiple sclerosis are characterized by heterogeneous T-cell infiltrates during active disease, a broad spectrum of superantigens could precipitate exacerbation. Organisms known to produce superantigens include streptococci, staphylococci, mycoplasma, toxoplasma, and retroviruses.

In addition, it has been suggested that unknown superantigens play a role in the induction of autoimmune disease by activating autoreactive T cells outside the target organs of autoimmunity. Because of their prior activation, these cells can enter body tissues such as the synovia of the joints and persist because of reactivation by autoantigens, ultimately resulting in a local inflammation in the joint and arthritis. Similarly, T cells reactive to antigens of the central nervous system could be activated by appropriate superantigens in the blood or gut of a patient. In an activated state, the lymphocytes can then cross the blood-brain barrier which functions as a gatekeeper that prevents most cells from entering the brain tissue. Once inside the central nervous system, autoreactive T cells could encounter an autoantigen such as myelin basic protein and produce multiple sclerotic lesions.

Cytokines. Soluble mediators of communication between cells, termed cytokines, play an important role in the regulation of immune responses. Certain patterns of cytokine production by lymphocytes, mainly T-helper lymphocytes, have been associated with autoimmunity. Infectious agents are able to regulate the direction of cytokine phenotype development, that is, a predominant expression of interleukin 2, interferon γ, and tumor necrosis factor α. This pattern of cytokine production by T cells is called the T helper 1 (TH1) phenotype. The preferential induction of TH1-type T cells with broad specificity by pathogens such as *Listeria monocytogenes* provides a mechanism for the modulation of immune responses by microorganisms. In clinical studies involving individuals with multiple sclerosis, it was shown that minor infections precipitate autoreactive responses to central nervous system antigens, possibly through elevated production of interferon γ and tumor necrosis factor, just before acute attacks. Pathogens associated with exacerbations include Epstein-Barr virus and herpes simplex virus.   *Stefan Brocke; Lawrence Steinman*

Defective apoptosis. During the early 1900s, it was recognized that the immune system had such an enormous capacity for antigen recognition that tightly controlled regulation was required to prevent recognition of self antigens. The term horror autotoxicus was coined to describe fear of the self-destruction that would ensue if this regulation were to fail. Although the necessity for such regulation was recognized, determination of the mechanism involved required a fuller understanding of the interactions involved in the immune system. Based on studies of graft rejection, F. M. Burnet formed the hypothesis that all progeny of those cells that were self-reactive were deleted, and thus established the concept of clonal deletion. Later, it was recognized that the immune system was composed of B cells and T cells which recognize antigens through the variable regions of T-cell receptors (TCRs). Improved methodologies, including the ability to produce hybridoma cells that produce monoclonal antibodies specific for a single epitope and radioimmune assay and enzyme-linked immunosorbent assay (ELISA), allowed the rapid and specific analysis of the production of autoantibodies by B cells and their role in autoimmune disease. Analysis of the T-cell compartment of the immune system was made possible by the molecular cloning of the various chains that form the T-cell receptors, notably the cloning of the β chain. The result was the production of T-cell receptor transgenic mice that expressed a T-cell receptor for a single, defined antigen on all T cells.

In the late 1980s, the first self-reactive T-cell receptor transgenic mouse was produced with specificity for the antigenic determinant that is defined by interaction of the major histocompatibility class I antigen complexed with the male (H-Y) antigen. In these mice, there was clonal deletion of self-reactive T cells that express the H-Y antigen in the thymus of male mice, but not in female mice, which do not express the antigen. At approximately the same time, it was demonstrated that clones containing certain self-reactive variable (V) regions of the β chain of the T-cell receptor (TCR Vβ) were eliminated in the thymus of mice treated with superantigens. Thus, it was clear that induction of clonal deletion was critical for achieving self-tolerance of T cells. Also in the late 1980s, it was established that certain autoreactive B cells and T cells could exist in a nonreactive (anergic) state. That is, they were autospecific but not autoreactive. For T cells, the interleukin-2 (*IL-2*) gene played an important role in the maintenance of T-cell anergy. In addition, transcription of interleukin 2 was associated with loss of tolerance and T-cell activation, whereas repression of the transcription of the interleukin-2 gene was associated with tolerance. Various molecules that are expressed on the T-cell surface have been shown to be of importance in the regulation of interleukin-2 gene transcription and the transcription of other cytokines. The cell surface molecules include CD28 and the CTLA-4 coreceptors, the CD40 ligand, and certain adhesion molecules.

Genetic basis of autoimmunity in mice. In 1992, the first autoimmune gene was identified and isolated in a mouse model for autoimmune disease. This gene, called lymphoproliferation (*lpr*), was found in mice that produced autoreactive T cells, autoreactive B cells, and autoantibodies. The mice developed lymphadenopathy and severe autoimmune disease resulting in early mortality. The *lpr* gene was a point mutation of the intracellular domain of the *fas* gene. This mutation of the *fas* apoptosis gene interrupted a programmed cell death signal that normally led to the death of lymphocytes. In 1994, it was demonstrated that the ligand for the Fas protein was mutated in a second strain of autoimmune mice known as generalized lymphoproliferative disease (*gld*) mice. Therefore, autoimmune disease in mice could be explained by mutation of either an apoptosis signaling molecule or its ligand.

Correction of autoimmunity transgenic mice. Although it was clear that autoimmune strains of mice had mutations in molecules that mediated programmed cell death, the mechanism by which these mutations led to autoimmune disease remained uncertain. One reason was that investigators were unable to detect defective apoptosis function in *lpr* or *gld* mice, despite the mutation. To directly demonstrate that a mutated Fas molecule can lead to autoimmunity, and conversely that correction of the defect prevents autoimmunity, the *fas* gene was used to produce Fas transgenic *lpr* mice. The transgenic mice expressed normal levels of nonmutated Fas, resulting in a complete correction of the lymphoproliferative disease, and a partial correction of autoantibody production and of the development of autoimmune disease. These results indicated that autoimmune disease could be caused by a mutation of the Fas apoptosis molecule. However, the apoptosis defect responsible for the manifestations of autoimmune disease can be difficult to detect as it is effective at specific developmental stages. Furthermore, it is apparent that autoimmune disease can be due to a relatively small leak of autoreactive T or B cells.

Regulation of apoptosis and implications. Although the genetic and molecular basis for autoimmune disease in mouse models of autoreactivity is being defined, the importance of apoptosis defects in human autoimmune diseases such as systemic lupus erythematosus and rheumatoid arthritis remains unclear. One mechanism for regulation of molecular interactions with the receptor is known to be the production of soluble receptors that block this interaction. It has been demonstrated that elevated levels of soluble Fas protein exist in the serum of patients with systemic lupus erythematosus. These soluble proteins were found to be normal receptors that are missing a single genetic exon (a portion of deoxyribonucleic acid) that encodes for the transmembrane anchoring region of Fas. The result is a soluble Fas molecule that can act as a competitive inhibitor for the ligand. Thus, one mechanism for the abnormal survival of unwanted self-reactive T or B cells in patients with systemic lupus erythematosus might be due to overproduction of this natural soluble inhibitor of the Fas-ligand interaction with Fas. Thus, therapeutic strategies might include use of a special plasmaphoresis column to remove soluble Fas from the serum of individuals with lupus. An alternative would be to interfere with the ability of the cells to produce soluble Fas. The activation of cells results in down modulation of soluble Fas and up modulation of surface-expressed Fas that is capable of mediating apoptosis. This fluctuation might explain the observed phenomenon of activation-induced apoptosis, and may be involved in homeostasis and elimination of cells after activation. Further understanding of this mechanism and possible defects in individuals with autoimmunity would lead to methods for increasing apoptosis after activation and consequently improving the elimination of autoreactive cells.

Other mechanisms of modulation of Fas-induced apoptosis include modulation of Fas molecule signaling. It has recently been shown that the presence of Fas and Fas-ligand is not sufficient for induction of apoptosis, and that the Fas molecular signaling pathway must be facilitated. One molecule important for Fas signaling appears to be hematopoietic cell phosphatase (HCP). Cell lines lacking hematopoietic cell phosphatase do not undergo Fas-mediated apoptosis, as was demonstrated in another mouse model of autoimmune disease, called the moth-eaten (me) mouse. Moth-eaten mice develop high levels of immunoglobulin and a very large spleen, and die at only a few weeks of age. These mice have been found to have a mutation of the *HCP* gene. After treatment with anti-Fas antibody, there was defective Fas-mediated apoptosis. Stimulation of the Fas pathway resulted in increased proliferation of cells, indicating that stimulation of the Fas molecule results in both a signal for proliferation and a signal for cell death. Thus, the hematopoietic cell phosphatase is critical for signaling cell death, and if it is mutated as in moth-eaten mice, autoimmune disease and overproliferation of immune cells can occur. Consequently, another potential mechanism for regulation of autoimmune disease in humans would be to augment the Fas signaling pathway to enhance or normalize apoptosis. In addition to hematopoietic cell phosphatase, Fas-mediated apoptosis can occur by activation of the enzyme sphingomyelinase, which converts sphingomyelin to ceramide. Regulation of Fas signaling may become an important mechanism for restoring normal apoptosis in human autoimmune disease.

Activation and apoptosis signaling. Activation-induced apoptosis is an important mechanism that allows the initial stimulation and proliferation of immune cells and the subsequent elimination of the cells according to the appropriateness of their response. Cells with too high or too low affinity for the interactive antigen, or cells with autoreactive potential, should be eliminated by activation-induced apoptosis. Determination of the underlying mechanism should lead to a better understanding of appropriate intervention for elimination of autoreactive T cells. Signaling through the T-cell receptor has been demonstrated to lead to proliferation and to apoptosis when the nuclear binding protein Nur77 is involved. Apoptosis mediated by this protein is critical for elimination of self-reactive T cells. In addition, appropriate engagement of costimulatory molecules such as CD4, CD28, CD40-ligand, and adhesion molecules appears to be critical to proper apoptosis signaling after T-cell activation. Thus, apoptosis signals can be regulated at the cell surface by inhibition of certain molecular interactions. This regulation has been demonstrated for the interaction of CD28/CTLA-4 with its counterreceptors B7-1 and B7-2. Autoreactive T and B cells can potentially be regulated by blocking signals that reduce T-cell apoptosis or by enhancing signals that increase T-cell apoptosis after activation, thereby leading to an overall decrease in the immune response and a decreased propensity toward development of autoimmunity.

For background information SEE ANTIBODY; ANTIGEN; AUTOIMMUNITY; GENE; HISTOCOMPATIBILITY; IMMUNITY in the McGraw-Hill Encyclopedia of Science & Technology.

John D. Mountz

Bacteria

Amber is fossilized tree resin that formed millions of years ago. Because of resin's sticky nature, a wide variety of plant and animal life forms were trapped and preserved in amber. Although amber has traditionally not been used extensively in the field of micropaleontology (the study of microscopic fossils), recent investigations have demonstrated that amber preserves many types of microorganisms in infinite detail.

Presence in amber. The presence of microorganisms in amber can occur by active or passive action, or simultaneously by both. Active action occurs when a resin flow actively covers microorganisms that are growing on the bark of a tree or thriving in stagnant water in bark crevices or tree holes. These actions normally preserve the vegetative or developing stages of microorganisms. Passive action occurs when microorganisms (including microspores such as pollen grains) are passively blown by the wind onto a mass of resin adhering to the bark of a tree. Passive action results in the preservation of resting or transport stages of microorganisms, such as spores and cysts.

Bacteria. Aside from bacterial spores which are blown by passive action into the sticky resin, vegetative cells and spores can occur inside the remains of insects and other forms of animal and plant life that have been entombed in amber. Such ancient bacteria can be observed directly in the amber with the aid of an electron microscope (**Fig. 1**). Although morphological features (shape, size, surface projections) can be useful in characterizing these bacteria and assigning them to modern-day genera, deoxyribonucleic acid (DNA) analysis is proving to be a more refined way to identify such fossil bacteria and compare them with present-day genera and species. Recent studies comparing a 546 base pair fragment of the 16S ribosomal ribonucleic acid (rRNA) gene from *Bacillus* species found in the abdomen of stingless bees in amber have shown how these fossil bacteria are related to forms found in extant tropical bees.

Sheathed bacteria similar to living representatives of the genera *Crenothrix* and *Spaerotilus* have been reported from amber dating back to the Triassic period. These filamentous gram-negative bacteria divide by binary fission within a mucilaginous sheath and commonly occur in aquatic habitats. Actinomycetes are gram-positive bacteria that form well-developed branching filaments. They have been found in amber from both the Tertiary and Cretaceous periods. Cyanobacteria, photosynthetic bacteria that release molecular oxygen as a consequence of their photosynthesis, have also been reported in amber. Forms resembling the extant genus *Scytonema* occur in Triassic amber.

Algae. A number of algal genera are aerial and grow on damp rocks and stones. Others are epiphytic on the bark of trees, and these latter have been found in amber. Large dense branching filaments

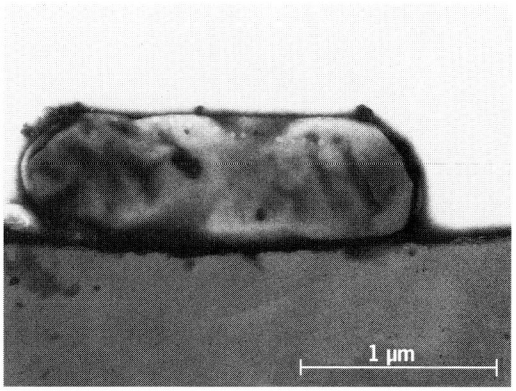

Fig. 1. Electron micrograph of a bacterium in Mexican amber (26 million years old). (*Courtesy of Roberta Poinar*)

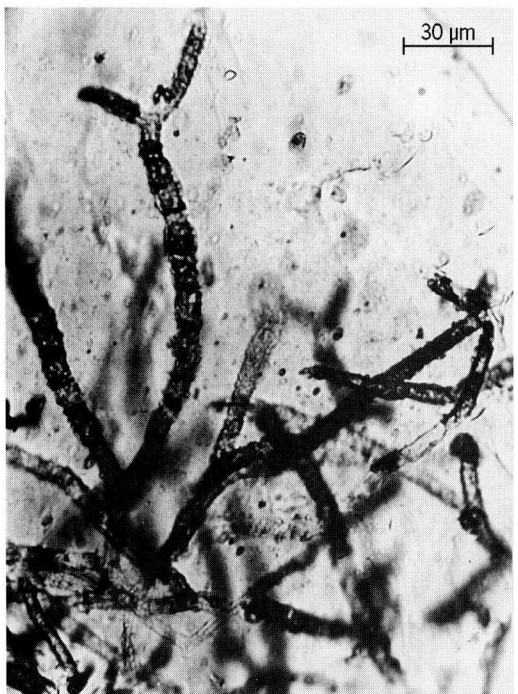

Fig. 2. Branching filaments of an aerial green algae similar to extant species of the genus *Trentepohlia* in Triassic amber (225 million years old).

of an aerial algae closely resembling extant members of the genus *Trentepohlia* have been discovered in Triassic amber (**Fig. 2**).

Slime molds. Biologists continue to have problems deciding how slime molds should be classified. Some place them in a separate kingdom, Protoctista, whereas others group them with the Protista. Since many slime molds grow on tree bark, they are sometimes covered and preserved in amber. A plasmodium resembling modern representatives of the order Physarales has been reported from Dominican amber, and the fruiting structure of a *Stemonites* was described from Baltic amber.

Protozoa. Although most protozoa are generally considered aquatic organisms, in some instances they occur in specialized habitats associated with cavities or crevices in trees. The delicate nature of amber preservation can be attested by the discovery of not just tests of protists, but developing stages of ciliates related to the modern genera, *Paramecium*, *Cyrtolophosis*, and *Nassula* in amber. Such discoveries represent the oldest known terrestrial fossils of these groups.

Perhaps the most spectacular find pertaining to fossil protozoa in amber was the discovery of a group of amebas with thick, irregular pseudopodia in Triassic amber. The activities of the amebas indicate that this small population was preserved instantaneously. Several of the organisms had their bodies extended as if they were moving along a water film (**Fig. 3**), another was in the process of dividing, and a third was ejecting from its body what appeared to be the indigestible portion of a diatom. These rare forms, named *Triassamoeba alpha* in recognition of being the earliest known terrestrial fossil amebas, give a snapshot view of life in an aquatic microhabitat that existed approximately 225 million years ago.

DNA studies. The ability of amber to preserve the internal DNA of entrapped microorganisms allows us to conduct detailed comparisons between fossil and modern-day forms of microbes. Such comparisons can be used to develop phylogenetic trees for evolutionary studies.

A recent study comparing DNA from a fossil microbe with DNA from a extant descendant showed that amber maintains the DNA integrity of microorganisms, in this case bacteria belonging to the genus *Bacillus*.

Certain species of *Bacillus* are known to occur in the digestive tract of stingless bees (Apidae: Hymenoptera) residing in tropical areas. Similar bees, but representing extinct species, occur in amber from the Dominican Republic. In order to determine if species of *Bacillus* also occurred in the digestive system of the extinct bees, 5–10 mg of the fossil bee's abdominal tissue were removed and the DNA was extracted. The very small amounts of DNA (2.5 nanograms/ml) were analyzed by utilizing the polymerase chain reaction (PCR). This revolutionary method enables the investigator to quickly obtain large quantities of specific nucleotide sequences (genes or portions of genes) directly, rather than attempting to estimate differences through DNA hybridization, electrophoretic analysis of isozymes, or restriction mapping. In the PCR

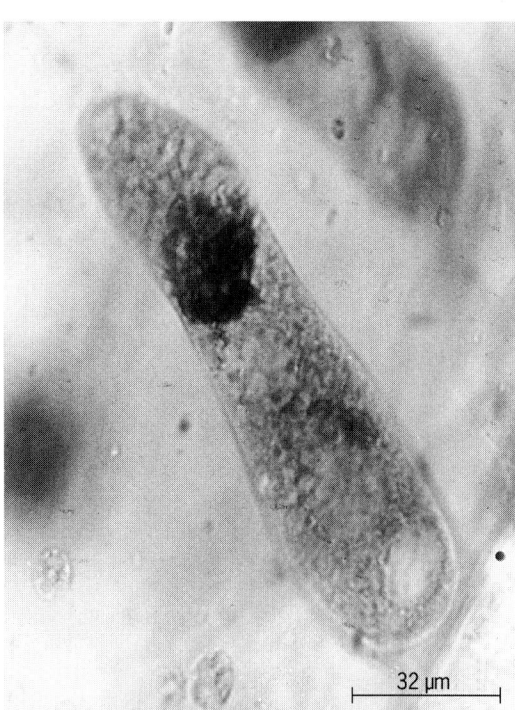

Fig. 3. Monopodial extension of *Triassamoeba alpha* in Triassic amber (225 million years old).

reaction primers or short nucleotide sequences match the portions of the gene that are to be amplified. The primers used in the bee *Bacillus* study were BCF1 and BCR2, which were obtained from the 16S rRNA sequence of *B. circulans* and amplify a 546 base pair region located between the 343 and the 889 base pairs of the gene. SEE CLINICAL PATHOLOGY.

The results of the DNA analysis of the abdominal tissue of the extinct bee showed that *Bacillus* DNA was indeed present. A comparison of this ancient DNA with corresponding sequences of modern DNA from the abdomens of extant bees showed a definite similarity. Although the DNA sequences originating from the extinct bees differed slightly from those in their modern-day descendants, the extinct sequences aligned or compared most closely with *B. circulans, B. pumilus, B. subtilis* and *B. firmus.* These bacterial species are associated with modern-day bees.

Thus, a relationship between the ancient and modern species could be established with the alignment of only 510 base pair sequences in the 5′ region of the 16S rRNA genes. Such molecular information can be used to establish relationships between ancient and modern bacteria, to show phylogenetic relationships, and to estimate the time needed for certain molecular changes. In this way, some idea of evolutionary change over time can be acquired.

For background information SEE ALGAE; AMBER; BACTERIA; DEOXYRIBONUCLEIC ACID (DNA); FUNGI; PROTOZOA; RIBONUCLEIC ACID (RNA) in the McGraw-Hill Encyclopedia of Science & Technology.

George O. Poinar, Jr.

Bibliography. R. J. Cano et al., *Bacillus* DNA in fossil bees: An ancient symbiosis, *Appl. Environ. Microbiol.,* 60:2164–2167, 1994; G. O. Poinar, Jr., *Life In Amber,* 1992; G. O. Poinar, Jr., B. M. Waggoner, and U. C. Bauer, Description and paleoecology of a Triassic amoeba, *Naturwissenschaften,* 80:566–568, 1993; G. O. Poinar, Jr., B. M. Waggoner, and U. C. Bauer, Terrestrial soft-bodied protists and other microorganisms in Triassic amber, *Science,* 259:222–224, 1993.

Barnacle

Barnacles have been described as the pinnacle of sessile evolution. Because most adult barnacles are sessile, the most critical point in their life cycle is in finding a suitable place to settle during the larval phase. This search is particularly important for species that are obligate cross fertilizers and that must, therefore, settle close to their own species in order to reproduce. Such settlement behavior is termed gregarious. Researchers have sought to understand, in particular, how the settlement-stage larva (the cyprid) is able to recognize its own species (conspecifics).

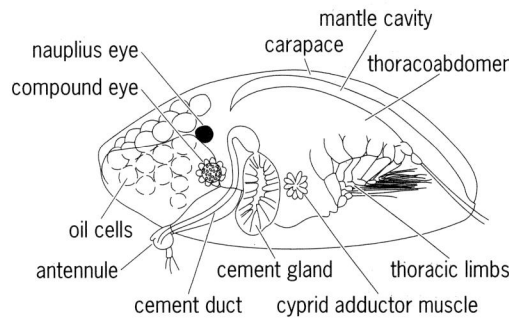

Main features of the barnacle cyprid larva. (*After L. J. Walley, Studies on the larval structure and metamorphosis of Balanus balanoides (L.), Phil. Trans. Roy. Soc. London, B256:237–280, 1969*)

Cyprid larva. The barnacle cyprid is a nonfeeding larva that is specialized for settlement. It is characterized by a bivalved carapace, compound eyes, lipid droplets (an energy store) and modified antennules, which hold the cyprid on to the substratum during searching behavior (see **illus.**). In effect, the cyprids "walk" with these antennules. Temporary adhesion is effected by an adhesive that is secreted onto the surface of the attachment disc by unicellular glands. Because the antennules are so closely associated with the substratum, they are believed to be the organs responsible for recognizing adult conspecific barnacles. Observations of the searching behavior of cyprids suggest that cyprids test the strength of adhesion between the temporary adhesive and the substratum; the cyprids seem to pull on their antennules. If a substratum is unattractive, the cyprid will swim off. This ability to discriminate between potential settlement sites decreases with time and is considered to be a response to depleted energy reserves.

Many factors influence the choice of settlement site, such as hydrodynamic conditions (passive settlement), and characteristics of the substratum that render it more or less attractive to the cyprid (active settlement). It is most likely that settlement is a combination of passive and active processes, the former acting on a large scale of the order of meters to kilometers and the latter on a relatively small scale of centimeters or less. Of the small scale factors, the biological character of the substratum is considered to be the most important.

Bacteria are usually (though not necessarily) the first living colonizers of a substratum. The bacteria produce exopolymers that entrap other bacteria, diatoms, and nonliving particles. The resulting biofilm has generally been considered favorable to barnacle settlement. However, recent research has shown that the result of the interaction between the cyprid and biofilm is dependant on the age of the biofilm: with increasing biofilm age, the effect on settlement reverses from inhibitory to facilitatory. However, the most potent modifier of barnacle settlement is the presence of conspecific barnacles.

Settlement pheromones. In the early 1960s it was shown that settlement of cyprids of one barnacle, *Semibalanus balanoides*, was induced by conspecific adults. The active factor (arthropodin) is a large glycoprotein that is believed to be a component of the barnacles' integument or cuticle. If water-soluble proteins are extracted from adult barnacles and applied to an inert substratum, they render that surface more attractive to cyprids with a corresponding increase in settlement over untreated surfaces. Cyprids of other species similarly respond to conspecific barnacles, but also to protein extracts of other species. The magnitude of the response is related to systematic affinity; that is, the more closely related the species, the greater is the inductive effect.

Pheromone recognition. Two theories have been advanced to explain how the cyprid detects the settlement pheromone. It is possible that the cyprid senses settlement pheromone through increased adhesion; measurements of the strength of cyprid adhesion to clean and arthropodin-treated surfaces support this theory. However, the finding that induction of settlement relates to systematic affinity supports the concept of a chemical sense rather than a physical (adhesion) mechanism of pheromone detection. Moreover, because those cyprids that have a prolapsed thorax (a condition which impairs both swimming ability and searching behavior) are still able to respond to settlement pheromone, increased adhesion is more likely a consequence than a cause of pheromone reception.

The alternative chemical sense theory of pheromone recognition was first suggested in the 1950s. However, recent evidence in support of this theory is suggestive rather than direct. For example, pharmacological studies reveal that some components of the signal transduction pathway of pheromone reception are common to other olfactory systems. Evidence is strongest for the involvement of cyclic adenosine monophosphate (cyclic AMP), although a G protein and calcium are also implicated in the pathway. Another finding indicative of a chemical sense is that the fourth antennular segments of the paired cyprid antennules are flicked through the water column while the cyprid is walking over the substratum. This behavior is reminiscent of decapod antennular flicking that increases stimulus access to the olfactory receptors. There are a number of bristly hairs (setae) on the fourth antennular segment that have at least an outward resemblance to crustacean aesthetascs, which are the crustacean olfactory receptors.

Pheromone reception needs to be examined with electrophysiological techniques. Recordings have recently been made from other, similarly sized, invertebrate larvae that have been challenged with chemical cues. The small size of the putative cyprid receptive structures should not, therefore, be an obstacle to their study.

Cyprid gregarious settlement. Cyprids are induced to settle by adult settlement pheromone, but gregarious settlement is still unexplained in the absence of similar adult species. One possibility is that among the cyprid progeny of a single adult, there is a population that is predisposed to the colonization of barnacle-free surfaces. Larvae of the tube worm, *Hydroides dianthus*, were recently shown to produce two types of larvae: one type colonizing uninhabited substrata, the other type settling in response to conspecifics. It is also possible that these substrata are colonized by old larvae that have lost their discriminatory powers. Finally, cyprid-cyprid interactions, expressed as a density effect in settlement assays, may be a factor. Cyprids are added to petri dishes with the compound of interest, and after a set period, settlement is quantified. For example, during the course of assays on the barnacle *Balanus amphitrite*, it was noted that dishes containing a relatively high number of cyprids had a greater percentage of settlements compared to dishes containing fewer cyprids. The effect is most marked with young cyprids.

The cyprid-cyprid interaction had previously been shown for the species *S. balanoides*. Although this species is more difficult to work with, researchers were able to show that substrata that had been in contact with cyprids were rendered attractive to other cyprids of the same species with a resultant induction of settlement. The cyprids were actually responding to so-called footprints of temporary adhesive left behind during cyprid searching behavior. Careful examination of glass slides stained with the protein dye reagent Coomassie blue revealed tracks of these footprints, which had a diameter and pace length that corresponded with predicted values. Subsequent studies with the more tractable species *B. amphitrite* have also confirmed the stimulatory effect of cyprid footprints on settlement of conspecifics.

One-half of a petri dish was treated with cyprid footprints. The dish was tilted so that the seawater meniscus spread halfway over the surface. Cyprids were then introduced and allowed to explore that half, and then removed; the dish was rinsed and then filled with seawater. When a single cyprid was then introduced into the dish, it could settle on either a footprint-treated or a surface free from the influence of other live cyprids. When a series of such dishes was set up with the appropriate controls, not only was the distribution of cyprids after a set period found to be biased to the footprint-treated side of the dish, but also settlement was found to occur exclusively on that side. That is, the footprints, by stimulating settlement, appeared to be acting as a pheromone.

The footprints of temporary adhesive may be a modified integumentary protein, perhaps similar in character to adult settlement pheromone. The respective proteins will first have to be isolated before this claim can be addressed.

Ecological implications. Footprints of *S. balanoides* have been found to persist on glass for more than 3 weeks, making possible a considerable buildup of temporary adhesive on a substratum with time. Substrata that are attractive to cyprids and stimulate exploration, perhaps by virtue of the presence of adult conspecifics, might be expected to gain a large number of footprints. A corresponding increase in the inductive stimulus could then lead to gregarious settlement. Certainly, the potential contribution of cyprid footprints warrants further investigation. Larval-larval interactions may well be found to be of more importance in the field than the intensively studied adult-larval interactions.

For background information SEE BARNACLE in the McGraw-Hill Encyclopedia of Science & Technology.

Anthony S. Clare

Bibliography. D. T. Anderson, *Barnacles*, 1994; A. S. Clare et al., On the antennular secretion of the cyprid of *Balanus amphitrite amphitrite*, and its role as a settlement pheromone. *J. Mar. Biol. Ass.*, 74:243–250, 1994; A. S. Clare and J. A. Nott, Scanning electron microscopy of the fourth antennular segment of *Balanus amphitrite amphitrite*, *J. Mar. Biol. Ass.*, 74:967–970, 1994; G. Walker, *Microscopic Anatomy of Invertebrates*, 1992.

Biological clocks

Although biological oscillations can range in period from milliseconds to years, a subset of these oscillations has evolved to match closely environmental periodicities. In this way, some endogenous oscillations have emerged as biological chronometers that allow organisms an opportunity to synchronize their physiology to environmental cycles in highly adaptive ways. The best known of these endogenous chronometers is the approximately 24-h biological clock. These clocks have been documented in most plants and animals, and even exist in primitive bacteria. They time a number of circadian phenomena ranging from the familiar daily activity cycles of many higher vertebrates to the time of emergence of adult *Drosophila* from their pupal cases, leaf movements in plants, and nitrogen fixation in some bacteria.

In addition to timing overt rhythmic behaviors, biological clocks play a more covert but critical role in Sun compass navigation by some birds and insects and also in the time measurement system that produces seasonal rhythmicity in reproductive cycles in plants and animals. Indeed, few processes within organisms escape the influence of the circadian timing system.

Environmental signals to set clocks. Before the 1950s there was serious debate over whether periodic behaviors in plants and animals are controlled by endogenous timers or by subtle geophysical cues within the environment that organisms are capable of detecting. Subsequently, sufficiently good scientific instrumentation was developed to confirm that when animals or plants are maintained under the constant conditions of steady temperature and continual darkness most behavioral rhythms do not occur in precise 24-h cycles but in slightly longer or shorter ones. The observation that rhythmicity was circadian, about 24 h but rarely exactly so, made implausible any exogenous timing hypothesis invoking precise 24-h signals derived from Earth's daily rotation.

The endogenous nature of the biological clock can easily be demonstrated by placing a rodent on a running wheel in which wheel movements are monitored by an event recorder. When the animal is placed in constant darkness, a rhythm in wheel-running activity is generated with a periodicity that, for nocturnal rodents, is typically less than 24 h. This periodicity, exhibited in constant conditions, is referred to as the free-running rhythm and reflects the natural periodicity of the organism's clock in the absence of environmental signals.

In order for the biological clock to be truly useful, it must be set to local time. This setting or synchronization process is referred to as entrainment, and leads to establishment of an exactly 24-h periodicity in the behavior. Entrainment occurs due to the action of periodic environmental signals that are involved in resetting the phase of the biological clock each cycle, much as the hands of a sluggish mechanical clock might be reset each day. An environmental cue that is effective in resetting the biological clock is referred to as a *Zeitgeber* (time giver). The entrainment process leads to control not only of the period of the biological clock, which is now forced to a 24-h cycle, but also of its phase. For example, when rodents are exposed to a 24-h light schedule (12 h of light and 12 h of darkness) not only is the wheel-running rhythm exactly 24 h but the phase of the activity rhythm (measured as the onset of wheel running activity) is also set by the entrainment process and begins at dusk (the precise phase varies as a function of the period of the light cycle and the duration of the photoperiod). Thus, entrainment controls exactly when the organism will be active and inactive during a day or night cycle.

Photoreceptors and pacemakers. Since the 1970s, much effort has been expended in attempts to identify the location of the biological clock in higher animals. Each clock system consists of at least three components, a photoreceptor system for entrainment, a circadian pacemaker, and effector processes that are directly responsible for generating the behavior being measured. In several mammals (and presumably in humans) the circadian pacemaker is located in the hypothalamus, among a group of brain cells referred to as the suprachiasmatic nucleus. This bilaterally paired brain structure, located just above the optic chiasm, is required for the generation of nearly all behavioral and neuroendocrine circadian rhythms. Recording electrodes placed inside the suprachiasmatic nucleus region of rodents reveals a circadian rhythm in electrical activity. It is possible to remove a brain slice containing the area that exhibits a circadian rhythm when maintained in tissue culture. It is also possible to restore circadian rhythmicity in rodents made arrhythmic with a suprachiasmatic nucleus lesion by implanting the fetal suprachiasmatic nucleus of another rodent. These results provide unambiguous evidence that this region is a circadian pacemaker in mammals.

In lower vertebrates the pineal gland appears to play the predominant role in circadian rhythm generation. For example, removal of the pineal gland from the house sparrow leads to a loss of circadian rhythmicity in perch-hopping behavior. Implantation of a pineal gland into a host whose pineal gland was previously removed restores behavioral rhythmicity. Furthermore, when removed from the bird and maintained in the laboratory in culture medium, the isolated pineal gland exhibits a circadian rhythm by releasing the hormone melatonin. In invertebrates circadian pacemakers have been localized to the brain, and in some cases, to the retina. In cockroaches and crickets the optic lobes of the brain contain the biological clock, whereas in some marine snails the clock is in the eye.

One surprising result regarding biological timekeeping is that photoreception for biological clock entrainment often occurs in structures other than the eyes. For example, blinded birds will entrain to light cycles, although entrainment can be blocked by preventing light from entering the skull. The eyes do apparently play a role, but there are extraretinal photoreceptors as well. Likewise, many arthropods do not require their eyes for rhythm synchronization, although some such as the cockroach do. Even in mammals whose eyes are clearly involved in entrainment, it is possible that a specialized set of photoreceptors is involved in this process.

Cellular and molecular mechanisms. Physiological study of model systems such as the opisthobranch mollusks *Aplysia* (the sea hare) and *Bulla* (the cloudy bubble snail) and molecular genetic studies of the fruit fly *Drosophila* and the bread mold *Neurospora* have yielded much information about the mechanisms responsible for generating circadian rhythms.

In *Aplysia* and *Bulla* the circadian pacemakers reside within the retina. The retina can be removed from the organism and maintained in tissue culture for several weeks. Under these conditions the retina generates a precise spontaneous rhythm in optic nerve impulse activity that is responsible (in the intact animal) for controlling the locomotor rhythm. In *Bulla* individual neurons have been removed from the retina and cultured. Electrophysiological recording from these isolated neurons, using intracellular microelectrodes, reveals circadian rhythms in membrane resistance. Thus, it has been suggested that circadian rhythms in the central nervous system of higher organisms are not due to neuronal circuits but rather are generated by cellular processes within individual neurons.

It has been possible to produce clock mutants through mutagenesis in *D. melanogaster* and *Neurospora crassa*. A hamster clock mutant, the *tau* mutant, and a mouse mutant, *clock,* have also been isolated. Chemically mutagenized *Drosophila* have been isolated with altered free-running periods in hatching and locomotor behavior. The changes in the circadian period are due to modification of a single gene, the period gene *per.* A second gene, timeless or *tim,* involved in the circadian timing system has also been identified. Similarly, in *Neurospora,* a single gene, the frequency gene *frq,* appears to play a critical role in circadian timing, and rhythmicity in the transcription of this gene appears critical for proper timing. In both *Drosophila* and *Neurospora,* clock genes have been identified, and are centrally involved in rhythm generation. In each case transcription of the gene is rhythmic, and in *Neurospora* rhythmicity of the ribonucleic acid (RNA) transcripts is definitely required for the circadian rhythm to be expressed. Thus, it is likely that a feedback loop centrally involving the *frq* gene, and other as yet unidentified genes, lies at the heart of the *Neurospora* biological clock.

For background information SEE BIOLOGICAL CLOCKS; BRAIN; GENE; PHOTOPERIODISM; PINEAL GLAND in the McGraw-Hill Encyclopedia of Science & Technology.

Gene D. Block

Bibliography. J. Aschoff, *Biological Rhythms: Handbook of Behavioral Neurobiology,* 1981; L. N. Edmunds, *Cellular and Molecular Bases of Biological Clocks: Models and Mechanisms for Circadian Timekeeping,* 1988; J. F. Hoffman and P. De Weer (eds.), Temporal organization: Reflections of a Darwinian clock-watcher, *Annu. Rev. Physiol.,* 55:17–54, 1993; M. C. Moore-Ede, F. M. Sulzman, and C. A. Fuller, *The Clocks That Time Us: Physiology of the Circadian Timing System,* 1982.

Biomass

Energy derived from biomass is receiving renewed interest worldwide. This interest is being driven by various concerns, including the continuing increase in worldwide fossil fuel consumption and its associated environmental impacts, energy security, and rural development. Biomass, in combination with advanced combustion and conversion technologies, has the potential to supply significant quantities of energy for transportation, power generation, and industrial processes.

It is estimated that biomass, consisting primarily of plant material, provides about 15% of the energy used worldwide, and an estimated 38% used in developing countries. In developing countries, it is mostly used for heating and cooking, which are inefficient means of converting fuel into useful energy. The increased use of biomass grown under sustainable conditions will improve rural development, enhance environmental quality, and reduce the net buildup of carbon dioxide (CO_2) in the

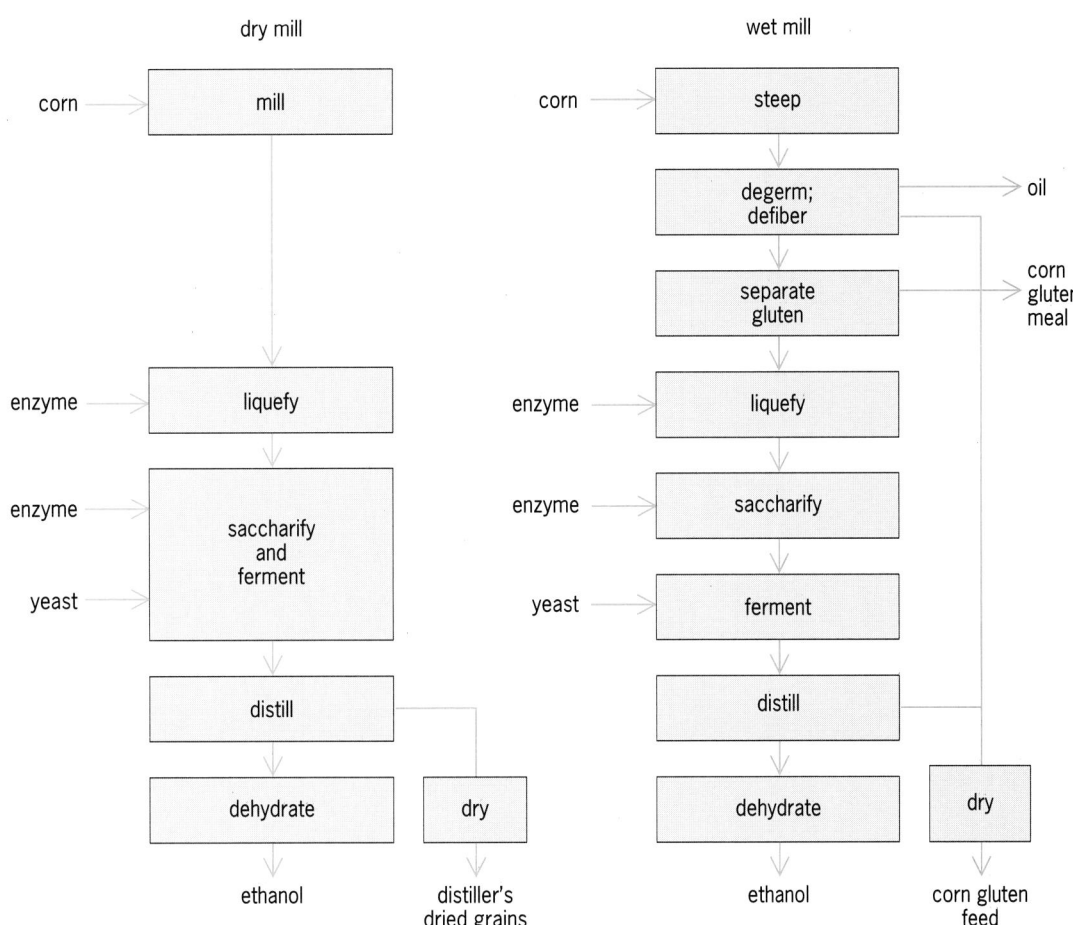

Fig. 1. Production of ethanol from corn by using wet- and dry-milling technologies.

atmosphere, because CO_2 released during combustion is partially offset by CO_2 extracted from the atmosphere during photosynthesis.

Some of the traditional technologies designed to convert biomass into energy involve direct combustion of wood for heating, cooking, generating electricity, and making ethanol from sugar and starch. Examples of other emerging technologies include converting lignocellulose material from agricultural and forestry crops and their residues into ethanol (CH_3CH_2OH), making biodiesel from oilseed crops and animal fats, and improving systems for generating electricity by using gasification (any chemical or heat process used to convert a substance into a gas) and advanced turbine technology.

Ethanol from grain. Ethanol is widely used in Brazil, where it is made from sugarcane, and to a lesser extent in the United States, where it is made primarily from corn because of corn's availability and high starch content.

Proven wet- and dry-milling grain-processing technologies are used to make ethanol from corn. Dry milling is based on traditional technology for the manufacture of potable alcohol, while the wet-milling process is based on refining of corn to starch and fructose, a very sweet sugar. Except for the initial separation process, the technology for the conversion of the starch to fuel ethanol is generally the same (**Fig. 1**).

Dry milling. Corn is milled and mixed with steam and enzymes to liquefy the starch component of the grain, which is converted to sugars by addition of other enzymes. The sugars are fermented to ethanol by using yeast. The mixture leaving the fermenter is distilled into 190-proof ethanol (95% ethanol and 5% water) and residues (whole stillage), and the ethanol is dehydrated to produce the 200-proof fuel grade. Water is removed from the whole stillage by mechanical and drying processes to produce the distiller's dried grains with solubles that are used as an animal feed. Current technology enables dry mills to produce approximately 2.6 gal (9.8 liters) of fuel-grade ethanol, 16.5 lb (7.5 kg) of distiller's dried grains with solubles, and 16.5 lb (7.5 kg) of carbon dioxide per bushel (35.2 liters) of corn.

Wet milling. In order to separate the corn into germ, fiber, gluten, and starch components, it is soaked in a mixture of water and sulfur dioxide. The soft kernels are then milled to separate the germ from the starch. The germ is dried, and the oil removed. The remaining slurry is screened to remove fiber, and then centrifuged to separate the gluten, which is used as a high-quality animal feed.

The starch is liquefied and converted to sugar, which is fermented, distilled, and dehydrated to produce fuel-grade ethanol. The thin stillage from distillation is combined with water used for steeping, and the solids are removed by evaporation. This mixture is added to the recovered fiber to produce corn gluten feed, which is used primarily for dairy cattle.

Ethanol yields from wet milling are slightly lower than yields from dry-milling plants. In addition to the 2.5 gal (9.5 liters) of ethanol per bushel (35.2 liters), there are about 1.7 lb (0.9 kg) of corn oil, 3 lb (1.4 kg) of corn gluten meal (60% protein), 13 lb (5.9 kg) of corn gluten feed (21% protein), and 17 lb (7.7 kg) of carbon dioxide.

Improvements in reducing the energy consumption for producing and processing corn into ethanol have resulted in a positive net energy balance. Reductions in energy use have resulted from efficiency improvements throughout the entire process, ranging from the cogeneration of electric power and steam to the use of corn grits and molecular sieves to remove the last 5% of the water in the ethanol. Improved enzymes for converting the starch to sugars permit a more efficient conversion at lower enzyme costs. New yeasts for fermentation have resulted in shorter fermentation times, higher levels of ethanol in the fermenter, lower residual sugars, and more sugar being converted. Higher ethanol concentrations have resulted in lower processing costs because less water needs to be removed.

One new approach for reducing the steeping time involves gaseous injection of sulfur dioxide. This treatment of the corn kernels facilitates the separation of starch and protein, reducing the steeping time from approximately 40 h to 8 h.

The current method for converting the grain starch to glucose sugar involves a series of processes using enzymes. New membrane technology to immobilize the enzymes will reduce the time from approximately 48 h to 5 h. By using biotechnology tools, new yeasts are being developed that are both more resistant to the concentration of ethanol in the fermenter and more heat-tolerant.

Ethanol from biomass. Ethanol is made from lignocellulose materials obtained from dedicated energy crops, agricultural and forestry residues, processing wastes, and domestic and animal wastes. Dedicated energy crops consist of annual and perennial plants and trees managed for high productivity using sustainable cultural practices. Annual and perennial crops are harvested once or twice per year, while the harvest interval for tree crops ranges from 5 to 12 years. Lignocellulose materials are composed primarily of cellulose, hemicellulose, and a binding material called lignin. In general, the cellulose component ranges from 30 to 50%, the hemicellulose from 25 to 35%, and the lignin from 10 to 30% of the biomass weight, depending upon the type of feedstock. The process for producing ethanol from lignocellulose materials (**Fig. 2**) is more complex than for corn or sugarcane, but the supply potential from dedicated energy crops is much greater.

Although both cornstarch and cellulose are composed of chains of glucose, the complex structure of cellulose makes the complete separation of all glucose molecules extremely difficult. There are two general methods for breaking the bonds between the glucose molecules. The first is acid-catalyzed hydrolysis of the biomass to sugars. The second method shown in Fig. 2 involves enzymatic hydrolysis that requires pretreatment of the feedstock to render the material more accessible to the enzymes. Combinations of mechanical and chemical pretreatments have been used. The enzymes break down the cellulose into sugars at ambient or slightly above ambient temperatures. Although reaction rates are slower than for acid hydrolysis and the raw material must be pretreated, enzyme hydrolysis gives greater sugar yields and results in fewer problems with waste product disposal.

Biodiesel. Biodiesel is a diesel-type fuel derived from biological feedstocks for use in unmodified compression ignition engines. Biodiesel is produced from vegetable oils (such as palm or soy), animal fats, used cooking oils, or combinations of all three sources. The raw oils and fats must be modified by

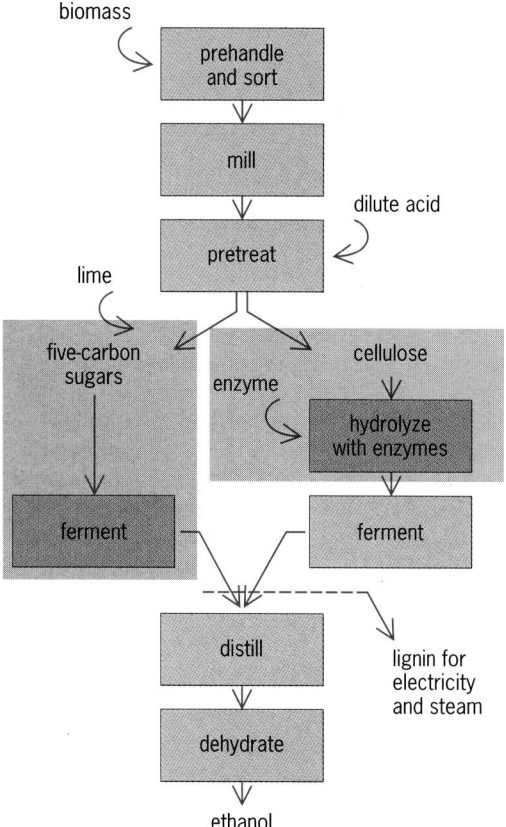

Fig. 2. Production of ethanol from lignocellulose material by using the enzymatic hydrolysis method.

chemical or thermal processes to reduce their viscosity and lower their boiling point. Currently, biodiesel manufacture in Europe and the United States involves a transesterification process long used in the fat and oil conversion industries to make soaps and cosmetics from triglycerides. The degummed oil is first filtered to remove water and other impurities, mixed with an alkali catalyst (either sodium or potassium hydroxide), and dissolved in an alcohol, usually methanol (CH_3OH). After the chemical reaction is complete, the glycerol is separated and the ester is washed with water or an acid to remove minute amounts of methanol, soap, and free fatty acids.

Neat (100%) biodiesel is biodegradable, nontoxic, and essentially free from sulfur and aromatics. Biodiesel and biodiesel blends reduce emissions of particulate matter, carbon monoxide, and total unburned hydrocarbons in proportion to the amount of esters; the higher the ester content, the greater the reductions.

Electricity. In the United States biomass is currently used to generate approximately 7100 megawatts of electricity. Recent estimates suggest that by using dedicated energy crops an additional 5 quads of generating capacity could be available by the year 2010. Wood is the largest source of fuel, followed by agricultural processing wastes. Roughly 68% of the biomass generating capacity occurs at cogeneration facilities (producing steam and electricity), or at paper and lumber mills. Approximately 28% of the capacity is at stand-alone nonutility biomass power plants, the remaining being obtained from plants operated by utilities using wood and wastes.

Advanced gasification technologies, developed for coal, to power a gas turbine offer greater potential. Biomass is more attractive than coal because it is easier to gasify and has a very low sulfur content, thus eliminating the requirement for expensive sulfur removal systems. Biomass integrated gasifier-gas turbine systems with efficiencies of 40% have already been demonstrated, and should be commercially available within a few years. They not only offer higher efficiencies but also lower capital costs at relatively modest scales of 100 MWe or less, even when higher cost biomass feedstocks are used. After the year 2000, conversion efficiencies as high as 57% may become feasible by using advanced fuel-cell technologies.

For background information SEE BIOMASS; BIOTECHNOLOGY in the McGraw-Hill Encyclopedia of Science & Technology.

W. Lamar Harris

Bibliography. L. Burnham, *Renewable Energy: Sources for Fuels and Electricity*, 1993; N. Hohmann and C. Rendleman, *Emerging Technologies in Ethanol Production*, USDA-ERS Agr. Info. Bull. 663, 1993; D. Klass, *Energy, Environment, Agriculture and Industry*, National Renewable Energy Laboratory, NREL/CP 200-5768, 1993; C. McGowin, *Strategic Benefits of Biomass and Waste Fuels*, Electric Power Research Institute, EPRI TR-103146, 1993.

Biosolids

Biosolids are organically derived solids that have been treated to ensure that their pathogenic content is undetectable or nominally acceptable, so that they can be safely recycled to the land for beneficial purposes.

The word biosolids is actually a new term for an old product. Residual solids remaining after the treatment of domestic wastewater have for many years been defined as sewage sludge. However, the term sludge conveys a negative image to a general public unable to differentiate between recyclable wastewater sludge and toxic sludges associated with hazardous wastes. To improve public acceptance of this product, the wastewater treatment industry and the U.S. Environmental Protection Agency (EPA) agreed to define more precisely recyclable wastewater sludge as biological solids, or biosolids. The latest available data indicate that more than 6.9×10^6 metric tons (7.6×10^6 tons) of dry biosolids are produced by all United States publicly owned treatment works annually. These totals are predicted to increase rapidly through the end of the twentieth century as a result of increasingly stringent standards for wastewater treatment and biosolids management. The new biosolids regulations that became effective in 1994 and the generally negative public reaction to proposed new or expanded biosolids application projects will only increase the challenge for biosolids management.

Anaerobic digestion. The proper treatment of wastewater in modern society results in a final liquid effluent that meets all applicable public health and environmental protection standards (see **illus.**). Each stage in the treatment process also results in an accumulation of solid organic material. In the primary stage, solids are settled out and concentrated prior to further processing. Wastewater effluent from this primary stage contains large quantities of soluble organic compounds that are converted in the secondary stage into a large mass of microorganisms. This biomass is settled and removed from the wastewater prior to discharge. Solid organic material from the primary and secondary treatment stages contains large quantities of biodegradable matter that is highly putrescible. Further treatment to produce a more stable, recyclable product is necessary prior to final use or disposal.

Although several physical, chemical, and biochemical alternatives are available to stabilize wastewater sludge, the most commonly used process is anaerobic digestion. Anaerobic digestion is the controlled breakdown of organic material by bacteria acting in an environment lacking in molecular oxygen. The microorganisms are mainly composed of facultative and anaerobic bacteria which derive their metabolic oxygen needs from chemically bound oxygen in the form of nitrates, sulfates, carbon dioxide, or organic compounds.

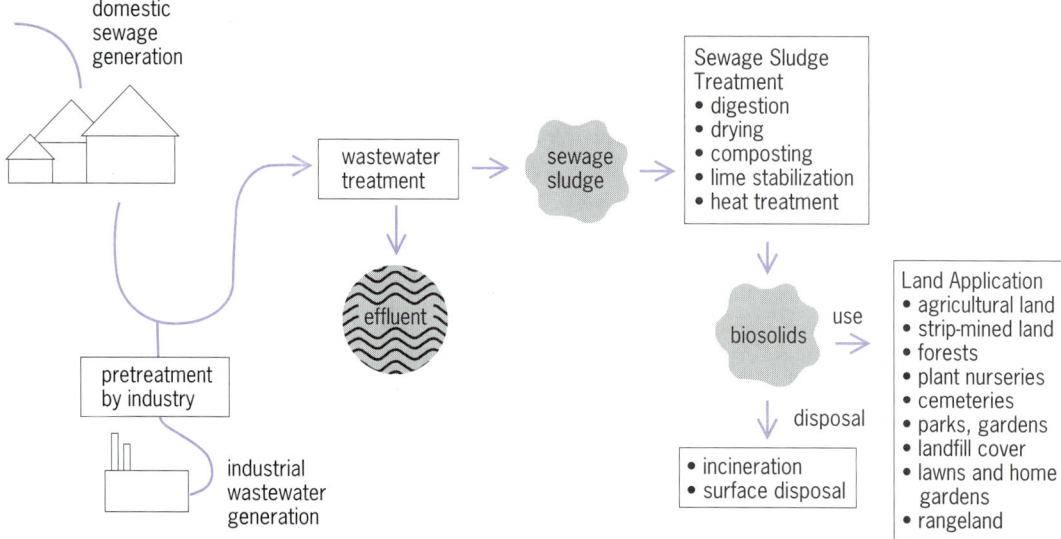

Generation, treatment, use, and disposal of wastewater sludge and biosolids. *(After U.S. EPA, Control of Pathogens and Vector Attraction in Sewage Sludge, EPA Pub. 625/R-92/013, 1992)*

The raw or unprocessed feed sludge is primarily composed of particulate matter that must be broken down into soluble end products before it can be utilized for microbial metabolism and growth. This multistage process begins with an acid fermentation stage, in which saprophytic bacteria attach to sludge solids and begin breaking down the organic solids by releasing extracellular enzymes. The first phase of digestion converts most of the solid organic matter into volatile organic acids, carbon dioxide, hydrogen, and bacterial cells. In the second phase, methane-forming bacteria begin converting hydrogen, carbon dioxide, and organic acids into methane gas. This phase is often referred to as methanogenesis, and it is responsible for significantly reducing the amount of organic matter and human pathogenic microorganisms within the total sludge mass. As with any functioning biological system, a wide range of physical-chemical conditions must remain within balance if the process is to continue efficiently. These key conditions include but are not limited to temperature, pH, alkalinity, nutrients, and the absence of significant concentrations of toxic compounds.

Anaerobic digestion has become the most commonly used sludge stabilization method because of its energy efficiency (the methane gas produced serves as the primary heating source for maintaining the required environment for mesophylic bacteria), its cost effectiveness, and its ability to reduce the amount of material that ultimately must be handled. Properly stabilized sludge meets the EPA definition for a process to significantly reduce pathogens (PSRP), and thus the material can be defined as biosolids.

Current challenges. In the past, biosolids were treated as a waste by-product, and the emphasis was on disposal rather than beneficial reuse. Most biosolids were thickened to a mudlike consistency (15–30% solids concentration) and dumped in sanitary landfills. Decreasing landfill disposal capacity is forcing many communities to divert significant quantities of waste materials out of the solid waste stream in order to extend the useful life of remaining landfills. At the same time, several large coastal cities, such as New York, Philadelphia, Boston, Seattle, and Los Angeles, can no longer discharge lower quality effluent and partially treated sludges directly to the ocean. Thus, the wastewater treatment industry has had to look for innovative strategies to manage increasing quantities of biosolids.

Recent innovations. In recent years, innovations in biosolids management have focused on two broadly defined approaches. The first approach involves processing the biosolids further in order to increase stability and decrease the concentration of human pathogens to nearly undetectable levels. Biosolids processed to this quality standard meet the EPA standard for a process to further reduce pathogens (PFRP) designation. The EPA currently approves seven different PFRPs, including composting, heat drying, heat treatment, thermophilic aerobic digestion, beta- and gamma-ray irradiation, and chemical pasteurization. Biosolids derived from these processes contain levels of pathogenic bacteria, enteric viruses, and viable parasitic worm ova that have been reduced to below detectability. The PFRP designation allows biosolids to be distributed and marketed with no significant limitations. Therefore, publicly owned treatment works can market their biosolids to the general public as a low-grade fertilizer or soil amendment. In addition, the PFRP designation allows unlimited use on agricultural lands that produce fruit and vegetable crops for direct human consumption, and use on public access areas such as parks and school grounds as a soil amendment.

The second approach involves identifying safe and acceptable uses for lower-quality PSRP biosolids. This distinction is frequently misunderstood by the general public. When EPA established new standards for biosolids use and disposal in 1994, the agency established standards of use for both PSRP and PFRP biosolids. PSRP classified biosolids have more limited acceptable applications and uses. However, if all controls and limitations are met the overall level of risk to public health and the environment is essentially the same as for the higher-quality PFRP biosolids.

Many publicly owned treatment works have found significant economic efficiencies associated with innovative new uses for PSRP biosolids. Two of these innovations involve silviculture and rangeland management.

Silviculture. The Municipality of Metropolitan Seattle, Washington, has been a leader in researching and establishing the benefits of recycling biosolids in forestry applications. Dewatered biosolids are trucked into forest application areas, where they are mixed with water to form a liquid slurry and then sprayed under high pressure onto groves of relatively young coniferous trees. Nitrogen and other trace minerals in the biosolids promote rapid tree growth, allowing the trees to be harvested in a significantly shortened time frame by the forest producer. A side benefit to this process is the equally enhanced growth of shrubs and other low-growing vegetation in the understory, thus reducing runoff and erosion from the sites.

Rangeland management. Another innovation generating great interest involves the application of PSRP biosolids to arid and semiarid rangelands in the western United States and Canada. Research and operating projects throughout the region have achieved significant improvements to overgrazed or otherwise degraded rangelands when biosolids are applied to the surface as a top dressing. The concept was developed and defined as part of a long-range soil erosion control project in the Rio Puerco Watershed in western New Mexico. U.S. Department of Agriculture (USDA) scientists concluded that plant biomass increased significantly for several years after the initial applications were made. The applications also increased the crude protein content of the grasses favored by grazing livestock, and decreased the amount of nonpalatable toxic range plants. Trace metals did not increase in the plant tissues, and no negative environmental impacts were identified.

Follow-up research conducted in the early 1990s focused on potential water quality impacts due to runoff from recently applied sites. USDA soil scientists conducted rainfall simulation experiments on small test plots located within the Sevilleta National Wildlife Refuge, south of Albuquerque, New Mexico. Simulated rainfall was applied at a rate of 8 cm/h (3 in./h), replicating the conditions of a severe thunderstorm. Runoff water was collected and analyzed for nitrate-nitrogen, copper, and cadmium concentrations during both natural and simulated rainfall events. All concentrations were within both state and federal water quality requirements.

The City of Fort Collins, Colorado, in conjunction with Colorado State University, EPA, and other agencies, has been studying the effects of rangeland biosolids application on a 10,500-hectare (26,000-acre) operating cattle ranch. Research has been conducted on the effects of biosolids application on soil, vegetation, surface water, ground water, and air. Vegetative test plots established early in the study received surface applications of dewatered biosolids at rates of 0, 2.2, 4.5, 11.0, 22.0, and 34.0 dry metric tons/ha (0, 1, 2, 5, 10, and 15 dry tons/acre). Equivalent rates of composted biosolids that met the PFRP designation were also applied to separate plots. Both sets of plots showed results similar to those found in the New Mexico research study, as was the case for a runoff study conducted on moderate to steeply sloped terrain (8 and 15% slopes). Additional studies at the site included an evaluation of respirable dust associated with applied biosolids; a wildlife and plant species survey, with an emphasis on threatened and endangered species; an analysis of wetlands and riparian areas; and an evaluation of existing and innovative grazing management practices.

The first research phase of the project was completed in 1994, and the project was expanded to a larger-scale demonstration phase. The City of New York has also become a major innovator in the application of biosolids to rangeland. New York was forced to halt dumping of biosolids into the Atlantic Ocean in the late 1980s and early 1990s. One of the short-term solutions was to privatize the transport and recycling of the biosolids to other states. One of the private contracts involves shipping large quantities of dewatered biosolids to eastern Colorado by rail, where a portion has been applied to degraded rangeland and other highly eroded areas. The other major contractor has purchased a 51,700-ha (128,000-acre) ranch in extreme west Texas, where large quantities of rail-shipped biosolids are being applied and monitored by researchers from Texas Tech University.

Other experimental rangeland biosolids applications and studies are under way near Portland, Oregon; Vancouver, British Columbia; Albuquerque, New Mexico; Denver, Colorado; and Bakersfield, California. All of these projects are located in arid to semiarid climates on overgrazed and eroded soils that are deficient in nutrients and organic matter. If research continues to support this application method it could be used to recover many thousands of acres of overgrazed western rangelands, significantly improving soil and water quality, and improving forage conditions and wildlife habitat.

For background information SEE MICROBIAL ECOLOGY; RANGELAND CONSERVATION; SEWAGE DISPOSAL; SEWAGE SOLIDS; SEWAGE TREATMENT; SILVICULTURE in the McGraw-Hill Encyclopedia of Science & Technology.

W. Thomas Gallier

Boring and drilling (prospecting)

In recent years a number of advances have been made in directional drilling. Drill holes seldom proceed straight to their target, generally wandering or migrating from their prescribed path. Thus, small down-hole survey devices [less than 4-in. (102-mm) diameter] are used to measure the orientation (azimuth and dip) of a drill hole systematically, so that it is possible to determine the direction and magnitude of the deviation. The challenge of directional drilling is to measure drill-hole deviation and employ down-hole tools to guide the holes to their targets. Directional drilling involves the careful aiming of boreholes through use of oriented wedges and down-hole motors whose attitude (dip angle and azimuth) can be controlled. In exploration and mining, mineral targets have become deeper because of exploitation of near-surface deposits; consequently deep drilling has created a need to control more carefully the direction of drill holes. In deep drilling at greater than 2000 ft (630 m), wedging and directional drilling can be used to intersect several targets so that the overall amount of drill footage is reduced—as opposed to starting a new hole for each target (**Fig. 1***a*).

Research in directional drilling for the minerals and mining industry is focused on the development of efficient small-diameter tools for surveying and guiding drill holes to their targets.

Standard technology. As a hole is drilled by using standard coring methods, the trajectory of the hole will normally deviate from a straight path because of several factors, including the fabric or layering of the rock, rotation of the drill rods, pressure on the bit, bit wear, and drill-rod flexibility. In mining and exploration drilling, it is general practice to penetrate mineral-bearing strata or veins at an angle perpendicular to the major planar direction, thus permitting a determination of the true grade and width parameters of mineralization. Because rock is often tilted, folded, or with a pervasive planar metamorphic fabric, the drill bit intersects the rock fabric at an angle. If the angle of intersection is acute, on the order of 15° or less, the drill bit will tend to seek a path parallel to the rock fabric. At higher intersection angles (15–90°; as in the majority of cases), the drill bit will tend to turn into the rock fabric in order to intersect the planar fabric at a right angle. In rock masses with homogenous fabrics such as granite or diorite, drill holes may wander from their desired path because of effects brought about by the rotation of the drill string. Thus, the earth scientist or driller is confronted with forces that turn the hole from its intended direction in all but a few special geometric cases.

Advanced techniques. Sophisticated directional drilling techniques, long employed by the petroleum industry, have recently become available to the mining and minerals exploration industry, in part because of need induced by deep drilling projects, and in part by development of small-diameter (less than 6 in. or 15 cm) down-hole motors. Typically, exploration holes in the minerals industry are much

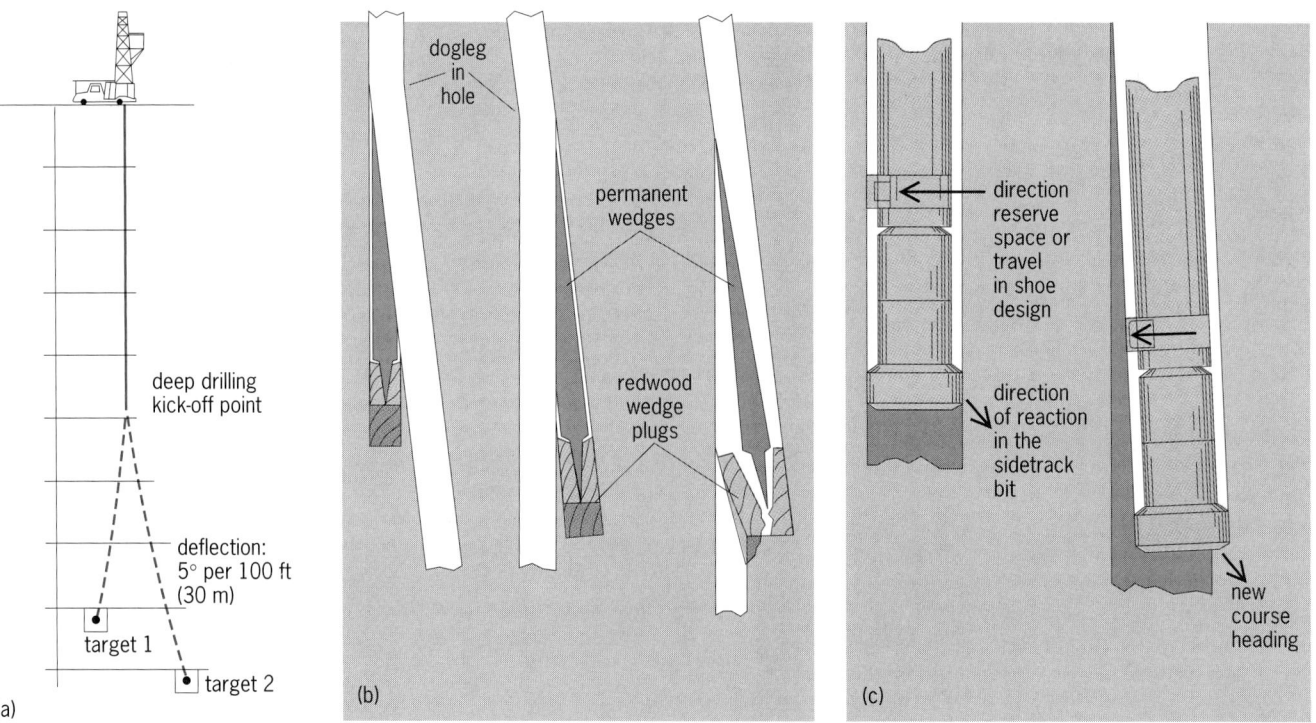

Fig. 1. Methods of wedging or kicking off a drill hole. (*a*) Multiple intercepts from one master hole. (*b*) Use of conventional steel wedges, illustrating wedge failure. (*c*) Addition of a deflection shoe when drilling off a cement plug by using a positive-displacement motor. *(Christensen Boyles Corp.)*

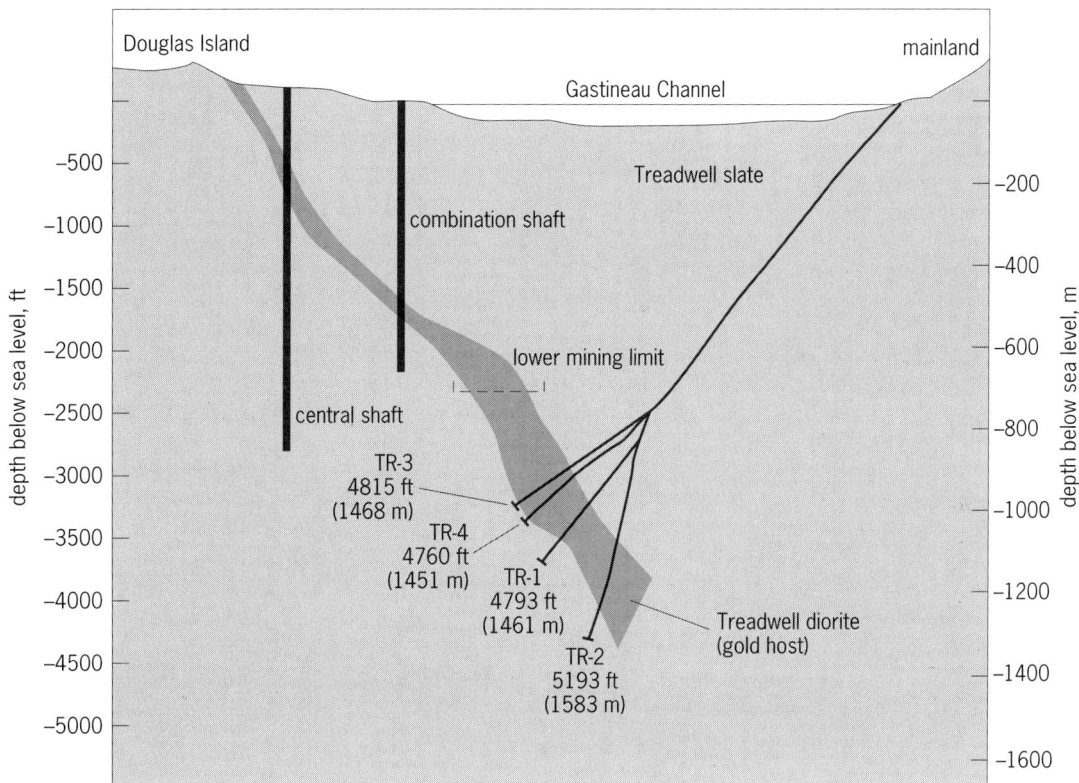

Fig. 2. Vertical cross section showing attitude of the Treadwell gold-bearing diorite and the trajectory of drill holes, TR-1–TR-4, from the deep drilling project. *(Echo Bay Mines)*

smaller in diameter than those used in petroleum exploration. For shallow drilling, where relative deviations are small, oriented steel wedges (Fig. 1*b*) are used; although a number of drill-string trips (5–6) are necessary to plug the hole, set the wedge, and drill from the wedge, each trip is not time consuming. In deep drilling, however, it becomes more efficient practice to use a down-hole motor to deviate the hole, since the rods can be left in the hole during most of the exercise. Use of a down-hole motor is also appropriate where a severe deviation is called for over a relatively short hole length. In such instances a large number of steel wedges would be required; if any were to come loose, the drill hole could be lost.

Steerable down-hole motors (also known as positive-displacement motors) transfer mechanical energy in the form of pressurized drilling fluid into rotational forces through the motor to turn a rotary bit at the drilling face. The drill rods (string) remain stationary, with the motor providing rotation of the drill bit only. The bit is offset from the drill string by using either a bent housing or offset stabilizer, causing the hole being drilled to curve (Fig. 1*c*). Down-hole surveys of the azimuth and dip angle of the hole path are regularly measured, and the hole is steered to the desired target or heading. Positive-displacement motors are powered by pressurized drilling fluid consisting of water, mud, or hydrocarbons, sometimes mixed, using a power-generation section consisting of a rotor and stator. Together they act as a system of gears with the rotor inside the stator. The metal rotor usually has one less gear (lobe) than the elastomer stator; under pressure the fluid applies a rotational force on the rotor. Torque produced by the motor is proportional to the differential pressure of the drilling fluid pumped through the motor (top to bottom), and the rotation speed is proportional to the volume of mud pumped through the positive-displacement motor. Because of the inverse relationship between torque and motor diameter, the slimmest positive-displacement motor currently in use has a diameter of 1.75 in. (45 mm). However, for most deep drilling applications, motors with a 2.375-in. (60-mm) or more likely 2.75-in. (70-mm) diameter are used. A typical small diameter positive-displacement motor operates at a rotational speed of 500–1000 rpm and a torque of 85–120 ft-lbs (115–163 joules), with a bit pressure of 85 lb/in.² (587 kilopascals); flow rates through the motor range from 30 to 50 gal/min (1.9 to 3.2 liters/s), producing a motor pressure differential of approximately 800 lb/in.² (5520 kPa). These motors are capable of effecting rates of curvature that, depending upon rock hardness, vary upward between 6.5 and 10° per 100 ft (30 m) of advance. Higher rates of curvature are also possible.

Directional drilling in minerals exploration. An example of directional drilling is the work done in the Treadwell mine, located 1 mi (1.6 km) from the mainland on Douglas Island in the Alexander

Archipelago in southeast Alaska. The gold deposit, roughly tadpole-shaped in plan view, was exploited from 1880 to 1917, yielding 3.2×10^6 troy ounces (100,000 kg) of gold. Recently, a deep directional drilling program was conducted at Treadwell to determine if the deposit continued uninterrupted down dip below the limit of previous mining. The deposit strikes northwest and dips northeast at an average of 53° from the horizontal (**Fig. 2**). Complications involved with drilling the target include (1) a 60° southeast inclination of the northern lode of the deposit, (2) lack of exposure of the hanging-wall rocks, and (3) limited drill-site selection caused by the proximity to Gastineau Channel, a fiord. Consideration of these elements and of the need to intersect the deposit at right angles, led to the selection of a drill site on the mainland side of Gastineau Channel approximately 1 mi (1.6 km) from the target (Fig. 2).

The drilling plan at Treadwell entailed drilling a master hole 3.75 in. (95.3 mm) in diameter to a target depth of approximately 4800 ft (1464 m) at a hole angle of 53° from the horizontal. Because of the fabric orientation of the hanging-wall Treadwell slate, the master hole began to deviate from its planned trajectory; this deviation required a correction at a depth of 1500 ft (458 m) and again at a depth of 3000 ft (915 m). At 1500 ft (458 m) the hole had flattened in dip by 5.5° (from −53 to −47.5°) and swung in azimuth from 226 to 222°; the correction required a steepening of the dip by 7.5° and a change in azimuth back to 225°. Both corrections were completed by using a positive-displacement motor that achieved a rate of curvature of approximately 2–2.5°/100 ft (30 m). Because of inflow of seawater, which adversely altered the chemistry of the drilling fluids, the hole was cased to a depth of 3000 ft (915 m), at which point it was reduced in diameter to 3.1 in. (79 mm). The master hole was completed at a trajectory depth of 4793 ft (1462 m), intersecting the target within 100 ft (30 m) of the planned intersection (well within specifications). Much greater accuracy could have been achieved, but was not required. After completion of the master hole, three additional drill intercepts were obtained by kicking off (wedging) at depths of approximately 3000 ft (900 m; Fig. 2). At each kick-off point the hole was wedged by using a directionally set steel wedge (1.5°) followed by use of a positive-displacement motor. All these kick-off holes were guided successfully to their respective targets with good success.

For background information SEE BORING AND DRILLING (GEOTECHNICAL); BORING AND DRILLING (MINERAL); OIL AND GAS WELL DRILLING; PROSPECTING in the McGraw-Hill Encyclopedia of Science & Technology.

Rick S. Fredericksen

Bibliography. Drilex Systems, Inc., *Motor Operations Handbook,* 4th ed., 1992; M. Stenberg, Directional drilling: Four case studies, *Eng. Min. J.,* 193:70–75, September 1992.

Carbon nitride

Developing new materials that possess useful mechanical, electrical, and optical properties is a central goal of solid-state chemistry and materials science. To realize this goal requires the development of synthetic methods for the assembly of atoms and clusters into specific three-dimensional arrays that provide the physical properties of interest. Traditional materials synthesis techniques, which require high-temperature conditions in order to facilitate diffusion of atoms in the solid state, generally preclude the rational assembly of atoms since the products are restricted to those phases stable at high temperature. For example, heating graphite in molecular nitrogen does not provide carbon nitride (C-N) solids. However, binary carbon-nitride solids are desirable materials to prepare and study since they may possess extreme hardness and high thermal conductivity. Extreme hardness is important in thin-film coatings of cutting tools used for machining and in bearing surfaces used in a variety of high-performance mechanical devices, whereas high thermal conductivity is important to the fabrication of advanced microelectronics devices. To overcome the limitations imposed by classical methodologies of solid-state chemistry and provide access to these potentially exciting materials requires better control of the reactants and reactions that might lead to carbon nitride solids. A new experimental approach that meets these requirements and has provided access to carbon nitride materials was discovered in 1993. This approach utilizes pulsed-laser evaporation of graphite to generate reactive carbon fragments and an atomic nitrogen beam as a source of nitrogen that can readily react with the carbon fragments.

Preparation of carbon nitride materials. The synthesis of covalent carbon nitride solids was explored previously, although with little success. For example, methane was decomposed in a radio-frequency plasma reactor containing molecular nitrogen (N_2). However, the thin films produced by this procedure were polymeric carbon-nitrogen-hydrogen (C-N-H) materials. Polymeric organic C-N-H solids were also obtained from the decomposition of nitrogen-containing organic precursors at elevated temperature and pressure. These results are understandable since C-H and N-H bonded products are thermodynamically more stable than the desired C-N and molecular hydrogen (H_2) products. Furthermore, these results suggest that hydrogen should be excluded from reactions designed to produce inorganic carbon nitride materials under equilibrium conditions. A very different approach that has been attempted in the preparation of carbon nitride involves shock-wave compression of organic C-N-H precursors. The rationale behind this approach is that dense, high-pressure phases can be obtained from shock compression of solids. Unfortunately, the only crystalline product observed in these shock

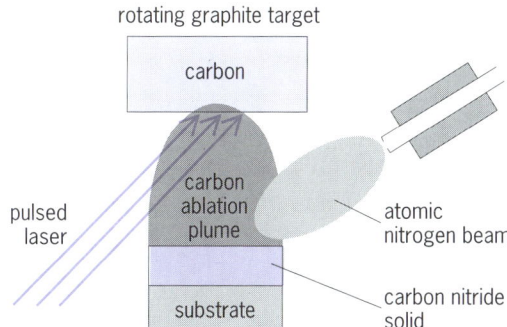

Fig. 1. Apparatus for the synthesis of carbon nitride.

compression experiments was diamond. This observation indicates that at ultrahigh pressure and high-temperature diamond and N_2 may be more stable than a covalent carbon nitride solid.

The new synthetic approach that has been developed is outlined schematically in **Fig. 1**. The apparatus contains a rotating graphite target that is irradiated with the focused output of a pulsed laser, and the resulting carbon ablation plume is directed toward the substrate. The atomic nitrogen beam intersects the carbon fragments at the substrate surface to produce the carbon nitride solid. The size and energies of the carbon fragments can be controlled through variations in the laser wavelength and power density, thus providing significant flexibility in the properties of the carbon reactants. Of perhaps greater importance is the choice of nitrogen reactant used in the synthesis: to avoid thermodynamic constraints imposed by molecular nitrogen or nitrogen-hydrogen compounds, atomic nitrogen is utilized. In the apparatus shown in Fig. 1, an atomic nitrogen beam is generated by using a radio-frequency discharge within an alumina nozzle through which N_2 seeded in helium (He) flows. This process can produce a very high flux of atomic nitrogen. Furthermore, by varying the N_2:He ratio and the radio-frequency power, it is possible to control both the flux and energy of this critical reactant. Hence, this synthetic approach provides a ready means of producing reactants that are free from impurities and that have controllable energies.

Composition. The synthetic control afforded by the new approach is clearly evident from compositional analyses of thin films prepared using varying fluxes of atomic nitrogen. The C-N composition in these materials was quantitatively determined by using Rutherford backscattering spectroscopy. Materials prepared by laser ablation of graphite at a 532-nanometer wavelength with the radio-frequency discharge off or in a background of 200 millitorr (27 pascals) of nitrogen consisted only of amorphous carbon; that is, molecular nitrogen does not appear to react with the carbon fragments produced by laser ablation. However, the C:N ratio systematically decreases (increasing nitrogen content) as the flux of atomic nitrogen is increased. The highest percentage nitrogen obtained initially in these studies by using a frequency doubled neodymium-yttrium-aluminum-garnet (Nd-YAG) laser (532 nm) was 40%, which corresponds to a stoichiometry of C_3N_2.

However, it cannot be inferred that there is a single carbon nitride phase of stoichiometry C_3N_4, since the composition measurements are averaged over the entire sample. Indeed, structural studies of these carbon nitride materials indicate that the solid product consists of both amorphous carbon and carbon nitride; thus, the carbon nitride phase may contain more nitrogen than expected from an average C_3N_2 stoichiometry. To assess the true composition of the carbon nitride phase, the effects of laser wavelength and power density on the observed composition of carbon nitride have been studied. Initial data show the percentage of nitrogen increasing as either the wavelength or power density decrease. The maximum nitrogen concentration observed after optimizing these experimental parameters is 50%. These results suggest that the carbon nitride composition could be CN.

Structure. The structural properties of the carbon nitride materials produced by the laser ablation approach have been investigated by using a variety of techniques, including x-ray photoelectron spectroscopy, infrared spectroscopy, and transmission electron diffraction. Photoelectron and infrared spectroscopies both show that carbon and nitrogen are bound covalently within this solid but cannot provide information about the three-dimensional structure. Structural information was obtained from electron diffraction. These latter studies show that the new carbon nitride films exhibit diffraction rings, although samples prepared by using only molecular nitrogen did not. Additional insight into the carbon nitride solid produced by this approach was provided by diffraction investigations of materials containing varying average nitrogen contents (determined by Rutherford backscattering spectroscopy). This work showed that the same diffraction ring pattern was obtained from carbon nitride samples with average nitrogen contents of 25–40%. These results suggest that a single crystalline carbon nitride phase was obtained by using the new synthetic strategy.

The structure of this crystalline carbon nitride phase has been inferred from comparisons to the diffraction patterns calculated for hypothetical carbon nitride structures. In one case, the six diffraction rings experimentally observed can be consistently indexed as the (101), (210), (320), (002), (411), and (611) reflections for a structure based on Si_3N_4 in which silicon is replaced by carbon (**Fig. 2**). These same six diffraction rings can also be indexed to a structure analogous to the high-pressure phase of germanium phosphide (GeP) in which Ge and P are replaced by carbon and nitrogen, respectively. Hence, additional experimental work is needed in order to identify unambiguously the structure of covalent carbon nitride.

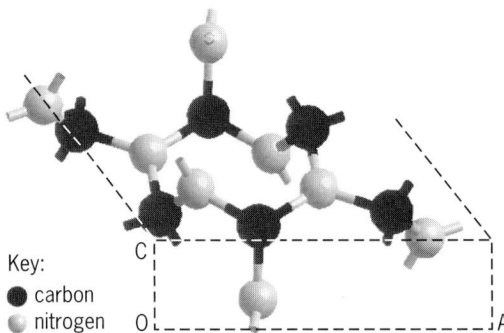

Fig. 2. Computer model of the carbon nitride solid C_3N_4, displaying one unit cell. Each sp^3 hybridized carbon atom is bonded to four nitrogen atoms in a distorted tetrahedral geometry, whereas each sp^2 hybridized nitrogen is bonded to three carbon atoms in a trigonal planar geometry. Broken lines correspond to the A and C axes of the crystal unit cell.

Other key properties. The new carbon nitride materials also exhibit interesting physical properties that may be attractive for high-performance engineering applications. Qualitative scratch tests indicate that carbon nitride materials produced by the new laser ablation technique are hard. For example, rubbing carbon nitride and hard amorphous carbon surfaces against one another produces damage in the carbon but not the carbon nitride material. Quantitative measurements of the microhardness are a much needed area of future work so as to allow quantitative comparison of the properties of carbon nitride to other hard materials such as diamond.

Very recent measurements of the electrical conductivity of carbon nitride are also quite exciting. Studies have shown that carbon nitride is an excellent electrical insulator, and that these electrical properties are stable to thermal cycling. Because carbon nitride is also expected to exhibit good thermal conductivity, it could be an attractive candidate for the dielectric in advanced microelectronics devices where thermally conducting electrical insulators are needed to enable further miniaturization of devices. Thus, future research probing these issues could result in significant scientific and technological payoffs.

For background information SEE CHEMICAL BONDING; LASER; SOLID-STATE CHEMISTRY; SPECTROSCOPY; X-RAY DIFFRACTION in the McGraw-Hill Encyclopedia of Science & Technology.

Charles M. Lieber

Bibliography. C. M. Lieber and Z. J. Zhang, Synthesis of covalent carbon nitride solids: Alternatives to diamond?, *Adv. Mater.*, 6:497–499, 1994; A. Y. Liu and M. L. Cohen, Prediction of new low compressibility solids, *Science*, 245:841–842, 1989; C. Niu, Y. Z. Lu, and C. M. Lieber, Experimental realization of the covalent solid carbon nitride, *Science*, 261:334–337, 1993; R. Riedel, Novel ultrahard materials, *Adv. Mater.*, 6:549–560, 1994.

Carpal tunnel syndrome

Carpal tunnel syndrome is caused by the thickening of ligaments and tendon sheaths at the wrist, with consequent compression of the median nerve at the palm. Affected individuals report numbness, tingling, and pain in the hand, which often becomes worse at night or after use of the hand. A physical examination of the injured hand during the early stages of the syndrome often reveals no abnormality. With more severe nerve compression, the individual experiences sensory loss over some or all of the digits innervated by the median nerve (thumb, index finger, middle finger, and ring finger) and weakness of thumb movement.

Assessment. Electrodiagnostic testing is important for an accurate diagnosis. The electromyographer uses sensory fibers to measure the nerve conduction velocity from the finger or the palm to the wrist and the motor conduction velocity from the wrist to the thumb muscles. Approximately half the individuals with carpal tunnel syndrome have abnormalities of the opposite median nerve. Electrodiagnostic values of these individuals therefore need to be compared with reference values obtained from normal subjects and with ulnar or radial nerve values. No other test has a higher diagnostic accuracy for individuals who have a final diagnosis of carpal tunnel syndrome, but to make a final diagnosis clinical data, including the individual's response to treatment, must be obtained.

False positive and false negative results are illustrated by the high rate of abnormalities in the opposite, nonsymptomatic hands of individuals with carpal tunnel syndrome. As testing becomes more complex and sophisticated, it becomes increasingly likely that results beyond the normal range (false positive) will be generated. With currently used criteria, the number of individuals with false negative results should be below 8%.

Thermography shows clear abnormalities in carpal tunnel syndrome, but an abnormal thermogram is found in many other conditions in which the pattern of blood flow to the hand is altered. Computed tomography and magnetic resonance imaging have not been widely used for carpal tunnel syndrome and have no role in management at this time.

Prevalence. Carpal tunnel syndrome is common. One study of the syndrome derived an incidence of 125 per 100,000 individuals affected from 1976 to 1980. In a survey of physicians it was estimated that 515 of every 100,000 patients sought medical attention for carpal tunnel syndrome in 1988; the syndrome in half of these individuals was thought to be occupational in origin.

The incidence of carpal tunnel syndrome is greater among electronic-parts assemblers, frozen-food processors, musicians, and dental hygienists. The use of highly repetitive wrist movements, vibrating tools, awkward wrist positions, and movements involving great force seems to be correlated with the disorder. Awkward and repetitive wrist motions occur in many office tasks, such as typing and word processing. In some occupations, such as shellfish packing, the incidence was more than 200 times higher than in the baseline data. The highest reported incidence of carpal tunnel syndrome in an industrial setting, based on the numbers of carpal tunnel releases performed, was 15% among a group of meatpackers.

Carpal tunnel syndrome probably accounts for a minority of the cases of overuse syndrome (cumulative trauma syndrome), which is a common problem in occupational settings. Overuse syndrome symptoms include muscle pain, tendinitis, fibrositis (inflammation of connective tissue in a joint region), and epicondylitis (inflammation of the eminence on the condyle of a bone). Although the causitive relationship between the two disorders has not been conclusively proven, the incidence of both carpal tunnel syndrome and overuse syndrome appears to increase in tandem in individuals who are at risk.

Pathophysiology. Under normal circumstances, the pressure within the tissues of a limb is 7–8 mmHg. In carpal tunnel syndrome, the pressure is often 30 mmHg, approaching the level at which nerve dysfunction occurs. With wrist flexion or extension, pressures may increase to 90 mmHg or more.

The increase in pressure within the carpal canal is usually caused by nonspecific inflammation of flexor tendon sheaths. Diabetes, pregnancy, rheumatoid arthritis, and hypothyroidism are the commonest medical conditions associated with carpal tunnel syndrome. Amyloidosis, acromegaly, and mycobacterial infections are important rare causes. It has been shown that the carpal canal is smaller in individuals with carpal tunnel syndrome than in control subjects, although not all investigators agree with this finding. A reduction in the flow of blood to the nerve can account for the intermittent tingling that occurs at night or with wrist flexion.

Treatment. Nonsurgical treatment includes avoidance of the use of the wrist, use of a splint to keep the wrist in a neutral position, and anti-inflammatory medications. These treatments are especially useful in individuals with an acute flare-up and in those with minimal and intermittent symptoms. Conservative nonsurgical treatment will not succeed for individuals over the age of 50, for those who have had the syndrome for more than 10 months, and for those having constant tingling.

Surgical treatment may be used if more conservative approaches fail. The procedure is usually done on an outpatient basis with prognoses of good to excellent in 80% of the cases. Although 40% of the individuals regain normal function, the condition of 5% may worsen. Most individuals return to an office job within a week of surgery, but it may be 4–6 months before carpenters, construction workers, or athletes are able to return to work. Many individuals with work-related carpal tunnel syndrome should consider changing jobs.

Surprisingly little information is available regarding the redesign of workstations and its effect on the reversal of symptoms or the prevention of carpal tunnel syndrome. Ergonomic redesign is widely practiced but rarely described in medical terms.

For background information SEE AMYLOIDOSIS in the McGraw-Hill Encyclopedia of Science & Technology.

David M. Dawson

Bibliography. D. M. Dawson, M. Hallett and L. H. Millender, *Entrapment Neuropathies*, 2d ed., 1990; R. B. Rosenbaum and J. L. Ochoa, *Carpal Tunnel Syndrome and Other Disorders of the Median Nerve*, 1993.

Cat scratch disease

Cat scratch disease in humans is typically a benign, mild, localized disease, involving an abnormal enlargement of the lymph nodes (lymphadenopathy), and results from dermal inoculation with *Bartonella* (formerly *Rochalimaea*) *henselae*. This bacterium has been isolated from immunocompromised patients with bacillary angiomatosis, a distinctive and potentially deadly vascular proliferative host response in skin, bone, or other organs associated with the presence of clumps of bacteria.

Although cat scratch disease was first described in 1950, the causative bacterial agent remained obscure until 1992, when *B. henselae* was implicated by serologic studies. In 1993 and 1994, major progress was made in the knowledge of cat scratch disease epidemiology. The cat seems to be the major reservoir of *B. henselae*; and the cat flea, *Ctenocephalides felis*, could be its vector of transmission from cat to cat.

Etiology. The cause of cat scratch disease has long been in question. It was initially considered a virus, then a gram-negative bacterium, and only recently was a specific organism identified. In 1983, a small bacillus was identified by Warthin-Starry silver deposition staining of lymph node biopsies from 39 individuals with cat scratch disease. In 1988, a pleomorphic, gram-negative bacterium was isolated from the lymph node of an infected individual and named *Afipia felis*. Its isolation was difficult, and the bacterium was limited to a few strains. In addition, serology was not highly specific and was difficult to standardize. For several years, *A. felis* was considered to be the most probable agent causing cat scratch disease. However, in the 1990s evidence was found clearly implicating *B. henselae* as the causative agent.

Bartonella henselae is morphologically very similar to *A. felis* when examined by Warthin-Starry staining, which may explain the previous confusion. Serological studies and isolation of the organism from lymph nodes of probable cat scratch disease cases substantiate the major role played by *B. henselae* in the etiology of the disease. *B. henselae* is a small, curved gram-negative rod which exhibits a twitching motility. Optimal growth on solid media is obtained by use of rabbit blood agar and incubation at 95°F (35°C) in 5% carbon dioxide (CO_2).

Epidemiology. There were an estimated 22,000 individuals with cat scratch disease in the United States in 1992, 2000 of these requiring hospitalization. Cat scratch disease occurs in immunocompetent individuals of all ages. However, a higher proportion of cases among children and teenagers than adults is reported (45–50% of infected individuals are less than 15 years old). Cat scratch disease is considered the most common cause of chronic benign adenopathy in children and young adults, but can easily be confused with neoplastic conditions. More than 90% of the individuals have a history of some type of contact with cats, and 57–83% have received a scratch from a cat in the past. Incidence varies by season; however, most cases are reported in fall and winter. More cases are observed in males than females.

A recent epidemiological study reported that infected individuals were more likely than healthy cat-owning control subjects to have at least one kitten 12 months of age or younger, to have been scratched or bitten by a kitten, and to have at least one kitten with fleas. Of 45 individuals observed, 38 (84%) had antibodies to *B. henselae* compared to 4 of 112 control subjects (3%). Interestingly, 81% (39 of 48) of the cats of the infected individuals also had antibodies to *B. henselae,* as compared with 11 of the 29 (38%) control cats.

Prevalence of *B. henselae* antibodies has been found to be common in cats from Baltimore (14.7%), especially feral cats (44.4%). Geographical variations of seroprevalence were also observed: of the cats tested in the southeast region of the United States 60% were seropositive but very low levels were observed in the Midwest, Alaska, and the Rocky Mountains. The presence of bacteria in the blood (bacteremia) is also common in cats. In the San Francisco area, of 61 cats 41% were found to be infected. Bacteremia can last several months in cats, but very high bacteremia levels do not last usually more than 4–6 weeks.

Clinical signs. In humans, 1–3 weeks elapse between the scratch (or bite) and the appearance of clinical signs. In 50% of the cases, a small skin lesion, often resembling an insect bite, appears at the inoculation site (usually on the hand or forearm) and evolves from a papule to a vesicle and partially healed ulcers. These lesions resolve within a few days to a few weeks. Inflammation of lymph nodes develops approximately 3 weeks after exposure, and is generally unilateral, commonly appearing in the epitrochlear, axillary, or cervical lymph nodes.

Swelling of the lymph node is usually painful and persists for several weeks to several months. In 25% of the cases, a discharge of pus occurs. A large majority of cases show signs of systemic infection, such as fever, chills, malaise, anorexia, and headaches. In general, the disease is benign and heals spontaneously without aftereffects. Atypical manifestations of cat scratch disease occur in 5–9% of the cases. The most common is Parinaud's oculoglandular syndrome (enlargement of the lymph node around the eye and conjunctivitis), but also encephalitis, degenerative bone lesions, and thrombocytopenic purpura (a bleeding disorder due to decreased platelet levels) may occur. Usually, complete recovery occurs with few aftereffects.

No clinical signs of cat scratch disease have ever been reported in cats. Suspicion of enlargement of the lymph nodes caused by a cat scratch disease–like organism has been reported. Whether members of *Bartonella* are pathogenic for cats or contribute to persistent lymphadenopathy in cats remains to be determined.

Diagnosis. For years, the diagnosis was based on clinical criteria, exposure to a cat, failure to isolate other bacteria, or histologic examination of biopsies of lymph nodes. A skin test was also developed and used. However, the antigen prepared from pasteurized pus from lymph nodes of patients with cat scratch disease was not standardized, and concerns were raised about the safety of such a product. In the past 2 years, a serological test and techniques to isolate the organism from human specimens have been developed. Additionally, the polymerase chain reaction is being used to confirm infection by *Bartonella.* At present, diagnosis is mainly based on a serological titer ≥1:64 (or a fourfold titer increase between early and late serum samples) and a history of cat scratch or bite. SEE CLINICAL PATHOLOGY.

In cats, isolation of the organism from blood samples is performed as in humans suffering bacillary angiomatosis. Blood samples are collected in pediatric lysis-centrifugation tubes and centrifuged; then, the pellet is resuspended in inoculation media and plated onto 5% rabbit blood agar.

Treatment. Most individuals with cat scratch disease experience mild illness and require minimal treatment. Usually antimicrobial therapy is not necessary since spontaneous resolution is common. In severe forms, especially in immunocompromised patients, antibiotics such as ciprofloxacin, rifampin, and gentamicin have been recommended. Erythromycin and doxyxycline are effective antibiotics in the treatment of bacillary angiomatosis and may be recommended in immunodeficient persons suffering from cat scratch disease.

For background information SEE CAT SCRATCH DISEASE; IMMUNOLOGICAL DEFICIENCY; MEDICAL BACTERIOLOGY in the McGraw-Hill Encyclopedia of Science & Technology.

Bruno D. Chomel

Bibliography. J. E. Childs et al., Epidemiologic observations on infection with *Rochalimaea* species among cats living in Baltimore, Md., *J. Amer. Vet. Med. Assoc.*, 204:1775–1778, 1994; M. J. Dolan et al., Syndrome of *Rochalimaea henselae* adenitis suggesting cat scratch disease, *Ann. Intern. Med.*, 118:331–336, 1993; J. E. Koehler, C. A. Glaser, and J. W. Tappero, *Rochalimaea henselae* infection: A new zoonosis with the domestic cat as reservoir, *J. Amer. Med. Assoc.*, 271:531–535, 1994; R. L. Regnery, M. Martin and J. G. Olson, Naturally occurring "*Rochalimaea henselae*" infection in domestic cat, *Lancet*, 340:557–558, 1992.

Cellular immunology

T-cell dependent humoral immune responses require interactions between antigen-specific T lymphocytes (T cells) and B lymphocytes (B cells), leading to B-cell antibody secretion of different isotypes with diverse effector functions. B cells proliferate and differentiate into antibody-secreting cells after they induce T-cell helper functions by acting as antigen-specific, antigen-presenting cells. T-cell help for resting B cells involves cell-cell mediated interaction, and the production of soluble cytokines that drive B-cell growth and regulate the immunoglobulin (Ig) switch from IgM to other isotypes. The identification of specific molecules on the B- and T-cell surface which regulate this process allows the development of therapies aimed at inhibiting or curing autoimmune disease, and has led to a better molecular understanding of certain immunodeficiencies. SEE AUTOIMMUNITY.

Antigen processing and presentation. Both B and T cells express antigen-specific cell surface receptors which give the immune system its specificity. The sequence of events leading to B-cell antibody production is initiated when an antigen binds to the clonotypic antigen-specific B-cell receptor. The B-cell receptor typically has very high affinities for its cognate antigen, and it is able to bind the antigen in its native configuration (see **illus.**). However, the antigen-specific T-cell receptor cannot recognize the antigen in its native form but only in the context of class II major histocompatibility complex (MHC) molecules expressed on the B-cell surface. The T-cell receptor recognizes only short peptides fragments that are generated intracellularly by B-cell proteases. These peptides bind to a groove in the class II MHC molecule that faces outward toward the T-cell receptor and can be recognized during antigen presentation. The interaction between antigen-specific B and T cells results in the clonal expansion of these cells, leading to immunity against only that particular foreign antigen and providing the basis for immune specificity. SEE ANTIGEN.

The requirement for contact between T cells and antigen-presenting B cells involves more than the recognition of peptide and class II MHCs on the B-cell surface by the T-cell receptor. T-cell receptor

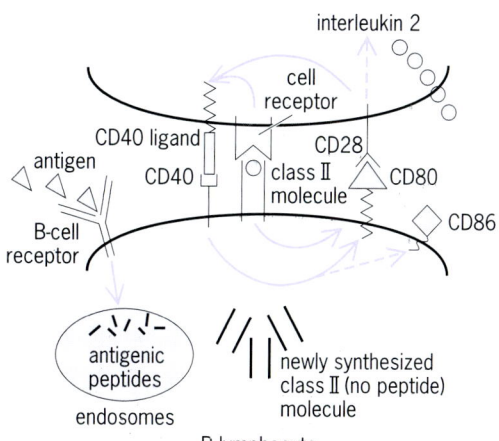

B cells bind and internalize a foreign antigen through the antigen-specific B-cell receptor. The foreign antigen is processed on endosomal compartments into small peptides which bind newly synthesized class II major histocompatibility molecules. Antigen presentation stimulates T cells by cross-linking the T-cell receptor, setting off a cascade of reactions that result in expression of inducible surface proteins and interleukin-2 production.

antigen recognition induces T-cell helper functions that include new expression of cell-surface proteins and soluble cytokines which are secreted locally into the external cellular milieu. The requirement for inducing T-cell help has been further elucidated by directing monoclonal antibodies against the T-cell receptor, thus mimicking the effects of T-cell antigen recognition. Fixed, preactivated T cells or membranes from activated T cells are able to induce B-cell activation, and in the presence of appropriate cytokines stimulate B-cell differentiation to antibody-secreting cells. Resting T cells or membranes from resting T cells do not stimulate resting B cells. Significantly, although antigen recognition and the initiation of T-cell helper functions is restricted to the MHC, the delivery of T-cell help is not. Once T cells are activated, newly expressed cell surface proteins stimulate B cells in a fashion that is not restricted by the MHC. The identification of T-cell surface proteins expressed after T-cell activation that induce B-cell responsiveness is paramount.

CD40 ligand. The major component of contact-dependent T-cell help has recently been identified as the counterreceptor for the B-cell surface antigen CD40. CD40 is a protein (40 kilodaltons) expressed on resting and activated B cells, as well as dendritic cells, activated macrophage, and follicular dendritic cells. It is a member of the tumor necrosis factor receptor family of genes, which include *CD27, CD30, fas* (a receptor whose engagement causes programmed cell death), and nerve growth factor receptor. CD40 ligand (CD40L) is a protein

(33 kDa) expressed predominantly on acutely, activated CD4[+] T cells, and shares homology with the tumor necrosis factor family, which includes CD27 ligand (*CD70*), CD30 ligand, and Fas ligand. The identification of CD40 ligand as the major stimulatory molecule on activated T cells fits well with previous studies using antibodies directed against B-cell CD40. Such antibodies are very active at low concentrations, are costimulatory with anti-immunoglobulin, and induce B cell differentiation in combination with various cytokines. *See* TUMOR.

Inhibition of CD40-CD40 ligand interactions has profound consequences on B-cell antibody secretion both in the laboratory and in living organisms. Chimeric fusion proteins consisting of the extracellular domain of CD40 block B-cell proliferation and antigen production in response to activated T cells under artificial conditions, whereas antibodies against CD40 ligand affect T-cell-dependent humoral immunity in living organisms. Mice treated with anti-CD40 ligand do not exhibit primary or secondary responses to either sheep red blood cells or protein antigens, and anti-CD40 ligand inhibits the induction of collagen-induced arthritis in susceptible mice. The efficacy of anti-CD40 ligand treatment in living organisms may provide a basis for the development of effective therapies aimed at controlling the onset and progression of certain autoimmune diseases such as type I diabetes and systemic lupus erythematosus.

The importance of CD40-CD40 ligand interactions for humoral immunity is underscored by the defect revealed in hyper IgM syndrome, a rare X-chromosome-linked immunodeficiency characterized by elevated levels of serum IgM. Afflicted individuals lack immunoglobulin of all other isotypes, and do not form germinal centers (where B cells grow) in response to foreign antigens. Although initially considered a B-cell abnormality, it was recently discovered that the genetic defect introduces point mutations into the CD40 ligand gene. These mutations alter T-cell CD40 ligand expression so that it no longer interacts with *CD40*, either by altering its three-dimensional structure or by precluding surface expression. B cells from these individuals still respond to anti-*CD40* or to normal, activated T cells in the laboratory, confirming that B-cell function is not impaired and that *CD40*-CD40 ligand conjugation in living organisms is necessary for secondary follicle generation and immunoglobulin isotype switch.

B- and T-cell costimulation and crosstalk. Although the series of events leading to B-cell antibody production is initiated by B-cell presentation of peptide in the context of class II MHC molecules, and T-cell receptor recognition, B and T cells need additional cell surface molecules for maximal stimulation and generation of an effective humoral immune response. Although some proteins function as adhesion molecules to stabilize the B-T cell interface, other proteins provide key costimulatory signals that optimize cell activation and proliferation. Depending on the state of the B or T cell involved (for example, resting versus activated) lack of costimulation may induce a state of nonresponsiveness or anergy in cells.

Optimal T-cell activation and interleukin-2 production requires additional costimulatory signals provided by antigen-presenting cells. One very important costimulatory molecule for T-cell activation is the CD28 antigen, whose ligands CD80 and CD86 are found on a variety of antigen-presenting cells, including activated B cells and macrophages, as well as dendritic cells (see illus.). Costimulation of T cells through the T-cell receptor and CD28 antigen augments T-cell proliferation and interleukin-2 production in living organisms and under particular conditions prevents nonresponsiveness. Signals transduced through CD28 also increase T-cell helper functions, an increase that correlates with a higher expression of cell-surface CD40 ligand. Functional CD80/86 ligand–CD28 binding in living organisms is critical for humoral immunity and T-cell activation, since blocking these receptor-ligand interactions blocks antibody production against protein antigens, and prolongs xenogeneic graft survival.

The commitment to T- and B-cell activation versus nonresponsiveness is apparently determined by cross talk between receptor-ligand pairs expressed on T and B cells. Cross-linking class II MHC molecules on B cells with antibodies induces CD80 ligand expression, which is mimicked by T-cell receptor recognition. The induction of CD40 ligand expression by T-cell receptor engagement would further T-cell activation since CD40-CD40 ligand interactions augment CD80/CD86 ligand upregulation on B cells, which would costimulate T cells through the CD28 antigen. This crosstalk between B and T cells leads to a highly regulated progression of intracellular signals derived from the interaction of specific receptor-counterreceptor pairs, and is controlled by the sequential expression of such molecules in a temporal fashion. Certain receptor-ligand pairs are functionally important early during B-T cell conjugation, whereas others are upregulated late, and operate at the distal stages of cell-cell interaction.

The amplification of T- and B-cell responses by inducible cell-surface proteins (CD40 ligand and CD80/CD86) that bind constitutively expressed molecules (CD40 and CD28) requires a means for downregulating lymphocyte responses. One newly identified pathway is the internalization of CD40 ligand that occurs after binding to CD40, which maintains transient CD40 ligand expression and could prevent nonspecific activation of bystander B cells. Also, the B-cell specific antigen CD22 binds a particular isoform of the T-cell CD45 molecule, termed CD45R0, which downregulates T-cell proliferative responses. The balance between these interactions makes it difficult to predict the intensity of cell stimulation or whether it will occur at all, and is further complicated by the different requirements of each cell, based on whether it has recently emerged from the bone marrow or thymus or is a memory cell which has previously recognized the antigen. Although recently identified proteins such as CD40 ligand and the ligands CD80 and CD86 have advanced our understanding of humoral immunity, predicting the outcome of B-T cell interactions with regard to particular B- and T-cell subsets will be critical for developing vaccines against different pathogens and for immunotherapies aimed at halting progression of disease.

For background information SEE ANTIBODY; ANTIGEN; AUTOIMMUNITY; CELL SURFACE INTERACTIONS; CYTOKINE; HISTOCOMPATIBILITY; IMMUNITY in the McGraw-Hill Encyclopedia of Science & Technology.

Stephen J. Klaus; Edward A. Clark

Bibliography. E. A. Clark and J. A. Ledbetter, Cell-cell interactions regulating T-cell-dependent B-cell maturation, Nature, 367:425, 1994; P. S. Linsley and J. A. Ledbetter, The role of the CD28 receptor during T cell responses to antigen, Annu. Rev. Immunol., 11:191, 1993; D. C. Parker, T cell-dependent B cell activation, Annu. Rev. Immunol., 11:331, 1993.

Clinical pathology

Since the mid-1980s scientists have made tremendous progress toward developing an understanding of the molecular biologic processes underlying many diseases. As a direct result of these investigations, new technologies have taken molecular genetic studies out of the research laboratory and into the diagnostic laboratory. These technologies are now being applied to the diagnosis and treatment of various forms of cancer, prenatal diagnosis of hereditary diseases, and the identification of infectious organisms.

Molecular diagnostic techniques. The deoxyribonucleic acid (DNA) molecule consists of two chains composed of four types of subunits. These subunits are deoxyribonucleotides that contain the bases adenine (A), cytosine (C), guanine (G), and thymidine (T). The two chains are arranged in a double helix so that each base pairs with a complementary base on the opposite DNA strand. Base pairing is specific in that A only pairs with T, and G only pairs with C. Thus, one DNA strand serves as a template for the other.

Hybridization techniques. These techniques take advantage of the formation of a duplex between two strands of complementary DNA. Typically, a short DNA strand (oligonucleotide) is designed to match a complementary DNA sequence in a gene of interest (**Fig. 1**). The target gene may be an normal gene, a known oncogene (a gene that initiates tumor growth), a viral gene, or any other gene of interest. The oligonucleotide strands used for hybridization are called probes.

Southern blotting is a hybridization technique in which spliced DNA extracted from an individual's sample is applied to a nylon membrane or nitrocellulose filter. A probe is then allowed to interact with the DNA adhering to the membrane. If the DNA sequence recognizable by the probe is present, the probe will bind to the sequence, and the interaction between the DNA probe and the target DNA in the sample can be visualized. (The probe is tagged with either a radioactive or an enzymatic label that can later be detected.) Southern blotting detects gene sequences present in any cell or organism in a sample but cannot localize the sequence to a specific cell type. In contrast, in in-place hybridization reactions the DNA probes are applied directly to tissue sections. This approach allows specific genes to be localized to specific cells in a tissue. In this way, the location and type of cells staining with the DNA probe can be seen directly.

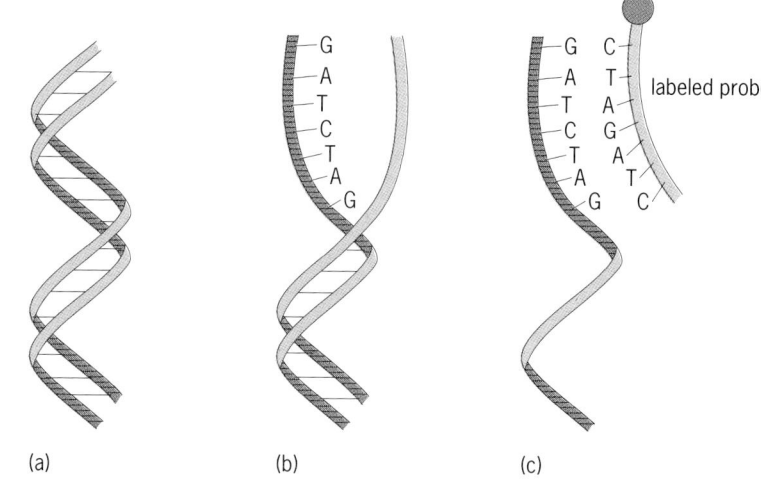

Fig. 1. Hybridization technique. (*a*) Normal double-stranded DNA. (*b*) DNA molecule denatured so that the two complementary strands separate from each other. (*c*) Application of oligonucleotide probe designed to recognize a specific DNA sequence.

Polymerase chain reaction. This extremely useful technique has revolutionized the field of molecular genetics (**Fig. 2**). With this method, the ability to amplify or to copy specific DNA sequences is possible, allowing the detection of a single copy of a DNA sequence among millions of other DNA sequences. As a result, the polymerase chain reaction markedly enhances the sensitivity of DNA testing procedures.

In the polymerase chain reaction, primers or oligonucleotide sequences are synthesized that are complementary to the DNA sequences on both ends of the fragment of DNA which needs to be identified and is destined to be copied. The DNA is heated to separate the complementary strands. The primer anneals, and the Taq polymerase enzyme adds nucleotides to the primer one by one by pairing them with the free nucleotides on the DNA template strand. The result is a new strand of DNA complementary to the strand under investigation. The DNA is denatured again by heating, the second primer anneals, and a new strand consisting of DNA of only the region of interest is synthesized by the Taq polymerase. The reaction is repeated to produce copies of the desired DNA sequence. With this technique, amplification serves as an indicator of the presence of the DNA sequence in question. If the sequence is not present, amplification will not take place.

Cancer. Cancers are groups of cells that have lost the ability to control normal cell growth. This loss of growth control often has a genetic basis, and, indeed, malignant cells frequently demonstrate genetic abnormalities. The range of abnormalities includes loss of whole chromosomes; chromosomal translocations, inversions, or deletions; and mutations limited to small regions of the DNA molecule. Specific molecular genetic alterations can be useful in characterizing the type of tumor a patient may have and predicting its behavior and response to treatment.

Cancer diagnosis. Molecular genetic techniques have been most successfully applied to the diagnosis of cancers of the hematologic (blood) system: leukemias (malignant proliferations of circulating white blood cells) and lymphomas (solid tumors of lymphocytes, cells which form a portion of the immune system).

Some forms of leukemia, such as chronic myelogenous leukemia, are consistently associated with specific chromosomal translocations. Cytogenetic analysis (the study of the mechanisms and behavior of chromosomes) of tumor cells from individuals with chronic myelogenous leukemia has revealed the presence of an abnormal chromosome referred to as the Philadelphia chromosome. Its abnormal appearance results from the displacement of part of chromosome 9 onto chromosome 22. In a process known as translocation, the long arm of chromosome 9 breaks in a specific region of a gene referred to as *abl*. The broken portion of the *abl* gene then fuses with the *bcr* gene located on chromosome 22. The newly created, abnormally fused *bcr-abl* gene can be specifically identified in blood samples taken from individuals suspected of having leukemia with the use of Southern blotting or the polymerase chain reaction. The identification of the *bcr-abl* gene establishes the diagnosis of chronic myelogenous leukemia.

Prognosis of malignant tumors. Molecular biologic techniques can also be invaluable in determining the future behavior, or prognosis, of a tumor. Molecular techniques are routinely applied in determining prognosis in breast cancer and in neuroblastoma, a malignant tumor containing embryonic nerve cells that affects infants and children.

Breast cancer is the most common form of cancer affecting women in the United States; it is estimated that 1 in 10 women will develop breast cancer. The decision of when to use chemotherapy in the treatment of cancer is critical because although treatment may mean prolongation of the individual's life, chemotherapy is also associated with numerous, often harmful, side effects. Reliable prognostic markers have now been developed that help clinicians determine which individuals will benefit from aggressive chemotherapy.

One such prognostic marker is a gene referred to as $ERBB_2$, or *HER-2/neu*. This gene normally encodes a growth factor receptor protein located within the cell membrane. In 25–40% of breast cancers, the number of copies of the $ERBB_2$ gene within the tumor cells is increased or amplified. Studies in breast cancer have shown that $ERBB_2$ amplification is strongly associated with an increased rate of tumor recurrence and with poorer survival. Therefore, individuals whose tumors demonstrate $ERBB_2$ amplification receive more aggressive treatment.

Neuroblastoma, a common childhood tumor, originates from cells of the peripheral nervous system. Approximately 50% of advanced neuroblastomas demonstrate amplification of the *N-myc* gene, an oncogene that plays a role in cell differentiation and proliferation. Individuals whose tumors show *N-myc* amplification do not live as long as individuals whose tumors have only one copy of the *N-myc* gene. As with $ERBB_2$ in breast cancer, amplification of *N-myc* can be used to define the most appropriate therapy.

Detection of minimal residual disease. If cancer treatment is to be successful, disease recurrence must be identified early. In this way, therapy can be instituted before the spread of the tumor from the site of its origin to numerous distant sites. Individuals with leukemia, for example, are treated with chemotherapeutic agents after the initial diagnosis. If all goes well, the chemotherapy destroys the malignant cells, and the individual enters a disease-free interval, or remission. Molecular genetic tech-

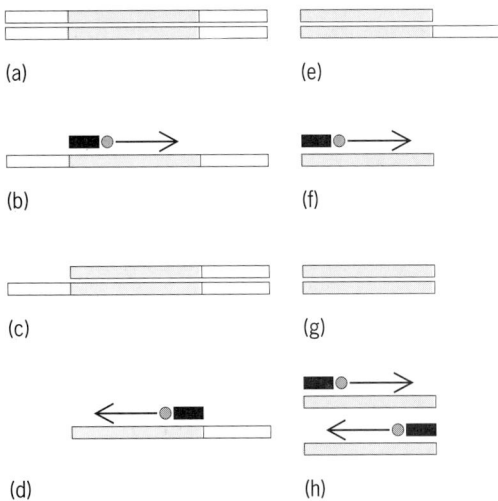

Fig. 2. Polymerase chain reaction. (*a*) Double-stranded sample DNA. The colored region represents the region to be amplified. (*b*) The DNA is heated to separate the complementary strands. The primer (black rectangle) anneals, and the Taq polymerase enzyme (gray sphere) adds nucleotide bases to it. (*c*) The result is a new strand of DNA complementary to the strand shown in *b*. (*d*) The DNA is denatured, the second primer anneals, and a new strand is synthesized by the Taq polymerase. (*e*) The result is a new DNA strand consisting only of the region of interest. (*f, g, h*) The reaction is repeated to produce copies of desired DNA sequence.

niques such as the polymerase chain reaction are now being used to monitor individuals for early disease recurrence. With this technique it is possible to identify a single malignant cell among 100,000 to 1,000,000 normal cells, a task not possible before. In this way, appropriate therapy can be instituted early in order to prolong individual's survival. The polymerase chain reaction can also be applied to screen bone marrow specimens of individuals with cancer for the presence of malignant cells before autologous transplantation.

Prenatal diagnosis of hereditary disease. In recent years, the molecular biologic changes underlying many genetic diseases have been elucidated. Currently, we know of more than 4000 hereditary disorders associated with an abnormality in a single gene. The genes responsible for many of these diseases have been identified, making it possible to screen for abnormal genes early in fetal development.

Some of the better characterized diseases include cystic fibrosis, Duchenne muscular dystrophy, Huntington's disease, myotonic dystrophy, and adult polycystic kidney disease. These diseases can be diagnosed prenatally in cases where there is a family history of the disorder. Prenatal diagnosis is achieved through chorionic villus sampling, a procedure that can be performed as early as weeks 10–11 of gestation. Chorionic villus sampling involves biopsy and evaluation of a small portion of placental tissue. Since the placenta contains tissue of fetal origin, it is genetically identical to the developing fetus. Molecular biologic techniques such as Southern blotting and the polymerase chain reaction are then applied to the chorionic villus sample to screen for the presence of the abnormal gene in question.

Diagnosis of infectious diseases. Until recently, the diagnosis of infectious disease depended on the identification of the responsible organism in culture, or on the identification of organism-specific antibodies in the blood of the affected individual. Because culture of some organisms, particularly fungi, viruses, and some bacteria, may take weeks, institution of appropriate therapy was sometimes delayed. In addition, in the case of organisms that cannot be grown in culture, diagnosis was presumptive, dependent on the concentration of antibodies, on symptoms, or on biopsy findings. The development of specific molecular biologic assays for many infectious organisms (see **table**) has solved some of these diagnostic problems.

The techniques most commonly employed for identification of infectious organisms are in-place hybridization and the polymerase chain reaction. These techniques can be performed rapidly enough that results are often available before the organism appears in culture. In addition, in-place hybridization and, especially, the polymerase chain reaction, are extremely sensitive and are capable of identifying organisms that are present in very small numbers, as is often the case with the bacteria that cause tuberculosis and the human immunodeficiency virus (HIV).

Infections for which molecular diagnosis is possible

Organism	Representative associated diseases
Viruses	
Cytomegalovirus	Multiple effects in immuno-suppressed patients
Herpes viruses	Genital herpes; fever blisters
Human papillomavirus	Venereal infections; warts; association with some forms of cancer
Epstein-Barr virus	Mononucleosis; association with some forms of cancer
Human immuno-deficiency virus (HIV)	Acquired immune deficiency syndrome (AIDS)
Hepatitis B virus	Hepatitis B
Hepatitis C virus	Hepatitis C
Measles virus	Measles
Rhinovirus	Common cold
Bacteria	
Mycobacteria	Tuberculosis; infections in immunosuppressed patients
Chlamydia	Venereal infections; eye infections
Helicobacter pylori	Gastritis; peptic ulcer
Neisseria	Gonorrhea; meningitis
Others	
Sporozoa	Malaria
Pneumocystis	Pneumonia in AIDS patients
Rickettsia	Rocky Mountain spotted fever; Q fever
Toxoplasma	Toxoplasmosis

For background information SEE ACQUIRED IMMUNE DEFICIENCY SYNDROME (AIDS); CANCER (MEDICINE); CLINICAL PATHOLOGY; DEOXYRIBONUCLEIC ACID (DNA); DISEASE; GENE AMPLIFICATION in the McGraw-Hill Encyclopedia of Science & Technology.

Cecilia M. Fenoglio-Preiser; Amy E. Noffsinger

Bibliography. M. J. Arends and C. C. Bird, Recombinant DNA technology and its diagnostic applications, *Histopathology,* 21:303–313, 1992; M. Loda, Polymerase chain reaction-based methods for the detection of mutations in oncogenes and tumor suppressor genes, *Hum. Pathol.,* 25:564–571, 1994; B. B. Rogers, Nucleic acid amplification and infectious disease, *Hum. Pathol.,* 25:591–593, 1994; J. D. Rowley, J. C. Aster, and J. Sklar, The clinical applications of new DNA diagnostic technology in the management of cancer patients, *JAMA,* 270:2331–2337, 1993.

Coal mining

For mining, teleoperation refers to the distant control of mining machinery from a protected operator compartment located out of the line of sight to the extraction site. Successful teleoperated mining systems have recently been developed to extract the extensive coal reserves in highwalls, that is, coal seams uncovered during surface strip mining operations or exposed in mountaintop ridges.

Highwall mining. The U.S. Bureau of Mines has estimated that there are 18,000 mi (29,000 km) of exposed highwalls in Appalachia and that a highwall coal reserve of 2×10^{10} tons (1.8×10^{10} metric tons) exists for a highwall mining system that can penetrate 850 ft (260 m). Significant highwall reserves consist of low-sulfur coal that is highly desirable for steam and metallurgical applications. Until recently, these reserves were left basically intact, because mining methods were unavailable or were hampered by inadequate safety, low productivity, and adverse environmental factors.

Highwall mining technology provides access to coal reserves at the interface between surface and underground mining. New highwall mining systems are safe, productive, and environmentally acceptable. These systems allow highwall penetration up to 1200 ft (366 m) and coal recovery ratios of about 60%. Much of the success of the new highwall mining systems is attributed to the technology of teleoperation.

Teleoperation. This technology allows equipment operators, who are not present at the area where the coal extraction occurs, to safely monitor and control the highwall mining systems.

In a typical teleoperated highwall mining system, coal is extracted by developing parallel entries in the exposed coal seam from the bench area adjoining the highwall (**Fig. 1**). The overburden above the coal seam is supported by pillars between the extracted entries. A teleoperator, located in a protected enclosure on the bench, controls the operation of the highwall mining system and is assisted by other workers. No workers are allowed in the developed entries, which increases safety, simplifies the mining process, and reduces costs. For example, barring workers from the extracted entry eliminates the requirement to install artificial roof supports in the strata above the entries. The standard operating procedure is to extract the coal and retrieve the equipment from the hole as quickly as possible. Concerns during highwall mining include ventilation to remove dust and explosive gases and procedures to extract the mining equipment in case of a breakdown.

A very successful design for a highwall mining system became available commercially in 1981. The original design was a hydraulic model that evolved to a more powerful and sophisticated electrical design introduced in 1985. An improved system was introduced in 1992, after having been systematically developed and tested in three Appalachian states. The new highwall mining systems have achieved success largely because of modern electronics, especially miniaturized cameras, and refinements in blasting and blast-hole alignment methods (used during the creation of the highwall) that increase the stability of the highwall.

Highwall mining system extraction. Highwall coal extraction is usually done with a continuous mining machine modified for distant remote control. A continuous mining machine, used extensively in underground deep mines, is a massive tracked machine equipped with a rotating drum that is fitted with tungsten carbide cutters. The position of the cutting drum is controlled by the operator, so that its penetration rate into the coal face and its cutting height can be varied to suit conditions. Continuous mining machines also include mechanisms to gather and direct cut coal onto a conveyor that transports it from the extraction face to the outby end of the machine (toward the shaft or entry of the mine).

Highwall mining systems use a variety of equipment to transport cut coal from the end of the continuous mining machine to the highwall bench. In some instances, continuous haulage systems have been developed for underground mining, with

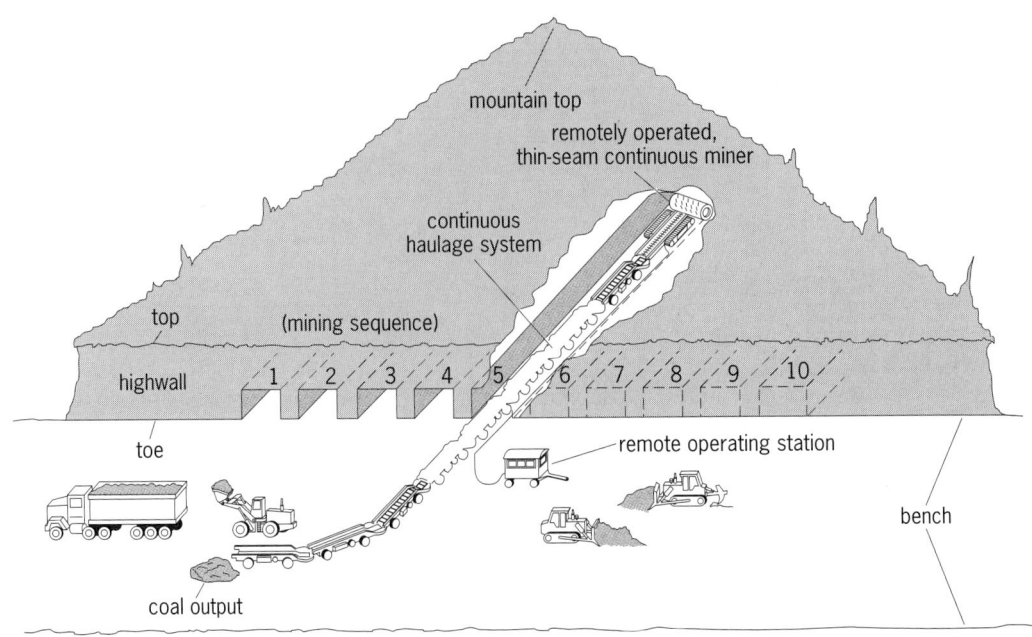

Fig. 1. Teleoperated highwall mining system.

joined vehicles carrying either conveyor chains or belts that have been adapted to this purpose. For other highwall mining systems, coal haulage systems have been developed specifically for the application; these use rigid pushbeam sections to convey the coal. Pushbeam sections are segmented structures, approximately 40 ft (12 m) long, that are designed both to contain the conveying mechanisms and to transmit compressive forces. To move the coal, one design uses internal augers and another uses conveyor belts. For both systems, bench-based machines progressively add the pushbeams until the desired depth of highwall penetration is obtained.

Teleoperator control. The teleoperator for the highwall mining system must be provided with sufficient information to control the mining operations, although the operator is located a great distance from the extraction face and cannot see it directly. Video systems are normally the operator's primary means of monitoring a highwall mining system in a teleoperated environment. Such video systems are usually multicamera, multimonitor, closed-circuit configurations. Typically, two or three cameras are mounted at strategic locations on the equipment to allow remote monitoring of critical operations and conditions, such as the roof type and condition, the boundary between the coal and rock above the seam, the cutter bit paths in the roof, and the color of the generated dust. Usually one video monitor is assigned to each camera. Sometimes, by using a video special-effects generator, simultaneous images from two cameras are presented on one monitor as a main and a smaller inset picture.

Sensors. Sound systems can extend the operator's sensory perception during teleoperation of a highwall mining system. Initial attempts using microphones mounted on the equipment were of little use, because loud noises generated by machines such as pump motors and conveyors masked other ambient sounds. The U.S. Bureau of Mines obtained much better results by using accelerometers (sensors that detect and produce electrical signals) coupled to the cutting head boom of the continuous mining machine. Teleoperators using this sound system reported that the accelerometer-based sound was very useful for determining coal/rock interfaces, aiding the operator in keeping the mining operations within the coal seam.

A variety of other sensors have been used with highwall mining systems to increase the information presented to the teleoperator. These sensors monitor operating parameters of the continuous mining machine and haulage equipment that include electrical motor currents, hydraulic fluid levels, the mine methane level, and the orientation (pitch and roll) of the continuous mining machine. Sensors are normally selected to furnish the operator with adequate information to control the mining system and to monitor the operational condition of its major subsystems.

Newer designs of highwall mining systems use computers to communicate operator control com-

Fig. 2. Workstation for a teleoperated highwall mining system.

mands and sensor data between the operator's control station and the mining equipment. Such systems usually provide the teleoperator with more detailed, perhaps processed, information on conditions within the mined entry and on important machine parameters. Some systems with computers incorporate or allow for automated cutting cycles. Continued advancements in computer systems and sensors, particularly coal interface detectors, are likely to result in future generations of highwall mining systems with more automation capabilities.

Ergonomics. The success of a teleoperated highwall mining system can depend on the operator-to-machine interface. Because the teleoperator is required to control the mining system for extended periods, it is important to engineer the workstation to accommodate the needs of the operator. **Figure 2** shows an ergonomically designed remote operator's control room developed by the U.S. Bureau of Mines for a teleoperated system. For this application, the operator controls the continuous mining machine, major functions of the haulage equipment, and some functions of highwall-based support equipment. Operator comfort is an important element of the design. The operator can change seating and body positions, while still having full access to control and monitoring equipment. The operator, in a semireclining position on a special, adjustable task chair, has access to clustered control and readout instrument panels and multiple video monitors. Sensory information from the continuous mining machine and haulage equipment is presented on color-coded bar-graph displays on an adjustable, swing-away console in front of the operator. Main controls for the continuous mining machine are grouped on a small hand-held module, while secondary controls are located on an adjustable floor-mounted panel next to the operator's task chair. Dual 19-in. (483-mm) color video monitors are housed in a custom cabinet at the optimum viewing distance. Aural information from the continuous mining machine is received over stereo headphones.

Advanced technology. The advent of advanced technology, particularly in sensor and control methodologies, is the basis for a successful application of the new teleoperated highwall mining systems. Because most new highwall mining systems use continuous mining machines to drive the extraction entries, no cutting power losses are associated with increased highwall penetration. For most systems, entry depth does not influence the power available for coal haulage at the face. However, having the power available to cut and haul from deep within the highwall is of little value without the capability to stay within the coal seam and to minimize ground falls. Advanced sensor and control subsystems allow a teleoperator to make informed decisions on what needs to be done and to control the equipment.

Prospects. Teleoperation has made deep-penetration highwall mining practical; the control method has been adopted by all the major manufacturers of highwall mining systems. Continued sensor, computer, and mining-specific advances should improve teleoperation technology, increasing its ability to mine coal safely and productively from highwall and other coal reserves.

For background information SEE COAL MINING; HUMAN-FACTORS ENGINEERING; INSTRUMENTATION; MINING in the McGraw-Hill Encyclopedia of Science & Technology.

August J. Kwitowski

Bibliography. J. Chadwick, Highwall mining, *Min. Mag.*, pp. 347–353, December 1993; A. J. Kwitowski, W. D. Mayercheck, and A. G. Mayton, Overview of highwall mining systems, *Session Papers from American Mining Congress Coal Convention,* pp. 25–40, June 1989; J. C. Volkwein, G. V. G. Goodman, and E. D. Thimons, *Ventilation of Automated Mining Systems*, SME Preprint 90-103, SME Annual Meeting, Salt Lake City, Utah, February 26–March 1, 1990.

Cognition

The origins of human cognitive abilities, such as perception, memory, thinking, and problem solving, have been the focus of much interest in psychology, biology, philosophy, and the emerging discipline of cognitive science. In the early 1960s, speculation, inferences from adult abilities, and a few observations of infants formed the bases of study of infant cognition. With the development of experimental methods for inferring what infants perceive and know, longstanding assumptions about the beginnings of thought have been challenged.

Methodological advances. Human infants do not crawl until about 6–7 months of age, reach effectively until 4–5 months, or speak until much later. The lack of observable responses makes studying infant abilities difficult, and has led observers to underestimate what infants know. Infants do actively fixate visual displays. Thus, a primary advance in studying infant cognition has focused on their visual attention span (length of looking time).

In the visual preference method, discrimination abilities and preferences of infants are tested by presenting two displays side by side and recording fixation preferences. In the habituation/dishabituation method, repeated presentation of one display leads to decreased attention. However, with a subsequent presentation of novel display attention rebounds. Even without habituation, presentation of an event that is perceived as bizarre or anomalous may lead to greater visual attention than a control event involving similar objects. By using these facts about infant attention many inferences about infant cognitive abilities can be made.

Infant perception. A classical view contended that the senses of newborns provide them only with sensations, particular qualities unique to each sense, such as brightness in vision or pitch in hearing. These sensations are due to activity in sensory receptors; by themselves they indicate nothing about objects or events in the outside world. Perception, or knowledge about the world, could be attained only by a long learning period, in which input from various senses and remembered sensations become associated with each other. Through such associative learning, the objects and events in the outside world come to be known. By this account, most of the first year of life is spent learning to perceive. As recently as the mid-1960s, many pediatricians believed that newborns were functionally blind.

Recent research presents a different picture. The senses of an infant place the infant in meaningful contact with the outside world from birth. Early sensitivity to properties of depth, three-dimensional objects, motion, and events can be demonstrated for infants as easily as sensitivity to sensory qualities such as color or pitch. Some limitations on the precision of perception are resolved by maturational processes that improve sensory resolution during the first several months of life, and others by the development of attentional skill in using perceptual systems. However, the general picture is consistent with an ecological view that the senses evolved to provide knowledge about the environment. Acquiring such knowledge does not depend on learning to interpret sensations but rather on brain mechanisms sensitive to spatial and temporal patterns of information in incoming stimulation. Some specific areas of early competence are described below.

Objects. A primary question concerning visual abilities involves the detection from reflected light of the unity and boundaries of physical objects in the world. Experiments suggest that some basic mechanisms for detecting object boundaries are in place well before infants acquire experience from crawling or grasping. Likewise, there is evidence for the perception of three-dimensional form of objects within the first several months of life. In both cases, infants seem to use only a subset of the information available to adults, that is, earliest competence depends on information carried by motion. An adult may perceive the boundaries of objects quite well in stationary scenes and also get information from which things move together, even when parts of the objects are hidden behind nearer objects. Infants are sensitive to the information carried by motion before they are able to detect object boundaries by shape and color differences alone. Although adults perceive overall three-dimensional forms either from certain single views or from changing projections given when an object moves, infants are initially sensitive to only the latter information. For example, in an habituation/dishabituation experiment, infants were shown successive single static views of an object until visual attention declined. Then, in alternating test trials new views of the same object or views of a similar, but different, object were shown. It was hypothesized that if infants detected that all the habituation views came from the same three-dimensional object, they would recover visual attention to the new object after habituation. However, results indicated no differential attention to the views of the two objects in the tests; infants did not detect three-dimensional structure from these multiple static views.

The same experiment was conducted by using the objects in continuous rotation. Research in adult vision indicates that the continuous projective changes given by rotation provide a special class of information for perception of three-dimensional structure. After habituation to one rotating object, infants were tested alternately with a novel object and an old object, both in a new axis of rotation. Looking time decreased during habituation and remained low when infants viewed the same object in the test trials. However, looking time rebounded when the new object was viewed. Thus, it was inferred that infants detect three-dimensional structure from motion.

Still other results indicate that infant perception of object properties such as size and shape are influenced by distance information in ways similar to adults. Overall, the results suggest that infants visually perceive objects and their properties by using certain classes of information from the earliest months of life.

Depth and space. A classic controversy is whether humans can perceive three-dimensional space from birth, or whether the world is initially a meaningless two-dimensional picture. Research supports the idea that infants, like adults, perceive a three-dimensional world. However, there are several categories of information about depth and space, and all do not develop in the same way. Motion-carried information includes various ways that spatial layout is indicated by changes in optical projections given to a moving observer. Although more evidence is needed, there are indications that this sort of information functions very early in life. Both performance tests and neurophysiological evidence suggest that stereoscopic vision, the use of differences in the two eyes to specify depth, emerges by maturation of the visual cortex around 16 weeks of life and rapidly attains adult levels. Oculomotor cues, use of focusing and convergence adjustments of the eye muscles, may provide information early in life, but they attain greater precision later. Finally, the pictorial cues to depth, such as linear perspective, arrive quite late, sometime in the latter half of the first year.

Motion. Infants detect and attend to moving objects from birth. Sensitivity to the paths of moving targets has been demonstrated, as has the role of motion in providing information about object structure and spatial relations.

Intermodal perception. It was once believed that sensations from the separate senses function independently until learning can relate them to each other. Evidence now indicates that much intersensory coordination is unlearned, and is instead based on temporal synchrony and patterns, such as the detection of object impacts simultaneously through vision and hearing.

Physical and social knowledge. What is known about early perceptual abilities gives a different view of infancy. Instead of learning how to perceive, infants are engaged in learning how the physical and social worlds work. Study of these processes is relatively new, but there are early indications of the tools that facilitate development. Newborns show some ability to imitate facial expressions, an ability that may imply surprisingly advanced perceptual and motor organization, and possibly rudiments of social knowledge. There is evidence that human infants enter the world with certain core beliefs about the physical world. Apparent violation of these beliefs evokes surprise, even when the observer has had little chance to learn these beliefs from experience. Core beliefs include the notions that two objects cannot occupy the same place at the same time, objects move in continuous paths through space and time, and objects cause others to move by contact. In contrast, notions of inertia and gravity do not appear to be known in the first year of life.

For background information SEE COGNITION; INFORMATION PROCESSING (PSYCHOLOGY) in the McGraw-Hill Encyclopedia of Science & Technology.

Philip J. Kellman

Bibliography. W. Epstein and S. Rogers (eds.), *Handbook of Perception and Cognition*, vol. 5: *Perception of Space and Motion*, 1995; C. Granrud (ed.), *Development of Perception: The 1989 Carnegie-Mellon Symposium on Cognition*, 1992; C. Rovee-Collier and L. Lipsitt (eds.), *Advances in Infancy Research*, vol. 9, 1995; E. S. Spelke et al., Origins of knowledge, *Psychol. Rev.*, 99(4):605–632, 1992.

Comet

The most spectacular astronomical event in recent history began on July 16, 1994, when Comet Shoemaker-Levy 9 plunged into Jupiter. For one week in

July 1994, every telescope on Earth (as well as in the vicinity of Earth) was pointed at Jupiter, as the comet's many fragments followed each other into the planet.

Comet Shoemaker-Levy 9. The comet had been discovered by the comet hunters Carolyn and Eugene Shoemaker and David Levy on March 24, 1993, using the 0.46-m (18-in.) Schmidt telescope at Palomar Observatory in southern California. At the time, the length of the comet train as projected on the plane of the sky was about 160,000 km (100,000 mi). It was soon determined that the comet fragments came from a single parent body that had been captured into an orbit around Jupiter. The parent comet had been in Jovian orbit for several decades, but in July 1992 it came close enough to Jupiter that the planet's tidal forces pulled it apart. The unusual circumstances surrounding Shoemaker-Levy 9 were of great interest. Comets have been observed to split before, but never have so many fragments been observed. When it was announced that the comet fragments were on a collision course with Jupiter, the largest observing campaign in astronomical history was organized.

The comet was heavily monitored throughout the months before the collisions (**Fig. 1**). Dust from each fragment streamed outward in broad tails, propelled by pressure from sunlight. The fragments moved away from each other on separate orbits determined by the gravitational pull of Jupiter and the Sun. By July 1993, the length of the comet train had increased by about 40%. While the comet length increased, the positions and appearances of individual fragments continued to change up until the time of impact. Images from the newly repaired Hubble Space Telescope showed that some fragments separated further into multiple components, whereas other fragments disappeared. The causes of the disappearances or splitting were not known. No gaseous emissions were detected in the fragments, raising the question whether the parent body was truly a comet or an asteroid. For the time being, the convention was to refer to Shoemaker-Levy 9 as a comet.

The motion of Shoemaker-Levy 9 was chaotic, meaning that it was impossible to deduce its complete orbital history; but accurate predictions of the final segment of the trajectory of the comet were possible. The comet fragment nearest to Jupiter at the time of breakup was the first to impact. The impacts occurred on the far side of Jupiter as seen from the Earth, near a latitude of 44° S and a longitude about 5° behind the limb. This distance was close enough to the limb that impact sites were carried by Jupiter's rotation into sight within half an hour after each impact. The impacts took place over 6 days, and the fragments approached the atmosphere at an angle of approximately 42° from the vertical.

In spite of the heavy monitoring of the comet, little was known for certain about the impacts before the events. The main problem was the unknown sizes of the fragments, which were crucial in determining the magnitudes of the impacts. Because of the coma surrounding each comet fragment, it was impossible to determine the fragment sizes. Furthermore, it was not ascertained whether each fragment was a cohesive, solid body or a rubble pile of smaller grains. What was known about Shoemaker-Levy 9 is summarized in **Table 1**.

Jupiter. As the largest planet in the solar system, the radius of Jupiter is 10 times the radius of Earth, and it is 300 times more massive. It is composed largely of hydrogen (~89%) and helium (~11%), which gives it a mean density of only about 1340 kg/m^3 (84 lb/ft^3, or 1.34 times the density of water). In comparison, the density of Earth is about 5200 kg/m^3 (325 lb/ft^3, or 5.2 times the density of water). There are also small amounts of gaseous ammonia, methane, water, carbon monoxide, hydrogen cyanide, and other more exotic compounds. The bulk of Jupiter rotates once in 9 h 55 min, although

Table 1. Properties of Comet Shoemaker-Levy 9

Property	Value
Length	22 or more nuclei
Size	100–1000 m (300–3000 ft)
Mass	4×10^9 to 4×10^{12} kg (9×10^9 to 9×10^{12} lb)
Impact velocity	60 km/s (37 mi/s)
Impact energy	7×10^{18} to 7×10^{21} joules (2×10^3 to 2×10^6 megatons of TNT)
Strength	$\leq 10^4$ N/m^2 (200 lbf/ft^2)

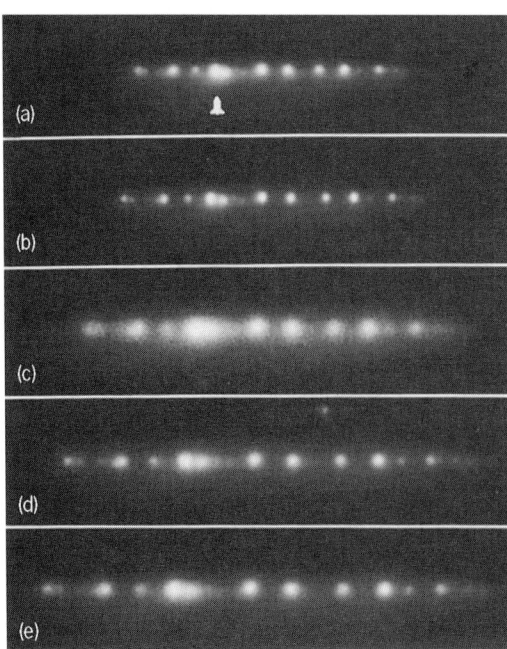

Fig. 1. Monthly photographs of Comet Shoemaker-Levy 9 during March–July 1993. Pictures were obtained with the University of Hawaii 2.2-m (87-in.) telescope, Mauna Kea, Hawaii. All photographs have the same scale. (*a*) March 27. (*b*) April 15. (*c*) May 21. (*d*) June 12. (*e*) July 17. *(Courtesy of D. Jewitt and J. Luu)*

Table 2. Properties of Jupiter

Property	Value
Radius	7.1×10^4 km (4.4×10^4 mi)
Mass	1.9×10^{27} kg (4.2×10^{27} lb)
Density	1340 kg/m^3 (84 lb/ft^3)
Composition	Mostly H_2 (89%) and He (11%)
Rotation period	9 h 55 min
Escape velocity from planet	60 km/s (37 mi/s)

the cloud features on Jupiter have slightly different rotation periods resulting from intrinsic cloud motion.

The visible surface of Jupiter is a deck of clouds, made of ammonia crystals. At the top of the cloud, the pressure is of the order of 1 atmosphere (100,000 pascals). Below the ammonia clouds, ammonium hydrosulfide clouds and water-crystal clouds are believed to be present, followed by clouds of liquid water. **Table 2** summarizes the Jovian properties most relevant to the comet collisions.

Impact phenomena. Comet Shoemaker-Levy 9 was predicted to plunge into Jupiter at a speed comparable to the escape velocity from the planet, 60 km/s (37 mi/s), or about 50 times the speed of sound in Jupiter's very light, mostly hydrogen atmosphere. At this velocity, the impact energy is much greater than the energy required to sublimate or vaporize the comets, so all the comet fragments should be vaporized. The atmosphere acts as a shield; it dissipates the momentum of the incoming projectile and tries to stop it, at which point the projectile should fragment. The fragmentation height depends on the velocity and strength of the projectile; the higher the incoming velocity and the weaker the projectile, the higher up it fragments. This fragmentation is the origin of atmospheric explosions; fragmentation causes the projectile to spread, leading to a deceleration by air friction, which can then lead to an explosion. If the projectile expands very slowly, it will experience little deceleration and will descend as a single body; no explosion will take place. However, if the projectile expands very rapidly, the deceleration will be dramatic and an explosion will occur.

Impact flashes. All the pieces of Shoemaker-Levy 9 were expected to fragment above the clouds, then explode either above or below the clouds, depending on the fragment size. Large fragments (radius equal to or greater than 1 km or 0.6 mi) were expected to explode below the clouds, whereas smaller ones (radius ~100 m or 300 ft) would explode above them. A bright flash lasting a few seconds was predicted for each comet nucleus passing through the Jovian stratosphere, much like the bright flash from a terrestrial meteorite entering the Earth's atmosphere. A second flash was expected to follow because of the explosion of the nucleus, analogous to the fireballs observed in atmospheric bomb tests on Earth. The brightness of both flashes would depend critically on the energy released at impact. The impacts were expected to occur on the far side of Jupiter, but close enough to the limb that impact flashes might be observable at the limb.

The explosion heights were not known for all impacts, but observations suggest that, indeed, some fragments exploded below the clouds. In the case of the strongest impacts (which yielded the most reliable data), two flashes were observed. The flashes were short (1–3 min), the second one lasting longer than the first. Probably the first flash corresponded to the entry of the fragment into the stratosphere, and the second to the explosion (the fireball itself). As the fireball cooled, the second flash faded. There was no evidence of the flashes being reflected by nearby Galilean satellites.

Atmospheric effects. The kinetic energy of a 1-km (0.6-mi) radius nucleus is 10^{21} joules, or the equivalent of 10^6 megatons of TNT. Deposited in the atmosphere, this energy should create a plume that will ascend. Thus, after the initial flashes, a hot plume of gas was predicted to rise in the atmosphere, lasting a few minutes and radiating mostly in the infrared. The plume would contain material from the comet as well as from Jupiter's atmosphere. The plume composition was thus of great importance as it was diagnostic of the depths to which the fragments penetrated, and of the composition of the Jovian atmosphere below the ammonia clouds. As the plume cooled, material in the plume might recondense and form clouds in the stratosphere. The motion of the clouds would help map the wind circulation in the Jovian stratosphere.

As predicted, a bright, rapidly expanding flare lasting ~10 min followed the second flash. At infrared wavelengths (for example, 2.3 microme-

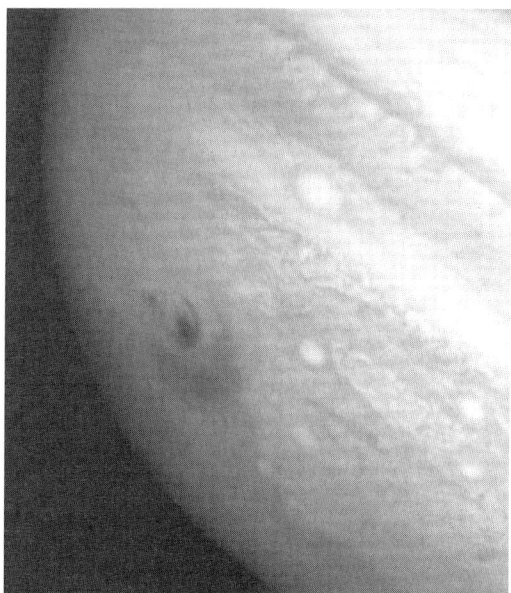

Fig. 2. Image of Jupiter from Hubble Space Telescope, 1 h 45 min after impact of fragment "G," the largest fragment of comet Shoemaker-Levy 9, on July 18, 1994. *(Hubble Space Telescope Comet Team and NASA)*

Fig. 3. Four photographs of Jupiter from the spacecraft *Galileo*, showing remnant cloud after the impact of fragment "W" on July 22, 1994. The cloud can be seen toward the bottom of the planet, near the South Pole. Images were taken 2.3 s apart, with a green filter. *(Galileo Science Team)*

ters), this flare even outshone the entire disk of Jupiter. This flare was best explained by a huge, hot plume of gas, located high above the explosion site and heated by the release of gravitational potential energy as the explosion ejecta fell back down in the atmosphere. As the plume expanded and cooled, it eventually turned into a dark dust cloud that lingered for weeks. **Figure 2** shows an image of Jupiter from the Hubble Space Telescope, 1 h 45 min after the impact of fragment "G," the largest fragment of Shoemaker-Levy 9, on July 18, 1994. The point of impact, the central dark spot, 2500 km (1550 mi) in diameter, is surrounded by a ring, 7500 km (4660 mi) in diameter, that may represent a sonic boom created by the exploding nucleus. The crescent-shaped object is ejecta from the fireball. The smaller feature to the left of the fragment "G" impact site was formed by the impact of fragment "D" about 20 h earlier. Another example of an impact cloud can be seen in **Fig. 3**, which shows four images of Jupiter after the impact of fragment "W."

Spectra of the impacts showed evidence of a large number of molecules, including most of the molecules known or expected to exist in Jupiter (such as ammonia, hydrogen cyanide, carbon monoxide, and methane), as well as unexpected ones. A considerable amount of water was also detected, supporting a cometary origin. The spectra are expected to yield valuable information on the composition and chemistry of the Jovian atmosphere when the data analysis is complete.

Unresolved questions. The challenge that now faces astronomers is to interpret the wealth of data that was collected during the impacts. As the data are examined, questions materialize more readily than answers. Some of the unresolved questions concern the nature of the parent body of Shoemaker-Levy 9, and, in particular, whether it was a comet or an asteroid, and the composition of the comet/asteroid projectile.

For background information *SEE* ASTEROID; COMET; JUPITER in the McGraw-Hill Encyclopedia of Science & Technology.

Jane Luu

Bibliography. J. K. Beatty and S. J. Goldman, The great crash of 1994: A first report, *Sky Telesc.*, 88(4):18–23, October 1994; J. K. Beatty and D. H. Levy, Awaiting the crash, *Sky Telesc.*, 87(1):40–44, January 1994, and 88(1):18–23, July 1994; R. A. Kerr, Second thoughts about Shoemaker-Levy impact, *Science*, 266:975, 1994; R. L. Newburn, Jr., *Periodic Comet Shoemaker-Levy 9 Collides with Jupiter: Background Material for Science Teacher*, NASA and Jet Propulsion Laboratory, JPL 400-520, March 1994.

Communications satellite

The most recent generations of INTELSAT satellites, which are the chief carriers of international satellite communications, continue to incorporate new technology and to increase in power and per-

Series*	Retired satellites	Operational satellites	Satellites to be launched	Total
INTELSAT I	1			1
INTELSAT II	4			4
INTELSAT III	8			8
INTELSAT IV/IVA	14			14
INTELSAT V	3	12		15
INTELSAT VI		5		5
INTELSAT K		1		1
INTELSAT VII		4	2	6
INTELSAT VIIA			3	3
INTELSAT VIII			4	4
INTELSAT VIIIA			2	2
Total	30	22	11	63

*As of the end of January 1995.

Fig. 1. INTELSAT communications satellite series. (*a*) INTELSAT VII. (*b*) INTELSAT VIII.

formance. Meanwhile, advanced communication technologies are being developed and tested by the Advanced Communications Technology Satellite (ACTS) program.

INTELSAT VII and VIII Spacecraft

Since its creation in 1964, INTELSAT has procured its geostationary telecommunications satellites in a succession of series (**Table 1**), for use in international, regional, and domestic communications systems. One exception to this multisatellite procurement rule is the single INTELSAT K. At the end of January 1995, INTELSAT operated 22 satellites in orbit, among which the most advanced belonged to the INTELSAT VII generation. The remaining INTELSAT VII and VIIA satellites are scheduled for launch by early 1996. The following generation, comprising INTELSAT VIII and VIIIA satellites, is under procurement for launches scheduled mainly in 1996 and 1997.

INTELSAT VII spacecraft. Of the nine spacecraft in the INTELSAT VII and VIIA generation (**Fig. 1***a*), six are called INTELSAT VII and three are called INTELSAT VIIA. The INTELSAT VII/VIIA series was designed to meet the unique requirements of the Pacific Ocean region, but it is a versatile spacecraft series that will be used throughout the INTELSAT system.

Some capabilities and characteristics of INTELSAT VII satellites are given in **Table 2**. INTELSAT VII and VIIA spacecraft provide peak power levels up to 41.8 decibels above 1 watt at C band and up to 51.4 dBW at Ku band for INTELSAT VII and 54.4 dBW for INTELSAT VIIA. Combin-

Table 2. Capabilities and characteristics of INTELSAT VII/VIIA and VIII/VIIIA spacecraft

Capabilities and characteristics	INTELSAT VII	INTELSAT VIIA	INTELSAT VIII	INTELSAT VIIIA
Two-way telephone circuits	18,000	22,500	22,500	14,000
Two-way telephone circuits with DCME*	90,000	112,500	112,500	70,000
Analog television channels[†]	3	3	3	3
Dry mass	1500 kg (3307 lb)	1850 kg (4079 lb)	1560 kg (3439 lb)	1600 kg (3527 lb)
Mass at launch	3700 kg (8157 lb)	4200 kg (9259 lb)	3600 kg (7937 lb)	3600 kg (7937 lb)
Total length of deployed solar array	22 m (72 ft)	27 m (89 ft)	24 m (79 ft)	24 m (79 ft)
End-of-life solar generator power	4 kW	5.3 kW	5.4 kW	5.4 kW
Total number of transponders	36	40	44	31
C band	26	26	38	28
Ku band	10	14	6	3

* DCME = digital circuit multiplication equipment.
[†] Greater television service is possible with digital compression.

Table 3. INTELSAT VII/VIIA and VIII/VIIIA launches

Satellite*	Series	Launch date	Launch vehicle
INTELSAT 701	INTELSAT VII	October 1993	*Ariane 44LP*
INTELSAT 702	INTELSAT VII	June 1994	*Ariane 44LP*
INTELSAT 703	INTELSAT VII	October 1994	*Atlas IIAS*
INTELSAT 704	INTELSAT VII	January 1995	*Atlas IIAS*
INTELSAT 705	INTELSAT VII	March 1995	*Atlas IIAS*
INTELSAT 706	INTELSAT VIIA	May 1995	*Ariane 44LP*
INTELSAT 707	INTELSAT VIIA	Fourth quarter 1995	*Long March 3B*
INTELSAT 708	INTELSAT VIIA	First quarter 1996	*Ariane 44LP*
INTELSAT 709	INTELSAT VII	Second quarter 1996	*Ariane 5*
INTELSAT 801	INTELSAT VIII	Second quarter 1996	*Ariane 44P*
INTELSAT 802	INTELSAT VIII	Third quarter 1996	*Ariane 44P*
INTELSAT 803	INTELSAT VIII	Fourth quarter 1996	*Ariane 44P*
INTELSAT 804	INTELSAT VIII	First quarter 1997	*Ariane 44P*
INTELSAT 805	INTELSAT VIIIA	Fourth quarter 1997	*Long March 3B*
INTELSAT 806	INTELSAT VIIIA	Third quarter 1997	*Long March 3B*

* Status at end of January 1995. Launches of *INTELSAT 705–709* and *801–806* are projected.

ing geographical separation and the two circular polarizations, INTELSAT VII/VIIA satellites reuse four times the bandwidth of 500 MHz allocated in the C band, and, through vertical and horizontal polarization, two times the 500 MHz allocated in the Ku band.

INTELSAT VII are 36-transponder spacecraft, with 26 in the C band and 10 in the Ku band; INTELSAT VIIA are 40-transponder spacecraft, with 26 in the C band and 14 in the Ku band. The 26 transponders in the C band are allocated to four types of coverage: 10 in two hemispheric coverages (hemi 1 and 2); 10 in four zone coverages (zones A, B, C, and D); and 6 in either global or C-spot coverages. The equivalent isotropic radiated power (eirp) is 33 dBW at the edge of the hemispheric and zone coverages, 26–29 dBW at the edge of the global coverages (depending on channel), and 33.3–36.3 dBW at the edge of the C-spot coverages. The 10 transponders in the Ku band for INTELSAT VII and the 14 transponders for INTELSAT VIIA are allocated to three Ku-spot coverages (S1, S2, and S3). Each spot coverage (C or Ku) is circular and steerable anywhere over the visible portion of the Earth.

INTELSAT VII are very flexible spacecraft whose capabilities include multiorbital location use in the three ocean regions (Atlantic, Indian, and Pacific), the ability to match east-west traffic imbalance patterns, improved radio-frequency performance with respect to earlier generations of INTELSAT satellites, spacecraft attitude inversion capability to improve coverages at some orbital locations, long in-orbit life (between 13 and 16 years depending on the launch vehicle used), and easier and more secure satellite control using on-board microprocessors.

Launches of INTELSAT VII/VIIA satellites are listed in **Table 3**.

INTELSAT VIII spacecraft. Of the six spacecraft in the INTELSAT VIII generation (Fig. 1*b*), four are called INTELSAT VIII and two are called INTELSAT VIIIA, or Land Mass satellites. Once again, INTELSAT VIII was designed with the specific needs of the Asia-Pacific region in mind for improved C-band coverage and service.

Some capabilities and characteristics of INTELSAT VIII satellites are given in Table 2. INTELSAT VIII will provide significantly more C-band capacity for public-switched telephony and business services, provide better quality for video services, and encourage new very small aperture terminal (VSAT) applications.

Combining geographical separation and the two circular polarizations, INTELSAT VIII will reuse six times the bandwidth of 500 MHz allocated in the C band, whereas INTELSAT VIIIA will reuse two times the extended bandwidth of 800 MHz allocated in the C band.

INTELSAT VIII are 44-transponder spacecraft, with 38 in the C band and 6 in the Ku band. The 38 transponders in the C band are allocated to three types of coverage: 12 in two hemispheric coverages (hemi 1 and 2), 20 in four zone coverages (zones 1, 2, 3, and 4), and 6 in global coverage. The six transponders in Ku band are allocated to two circular spot beams steerable anywhere over the visible portion of the Earth.

INTELSAT VIII are very flexible spacecraft whose capabilities include multiorbital location use in the three ocean regions (Atlantic, Indian and Pacific), improved radio-frequency performance with respect to INTELSAT VII (36-dBW equivalent isotropic radiated power at the edge of hemispheric and zone coverages, compared with 33 dBW for INTELSAT VII), spacecraft attitude inversion capability, use of arcjet thrusters for north-south orbit correction with improved performance, long in-orbit life (more than 15 years), and use of only solid-state power amplifiers for C-band transponders.

INTELSAT VIIIA use the same platform as INTELSAT VIII, but their payloads are devoted to specific communications functions, mainly in the C band, with a land-mass coverage tailored to the needs at specific orbital locations. One of the two INTELSAT VIIIA satellites under procurement, *INTELSAT 805*, is for the Asia-Pacific region, and *INTELSAT 806* is mainly for South America.

The INTELSAT VIIIA use the full 800-MHz bandwidth in the C band allocated to fixed satellite

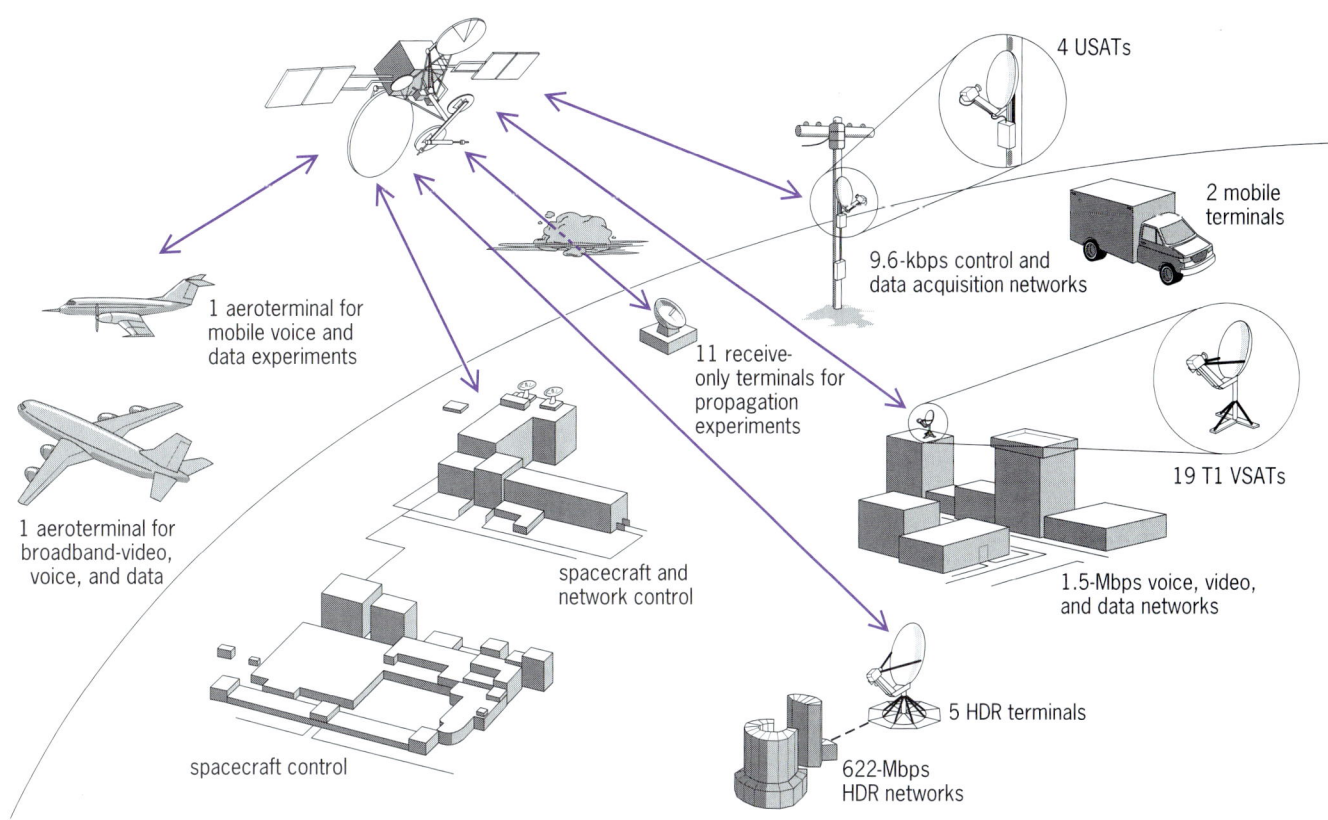

Fig. 2. Advanced Communications Technology Satellite (ACTS) system, showing operations of the satellite, control centers, and user terminals. The abbreviations are explained in the text.

services, with 28 C-band transponders in one composite C-band coverage. The high equivalent isotropic radiated power of 36 dBW is obtained by using 55-watt traveling-wave-tube amplifiers. In addition, INTELSAT VIIIA have three high-power Ku-band transponders (115-W linearized traveling-wave-tube amplifiers) allocated to a steerable coverage. Launches of INTELSAT VIII/VIIIA satellites are listed in Table 3.   *Pierre J. Madon*

ACTS Program

The Advanced Communications Technology Satellite program of the National Aeronautics and Space Administration (NASA) was intended to develop advanced technologies and prove them in actual user trials. *ACTS,* which was launched in September 1993, has created technologies such as high-gain hopping spot beams to small-aperture terminals, on-board circuit switching, on-demand circuit assignment, Ka-band (with an uplink frequency of 30 GHz and a downlink frequency of 20 GHz) radio-frequency transmission, and wide-bandwidth (900-MHz) channels, as well as new services.

ACTS system. The ACTS system (**Fig. 2**) is composed of the satellite, control centers, and more than 50 user terminals. **Figure 3** gives the information data rates for each type of user terminal. The communications services provided include T1(1.5 megabits per second), very small aperture terminal (VSAT; 1.2 m or 3.7 ft) links to customer premises, including integrated services digital network (ISDN) services, for compatible satellite-terrestrial communications; high-data-rate (HDR; 622 Mbps) terminal (aperture of 3.4 m or 11.2 ft) networks for compatible satellite-terrestrial communications; broadband data communications (up to 1.5 Mbps) to aircraft for video and data services, using a 15-cm (6-in.) aperture reflector terminal; voice and low-rate (64 kilobits per second) data communications to aircraft using phased-array patch antennas; terrestrial mobile voice and low-rate (64 kbps) data communications; and supervisory control and data acquisition (SCADA) networks using an ultra-small-aperture terminal (USAT, 36 cm or 14 in.) for low-data-rate (9.6 kbps) communications.

User program. The true value of the ACTS system is to be assessed by the user trials that are in process and planned to continue through 1998. **Table 4** gives the categories of uses and the number of users in each category.

The user trials have shown that the ACTS system meets or exceeds its performance requirements. Of the many T1-VSAT user trials, some of the more significant are as follows: A bank used ACTS to connect its branch offices to main computers without terrestrial interconnects, and concluded that an on-demand satellite system would meet all its needs for network restoral and backup. ACTS has been used in telemedicine experiments where the services of a specialist are made available to patients in rural or other remote settings by means of two-way video

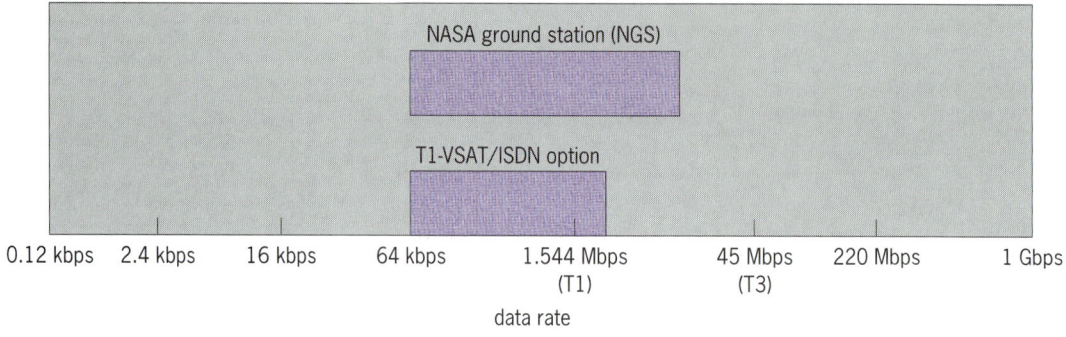

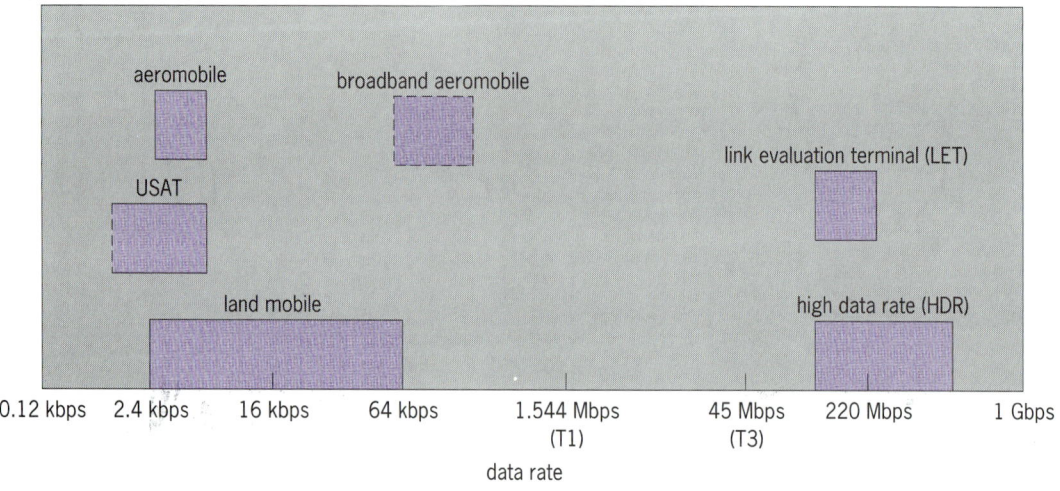

Fig. 3. ACTS user terminal types, for operation with ACTS in (*a*) baseband processor mode and (*b*) microwave switch matrix mode. The abbreviations are explained in the text.

communication. The U.S. Army has gained a great deal of experience using ruggedized units that are transported into the field and set up as needed to support operations in a given area. During 1994, the U.S. Army used the ACTS capability to support its field operations and found that the ACTS high-quality videoconferencing capability significantly enhanced its command-and-control function.

Future field trials. The wide-bandwidth channel capabilities of ACTS (900 MHz) offer a unique and unparalleled opportunity to deliver multimegabyte services through a compatible interface with the terrestrial network. NASA and the Advanced Research Projects Agency (ARPA) are sponsoring a joint development of Earth stations with 3.4-m (11.2-ft) apertures; these stations will provide OC-3 (155-Mbps) or OC-12 (622-Mbps) services using the synchronous optical network (SONET) physical-layer protocol. Three Earth stations will be connected into local-area fiber networks at Washington, D.C., Kansas City, and Hawaii. Two more Earth stations will be placed at other locations, depending on the operations to be performed.

Users will pioneer a wide variety of applications with this high-data-rate network. The applications, many of which will be performed by using asynchronous transfer mode (ATM) cell transmissions, include telemedicine; parallel supercomputing; remote calculations, visualization, and control; high-definition-television (HDTV) video distribution; and network investigations.

One rapidly expanding application is commer-

Table 4. ACTS user categories

Use category	Number of users
Business networks	9
Medical	7
Integrated services digital network (ISDN)	5
High-definition television (HDTV)	1
Supervisory control and data acquisition (SCADA)	1
Public switched network restoral	3
Land and aeronautical mobile	9
Educational	7
Gigabit network	6
Video conferencing	1
Department of Defense strategic/tactical	2
Science networks	3
Network protocol	4

cial communications for passengers aboard airliners. ACTS will be used to test the ability to install a so-called office-in-the-sky and to receive video. A two-axis-gimbal reflector of approximately 16 cm (6.3 in.) will be installed on the aircraft. Services to be tested include communications at data rates up to 1.544 Mbps from the ground to the aircraft and 386 kbps from the aircraft to the ground.

New systems. Several systems that are now either under development or under construction will utilize ACTS-type technologies. One is Iridium, a global personal communications system using 66 low-Earth-orbit (LEO) satellites. This system will provide voice and low-data-rate communications between users of hand-held telephones located anywhere on the globe. As in ACTS, Iridium uses an on-board baseband switching system. Instead of routing individual circuits between many ACTS spot-beam locations, the Iridium system will route circuits between individual satellites. In addition to following ACTS in using on-board baseband switching, the Iridium system will use the Ka band for the cross-links between satellites and for communications to the gateway stations. SEE DATA COMMUNICATIONS.

The advantage of high-gain spot beams is the ability to use smaller-aperture, inexpensive Earth stations. Thus, a proposal has been filed with the Federal Communications Commission to place in orbit a 48-fixed-spot-beam Ka-band geostationary satellite called *Spaceway* to provide inexpensive, ubiquitous, high-speed, switched data, video, and videotelephone communications services to both residential and commercial users. The spot-beam network would use Earth stations approximately 66 cm (26 in.) in aperture to provide transmission rates of 384 kbps, which would bring two-way picturephone and data services into the home. Nine satellites with cross-links are planned to provide global communications.

The concept of the information superhighway is one in which users will be able to call up, on demand, wide-bandwidth transmission capabilities over a global network relatively inexpensively. ACTS technologies or concepts, such as Iridium and Spaceway, hold the potential for realizing this concept. The ACTS user program will provide additional information for the communications carriers and suppliers to further assess the ACTS technologies and their potential use in new systems for the information superhighway. With the availability of ACTS technologies and of those developed by the U.S. Department of Defense and industry, future communications satellites will be substantially different from those of the past.

For background information SEE COMMUNICATIONS SATELLITE; DATA COMMUNICATIONS; INTEGRATED SERVICES DIGITAL NETWORK (ISDN) in the McGraw-Hill Encyclopedia of Science & Technology.

Richard T. Gedney

Compact disk

A new pathway for increasing the storage capacity of optical disks, called multilayer optical storage, has been developed. It extends optical storage to the third dimension by stacking data layers one above each other, with each layer separated by a spacer region (**Fig. 1**). This approach takes advantage of a unique property of electromagnetic radiation: light can be selectively focused to a distinct layer throughout the volume of an optical storage medium. Such an extension in the third dimension would obviously lead to substantial improvements of the volumetric storage capacities of an optical disk.

The concept is in principle applicable to all forms of read-only and writable optical disks. It can be used for relatively low-performance systems such as CD-ROM (compact disk, read-only memory) and for high-performance computer data applications. An important attribute is its ability to readily be implemented to allow an optical drive to be backward-compatible for reading and writing existing single-layer disks. In addition, the multilayer technology can be combined with other methods of increasing the areal density of an optical disk such as shorter-wavelength lasers and new methods of encoding data.

Optical storage. A number of attributes make optical storage useful for many applications of data storage. Read-only memory (ROM) disks allow large quantities of data to be distributed very inexpensively. Writable media allow archiving of data safely and inexpensively for decades. The optical disks can be removed from the drives, which allows cost-effective extension of the storage capacity by simply adding more disks. Optical libraries using robotics systems to implement automated disk exchanges exploit this removability to provide tens or hundreds of gigabytes of data on line at low cost (in terms of dollars per megabyte). Implementation of multilayer disks can multiply these capacities severalfold.

Drive system. Optical drives contain several key components for multilayer optical storage. A laser light source is focused by a high-quality objective lens to form a very small spot on the optical disk. This lens is mounted on an actuator that allows for

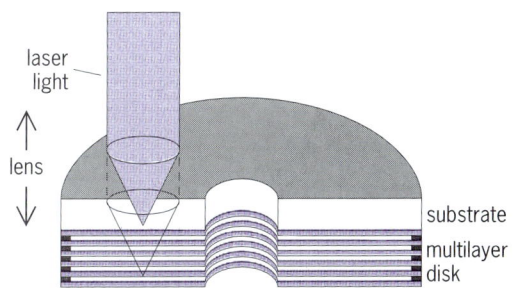

Fig. 1. Cross section of a multilayer optical storage disk. The objective lens is moved up and down to focus on the desired layer for writing and reading.

rapid adjustment of the focused spot in two directions. Movement perpendicular to the disk ensures that the laser beam maintains focus on the disk even if the disk is highly warped. Movement radially along the disk surface ensures that the focused beam remains aligned in the center of the track for reliable reading and writing of data, or for providing fast access to all other tracks on the disk surface. All motions of the lens are controlled precisely by closed-loop servo systems that sense when the lens is wandering from the desired position and correct that wandering. Separate servo systems control the laser. For reading data, the laser power is kept constant. During writing, the laser is pulsed rapidly between high- and low-output power to form marks corresponding to encoded digital data on the disk.

Storage capacity. In optical storage, improvements of storage capacity have traditionally been achieved by increasing the areal density, the amount of information stored in a unit area. This increase can be accomplished by decreasing the size of the written marks and placing them closer together. Light from near-infrared semiconducting lasers with wavelengths of 780 nanometers are currently focused to a spot size that is slightly less than 1 micrometer in diameter. With such a small spot, which is less than 1/50 the diameter of a human hair, 650 megabytes of digital data can be stored on each side of a 130-mm-diameter (5¼-in.) disk at current densities. In the future, as shorter-wavelength recording technology becomes available, the areal density can improve quadratically with the wavelength reduction.

Avoidance of crosstalk. Many technical issues are related to the implementation of multilayer volumetric storage, but a few stand out as key concerns. First, partially transmissive recording layers have to be made with a transmission that is high enough to allow light to reach the deepest layers and yet reflective enough to provide an adequate signal to the data detectors. It is also necessary that data present on one layer not cause false signals, known as crosstalk, while reading or writing on another layer.

Reading crosstalk could arise in principle from data on the out-of-focus layers affecting the beam while it is focused on another layer. However, detailed modeling and experiments show that, with reasonable layer spacings, the laser read beam is so large on the out-of-focus layers that it is only modulated by small edge effects and the reading crosstalk is insignificant. Similarly, crosstalk into the servo detectors is reduced significantly with reasonable layer spacings. In practice, writing crosstalk does not occur because the beam is spread out when it passes through the out-of-focus layers so that it does not heat any of the recording material above its recording temperature. Writing can occur only when the laser beam is focused to a small spot directly on the recording layer.

Read-only disks. The most popular form of optical storage is the compact disk, which is an example of read-only media. The compact disk is manufactured by first making a precision stamper that contains bumps corresponding to recorded data. Molten polycarbonate is forced around the bumps by an injection molding machine to make a disk which contains pits. Since many copies of the same disk are made from one stamper, the cost is quite low. Normally, a CD-ROM has an aluminum alloy coating deposited by vacuum deposition on the surface to raise the reflectivity above 70% and hence give a strong read-back signal.

A two-layer volumetric version of the compact disk can be made by suitably modifying the basic compact-disk design. The simplest implementation has two disks with the data layers facing each other across an air gap. The spacing between the disks is maintained by a set of spacer rings located at the inner and outer diameters of the disk. Optical interference considerations require that the depth of the pits on layer 2 be increased by 50% to compensate for the laser light being incident from the air rather than through the substrate. Because the presence of the aluminum layer on layer 1 would make the layer nontransmissive, it is not applied. Instead, a partially transmissive coating is applied which has sufficient reflectivity for reliably reading the data once the detected signal is boosted by a low-noise amplifier. Various implementations of such two-layer compact disks have been made, including both audio and video playback. Since each layer can store 70 min of compressed video data, a full-length movie of more than 2 h could be stored on such a two-layer disk.

Read-only volumetric optical storage disks can be made with many more than two layers. Laboratory experiments have demonstrated extensions of this concept and data from up to six layers have been reliably read. A mockup of a 10-layer multilayer volumetric disk is shown in **Fig. 2**. The drive implementation will be more complex than for a two-layer system since it is necessary to correct for distortion of the optical beam caused by its transiting several substrates. The maximum number of layers which can be achieved is not known yet. Basic physics suggests that more than 10 layers can be used.

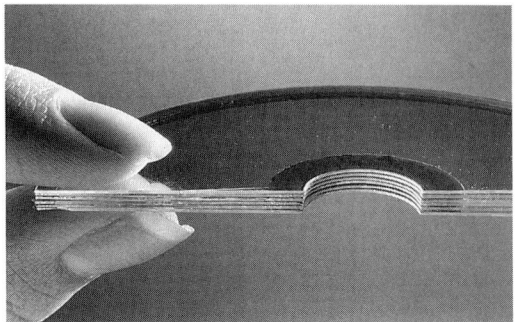

Fig. 2. A 10-layer read-only optical disk.

Writable optical disk. Writable versions of multilayer volumetric disks have been fabricated. The requirements for the media design are more stringent for this implementation since the writable layer must have several new attributes not needed for read-only media. It must absorb enough laser light so that a spot can be heated to a high temperature locally to form a written mark. The absorption of the layer must not be so great as to prevent the transmission of light to a deeper layer in the structure for reading and writing on that layer. These considerations will limit the number of writable layers to be less than the number of layers that can be implemented on read-only disks. However, even with these limitations the capacity of writable optical disks can be significantly increased by using this multilayer approach. Two-layer and four-layer writable memories have been demonstrated with good writing and reading performance. The development of writable memories with more layers will require continued research on new materials systems.

For background information SEE COMPACT DISK; COMPUTER STORAGE TECHNOLOGY; OPTICAL RECORDING; VIDEO DISK RECORDING in the McGraw-Hill Encyclopedia of Science & Technology.

Kurt Rubin; Wayne Imaino; Hal Rosen

Conodonts

Conodonts are a group of extinct marine animals that were entirely soft-bodied except for their bonelike teeth, which formed a complex oral apparatus. The teeth were usually dispersed across the sea floor after the death and decay of the parent animal and are common fossils in strata of Late Cambrian to Late Triassic age. Each tooth is normally microscopic, 0.008–0.08 in. (0.2–2 mm), in size, but rare larger examples up to 0.8 in. (20 mm) occur. They display a range of shapes from simple cones through denticulated bars and blades to highly ornamented plates.

Recent discoveries of fossilized soft-bodied animals and new research on the internal microstructure of conodont elements have provided evidence that the conodonts were primitive vertebrates. They were probably the first animals in the vertebrate lineage to secrete biomineralized phosphatic elements, suggesting that the vertebrate skeleton arose as a toothlike feeding device in an active predatory or scavenging animal.

Morphology. Fossils of soft-bodied organisms are very rare. For this reason, the nature of the conodont animal was totally unknown until 1982 when a nearly complete fossilized specimen was found in a museum drawer. The specimen had been collected early in this century from Lower Carboniferous rocks near Edinburgh, Scotland, but its significance had not previously been appreciated. Subsequently, nine additional specimens were found at the original locality, and in 1994 a single soft-bodied animal was discovered in Ordovician shales north of Cape Town, South Africa.

The features of the soft parts of the Scottish animals are preserved as a calcium phosphate film. These animals were elongate, 1.6–2.2 in. (40–55 mm) in length and less than 0.08 in. (2 mm) in height, and had a short head and large eyes. The preserved body traces show a long trunk, along which runs a pair of axial lines representing the decayed notochord, a precursor of the backbone. V-shaped muscle blocks are evident along the entire length, and on two of the specimens an apparent ray-supported caudal fin is present at the posterior end. Some of the muscle blocks and the notochord on one specimen show a fibrous microstructure, which may reflect original muscle fibers. Two specimens show a possible dorsal nerve cord, and one has evidence of possible auditory capsules and branchial structures behind the eyes. The only skeletal structure of the animals, the feeding apparatus, lies ventrally and immediately posterior to the eyes. There is no evidence of any tissue surrounding this skeleton, indicating that the soft parts are incompletely preserved, at least in the ventral anterior portion of the specimens.

The single specimen from South Africa is much larger; it belongs to a different order of conodonts but shows comparable soft-tissue features. Only the anterior end has been found; it displays eyes, a feeding apparatus, and 4 in. (100 mm) of preserved trunk. The skeletonized feeding apparatus is positioned below and just behind the eyes and is composed of 19 individual elements up to 0.4 in. (10 mm) long. The body shows a pattern of V-shaped muscle segments similar to the Scottish specimens, and the line of the notochord is represented by a longitudinal gap in mineralization at the apices of the muscle chevrons. The body trace is nearly 0.8 in. (20 mm) across, suggesting that the entire animal may have been up to 16 in. (400 mm) in length. However, the eyes are only a little larger than those of the Scottish animals.

By using evidence from the Scottish and South African specimens, the morphology of a living conodont animal can be broadly reconstructed (see **illus.**). A single poorly preserved specimen of a con-

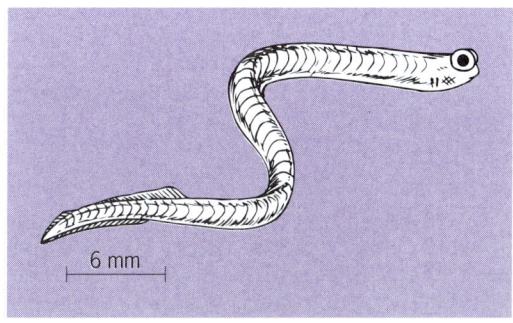

Restoration of the living conodont animal, with proportions based on the Scottish specimens.

odont animal with an apparatus of cone-shaped elements has also been reported from the Silurian of Wisconsin. Its indistinct features suggest that the bodies of some conodonts may have been broader than the Scottish and South African animals.

Biological affinities. The soft-part morphology of the conodonts indicates that they belong to the chordates: the V-shaped muscle blocks, the notochord and the dorsal nerve cord are all chordate features. Gill slits have not yet been recognized, but they have a low preservation potential and would lie in the anteroventral region where soft-tissue traces are absent in the fossils. Within the chordates, the presence of hardened cartilages that supported large eyes suggests a substantial degree of encephalization (the developmental process by which the cerebral cortex has taken over the functions of the lower, or spinal, centers), characteristic of the vertebrates. The termination of the notochord anteriorly before the feeding apparatus is also a vertebrate feature, which contrasts with the more extensive notochord of the more primitive cephalochordates, such as the lancelet *Branchiostoma*.

The vertebrate nature of the conodonts is corroborated by histological work on the teeth undertaken by using Nomarski differential interference contrast microscopy and scanning electron microscopy of polished and etched slices. This research has demonstrated the presence of hard tissues similar to cellular bone, enamel, and dentine in conodont elements of various types. The enamel forms translucent lamellar layers in the element crowns, which also show areas of opaque white matter that display the cell lacunae and small channels (canaliculi) typical of vertebrate cellular bone. Tubules comparable with those found in dentine have been recognized in the basal bodies of some Ordovician conodont elements; basal bodies of some other species contain globular calcified cartilage.

The absence of skeletonized vertebral elements along the notochord and the lack of phosphatic skin armor point to a primitive position for the conodonts in the vertebrate clade. There are many similarities to the myxinoids, or hagfishes, including the possession of a bilaterally operative feeding apparatus. However, the hagfishes have no phosphatic elements, possessing a lingual apparatus of simple teeth composed of organic keratin and without skeletal supports. The conodonts, therefore, may have been the first vertebrates to develop a biomineralized phosphatic skeleton. Thus, they are placed as a sister group to all other vertebrates, apart from the myxinoids.

The identification of the true affinities of conodonts pushes the vertebrate record back into the Late Cambrian. Possible ancestors to the unequivocal conodonts (euconodonts), represented by microscopic phosphatic cones referred to as paraconodonts also occur in Lower and Middle Cambrian strata. If the relationship of paraconodonts to the euconodonts is histologically confirmed, the record of vertebrate ancestry would be extended back to the Early Cambrian metazoan radiation.

Architecture and function. The earliest euconodont apparatuses were composed of an array of simple conical teeth that were most likely used to grasp small prey. In the Ordovician, a much more complex apparatus rapidly evolved, incorporating a variety of denticulated, serrated, and platformlike elements in a bilaterally symmetrical arrangement. Detailed studies of exceptionally preserved complete skeletons of giant conodonts from the Ordovician of South Africa have enabled reconstruction of three-dimensional models of their feeding apparatus. In these animals, 11 elongate, ramiform elements, each with 2–4 posteriorly directed denticulated processes, appear to have formed a grasping array for impaling prey and passing it backward; 4 opposing pairs of arched and platform elements above and behind probably served to shred and crush the captured food. Other types of apparatus, common from the Silurian onward, were simpler but had a comparable set of 11 elongate grasping elements, each with 2–3 processes of which one was long, barbed and posteriorly directed; directly behind this array was a scissorlike pair of slicing elements followed by a pair of shearing or grinding pectiniform elements. This type of apparatus was possessed by the Scottish conodont animals.

One theory is that the conodont elements did not serve as teeth but as supports for a soft, ciliated filter-feeding device. However, the arrangement of the apparatus in the preserved conodont animals, with the more delicate, denticulated ramiform elements in front of the robust serrated and platform-bearing elements, is evidence against such a passive function. This evidence is corroborated by studies of ontogenetic changes in the relative size and complexity of conodont elements within the preserved apparatuses. These analyses show that the conodont elements did not increase in size sufficiently quickly to provide enough food for the growing conodont animal through passive trapping of material in suspension. Some evidence of wear on the working surfaces of conodont elements also attests to an active food-processing function.

For background information SEE CHORDATA; CONODONT; VERTEBRATA in the McGraw-Hill Encyclopedia of Science & Technology.

Richard J. Aldridge

Bibliography. R. J. Aldridge et al., The anatomy of conodonts, *Phil. Trans. R. Soc. Lond. B*, 340:405–421, 1993; M. A. Purnell, Feeding mechanisms in conodonts and the function of the earliest vertebrate hard tissues, *Geology*, 21:375–377, 1993; I. J. Sansom et al., Dentine in conodonts, *Nature*, 368:591, 1994; I. J. Sansom et al., Presence of the earliest vertebrate hard tissues in conodonts, *Science*, 256:1308–1311, 1992.

Crustacean

Insects and crustaceans are closely related arthropod groups that have very similar life styles because growth and development are achieved by molting and metamorphosis. Molting is a process where the old exoskeleton is shed and replaced with a larger one. In both groups, it is regulated by the molting hormone 20-hydroxyecdysone. Metamorphosis takes place as the larval forms progressively molt to the adult forms. The juvenile hormones are responsible for the postmolt morphology of insects. Some insects (termed holometabolous), undergo complete metamorphosis, and the life-cycle stage following a molt (larva, pupa, or adult) is determined by the presence or absence of juvenile hormone. The absence of juvenile hormone during a critical period prior to the molt in a larva initiates differentiation to the pupa. Conversely, the presence of juvenile hormone prevents differentiation and the larva molts to larger larval stage. In addition to metamorphosis, the juvenile hormones are important for reproduction in insects, for example, blood levels change in females at various stages of egg development. In male insects, the presence of juvenile hormone in the blood is necessary for mating behavior.

The evolution of male polymorphism in crustaceans and insects has resulted in the development of alternative mating strategies for those individuals less able to compete with larger more aggressive opponents. Both the polymorphism and the reproductive behavior appear to be influenced by chemically similar hormones, methyl farnesoate in crustaceans and juvenile hormone in insects.

Crustacean juvenile hormone. The crustacean juvenile hormone, methyl farnesoate (MF), which is an unepoxidated form of insect juvenile hormone III, was identified in 1987. It is synthesized by the mandibular organs, which are regulated by secretions from the sinus glands located in the eyestalks. Methyl farnesoate was initially detected in the blood of the common east coast spider crab (*Libinia emarginata*) and since then it has been detected in 25 species of Crustacea.

The role of methyl farnesoate in crustaceans appears to be similar to that for juvenile hormone in insects. It is associated with reproduction and may also control morphogenesis in larvae, juveniles, and adults. In female spider crabs, methyl farnesoate blood levels and its synthesis by the mandibular organs under laboratory conditions is correlated with the ovarian cycle. Methyl farnesoate synthesis is lowest in juveniles of both males and females, and highest during yolk production.

Male morphotypes in spider crabs. Male spider crabs are markedly polymorphic regarding the length of the claws compared to the length of the carapace and the condition of the epicuticle covering the exoskeleton (see **illus**.). They also differ with respect to the size of their reproductive

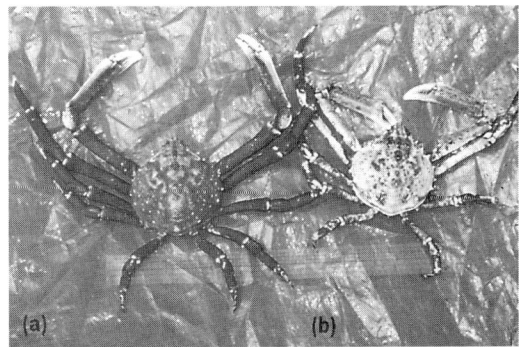

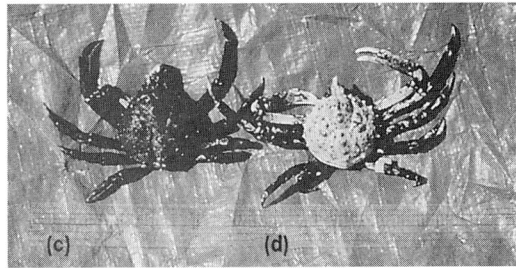

Male morphotypes of the spider crab *Libinia emarginata*. (*a*) Long-clawed unabraded. (*b*) Long-clawed abraded. (*c*) Short-clawed unabraded. (*d*) Short-clawed abraded.

systems, blood levels of methyl farnesoate, and mating behavior.

Newly molted crabs are completely covered with a very soft epicuticle, which wears off in approximately 18–24 months. Crabs with an intact epicuticle are referred to as unabraded, and crabs with completely bare exoskeletons are termed abraded. Among the morphotypes, there are unabraded, abraded, and incompletely abraded large-bodied males with long claws. Additionally, there are large-bodied, unabraded males with short claws, and small-bodied, short-clawed males, both unabraded and abraded.

Unabraded males, regardless of body or claw size, have the smallest reproductive systems (less than 1% of their body weight) and lowest levels of methyl farnesoate. The abraded males, both large-bodied, long-clawed and small-bodied, small-clawed, have the largest reproductive systems (greater than 2% of the body weight, and perhaps as much as 6%). The highest levels of methyl farnesoate in the blood occur in the abraded males with the largest reproductive systems. The incompletely abraded large-bodied, long-clawed males have reproductive systems that are somewhat smaller (1–2% of body weight) than those that are completely abraded, and their methyl farnesoate levels are half or two-thirds those of completely abraded morphotypes.

Reproductive behavior of male morphotypes. The abraded large-bodied, large-clawed morphotype is the primary reproductive that aggressively competes for females. After copulation, females are carried and guarded until their fertilized eggs are released.

The incompletely abraded long-clawed males attempt copulation if they are isolated with a receptive female, but they do not compete with completely abraded ones. The presence of a partially worn epicuticle indicates that the male has undergone its final molt fairly recently, and it is probably participating in the mating season for the first time.

Small-bodied, small-clawed abraded males are approximately the same size as mature females. Passive in the presence of dominant large-clawed abraded males, they have been observed being carried like a female by primary reproductives. Thus, small males appear to exhibit female mimicry. Although they do not compete with the primary reproductives for females, they can increase their mating opportunities by so-called sneak mating, that is, by attempting to copulate with a receptive female when the primary reproductives are distracted.

None of the unabraded males exhibit mating behavior. Small-clawed unabraded crabs are still growing, not having undergone their molt to the long-clawed form, and their methyl farnesoate production is low; thus mating is delayed. The low methyl farnesoate production may be an adaptation to redirect energy investment into growth, resulting in a larger body and longer claw size that ultimately enhance mating success. In contrast, the large unabraded males have reached their maximal size externally, but their internal organs are approximately the same size they were prior to the molt. Thus, energy is directed toward growth of the reproductive system prior to beginning the energetically expensive effort of competing for females and mating.

The lack of mating behavior observed in unabraded males can be regarded as a state of reproductive diapause. In insects, this physiological state is found in prereproductives and in adults during the nonbreeding season that is characterized by low blood levels of juvenile hormone and small reproductive systems.

For background information SEE CRUSTACEA; ENDOCRINE SYSTEM (INVERTEBRATE) in the McGraw-Hill Encyclopedia of Science & Technology.

Hans Laufer; Jonna S. B. Ahl

Bibliography. R. G. H. Downer and H. Laufer (eds.), *Endocrinology of Insects,* 1983; H. Laufer et al., Identification of a juvenile hormone-like compound in a crustacean, *Science,* 235:202–205, 1987; H. Laufer and R. G. H. Downer (eds.), *Endocrinology of Selected Invertebrate Types,* 1988; A. Sagi, E. Homola, and H. Laufer, Distinct reproductive types of male spider crabs *Libinia emarginata* differ in circulating and synthesizing methyl farnesoate, *Biol. Bull.,* 185:168–173, 1993.

Data communications

The ever-increasing miniaturization of electronic circuits, especially those passing digital information, has made it technically possible to build very inexpensive data-communications systems in which battery-powered, hand-held transmit-receive units can communicate directly with small satellites orbiting Earth at a low altitude, generally below 1000 km (625 mi). Such satellites are known as Little LEOs (low Earth orbits), distinguishing them from the well-established communication satellites at the geosynchronous orbit altitude of 35,786 km (22,236 mi), the geosynchronous (GEO) satellites. So-called Big LEOs will also orbit at low altitudes, but are being designed to carry voice traffic and as a result are much larger and more expensive. Big LEOs will not be available for several years. The advantage in communications capability to be derived from using satellites instead of transmitter towers (as in terrestrial cellular phone systems) is that complete global coverage can be provided. The small, battery-powered remote units add mobility to the Little LEOs' communications capability characteristics.

ORBCOMM. The first of the Little LEO systems, called ORBCOMM, is scheduled for use in 1995 in the United States and some Latin American countries. It consists of a constellation of 26 satellites, which orbit Earth every 100 min in circular orbits at an altitude of 785 km (490 mi). Since Earth is also rotating beneath them every 24 h, each of the satellites overflies all the world (excepting the polar regions). Hence, the entire constellation of satellites is available for use with networks on the ground worldwide. Generally, a network will use only one of the satellites at any time, and will switch from one setting below the horizon to another appearing above the horizon in another direction. Some of the larger networks, however, such as those in the United States, will use three or even four satellites at a time.

Satellites. Since the first geosynchronously orbiting communication satellites were launched in the 1960s, they have been steadily growing in size, mass, complexity, capability, and cost. By comparison, the Little LEO satellites are of small dimensions and low mass, and are considerably cheaper to build and launch. They are nevertheless quite complex. When the antenna boom and solar arrays of an ORBCOMM satellite are stowed for launch, it has the shape of a disk about 1 m (40 in.) in diameter and about 10 cm (4 in.) thick, and weighs only 40 kg (88 lb). It is launched by the technically innovative Pegasus rocket, which is air-dropped from a modified commercial aircraft and then ignited to propel the satellites into orbit. Each Pegasus can place eight of these satellites in orbit, allowing major launch-cost savings. The constellation has eight satellites equidistantly placed in the same circular orbit, and three such orbits in different orbital planes. These orbit planes are inclined at 45° to Earth's equatorial plane, and the ground tracks of the satellites oscillate between the latitudes of 45° north and south. In addition to these 24 satellites, 2 in a highly inclined orbit plane (70°) provide less continuous coverage of the polar regions.

In orbit, the faces of the disk are opened. Solar cells are mounted on the faces (called solar arrays), which are pointed toward the Sun so that the solar cells can convert the Sun's radiation into electrical current. The antenna boom is folded and stowed inside the disk during launch. When it is deployed in orbit, it is pointed down toward the center of Earth. The satellite senses its attitude and keeps the antenna boom in the correct direction. Because the antenna beams are very broad (with about a 62° half-cone angle), extremely precise attitude control, such as is needed with a GEO satellite, is not required. Hence, instead of using complex gas propulsion systems for attitude control, these LEO satellites use passive devices (gravity gradients and magnetic torquers). Also, exact positioning of the satellites relative to Earth is not required; only their positions relative to each other must be maintained. Thus, instead of station-keeping subsystems, the satellites need only the simpler formation-keeping devices. However, although the so-called satellite bus can be kept quite simple, these satellites have very complex and capable communication payloads. All received signals are demodulated from the radio carrier down to baseband (the information 1's and 0's) and are stored, processed, and rearranged in the satellite before being remodulated for the downlink transmission.

Networks. Within a few years, nearly every country in the world will have its own ORBCOMM network, or will join with other neighboring countries in a regional network. The network owner-operator will have a network control center and one or more gateway Earth stations, and will provide data communications services to its customers, who will have purchased small communicator units from one of several manufacturers of these devices. The network control center will consist of computers with large, fast-retrieval memories, and displays to show the operators how the system is operating. Human intervention will be required only if there is a fault to be diagnosed and repaired. The gateway Earth stations have large antennas pointed at one of the satellites and follow that satellite across the sky while it is being used by that network. Most networks will be serving countries that are small enough in area that only one gateway will be needed. The gateways will be located in remote areas of the country, and will be unattended. They will be monitored and controlled by the network control center.

Signal routing. Most users of ORBCOMM will be organizations rather than individuals, such as commercial companies and government agencies. They will have a fixed base station (often no more than a desktop computer connected through a modem to the public switched telephone network), and several remote units, which may be fixed in location or mobile. They will send messages from the base station outbound to one of the remotes, or to the base station inbound from a remote. In the first case, the routing is from the base station to a network control center over the public switched network, then to one of the gateways via dedicated circuits (usually leased from the public switched network provider), then up to one of the satellites on a very high frequency (VHF) radio link, and finally down to the remote on another VHF radio link. The inbound routing is the same but in reverse (see **illus.**). The VHF frequencies used are between 137 and 138 MHz for signals transmitted by the satellites, and between 148 and 150.05 MHz for those transmitted to the satellites. These frequencies are also used by ground-based communication

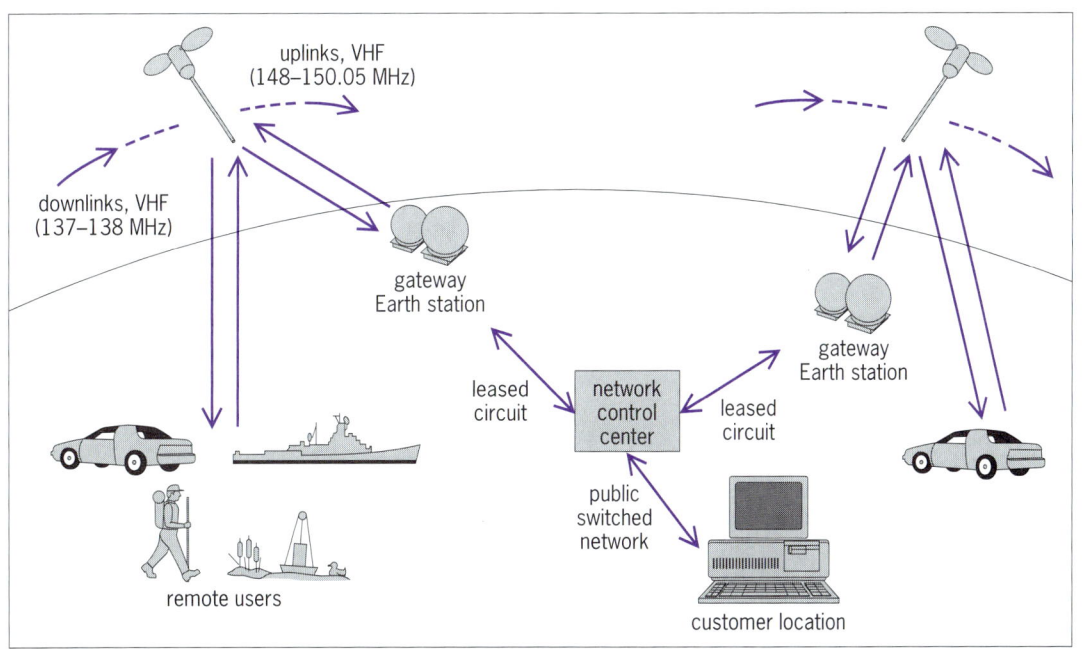

ORBCOMM network, showing signal routings between the base station and a remote communicator unit.

systems, generally push-to-talk mobile radios and one-way pagers, so that special techniques must be used to avoid interference between them and ORBCOMM facility transmissions. ORBCOMM uses very narrow channel bandwidths (10 and 15 kHz) that jump from one part of the band to another to avoid channels that are being used by terrestrial radio transmitters. Another technique to mitigate interference available to Little LEOs is to spread the energy out by using spread-spectrum modulation over most of the radio-frequency band so that it is very low at any narrow channel being used by another system.

Position determination. Although ORBCOMM is primarily designed to communicate data, either outbound to a remote at the rate of 4800 bits/s, or inbound from a remote at 2400 bits/s, it is also capable of position determination. Each of the remote communicator units can determine its location by internally processing its measurements of the frequency shift of the downlinked signals caused by the relative motion between the satellite transmitting the signal and the unit receiving it (the Doppler shift). The accuracy of this position will vary from about 100 m (300 ft) to about 1 km (0.6 mi), but for many applications this degree of accuracy is sufficient.

Applications. The applications for Little LEO systems are many and varied. They include the entire family of supervisory control and data acquisition (SCADA) uses, in which the status and performance of remote, unattended machinery can be monitored at a base station (by using the inbound communication capabilities); commands to turn the machinery on or off, or to change its operating configuration, can be sent by using the outbound capabilities. This application is typical for oil wells, pumping stations, and pipelines in the energy industry, and for center-pivot irrigation devices of increasing use in agriculture.

Another class of application is the tracking and monitoring of trucks and, particularly, refrigerated containers. The remote communicator unit, with its self-contained battery, is mounted on the outside of the container. An outbound message can be sent to the unit, commanding it to respond with its position and status (such as the temperature inside the container). Importantly, if the temperature rises above some preset value because the refrigeration unit is failing or some other parameter exceeds its allowable range, the communicator unit can automatically alert the base station so that the container can be located and repaired before its contents are ruined. There are also a whole range of applications in which the remote unit is carried by a person, such as a surveyor working in the field.

One of the applications that could well become the most common is emergency reporting from passenger cars. Should a car be involved in an accident that triggers the deployment of the air-bag restraint system, an ORBCOMM unit mounted in the car and connected to the standard AM-FM radio whip antenna would send out an emergency alert, identifying the car and its location.

For background information SEE COMMUNICATIONS SATELLITE; DATA COMMUNICATIONS; SATELLITE NAVIGATION SYSTEMS; SPREAD SPECTRUM COMMUNICATION in the McGraw-Hill Encyclopedia of Science & Technology.

Martin Deckett

Deoxyribonucleic acid (DNA)

Deoxyribonucleic acid (DNA), as it occurs in nature, possesses a double helix structure, consisting of millions of nucleic acid base pairs. This structure has the appearance of a spiral staircase whose steps encode the genetic information that controls the development of all living organisms. The double helix structure is stabilized by interactions between two individual strands of nucleic acid bases and leads to stacking of the base pairs over one another. This stacking, in turn, causes the outside of the DNA double helix to have a characteristic shape that includes so-called pockets or grooves. Small molecules possessing a shape complementary to these grooves can bind to DNA through a process by which a portion of the bound molecule structure injects itself between two of the base pairs forming the double helix. This mode of binding is called intercalation, to emphasize the penetration of the molecular structure between the base pairs.

DNA as a molecular wire. The electrons of the base pairs in the DNA double helix are strongly coupled because the double helix structure jams flat-shaped base pairs against one another like cards in a deck. A strongly bound, intercalated molecule of the correct structure may be considered as part of the DNA structure in the sense that the host DNA and bound guest may be strongly coupled electronically. That is, the electrons of the DNA and intercalated molecule are strongly interacting and in the limit may be viewed as part of the same extended electronic system. Such strong coupling elicits a metaphor of DNA as a molecular wire capable of transmitting an electric current along the double helix. It has recently been shown that when certain molecules possessing metal centers intercalate to DNA structures and are excited with light, electrons can be transferred from one bound molecule to another, with the DNA structure assisting in the electron transport; that is, in these electron-transfer reactions, DNA truly may behave as a molecular wire that conducts electric current.

Photoinduced electron transfer. Electron transfer between metal centers is an important reaction in many chemical and biochemical systems. It is of great interest, therefore, to examine the factors that control the efficiency of electron transfer between metal centers and especially to search for mechanisms that allow for specificity in the transport of electrons between two points in space or that allow for specific pathways for electron transfer. A particularly intriguing issue is whether electrons can be transferred over large distances on the molecular and atomic scale, and by which mechanisms electron transfer over large distances might occur.

When light is absorbed by any molecule, electrons are displaced from orbitals that are close to nuclei to orbitals that are farther away. Electrons that are farther removed from nuclei are more loosely bound, and therefore they are more readily lost or transferred. Thus, it is not surprising that many metals which already have an inherent tendency to lose electrons have this tendency enhanced when the metal absorbs light. For example, if a certain ruthenium metal complex is photoexcited in a solution containing a specific rhodium complex (**Fig. 1**), electron transfer from the excited ruthenium complex to the rhodium complex occurs upon collision of the two species. Since the electron transfer requires light absorption, it is termed photoinduced electron transfer.

Molecular light switch. In aqueous solution, the ruthenium complex shown in Fig. 1 does not emit light after it is photoexcited. However, in the presence of double helical DNA, the complex emits strongly. Thus, DNA may be viewed as a molecular light switch that serves to report the binding of the metal complex to the DNA. It has been further shown that the light-switch effect operates only when the complex is intercalated into the DNA structure.

A special DNA has been synthesized that consists of two independent strands of a short piece of DNA, with a metal atom of rhodium attached to the end of one strand and a metal atom of ruthenium attached to the end of the other. Each strand contains 15 nucleic acid bases (**Fig. 2**). When these two DNA strands are allowed to interact in a water solution, they spontaneously combine to form a DNA double helix that contains a ruthenium atom at one end and a rhodium atom at the other end. When a double helix is prepared with just a ruthenium atom (the second strand not possessing a rhodium atom) the DNA material glows brightly when excited with light; as expected, the DNA serves as a light switch to turn on the ruthenium

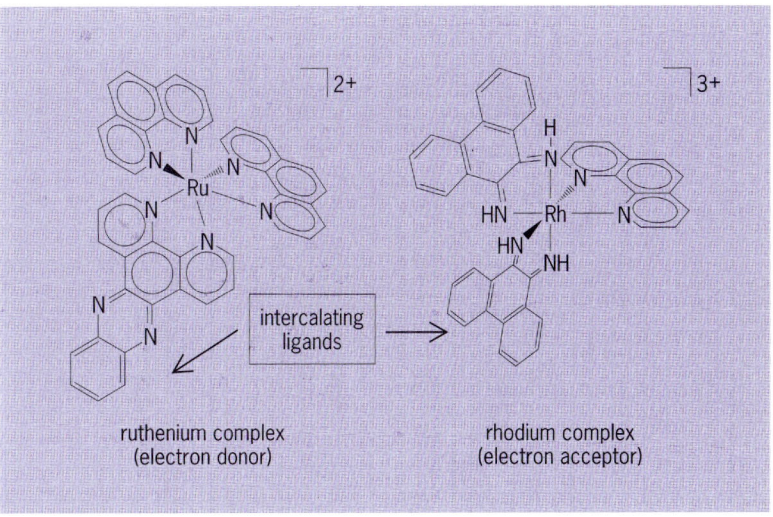

Fig. 1. Photoinduced electron transfer from a ruthenium metal complex [Ru(dppz)(phen)$_2^{2+}$] to a rhodium π metal complex [Rh(phi)$_2$(phen)$^{3+}$]. When the ruthenium complex is photoexcited in the presence of the rhodium complex, electron transfer occurs from the photoexcited ruthenium complex to the unexcited rhodium complex.

emission. However, when the double helix contains both a ruthenium and a rhodium atom at opposite ends, the glow is extinguished.

The glow is extinguished because an electron is transferred from the ruthenium to the rhodium. It seems that the electron passes through the DNA double helix. From molecular models, the distance between the ruthenium and rhodium centers for a piece of DNA that has 15 bases in each strand is around 4 nanometers. The size of the bound molecules and the bases are of the order of a fraction of a nanometer. Thus, a number of base pairs must intervene between the metals involved in the photoinduced electron transfer. In other words, the DNA double helix serves as an incredibly small wire that conducts electric current from the photoexcited ruthenium atom to the rhodium atom. In effect, a so-called molecular light switch for determining changes that occur in DNA has been developed.

Whether and where the ruthenium light is turned on or off can be used to watch changes in DNA to

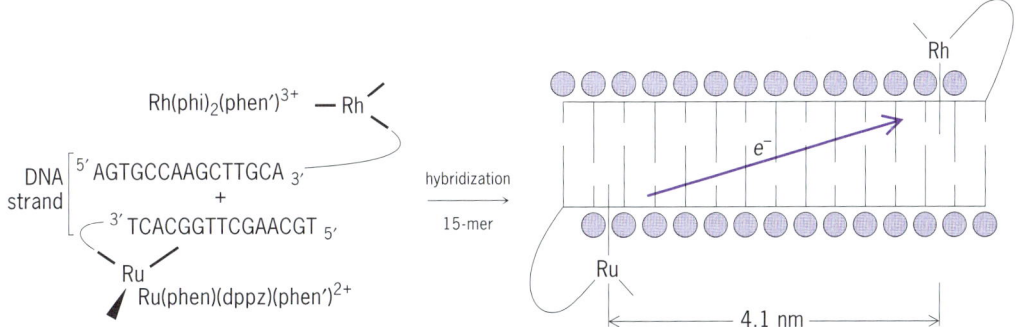

Fig. 2. Electron transfer through DNA. In a long DNA strand containing the bases adenine (A), cytosine (C), guanine (G), and thymine (T), metal complexes intercalate into the DNA double helix, and photoinduced electron transfer occurs.

determine, for example, whether the DNA becomes damaged, and even whether specific sequences of DNA are present or absent.

Biological relevance. Since light has been shown to activate the flow of electric current through a piece of DNA, which serves as a molecular wire through which electrons flow, it is possible to speculate on the potential importance and applications of these observations. On the one hand, these results provide a starting point for basic research for a new range of studies into biological electron transport systems; on the other hand, they have the potential to be developed into practical innovations such as sensitive agents for diagnoses of DNA-based diseases or the construction of nanoscopic electric circuits and biosensors.

For several decades, scientists have speculated whether some cancers might have involved damaging DNA through reactions involving electron flow through the DNA. This speculation may now be tested by investigations that employ the sensitive glow of the ruthenium as a sensitive reporter of the damage. A knowledge of how DNA operates as a wire is important for the understanding of how electric current may be directed in cells in the body. In many critical biological activities, electrons must travel over large distances (compared to the size of atoms) to move across membranes in a cell or to move from one protein to another. The results provide a working model for very long range electric current flow through the DNA double helix and suggest how nature may direct the flow of electrons.

The observation that the DNA helix is such an efficient molecular wire also opens up two completely new areas for investigation. Scientists may now begin to take advantage of the natural DNA wire in constructing exceedingly tiny nanosized circuits and biosensors. In addition, the efficiency of the DNA wire presents the possibility that nature may already employ the DNA double helix in reactions involving the flow of electrons in the cell, and a search can be conducted to confirm this possibility.

For background information SEE DEOXYRIBONUCLEIC ACID (DNA); ELECTRON-TRANSFER REACTION; NUCLEIC ACID in the McGraw-Hill Encyclopedia of Science & Technology.

Nicholas J. Turro

Bibliography. A. E. Friedman et al., Molecular "light switch" for DNA: $Ru(bpy)_2(dppx)^{2+}$, *J. Amer. Chem. Soc.*, 112:4960–4962, 1990; C. J. Murphy et al., Long-range photoinduced electron transfer through a DNA helix, *Science*, 262:1025–1029, 1993; W. Saenger, *Principles of Nucleic Acid Structure*, 1984; N. J. Turro, J. K. Barton, and D. A. Tomalia, Molecular recognition and chemistry in restricted reaction spaces: Photophysics and photoinduced electron transfer on the surfaces of micelles, dendrimers, and DNA, *Acc. Chem. Res.*, 24:332–340, 1991.

Dinosaur

Although the scientific concept of dinosaurs dates back to 1842, since the 1970s there has been a renewed scientific interest in these long extinct reptiles. New ideas, data, and analyses about evolutionary relationships, the earliest dinosaurs, distribution, bird origins, nests and hatchlings, metabolism, footprints, extinction, and bone morphology have fueled a vigorous renaissance of dinosaur studies.

Evolutionary relationships. Phylogenetic systematics, or cladistics, has modified some and upheld other older views of the evolutionary relationships among dinosaurs. Cladistics identifies groups of organisms that share evolutionary novelties as closely related. It does not establish relationship by overall similarity or proximity of organisms in geological time. Dinosaurs can be identified as a group of archosauromorph reptiles having a single, common ancestry by their evolutionary novelties, which include a shoulder joint that faces backward, three or fewer bones in the fourth finger of the hand, an open hip socket, and a distinct neck-and-ball joint offset from the shaft of the thigh bone.

Among dinosaurs, cladistics upholds the long-identified separation into two groups, the saurischian (lizard-hipped) and ornithischian (bird-hipped) dinosaurs. Nevertheless, some of the earliest dinosaurs, for example, Late Triassic *Herrerasaurus* from Argentina, cannot be assigned to either group. Indeed, their status as dinosaurs has been recently challenged by some paleontologists.

The saurischians consist of the theropods and sauropodomorphs. The theropods (meat-eating dinosaurs) can be divided into three evolutionary branches: the ceratosaurs, characterized by fusion of the ankle and hip joints; the small and very birdlike coelurosaurs; and the giant meat eaters, the carnosaurs. Sauropodomorphs include the prosauropods, the earliest great group of dinosaurian plant eaters and the sauropods (brontosaurs of popular terminology), the largest of all dinosaurs and largest land animals of all time.

Ornithischians consist of the thyreophorans (shield bearers), which include the heavily armored ankylosaurs and the stegosaurs; the ornithopods, which include the duck-billed dinosaurs; and the marginocephalians (edge-headed dinosaurs), which include the horned dinosaurs (ceratopsians) and the bone-headed dinosaurs, the pachycephalosaurs. Nevertheless, a close relationship of ceratopsians and pachycephalosaurs based largely on the so-called incipient frills on the skulls of the latter is not probable.

Earliest dinosaurs. New discoveries of some of the earliest dinosaurs, especially in Argentina, have revolutionized understanding of dinosaur origins. Examination of isolated teeth and jaw fragments from North America has led to the designation of several Late Triassic ornithischians: *Lucianosaurus*, *Revueltosaurus*, *Technosaurus*, and *Tecovasaurus*.

Shuvosaurus, from the Late Triassic of Texas, is a toothless reptile identified as a theropod closely related to the much younger ornithomimids. The newly discovered theropod *Eoraptor* and recently described specimens of *Herrerasaurus* from the Argentinian Late Triassic augment one of the most extensive records of Late Triassic dinosaurs. *Rioarribasaurus* has been introduced as the correct name for the dinosaur long referred to as *Coelophysis,* which is the best-known Late Triassic theropod and is represented by hundreds of skeletons from Ghost Ranch in northern New Mexico.

The earliest dinosaurs were diverse, and some kinds were surprisingly specialized. For many years, paleontologists believed that the oldest dinosaur fossils came from Brazil and Argentina, and thus that South America was the locus of dinosaur origins. However, recent reassessment suggests that the South American dinosaurs are the same geological age as the oldest dinosaurs known from North America, North Africa, and India, implying that dinosaurs appeared suddenly over a broad geographic area, became diverse very early, and evolved into some specialized forms. If the period of dinosaur origins was not a time of very rapid evolution, much older (and more primitive) dinosaurs remain to be discovered; possibly both assumptions are valid.

Distribution. Paleontologists are now exploring the globe for new dinosaurs. Recent discoveries of poleward dinosaurs (on Alaska's north slope and in New Zealand, Australia, and Antarctica) are particularly intriguing. They establish a dinosaur presence on all the continents and in latitudes that, even when adjusted for the effects of continental drift, place dinosaurs in locations with winter months of virtual darkness. Thus, some researchers advocate that dinosaurs were cold-blooded and suggest regular, annual migrations by dinosaurs to a warmer, sunnier climate.

New discoveries have significantly augmented the known diversity of dinosaurs, especially of theropods. Recently discovered theropods of post-Triassic age include *Cryolophosaurus,* a crested form from the Jurassic of Antarctica (as late as the mid-1980s no dinosaurs were known from the South Pole region), *Monolophosaurus* from the Jurassic of China, *Alxasaurus* from the Early Cretaceous of China, and *Afrovenator* from the Early Cretaceous of Niger.

Bird origins. A dinosaurian ancestry of birds from a small theropod to Late Jurassic *Archaeopteryx* (the first bird) has met several unconvincing challenges. *Protoavis,* a supposed Late Triassic bird from west Texas nearly as old as the oldest dinosaurs, and much older than *Archaeopteryx,* is not a bird. Most of the fossil indicates a small theropod; the rest is unrecognizable.

Diverse and advanced birds from the Early Cretaceous that were capable of sustained, powered flight are only a few million years younger than *Archaeopteryx. Cathayornis* and *Sinornis* from northeastern China, *Ambiortus* from Mongolia, and *Iberomesornis* from Spain indicate either that the early evolution of birds was very rapid or that *Archaeopteryx* is not the first bird. *Mononykus* from the Late Cretaceous of Mongolia is a bizarre animal that is either a flightless bird or a birdlike dinosaur; the former possibility underscores the early diversity of primitive birds. Yet, despite these new discoveries, *Archaeopteryx* still is considered the first bird, a so-called feathered dinosaur descended from a small theropod.

Nests and hatchlings. Nests and hatchling ornithopods from Montana provide rich insight into the reproductive behavior and early life stages of some dinosaurs. Among other things, they provide evidence that some dinosaur hatchlings (for example, those of the hypsilophodontid *Orodromeus*) were relatively mature at birth, whereas others (for example, those of the hadrosaurid *Maiasaura*) were relatively immature at birth. The immature hatchlings must have received some parental care. Indeed, analysis of the rich record of dinosaur eggs and babies suggests remarkably birdlike modes of reproduction and parenting among many kinds of dinosaurs.

In southeastern China, geochemical analysis of dinosaur eggs laid just before the era of dinosaur extinction has implicated concentrations of heavy metals in the eggshells as having led to the shell thinning and loss of viability that was the cause of dinosaur extinction. However, the analyses failed to acknowledge the possibility of enrichment of heavy metals during fossilization, and therefore did not establish a definitive link between egg chemistry and extinction.

Metabolism. The idea that all dinosaurs were warm-blooded, which was strongly advocated by some paleontologists during the 1970s, has not withstood recent analysis. Particularly important has been recognition of the significance of gigantothermy, the idea that any vertebrate animal weighing more than 1000 kg (2200 lb) maintains a nearly constant body temperature by virtue of its great mass and relatively small surface area (thermal inertia). Thus, most adult dinosaurs had a constant body temperature regardless of their metabolic rate, confounding efforts to determine that rate. Nevertheless, the metabolic rate of juvenile (and hatchling) dinosaurs remains an open question, leading to the suggestion that most dinosaurs may have been heterotherms whose metabolic rates changed throughout their life cycle. Some degree of warm-bloodedness in the small theropods is well supported by diverse evidence.

Footprints. Long regarded as curiosities, dinosaur footprints are now being subjected to rigorous and extensive analysis. They provide direct evidence of dinosaur modes of walking; and they are the only reliable way to estimate speed, which turns out to have been mostly a slow walk, although a few fleet-footed dinosaurs ran as fast as 43 km/h (27 mi/h). In many rock formations where bones are lacking, the footprints are the only evidence of dinosaurian presence. Furthermore, social groupings of dinosaurs and dinosaurs feeding on particular plant types have been inferred from footprints.

Extinction. The evidence for a bolide (a comet or meteor that explodes when striking Earth) impact 66 million years ago at the time of dinosaur extinction is incontrovertible. On the northern tip of Mexico's Yucatan Peninsula, the Chicxulub impact structure has been estimated to be 300 km (180 mi) across. Debate about its age and origin has been fierce, but the preponderance of evidence identifies Chicxulub as the crater formed by the bolide that impacted Earth at the end of the Cretaceous.

However, proposing this bolide impact as the cause of dinosaur extinction is still hotly debated. Most of the relevant evidence comes from Montana, where a wealth of terrestrial and fresh-water animals (more than 51% of the total) survived the impact unscathed. Here, patterns of dinosaur extinction have been interpreted as either gradual or abrupt, but most vertebrate paleontologists favor the gradual extinction of dinosaurs over an interval of 3 million years. They see it as part of a complex ecological collapse of the terrestrial biota in response to changing landscapes and deteriorating climate. It is important to note that almost all the useful data on dinosaur extinction comes from Montana and nearby states, and may not reliably indicate the global pattern. *Spencer G. Lucas*

Bone. A wealth of information about the diversity of dinosaurs and various aspects of their biology is derived directly from their fossilized bones. Soft tissues of the body, including organic material such as the collagen within bones, generally decompose shortly after death. The inorganic mineral phase of bone, hydroxyapatite, a calcium phosphate, is much more resistant to decomposition and undergoes only minor change. It is the preservation of this mineral component, and in particular its microscopic organization, that permits histological analysis of fossil bone.

As bone forms, processes affecting bone growth are recorded and the rate of bone deposition is indicated. Studies of dinosaur bone microstructure provide a direct assessment of bone growth and overall growth strategy, and permit indirect inferences regarding the physiology of these extinct animals.

General histology. Macroscopic examination of transverse and longitudinal sections of any fossil long bone shows a distinct bone wall consisting of a dense bone (compact bone) that encloses a central medullary cavity. A porous tissue (spongy bone or cancellous bone) is generally restricted to the distal or proximal ends but is occasionally distributed in the bone's perimedullary region.

Although a variety of diagenetic factors affect the history of a bone, the microstructure of fossil bone is usually well preserved. Fungal damage to fossil bone or extensive chemical alteration causing total destruction of the microstructure has been reported rarely.

By examining a polished thin section of fossil bone (about 30 micrometers thick) under a light microscope, the microstructure (histology) of the bone is revealed. The cavities that once housed the blood vessels are readily recognizable, as are the lacunae that the bone cells (osteocytes) occupied in life. Even the branching canaliculi of the lacunae that allowed communication among neighboring cells in the living bone are distinctly visible. All the natural cavities in the fossil bone, as well as those caused by fracturing during fossilization, are generally filled by sediments or minerals derived from the environment.

Compact bone can be either primary or secondary in origin. Primary compact bone is generally formed under a fibrous membrane that envelopes bone (periosteum) during the growth of new layers of bone tissue. Secondary compact bone involves the resorption of primary bone and subsequent redeposition of bone tissue. Both primary and secondary compact bone tissues are readily observed in dinosaur bone. The organization of the collagen fibrils that were once present may be determined by studying the organization of the apatite (a group of phosphate minerals), which during life closely followed the collagen fibrillar matrix arrangement. Thus, the gross textural organization of the apatite provides insight into the rate of bone formation. Two fundamental types of primary compact bone tissues are recognized on the basis of fibrillar arrangement. Bone deposited quickly results in fibro-lamellar bone characterized by a fibrillar matrix with a relatively haphazard, woven arrangement. Lamellar bone results when bone is formed slowly, and typically has a more organized fibrillar matrix.

Compact bone, regardless of whether primary or secondary in origin, is either vascular or avascular. If the bone is vascular, the orientation of the vascular canals can provide useful information for further classification of the tissue. The vascular canals are classified into simple vessels, primary osteons, and secondary osteons. Simple vessels and primary osteons are formed during primary bone formation. The simple vessels do not interrupt the structure of the bone surrounding them, whereas the primary osteons are surrounded by a centripetal deposit of bone (**Fig. 1**). Secondary osteons, or Haversian systems, are formed as a result of secondary or Haversian reconstruction that involves resorption and subsequent redeposition of bone. Secondary osteons are easily distinguished from primary osteons by the presence of a distinct cement line which marks the furthest extent of bone resorption (Fig. 1). Bone around each secondary osteon is initially primary in origin; however, as the process of secondary reconstruction continues, the number of secondary osteons increases, eventually resulting in dense Haversian

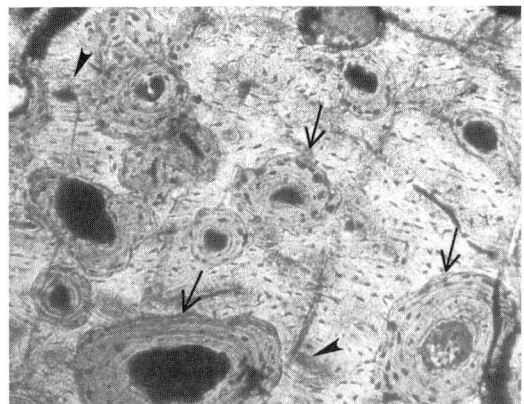

Fig. 1. Transverse section of a femur of the Early Jurassic dinosaur *Syntarsus rhodesiensis*, showing primary osteons (arrowheads) and secondary osteons (arrows). Note the cement line around the secondary osteons.

bone in which even the interstitial bone consists of remnants of older secondary osteons.

Dinosaur bone is generally very highly vascularized by primary osteons, which are embedded in the woven matrix of the fibro-lamellar bone. The number of secondary osteons formed is highly variable. It ranges from fairly limited occurrences, with just a few isolated secondary osteons, as in *Massospondylus* and *Dryosaurus*, to extensive occurrences of dense Haversian bone, as in *Brachiosaurus* and *Pachyrhinosaurus*.

The following bone tissues are also fairly well known in dinosaurs: endochondral bone, resulting from replacement of cartilage; compacted coarse cancellous bone, formed by endosteal infilling of cancellous spaces; metaplastic bone, resulting from ossification of ligaments or tendons; and pathological bone. However, the nature of the compact bone has provided the greatest insight for interpreting the rate of growth of the animal, and has permitted inferences about their physiology.

Significance. Early studies of dinosaur bone histology recognized the frequent occurrence of fibro-lamellar bone and Haversian bone. These histological characteristics bore a striking resemblance to features of mammalian bone, and were later cited as evidence of warm-bloodedness in dinosaurs. Although fibro-lamellar bone is typical of warm-blooded mammals, more recently it has been recognized in known cold-blooded animals, such as juvenile crocodiles, and in mammallike reptiles from the Permian and Triassic, which are not considered on anatomical or phylogenetic grounds to have been warm-blooded animals. Secondary reconstruction is also recognized in reptiles and is extensively developed in turtles. Thus, Haversian bone is now not considered a unique characteristic of warm-blooded animals.

More recent work on dinosaur bone histology has indicated that at least some taxa form bone with distinct interruptions that are manifest as lines of arrested growth in bone deposition, or as lamellated bone, or as both. A distinctly stratified compacta, termed zonal bone, results from such alternate periods of slowed and rapid growth (**Fig. 2**). Isolated bones of various dinosaur taxa examined (for example, *Rhabdodon*, *Baryonyx*, *Allosaurus*, *Tyrannosaurus*, and *Protoceratops*) have demonstrated that zonal bone is fairly widespread in the Dinosauria.

Such a cyclical pattern of bone formation is characteristic of modern reptiles. Experimental analysis using fluorescent dyes on animals of known age has identified such cycles as annual, resulting from an innate biological rhythm synchronized by seasonality but not necessarily controlled by it. It cannot be proven that the cycles evident in dinosaur bones are annual, but since no other cycle in living animals is known to form such rings, they are assumed to be annual. Furthermore, the examination of growth rings in various specific dinosaurian taxa has shown that the number of rings increases with the size of the animal. In the small theropod *Syntarsus*, at least eight cycles of bone deposition can be recognized. If these cycles are annual, it took about 8 years for this animal to reach mature body size. Another smaller theropod, *Troodon*, is considered to have taken about 3–5 years to reach maximum body size. The prosauropod *Massopondylus* has 15 growth rings in its bone, but no examined specimen of this group appears to have reached maximum body size. Even if the growth rings were not annual, the occurrence of periodic pauses in the rate of bone deposition suggests that these animals were growing significantly more slowly than modern mammals and birds, and that they were unable to maintain a sustained high rate of bone deposition. Lines can result in the bones of mammals because of trauma, but such lines are infrequent and are not an integral part of normal growth of immature animals. As in mature mammals, mature individuals of both *Troodon* and *Syntarsus* exhibit closely spaced rest lines in the peripheral region of their bones, which suggest that they have reached mature body size and hence have a determinate growth pattern. The largest individuals of *Massospondylus* and *Dryosaurus* do not show these characteristics and

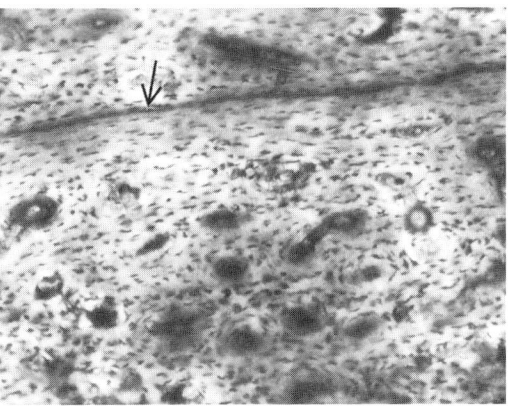

Fig. 2. Transverse section of a femur of *Patagopteryx deferrariisi*, a Late Cretaceous bird, showing zonal bone. The arrow indicates the line of arrested growth.

are considered to have an indeterminate growth strategy similar to modern reptiles.

Not all dinosaurs have zonal bone. A study of 13 femora of *Dryosaurus* (a medium-sized ornithopod) showed no growth rings, and therefore no evidence that growth slowed down or even stopped during the life-span of the individual (ontogeny). The *Dryosaurus* study is the only growth series of a dinosaur yet described that apparently grew rapidly throughout ontogeny. It also appears to have had an indeterminate growth strategy, since there is no evidence of growth having stopped in the largest specimen studied. *Dryosaurus* was relatively small; therefore mass effects could not have played a major role in maintaining a high stable body temperature. Modern reptiles living in aseasonal environments still show growth rings. Thus, the absence of any growth rings in the bones of *Dryosaurus* indicates that it was physiologically capable of growing at a high, sustained rate without any interruptions in the rate of bone deposition. In fact, of all dinosaurs *Dryosaurus* most closely resembles a warm-blooded mammalian growth pattern. However, the absence of rest lines in the largest individual seems to indicate a reptilian pattern of indefinite or indeterminate growth.

A combination of both reptilian and mammalian characteristics is thus distinctly evident in dinosaur bone. Thus, these organisms were neither typical warm-blooded nor typical cold-blooded animals. It is quite possible that they represent some intermediate position along the continuum of physiological strategies or were perhaps specialized cold-blooded or specialized warm-blooded animals. Their bone histology does not reveal which possibility is more likely.

It is widely held that birds evolved from avian theropod dinosaurs. The histological changes that occurred through this transition were recently explored by examining the bone microstructure of three Late Cretaceous birds from Argentina. It was found that the early birds *Patagopteryx* and two species of Enantiornithines resembled their theropod ancestors in having growth rings in the compacta of their bones. This characteristic sets them apart from their modern descendants, which grow rapidly without any interruptions in the deposition of bone. These basal birds were physiologically different from modern birds and had slower growth rates. Thus, even halfway through the evolutionary history of birds, these basal birds, like their dinosaurian ancestors, were not classic warm-blooded animals.

For background information SEE BONE; DINOSAUR; ORNITHISCHIA; PALEONTOLOGY; SAURISCHIA; TAXONOMY in the McGraw-Hill Encyclopedia of Science & Technology.

Anusuya Chinsamy

Bibliography. K. Carpenter, K. F. Hirsch, and J. R. Horner (eds.), *Dinosaur Eggs and Babies,* 1994; A. Chinsamy, Bone histology and growth trajectory of the prosauropod dinosaur *Massospondylus carinatus* Owen, *Mod. Geol.,* 18:319–329, 1993; A. Chinsamy, L. Chiappe, and P. Dodson, Growth rings in Mesozoic avian bones: Physiological implications for basal birds, *Nature,* 368:196–197, 1994; A. P. Hunt, The early diversification pattern of dinosaurs in the Late Triassic, *Mod. Geol.,* 15:43–60, 1991; M. Lockley, *Tracking Dinosaurs: A New Look at an Ancient World,* 1991; S. G. Lucas, *Dinosaurs: The Textbook,* 1994; R. E. H. Reid, Zonal "growth rings" in dinosaurs, *Mod. Geol.,* 15:19–48, 1990.

Drug design

Chemotherapy and chemoprophylaxis (the use of drugs to prevent the development of infectious disease) of a number of viral-induced diseases, for example, influenza and acquired immune deficiency syndrome (AIDS), have been largely dependent on compounds that have undesirable side effects or low efficacy. The nature of the infection and of the virus has presented significant difficulties in finding alternative approaches toward the discovery of clinically useful drugs. Traditionally, new drug entities have been identified by a long and tedious process of fine-tuning the appropriate receptor models through the biological evaluation of a large array of structural analogs of a lead compound. However, recent advances in computational chemistry, x-ray crystallography, and nuclear magnetic resonance spectroscopy have provided new insight to the rational design of novel drug entities, in particular, antiviral drugs. One of the most significant advances is the substantial increase in the user-friendly nature of computer hardware and software, which has made computational chemistry accessible to beginners in the drug design field. Rational drug design methods have been successfully employed in the discovery of potent and specific antiviral drugs, in particular, potential anti-influenza drugs.

Computer-assisted design of antiviral drugs. An array of proteins associated with a number of pathogenic viruses have been isolated and characterized, and their three-dimensional structure determined. Many of these proteins have been implicated as essential components of the infection process. Therefore, the opportunity for intervention in this process would involve preparing molecules that would interact with these proteins and prevent the normal function of the relevant protein. The computer-assisted drug design process allows examination of the interactions between bioactive molecules and the target protein. For example, one type of computer software makes possible the evaluation of a predetermined region of a protein, perhaps an active site on an enzyme, in terms of high-energy interaction sites for a variety of functional group probes. The result of these calculations provides a map of so-called hot spots for the potential introduction of new functional groups, which can be further explored by using sophisticated computer graphics on any given molecular template (framework). These new functional groups should add to the overall binding energy, and hence increase the affinity of the original chemical structure for its corresponding protein.

Anti-influenza drugs. Computer-assisted drug design, based on the x-ray crystallographic structure of sialidase, the surface glycoprotein of the influenza virus, was recently utilized to design potential anti-influenza drugs.

Mechanism of infection. Influenza virus sialidase is a sialic acid–specific enzyme (an exoglycohydrolase) that is required for the mobility of the virus particle in the upper respiratory tract, and the dispersion of virion (the extracellular infective form of a virus) progeny from the site of initial infection. The enzyme enables the virion progeny to infect new cells in the upper respiratory tract; without it, the infection would not significantly advance. At a molecular level, sialidase acts as a pair of biological scissors, cleaving terminal-linked 2 → 3 α-ketosides of the carbohydrate N-acetylneuraminic acid. N-Acetylneuraminic acid [structure (I), where R′ = NHCOCH$_3$] is considered to be one of the most significant naturally occurring sugars in the biological system and appears to play an essential role in a number of important bioprocesses. As the virion progeny are assembled at the surface of the infected cell, the budding process is facilitated by the removal of N-acetylneuraminic acid from the surface glycoproteins on the freshly synthesized virion progeny and from any glycoconjugates in the immediate vicinity. If this event did not occur, the hemagglutinin on the virion progeny would recognize the sialic acid–containing glycoconjugates as ligands, resulting in the clumping of the progeny on the surface of the infected cells. This clumping (agglutination) effect would give the immune system enough time to mount a response and dispose of the invading microorganism.

Molecular modeling. The crystal structure of influenza virus sialidase provided a focus for designing small high-affinity molecules that would intervene in the normal biological function of the protein by effectively plugging the active site of the enzyme. In this example, two functional group probes, namely a carboxylic acid and an amino group probe, suggested that there were a number of potential high-interaction energy sites.

Modeling began with N-acetylneuraminic acid (Neu5Ac), structure (I), where R′ = NHCOCH$_3$. Its naturally occurring analog, Neu5Ac2en, where R = OH and R′ = NHCOCH$_3$, is shown as structure (II).

If the carboxylic acid group of this original template (II) is in the correct orientation, the introduction of an amino group at the carbon-4 position can complete the suggested functionalization. The theoretical enhancement in affinity to the enzyme on going from Neu5Ac2en (I) and the original template (II) to the 4-amino-Neu5Ac2en analog where R = NH$_2$ and R′ = NHCOCH$_3$ was estimated to have increased from micromolar to approximately 10^{-8} molar affinity. Indeed, this increase was observed when the compound was evaluated in the appropriate enzyme bioassay: the 4-amino-Neu5Ac2en analog was found to be 100 times more active than the original template.

Further molecular modeling indicated that replacement of the amino group with the bulkier and more basic guanidinyl group at the carbon-4 position would lead to an even more potent inhibitor. This prediction led to the synthesis of the final compound 4-guanidino-Neu5Ac2en, where R = [(NH$_2$)$_2$NH]$^+$ and R′ = NHCOCH$_3$, which binds with greater than nanomolar affinity and is the most potent influenza virus sialidase inhibitor.

The compound 4-guanidino-Neu5Ac2en has been evaluated in a number of animal models and has shown high levels of activity in organisms. Indeed, recent results from a trial experimental infection suggest that the compound is also active in humans; the compound has significant preventative and therapeutic antiviral effects. Full clinical trials are currently in progress to assess the potential of this drug candidate.

Future developments. Rational drug design based on structural information (receptor-based design) is currently being developed by using other viral proteins such as human immunodeficiency virus (HIV) protease. The potential for the development of an anti-AIDS drug based on this process is being considered by both research institutes and industry-based laboratories. The potential of carbohydrates as a new class of compound that acts as a key to the discovery of new drug entities has led to an industry termed glycotechnology.

The future for rational drug design appears promising and offers the possibility of tackling viruses which have been traditionally difficult to control because of their ability to either escape or modulate the response of the immune system. As further advances in computational chemistry are made, the analysis of structural information will become routine, with the slow step being the acquisition of the actual protein structures by either x-ray crystallography or nuclear magnetic resonance spectroscopy.

For background information SEE ACQUIRED IMMUNE DEFICIENCY SYNDROME (AIDS); COMPUTER-AIDED DESIGN AND MANUFACTURING; COMPUTER GRAPHICS; INFLUENZA in the McGraw-Hill Encyclopedia of Science & Technology.

Mark von Itzstein

Bibliography. S. Borman, New 3-D search and de novo design techniques aid drug development, *Chem. Eng. News*, 70(32):18–26, 1992; T. J. Perun and C. L. Propst (eds.), *Computer-Aided Drug-Design: Methods and Applications*, 1989; M. von Itzstein et al., Rational design of potent sialidase-based inhibitors of influenza virus replication, *Nature*, 363:418–423, 1993.

Earth

Most minerals present at the Earth's surface are dramatically transformed under the conditions of high pressures and temperatures existing deep inside the planet. Perhaps the best-known example is carbon: usually found as a soft, dark mineral graphite, it converts into hard and clear diamond when taken to pressures exceeding 5 gigapascals (50,000 times atmospheric pressure) at temperatures above 1000°C (1830°F; **Fig. 1***a*). Such conditions are achieved at about a depth of 150 km (90 mi).

Crustal and mantle minerals. By understanding the ways in which minerals transform at depth, geologists are able to decipher much about the conditions existing within Earth. Indeed, only through laboratory experiments and theoretical investigations of material properties at high pressures and temperatures can the measurements geophysicists make on Earth's interior be interpreted. The reason is that samples of the material existing at great depth cannot be found at the Earth's surface. In fact, natural diamonds have the deepest known origins of any minerals found at the surface, having been rapidly ejected from a depth of ~150–200 km (90–120 mi) by means of exceptionally violent volcanic eruptions. It is speculated that some diamonds may even come from a few hundred kilometers deeper, but in view of the 2890-km (1850-mi) thickness of the outer rocky shell (the mantle and crust) or the 6370-km (3950-mi) distance from the surface to the center of the planet, it is clear that geologists can obtain samples only from the outermost portions.

By examining rocks brought to Earth's surface, whether by volcanic eruptions or by other processes, geologists are able to decipher the nature of the crust and the uppermost mantle. Thus, it is known that these regions consist primarily of silicate minerals, that is, silicon-oxygen (Si-O) compounds. Indeed, oxygen constitutes more than 50% of the atoms typically making up rocks.

The crust, which is about 5–10 km (3–6 mi) thick under oceans and about 20–70 km (12–40 mi) thick under continents, is somewhat distinct in composition, on average containing slightly more silicon and aluminum (Al) than the underlying mantle. Feldspars ($NaAlSi_3O_8$-$KAlSi_3O_8$-$CaAl_2Si_2O_8$) and quartz (SiO_2) are among the most common minerals of the crust, whereas olivines [$(Mg,Fe)_2SiO_4$], pyroxenes [$(Mg,Fe,Ca)SiO_3$], and garnets [$(Mg,Fe,Ca)_3Al_2Si_3O_{12}$] make up the topmost mantle (**Fig. 2**).

Minerals are crystalline compounds in which the atoms are arranged in a highly regular order. Therefore, a mineral name refers both to a chemical composition or formula and to a particular crystal structure, that is, a particular way in which the atoms are arranged. For example, diamond and graphite possess two different crystalline arrangements of carbon (Fig. 1*a* and *b*), and quartz (Fig. 1*c*) is SiO_2 with a particular crystal structure.

Unfortunately, the structures of most minerals in the crust and mantle cannot be easily pictured because they are far more complex than the crystalline structures of graphite and diamond. A common theme, however, is that the silicon in most minerals at Earth's surface is surrounded by four oxygen atoms to form the tetrahedral unit SiO_4. In some cases, such as quartz and the pyroxenes, these SiO_4 units are linked together to form chains or spirals that can be thought of as the backbone of the mineral's crystal structure (Fig. 1*c*).

Experimental mineral transformations. The most direct method for deducing what happens to the crustal and mantle minerals at greater depth inside Earth is to simulate the conditions existing deeper in the mantle by squeezing and heating minerals in the laboratory. One of the most versatile instru-

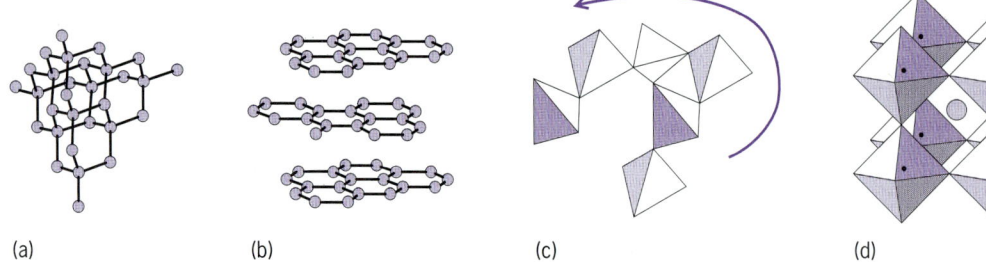

Fig. 1. Crystal structures of some crustal and mantle minerals. (*a*) Diamond. (*b*) Graphite. (*c*) Quartz (part of structure). (*d*) Perovskite. Quartz and perovskite are composed of SiO_4 tetrahedra and SiO_6 octahedra, respectively.

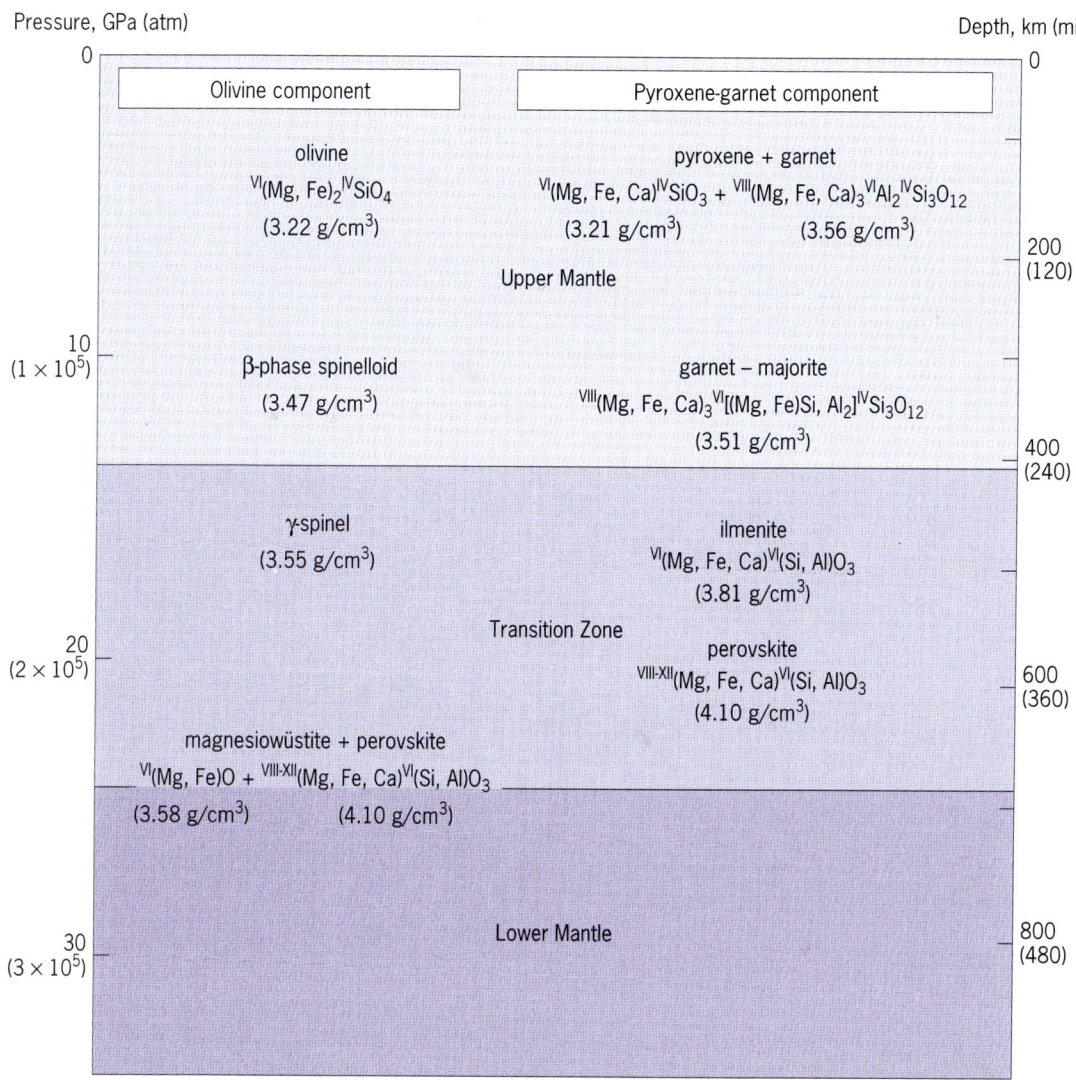

Fig. 2. Phases of mantle minerals. Each name refers to a specific crystal structure existing at a given pressure or depth within the Earth. Roman numeral superscripts in the chemical formulas indicate coordination of oxygen (O) around silicon (Si) and other cations. Density values given in parentheses are for the Mg compositions (end members) at ambient conditions.

ments available for such experiments is the diamond cell, in which a tiny specimen is compressed between the tips of two gem-quality diamonds.

Because the surface area of each diamond tip is so small, and pressure is given by force per unit area, very high pressures—in excess of 500 GPa (5×10^6 atm)—can be achieved with relatively modest forces pushing the diamonds together (for comparison, the pressure at the center of Earth is only 365 GPa or 3.65×10^6 atm). Because diamond is the strongest material known it is an ideal anvil for exerting ultrahigh pressures on mineral samples. However, it is perhaps even more significant that diamond is transparent; that is, the sample can be observed right through the diamond anvil, while it is at high pressures. In this sense, the diamonds might better be referred to as windows than anvils.

Moreover, a high-power laser beam can be focused directly through the diamonds, thereby heating the sample to several thousand degrees while it is under pressure. In this way, the conditions existing throughout Earth's mantle have been recreated in the laboratory, and it has been possible to determine what happens to minerals as they are taken to great depths inside the planet.

The results of many high-pressure experiments obtained with diamond cells, as well as with other high-pressure instruments, show that the minerals of the outermost Earth all transform to new mineral forms when taken to the conditions of the deeper mantle. As shown in Fig. 2, $(Mg,Fe)_2SiO_4$ olivine transforms to the spinel-type crystal structure at depths of 400–650 km (240–390 mi), for example, whereas $(Mg,Fe)SiO_3$ pyroxene takes on the garnet and ilmenite-type structures at similar depths.

Ultimately, all the important minerals of the outermost mantle—olivines, pyroxenes, and garnet—transform to a very dense mineral having the perovskite structure. According to the experiments, the perovskite form of $(Mg,Fe)SiO_3$ appears at pressures of ~20 GPa (2×10^5 atm; corresponding to a depth of 650 km or 390 mi) almost independent of composition, although a small amount of the oxide mineral (Mg,Fe)O magnesiowüstite is also formed in the high-pressure transformation of spinel (high-pressure form of olivine).

Perovskite mineral structures. It is remarkable that the one mineral (Mg,Fe)SiO$_3$ perovskite takes over for nearly all the upper-mantle minerals when these are taken to the conditions of the lower mantle (below 650 km or 390 mi). Even more interesting is that once it is formed, the silicate perovskite (and coexisting magnesiowüstite) is stable all the way down to the bottom of the mantle, where pressures reach 135 GPa (1.35×10^6 atm). The ability of this perovskite mineral to remain stable over such a broad range of pressures [~20–130 GPa (2×10^5–1.3×10^6 atm) or more] is no doubt related to its crystal structure (Fig. 1d), in contrast to the stability range of olivine, which is limited to 0–12 GPa (0–1.2×10^5 atm).

Hundreds of different crystalline compounds having perovskitelike structures are known, and there is widespread interest in understanding the source of their stability. These are all named after the mineral CaTiO$_3$ perovskite, which is present in Earth's crust and has the same crystal structure as the high-pressure (Mg,Fe)SiO$_3$ perovskite. Numerous materials that are enormously important for technology have perovskite-related crystal structures, including the high-temperature oxide superconductors discovered in the 1980s.

One of the major differences between the crystal structures of the outermost-Earth and the perovskite-type silicates is that each silicon atom is surrounded by six oxygen atoms in the high-pressure form, rather than the four oxygen neighbors typical of the low-pressure minerals (Fig. 1). Numerous studies of silicate crystals have shown that fourfold coordination (numbers of oxygen atoms) around silicon is transformed to sixfold coordination at high pressures (Fig. 2). The transformation from tetrahedral SiO$_4$ to octahedral SiO$_6$ arrangements is thought to alter the nature of the chemical bonding between the atoms, from relatively covalent at low pressures to more ionic in the high-pressure form.

Significantly, the effect of increasing coordination is to pack the atoms more tightly together; it is a characteristic effect of pressure to transform the low-pressure crystals to more tightly packed structures. Thus, for a given composition, high-pressure crystals are always denser (packed into a smaller volume) than the low-pressure forms: for example, the perovskite form is denser than the ilmenite and pyroxene forms of (Mg,Fe)SiO$_3$ (Fig. 2).

Extrapolation and interpretation. The changes in density and other physical properties due to high-pressure mineral transformations are important

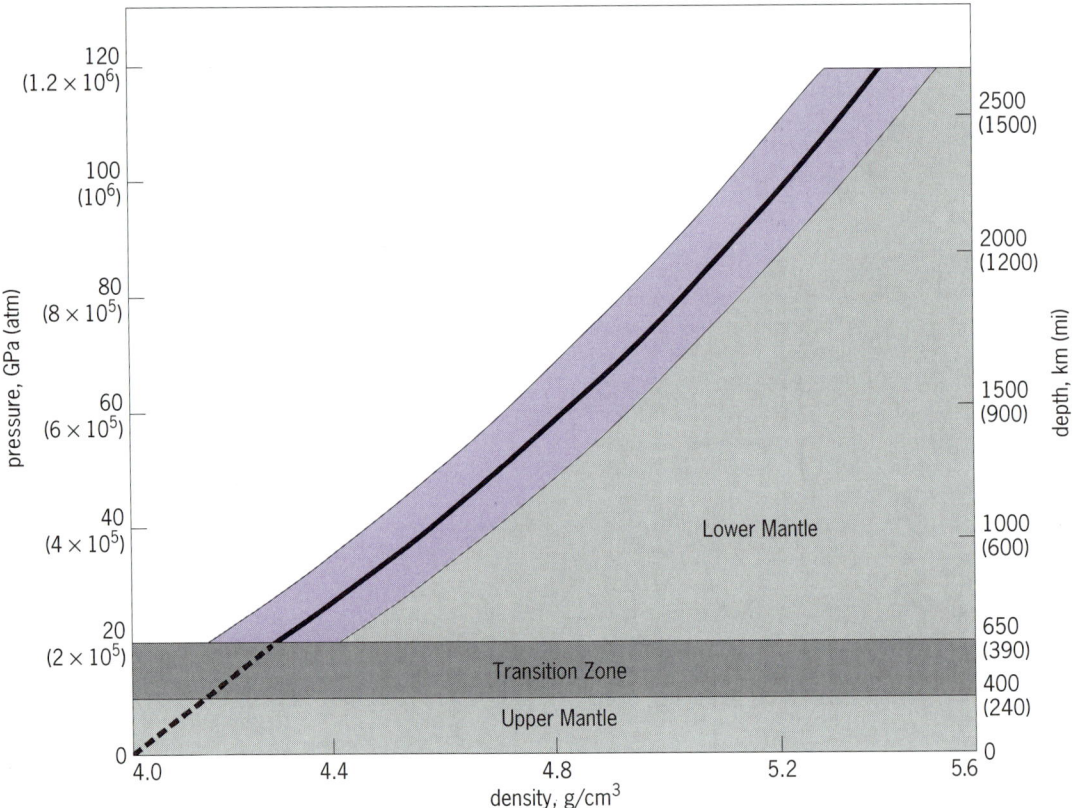

Fig. 3. Comparison of the density of (Mg,Fe)SiO$_3$ perovskite + (Mg,Fe)O magnesiowüstite at high pressures and at temperatures of 1000–2000°C (1830–3600°F; tinted band) with the geophysically determined density profile of the lower mantle (heavy line).

Earth interior

Small but abrupt changes in seismic velocity are observed at depths of about 410 and 660 km (250 and 410 mi) beneath Earth's surface. These velocity changes are most likely caused primarily by changes in the crystal structure (phase changes) in the minerals that make up mantle rocks, a result of the increasing pressure at depth. The properties of these abrupt changes or discontinuities provide valuable clues regarding the composition, temperature, and long-term deformation processes within the mantle. However, they are difficult to observe since the velocity changes are only about 5% in amplitude; therefore, the seismic signals they generate are relatively weak. Recently, seismologists are using special processing techniques to enhance the visibility of energy coming from the discontinuities; as a result they are learning much more about these interfaces and the properties of the surrounding rock.

Velocity discontinuities. Throughout most of Earth's mantle (the solid outer 45% that surrounds the liquid outer core), seismic velocities increase fairly gradually (**Fig. 1**). However, at depths between about 400 and 700 km (240 and 430 mi), an interval referred to as the transition zone, velocities increase much faster than can be explained from the increasing pressure alone, suggesting changes in composition or crystal structure. Rapid changes in velocity observed near 410 and 660 km (250 and 410 mi) are consistent with high-pressure laboratory measurements that indicate phase changes in

for geophysics because measurements on Earth's interior show that such changes take place inside our planet. Discontinuous jumps in density and elastic properties observed at depths of 400 and 650 km (240 and 390 mi) define the transition zone of the mantle; it is at these depths that, according to the high-pressure experiments, an upper-mantle rock containing olivine, pyroxene, and garnet would be expected to transform, first to spinel-type structures and then to the dense silicate perovskite.

Actually, laboratory experiments show that the pressures at which these transformations take place depend somewhat on temperature. In order to exactly match the pressures corresponding to the geophysically observed depths of the transitions, temperatures of 1400 and 1700°C (2550 and 3090°F) are required at 400 and 650 km (240 and 390 mi) depth, respectively. In short, by comparing laboratory measurements on mineral transformations with geophysical observations on the depths at which these transformations take place, the temperatures existing deep within the mantle can be determined.

It can be confidently stated that these high-pressure mineral phases exist deep inside Earth, despite the fact that they are never brought to the surface where they can be collected (the high-pressure minerals are too unstable to survive near the surface, converting back to low-pressure forms on their way up from depth). Not only do experiments show that upper-mantle minerals transform at just the right pressures corresponding to the geophysically observed jumps in properties at depth, but the properties of the high-pressure minerals match those of the transition zone and lower mantle.

Thus, it is known that all rocks bearing any resemblance to the fragments brought up from the uppermost mantle transform first to spinel and then to perovskite minerals with increasing depth (Fig. 2). Furthermore, the properties of the entire lower mantle, the single largest region of the planet, are in good agreement with measurements on the silicate perovskite at high pressures (**Fig. 3**).

Therefore, it can be safely concluded that this one mineral, $(Mg,Fe)SiO_3$ perovskite, makes up more than 70% of the lower mantle, or 40% of the entire Earth. Although a total of only a few grams have been made in laboratories at the surface, more than 2×10^{21} tons (2×10^{24} kg) of this mineral exist within the planet. The properties of this single most abundant mineral of Earth have surely played a crucial role in controlling the ways in which the planetary interior has evolved over geological time.

For background information SEE DIAMOND; HIGH-PRESSURE MINERAL SYNTHESIS; PEROVSKITE; SILICATE MINERALS in the McGraw-Hill Encyclopedia of Science & Technology.

Raymond Jeanloz

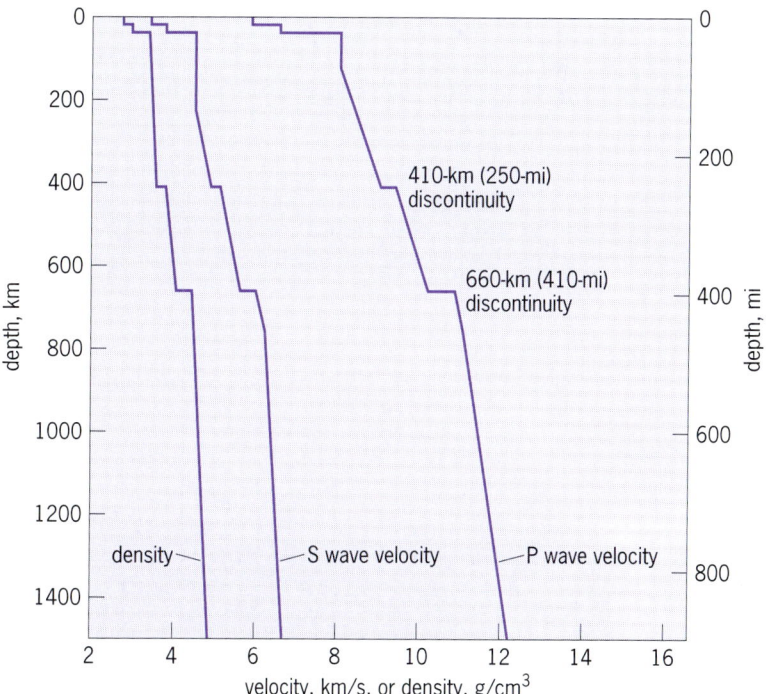

Fig. 1. Compressional (P) and shear (S) wave velocities and density versus depth in Earth's upper mantle. Discontinuous changes are observed near 410 and 660 km (250 and 410 mi).

olivine, the dominant mineral in the mantle, at analogous pressures. Seismologists discovered these velocity discontinuities several decades ago by noting anomalies in the travel times from earthquakes to seismographs at different distances. However, with this type of analysis it is difficult to determine the exact depth of these features, how truly discontinuous the velocity changes are, and whether there is any topography on the interfaces. Higher resolution can be achieved by studying the relatively weak seismic arrivals that result from direct reflections off either the top or the bottom of the velocity discontinuities, or that are due to conversions between compressional and shear waves at the interface.

Observation methods. Increasing numbers of observations of these secondary phases have been made in the last few years, a result both of expanded global seismic data sets and of better processing techniques. The processing methods combine data from hundreds of seismograms in order to suppress the effects of incoherent noise and enhance the visibility of weak seismic arrivals. These procedures are termed stacking; and they are similar in many respects to reflection seismic techniques used in the oil industry to image shallow crustal structure. However, the new results extend much deeper, and they map reflections from internal discontinuities within the upper 700 km (430 mi) of the mantle. Long-period (low-frequency) seismic waveforms tend to be simple and globally coherent, making them ideal for stacking methods. Results from long-period studies indicate that depths to the discontinuities vary by up to 40 km (24 mi) between different regions.

Mantle models. The amplitude of this topography has important implications for an ongoing controversy regarding the nature of flow within the mantle. The rocks in the mantle, although solid, gradually deform and flow at the high temperatures and pressures that exist in Earth's interior. Over millions of years this process results in a gradual overturn, or convection, in the mantle rocks, with warm material tending to rise and cold material to sink. The 660-km (410-mi) discontinuity has long been recognized as a potential barrier to mantle convection, which may separate the mantle into two distinct layers. In one model, termed layered convection, material does not flow across the 660-km (410-mi) interface, but convects separately within two layers of different composition. In an alternative model, termed whole mantle convection, convection involves an overturn of material throughout the mantle with significant flow across the 660-km (410-mi) discontinuity. Topography on the 660-km (410-mi) discontinuity can act to oppose flow through the interface, because of the density jump that occurs across the boundary. However, it appears that the relatively small relief on this discontinuity seen in the recent observations is insufficient to halt vertical flow through it, thus favoring models of whole mantle convection. The 660-km (410-mi) discontinuity is mainly caused by a mineralogical phase change, but controversy exists as to whether a small compositional change is also present. The deflections in the 660-km (410-mi) discontinuity are much less than those predicted for a purely compositional boundary, supporting the view that the bulk of the velocity jump at 660 km (410 mi) is due to a mineral phase change.

Seismograph networks. To obtain higher resolution images of discontinuity structure, short-period records from deep earthquakes can resolve features in the immediate vicinity of the events. Deep earthquakes tend to occur in subduction zones (regions where the surface plates are plunging down into the mantle), where it might be expected that the 660-km (410-mi) discontinuity is deflected downward in response to the sinking material. Since short-period seismic data are less coherent than long-period data, it is often necessary to stack data from arrays of instruments to produce usable signals. The regional networks of seismographs in the western United States, originally deployed to study local earthquakes, have proven to be particularly useful for this type of work, and are now being used to reveal details in the structure of subduction zones thousands of miles away (**Fig. 2**). Results of these and other studies suggest that in the vicinity of some subducting slabs the 410-km (250-mi) discontinuity is elevated by about 15 km (9 mi) and the 660-km (410-mi) discontinuity is depressed by 20–50 km (12–30 mi) compared to recent global averages of discontinuity depths.

Thermal models. The depth at which mineral phase changes occur depends upon the temperature; so, in principle, observations of discontinuity depth variations can be used to provide in-place mantle temperatures. The observed elevation of the 410-km (250-mi) discontinuity and the depression of the 660-km (410-mi) discontinuity are consistent with the effect of a cold subducting slab on the appropriate phase boundaries but require significant broadening of the slabs near 660 km (410 mi). Thermal models of subducting slabs that pass cleanly through the 660-km (410-mi) discontinuity into the lower mantle predict greatly reduced mantle temperatures in a narrow zone within the subducting slab itself but do not explain a broad regional depression in the 660-km (410-mi) discontinuity

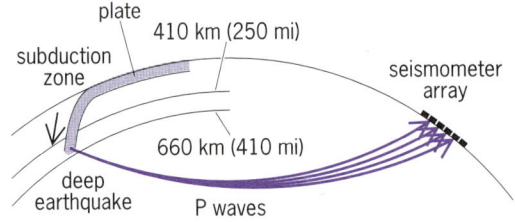

Fig. 2. Seismic waves (compressional or P waves) from deep earthquakes in subduction zones can be recorded by seismometer arrays thousands of miles away to reveal details of the mantle discontinuities near the earthquakes.

seen in long-period data for some subduction zones. The most likely explanation for this anomaly is that the slab broadens or turns near 660 km (410 mi), thus spreading the cold slab temperatures into the adjacent mantle. This model is supported by recent seismic velocity inversions for northwest Pacific subduction zones, which image horizontal extensions of the slabs just above 660 km (410 mi). The apparent deflection of the slabs could result from resistance to slab penetration through the 660-km (410-mi) phase change, as some models of mantle convection have indicated. However, the observed depression in the 660-km (410-mi) discontinuity is too small to prevent the slab material from eventually sinking into the lower mantle.

Seismic energy frequency. The frequency content of reflected seismic energy can provide information regarding how abruptly velocity changes with depth within the apparent velocity discontinuities. High-frequency seismic energy can be reflected only by relatively sharp interfaces, whereas low-frequency arrivals can be generated even by a fairly gradual change in properties. Short-period observations of P waves that have reflected off the bottom of the upper mantle discontinuities suggest that a large part of the velocity change near 410 and 660 km (250 and 410 mi) occurs within a depth interval of 5 km (3 mi) or less. Thus, a constraint is provided for the chemical composition of the mantle rocks.

Other anomalies. In addition to the discontinuities at 410 and 660 km (250 and 410 mi), which are strong global features, there are suggestions in many of these studies of additional reflectors at other depths, including anomalies near depths of 220 and 520 km (140 and 320 mi). These apparent discontinuities are often seen only in particular data sets, and their global distribution has not yet been established. However, reflection seismic images of the mantle will continue to improve as additional seismic stations and data become available. Increasingly more detailed maps of the structure of both major and minor mantle discontinuities are likely in the future.

For background information SEE EARTH INTERIOR; PLATE TECTONICS; SEISMOLOGY in the McGraw-Hill Encyclopedia of Science & Technology.

Peter M. Shearer

Bibliography. J. Phipps Morgan and P. M. Shearer, Seismic constraints on mantle flow and topography of the 660-km discontinuity: Evidence for whole-mantle convection, *Nature,* 365:506–511, 1993; P. M. Shearer and T. G. Masters, Global mapping of topography on the 660-km discontinuity, *Nature,* 355:791–796, 1992; J. E. Vidale and H. M. Benz, Upper-mantle discontinuities and the thermal structure of transition zones, *Nature,* 356:678–683, 1992; C. W. Wicks and M. A. Richards, A detailed map of the 660-kilometer discontinuity beneath the Izu-Bonin subduction zone, *Science,* 261:1424–1427, 1993.

Earthworm

Earthworms are invertebrates common in soils in most parts of the world. They are extremely important in breaking down organic matter and releasing its plant nutrients, as well as in turning over large quantities of soil, thereby improving soil fertility, structure, aeration, and drainage. In recent years, considerable advances have been made toward using earthworms in environmental management by manipulating earthworm populations, introducing them to new areas, and promoting their activities in soil and organic wastes so as to improve the environment and decrease the need for environmentally polluting inorganic chemicals.

Charles Darwin was the first to point out the great importance of earthworms in soil formation, turnover, and maintenance. He outlined their major role in combining mineral fragments with organic material to form relatively water-stable soil aggregates and in maintaining the fertility of the upper layers of soil. A large body of scientific evidence now supports and extends his conclusions.

There are about 1800 species of earthworms distributed around the world, and they differ greatly in size, number, life cycle, habits, and behavior. Some species are common all over the world, while others are confined to either temperate or tropical soils. One main family of earthworms, the Lumbricidae, have migrated to and become established in most parts of the world. In recent years they have been used extensively in environmental improvement and management.

Role in organic waste management. Earthworms consume dead or dying plant organic matter, which they grind in their gizzards, thereby increasing its surface area and promoting microbial activity as it passes through their intestines. They obtain their nutrition from the microorganisms, and this synergy between earthworms and microbes releases carbon, nitrogen, and other essential plant-nutrient elements from dead plant material, combining them into compounds that can be readily taken up by living plants, thereby completing the soil-nutrient cycle.

The species used in waste management are those that invade compost heaps and help to decompose decaying organic matter. However, the overall role of earthworms in compost heaps is relatively limited, because the microbial activity heats up the center of the compost heap to temperatures too high to allow earthworms to survive; hence they are confined to the outer layers of the heap.

Historically, earthworms have been bred in organic wastes for use as fish bait. However, intensive research since the mid-1970s in the United States, England, and elsewhere has harnessed the ability of earthworms to break down organic wastes into useful materials. Interdisciplinary research has developed low-, intermediate-, and high-technology methods of organic waste management using earthworms. The species used most commonly is *Eisenia fetida,* although several other species are suitable and may be more effective under tropical conditions.

To maximize the productivity of earthworms in waste management, it is important to provide optimal environmental conditions in the waste. Each species has different requirements, but temperatures of 15–25°C (59–77°F), moisture contents of 80–90%, ammonia contents below 0.5 mg/g (0.038 oz/oz), salt contents below 0.5%, and a pH of 5.0–9.0 is a general requirement. Earthworms in organic wastes need aerobic conditions, so they are mostly confined to the top 10–15 cm (3.9–5.9 in.) of a waste heap. The majority of efficient systems of organic waste management using earthworms is based on the regular and frequent addition of thin layers of waste to a base layer of waste that has been inoculated with a large population of an appropriate species. The waste does not heat up, and the earthworms remain near the surface, breaking down the wastes extremely rapidly and converting them into finely divided particulate materials resembling peat; these materials are rich in available plant nutrients and have considerable commercial potential as horticultural plant growth media.

Initial research in the United States in the late 1970s, involving the utilization of earthworm breakdown of organic wastes, was targeted at processing human sewage. It was found that sewage solids and sludges were broken down rapidly by some species of earthworms, and pilot plants using earthworms were tested in the late 1970s and early 1980s in Texas, Maryland, Washington, Utah, and Florida. However, this technology was not developed further because the costs of processing sewage with available methodologies were greater than those of conventional sewage waste treatments. There was also concern that chemicals toxic to earthworms deposited into the waste stream might occasionally kill many of them.

Later research in the United Kingdom demonstrated, on a commercial scale, that several species of earthworms can be used to break down a wide range of organic wastes, including cattle, horse, pig, and poultry waste, as well as industrial wastes from the paper, brewery, mushroom, preprocessed potato industries, and restaurant and supermarket organic wastes. These wastes differ in the pretreatment required (such as chopping and mixing, precomposting, and elimination of excessive levels of ammonia and salts) to ensure the maximum activity and productivity of the earthworms.

Since most earthworm-processed organic wastes can be reclaimed and utilized as commercially viable plant growth media, the economic returns on these technologies have proved to be excellent. The low-technology methods tend to require large areas of land and are labor-intensive, whereas the intermediate- and high-technology methods, using automated continuous reactors, require relatively small areas of land and have low labor requirements; both methods are viable commercially and have been used in various parts of the world for organic waste management. There is a valuable by-product when surplus earthworms are produced: earthworms are about 60–70% high-grade protein, up to 10% fat, and about 20% carbohydrate, and they contain essential minerals and vitamins. They can be processed through drying or ensiling into an excellent feed protein for fish, poultry, and pigs.

Soil amelioration and land reclamation. The clear link between overall soil fertility and the size of earthworm populations has resulted in many attempts to introduce earthworms into soils that contain no earthworms or small populations. The first essential step is to provide soil conditions suitable for earthworm survival and multiplication. Extensive research in the United Kingdom and other parts of Europe has shown that earthworm soil requirements involve adequate supplies of organic matter; conservation measures; minimal or no tillage cultivation; a pH that is not excessively acidic or alkaline; and reasonable aeration and plant cover. Given suitable conditions, populations of up to 500/m^2 (47/ft^2) are possible in woodlands, orchards, and pastures and up to 400/m^2 (37/ft^2) in arable land, although 100–150/m^2 (9–14/ft^2) are common most parts of the world.

Introduction of typical populations of deep and shallow-working earthworms into soils with few or no earthworms has produced significant increases in yields of many crops. This result has major environmental relevance, since use of earthworms can minimize the need for inorganic fertilizers and synthetic pesticides, which may disrupt dynamic soil biological processes and contaminate surface and ground waters.

The most successful introductions of earthworms into new areas has been in New Zealand, which lacks native lumbricid earthworms. Introductions of several species into agricultural fields has increased grass yields greatly, sometimes even doubling them, with a significant decrease in the demand for inorganic fertilizers. It takes 6–7 years for a field with no previous population of earthworms to become completely colonized and for maximum effects on fertility to be achieved. In West Africa (Nigeria and Ivory Coast) and Central America, a group of French researchers has successfully demonstrated considerable increases in yields of a range of crops after the introduction of earthworms into some of these more fragile tropical soils.

As more such successes are reported, there is increasing research and interest in many countries into the environmental benefits of earthworms for their poorer soils. Experimental introductions of earthworms have been made into poor upland soils, urban areas, sports grounds, gardens, and newly capped landfills. Many of these introductions accelerated the reclamation and improvement of these soils. Considerable research involving the introduction of earthworms into land newly reclaimed from the sea, such as the polders in the Netherlands, has resulted in a great acceleration of the rates of reclamation. If successful methods of cultivating soil-inhabiting earthworms can be developed, such inoculations will be facilitated greatly.

Bioindicators of environmental contamination. There is increasing evidence that a good earthworm population, which depends heavily upon microbiological activity in soils, is one of the best bioindicators of both soil health and freedom from chemical contamination. Research in the United Kingdom has enabled development of a standardized ecotoxicological test, using *Eisenia fetida* exposed to potentially toxic chemicals in a defined artificial soil (69% industrial quartz sand, 20% kaolinite clay, 10% sphagnum peat, 1% calcium carbonate). This test has been adopted globally by the Organization for Economic Cooperation and Development and the European Union as a standardized test for environmental contamination. This test protocol has been used extensively in recent years by pesticide registration authorities and by the U.S. Army Corp of Engineers in assessing the suitability of contaminated materials dredged from rivers for land disposal in the United States.

For background information SEE LAND RECLAMATION; OLIGOCHAETA; SEWAGE SOLIDS; SOIL MICROBIOLOGY in the McGraw-Hill Encyclopedia of Science & Technology.

Clive A. Edwards

Bibliography. C. Darwin, *The Formation of Vegetable Mould through the Action of Worms: With Observations on Their Habits,* 1881, reprint 1985; C. A. Edwards and J. E. Bater, The use of earthworms in environmental management, *Soil Biol. Biochem.,* 24(12):1683–1690, 1992; C. A. Edwards and P. J. Bohlen, *The Biology and Ecology of Earthworms,* 1995; C. A. Edwards and E. F. Neuhauser (eds.), *Earthworms in Waste and Environmental Management,* 1988; P. W. Greig-Smith et al., *Ecotoxicology of Earthworms,* 1992.

Electric power systems

Synchronization of power-system measurements with the help of high-precision timing pulses such as those provided by the Global Positioning Satellite (GPS) system will have far-reaching impact on the fields of power-system monitoring, protection, and control. Several experimental installations of these systems have already taken place, and more are planned. With increasing use of such measurement systems, operation of the electric power networks should be dramatically improved. SEE APPLICATION SATELLITES.

Role of measurements. Electrical power networks are perhaps the largest and most complex engineering systems of modern times. Large generators separated by hundreds of miles spin in synchronism and at the same time are interconnected by continent-wide networks that supply industries, cities, and consumers. Measurements of network electrical parameters play a vital role in making this power supply reliable and of high quality. Measurements serve a variety of crucial functions: providing a feedback which ensures that the system is providing power at correct voltage and frequency levels; providing inputs to protective elements which respond automatically to emergencies created by abnormal circumstances such as tornadoes, lightning strikes, or earthquakes; informing power-system operators on the network if the system is able to withstand the next contingency; and finally, serving as the basis for charging the customer for the use of electrical energy.

Positive-sequence measurements. Modern ac power systems are three-phase systems. Under normal balanced conditions, the three phases carry voltages and currents that are equal in magnitude to each other, and have a mutual phase-angle difference of ±120°. Every attempt is made to balance the three phases. It is thus advantageous to measure the principal (balanced) portion of current and voltage distribution on the power network. This measurement is called the positive-sequence measurement. It is an average of the three phase measurements, with appropriate phase shifts introduced in the averaging process to account for the ±120° angle shift between the three phases.

Synchronized phasor measurements. Such measurements are made possible by the relatively recent availability of precise timing pulses from the Global Positioning Satellite (GPS) system. From this system, pulses at 1 pulse per second can be derived at any point on Earth at all times, with a precision of 1 microsecond or better. This accuracy corresponds to an accuracy in phase angle of 0.022° at the power-system frequency. The synchronizing pulse received from the GPS system is utilized to produce phase-locked sampling pulses, which are used to sample the power-frequency voltage and current waveforms in synchronism (**Fig. 1**). The

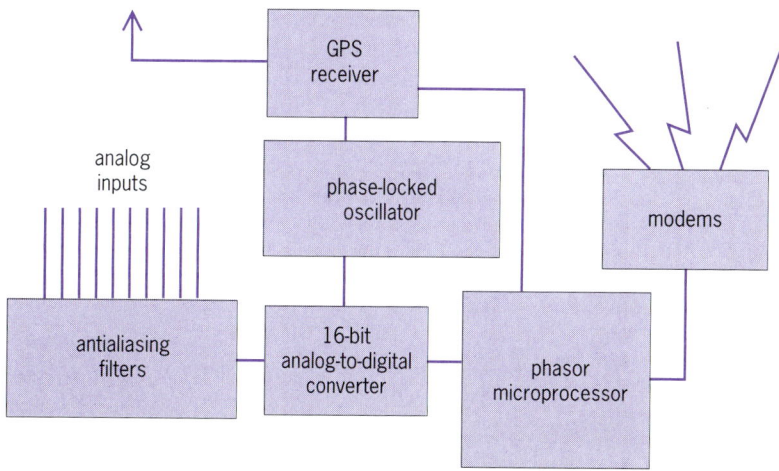

Fig. 1. Synchronized phasor measurement system using pulses from the Global Positioning Satellite (GPS) system.

sampling rates are modest, varying between 480 and 1920 Hz, with the most commonly used rate being 720 Hz. Because the data samples are given time tags, which describe the instant when the measurement was made with a precision of 1 μs, any parameters determined from these samples also carry the same precision once the precise sample numbers (and the rest of the time tag) are permanently attached to the measurement. It has been shown that the positive-sequence measurement from a three-phase input signal can be computed with a linear transformation. In computer-relaying algorithms, fractional-cycle data windows (that is, measurements of power-system quantities taken over a total duration of less than one period of the fundamental power-system frequency) are of interest for achieving high-speed relaying, whereas in most of the applications involving synchronized phasor measurements, one- or multiple-cycle windows (measurements taken over one or more such cycles) are commonly used.

Communication to central site. The phasors and their attached time stamps are sent to a central site for further processing. When the phasors serve as vehicles for recording system transient events for post-mortem analyses, the communication to a central site need not be in real time. Several applications of the phasors envision using them in a real-time control loop, for example, for advanced protection and control functions. In such cases, it is necessary to transmit the phasors on dedicated communication lines to the site where the control or protection function resides. Communication links with speeds of 4800 or 9600 bauds are able to sustain a data rate of about one transmission every 2–5 cycles of the fundamental power-system frequency.

Applications. Since the evolution of the synchronized phasor-measurement technology is still in its early stages, not all the applications of this technology are yet identified. However, several applications appear to be promising.

State estimation. State estimation is a monitoring function in power-system control centers, through which the state of the power system is determined from a set of measurements. In current practice, the measurements are not made simultaneously, and consequently the state of the system (defined as the collection of positive-sequence voltages at all network buses, where transmission lines and transformers are connected together) is at best a crude average state. Furthermore, the measurements made are nonlinear functions of the system state, so that an iterative state-estimation calculation must be performed. The state estimate obtained in the present technology is therefore an approximate quasi-steady-state picture of the system.

The synchronized phasor-measurement system completely changes this scenario. The positive-sequence voltage at a network bus is an element of the system state vector, and if all these voltages are measured and are synchronized, no further processing is needed. Thus, there is a progression from the technology of state estimation to the technique of direct measurement of the state. The measurements are truly simultaneous, so that they reflect the true state of the dynamic system. Thus, for the first time, it is possible to deal with dynamic problems at the control center. As an added benefit, a true static-state measurement is also obtained. Positive-sequence currents are measured, as well as voltages, providing redundancy in the measurement set.

Adaptive relaying. Relaying systems are designed to respond quickly and accurately to faults on a power network, taking action to isolate the faulted portion of the power system. Because of the speed of response required of protective relays (of the order of tens of milliseconds after the occurrence of a fault), the relays depend upon built-in logic with preprogrammed response modes. In recent years, it has been noted that some of these preprogrammed responses, although correct when originally determined, are no longer appropriate. This condition is not universal but occurs with sufficient regularity to cause major complications when the power system is in danger of going into a blackout. With the advent of computer-based relays it seems appropriate that the settings of the offending relays be revised from time to time automatically, as some crucial conditions on the power network change, a process known as adaptive relaying. Synchronized phasor measurements are found to play a significant role in many adaptive relaying functions. An example is the out-of-step relay that predicts the outcome of a system electromechanical swing. The synchronized phasor measurements bring back information about the swing as it evolves, and by using one of several prediction techniques, it is often possible to determine the outcome of the swing in real time, based upon actual system conditions. Experimental systems of this type are currently under investigation in a field installation.

Advanced control. Many of the control systems employed on modern power systems use feedback signals derived from local measurements made at the site where the controlled element resides. Examples of large controllable elements include excitation systems of large generators, high-voltage dc converters, static var converters, and variable series capacitors. The objective function of the control process is often to improve some overall performance index on the power system, and involves controlling certain remote variables to within desired bounds. For example, a controller for the high-voltage dc terminal may be responsible for maintaining power flows on a parallel ac network within bounds. In all these cases, since the feedback signal is derived locally but the objective of the control process is expressed in terms of remote parameters, the controller is generally based upon a mathematical model of the connection between the local measurement and the remote parameter, and this often leads to unsuitable controller design.

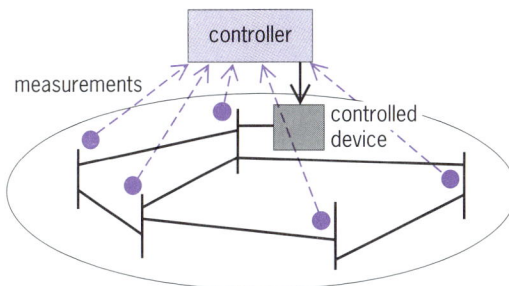

Fig. 2. Power-system control element with feedback control by remote variables (positive-sequence phasors).

Time-tagged synchronized phasors communicated to the controller from remote sites provide a very powerful means of improving the efficacy of the control process, since there is less dependence on the mathematical model of the power-system behavior and more direct feedback from the power system itself (**Fig. 2**).

For background information SEE ALTERNATING CURRENT; CONTROL SYSTEMS; ELECTRIC POWER SYSTEMS; RELAY; SATELLITE NAVIGATION SYSTEMS in the McGraw-Hill Encyclopedia of Science & Technology.

Arun G. Phadke

Bibliography. C. T. Leondes (ed.), *Advances in Electric Power and Energy Conversion System Dynamics and Control,* vol. 43 of *Control and Dynamic Systems,* 1991; A. G. Phadke and J. S. Thorp, *Computer Relaying for Power Systems,* 1990.

Electric protective devices

The development of the very large scale integrated circuits (VLSI) technology has changed the economics of using computers to protect components of electric power systems. Microprocessor relays have become as economically viable as previous designs manufactured with electromechanical and solid-state (analog electronic) components. VLSI technology has also made possible the development of stand-alone relays that can perform either a single function or multiple functions.

This activity started with the development and marketing of inverse-time overcurrent relays; within a short period of time a large variety of relays were offered for protecting generators, transmission lines, transformers, and other components of power systems. The initial designs used 8-bit microprocessors. As further developments in the area of microprocessors took place, designs were upgraded, using 16-bit and later 32- and 64-bit microprocessors and digital signal processors (DSPs). Although the processing devices now used in relays are of the 16-, 32-, and 64-bit types, analog-to-digital converters with only 12 to 16 bits continue to be the devices of choice. There are three reasons for not using more bits in analog-to-digital converters: (1) the low-order bits of an analog-to-digital converter could represent noise rather than the true strength of a signal; (2) analog-to-digital converters with more bits cost substantially more and, therefore, their use would increase the production costs of the relays; and (3) the benefits of using more bits can also be derived with range switching, which is more economical.

In addition to the traditional protection functions, considerable attention has been given to including features that assist the electric power utilities in monitoring their equipment more effectively. For example, a microprocessor relay could keep track of the interrupting duties performed by a circuit breaker and alert an operator when maintenance work is needed. Another important feature provided by these relays consists of saving the data from disturbances experienced on power systems. The data can be later checked and analyzed by a relay engineer. These and many other innovative features included in microprocessor relays have become important issues for product selection.

Although commercial microprocessor relays have become permanent components of electric power systems, research and development continues in several directions. Most recent activities have been in the areas of adaptive relay settings, testing of microprocessor relays in the laboratory and at site, protecting three-terminal lines, detecting downed conductors, system-wide synchronized sampling of signals, and so-called open-system relaying. Some of these areas of activity will be discussed. SEE ELECTRIC POWER SYSTEM.

Adaptive relaying. Relays have always been adaptive in some sense because their performance is always dictated by the prevalent operating state of the system. For example, the operating time of an inverse-time overcurrent relay depends on the fault current sensed by the relay.

The present interpretation of adaptive relaying is somewhat different than the adaptive performance which relays have always provided. A recent development in this area is the use of adaptive settings on distance relays to protect three-terminal transmission lines. The possibility was investigated of changing the settings of the relays as the amplitudes and phase angles of voltages at the three terminals change. Although the analytical developments have been reported, the work does not seem to have been implemented on an actual line.

Making relay settings adaptive for detecting high-impedance faults and for shedding loads, when generation-load imbalance is experienced, is an interesting and useful application; however, most benefits are likely to be achieved by making the time delays of relays adaptive to the state of the power system they protect. The levels of currents during faults depend on the capacities of generators, their locations, and the configuration of the transmission system up to the location of the fault. The fault levels are high during peak load times but low at off-peak load times. In present practice, the inverse-time overcurrent relays are set to maintain proper coordination at peak generation (and peak load) time. As a consequence, the tripping times of the relays are substantially larger than necessary during low-generation (low-load) periods. This practice unnecessarily subjects the system to prolonged shocks.

Adapting the relay settings to the prevailing system operating state could reduce the durations of disturbances and, as a consequence, prolong the life of the equipment. The time delays of relays, especially overcurrent relays, can be optimized so that they are kept at the minimum possible levels for all operating conditions.

To achieve this objective, several technologies have to come into play. The relays must be able to change their settings on receiving commands from a remote location. In addition, the system state must be estimated. Estimation requires that data be acquired from all over the system and brought to a central location. From these data, it must be determined which generators and lines are in service and how much load is being supplied. Currents at the relay locations in the event of faults at key locations, at least two on each line, must be calculated and then the settings of relays which are optimum for the prevailing state must be determined. All this activity requires extensive communication facilities to bring the required information from remote locations to a control center, and the use of advanced analytical techniques. After the optimum settings are determined, they must be communicated to the relays, which are dispersed throughout the system. On receipt of instructions, all relays must change their settings. Although some of these activities are already performed in power systems, many others are not. An experimental setup demonstrating the feasibility of achieving these activities has been reported.

Relay testing. Relay testing techniques have also changed with the introduction of microprocessor relays. It has become essential to test the hardware as well as the software of the relay. For example, calibration of analog-to-digital converters can be checked by using a signal of known magnitude and processing it through the software while the relay is in operation.

An elementary form of testing, which can check most blocks of the software, could take the form of providing data from a file and having it processed by the relay software. Fortunately, the testing technology has also advanced substantially. The relay test sets can now simulate the transient performance of power systems of moderate size. The results from these studies drive digital-to-analog converters, which provide analog signals to the relays being tested. Depending on the sophistication of the analysis software used and the response of the digital-to-analog converters, the testing now is much more realistic than testing the relays with the fundamental-frequency voltages and currents, as was done in the past. The size of the test equipment is not prohibitively large and testing of the relays at site is also possible.

Open-system relaying. Although other activities are likely to advance the state of relaying, open-system relaying is a major challenge to the way relays have been designed in the past. The basic idea is that the manufacturers provide the hardware suitable for use on a power system as well as the tools for writing the software, using a package of programs of individual functions provided with the relay. Although relay engineers would have the opportunity to put together relay designs that match their particular needs, they would also assume considerable responsibility, because they would have to be almost as well informed as the manufacturers are now.

For background information SEE ELECTRIC POWER SYSTEMS; ELECTRIC PROTECTIVE DEVICES; INTEGRATED CIRCUITS; MICROPROCESSOR; RELAY in the McGraw-Hill Encyclopedia of Science & Technology.

M. S. Sachdev

Bibliography. B. Chattopadhyay et al., Protection of a distribution network: An adaptive approach, *Can. J. Elec. Comp. Eng.*, 19(3):103–112, 1994; P. G. McLaren et al., On site relay testing for a series compensation upgrade, *IEEE Trans. Power Deliv.*, 9(3):1308–1315, 1994; P. G. McLaren et al., Open systems relaying, *IEEE Trans. Power Deliv.*, 9(3):1316–1324, 1994; M. S. Sachdev (ed.), *Microprocessor Relays and Protection Systems*, IEEE Tutorial Text 88EH0269-1-PWR, 1988; Y. Q. Xia et al., High-resistance faults on a multi-terminal line: Analysis, simulated studies and an adaptive distance relaying scheme, *IEEE Trans. Power Deliv.*, 9(3):1308–1315, 1994.

Electric utility industry

In 1994, the electric utility industry once again experienced significant change: California and other states began considering retail wheeling. The U.S. Federal Energy Regulatory Commission (FERC) continued implementing changes as a result of the passage of the Energy Policy Act of 1992. Two regional reliability councils filed with FERC to become regional transmission groups (RTGs), and FERC conditionally approved these filings. Having exercised its newly acquired authority to order transmission service in 1993, it followed in 1994 by imposing a solution on the two utilities in the case. The year began with rolling blackouts in the mid-Atlantic states and Virginia on January 19 because of a severe cold wave in much of the eastern two-thirds of the United States and Canada. Other significant activities included continuation of merger activity, environmental concerns, including nitrous oxides, electric and magnetic fields, alternate energy applications, and cancellation of five partially completed nuclear units.

Retail wheeling. The California Public Utilities Commission (CPUC) proposed an ambitious plan to reform the way it regulates electric service by investor-owned electric utilities in the state. The proposal relies on competition and the discipline of markets to replace often burdensome administrative regulatory approaches. In essence, the proposal allows retail wheeling by permitting all customers of investor-owned utilities to shop for electricity supplies among electric utilities, independent power producers (IPPs), and power marketers. (Independent power producer, as used here, is a generic term that includes all nonutility generators of electricity, such as cogenerators, qualifying facilities, exempt wholesale generators, and independent power producers that sell their output to electric utilities. Power marketers are those entities approved by FERC that buy electricity from an electric utility or independent power producer and sell it to another electric utility.) The second part of the CPUC proposal relies on a performance-based regulatory approach that rewards a utility for efficient operations, management, and investment. The commission anticipates the latter approach will better equip utilities to develop the tools to make the transition from a regulated to a market-driven electricity market. Three of the four investor-owned utilities in the state have a performance-based proposal before the commission. The fourth such utility, Pacific Power & Light Company, had a performance-based proposal approved earlier in 1994.

Although the CPUC has jurisdiction over only investor-owned utilities, these utilities have included on a voluntary basis noninvestor-owned utilities in the work groups and committees that are developing the restructuring proposals. Noninvestor-owned utilities account for about 25% of the demand in California.

Beginning in 1996, the CPUC would allow the state's largest electric customers, those served at a transmission level of 50 kV or higher, to tap the competitive markets instead of relying on the local investor-owned electric utility. Beginning in 1997, customers receiving service at the primary voltage level would be eligible. By the start of 1999, all commercial customers would be eligible. Finally, starting in 2002, all remaining customers, that is, residential customers, would be eligible.

There was considerable opposition to the restructuring proposal. Congressman R. Lehman, Chairman of the House Natural Resources Subcommittee on Energy and Mineral Resources, questioned the CPUC's authority to implement the proposal. The U.S. Department of Energy expressed concern that this proposal, which could spawn similar proposals in other states, would threaten the Administration's climate control goals by forcing utilities to cut demand-side management and other energy-efficiency programs. The Natural Resources Defense Council said the proposal requires an environmental impact statement, discourages energy efficiency, and would raise electricity bills for small customers.

The Michigan Public Service Commission approved an experimental 5-year retail wheeling plan for customers of Consumers Power Company (CP) and Detroit Edison Company (DE). The experiment would be limited to customers that take service at transmission or subtransmission voltages and have an electrical demand of 60 MW for CP customers and 90 MW for DE customers. The experiment was challenged in federal court by DE, which claimed that it impinges on FERC authority over interstate electricity flows.

Connecticut, New Jersey, and New York commissions determined that retail wheeling is not in the best interest of their states.

RTG formation. Pushed by FERC to form RTGs so transmission access by transmission-dependent utilities would be enhanced, the Western Regional Transmission Association (WRTA) filed in May for FERC approval, followed in June by the Southwestern Regional Transmission Association (SWRTA) and in October by the Northwest Regional Transmission Association (NWRTA).

FERC approved the WRTA and SWRTA filings in October, but placed two significant conditions on the approval. FERC required the RTGs or the transmission-owning members to file tariffs offering conditions of service and pricing to users comparable to what the owners charge themselves. It also ordered each group's members to file a single regional transmission plan instead of each member planning its own system. Although the FERC conditions gave WRTA and SWRTA reason to reconsider their actions, both groups decided to proceed with formation of the groups. The WRTA group, as filed, consisted of 30 entities and SWRTA of 18.

Comparability. FERC stated early in the year that it would require utilities to provide their competitors with transmission services comparable in quality and price to what the utilities provided themselves. The new standard drastically altered the commission's definition of what constituted undue discrimination and anticompetitive practices by transmission-owning utilities under the Federal Power Act. Previously, utilities were barred from discriminating among third parties by offering some customers better transmission deals than others. FERC's new policy is intended to prevent utilities from discriminating in favor of themselves. FERC's new policy found great interest among transmission-dependent utilities and power marketers. They filed with FERC for transmission access and comparable standing with transmission-owning utilities in the use of the transmission systems.

In midsummer, FERC expanded use of the comparability rule to include power-marketing affiliates of electric utilities. The first such affiliate to be affected by the rule was Heartland Energy Services, Inc., an affiliate of Wisconsin Power and Light Company. Later in the year the rule was expanded again in a case involving Pacific Gas and Electric Company to include the electric utility parent of the affiliate. Under the expanded rule, an electric utility parent organization must have comparable tariffs on file with FERC before blanket approval of market-based rates is given for the affiliate.

Power marketers. The number of power marketers approved by FERC at the beginning of 1995 reached 57 with 14 additional applications pending. The actual number of power marketers doing electricity transactions is still small, however. During the first quarter of 1994, only two power marketers filed the required activity reports. In the second and third quarters, the numbers rose to five and seven, respectively. Brokers are not required to obtain FERC approval to negotiate electricity transactions, nor are they required to file reports. They also were not involved in many transactions during 1994. (Brokers arrange for the sale of electricity between two electric utilities or independent power producers. Unlike power marketers, they do not take ownership of the electricity that is transferred.) Although power marketer and broker activity was low in 1994, it is expected to grow significantly.

Integration of IPPs. In fall 1993, the Board of Trustees of the North American Electric Reliability Council (NERC) recommended sweeping changes to the way NERC and the regional reliability councils would operate. One of these changes dealt with membership. The board recommended that all regional reliability councils expand their membership to include independent power producers and other entities with an interest in the electricity supply system. The board recognized that these new entities had to participate in the regional reliability councils and help make the rules for operating the electricity supply systems, if they were expected to follow those rules.

As of the beginning of 1995, 29 independent power producers, 15 power marketers, and 1 broker were members of the nine regional reliability councils. Because some of these entities participate in more than one council, the actual number of individual entities participating is lower: 20 independent power producers and 8 power marketers.

Division by plant and utility type. The division of generating capacity by plant type in 1993, along with other United States electric utility industry statistics, is given in **Table 1**. The historical breakdown and a forecast of these figures are shown in the **illustration**. Generating capacity is based on net output rather than gross (nameplate) capacity. Also, for consistency net energy for load (the amount of energy available to serve customers plus losses) is shown, rather than energy production. Capacity additions in 1994 and a comparison of capacity in 1993 and 1994 are given in **Table 2**.

The division of capacity among utilities with various types of ownership at the end of 1994 was as follows: investor-owned utilities, 581,498 MW (77.4% of the total); cooperatives, 26,474 MW (3.5%); federal agencies, 66,190 MW (8.8%); municipal utilities, 42,423 MW (5.6%); and state and public power districts, 35,116 MW (4.7%).

Transmission case settled. In fall 1993, the FERC ordered Florida Power and Light Company

Table 1. United States electric power industry for 1993

Parameter	Amount	Change compared to 1992, %
Net generating capacity, MW		
Hydroelectric	84,890 (11.9%)	−1.48
Fossil-fueled steam	433,639 (60.9%)	−0.30
Nuclear steam	100,071 (14.1%)	2.16
Combustion turbine, internal combustion	63,446 (8.9%)	5.15
Other utility*	572 (0.1%)	28.54
Independent power producers[†]	29,339 (4.1%)	8.53
TOTAL	711,957	0.58
Transmission, circuit miles		
Alternating current, 230–765 kV	148,527	1.32
Direct current, ±250–500 kV	2,426	0.00
Noncoincident demand,[‡] MW	580,753	5.84
Net energy for load, TWh[§]	2,882.5	4.43
Energy sales, TWh		
Residential	990.1	6.55
Commercial	781.8	3.45
Industrial	963.7	1.52
Miscellaneous	100.4	−0.33
TOTAL	2,836.0	3.70
Revenues, total, 10^9 dollars	196.6	4.89
Capital expenditures, total, 10^6 dollars[¶]	25,209	3.97
Customers, 10^3		
Residential	101,126.1	1.50
TOTAL	114,796.2	1.52
Residential usage, kWh/(customer) (year)	9,874	5.13
Residential bill, cents/kWh (average)	8.32	1.22

* Includes solar, wind, biomass, and so forth.
[†] Includes independent power producers, exempt wholesale generators, qualifying facilities, cogenerators, and other nonutility sources of generation.
[‡] Noncoincident demand is the sum of the peak demands of all individual electric utilities, regardless of the day and time at which they occurred.
[§] 1 TWh (terawatt-hour) = 10^{12} watt-hours.
[¶] Investor-owned electric utilities only. Does not include cooperative or public power utilities, estimated at $3.2 billion.
SOURCE: North American Electric Reliability Council, *Electricity Supply and Demand 1994–2003,* 1994; Edison Electric Institute, *Statistical Yearbook of the Electric Utility Industry,* 1993.

Table 2. Generating capacity added in 1994, and comparison of capacity in 1994 and 1993*

	Capacity added in 1994		Capacity, MW		
Plant type	Power, MW	Number of units	1993	1994	1994 total capacity, %
Fossil-fueled steam	2,279	42	433,639	435,918	60.3
Nuclear steam	150	11	100,071	100,221	13.9
Combustion turbines†	3,984	56	63,446	67,430	9.3
Hydroelectric	144	17	84,890	85,034	11.7
Other‡	134	2	572	706	0.1
Independent power producers§	4,489	84	29,339	33,828	4.7
TOTAL	11,180	212	711,957	723,137	

* Not adjusted for retirements.
† Includes diesel units.
‡ Includes solar, wind, biomass, and so forth.
§ Includes independent power producers, exempt wholesale generators, qualifying facilities, cogenerators, and other nonutility sources of generation.

(FP&L) to provide network transmission service to the Florida Municipal Power Agency (FMPA). This order was the first use by FERC of its newly acquired power to order transmission service under the Energy Policy Act of 1992. FERC's order gave the two utilities 60 days to negotiate rates and terms and conditions for services before FERC would render its decision.

The two utilities were unable to resolve their differences and turned to FERC for a decision. In May 1994, FERC issued a final order that granted FMPA the network service it requested. However, FERC agreed with FP&L on how that service should be priced. The order also clarified a provision of the Federal Power Act that prohibits FERC from ordering transmission service when a customer could use that service to replace electricity that it is under contract to buy. Parties under contract for electricity requirements can request a transmission service order, which FERC can issue, but that order cannot be effective before the end of the existing contract. Thus, customers can line up alternative suppliers and obtain the necessary orders while still under contract with their existing supplier.

Rotating blackouts. Much of the Eastern Interconnection (the area of the United States and Canada east of the Rocky Mountains excluding about 75% of Texas and the province of Quebec) was subjected to very cold temperatures and waves of snow and ice storms during January 1994. The severe weather reached a peak during the week of January 16. Faced with customer demands that far exceeded expectations and cold-weather-related problems with their generators and fuel supplies, electric utilities imported large blocks of electricity from throughout the Eastern Interconnection. For almost 4 h during the morning and early afternoon of January 19, utilities in the Mid-Atlantic Area Council (MAAC, which encompasses most of New Jersey, the eastern 75% of Pennsylvania, Delaware, the eastern half of Maryland, the District of Columbia, and a small portion of northern Virginia) instituted a total of 1500 MW of rotating blackouts and Virginia Power instituted 800 MW of rotating blackouts to maintain the integrity of the electricity supply system.

Transmission systems in many parts of the Eastern Interconnection were loaded to their limits much of the time. On January 18 and 19, utilities in Georgia and Alabama allowed overloads on their transmission facilities to permit higher than normal electricity transfers to neighboring utilities.

The rotating blackouts were the subject of many investigations by NERC, FERC, the U.S. Department of Energy, and various state governments and regulatory agencies. Although all the investigations developed recommendations to mitigate similar future situations, they all concluded that MAAC and Virginia Power acted properly by instituting the blackouts.

Mergers. At least 18 mergers were under review in 1994. Anticipating more mergers as competition increases, FERC took the opportunity to close a

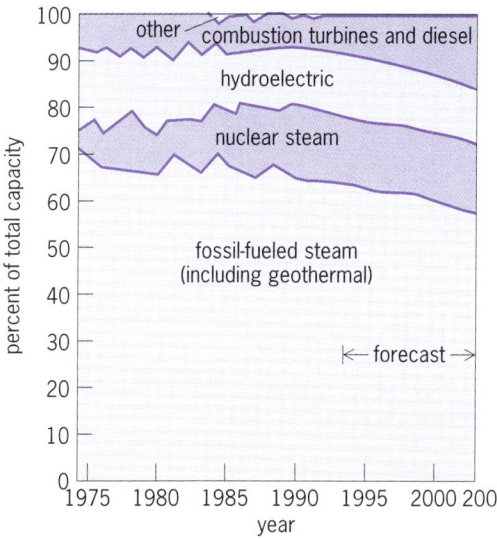

Probable mix of net electric utility generating capacity. "Other" includes solar, wind, and other nonconventional generating technologies. (*After North American Electric Reliability Council, Electricity Supply and Demand 1994–2003, 1994*)

loophole in its merger proceedings. Under its new policy utilities cannot avoid FERC scrutiny by first merging their holding companies, which does not require FERC review, and later seek approval to merge the subsidiary operating companies. The new policy was prompted by the mergers of utilities in Iowa in 1990 and 1991, which took this approach.

Even though Entergy Corporation completed its merger with Gulf States Utilities, unresolved issues remained at year end. PSI Energy and Cincinnati Gas & Electric Company, having obtained FERC approval to merge, still needed approval from the Securities and Exchange Commission. The proposed merger between Central and Southwest Corporation (C&SW) and El Paso Electric Company (EPE) continued to encounter obstacles. Las Cruces, Texas, which accounts for 8% of EPE's customer base, sought to establish a municipal utility that's supplied electricity by another entity.

Several new mergers were announced, including that of Washington Water Power Company and Sierra Pacific Power Company to form Resources West Energy Corporation, which would serve customers in five states. The proposed merger of Midwest Resources Inc. and Iowa-Illinois Gas & Electric Company would form the MidAmerican Energy Company, the largest combination utility in Iowa.

Nitrous oxide regulations. Late in 1994, a three-judge panel of the U.S. Court of Appeals for the District of Columbia threw out a U.S. Environmental Protection Agency (EPA) nitrous oxide rule. That rule required electric utility generating plants to use so-called overfire air technology to meet the Clean Air Act of 1990 acid-rain program's nitrous oxides emission limit. The court ruled in a suit brought by Alabama Power Company and the Utility Air Regulatory Group that EPA exceeded its authority. The court also ruled that electric utilities would not have to meet a January 1995 deadline for nitrous oxide compliance.

In 1992, EPA proposed regulations to control nitrous oxide emissions from about 200 coal-fired electric utility boilers (tangentially fired and dry bottom). The EPA was almost 2 years late in issuing these regulations. That delay allowed electric utilities only about 8 months to comply with the regulations; the Clean Air Act allowed 30 months. In issuing the successfully challenged regulation, EPA ignored a nitrous oxide control proposal developed by the Acid Rain Advisory Committee formed by EPA to help implement the rewrite of the Act. A majority of the committee concluded that Congress did not intend requiring use of overfired air technology to meet emission limits.

Electric and magnetic fields. Electric and magnetic fields continue to draw the interest of government and the media. Current research requests focus on exposure assessment and field management. Studies of the relationship between electric and magnetic fields and various types of cancer continue to show no uniformity of results.

A jury in Georgia found that electric and magnetic fields from electric lines of the local utilities did not cause cancer in a woman living near the lines. In Washington a state administrative law judge upheld a decision by the Washington State Department of Labor and Industries denying compensation to a Seattle City Light worker who claimed he developed leukemia from exposure to electric and magnetic fields during his work for the utility. These decisions follow a 1993 decision in San Diego in which a jury failed to find a link between electric and magnetic fields and a child's kidney cancer.

Alternate energy applications. Plans to generate electricity from several wind-powered generators were announced during the year. If the full scale of these agreements come to fruition, the capacity will amount to about 150 MW in 2–3 years and exceed 1300 MW within a decade. These projects are in the Pacific Northwest, southern California, Texas, West Virginia, and New England.

Two California electric utilities have entered the photovoltaic market. The Sacramento Municipal Utility District is using so-called green pricing to encourage a sustained orderly development of this market. This program relies on a voluntary rate premium of 15% to help finance the program. The utility's goal is to accumulate 50 MW of photovoltaics by 2000. Southern California Edison launched a 3-year photovoltaic program to serve customers in isolated areas where electricity is not available.

A joint venture of Amoco and Enron Solar Power Development will build utility-scale photovoltaic plants. The joint venture will run a photovoltaic module manufacturing business previously operated by Amoco's subsidiary, Solarex. The company plans on building 10 MW of photovoltaic modules annually.

A 19.9-MW (net) waste-tire-to-energy plant is scheduled for a Chicago suburb. An unregulated subsidiary of Houston Industries, Inc. and Chewton Glen Energy will build the facility and sell the output to utilities. It would be fueled by 65 tons (59 metric tons) of scrap tires per year, about 30% of the area's scrap tire production. It is the third waste-tire-to-energy plant in the United States.

Nuclear cancellations. Two electric utilities agreed to discontinue construction of nuclear facilities. The Washington Public Power Supply System (WPPSS) Board of Directors voted to terminate the Hanford No. 1 and Satsop No. 3 units, which total about 2500 MW. Construction of the units stopped in 1983, and they have been preserved since then. This decision stops any further involvement in the units by more than 100 public utilities and the Bonneville Power Administration. The board did leave open the possibility an independent developer or a consortium might buy one or both units.

The Tennessee Valley Authority (TVA) said it will not complete by itself three nuclear plants because of the high cost of completion. The three units with a total capacity of 3940 MW are the Watts Bar No. 2, and Bellefonte Nos. 1 & 2. TVA agrees that there is little chance of finding a partner to complete the plants.

For background information SEE ELECTRIC POWER GENERATION; ELECTRIC POWER SYSTEMS; ENERGY SOURCES; SOLAR ENERGY in the McGraw-Hill Encyclopedia of Science & Technology.

Eugene F. Gorzelnik

Bibliography. Edison Electric Institute, *Statistical Yearbook of the Electric Utility Industry*, 1993; North American Electric Reliability Council, *Electricity Supply and Demand 1994–2003*, 1994; North American Electric Reliability Council, *Reliability Assessment 1994–2003*, 1994.

Electron holography

Electron holography is an imaging technique using the wave nature of electrons and light. The microscopic world seen with the eye of the extremely short wavelengths of electrons is enlarged and observed as an optical image. The development of an extremely coherent field-emission electron beam in 1979 together with subsequent progress in reconstruction techniques opened the way to applications of this method. In addition to the original objective of high-resolution microscopy, methods for displaying the phase distribution of an electron beam were developed, thus making it feasible to observe and measure microscopic objects and fields which are inaccessible to electron microscopy.

Experimental method. The principle behind electron holography was devised by D. Gabor in 1949 to overcome the resolution limit of an electron microscope by compensating for the inevitable aberrations of the electron lens. However, because of the unavailability of coherent waves, the intrinsic value of this principle was not fully recognized until 1960 when coherent light was first produced and holography was developed in the field of light optics, in contrast to Gabor's intention. Electron holography, using a coherent field-emission electron beam, is based on similar principles and has become applicable to practical problems. This beam is produced by applying a voltage of a few kilovolts to a sharp tungsten needle 100 nanometers in radius, and has a bright point source: the current density at the tip surface is 100,000 times higher than that of a conventional thermionic electron beam, and the virtual source is as small as 5 nm. In addition, the energy spread of the electron beam is as narrow as 0.3 eV since the cathode can be operated at room temperature. Because of these features of high brightness and monochromaticity, the total number of electron interference fringes recordable on film reaches 3000, and the fringes can be seen with the unaided eye on a fluorescent screen when the fringe number is less than 100.

Electron holography consists of two steps: An interference pattern between an object wave and a reference wave is formed using an electron beam and is recorded on film as a hologram. Then, the image of the original object is reconstructed by illuminating a light beam equivalent to the reference wave onto the hologram.

Optical reconstruction using light is simple, but is off line because of the time consumed in developing hologram film. On-line reconstruction techniques are being developed by using computers and optical devices. For example, an image can be numerically reconstructed from a hologram recorded on a charge-coupled device (CCD) attached to an electron microscope, and a reconstructed image can be displayed. Images can be obtained in a fairly short time depending on the performance of the computer used, but not yet in real time. A real-time method using a liquid-crystal panel as a phase hologram has also been developed. The video signal for a hologram detected by a television camera attached to an electron microscope is transmitted to the liquid-crystal panel, where the intensity distribution of the hologram is transformed into the phase-shifting function for an incident light beam. Images are instantly reconstructed by illuminating a laser beam onto the panel. The time resolution of these images is 1/30 s. This method thus allows dynamic phenomena to be observed in real time.

The image reconstructed contains complete information about the object wave, that is, the phase and the amplitude. Therefore, images under arbitrary focusing conditions can be obtained from a single hologram, and a disturbed image due to the aberrations of an electron lens can be restored in the reconstruction stage. In addition, the phase distribution of an electron beam transmitted through an object can be obtained as an interference micrograph. Applications of electron holography are in the areas of high-resolution microscopy and interference microscopy.

High-resolution microscopy. The spatial resolution of an electron microscope is determined by the spherical aberration of the objective lens, which is far from the fundamental limit of the wavelength. This aberration can now be optically or numerically compensated for in the image reconstruction stage of electron holography to improve the resolution. Since the lens aberration influences an image in such a way that the corresponding phase shift is added to the Fourier transform of the object wave, the effect of the aberration can be eliminated by subtracting the phase shift at the Fourier-transform plane of the reconstructed image. However, this method has not yet been developed to the point where an object which cannot be observed by electron microscopy can be seen by electron holography. Ultrahigh resolution will be achieved in the future with a higher-energy electron beam.

Interference microscopy. Electron interferometry has been carried out with an electron microscope equipped with an electron biprism. The phase distribution of an electron beam transmitted through an object is measured as an interferogram

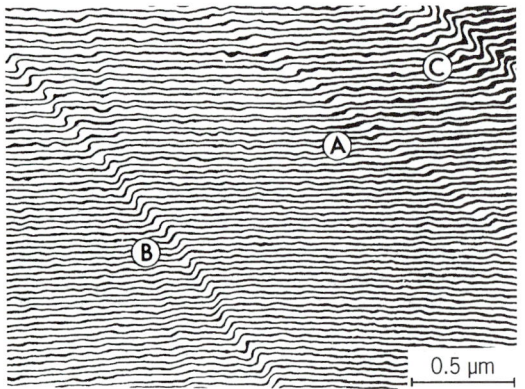

Fig. 1. Interference micrograph of molybdenite thin film, phase-amplified 24 times. Deviations from regular fringes, labeled A, B, and C, indicate thickness changes. Step A corresponds to a monoatomic step of 0.6-nanometer height.

by using the biprism to overlap the transmitted beam and an electron beam passing through outside the object. The precision in phase measurement is $\lambda/4$, where λ is the wavelength.

The developments of both electron holography and a coherent field-emission electron beam have improved the performance of electron interferometry. The precision in phase measurement has increased to $\lambda/100$, and real-time observation has become feasible. This technique has been used to observe thickness distributions in atomic dimensions, electric and magnetic fields, and vortices inside superconductors.

Thickness distributions. A thin-film specimen made of uniform material can be regarded as a space in which an electrostatic potential is higher than that of a vacuum by an inner potential V_0. Therefore, electrons are accelerated by V_0 when they enter the specimen, and the wavelength becomes slightly shorter. The phase shift of an electron beam having passed through the specimen is proportional to the film thickness.

An interference micrograph of a molybdenite thin film is shown in **Fig. 1**. The phase distribution is displayed here as deviations from regular fringes. Since this interferogram is phase-amplified 24 times, one-fringe displacement corresponds to a phase shift of $\lambda/24$, or a thickness change of 1.3 nm. Steps A, B, and C in the micrograph correspond to one, three, and five layers of surface atomic steps, respectively. The thickness change at step A is 0.6 nm, and corresponds to a $\lambda/50$ electron phase shift.

In the transmission mode of electron holography, surface topography can be investigated only through thickness measurements. However, a reflection mode using a Bragg-reflected electron beam was demonstrated to be feasible and the surface topography was measured to a precision as high as 0.01 nm without any phase amplification. This mode was used to quantitatively measure the surface undulation due to a single dislocation.

Electric and magnetic fields. When the observed object is an electric field or a magnetic field, the contour fringes in the interference micrograph directly indicate projected equipotential lines of an

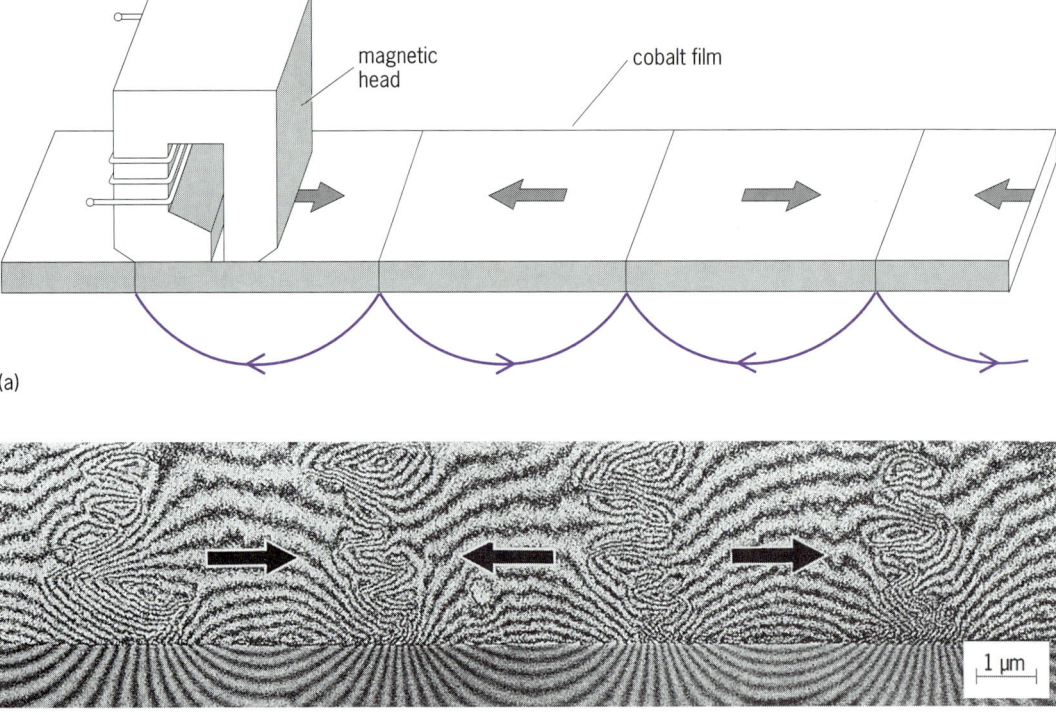

Fig. 2. Observation of magnetic lines of force on a recorded magnetic tape. (*a*) Recording method. (*b*) Interference micrograph.

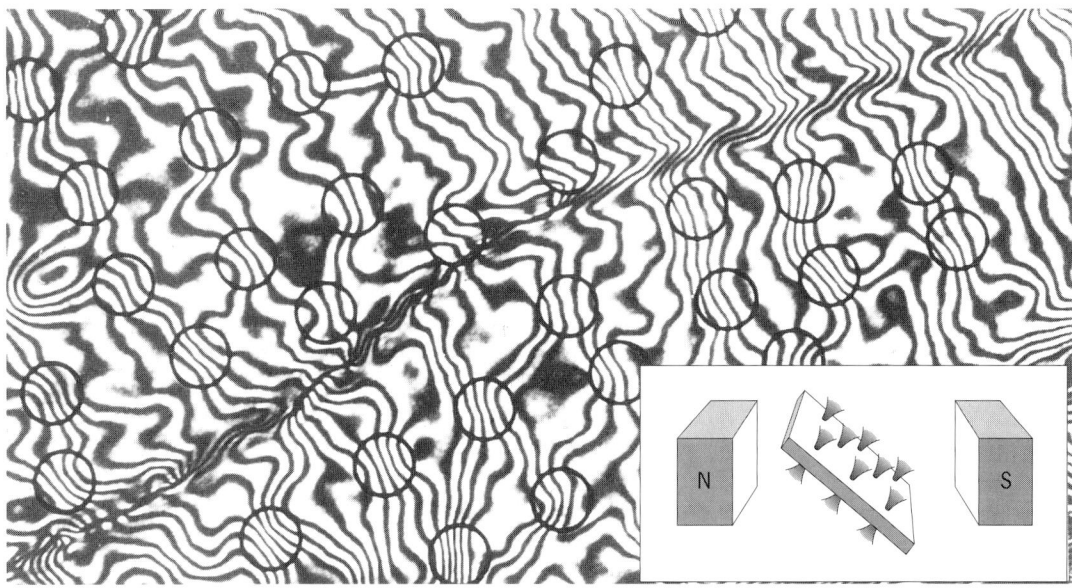

Fig. 3. Interference micrograph of a superconductive niobium thin film at 4.5 K (−452°F) and 0.01 tesla (100 gauss), phase-amplified 16 times. Magnetic lines become dense in encircled regions, which correspond to vortices. Inset shows configuration of film, with vortices, and north (N) and south (S) poles of magnet.

electric field or projected magnetic lines of force. Especially in case of a magnetic field, a constant magnetic flux of h/e (where h is Planck's constant and e is the electron charge) flows between adjacent contour fringes. An observation of a recorded magnetic tape is shown in **Fig. 2**. A magnetic head was moved on the surface of a cobalt thin film, and a bit pattern was recorded on the film. Magnetic lines of force can be directly observed as contour fringes in the micrograph. A constant flux of h/e flows between two adjacent fringes. The magnetization recorded inside the film, the direction of which is indicated by arrows, can be observed, as well as the magnetic fields leaking out from the film. Two oppositely directed streams of magnetization collide head on with each other at the boundary, and produce vortices similar to those produced by streams of water.

Vortices in superconductors. A vortex has the minimum quantum of magnetic flux in a superconductor, and is shaped like an extremely thin filament. Therefore, it has evaded direct observation. However, such tiny vortices play an important role in the practical applications of superconductors. For example, the critical current of a superconductor is determined by the dynamic behavior of the vortices. When a current is applied to a superconductor, a Lorentz force proportional to the current is exerted on the vortices, and superconductivity breaks down if the vortices are not pinned at defects in the superconductor but instead begin to move. SEE SUPERCONDUCTIVITY.

Interference microscopy has made it possible to directly observe vortices in superconductors. In the example shown in **Fig. 3**, a niobium thin film is tilted 45° with respect to both the electron beam and the magnetic field. Projected magnetic lines of force can be directly observed as contour fringes. Magnetic lines flow generally in the direction of the applied field but become locally dense in the circled regions. Each such region corresponds to a vortex. The temperature dependence of the magnetic-field radius of a single vortex was measured with this technique.

Although interference microscopy provides high resolution and quantitative information about the magnetic-field distribution of vortices, Lorentz microscopy is more suited for dynamic observation of vortices. A Lorentz micrograph can be obtained simply by greatly defocusing the electron micrograph under the illumination of a collimated electron beam, and therefore can be seen even if it is moving. A vortex can be seen as a spot consisting of a pair of regions, one bright and one dark, in contrast. The dynamic interaction of vortices with pinning centers has been investigated with this technique.

For background information SEE ELECTRON MICROSCOPE; HOLOGRAPHY; INTERFERENCE MICROSCOPE; SUPERCONDUCTIVITY in the McGraw-Hill Encyclopedia of Science & Technology.

Akira Tonomura

Bibliography. A. Tonomura, *Electron Holography*, 1993; A. Tonomura et al. (eds.), *Proceedings of the International Workshop on Electron Holography, Knoxville, August 29–31, 1994*, 1995.

Electronic displays

Recent advances in electronic display technology include the development of ferroelectric liquid-crystal displays and volumetric three-dimensional displays.

Ferroelectric Liquid-Crystal Displays

Liquid-crystal displays (LCDs) are commonplace in laptop computers, portable television receivers, camcorders and numerous other portable equipment applications where flat-panel displays are needed. The LCDs used in these applications are usually either the active-matrix LCD or the supertwisted nematic LCD. A new type of LCD, the ferroelectric liquid-crystal display (FLCD), is not yet commercially available but offers the unique property of being able to store images without any power, and can be viewed from any angle, which is not usually possible with other LCDs.

Comparison of LCD types. In order to appreciate the significance of the properties of FLCDs, it is worth reviewing the two main LCD types: the active-matrix LCD and the supertwist LCD. The active matrix has an active element, such as a transistor or diode, on every picture element (there are more than 900,000 in a standard color computer display), which serves to store charge on that element while all the other elements in the display are being written and also to provide a sharp voltage threshold for switching the element, to enable the elements to be individually selected. Because this charge gradually leaks away, the active-matrix display needs to be constantly refreshed, and the total number of lines that can be addressed before a refresh is necessary is limited.

In contrast, supertwisted nematic LCDs respond to the average (root-mean-square) voltage that is applied to the picture elements to switch the liquid crystal, and depend on their own intrinsic very steep threshold to enable the elements to be individually selected. As the number of lines that are successively written is increased, the root-mean-square voltage applied to any one particular element becomes lost in the average of the root-mean-square voltages applied to all the other rows of picture elements, and the brightness of that element becomes indistinguishable from the brightness of the other elements. Thus, even with the sharp intrinsic threshold, there is a limit to the number of lines that can be written (typically 300).

Ferroelectric liquid-crystal displays, however, have the unique property that once they are switched into one state they remain in that state indefinitely, even if the applied voltage is removed or is a low (up to 10 V) alternating-current voltage. Thus, in theory an infinite number of lines can be written without the need to refresh the display. In practice, a limit is imposed, depending on how often the display is required to be refreshed. For example, the display needs refreshing at 25–50 Hz to show moving images. The important difference with FLCDs is that the display does not flicker even when refreshed at frequencies as low as 10 Hz, because there is no decay of the light output with time. At the same time, the fast optical response means there is no blurring of moving objects such as a mouse cursor. If the display is used in a reflective mode (no backlight), power is consumed by the display only when the image changes, making it attractive for portable applications. The main differences of FLCDs from the two major LCD types are summarized in the **table**.

Operating principle. Like other LCDs, the FLCD modulates the intensity of polarized light passing through it. The basic principle of operation (**Fig. 1**) is that of the classical half-wave plate of polarized optics, with an optic axis that can reorient under the applied field. Light polarized by an entrance polarizer that enters the FLCD, with the polarization direction either parallel or perpendicular to the optic axis, passes through unchanged and cannot emerge from the exit polarizer, as it is perpendicular to the entrance polarizer (Fig. 1a). After the application of a suitable polarity pulse the ferroelectric liquid crystal molecules reorient (**Fig. 2**) so that, ideally, the optic axis now makes an angle of 45° to the incoming polarized light (Fig. 1b). The light is then transmitted according to the equation for polarized light traveling through a half-wave plate. A pulse of opposite polarity would switch the FLCD back to the original state. The optical response to pulses of alternating polarity is shown in Fig. 1c.

Although Fig. 1 is an idealized model of the FLCD, it represents a sufficiently good approximation for considering display applications.

Color and shading. As with other LCDs, color is introduced by using red, green, and blue filters over the top of the picture elements, so that color shading can be introduced by varying the relative brightness of these primary colors. Varying the brightness of the picture elements is complicated for the FLCD, because it is inherently binary. In order to introduce shading into such a binary display a number of techniques can be employed. The

Comparison of liquid-crystal display (LCD) types

LCD types	Viewing angle	Optical response time (room temperature)	Liquid-crystal phase	Cell spacing	Relative cost	Shading
Active matrix	Good	25 ms	Nematic	5–8 µm	High	Analog
Supertwisted nematic	Poor	50 ms	Nematic	4–6 µm	Medium	Analog/binary
Ferroelectric	Very good	30 µs*	Smectic C	1.3–2.0 µm	Medium	Binary

*This number can vary by several orders of magnitude depending on the surface-alignment conditions and the magnitude of the spontaneous polarization.

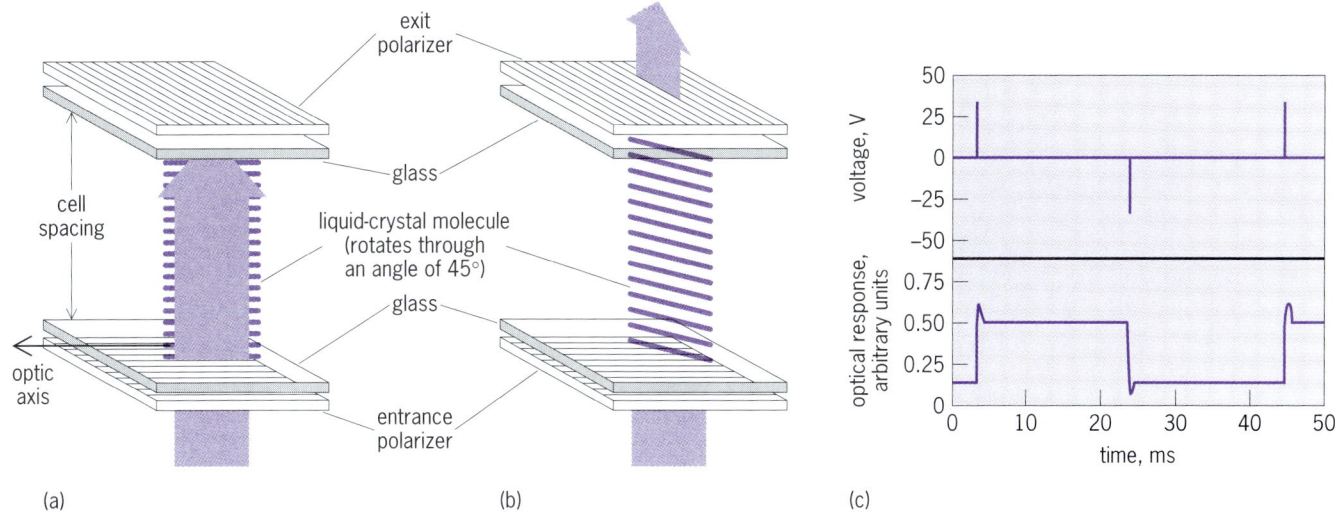

Fig. 1. Operating principle of the ferroelectric liquid-crystal device (FLCD). (*a*) Device in dark or off state, and (*b*) light, bright, or on state. The liquid-crystal molecule rotates through an angle of 45°. (*c*) Measured optical response to applied pulses. (*After P. W. H. Surguy and L. K. M. Chan, Changing the face of flat displays: The ferroelectric LCD, IEE Rev., pp. 249–252, November 1993*)

most commonly used techniques are spatial dither, temporal dither, and variable threshold.

Spatial dither simply means that every picture element is divided into subelements, with differing size. These subelements are sufficiently small that when the display is viewed at the normal viewing distance they are not perceived; only an average of them is perceived, so that the brightness varies according to the number of subelements that are switched on. The disadvantage of spatial dither is that the number of elements, and hence interconnections and drive electronics, is increased.

Temporal dither introduces brightness shading by switching each picture element on and off several times in a frame time (that is, in several subframes) at a sufficiently fast rate that no flicker is perceived. The brightness is then determined by the relative amount of time that each element spends in the on state. Temporal dither makes use of the fast switching time of the FLCD, but at present the amount of brightness variation possible with this method is limited by the FLCD switching time.

A different approach from either of these dithering techniques relies upon introducing a spread in the voltage threshold needed for switching in each picture element. Thus, when the applied voltage is increased more and more of the picture element switches so that the brightness of the element is governed by the size of the applied voltage.

Applications. The advantages of low power arising from the bistability of FLCDs means that they may be used in portable devices such as personal digital assistants and portable computers. However, since there is no limitation on the number of lines that can be written in an FLCD, they will probably be used for high-resolution displays (greater than 1000 lines), desktop computers (15-in. or 38-cm diagonal or greater), and high-resolution projectors, where active matrix and supertwisted nematic displays cannot be used.

Because of their fast switching speed, FLCDs can be used as optical correlators for pattern recognition and as optical crossbar switches for controlling the interconnection of light beams in optical-fiber telecommunications systems.   *Paul W. H. Surguy*

Volumetric Three-Dimensional Displays

Display systems able to faithfully reproduce the spatial structure of three-dimensional images have long been the subject of scientific investigation. The human visual system perceives spatial relationships by the often subconscious interpretation of a variety of depth cues, the relative emphasis placed upon each often being determined by the type of scene under observation. A variety of three-dimensional displays have been developed in recent years, each of which is useful for a limited range of applications; these displays are classified according to their general operating principles. One such class is known as volumetric display sys-

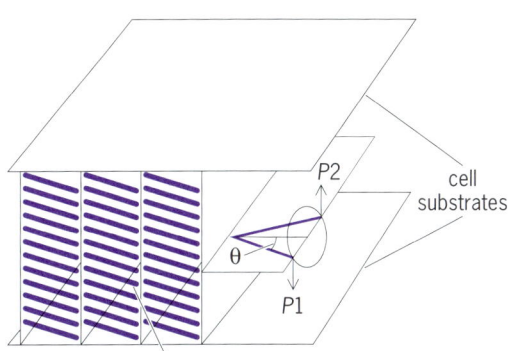

Fig. 2. Switching of molecule between two states defined by the surface of a cone in the smectic C liquid crystal. The angle θ is the molecular tilt angle of the liquid crystal; *P*1 and *P*2 show the two possible orientations of the ferroelectric dipole. (*After P. W. H. Surguy and L. K. M. Chan, Changing the face of flat displays: The ferroelectric LCD, IEE Rev., pp. 249–252, November 1993*)

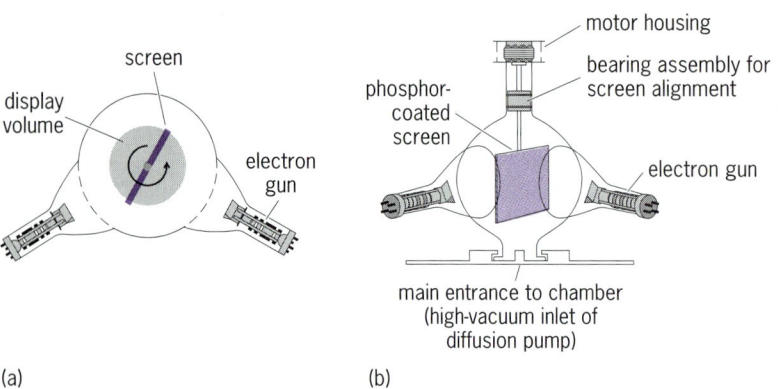
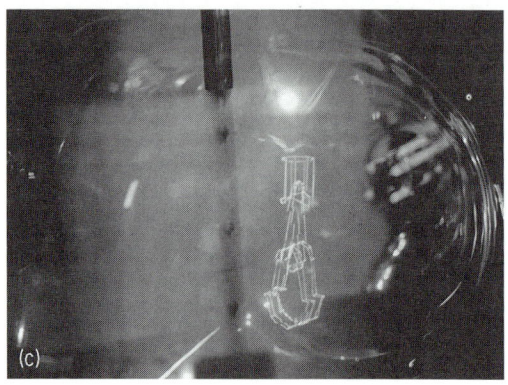

Fig. 3. Swept volume volumetric display. The rotating target screen is addressed by two electron guns, positioned 120° apart. The spherical vessel contains a vacuum and permits viewing from practically any angle. (*a*) Plan view. (*b*) Side elevation. (*c*) Volumetric image of piston, crankshaft, and connecting rod displayed by prototype swept-volume system.

tems. These systems are three-dimensional displays which permit images to be drawn within a volume rather than upon a stationary surface. Because these images occupy three physical dimensions, a number of depth cues are automatically satisfied.

Although there are many possible methods for implementing such a system, fundamental to each is the creation of a transparent three-dimensional display space within which images may be drawn. If this space is achieved by the rapid periodic motion of a two-dimensional surface, the display may be referred to as a swept volume system. If the display volume does not include a moving component, the display is known as a static volume system.

Swept volume systems. The most common approach to the creation of the display volume involves the use of a rotating or reciprocating target screen. In the former method (**Fig. 3**), a planar screen is rotated at speeds above 900 revolutions per minute so as to sweep out a cylindrical volume. The generation of voxels (volume elements, three-dimensional analogs of pixels) within this volume may be achieved either by equipping the screen with a matrix of individually addressable light sources or by addressing it with external laser or electron beams. In the latter case, the surface of the screen is coated with a phosphorescent material which emits light when impinged upon by the electron beam. Images are constructed by the careful synchronization of the beam deflection and intensity with the screen position. Other configurations under development involve the use of a helical-shaped screen, and a planar screen in conjunction with corotating beam sources.

Static volume systems. Significant work has also been undertaken in developing static volume displays. In one such technique, the display volume is composed of a material (a gas or a transparent solid medium doped with suitable ions) in which voxels may be created by a stepwise excitation of fluorescence, a process also known as resonant upconversion (**Fig. 4**). At the intersection of two infrared (nonvisible) laser beams, each having a frequency resonant with one of the excitation transitions, a voxel is created by the emission of visible light when the material decays to the ground state. This approach necessitates the identification of a material with precise spectroscopic properties. These properties determine both the nature of the required excitation sources and the color and intensity of the voxels generated. Although a number of gases (for example, mercury vapor and iodine monochloride) have been investigated, it has proven difficult to identify one with all the required attributes.

Recent work has concerned rare-earth ions doped into glass. This avenue appears to be encouraging, although a solid display volume may, in view of weight considerations, be limited in size.

Current volumetric techniques generate only translucent voxels, whereas the ideal system would also permit opaque voxel generation, enabling the depth cues of shading and occlusion to be satisfied. One possible means by which opaque voxels may be generated is via a photochromic or thermochromic mechanism at the intersection of two nonvisible laser beams. The process is analogous to the two-step excitation technique mentioned above. The difference is that, when excited, the voxels reflect ambient visible light rather than acting as light sources.

Another technique for implementing a static display volume is to comprise the volume from a three-dimensional array of physically discrete, individually addressable voxel elements. The difficulty with this approach is to ensure that both the voxel elements not being addressed and the physical means by which they are addressed are not visible and therefore do not interfere with the transparent display environment. One current embodiment uses optical fibers to address each element.

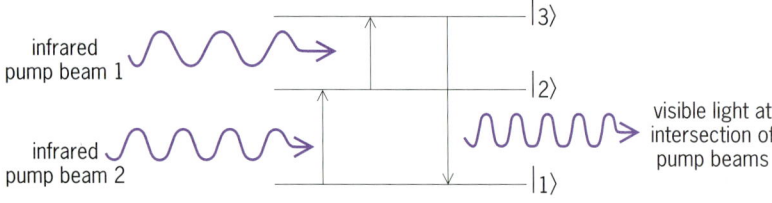

Fig. 4. Three-energy-level system, illustrating the concept of two-step excitation of fluorescence.

Nonuniformities. The ideal display volume would behave as a three-dimensional cartesian space, thus ensuring that images appear the same, independent of their location in the display space. This behavior is seldom an intrinsic property of a display volume, but may generally be provided via external manipulation of the data to be displayed. The characteristics in which nonuniformities are often found are voxel density, voxel size, and voxel positioning accuracy. An example of nonuniformity is provided by the case of voxel density in a system incorporating a rotating planar screen. Each voxel requires a certain length of time to be created, during which the screen moves through a small angle. Thus a limitation is provided on the angular density of voxels that can be supported by the display, and the maximum tangential voxel density therefore decreases with increasing distance from the rotation axis.

Status. Recent advances in computer performance have brought swept volume display systems close to realization as workstation graphics peripherals with a limited color capability. By eliminating the mechanical mechanism by which the display volume is created, static volume systems are likely to provide a more elegant implementation of the volumetric display concept. These systems still require considerable development and for systems reliant upon two-step excitation and photochromic techniques suitable materials (or material mixtures in the case of color versions) have to be identified. The results that are currently being achieved by experimental swept volume displays (Fig. 3) indicate that volumetric systems will be able to greatly assist in areas such as medical and educational visualization. Their ultimate form and performance limitations remain unknown.

For background information SEE ELECTRONIC DISPLAY; LIQUID CRYSTALS in the McGraw-Hill Encyclopedia of Science & Technology.

B. G. Blundell; A. J. Schwarz

Bibliography. B. Bahadur (ed.), *Liquid Crystals: Applications and Uses,* vol. 1, 1991; B. G. Blundell, A. J. Schwarz, and D. K. Horrell, Volumetric three-dimensional display systems: Their past, present and future, *IEE Sci. Eng. Educ.,* 2(5):196–200, October 1993; J. A. Castellano, *Handbook of Display Technology,* 1992; T. E. Clifton and F. L. Wefer, Direct volume display devices, *IEEE Comput. Graph. Appl.,* 13(4):57–65, 1993; R. B. Meyer, Ferroelectric liquid crystals: A review, *Molec. Crys. Liq. Crys.,* 40:33–38, 1976; A. J. Schwarz and B. G. Blundell, Considerations regarding voxel brightness in volumetric displays utilizing two-step excitation processes, *Opt. Eng.,* 32:2818–2823, 1993.

Elementary particle

Diffractive scattering events are of particular interest in the study of hadrons and of the strong nuclear force holding them together. In the currently accepted theory of the strong nuclear force, quantum chromodynamics (QCD), the scattering of two hadrons involves the interactions of quarks and gluons. The quarks and gluons carry color, the charge of the strong interaction. However, no particle carrying color has been observed in nature, probably as a result of a special property of the strong force: it increases in strength as the separation between two colored particles increases. As the two colored particles separate, new quarks and antiquarks are created and combine with the original quarks and gluons to form known color-neutral hadrons. The initial hadrons that took part in the scattering are thereby transformed into a new set of particles. However, in diffractive scattering one of the original hadrons remains intact, even at very high energies. Thus, it is implied that a colorless object, the pomeron, and not a simple quark or gluon, has taken part in the scattering process. The nature of the pomeron and its relationship to the quarks and gluons of quantum chromodynamics are of concern to experimentalists and theorists alike. A complete theory of the nuclear force, the strongest force known in nature, should provide a unified description of every aspect of the strong interactions. As a result of recent experimental data and theoretical activity, diffractive scattering is better understood.

Diffraction. Diffraction is a commonplace phenomenon, familiar from everyday experience. The effects of edges and pinholes on light are seen as patterns of bright and dark fringes at the edges of the so-called geometrical shadow. These effects represent deviations from the ray representation of light, and indicate its wave nature. When the trajectories of light waves are perturbed, constructive and destructive interference occurs. Interference patterns become visible even though in other respects the light behaves as a particle (photon).

In the scattering of highly energetic particles, diffraction is said to have occurred if one of the particles remains intact. For example, if two protons collide and only one breaks up, the event is labeled as diffractive. One proton (the one that breaks up) has diffracted around the other proton, which has effectively acted as an obstruction. If both remain intact and no other particles are created in the scattering, the event is called elastic, and is analogous to the billiard-ball scattering studied in introductory physics courses.

In electron-proton scattering, the electron does not break up, thus providing evidence that it has no substructure. The scattering is said to be diffractive if the proton remains intact while other particles are created, for example, in the process $e + p \rightarrow e + p + X$, where X represents any particles.

Hadron structure. In the series of experiments carried out by E. Rutherford and his collaborators in 1909, alpha particles (helium nuclei) were observed to scatter from thin metal foils. The unexpectedly large scattering angles measured indicated that the atoms in the foil had small, charged nuclei. This discovery helped set the stage for the current picture of

98 Elementary particle

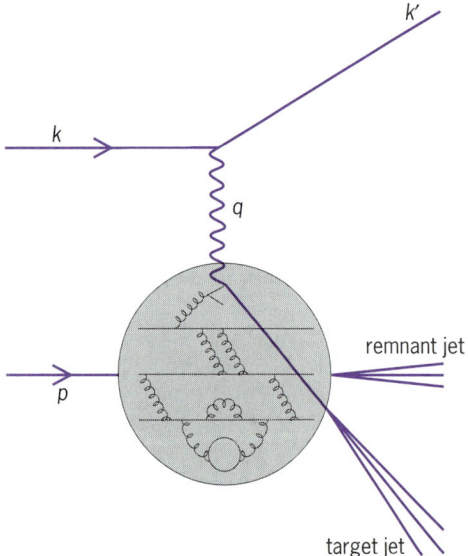

Fig. 1. Feynman diagram of an electron-proton interaction at high energy. The virtual photon (labeled q) is emitted from the electron (whose momentum changes from k to k') being absorbed by a quark in the proton (p).

atoms as composed of a central nucleus with dimensions much smaller than those of the atom itself. About 60 years later, a very important series of experiments essentially repeated the Rutherford program at higher energy. In these experiments, an electron beam probed the structure of the proton. Once again, substructure was found. The proton could be described as consisting of smaller constituents. Some of these smaller constituents, called partons, were identified with the quarks introduced in prior theories. The proton was understood as containing three quarks, with fractional electric charge, together with further quark-antiquark pairs and gluons. In quantum chromodynamics, the quarks also have strong, or color, charges, in addition to their electric charge. Strong nuclear interactions between quarks occur in quantum chromodynamics through the exchange of electrically neutral colored gluons, analogously to the exchange of photons in electromagnetic interactions. It has been determined experimentally that the combinations of partons making up the hadrons must be such that the hadrons are color neutral.

High-energy scattering. According to the very successful standard model of particle interactions, the forces of nature are mediated by the exchange of particles called gauge bosons. These gauge bosons interact with particles carrying the charge associated with a particular force. For example, the electromagnetic force is mediated by photons, and the photons interact with only particles possessing an electric charge. The strong force that holds the hadronic particles together is mediated by the exchange of gluons, and these gluons interact with only quarks and other gluons, which themselves carry color charge.

In the scattering of two hadrons at high energy, the distances probed are so small that the scattering is best regarded as occurring between the constituents of the hadrons, and not by the hadrons as a whole. A strong interaction takes place between a quark or gluon of one hadron and a quark or gluon of the other hadron. As the parton is knocked out of the hadron, it takes away color. No isolated quarks or gluons have been observed in nature; the parton must be "dressed" with more quarks and antiquarks (created from the vacuum) as it separates from the hadron remnant. At the end of this process, all particles must be color neutral. The result is that jets of particles are produced in the direction of the incoming hadrons, as well as in the directions of the scat-

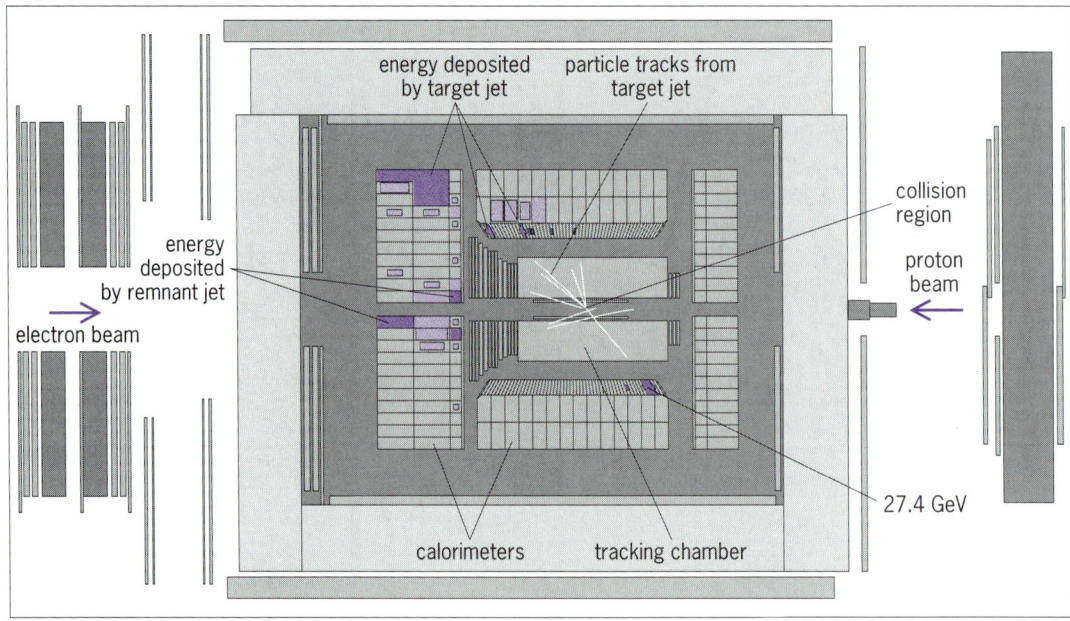

Fig. 2. Event recorded with the ZEUS detector at HERA. The electron beam circulates from left to right, and the proton beam from right to left. In this event, the 27.4-GeV electron has been scattered by about 70°, and is labeled by the marker.

tered partons. In the typical high-energy scattering process, the hadrons do not remain intact.

In electron-proton scattering, the scattering of the electron from the proton proceeds via the exchange of either a photon or a weak boson (W^+, W^-, Z^0). The rate for scattering is determined primarily by the photon exchange contribution (**Fig. 1**). At high energies, the wavelength of the photon is so short that it probes inside the proton and scatters from individual partons. The photon can interact with only a parton having electric charge, that is, a quark, which is then knocked out of the proton. Since the proton remnant and the struck quark have net color, they must become dressed. The result is jets of particles being produced in the direction of the proton (the remnant jet) and in the direction of the struck quark (the target jet).

In both of the processes described above, the hadrons break up and the scattering is not diffractive. For diffractive scattering to occur, a colored object cannot be removed from the hadron without it breaking up. The name pomeron exchange is used to describe diffractive scattering. In the pomeron exchange picture, the scattering does not take place from a quark or gluon in the hadron but rather from the color-neutral pomeron associated with the hadron. The hadron therefore does not necessarily break up, since it remains color neutral. The pomeron was originally introduced before the discovery of quarks and gluons to explain the behavior of hadron-scattering cross sections. Many of its properties were discussed in the framework of Regge theory, an older, semiphenomenological theory of strong interactions. It is now recognized that the pomeron may just be shorthand for describing multiparton exchange resulting in no net color being removed from the hadron.

Experimental data. In hadron-hadron scattering, such as that recorded experimentally by the UA8 collaboration operating at the CERN (Centre Européen pour la Recherche Nucléaire) laboratory in Geneva, diffractive scattering is recognized by the presence of one of the original hadrons among the final-state particles. This type of experiment has been the main source of information on diffractive scattering and the nature of the pomeron, including the evidence for the possible presence of gluons in the pomeron presented in 1988 by the UA8 collaboration. Further investigations into the nature of the pomeron have been carried out in Hamburg at the accelerator facility HERA (Hadron Elektron Ring Anlage). Two international collaborations, H1 and ZEUS, operate large, multipurpose detectors to record events at HERA, where 30-GeV electrons collide with 820-GeV protons.

In the ZEUS detector (**Fig. 2**), tracking chambers measure the trajectories of charged particles at the center of the detector, and calorimeters farther out measure total energy. In the event shown, the electron is scattered, through an angle of approximately 70°, into the detector. The proton produces a jet of particles as it breaks up (the remnant jet), thereby depositing energy in the calorimeter near the beam line. The remaining energy in the event is from the jet of particles from the quark struck by the photon (the target jet), which balances the transverse momentum of the electron. In this typical scattering event, the proton has broken up due to the knocking out of a colored quark.

However, a sizable fraction of the events show no signs of the proton breaking up. For example, in the event seen in **Fig. 3**, the jet formed from the struck pomeron is visible but no energy is seen in the

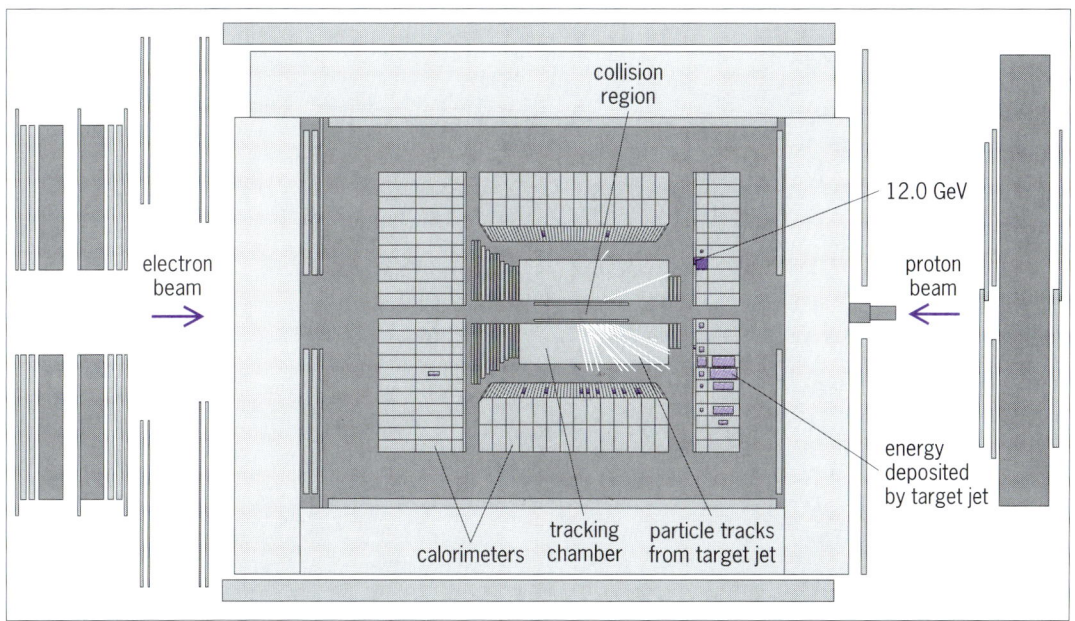

Fig. 3. Event recorded with the ZEUS detector, in which the 12-GeV electron has been scattered by about 30°, while the jet from the struck quark (target jet) produces many particles in the backward region relative to the proton beam direction. The lack of a proton remnant jet indicates a diffractive scattering.

direction of the proton, suggesting that diffractive scattering has taken place. At HERA, diffractive scattering appears to account for about 10% of the events where a large (4)momentum has been transferred from the electron. There is evidence for partons as components of the pomeron. Ongoing work includes comparisons of the experimental data with theoretical calculations, and the evaluation of specific classes of diffractive events.

Implications for the strong force. The issue being addressed in the study of diffractive scattering is the nature of the pomeron, and whether it is just a shorthand for a complicated interaction involving several partons (quarks or gluons) or is something different. Physicists believe that quantum chromodynamics is complete, and should in principle be able to explain all strong interactions. It would follow that diffractive scattering events, and the nature of the pomeron, should be explained from first principles within quantum chromodynamics. Because of the complexity of quantum chromodynamics, this explanation is not yet possible. New insights are needed for further progress. Recent data suggest that the pomeron is indeed composed of strongly interacting partons, and there has been much progress in describing diffractive scattering events theoretically within quantum chromodynamics. The continued study of these events remains a top priority for experimenters and theorists interested in the strongest force known in nature.

For background information SEE DIFFRACTION; ELEMENTARY PARTICLE; PARTICLE ACCELERATOR; PARTICLE DETECTOR; QUANTUM CHROMODYNAMICS; QUARKS; REGGE POLE; STANDARD MODEL in the McGraw-Hill Encyclopedia of Science & Technology.

Allen C. Caldwell

Bibliography. R. N. Cahn and G. Goldhaber, *The Experimental Foundations of Particle Physics,* 1989; F. Halzen and A. D. Martin, *Quarks and Leptons,* 1984.

Embryogenesis

Embryogenesis is the study of the development and transformation of embryonic plants into seedlings and mature plants and of embryonic animals into fetuses or larvae and adult animals. Among its many topics is the study of the formation of different shapes of embryos (morphogenesis or pattern formation). Two aspects of pattern formation in animal embryos are discussed here: the transformation of radially symmetrical eggs into bilaterally symmetrical embryos, and the recognition of the segmental blocks (somitomeres) found in the heads of developing embryos.

Elongate eggs. Although eggs of most animals are spherical, many insects, such as the common fruit fly *Drosophila,* have elongate eggs, with a long and a short axis. Typically, the embryo develops along the long axis, with the head at one end and the tail at the other. Bilateral symmetry, where an object can be bisected into equal parts, is therefore evident before such an egg begins to develop, even before fertilization. Although elongate eggs have an obvious and intrinsic bilateral symmetry, it is not obvious which end will become the head (anterior) and which the tail (posterior). Determination of the body axis lies in the functioning of sets of genes localized in the egg cytoplasm and inherited from the mother. Maternal gene activity at or soon after fertilization sets off a series of gene activations that progressively specify increasingly localized regions of the embryo. Such maternal gene activity specifies two axes of the embryo, the anteroposterior (head to tail) axis, and at a right angle the dorsoventral (back to belly) axis. Specification of these two axes automatically sets the third embryonic axis, the right-left axis.

The anteroposterior and dorsoventral axes in *Drosophila* are not each specified by only a single gene. At least 10 maternal effect genes have been identified and assigned names such as *bicoid, oskar, caudal, bicaudal,* and *nanos* for genes specifying the anteroposterior axis, and *snake, toll,* and *cactus* for genes specifying the dorsoventral axis.

The combined action of genes localized at one or the other end of the egg establishes the anterior and posterior ends of the body and the primary body axis. This process is often referred to as establishing the body plan. Two major lines of evidence indicate that maternal effect genes specify embryonic axes and establish bilateral symmetry. First, gene products, usually messenger ribonucleic acid (mRNA) but sometimes protein, can be visualized in the egg by using probes directed against sequences in individual genes. Second, mutations in these genes can duplicate or delete the anteroposterior or dorsoventral axes.

The *bicoid* gene has its highest concentration at one end and diffuses toward the other, establishing a concentration gradient. *Nanos,* which is involved in specifying the posterior end of the embryo, has its peak concentration at the opposite end of the egg. *Nanos* blocks transcription of *hunchback,* a gene which is activated by *bicoid,* an anterior gene. Such inactivation of anterior genes in the posterior region of the egg further enhances anteroposterior regionalization. Thus, in the future posterior region positive control from the presence of posterior genes and negative control from inhibition of anterior genes in the posterior region are contributing mechanisms for bilateral symmetry of the embryo.

Nanos is produced by nurse cells located at one end of the egg, thereby specifying the egg's posterior end. The factor determining the end of the egg for development of the nurse cells is not known. Indeed, the resolution of mechanisms of pattern formation can become a problem requiring evolutionary explanations.

Spherical eggs. Determination of axes is not quite so obvious in species with round or spherical radially symmetrical eggs, which usually produce bilaterally symmetrical embryos. Superficially, such eggs appear to have no preferred axis of symmetry (a requirement for bilateral symmetry) but subtle signs of bilateral symmetry are present even in the unfertilized spherical egg. The presence of more yolk at one end than at the other establishes two poles, but only one region on the surface may permit sperm to enter the egg—another sign of bilateral symmetry.

More subtle than these structural differences are the functional differences in internal organization of the egg. In frog eggs, the area of cytoplasm from which the future skin and nervous system will develop is already specified as different from the area from which the future alimentary canal will develop. These two areas, or germ layers (future ectoderm and endoderm), are organized as two halves of the egg on either side of an equator, thus establishing bilateral symmetry.

In frog eggs, the rotation of the superficial (cortical) cytoplasm over an arc of about 30° occurs at fertilization, in a precise relationship to the point of sperm entry. This rotation specifies the dorsal side of the future embryo; therefore the dorsoventral axis is prepatterned. The area of yolk-rich cytoplasm in frog eggs (more or less equivalent to future endoderm) divides more slowly and so produces larger cells at one pole than the less yolky cytoplasm does at the other. Future ectoderm and endoderm interact at the equator to specify the third germ layer, the mesoderm, with its dorsal and ventral sides. SEE FERTILIZATION.

In the embryos of some species, such as the tunicate, *Styela,* future germ layers are distinguished by color. Although mixed before fertilization, the cytoplasm rearranges itself immediately after fertilization. The yellow endoderm relocates to one pole, and the clear cytoplasm (future mesoderm) accumulates along the equator at one hemisphere as cytoplasmic constituents for individual regions establish symmetry by aggregating and relocating.

Thus, spherical eggs are not necessarily radially symmetrical but have intrinsic axes of symmetry. Indeed, there may be no such thing as a truly radially symmetrical egg.

Somitomeres. Many vertebrate organs are patterned, and they are often segmented. This patterning of internal organs is evident in vertebrate embryos. Nerves emerge segmentally from the developing spinal cord as pairs to the right and to the left. Between them lie pairs or blocks of mesoderm, again one on the right and one on the left. These are the somites, forerunners of the vertebrae, muscles, and connective tissue associated with the axial skeleton. Right and left symmetry is therefore evident in the trunk, but in the head, where somites do not exist, segmentation is much less evident.

The difficulty of seeing segmental organization in the head may be explained by the fact that much of the head does not arise from mesoderm but from the neural crest. Neural crest cells migrate away from the developing brain, divide rapidly, and fill available spaces. Although segmented when they leave the neural crest, these cells quickly mingle, obscuring segmented organization. Mesoderm makes a much smaller contribution to the head than to the trunk, forming muscle but little skeleton or connective tissue. A small mesodermal component is swamped by a much larger neural crest component.

In addition, the segmentation in the head is not characterized by somites but by somitomeres—patterned aggregations of mesodermal cells and the precursors of somites in the trunk—that merge together in the head. Although the somites in the trunk start out as somitomeres, they rapidly transform into easily recognizable somites. In the head, somitomeres are transitory, do not transform into somites, and so are difficult to see. Trunk somites contribute to individual vertebrae and muscles, a feature that can be seen without experimental manipulation. The contribution of head somitomeres to segmental organization is much more subtle because they require transplantation or replacement of individual somitomeres with labeled cells.

Somitomeres were first identified in chick embryos by using stereo pairs of scanning electron microscope images. They have now been identified in representatives of all six classes of vertebrates. Strangely, the number of pairs of somitomeres in the head differs between the vertebrate classes. Newts, frogs, and sharks have four pairs, whereas mice, chicks, turtles, and fish have seven. It is clear that somitomeres occur along the body axis, and in the head and trunk somitomeres and somites contribute to embryonic patterning by producing segmented structures.

For background information SEE CELL POLARITY; CLEAVAGE (EMBRYOLOGY); EMBRYOGENESIS; EMBRYOLOGY in the McGraw-Hill Encyclopedia of Science & Technology.

Brian K. Hall

Bibliography. S. Gilbert, *Developmental Biology,* 4th ed., 1994; N. Maclean and B. K. Hall, *Cell Commitment and Differentiation,* 1987; J. M. W. Slack, *From Egg to Embryo: Determinative Events in Early Development,* 1985; R. Wall, *This Side Up: Spatial Determination in the Early Development of Animals,* 1990.

Endometrial cups

Endometrial cups are areas of specialized glandular tissue, unique to pregnant mares, that are located on the interior (endometrial) surface of the gravid uterine horn. When the endometrium contacts the fetal membranes, specifically the area associated with the chorionic girdle, fetal trophoblastic cells in this region are stimulated to form temporary attachments to the endometrium. Trophoblast cells migrate from the fetal membranes to the adjacent endometrial epithelial layer and intrude into the endometrial stroma where they become permanently attached, growing into mature cup cells. Cell migration and cup formation begin at approximately day 35 of gestation.

Structure. The cups are visible from day 40 of gestation as pale, discrete, slightly swollen, raised endometrial growths. The surface of each structure is depressed in the center and may be round, oblong, or linear in shape, measuring from a few millimeters to several centimeters in length or diameter. The cups show a roughly circular pattern, whose orientation depends on the location of the fetus and the plane of the fetal chorionic girdle. The number of cup structures and volume of tissue is variable. Ultrastructurally, cup cells are large, binucleated, and epithelial in nature. The presence of large nucleoli and excessive amounts of endoplasmic reticulum suggests that the cup cells synthesize and secrete proteins.

Function. Endometrial cups produce a unique glycoprotein hormone, equine chorionic gonadotropin (eCG), formerly referred to as pregnant mare serum gonadotropin (PMSG). Detectable in equine plasma and urine at day 35–40 of pregnancy, eCG reaches maximal plasma concentrations at 55–70 days and slowly declines to nondetectable concentrations by 120–150 days. A reduction in the concentration of eCG is associated with cup degeneration, which is initiated at 70 days and completed by 130–150 days. Cup degeneration and eventual sloughing into the uterine lumen relates to maternal cell-mediated response to foreign antigens expressed by the fetal cup cells. Maternal leukocytes (lymphocytes, plasma cells, and eosinophils) accumulate in the stroma around the periphery of the cup, and as the cup cells begin to degenerate (about day 70), actively invade and destroy the fetal-derived tissue.

Peak plasma eCG concentrations vary tremendously between individual mares, and appear to be proportional to the amount of mature cup tissue. Fluctuation in the amount of cup tissue present is related to maternal size, parity, and breed as well as number and genotype of the fetus. Detection of this hormone in the serum or urine of horses is possible by utilizing enzyme-linked immunosorbent assay, and quantification is possible with radioimmunoassay. The half-life of eCG in the horse is reported to be 6–7 days.

Role of eCG. It is possible that eCG stimulates the production of progesterone and estrogen in the ovaries until the developing fetoplacental unit is able to provide amounts adequate for pregnancy. Chorionic gonadotropin may initiate production of the ovarian steroids, progesterone and estrogen, by stimulating primary corpus luteum resurgence and formation of secondary corpora lutea from ovulatory and anovulatory follicles. However, the role of eCG in this capacity may be secondary, as detectable levels of eCG are not absolutely necessary for successful pregnancy. It is likely that the primary function of this hormone has not been elucidated, but current theories suggest that its function may relate to an immunoprotection role for the developing fetus.

Pregnancy loss. The termination of a pregnancy after day 37–40 results in retained cup cell function, continued secretion of eCG, and maintenance of detectable eCG levels up to day 150. Elevated eCG levels interfere with normal endocrine events responsible for follicular growth, ovulation, and maintenance of corpus luteum function. Mares fail to return to normal fertile estrous cycle activity and frequently exhibit estrus, where ovulation is suppressed, eCG-dominated pseudopregnancy or prolonged luteal function, ovulation without estrus, or anestrus with small nonfunctional ovaries. A lack of regular, fertile estrous cycles after an embryonic loss usually delays the next conception for an extended period. Depending on the stage of pregnancy at the time of fetal loss, infertility may exist for 90–100 days. This delay and the fact that horses are seasonal breeders results in lost breeding potential.

Research. Concurrent with ongoing research to further characterize the primary role of cup tissue and eCG, methods are being evaluated that allow earlier rebreeding of temporarily infertile mares that have experienced fetal loss after formation of endometrial cup tissue. It is generally accepted that a return to fertility is inversely related to eCG concentration, and efforts are directed to reducing the hormone level below threshold to allow conception to occur. The enhancement of the normal immunological destruction of endometrial cup tissue by stimulating endogenous antibody production to the cup tissue is being evaluated. However, such a method of control must not inhibit desired cup formation in subsequent pregnancies.

Monoclonal antibody to eCG has reduced measured concentrations of this hormone in the serum of pregnant ponies, but the amount of antibody required and the frequency of administration render this approach impractical as a means of inhibiting cup tissue production.

Surgical removal of all visible endometrial cup tissue results in rapid reduction of eCG levels and subsequent conception within the same breeding season.

For background information SEE ESTROGEN; HORSE; OVARY; PROGESTERONE in the McGraw-Hill Encyclopedia of Science & Technology.

Michael J. Huber

Bibliography. W. R. Allen, D. W. Hamilton, and R. W. Moor, Origin of equine endometrial cups, II: Invasion of the endometrium by the trophoblast, *Vet. Rec.*, 177:485–502, 1973; W. R. Allen and R. M. Moor, The origin of the equine endometrial cups, I: Production of PMSG by the fetal trophoblastic cells, *J. Repro. Fert.*, 29:313–316, 1972; O. J. Ginther, *Reproductive Biology of the Mare*, 2d ed., 1992.

Environmental pollution

The adverse human health effects of exposure to mineral dusts, especially asbestos and crystalline silica dusts, have received much public scrutiny in recent years. The high incidence of cancer and asbestosis among professional asbestos workers and silicosis among sandblasters was particularly important in promoting this concern. The fact that asbestos causes cancer among heavily exposed workers, coupled with the hypothesis that there is no known threshold for the induction of a tumor by a carcinogenic substance, was the justification for a multibillion dollar program of asbestos removal in commercial buildings and schools. A further concern for schools was the idea that being exposed to asbestos fiber, even in minute quantities, as a child could cause later development of cancer.

In 1990 the U.S. Environmental Protection Agency (EPA) published an advisory document concerning managing asbestos in place. This advisory included the following information: (1) Asbestos hazard is dependent on dose; therefore a low dose means low risk. (2) On the basis of available data, fiber levels in buildings are low; accordingly, health risk to occupants also appears to be low. (3) Removal is often not the best course of action; an improper means of removal can create a dangerous situation where none existed previously. Despite this advice by the EPA, the removal of asbestos from school and other buildings continues.

Asbestos minerals and disease. Of the six types of asbestos (**Table 1**), only three have been used to any significant extent in commerce: chrysotile (white), crocidolite (blue), and amosite (brown). Chrysotile, a member of the serpentine mineral group, is distinctively different in chemical composition and atomic structure from crocidolite and amosite, which belong to the amphibole group of minerals. Between 1870 and 1992 approximately 165 million tons (150 million metric tons) of asbestos were mined worldwide, 90% being chrysotile. Combined production of crocidolite and amosite amounted to approximately 5% of the total asbestos production. Photomicrographs of chrysotile and crocidolite are shown in the **illustration**.

The three principal diseases caused by inhalation of asbestos dusts are lung cancer; mesothelioma, a cancer of the pleural and peritoneal membranes that invest the chest and abdominal cavities; and asbestosis, a condition in which the lung tissues become fibrous, thus losing the ability to function.

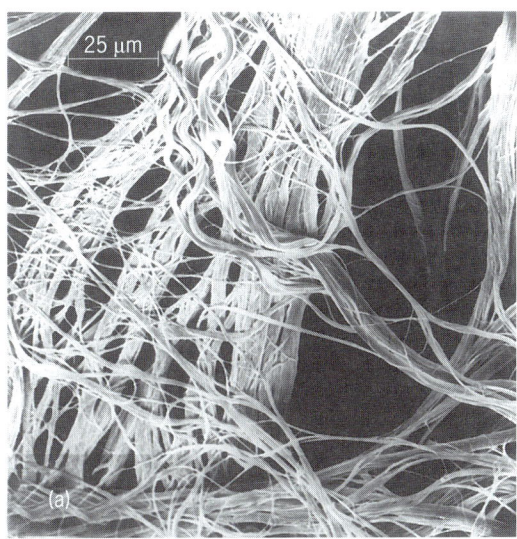

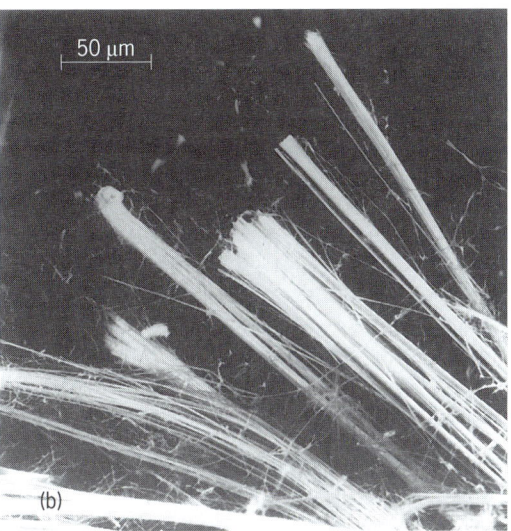

Light optical photomicrographs of (a) chrysotile asbestos and (b) crocidolite asbestos. Note the curly and ropelike nature of the chrysotile fiber bundles versus the straight and splintered nature of the crocidolite fiber bundles. (*U.S. Geological Survey*)

Table 1. Asbestos and crystalline silica minerals

Name	Chemical formula
Asbestos minerals	
Chrysotile	$Mg_3Si_2O_5(OH)_4$
Amosite	$(Fe,Mg)_7Si_8O_{22}(OH)_2$
Crocidolite	$Na_2(Fe,Mg)_5Si_8O_{22}(OH)_2$
Anthophyllite	$Mg_7Si_8O_{22}(OH)_2$
Tremolite	$Ca_2Mg_5Si_8O_{22}(OH)_2$
Actinolite	$Ca_2(Mg,Fe)_5Si_8O_{22}(OH)_2$
Crystalline silica minerals*	
Quartz	SiO_2
Cristobalite	SiO_2
Tridymite	SiO_2

*The four additional minerals in this group are rare and need not be considered as possible health hazards.

These three diseases are not equally prevalent in the various groups of asbestos workers that have been studied: the extent and type of disease depends on the duration and intensity of exposure and particularly on the types of asbestos to which the individual has been exposed. Lung cancer is caused by exposure to chrysotile, crocidolite, and amosite; however, increased risk of disease is found in those who smoke. Asbestosis is caused by all three forms of asbestos, whereas mesothelioma is caused principally by exposure to crocidolite asbestos. There is good evidence that chrysotile asbestos does not cause a significant risk for mesothelioma, even after long periods of intense exposure. The ingestion of asbestos has not been proven to cause disease in humans or animals; thus the EPA has determined that asbestos is carcinogenic only when inhaled, not when ingested. Tremolite, actinolite, and anthophyllite asbestos are rare (Table 1); however, excessive exposure to these types of asbestos is also known to cause disease. The common nonasbestiform varieties of tremolite, actinolite, and anthophyllite, which are sometimes defined (incorrectly) as asbestos for regulatory purposes, have not been shown to be harmful.

School building exposures. A recent review document by the Health Effects Institute-Asbestos Research (HEI.AR) summarized the asbestos air concentrations measured in United States and Canadian schools. In 48 school buildings containing asbestos, the average fiber level was 0.00051 fiber per milliliter of air. Examination of the 398 air samples showed that the highest value (0.02 f/ml) was obtained in a janitor's closet that contained a spray-on asbestos-containing product. Additionally, 171 school buildings containing asbestos were evaluated for litigation purposes. Within these buildings, 1008 air samples produced a mean average fiber level of 0.00011 f/ml. In the absence of data to the contrary, it is reasonable to assume that these 219 buildings are representative of the New York and other 31,000 nationwide schools thought to contain asbestos. The conditions of the asbestos materials in New York City schools, as described in the media, do not significantly differ from conditions in schools described elsewhere. Many of the 219 buildings in the study contained surface materials in poor and damaged condition, and were studied specifically for suspected airborne fiber. Therefore, the 1406 data points may actually represent a worst case scenario. The air data indicate that children who continuously attend school between the ages of 5 and 18 years are subjected to very low risk. At the average concentration value of 0.00022 f/ml, using the most pessimistic methods for calculating risk (the effect is proportional to dose with no threshold, all fiber types give the same risk for disease, and everyone smokes cigarettes), the calculated risk is 3 excess cancer deaths per 10^6 lifetimes. Estimates of risk from asbestos exposure in schools in comparison to other risks found in United States society are shown in **Table 2**. The air-sampling data collected in schools do not support the concept that low-level exposure to asbestos in buildings is a significant risk. Experts in public policy have suggested that unwarranted and poorly controlled abatement results in unnecessary risks to removal workers.

Crystalline silica and disease. There are seven naturally occurring crystalline silica minerals, but only three (quartz, cristobalite, tridymite) need be considered as possible health hazards (Table 1). Quartz is by far the most common of the three, composing a large percentage of many types of rocks and almost 100% of beach sands. Inhalation of fine quartz dusts in significant amounts over extended periods causes silicosis, a condition in which the lung develops fibrotic (scar) tissue and loses its air-exchange capacity. Some epidemiological data suggest that those who develop silicosis may be at additional risk for lung cancer. In 1987 the International Agency for Research on Cancer (IARC) des-

Table 2. Estimates of risk of death per 10^6 lifetimes* from asbestos exposure in schools in comparison to other risks in United States society

Nature of risk	Number of deaths
Smoking (all causes)	210,000
Motor vehicle accidents	17,000
Home accidents	8,400
Falls	5,600
Drowning	2,100
Frequent airline travel	1,000
Drinking one pint of milk per day (aflatoxin)	700
Consuming one diet drink per day (saccharin)	700
Living in a brick building (radiation)	350
Electrocution	350
Tornadoes	49
Lightning	35
Hurricanes and tropical cyclones	28
Exposure to asbestos in schools for 5 years at 0.00022 f/ml	3

* A lifetime = 70 years.

ignated quartz as a carcinogen; this decision was based on experiments in which cancer tumors were produced in rats exposed to airborne quartz dusts. The rules of the U.S. Occupational Safety and Health Administration (OSHA) Communication Standard were automatically invoked by the IARC advisory. This OSHA standard requires that any United States product that contains more than 0.1% free silica (quartz, tridymite, and cristobalite) must display hazardous warning signs. For example, in California, bags of play sand and other quartz-bearing products such as pottery clay are labeled as containing a possible human carcinogen. When one such bag of clay recently fell off a truck on a Sacramento freeway, traffic was tied up for hours while the material was carefully removed by hazardous material experts wearing special protection gear.

Implications. Common sense must prevail with regard to regulation of carcinogens; otherwise quartz-bearing beaches, dirt roads, farmlands, and thousands of commercial products will be designated as possibly carcinogenic to humans. The "no threshold" hypothesis of cancer induction may be invalid, and thus other criteria should be developed to control substances that may cause disease. A review of many epidemiological studies of those exposed to mineral dusts indicates that if there is no evidence of nonmalignant chronic disease in humans, such as asbestosis or silicosis, then there is no justification for predicting an increased cancer risk. In order to prevent cancer, the object of regulatory control should be to prevent chronic cell damage, where thresholds can be clearly defined.

For background information SEE AMPHIBOLE; ASBESTOS; MUTAGENS AND CARCINOGENS; PUBLIC HEALTH; SERPENTINE; SILICA MINERALS in the McGraw-Hill Encyclopedia of Science & Technology.

Malcolm Ross

Bibliography. Health Effects Institute-Asbestos Research, *Asbestos in Public and Commercial Buildings: A Literature Review and Synthesis of Current Knowledge,* 1991; B. T. Mossman et al., Asbestos: Scientific developments and implications for public policy, *Science,* 247:294–301, 1990; P. Slovic, Informing and educating the public about risk, *Risk Anal.,* 6:403–415, 1986; H. C. W. Skinner, M. Ross, and C. Frondel, *Asbestos and Other Fibrous Materials,* 1988.

Enzyme

Methanotrophic bacteria, which utilize methane (CH_4) as their sole source of carbon and energy, have evolved a complex multienzyme system to activate molecular dioxygen (O_2) reductively in the first step of their metabolic pathway. This enzyme system catalyzes the conversion of methane to methanol (CH_3OH), as in the reaction below, where

$$CH_4 + NADH + H^+ + O_2 \longrightarrow CH_3OH + NAD^+ + H_2O$$

NADH represents the reduced form of nicotinamide adenine dinucleotide. In this reaction, one of the two atoms of molecular oxygen is incorporated into methane, and the other is reduced to water (H_2O). The enzyme system responsible is therefore referred to as a monooxygenase.

Methane monooxygenase system. Three distinct proteins make up the complete monooxygenase system: a multisubunit $\alpha_2\beta_2\gamma_2$ (composed of two copies each of three different polypeptide chains designated α, β, and γ); nonheme iron hydroxylase (molecular weight of 251,000), which is the site of methane oxidation; a two-iron two-sulfur (Fe_2S_2) reductase (molecular weight of 39,000), which accepts electrons from NADH and transfers them through its own flavin adenine dinucleotide (FAD) moiety to the hydroxylase; and a regulatory protein (molecular weight of 16,000), which has no prosthetic groups but is thought to regulate the electron transfer between the other two components.

The methane monooxygenase (MMO) system is of general interest for several reasons. First, current industrial processes for the conversion of methane to methanol are carried out at high temperature and pressure with a low efficiency. An understanding of the biological oxidation of methane to methanol might make it possible to improve these processes and to use the world supply of natural gas more effectively. Second, methane monooxygenase catalyzes the oxidation of a variety of hydrocarbons in addition to methane. This property of the enzyme has led to the use of methanotrophic bacteria in cleaning up land contaminated by oil spills such as the 1989 Exxon *Valdez* spill and in the oxidative removal of hydrocarbon contaminants such as trichloroethylene from drinking water. Finally, methane is a greenhouse gas, and methanotrophic bacteria help to control the flux of methane in the atmosphere.

Research efforts aimed at understanding the mechanism of hydrocarbon oxidation by methane monooxygenase have focused on a catalytic dinuclear iron core found in the hydroxylase protein. Extensive biochemical and spectroscopic analyses have been carried out in several laboratories on the hydroxylase components from two different types of methanotrophs. Recently, the three-dimensional structure of the hydroxylase has been determined from the methanotroph *Methylococcus capsulatus* by means of x-ray crystallography.

Related iron-containing proteins. The MMO hydroxylase belongs to a large and diverse family of proteins that have iron atoms in their active sites. The best-studied examples are the hemoglobins, in which oxygen is transported by binding to a heme group, whose structure consists of a porphyrin ring and one iron atom. Oxygen activation in hemoglobin is accompanied by significant conformational changes in the orientation of the iron atom with respect to the porphyrin ring as well as in the position of an adjacent helix that donates a histidine ligand to the iron atom. Unlike hemoglobin, the hydroxylase does not contain a heme group; instead it contains a dinuclear iron center, in which the two iron atoms are linked by several exogenous and endogenous ligands. The hydroxylase, therefore, belongs to a growing class of functionally diverse diiron proteins, which includes the crystallographically characterized proteins hemerythrin and ribonucleotide reductase. The diiron center in hemerythrin functions as an oxygen carrier, reversibly binding dioxygen. In ribonucleotide reductase, the diiron center utilizes dioxygen to generate a tyrosyl radical as an intermediate step in the reduction of ribose moieties for synthesis of deoxyribonucleic acid (DNA). Both the heme and the nonheme proteins use helical conformations to stabilize the iron centers, either by supporting the heme pocket or through direct coordination of the iron atoms by neighboring amino acid side chains.

In terms of its function and wide substrate specificity, the MMO hydroxylase is most similar to the heme enzyme cytochrome P450. Cytochrome P450 can function with a variety of substrates, depending on the particular cell type. For example, in the adrenal cortex there are primary substrates for conversion of steroids to adrenal cortical hormones, whereas in the liver the detoxification of lipid soluble foreign substances requires that this enzyme recognize many chemical targets. The MMO hydroxylase also incorporates oxygen into a range of hydrocarbon substrates, including aromatic and alicyclic hydrocarbons, phenols, alcohols, amines, chlorinated hydrocarbons, and (uniquely) methane. In the liver, cytochrome P450 must be able to respond to whatever new toxin is present. Similarly, in methanotrophs, the MMO hydroxylase might have evolved to enable the bacteria to survive in environments with a range of natural and synthetic pollutants.

Chemical analysis of the diiron center. In the absence of detailed structural information on the MMO hydroxylase, a number of different models were proposed for the coordination environment of the iron atoms. Most of these models were obtained by the comparison of spectroscopic properties of the diiron center in the hydroxylase with spectroscopic properties of a number of small, chemically synthesized model compounds, for which the three-dimensional structure was known by x-ray crystallography. From these studies, which included extended x-ray absorption fine structure (EXAFS), Mössbauer, visible absorption, electron paramagnetic resonance (EPR), and electron nuclear double magnetic resonance (ENDOR) spectroscopic measurements, it was possible to draw preliminary conclusions regarding the geometry of the active site. The two iron atoms were predicted to be 0.34 nanometer apart, linked by one bridging hydroxide ion and one or two bridging carboxylate ligands derived from the protein. The bridging hydroxide distinguished the hydroxylase from hemerythrin and ribonucleotide reductase, both of which contain bridging oxo groups. In addition, these studies predicted the presence of a terminal water or hydroxide ligand and a greater number of oxygen than nitrogen nonbridging protein ligands. Thus, the hydroxylase core was rendered more similar to the ribonucleotide reductase diiron core, which has mostly oxygen ligands, than to the hemerythrin core, which has all nitrogen ligands.

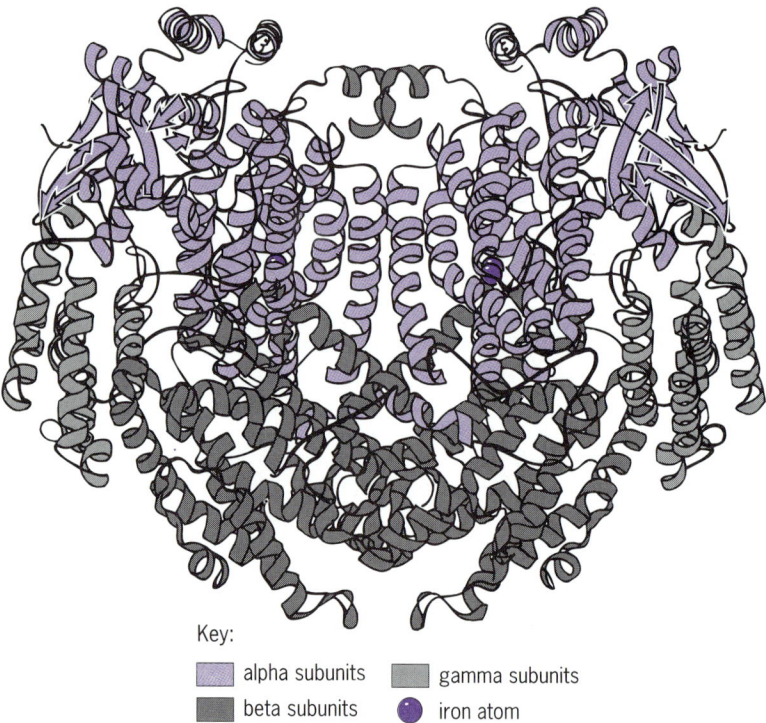

Fig. 1. Overall structure of the methane monooxygenase hydroxylase. A noncrystallographic twofold axis runs vertically, relating the two $\alpha\beta\gamma$ protomers. Helical secondary structure is represented by coils, beta sheet structure is represented by arrows, and random loops are represented by thin lines.

Key:
- alpha subunits
- beta subunits
- gamma subunits
- iron atom

Structure of the hydroxylase. The hydroxylase consists of three individual polypeptide chains of molecular weights 60,600, 45,000, and 19,800, respectively, that associate in an $\alpha_2\beta_2\gamma_2$ complex that has an apparent molecular weight of 251,000. The 0.22-nm-resolution three-dimensional molecular structure of the hydroyxlase protein has been determined by x-ray crystallography. The hydroxylase protein was purified directly from the methanotrophic bacteria *M. capsulatus* and crystallized by the method of vapor diffusion. The crystals diffract to very high resolution, and they contain one $\alpha_2\beta_2\gamma_2$ complex in the asymmetric unit. Therefore, the solved structure corresponds to a dimer that is the active form of the enzyme. The high resolution of the structure determination provides a detailed picture of the active site. In addition, knowledge of the overall shape of the molecule permits speculation concerning interaction of the other two proteins with the hydroxylase to form the complete active complex.

The structure of the hydroxylase is shown in **Fig. 1**. The hydroxylase secondary structure is mostly helical with only one small region of beta structure in the carboxy terminal segment of the alpha subunit. The large domain of this subunit is made up of three layers of helices, including six very long helices and six shorter ones. The diiron center is entirely contained within this domain and is surrounded by four long alpha helices, two each from adjacent antiparallel helical motifs. Each iron atom has one histidine ligand (His) from an adjacent helix as well as at least one monodentate carboxylate ligand, consistent with the spectroscopic observations (**Fig. 2**). The other ligands are a semibridging carboxylate (a glutamic acid residue; Glu 144) and a terminal water molecule or an additional monodentate carboxylate, respectively, for the two individual iron atoms. The two nonprotein bridges between these iron atoms, which are 0.34 nm apart, have been modeled as monodentate hydroxide and bidentate acetate ions.

The second domain of the alpha subunit consists of eight additional helices and two beta hairpins that lie close to the enzyme surface (Fig. 1). The beta subunit of the hydroxylase is composed solely of alpha helices, 10 of which have an almost identical three-dimensional arrangement to the corresponding helices of the alpha subunit. This similarity is remarkable considering the absence of any detectable primary sequence homology between these two polypeptide chains. Finally, the smaller gamma subunit consists of two clusters of four helices each. All the intersubunit contacts within the hydroxylase dimer are made by the alpha and beta subunits (Fig. 1). These two chains are arranged around a local twofold symmetry axis, relating the two alpha-beta-gamma protomers (subunits) such that the overall shape resembles a flattened toroid. The gamma subunits lie at the edge of this complex approximately along the central line of the alpha-beta interface within each protomer.

Implications of the structure. The crystallographic analysis permits speculation concerning the chemical mechanism of the hydroxylation reaction. The presence of a cysteine residue (Cys 151, **Fig. 3**) in an analogous location to the reactive tyrosyl radical in ribonucleotide reductase suggests that this sulfhydryl may play an oxidation-reduction role during catalysis. One additional protonated residue, a threonine near the active site (threonine residue; Thr 213; Fig. 3), may also provide a source of protons during catalysis. The sites of water and acetate ligands might represent the binding sites for dioxygen to the reduced protein as well as oxo intermediates and oxidation products of various substrates. In addition, a number of hydrophobic protein side chains adjacent to the diiron center provide a pocket for methane and other substrates near the active diiron center. Finally, the second coordination sphere of the dinuclear iron center contains two aspartic acid residues (Asp) hydrogen-bonded to the two histidine iron ligands (Asp 143 and Asp 242, Fig. 3). These negatively charged aspartic acid residues may increase the affinity of the bridging oxygen for protons, explaining why the hydroxylase contains a bridging hydroxo group rather than a bridging oxo group.

The relative positions of the different protein subunits within the hydroxylase dimer illuminate several other aspects of the activity of the enzyme and its modulation by the presence of the other protein components. First, there are several possible routes for a substrate to access the active site and for a product to leave the active site. Conformational changes in the helices that contain the coordinating ligands might open a route from the diiron center directly to the surface. There are also several hydrophobic pockets between the active site and the smaller domain of the alpha subunit. Protein conformational changes that could be induced by the binding of the other proteins to the hydroxylase might facilitate the movement of substrate molecules along these channels. It has been shown by spectroscopy that the regulatory protein directly interacts with the diiron center, influencing the rate and regioselectivity of substrate oxidation. If this component were to bind to the extended region formed between the alpha-beta pairs of the dimer, interactions through the helices at the interface could affect the coordination at the iron center. The reductase does not alter the spectroscopic properties of the iron center, but it does interact with both the hydroxylase and the coupling protein. A possible binding site for the reductase would also be along the extended region between alpha-beta-gamma protomers, but not in direct contact with the helices that provide the iron ligands.

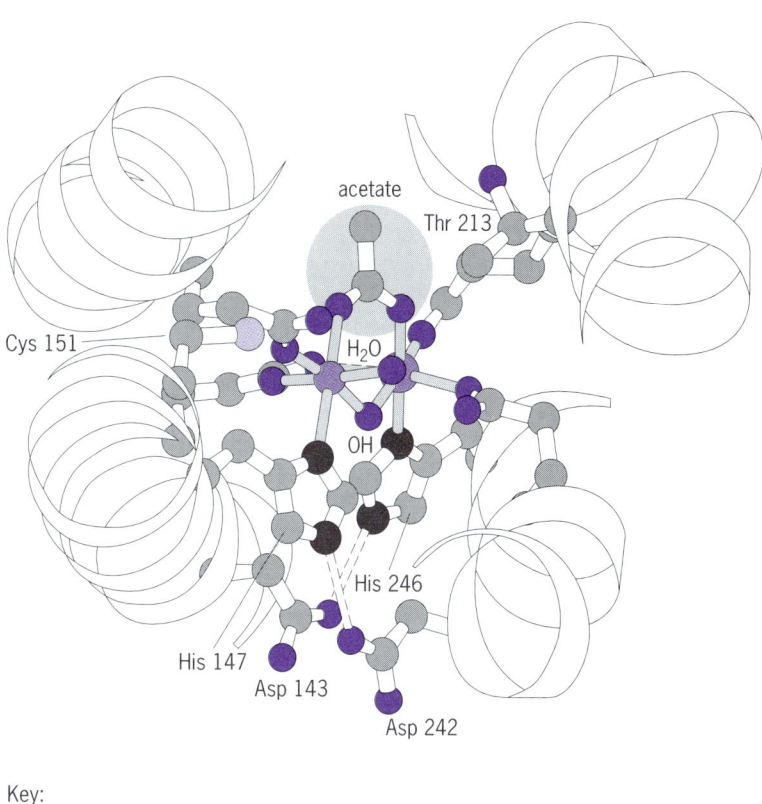

Fig. 3. Active site of the methane monooxygenase hydroxylase. A histidine residue (His 147) is hydrogen-bonded to aspartic acid residue (Asp 242), and His 246 is hydrogen-bonded to Asp 143. Broken-line bonds are hydrogen bonds. OH indicates an oxygen atom bonded to one hydrogen atom, and H₂O indicates an oxygen atom bonded to two hydrogen atoms. Thr = threonine residue. Cys = cysteine residue.

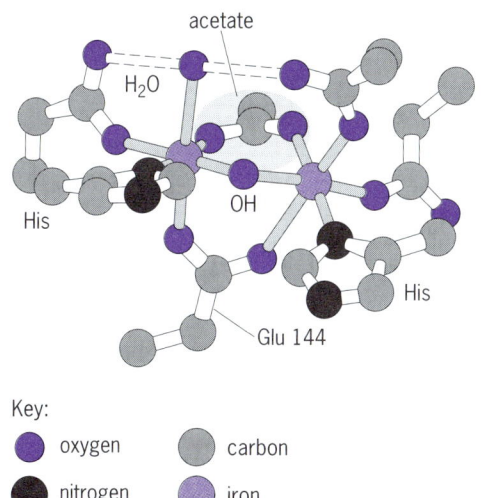

Fig. 2. Coordination of the diiron center in the methane monooxygenase hydroxylase. Broken-line bonds are hydrogen bonds. OH indicates an oxygen atom bonded to a hydrogen atom, and H₂O indicates an oxygen atom bonded to two hydrogen atoms. Glu = glutamic acid residue. His = histidine residue.

For background information *SEE* AMINO ACIDS; COORDINATION CHEMISTRY; CYTOCHROME; ENZYME; HEMOGLOBIN; PORPHYRIN; SPECTROSCOPY in the McGraw-Hill Encyclopedia of Science & Technology.

Amy C. Rosenzweig; Christin A. Frederick

Bibliography. A. L. Feig and S. J. Lippard, Reactions of non-heme Fe(II) with dioxygen, *Chem. Rev.*, 94:759–805, 1994; B. G. Fox et al., Complex formation between the protein components of methane monooxygenase from *Methylosinus trichosporium* OB3b, *J. Biol. Chem.*, 266:540–550, 1991; A. C. Rosenzweig et al., Crystal structure of a bacterial non-haem iron hydroxylase that catalyses the biological oxidation of methane, *Nature*, 366:537–543, 1993; A. C. Rosenzweig and S. J. Lippard, Determining the structure of a hydroxylase enzyme that catalyzes the conversion of methane to methanol in methanotrophic bacteria, *Acc. Chem. Res.*, 27:229–236, 1994.

Evolution

The conventional view in evolutionary biology is that complexity increases in evolution; but little solid evidence exists to support such a trend because the complexity of organisms has been difficult to define and measure. However, in recent years the problem has received increasing attention and novel methods have been devised for measuring complexity and analyzing trends. In this article, only complexity of form, or morphological complexity, is considered. The relationship between morphological complexity and genetic, developmental, and ecological complexity remains unclear, but as robust evolutionary patterns emerge from morphological studies the prospects for studying that relationship, and ultimately for synthesis, can be expected to improve dramatically.

Complexity and evolutionary development. Some compelling anecdotal evidence has sustained the conventional view, in particular a small number of familiar and salient cases of apparent increase in complexity. Examples include the transitions from single-celled organisms to multicelled invertebrates and then to vertebrates, culminating in human beings. The use of such cases as evidence for a general trend has at least three problems. First, although the transitions to multicellularity seem to be increases in complexity, it is not obvious that complexity has increased within multicellular groups. Modern clams, for example, seem no more complex than ancient clams.

Second, these cases may not be representative. Skeptics can cite an equal number of cases of complexity reduction, such as the decrease in number of skull bones in vertebrates, the decrease in differentiation of the vertebral column in the evolution of aquatic mammals, and the apparent decrease in overall complexity in the evolution of many parasites.

Third, and most important, the sense in which a modern fish or even a human is more complex than a typical invertebrate, modern or ancient, is unclear. For example, a lobster is a fully competent animal with a complex internal anatomy and external structure. By what standard is that anatomy and structure less complex than that of a human?

Definitions of complexity. There is some agreement now among biologists that the complexity of an organism, or any object, is a function of the number of different parts, or the degree of differentiation among parts, and the irregularity of their arrangement. Thus, heterogeneous and irregularly configured systems, such as organisms, automobiles, and junk heaps, are complex, whereas homogeneous and regular systems, such as crystal lattices, picket fences, and wallpaper patterns, are simple. In this scheme, complexity and simplicity define the extremes of a scale that refers only to structure.

This scheme sounds counterintuitive, because it seems to confound the very specific and functional heterogeneity of an organism with the random and (ordinarily) functionless heterogeneity of a junk heap. This objection reflects the fact that in common usage complexity includes some notion of functionality or efficacy. Unfortunately, measures of functionality have not been developed, and the functions of particular structures are often unknown, especially in fossil organisms. For example, the massive antlers of the extinct Irish elk may have been an adaptation for use in sexual display or may have had no function, arising merely as a side effect of a size increase in the animal. Conversely, an apparently functionless junk heap may actually be functioning as a work of art, perhaps one requiring a very specific arrangement of parts. With a definition based purely on structure, the complexity of these systems can be measured, at least in principle, without any knowledge of function.

Evolutionary trends. A new conceptual scheme has been developed for understanding evolutionary trends, including trends in complexity. The **illustration** demonstrates this scheme in a series of computer-generated evolutionary systems. In each system, a group of organisms begins at the bottom as a single lineage or species, and as time passes, new species arise and complexity changes in each lineage. Trends occur in illus. *a*, *c*, and *d*, and initially in *b*.

In illus. *c*, the system is driven by a bias so that most changes are increases in complexity, and lineages tend to move to the right. The system in illus. *d* is also driven, but weakly, so that most changes are also increases in complexity but change of any kind is rare. For organisms, various sources of bias have been proposed. For example, it is possible that more complex organisms have more parts and therefore greater division of labor among parts, resulting in greater efficiency. If so, increase in complexity would be favored by natural selection in most cases, and on average increases should occur more often than decreases.

Passive systems. The systems in illus. *a* and *b* are passive, that is, no bias or driving force is present and increases and decreases are equally likely. In illus. *a*, mean complexity increases without a

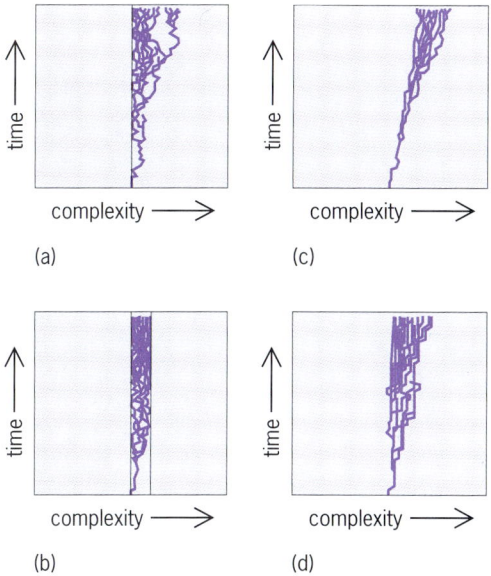

Computer simulation of the diversification of life (or of smaller groups of organisms). (*a, b*) Passive systems in which increases and decreases in complexity are equally likely but a lower bound (vertical line) limits decrease. (*c, d*) Driven systems in which increases in complexity are much more likely than decreases when change occurs.

driving force, the system spreading passively to the right. The spread is asymmetrical, because movement to the left is blocked by a lower bound, located at the thin vertical line. In evolution, that line might represent a lower limit on complexity, or the complexity of the simplest possible organism. If the first organisms originated near such a lower limit, mean complexity could only have increased as life diversified.

Illustration *b* shows a system where upper and lower bounds are present, resulting in an initial increase in mean complexity, but not in continued increases. Possibly the complexity of certain dynamical systems, including organisms, may be limited in this way. An organism can be understood as a network of interacting components. Computer simulations of such networks seem to show that they behave chaotically when their complexity exceeds certain critical values. This finding supports the views of a small minority of evolutionists who have long argued that complexity increased until the initial diversification of multicellular organisms in the Cambrian, but that a limit was reached, increases since then being modest or nonexistent.

Historically, a distinction has been made between necessary and occasional increase in complexity. Proponents of a trend argue that increases in complexity do occur, but only occasionally, in complexity revolutions (for example, the origin of mammals). Further, increase in complexity is not necessary, so simple forms persist. In the current conceptual scheme, the distinction is not especially useful because complexity increases occasionally in all systems, even in a system in which no long-term trend occurs (illus. *b*). Also, increase in complexity is not necessary, that is, simple species persist, in all but the strongly driven system.

Passive-driven distinction. The passive-driven distinction is more significant. It refers to the nature and distribution of the basic forces acting on complexity: whether they act only locally, at boundaries (passive), or globally, affecting all lineages (driven).

A number of tests have been devised to distinguish passive and driven concerns, using data from fossil and recent organisms. For example, in a driven trend a bias should be found in a random sample of ancestor-descendant pairs, with increases occurring more often than decreases (illus. *c* and *d*). In a passive trend (illus. *a*), increases and decreases should occur about equally often. Another test is the minimum test. In the strongly driven trend (illus. *c*), the complexity of the simplest species increases over time. However, for the passive trend (illus. *a*), the minimum is nearly constant, staying close to the boundary. But the minimum also changes little in the weakly driven system (illus. *d*), so that the minimum test is asymmetrical. That is, if the minimum increases, a trend is probably driven; if not, it could be either passive or driven.

Empirical studies. Applying even a purely structural definition to organisms is not straightforward because parts in organisms are not well defined. Whether a heart should be considered a part and one of its components, such as a valve, also a part is problematic. Identifying comparable parts among organisms is usually difficult, especially for organisms with radically different designs or body plans, such as a rose and an elephant.

One solution is to use cells as parts and to measure complexity as number of cell types. Counting cell types is a slow, laborious process; so far, reliable counts have been completed for only a dozen or so species, including a number of invertebrates, such as sponges (5 cell types) and arthropods (55), and vertebrates, such as lampreys (65) and humans (210). By this measure vertebrates are more complex, but more data are needed to document a trend. These estimates are based on cell-type counts for only a few species within each group, and the possibility remains that some arthropod species not yet studied will, for example, have many more than 55 cell types. Also, the figure for humans may not be comparable to the others, because human cells have been studied far more intensively, and finer distinctions among types have doubtless been made. Finally, it has not been determined whether the trend (if there is one) is passive or driven. Minimum complexity probably has not changed, because many groups with few cell types still survive and prosper, but this finding is consistent with either a passive or driven mechanism.

Measuring complexity within body plans and within single substructures or organ systems where parts are comparable is an alternative strategy. Each substructure might represent a random sample of the organism, with any pattern of change in the organism as a whole likely occurring also in its component substructures. Three studies of substructure complexity have been done: limb pairs, vertebral columns, and ammonoid sutures.

Limb pairs. In some substructures, the parts are well defined, and complexity can be measured simply by counting the number of different types. For example, aquatic arthropods are composed of a series of segments, each with a distinct leg-pair type. In the ancient trilobites, most leg pairs were very similar, whereas in modern lobsters, leg pairs in the midsection have been modified for walking, a forward pair has been enlarged and modified for grasping, and others have become specialized as mouth parts. For the entire group, from the Cambrian to the present, the data clearly show an increase in mean complexity. The trend appears to be driven, because minimum complexity increased, although the increase is mainly a consequence of the extinction of a single group (the trilobites) about 235 million years ago.

Vertebral columns. Counting parts is not possible in substructures where variation is continuous, such as vertebral columns. In mammals, for example, the thoracic vertebrae (which support the rib cage) differ from the lumbar vertebrae (near the hip), but the change from one region to the next is gradual and without clear boundaries. In such cases, complexity can be measured as degree of differentiation among parts. For example, the complexity of a column can be measured as the average absolute difference between adjacent vertebrae in some vertebral dimension, say vertebral length. In a typical fish, all vertebrae are about the same length over much of the length of the column, so the average difference between adjacent vertebrae is very low. In mammals, vertebral length varies considerably throughout the column, and average differences are therefore much greater. One study has shown that mean column complexity increased in the vertebrates in the transition from fish to mammal, but in ancestor-descendant comparisons within mammals increases occurred as often as decreases. For example, column complexity increased during the evolution of ruminants with the lengthening of the neck, but decreased in the evolution of the whale; the column became simpler and more fishlike as it was modified for life in the water. The absence of a bias in mammals suggests that the trend in vertebrates as a whole was passive.

Ammonoid sutures. The fractal dimension can be used as a measure of hierarchical complexity, that is, the degree to which an organism is composed of nested parts within parts. The fractal dimension has been used to study the complexity of sutures in ammonoids, an extinct group related to the chambered nautilus. Ammonoids secreted a coiled, chambered shell as they grew, and septa separating the chambers are visible externally in fossil specimens as curvy lines or sutures. In early ammonites, the sutures were fairly straight, but in many later species, the sutures became quite complex, showing sharp curves and curves within curves. Over most of the entire history of the group, from 400 to 65 m.y.a. (in the same mass extinction that eliminated the dinosaurs), mean fractal dimension increased. The minimum increased initially but remained roughly constant later on, and ancestor-descendant pairs show no bias, a pattern consistent with a passive mechanism.

Implications. The trend in aquatic arthropods appears driven, but those in the vertebral column and in ammonoid sutures are probably passive. However, no general conclusions about either trends or trend mechanisms for cell types follow from these few cases. If general tendencies exist, they can be expected to emerge only as statistical regularities in large samples. Many more quantitative studies, covering a wide range of organisms and using a variety of metrics, are needed.

For background information SEE AMMONIDEA; CELL DIFFERENTIATION; DEVELOPMENTAL BIOLOGY; FRACTALS; NAUTILUS in the McGraw-Hill Encyclopedia of Science & Technology.

Daniel W. McShea

Bibliography. G. Boyajian and T. Lutz, Evolution of biological complexity and its relation to taxonomic longevity in the Ammonoidea, *Geology*, 20:983–986, 1992; J. L. Cisne, Evolution of the world fauna of aquatic free-living arthropods, *Evolution*, 28:337–366, 1974; D. W. McShea, Complexity and evolution: What everybody knows, *Biol. Philos.*, 6:303–324, 1991; D. W. McShea, Evolutionary change in the morphological complexity of the vertebral column, *Evolution*, 47:730–740, 1993.

Evolutionary developmental biology

Evolution is change over time and across generations; development is change within an individual's lifetime. Evolutionary developmental biology is the study of how alterations in embryonic development bring about evolutionary change, especially change in morphology or phases of the life history. Although sometimes described as a new field of study, philosophers who lived as long ago as Aristotle and many nineteenth-century natural philosophers devoted major efforts, even their entire lives, to studying interactions between embryonic development and evolutionary change. A brief history of the field is followed by a discussion of late twentieth century evolutionary developmental biology.

History. At one level, the link between evolution and development lies in the meaning of the two words. "Evolution" was used in 1774, and increasingly in the eighteenth and first half of the nineteenth centuries, to denote embryonic development. Thus, in 1878, Thomas Henry Huxley could write an essay entitled "Evolution in Biology," and devote it entirely to embryonic development. "Evolution" here was used for a particular type of development known as preformation. According to this scheme all the parts of an organism are preformed or predestined in eggs or in sperm, and individuals simply unfold or evolve.

By the early nineteenth century, evolution was beginning to be used for the origin of new groups of organisms, for gradual improvement in organisms over time, and for transformation of aquatic to land-dwelling animals. It was in this sense that Charles Darwin in 1859 used "evolved" as the last word in the *Origin of Species:* "endless forms... have been, and are being, evolved."

Evolution. Evolution has now taken on a multitude of meanings reflecting its hierarchical nature; evolution operates at various biological levels. Evolution is used to describe changes in gene frequencies, the appearance of new structures, or the appearance, change, and spread of new species. Microevolution is change within species, macroevolution is speciation itself, and megaevolution is change within the origin of groups above the species level, groups such as families, orders, classes, and phyla.

Microevolution can occur quite rapidly and, therefore, may be studied in the laboratory, and with selection and breeding experiments. It is therefore relatively easy to investigate the developmental basis of microevolutionary change.

However, macroevolution is a much longer process and can be studied only indirectly. Even the fastest evolving animals, such as Hawaiian fruit flies or South African cichlid fishes, take several hundred years to produce new species. Most speciation events take hundreds of thousands or millions of years.

Some biologists maintain that megaevolution is over. Although the phyla into which organisms are classified evolved some 5×10^8 years ago, some phyla were discovered only this century. For example, the deep-sea red tube worms (at a depth of 8202 ft or 2500 m) associated with hydrothermal vents were assigned to a separate phylum (Vestimentifera) as late as 1985. Although new phyla are still being discovered (five so far this century), it seems highly unlikely that new phyla are still evolving. The reasons are partly ecological and partly developmental. Understanding why additional phyla are unlikely to evolve is an area in which evolutionary developmental biology makes a unique contribution.

Phyla. A resurgence of study of the Burgess Shale fossils of the Cambrian of British Columbia has rekindled interest in phyla. For example, there are questions as to the number of phyla represented and their origin and development.

Phyla are distinguished from one another on the basis of fundamental differences in how their members are constructed—their body plans. One feature used to distinguish body plans among phyla is symmetry, which can be bilateral or radial. A second feature is the number of basic layers that compose the animals. The numbers can vary from one, as in protozoa, to three as in humans and many other multicellular animals. A third feature is the organization and development of the body cavity.

These are such fundamental features of early embryonic development that it seems unlikely that new schemes could evolve. Embryonic development is a series of highly integrated processes, integrated in space and through time. Features distinguishing major groups such as phyla appear very early in embryonic life. Evolving a new scheme, such as an organism with wheels instead of legs, would entail such major restructuring of the embryonic body plan that it is highly unlikely such an embryo would survive long enough to grow into adulthood and pass on the new trait.

Restructuring early embryonic development can occur, and can produce major alterations in embryonic or larval structure without changes necessitating the placement of organisms into a new phylum. However, such changes are often, and perhaps usually, associated with the origin of new species. A good example of restructuring and of the role of evolutionary developmental biology in understanding such developmental and evolutionary change is the transition from indirect to direct development in animals such as sea urchins or frogs.

Indirect to direct development. Sea urchins and frogs are characterized by having life cycles in two major stages, which are separated by a metamorphosis that converts one stage into the other. In both groups embryos develop into larvae or tadpoles, respectively, that are structurally and functionally distinct from the adults. Moreover, tadpoles inhabit a different environment from frogs, an aquatic one. Being able to colonize different environments is one of the advantages of having two phases in the life cycle.

Such patterns of embryonic development with larvae and metamorphosis are called indirect because the embryos do not directly transform into adults. The newly hatched tadpole does not look like a small frog. There are famous cases of tadpoles and frogs being assigned to different species before it was realized that one transformed into the other. Animals such as humans have direct development: the newborn baby looks like a small adult.

Indirect development is the primitive or ancestral condition for sea urchins and for frogs. It is surprising that 180 out of 900 species of sea urchins and some 600 out of 4000 species of frogs have lost the larval or tadpole stage. In one genus of frogs, *Eleutherodactylus,* more than 500 species display direct development. In direct development, frog embryos hatch as small frogs, not as tadpoles. They have switched or evolved from indirect to direct developers, forming new species in the process. Even though profound embryonic changes occur during this evolutionary switch, it is impossible to tell from looking at the adult whether its embryonic development was direct or indirect. There are alternate developmental ways of attaining the final form. Evolutionary developmental biology attempts to explain how these developmental processes and mechanisms have evolved and how evolution is related to micro- and macroevolution.

Discovering the lack of a correlation between type of embryonic development—whether direct or indirect—and adult structure has been an important recent contribution from evolutionary developmental biology. During the nineteenth century it was believed that early embryonic development was not subject to evolutionary change, or, that even if changes in early development did occur, they would be lethal. It is now known from studies of direct developing sea urchins and frogs that profound changes in early embryonic development can occur.

In both sea urchins and frogs, direct development is associated with vast increases in egg size. The diameter of eggs of direct developers can be 10 or more times larger than those of indirect developers. In sea urchins particular embryonic cells in indirect developers exclusively produce larval structures such as the skeleton. Other embryonic cells exclusively produce adult structures. In direct developing species the larval cell lineages switch from producing larval structures to producing adult structures, often for different parts of the organism (for example, cells for the larval skeleton might form adult muscle). Thus, embryos are restructured, not with the loss of larval cell lines but with their reprogramming for adult structures.

The timing of development may also be altered in direct developers, adult structures appearing earlier in embryonic life than in indirect developers. A change in the timing of development as a mechanism of evolutionary change, a process known as heterochrony, is another important area of investigation in evolutionary developmental biology.

For background information SEE DEVELOPMENTAL BIOLOGY; EMBRYOGENESIS; EMBRYOLOGY; INVERTEBRATE EMBRYOLOGY in the McGraw-Hill Encyclopedia of Science & Technology.

Brian K. Hall

Bibliography. S. Gilbert, *Developmental Biology*, 4th ed., 1994; B. K. Hall, *Evolutionary Developmental Biology*, 1992; R. A. Raff and T. C. Kaufmann, *Embryos, Genes, and Evolution*, 1983.

Extremely high heat flux

Current technologies operate at increasingly higher power densities, creating a need for cooling systems that can remove increasingly large heat fluxes. The design of high-heat-flux systems has generally been limited to about 10^7 W/m^2, because of restrictions imposed by the coolant flow. Impinging liquid jets may offer a means of cooling at high heat flux that does not suffer from the usual limitations. Experiments with such jets have produced heat fluxes substantially higher than those at the surface of the Sun, and if jets are combined with properly chosen materials, even higher heat fluxes may be possible.

Heat fluxes. Heat flux (q) is the heat flowing through a surface per unit area per unit time [in joules/(s · m^2) or W/m^2; 1 W/m^2 = 0.317 Btu/h · ft^2)]. In the present context, heat flux measures the rate of heat loss from an object to a surrounding medium that cools it or into a passage that carries cooling fluid through the object. The integral of the heat flux over the entire surface of an object equals the power dissipated or absorbed within the object (in joules per second or watts).

For an object of a fixed size, an increase in the power dissipation causes an increase in the heat flux through the object's surface. For example, when more junctions are incorporated into an integrated circuit package (chip) of a given size, the heat flux that must be removed from the chip's surface will be higher. Electronics are one typical technology that demands cooling at increasingly high heat fluxes.

In day-to-day experience, heat flux is generally fairly low (**Fig. 1**). The radiant heat flux from bright sunshine is about 10^3 W/m^2; the heat flux into a boiling pot of water is about 2.5×10^4 W/m^2. Industrial fluxes are often larger. The heat flux from an advanced digital logic chip may exceed 10^5 W/m^2. The flux from the fuel rods of a nuclear reactor may run from 2×10^5 to 5×10^6 W/m^2.

For systems that run in a steady state, heat flux rarely exceeds 10^7 W/m^2 because maintaining a sufficiently low temperature at such high fluxes can be quite difficult. Limiting the temperature at high flux requires a high-performance cooling system, and cooling technologies for very high flux are scarce. However, for unsteady heating such as used in welding or cutting, the flux can be well above 10^7 W/m^2. An arc welder produces fluxes of 10^7–10^9 W/m^2, and lasers used for cutting materials can produce fluxes well above 10^9 W/m^2. In contrast, the heat flux at the surface of the Sun is only 6.3×10^7 W/m^2.

Trends toward higher power density are increasing the need for steady-state high-heat-flux systems. Examples may be found in densely packed electronic chips, proposed fusion reactors, and high-power laser-diode arrays. In some synchrotron x-ray mirrors, fluxes of 2×10^7 W/m^2 must be removed from delicate mirrors while holding the temperature within a tightly constrained range. The development of cooling schemes for such systems has received much attention.

Limitations on heat flux. The heat flux from an object is related to the surface temperature of that object through the heat-transfer coefficient, h, of

Table 1. Typical values of thermal conductivity (k)

Material	Thermal conductivity (k),* W/(m · K) [Btu/(h · ft · °F)]	Melting point,† °C (°F)	Tensile strength,* MPa (ksi)
Copper	400 (230)	1085 (1985)	~200 (29)
Molybdenum	140 (80)	2610 (4730)	~400 (58)
Tungsten	190 (110)	3410 (6170)	~400 (58)
Steel (1018)	30 (17)	1370 (2500)	~440 (64)
Diamond	2100 (1200)	4300 (7770)	>3000 (>435)

* Values vary significantly with processing, alloying, and temperature.
† Maximum service temperatures are significantly lower.

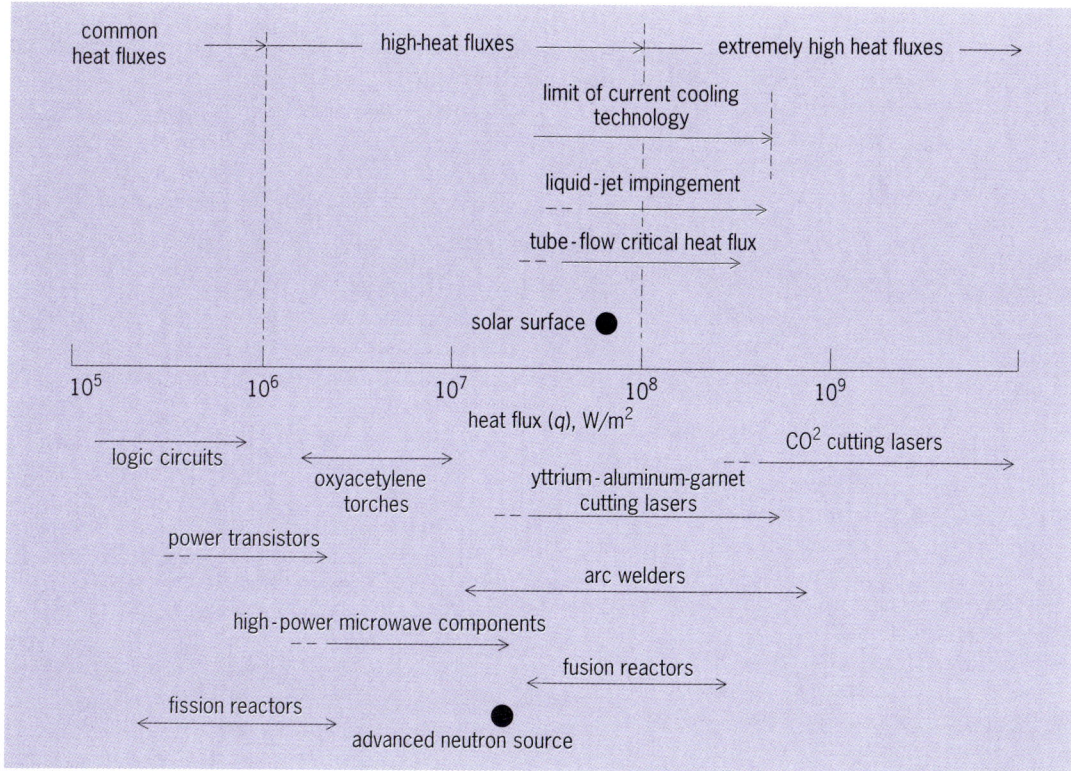

Fig. 1. Magnitudes of heat flux for heat sources associated with various technologies (below the scale) and cooling technologies (above the scale). Heat flux at solar surface is shown for comparison.

the cooling fluid, given by Eq. (1), where $T_{surface}$ and

$$q = h(T_{surface} - T_{fluid}) \quad (1)$$

T_{fluid} are the temperatures of the surface and fluid. A higher value of h leads to a cooler surface temperature for a given q. The value of h depends on the cooling fluid, its flow speed, and the geometry of the flow. The surface temperature of the object is related to the heat flux and the internal temperature of the object (T_{inside}). For a plane slab, this relation is given by Eq. (2), where k is the slab's thermal

$$q = \frac{k}{t}(T_{inside} - T_{surface}) \quad (2)$$

conductivity and t is its thickness. At a given heat flux and surface temperature, a thinner slab of a higher-conductivity material will become less hot inside. Some typical values of thermal conductivity and the heat-transfer coefficient are given in **Tables 1** and **2**.

For a high heat flux, the internal temperature can be minimized by maintaining a high heat-transfer coefficient and using a material with high thermal conductivity. The thickness of the object should be minimized, but requirements for mechanical strength usually prevent t from being too small. In fact, the choice of material for the object being cooled will depend not only on the need for high thermal conductivity, but also on considerations related to thermal stress, material strength, and melting or softening temperatures. Thus, although copper has a very high thermal conductivity, it may sometimes be less suitable than lower-k materials such as tungsten or molybdenum, which are significantly stronger and can be used at higher temperatures. In some cases, the maximum allowable internal temperature will be fairly low; in a semiconductor laser array, it must not exceed 150°C (300°F).

The other element of achieving a high heat flux while maintaining a low internal temperature is producing a large heat-transfer coefficient. The heat-transfer coefficient is typically 10–100 times larger for a liquid flow than for an otherwise identical gas flow, and consequently only liquid coolants are used in high-heat-flux cooling. In electronics applications, highly dielectric liquids, such as fluo-

Table 2. Typical values of heat-transfer coefficient (h)

		Heat-transfer coefficient (h), W/(m² · K) [Btu/(h · ft² · °F)]	
Fluid	Speed, m/s (ft/s)	Tube flow*	Impinging jets†
Air	10 (30.5)	60 (10)	160 (28)
FC-77‡	10 (30.5)	5,100 (900)	8,800 (1500)
Water	10 (30.5)	28,600 (5000)	49,300 (8700)
Water	30 (100)	69,000 (12,000)	98,800 (17,400)
Gallium	10 (30.5)	106,000 (18,700)	270,000 (47,500)

*1-cm (0.4-in.) diameter, nonboiling, fully developed, near 30°C (86°F).
†1-cm (0.4-in.) diameter, nonboiling, near 30°C (86°F), nozzle near target.
‡Fluorinated hydrocarbon liquid, commonly used to cool electronic components.

rocarbons, must be used. When electrical problems are not at issue, water is a more effective coolant. When the highest values of h are required, liquid metals (such as gallium or a mixture of sodium and potassium) will produce heat-transfer coefficients 3–8 times higher than water in an otherwise identical configuration.

The geometry of the coolant flow may vary significantly. Usually, flows for high heat flux are chosen so as to inhibit the growth of the slow fluid layer near the wall, the so-called boundary layer. This slow layer results from the viscous drag of the wall on the fluid. If the boundary layer grows, the resistance to heat transfer grows and the heat-transfer coefficient decreases. For turbulent flows inside tubes, a thin boundary layer is effectively maintained along the entire axis of the tube, owing to the confinement of the flow and the pressure gradient along the tube. Very thin boundary layers also occur in regions where a flow impinges on the outer surface of an object, such as at the point of impact of a liquid jet striking a target.

In the flow configurations typical of such cooling systems, the heat-transfer coefficient increases strongly with increasing flow speed, V. For example, for a coolant flowing through a tube, the heat-transfer coefficient increases roughly as $V^{0.8}$. High speeds are essential in these systems. Historically, high-speed tube flows of water or liquid metals have been the principal means of cooling at very high heat fluxes.

Liquid coolants have the added advantage that they can be boiled. Boiling increases the heat-transfer coefficient because the agitation of the flow by bubbles tends to reduce the boundary-layer resistance to heat transfer. (The fact that heat is absorbed by the latent heat of vaporization is often a secondary factor.) However, when the coolant flows through a tube the formation of vapor during boiling can obstruct the flow of liquid to the tube wall. If the heat flux becomes too high, the accumulated vapor at the wall will tend to insulate it, resulting in a dramatic increase in wall temperature that can actually melt the tube or the object being cooled. This event is called critical heat flux, and it is the usual obstacle to designing systems for extremely high heat flux. Dedicated efforts to maximize critical heat flux in tube flow have generally led to critical fluxes of 1.7–3.5×10^8 W/m^2, over small lengths of tubing a few millimeters in diameter.

Liquid-jet impingement. Impinging liquid jets have recently been put forward as an alternative to tube flows in extremely high-heat-flux systems. In **Fig. 2**, a laminar water jet strikes a surface and spreads into a radially flowing film. This particular film is initially laminar and becomes turbulent farther downstream; the film is terminated by a hydraulic jump caused by a raised edge on the target surface. Near the point of impact, jets like these produce a heat-transfer coefficient that may be three times larger than for a tube flow of the same speed and diameter. More importantly, near the point of impact, jets are less suscepti-

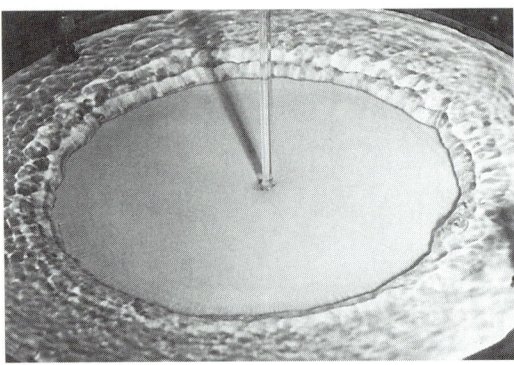

Fig. 2. Impinging water jet. *(After J. H. Lienhard V, X. Liu, and L. A. Gabour, Splattering and heat transfer during impingement of a turbulent liquid jet, J. Heat Trans., 114:362–372, May 1992)*

ble to critical-heat-flux phenomena than tube flows because vapor is easily removed by the outward radial flow while liquid continues to reach the wall from above. The region of high heat-transfer coefficient has a diameter roughly 1.5 times that of the jet.

Experiments were performed in 1992 to probe the ultimate performance of jet impingement cooling. A 1.9 mm (0.075 in.) water jet at speeds of up to 134 m/s (440 ft/s) was used to cool various metal targets with a thickness of 1–2 mm (0.04–0.08 in.). These targets were heated from the rear by a plasma arc (**Fig. 3**). The target would often melt partway through on the plasma side, but the cooling capability of the water jet was sufficient to prevent complete melting and to remove the heat flux from the plasma. The highest fluxes were obtained by using molybdenum targets; other materials (including steel, copper, tantalum, and tungsten) exhibited failures by brittle fracture and plastic yielding. Heat fluxes of up to 4×10^8 W/m^2 were sustained on molybdenum targets without evidence of critical heat flux. (In fact, the only limitation to higher flux in that experiment was an insufficiently powerful plasma torch.) This flux is more than six times that at the surface of the Sun.

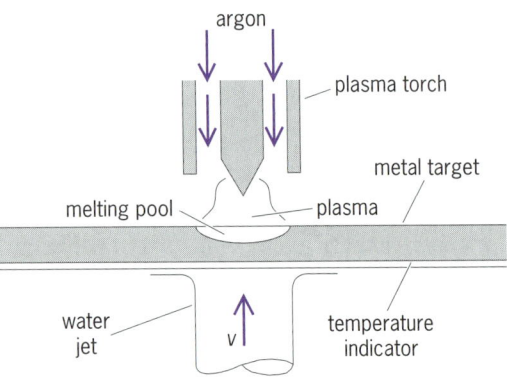

Fig. 3. Water-jet experiment configuration. High cooling is confined to circle about twice the jet diameter. Temperature indicator is used to calculate heat flux. v = velocity of water jet. *(After X. Liu and J. H. Lienhard V, Extremely high heat fluxes beneath impinging liquid jets, J. Heat Trans., 115:472–476, May 1993)*

Prospects. These experiments suggest that liquid-jet impingement cooling can reach substantially higher heat fluxes and that the principal limitation to reaching higher fluxes is the wall material. Wall materials are needed that have higher resistance to thermal stress and better ability to sustain use at high temperature.

Apart from metals, synthetic diamond surfaces show particular promise for high-heat-flux applications. Polycrystalline diamond films can now be produced in reasonable quantity and at a reasonable cost, including single-crystal diamonds of up to 1 cm^2 (0.16 in.^{2}). Diamonds offer several features desirable in high-heat-flux cooling. The conductivity is five times that of copper, the thermal expansion coefficient is very low, and the tensile strength is high. Estimates suggest that if a diamond surface were cooled with the water jet used in the experiments discussed above, fluxes of 5×10^8 W/m^2 or more could be reached with no danger of mechanical failure.

The use of liquid-metal jets may further enhance the cooling capability of jet-impingement systems. For example, gallium liquifies at 29.8°C (85.7°F) and could be used in a relatively low-temperature jet-cooling system. In comparison to a water jet of identical size and speed, estimates show that a gallium jet would produce a heat-transfer coefficient more than five times greater than the water jet. The full potential of jet impingement cooling systems remains to be determined.

For background information SEE BOUNDARY-LAYER FLOW; CONDUCTION (HEAT); CONVECTION (HEAT); HEAT TRANSFER; JET FLOW in the McGraw-Hill Encyclopedia of Science & Technology.

John H. Lienhard V

Bibliography. A. M. Khounsary (ed.), *High Heat Flux Engineering II*, SPIE, 1997:29–43, 1993; J. H. Lienhard V, Liquid jet impingement, *Annu. Rev. Heat Trans.*, vol. 6, 1995; X. Liu and J. H. Lienhard V, Extremely high heat fluxes beneath impinging liquid jets, *J. Heat Trans.*, 115:472–476, 1993.

Fertilization

Fertilization is the meeting and union of gametes that leads to the formation of a new individual of a species. Fertilization has enormous significance as a biological event, representing a point of no return: once joined together at fertilization, neither sperm nor egg can separate. Fertilization is the culmination of sexual reproduction, the end to which all the specializations of sexual adaption are directed. Fertilization has two main purposes: it generates a unique genome, and it triggers the start of the developmental program that forms the individual during embryogenesis. The egg and sperm find one another by chemotaxis, choosing one another through species-specific gamete recognition. They unite their genomes by cell-cell and nuclear fusion, and trigger the start of development by abrupt activation. The arrival of the sperm at fertilization sets off a calcium wave that sweeps through the egg's volume from the point of sperm entry, and is the means by which the small sperm can activate the entire egg.

Gametes. A generalized picture of fertilization would show a small and motile spermatozoon swimming toward a larger, spherical egg. The picture is a good representation of fertilization in animals and even in some plants, for example, the alga *Fucus serratus*. It is less applicable to higher plants, where the pollen tube releases a pair of nonmotile sperm very close to their targets, the egg cell and central cell. The egg cell becomes the plant embryo and the central cell, the nutritive endosperm. Very little is known about the mechanisms of fertilization in plants, so for the most part, the mechanisms of fertilization in animals will be discussed below.

Spermatozoa are highly specialized cells adapted to motility. Their genetic information is very tightly packed into the sperm head. The sperm head generally has a long tail (flagella) attached. The tail beats with a whiplike action which propels the sperm forward. At the base of the tail, one or several mitochondria provide the chemical energy that powers the tail beat. At the tip of the head is the acrosome, a granule whose contents are discharged as the sperm meets the egg. The granule contains digestive enzymes that break down the coatings that surround the egg.

Eggs are also highly specialized cells adapted to undergo rapid division, and store the information and nutrients that are required during the early stages of development. By the standards of cell biology they are large, with a median size of around 0.1 mm. Frog and fish eggs are 10 times bigger and can easily be seen with the unaided eye. Eggs of birds and reptiles are very large, consisting mostly of yolk (nutrient) and a protective fluid. Even in these eggs, the oocyte (the part that is fertilized) is small.

In aquatic environments, fertilization occurs in the water column. In land animals, fertilization usually occurs within the female reproductive tract. Shelled eggs are fertilized at an early stage, before being packaged into the shell. Most is known about fertilization in aquatic animals, because fertilization in water is easiest to study in the laboratory. More recently, the conditions inside the mammalian reproductive tract have become sufficiently well understood to permit the preparation of solutions allowing fertilization between mammalian (including human) gametes under laboratory conditions.

Chemotaxis. There may be selective advantage in chemoattraction between gametes that can increase the rate of fertilization. Substances released by the egg modify the swimming pattern of the sperm. The best-studied case involves the peptide resact, which is released from sea urchin eggs and causes the sperm to swim in tighter circles. This modification of behavior traps the sperm in the vicinity of the egg, and is thus the first of a series of signal transduction events that control the behavior of the gametes during fertilization. In this case, the resact peptide binds to a receptor on the sperm tail. The receptor spans the plasma membrane of the sperm tail and generates cyclic guanosine 5'-monophosphate (cGMP), a cell messenger which changes the characteristics of the tail beat, probably through phosphorylation of the motor proteins responsible for the movement.

Species-specific recognition. Individual species can remain distinct from one another only by maintaining a genetic isolation from other species. Various factors favor genetic isolation, the simplest being geographic isolation. For cohabiting species, behavioral mechanisms governing mating behavior are an adequate barrier. Nonetheless, it is rare for fertilization to occur between gametes of different species, even in the test tube. There must also be cell-cell recognition mechanisms that operate at fertilization.

One recognition mechanism involves the reaction of the acrosome (anterior body of the sperm). Most eggs have coats or layers that the sperm must penetrate to reach the egg surface. The enzymes in the acrosome eat a hole through these coats, creating a path for the sperm. The enzymes are released close to the surface of the coats and are triggered by coat components. There is a receptor for the coat components in the sperm head. When the receptor binds the component (for example, ZP_3 in mice), channels open in the sperm head and calcium floods in. The increased calcium concentration in the sperm causes the acrosome reaction: the fusion of the acrosomal vesicle with the plasma membrane. This action releases the acrosome contents in a way analogous to the release of neurotransmitters by nerve cells. Coat components from one species are often completely ineffective in triggering the acrosome reaction in sperm of another species. If the coats are removed from the egg, and the acrosome reaction of the sperm is induced artificially by raising the calcium levels (for example, by using a calcium ionophore), interspecific fertilization can occur in many cases. However, in some species, it does not, implying that a second recognition mechanism is at work.

The second recognition mechanism involves the interaction between the sperm and egg plasma membranes. Molecules on the sperm head interact with molecules in the egg plasma membrane. In mammalian gametes, these molecules are related to cell adhesion molecules known as integrins. Integrins are known to be involved in cell adhesion in mechanisms in many different sorts of cells, as well as at fertilization.

The purpose of these cell-recognition molecules is not to prevent interspecific fertilization, since the chances of egg and sperm of different species coming together by chance is very low; in most instances, physical or behavioral barriers prevent it. Rather, these recognition steps trigger a fertilization event at just the right time. Fertilization is most efficient when the acrosome reaction occurs just as the sperm reaches the egg surface, so it is triggered only by a component in egg jelly. Thus, when the plasma membranes of egg and sperm come together, fusion of sperm and egg and egg activation occur. The fact that the triggering molecules of one species are not recognized by the gametes of another is due simply to genetic drift. The same protein in different species has slightly different forms, and as evolutionary time goes by, the differences increase randomly. Thus, there comes a point when proteins that once fitted together no longer recognize one another.

In some situations, closely related species share the same habitat and breeding season. If fertilization occurs externally (for example, in seawater), the mechanisms of gamete recognition become an important matter of survival of the species. The best-studied example of gamete recognition is in abalone. The abalone sperm releases a lysin from its acrosome that punches a small hole in the protective coat surrounding the egg. The lysin from one species is ineffective against the coat of another, thereby preventing cross-species fertilization. These lysin proteins vary enormously in certain regions, far more than can be explained by genetic drift. In fact, they are among the fastest-evolving protein sequences discovered. There must be a very strong positive selection pressure on these proteins that is maintaining speciation in abalone, although the source of the pressure is unknown.

Fusion and activation. Sperm and egg must fuse so that the genetic material deoxyribonucleic acid (DNA) can be transferred from egg to sperm. Sperm-egg fusion is the moment that fertilization occurs, but what induces the sperm and egg to fuse is not yet clear. Perhaps sperm and egg plasma membrane fusion is induced by the the recognition molecules that bind the sperm and egg just prior to fusion. The analogy here is with molecules in virus envelopes that are known to induce fusion of the virus with cell membranes. It is also possible that the recognition molecules trigger a chemical reaction within the egg that induces fusion.

Immediately after the sperm and egg fuse, the egg is activated. Eggs lie in a quiescent state awaiting the sperm. Egg activation stimulates the egg to resume cell division in order to generate the cells that will form the embryo.

Calcium waves. In mammals and in some invertebrates (echinoderms), the activating signal is a wave of calcium that starts at the site of sperm-egg fusion and sweeps across the egg. In fact, calcium in the egg can be increased artificially in various ways (for example, by using a calcium ionophore) and the egg will act as though it is fertilized. It begins to grow and divide, although it lacks the sperm DNA that is the entire point of sexual reproduction. Usually, such parthenogenetic embryos do not survive for long because of accumulation of recessive lethal alleles within a single genome. However, in some insects, parthenogenesis is the norm at times, and fertilization (and the male) is dispensed with altogether.

Calcium ions are present at very low concentrations even during the peak of the calcium wave. At the peak, calcium can reach 10 micromolar; in the resting cell it is 100 times lower. The major advance since the 1970s has been that these changes can be visualized. Calcium was first seen by using a calcium-sensitive protein from jellyfish. The jellyfish uses the protein to glow gently and activates the light emission using calcium at similar concentrations to those found at fertilization. If the protein is extracted from jellyfish and microinjected into a fertilizing egg, using a fine glass micropipette, it glows more brightly as calcium concentrations increase. By using image intensification techniques, the calcium wave was first seen as a band of light that spread through a fish egg at constant velocity. More recently, fluorescent indicator dyes that change color or intensity as calcium changes are used. The principle is much the same as that of litmus paper of other pH indicators, but the dyes are sensitive to calcium ions rather than hydrogen ions. Calcium waves have now been seen in frog, sea urchin, starfish, rabbit, mouse, hamster, and even human eggs by microinjecting a calcium-sensitive fluorescent dye into the egg and using sensitive video cameras or confocal microscopes to generate an image. The properties of such waves are very similar wherever they are seen. They travel at a velocity of 5–10 micrometers/s through the egg cytoplasm, starting at the point of sperm entry (see **illus.**).

Generation. Calcium signals like the calcium wave generated at fertilization are produced by shifting calcium very rapidly between cell compartments. The process was first understood by studying the process of muscle contraction, a mechanism controlled by similar rapid changes in intracellular calcium levels. Calcium is stored at high (millimolar) concentrations in an intracellular compartment known as endoplasmic reticulum, which is separated from the cytoplasm by a lipid bilayer membrane that at rest is a barrier to calcium diffusion. The endoplasmic reticulum pervades the cytoplasm in an extensive three-dimensional network. Calcium ions are constantly scavenged from the cytoplasm by protein pumps in the bilayer membrane, which pick up a calcium ion by binding it at the cytoplasmic surface of the endoplasmic reticulum and releasing it inside. Because these protein pumps are moving calcium up a chemical gradient, they need a source of energy and consume adenosinetriphosphate (ATP).

Calcium increases occur when protein channels permeant to calcium ions in the endoplasmic reticulum open transiently. Ions can move much more quickly through channels than through pumps (the pumping mechanism is overwhelmed), and calcium rises very rapidly in the cytoplasm, increasing from 10 to 100 times in hundredths of a second. The channels then close, partly through a protective and self-limiting mechanism; high calcium in the cytoplasm inactivates them. With the channels closed, the calcium pumps are able to return the cytoplasmic calcium concentration to resting levels over a matter of a second or less.

The question is what opens the calcium channels. The answer has been determined by isolating the channel proteins and studying their behavior in artificial lipid bilayer membranes, where conditions can be tightly controlled. Calcium ions themselves can cause the channels to open. As calcium concentrations are increased from resting levels, the channels open more and more, up to a certain level, after which increasing calcium causes them to close progressively. During a calcium wave, calcium diffuses from areas of high calcium, where calcium channels have opened, to areas of the cytoplasm in front of the wave. Here, calcium ions open more calcium channels in the endoplasmic reticulum, and its concentration rises rapidly; the process is repeated by further diffusion. Behind the wavefront, calcium concentrations are so high as to bring about the protective calcium channel closure. This type of mechanism can also be observed in chemical reactions in solution, where it is called a reaction-diffusion process. By carefully observing the wavefront using calcium indicator dyes, it has been noted that the calcium release channels are separated by distances of several micrometers. Furthermore, individual channel openings can be observed. The wave appears to jump from one channel to the next, only later infilling the zone of propagation.

Activation. The endoplasmic reticulum and the its calcium channels provide a self-contained mechanism for spreading a signal from the sperm throughout the egg. At fertilization, the sperm sets the calcium wave off. Ideas about how this process happens differ, and there is no clear consensus. However, one prevalent idea is that the sperm behaves much like a hormone molecule or neurotransmitter. Hormones and neurotransmitters can cause a calcium release inside cells by a signal transduction

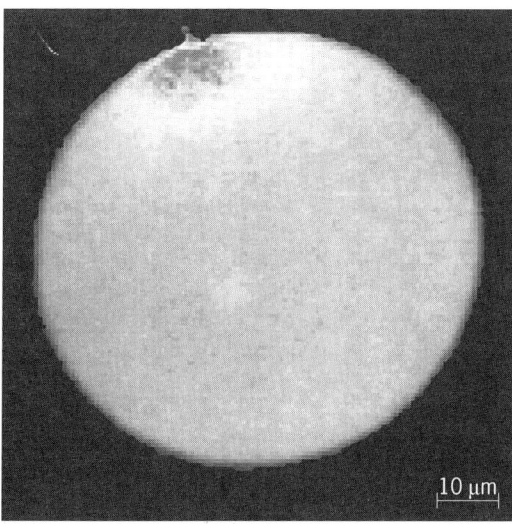

Moment of fertilization. The sperm has fused with the egg and triggered the calcium wave. The head of the sperm is the dot outside the egg.

mechanism. The hormone (the primary messenger) binds to a protein receptor molecule that crosses the cell's plasma membrane. Binding of the hormone outside the cell generates a conformational change in the protein receptor molecule that is transmitted through the plasma membrane by the protein, altering its conformation at the cytoplasmic face. This change in shape activates associated protein molecules, which then generate a second messenger. One such second messenger ($InsP_3$) can diffuse away from the membrane, bind to the calcium channels of the endoplasmic reticulum, and open them. Calcium waves are certainly generated this way in many cells. Whether this process happens at fertilization is less clear. Although there is a receptor for sperm on the egg membrane, it is not of the molecular type that generates $InsP_3$. Nonetheless, microinjecting $InsP_3$ into eggs activates them, and the egg receptor for the sperm may generate $InsP_3$ in a novel way not yet understood.

Another idea is that activation is linked to sperm-egg fusion. Sperm and egg must fuse in order for the sperm nucleus to enter the egg and make its contribution to the embryo's genome. Once the sperm cytoplasm is confluent with the egg cytoplasm after fusion, an activating molecule would diffuse from egg to sperm. Sperm-egg fusion is the first detectable event at fertilization. The sperm head contains a number of likely candidates, including $InsP_3$, calcium itself, and another second messenger, cyclic guanosine 5′-monophosphate, although it is unclear whether sufficient amounts are present to activate the egg, given how small the sperm is. However, it has been shown that hamster sperm contain sufficient protein to set off the calcium wave if it is isolated and microinjected into hamster eggs. The idea of an activating, diffusing protein is a novel one in the field of cell signaling. However, studying the fertilization calcium wave has revealed much about calcium waves in other cell types, so it is possible that fertilization may also offer the first instance of a new cell signaling mechanism.

For background information SEE CALCIUM; CELL MEMBRANE; CYTOPLASM; DEVELOPMENTAL BIOLOGY; ENDOPLASMIC RETICULUM; FERTILIZATION; SPERM CELL in the McGraw-Hill Encyclopedia of Science & Technology.

Michael Whitaker

Fertilizer

Fertilizers are important in modern agriculture. Potash and phosphate ores are the sources for potassium and phosphorus, whereas manufactured ammonia, ammonium nitrate, and urea are the sources for nitrogen in the commercial preparation of fertilizers.

Potassium and Phosphorus

Potassium and phosphorus are 2 of 16 elements essential for plant growth. Research documents that potassium influences more than 60 enzyme systems, regulates photosynthesis and plant respiration, improves plant disease resistance, increases tolerance to stress, and improves crop market quality. Phosphorus is known to stimulate root development, store and transfer energy for plant growth, improve crop maturity, promote seed development, enhance photosynthesis, and combine with potassium to improve nitrogen use efficiency. A shortage of either nutrient adversely affects plant growth and use efficiency of other production inputs.

Both nutrients have been used to improve crop growth for hundreds of years. Wood ashes were used as a source of potassium by early American pioneers. In fact, the first United States patent is for the production of potassium from wood ashes. Animal manures, fish, and other such products served as early sources of phosphorus. However, these products could not supply the need as cropland acreage increased.

Naturally occurring deposits of potassium or phosphorus materials exist in many regions of the world. Commercial deposits of phosphorus exist on most continents with major production in countries such as the United States, North Africa, the Commonwealth of Independent States (C.I.S.), and China. Commercial deposits of potassium are now in production in Canada, Germany, the C.I.S., the United States, Israel, Jordan, France, Brazil, and other nations. The quantity and quality of the ore as well as the distance of the ore from the production site to the region of use influence the commercial feasibility of an ore body.

Potassium. Potassium is mined commercially from naturally occurring salt brines and from deep underground crystalline mineral ore deposits. In arid regions, evaporation concentrates lake water into brine that contains elevated concentrations of salts, including potassium. The Dead Sea provides brine to the potassium industries in Israel and Jordan, while the Great Salt Lake in the United States is a major source of brine for potassium fertilizer production. The brines are pumped into vast ponds for solar evaporation. The result is the selective crystallization or so-called salting out of potassium minerals. These mineral salts become the raw material for developing processed potassium fertilizer.

World need for potassium, approximately 2×10^7 metric tons (2.2×10^7 tons) per year, is supplied mainly by potassium mined from mineral ore bodies buried deep beneath the soil surface (**Table 1**). The world potassium reserve base that is currently economical or believed to be economical with future technology has been estimated at 3.24×10^{10} metric tons (3.56×10^{10} tons).

Solution and shaft mining procedures are used to extract potassium from subsurface mineral ore bodies. In solution mining, potassium and sodium chlorides are dissolved from the ore vein by injecting water into the vein and then pumping the solution containing dissolved potassium to the surface, recrystallizing the dissolved potassium salt, and

Table 1. Primary mineral sources of potassium

Mineral	Composition	Potassium oxide (K_2O), %	Potassium (K), g/kg (oz/lb)
Sylvite	KCl	63.1	524 (8.38)
Carnallite	$KCl \cdot MgCl_2 \cdot 6H_2O$	17.0	141 (2.24)
Langbeinite	$K_2SO_4 \cdot 2MgSO_4$	22.7	188 (3.01)
Brine	Great Salt Lake, Utah	2.4	20 (0.32)
	Searles Lake, California	8.5	71 (1.1)

Table 2. Nutrient content of potassium fertilizers

Source	Formula	Potassium oxide (K_2O), %	Potassium (K), g/kg (oz/lb)
Muriate of potash	KCl	60–62.5	498–519 (8.00–8.30)
Sulfate of potash	K_2SO_4	50–52	415–432 (6.64–6.91)
Sulfate of potash magnesia	$K_2SO_4 \cdot MgSO_4$	22	183 (2.93)
Potassium nitrate	KNO_3	44	365 (5.84)

processing the crystals into fertilizers. The more common shaft-mining procedure involves the construction of shafts down to the ore body, removal of the ore with high-technology continuous mining equipment, and lifting of the potassium ore to the land surface for specialized processing and development into the different potassium fertilizer materials (**Table 2**).

Phosphorus. World consumption of phosphorus fertilizer during 1993 has been estimated at approximately 1.31×10^7 metric tons (1.44×10^7 tons), down from a high of 1.64×10^7 metric tons (1.8×10^7 tons) in 1988. Commercial production of phosphorus to supply this need occurs from deposits of naturally occurring phosphate minerals (**Table 3**). World reserves of phosphate rock with potential for commercial production are estimated at approximately 4×10^{10} metric tons (4.4×10^{10} tons). Such reserves exist as several different apatite minerals. In addition, there are high-phosphorus-containing ore bodies that were created over centuries as marine deposits accumulated on the ocean floor. When the oceans receded, the phosphorus ore bodies became inland deposits, such as those supplying commercial fertilizer manufacturers in Florida and North Carolina. Removal of a shallow overburden of topsoil by using strip mining techniques and technology exposes these deposits for commercial development.

Availability of phosphorus to plants from unprocessed naturally occurring ore (rock phosphate) varies between ore deposits, but in general it is very low. Availability is improved by using procedures in which rock phosphate is reacted with sulfuric acid, phosphoric acid, or other specialized chemical compounds; also the reactants may be heated. These treatments result in the formation of phosphorus fertilizer products (**Table 4**) that contain higher concentrations of phosphorus in a more readily available form for plant root absorption. Other treatments result in the formation of products with high concentrations of phosphoric acid that are suitable for industrial use and the formation of specialty phosphate fertilizer materials.

Worldwide requirements. The need for fertilizers containing both phosphorus and potassium is expected to continue to increase as the world population grows at a rate of nearly 9×10^7 more peo-

Table 3. Classes of phosphate-containing minerals in commercial phosphate deposits

Mineral	Composition
Iron-aluminum phosphates	
Wavellite	$Al_3(PO_4)_2(OH)_3 \cdot 5H_2O$
Variscite	$AlPO_4 \cdot 2H_2O$
Strengite	$FePO_4 \cdot 2H_2O$
Calcium-iron-aluminum phosphates	
Crandallite	$CaAl_3(PO_4)_2(OH)_5 \cdot H_2O$
Millisite	$(Na,K)CaAl_6(P_2O_4)_4(OH)_9 \cdot 3H_2O$
Calcium phosphates	
Fluorapatite	$Ca_{10}(PO_4)_6F_2$

Table 4. Nutrient content of phosphate fertilizers

Source	Phosphate (P_2O_5), %	Phosphorus (P), g/kg (oz/lb)
Ordinary superphosphate	18–20	78–87 (1.3–1.4)
Triple superphosphate	45–46	195–200 (3.12–3.20)
Diammonium phosphate (DAP)	46	200 (3.20)
Monammonium phosphate (MAP)	52–55	225–240 (3.60–3.84)
Phosphoric acid	52–54	225–235 (3.60–3.76)
Superphosphoric acid	76–83	330–360 (5.28–5.76)
Ammoniated polyphosphates	55	240 (3.84)
Urea-ammonium phosphates	28	120 (1.92)
Potassium phosphate	25	110 (1.76)

ple each year and while the area of arable and productive land remains essentially constant. These fertilizer products are essential for the continued production of food, feed, fuel, and fiber worldwide. Such production systems will continue to require the best scientifically developed management practices to continue to provide an abundance of safe, high-quality, low-cost food and fiber. Such systems must also be capable of sustaining optimum levels of production that are profitable to the farmer and at the same time compatible with the environment.

Noble R. Usherwood

Nitrogen Fertilizers

It has been estimated that as much as 20–25% of the world's food supply is directly due to the use of commercial fertilizers. Nitrogen fertilizers probably contribute more to the increased food supply than any other single input.

Types. Anhydrous ammonia, with few exceptions, forms the basis of nitrogen manufacturing throughout the world. It is synthesized from hydrogen and nitrogen in a high-pressure and temperature reaction known as the Haber-Bosch process. The hydrogen may be obtained from a number of sources, including water, coal, and oil; the most important source of hydrogen is natural gas. Consequently, nitrogen fertilizer plants are located close to natural gas fields or along major natural gas pipelines. The nitrogen used comes from the atmosphere.

Nitrogen fertilizers are widely used on agronomic, vegetable, horticultural, ornamental, and greenhouse crops. In general, there are no distinct agronomic advantages of one fertilizer compared to another for use on most crops.

Anhydrous ammonia (82 wt % nitrogen) is used for direct fertilizer application and also for the production of dry nitrogen fertilizers, including ammonium nitrate (34 wt % nitrogen), ammonium sulfate (21 wt % nitrogen), and urea (46 wt % nitrogen), and liquid nitrogen fertilizers such as urea-ammonium nitrate solution (32 wt % nitrogen), aqua ammonia (20 wt % nitrogen), and calcium ammonium nitrate solution (17 wt % nitrogen). Ammonia is also used to produce nitrogen-phosphorus fertilizers such as monoammonium phosphate (11 wt % nitrogen), diammonium phosphate (16–18 wt % nitrogen), and ammonium polyphosphate (10 wt % nitrogen).

Production. Fertilizer manufacturing is energy intensive. Production of ammonia requires 57 gigajoules per metric ton (4.9×10^7 Btu per ton) of nutrient, whereas ammonium nitrate, ammonium sulfate, and urea require 78, 73, and 61 Gj per metric ton (6.7×10^7, 6.3×10^7, and 5.3×10^7 Btu per ton) of nutrient, respectively. These estimates are representative of North American producers, but they may not be typical of producers in other regions because of differences in technology, process management, and efficiency.

Total worldwide nitrogen production was about 8×10^7 metric tons (8.8×10^7 tons) in 1992–1993, the latest year for which data are available. (Fertilizer production and consumption are reported on a fiscal-year basis.) This number is up from about 6.4×10^7 metric tons (7×10^7 tons) in 1982–1983. Nitrogen-fertilizer production peaked in the late 1980s at about 8.5×10^7 metric tons (9.4×10^7 tons) per year. The collapse of the Soviet Union has led to a precipitous decline in the production of nitrogen fertilizer in eastern Europe and the C.I.S. Production has fallen about 35% and is expected to decline further before it stabilizes. Asia is the world's largest nitrogen-producing region; of the more than 3.3×10^7 metric tons (3.6×10^7 tons) produced in 1993, the majority was produced in China, which is also the world's largest consumer of nitrogen fertilizer. From 1982 to 1993, production increased dramatically in Africa and Asia, while increasing modestly in North America. During the same period, Africa experienced the greatest growth with an increase of about 75%. Latin American production peaked in the late 1980s at about 3×10^6 metric tons (3.3×10^6 tons) and has fallen about 10%. At the end of 1993, production in western Europe had fallen almost 25% from its peak in 1984–1985.

The proportion of the different nitrogen fertilizers manufactured varies widely depending upon the crops grown, the infrastructure and technology in place, the historical use patterns, and the political and economic structure. Europe uses relatively more ammonium nitrate than any other region, while North America uses more anhydrous ammonia for direct application than any other region. Urea is the predominant fertilizer in Latin America, Asia, and the Middle East.

Urea usage increased markedly during the period from 1973 to 1991, the most recent period for which data are available (see **illus.**). Urea consumption increased more than 350% during this period whereas ammonium nitrate consumption increased only 13%. Values for ammonium nitrate include both pure product and ammonium nitrate with lime, a mixture of ammonium nitrate and calcium carbonate that has less nitrogen but that cannot be made explosive. Ammonium sulfate consumption actually fell 6% whereas direct application of ammonia increased 42%. All values are given in terms of actual nitrogen.

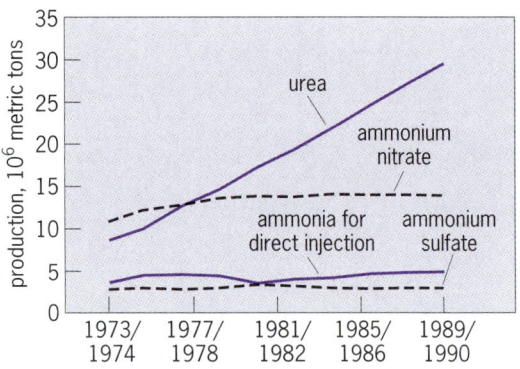

World nitrogen fertilizer production, plotted by product, 1973/1974–1989/1990.

Urea consumption has increased for several reasons. Production of urea is easily integrated with ammonia production; the carbon dioxide that is removed from the synthesis gas step is utilized in the production of urea. Urea has the highest nitrogen content of any dry nitrogen fertilizer, so shipping and handling costs per unit of nutrient are lower. It is also less corrosive than other dry nitrogen fertilizers and safer to ship and store than ammonium nitrate.

Consumption of nitrogen solutions such as urea-ammonium nitrate solution, urea solution, and ammonium nitrate solution has roughly doubled from about 2×10^6 metric tons (2.2×10^6 tons) in 1973–1974 to about 4.5×10^6 metric tons (5×10^6 tons) in 1989–1990. The United States consumes about half of the world's total; liquid fertilizers offer easier handling, greater convenience, and the opportunity to apply fertilizer in irrigation water (fertigation). The liquid fertilizers tend to be more expensive per unit of nutrient compared to dry fertilizers.

Consumption of nitrogen fertilizers is expected to grow only modestly in the near future. Nitrogen fertilizer manufacturing is closely tied to the availability of natural gas; world natural gas supplies are adequate to meet the demand for nitrogen fertilizer for at least 30–40 years.

For background information SEE FERTILIZER; NITROGEN FIXATION; PHOSPHATE MINERALS; PLANT MINERAL NUTRITION; UREA in the McGraw-Hill Encyclopedia of Science & Technology.

Steven E. Petrie

Bibliography. D. Armstrong (ed.), *Phosphorus for Agriculture*, 1990; D. Armstrong (ed.), *Potassium for Agriculture*, 1990; B. L. Bumb, *Global and Regional Data on Fertilizer Production and Use 1959/60–1992/93*, 1994; O. P. Englestad (ed.), *Fertilizer Technology and Use*, 3d ed., 1985; G. T. Harris et al., *World Fertilizer Production*, 1994; R. Manser (ed.), On course for recovery, *Phosph. Potas.*, no. 192, 1994.

Flow of solids

Industries as varied as chemical processing, food preparation, pharmaceuticals, electric utilities, metals, and consumer products handle millions of pounds of materials every year. Bins, silos, and hoppers used to store these materials vary in capacity from a few pounds of material (such as a press feed hopper used in a pharmaceutical facility) to multi-thousand-ton capacity vessels (such as train-loading silos at a coal preparation plant). However, in spite of the vast quantities of bulk solids handled, this field is not well understood by most engineers.

Start-up delays and ongoing inefficiencies are common in solids processing plants. The most prevalent cause of these problems is the improper design of bulk solids handling equipment, specifically, bins, hoppers, and feeders. Problems are also common in processing vessels that are used to purge, dry, heat, cool, or condition materials.

Flow problems. Typical flow problems include lack of flow, segregation, flooding, structural failure, and nonuniform flow in solids processing vessels.

No flow. Lack of discharge can be attributed to the formation of a stable arch over the outlet, or a stable vertical cavity called a rathole (**Fig. 1***a*). The use of sledgehammers or vibrators, while sometimes useful, causes flow to be erratic at best. In addition, it may compact the material, compounding the flow problem instead of solving it.

Segregation. Many materials experience separation of fine and coarse particles (Fig. 1*b*). Such segregation can significantly compromise the quality of the final product and the efficiency of the process. The result can be considerable expense when the product must be discarded or reprocessed.

Flooding. The collapse of a rathole in a bin that contains fine powder can cause the powder to exhibit fluidlike characteristics. The result is uncontrollable flow of material, loss of product, and clouds of dust.

Structural failure. Each year, hundreds of silos, bins, and hoppers fail in North America alone. Most of these failures could have been prevented if the

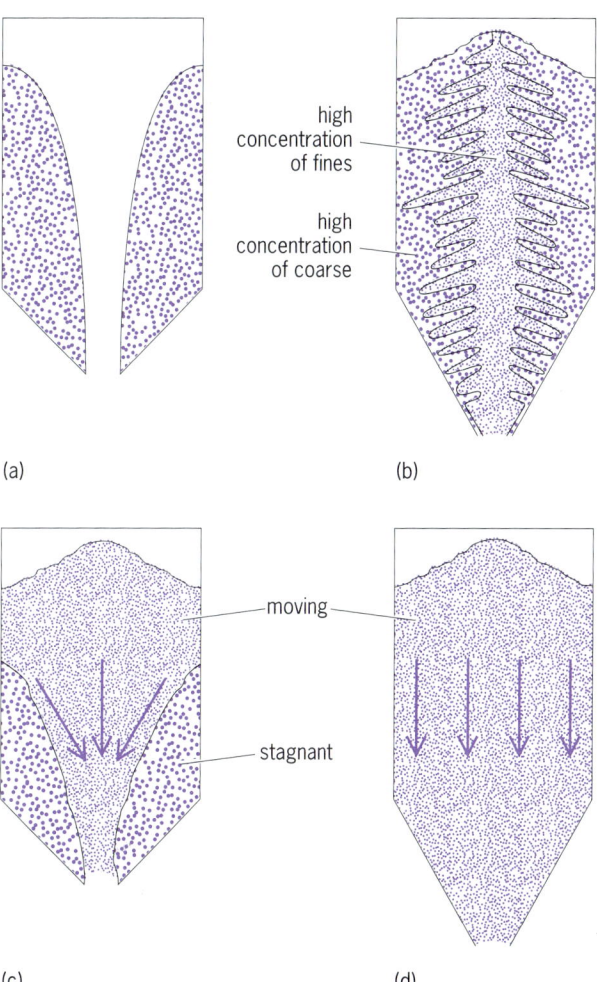

Fig. 1. Flow patterns and flow problems. (*a*) Stable rathole. (*b*) Sifting segregation during filling. (*c*) Funnel flow. (*d*) Mass flow.

designer had properly considered the loads imposed by the bulk solid being stored.

Nonuniform flow. Contact-bed–moving-bed processing vessels are used to dry, heat, purge, or condition bulk solids. The lack of uniform and reliable flow of the bulk solid in these vessels can drastically undermine the efficiency and cost-effectiveness of a process.

Impact of flow patterns. The type of flow pattern that develops in a bin, silo, or hopper significantly affects how this equipment performs. With a funnel-flow pattern (Fig. 1c), some of the bulk solid remains stationary while the rest moves. If the material being stored is subject to caking, spoilage, or oxidation, this first-in–last-out sequence is unacceptable. Ratholing (Fig. 1a) is likely, and fine powders will have a tendency to flood through the outlet when a rathole collapses. Funnel-flow designs can be cost-effective, but only if segregation is unimportant and the bulk solid is relatively coarse, free-flowing, and nondegrading.

The alternative to funnel flow is mass flow (Fig. 1d). Here, relying only on gravity, particles flow evenly along the hopper walls with no stagnant regions. Ratholing is eliminated and a first-in–first-out flow sequence is established. In addition, mass flow can prevent segregation and ensure uniform flow through the outlet.

Measures of flowability. To assess flowability, at least three key parameters need to be measured: cohesive strength, wall friction, and compressibility.

Cohesive strength. By using a direct shear tester, a sample of the bulk solid is consolidated by applying both compressive loads and shear, simulating flow conditions in a bin. Once the sample is consolidated, its strength is measured by shearing it to failure. The resulting relationship of cohesive strength versus consolidating pressure is called a flow function. Variables that can affect the cohesiveness of a bulk solid are moisture content; particle size and shape; temperature, both absolute temperature and temperature changes; storage time at rest; and chemical additives.

Once a material's flow function has been determined, minimum dimensions necessary to prevent flow stoppages due to arching and ratholing can be calculated. These dimensions determine not only the minimum size of the hopper outlet but also the size of the feeder used to control flow.

Wall friction. Expressed as either a wall-friction angle or coefficient of sliding friction, wall friction can be measured by sliding a sample of the bulk solid across a stationary surface representative of the vessel wall. The ratio of the shear force to cause sliding of the bulk solid to the applied load acting normal to the plate surface is the coefficient of sliding friction. The arc tangent of this coefficient is the angle of wall friction. Variables that can affect the frictional values of a bulk solid are similar to those which affect cohesiveness.

Two key uses of wall-friction data are mass-flow hopper design and evaluation of bin loads. By using values of both the wall-friction angle and the effective angle of internal friction, along with design charts, allowable hopper angles for mass flow can be determined. The lower the wall-friction angle, the less steep the hopper wall need be to provide mass flow. Wall-friction properties also are used in the evaluation of loads that are applied to the walls of a bin.

Compressibility. A solid's bulk density varies continuously as a function of the consolidating pressure that is acting on the material. For many bulk solids, this density-pressure relationship can be expressed as a straight line on a log-log plot, the slope of which is called the material's compressibility. Compressibility data can be used to determine minimum dimensions to overcome arching and ratholing, mass-flow hopper angles, bin loads, and feeder loads.

Feeders. These devices are used to control the rate of material discharge from a hopper outlet. They differ from conveyors, which simply transport material from one location to another without any control of the flow rate. Common mechanical feeding devices include screw, belt, chain, rotary vane, rotary plow, rotary table, vibrating pan, and vibrating louver types. These feeders are most commonly used in volumetric mode wherein the volume of material being discharged per unit time is varied by changing feeder speed, amplitude, or frequency. Several of these feeders can be designed to operate in a gravimetric mode wherein the mass flow rate of material is measured and controlled.

Screw- and belt-type feeders are commonplace because of the many competitive sources of supply, their ability to handle a wide range of bulk solids, and their design flexibility with respect to bin outlet dimensions and location of feeder discharge. Because of this flexibility, most units are custom designed. Unfortunately, few designers understand material flow properties and the mechanisms of interaction between a bin and a feeder. As a result, unnecessary and avoidable problems occur, such as poor bin flow, discharge rate fluctuations, excessive power consumption, accelerated feeder wear, and particle attrition.

Bin inserts. An insert is a static device placed within a bin (usually within its hopper section). Its purpose may be to expand the size of the active flow channel in a funnel-flow bin (thereby approaching mass flow), relieve pressure at the outlet region, control the solids velocity pattern, or allow gas introduction. Inverted cones and pyramids have been used for years in this regard but with limited success. A new concept in flow pattern modification was developed in the 1980s.

This device consists of a hopper within a hopper (**Fig. 2**). Material flows through the inner hopper as well as through the annulus between the inner and outer hoppers. By proper proportioning of the inner and outer hopper geometries as well as the materials of construction, it is possible to use this device to develop a mass-flow pattern in a bin that

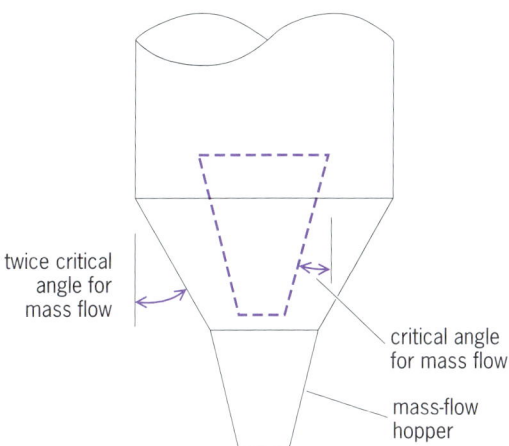

Fig. 2. Bin insert system to convert funnel flow to mass flow.

would otherwise exhibit funnel flow, thereby eliminating common problems such as ratholing and flooding of fine powders. In other applications, this device can be used to achieve a completely uniform velocity profile, thereby providing an absolute minimum degree of particle segregation. This application is particularly useful for products such as pharmaceutical powders in which segregation problems are extremely costly. A third use of this device is as an in-bin blender.

Purge and conditioning vessels. Vessels designed for processing solids are often adaptations of conventional storage bins to achieve the desired process activity. A wide variety of solids including chemicals, plastics, and sugar are processed in this way. Some of the processes carried out include heating, cooling, polymeric phase transformation, drying, curing, and suppressing or enhancing a particular chemical reaction. Purge and conditioning vessels often exhibit one or more of the following problems: nonuniform purge or condition, cross-contamination, lack of flow, erratic flow, and segregation. The key to solving these problems is to design the vessel for a mass-flow solids flow pattern, where all the material is in motion whenever any is withdrawn.

Tumble blending. Tumble blending by using portable containers (**Fig. 3**) has a number of advantages over other common blending techniques. There is no need to discharge the blender into the portable container, because at the end of the blending cycle the material in the container is blended. The segregation that can occur during filling of the portable container is thereby avoided.

Tumble blending also eliminates the costly step of blender cleaning between batches. All of the material is confined within the portable container; when cleaning is required, only that container needs to be cleaned, not the stationary blender.

Finally, tumble blending eliminates the downtime required to fill and empty a stationary blender, because the tumble station is ready to accept a new container as soon as the previous one has finished tumbling.

Design of the portable containers used in a tumble blending system is crucial to achieving the maximum benefits possible with such a system. Flow problems can be minimized, if not eliminated, by using containers designed for mass flow as opposed to funnel flow. The bin insert containers discussed above (Fig. 2) have particular advantages in terms of nonsegregated discharge.

Quality-control testers. Conventional testing equipment is not suitable for most quality-control applications because it requires considerable training and skill to operate, and consumes valuable production time. Quality-control testers overcome

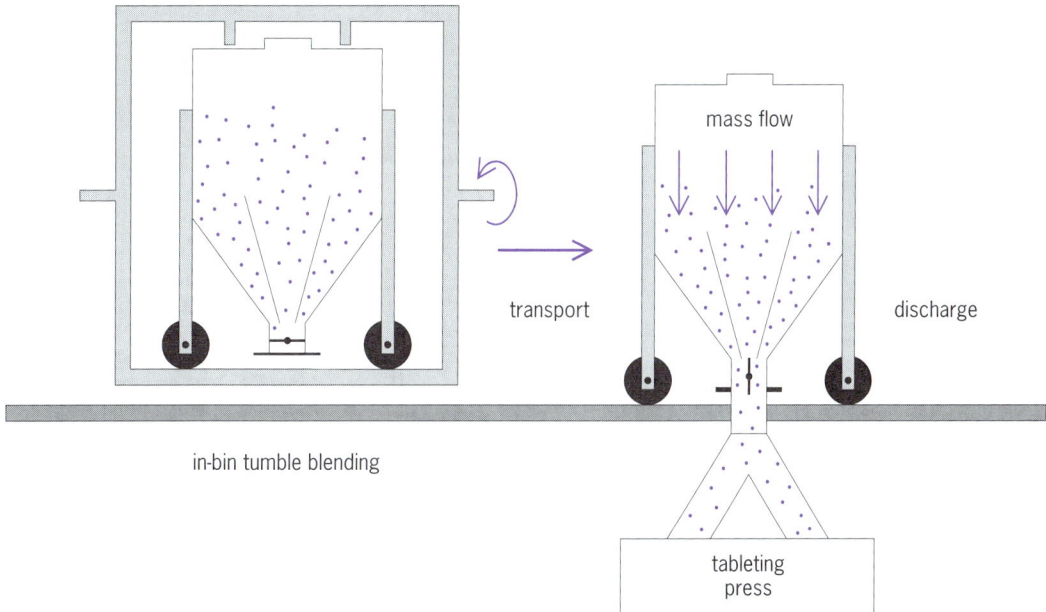

Fig. 3. Tumble blending operation using portable bin insert containers. The operation consists of three steps: in-bin tumble blending, transport, and discharge.

these disadvantages by quickly, simply, and reliably measuring a powder's relative flowability. Typical applications are to powder production or processing lines, incoming powders, and maintaining the consistency of packaging and die-filling machines.

These testers measure a powder's cohesive strength, which, as discussed above, indicates the potential for flow problems such as arching and ratholing. A simple accept-reject criterion is established for a given material. The test technique is simple and straightforward: one test takes about 5 min compared to several hours using conventional shear-test equipment.

For background information SEE BULK-HANDLING MACHINES; FLOW OF SOLIDS; FRICTION; QUALITY CONTROL in the McGraw-Hill Encyclopedia of Science & Technology.

John W. Carson

Bibliography. J. W. Carson and J. Marinelli, Characterize bulk solids to ensure smooth flow, *Chem. Eng.*, pp. 78–89, April 1994; J. W. Carson, T. A. Royal, and D. J. Goodwill, Understanding and eliminating particle segregation problems, *Bulk Solids Hand.*, 6:139–144, February 1986; J. Marinelli and J. W. Carson, Solve solids flow problems in bins, hoppers and feeders, *Chem. Eng. Prog.*, pp. 22–27, May 1992; J. Marinelli and J. W. Carson, Use screw feeders effectively, *Chem. Eng. Prog.*, pp. 47–51, December 1992.

Flow visualization

Computational flow imaging (CFI) is a technology for generating digital images of theoretical fluid dynamic phenomena in optical formats that mimic real observations of the corresponding real flow fields. Constructing the optical simulations requires the use of flow-field theory, principles of wave optics based on electromagnetic wave theory, treatment for sensor responses and background noise, sometimes principles of spectrophysics (theories for absorption, scattering, illuminance, spectroscopy, chemical gas kinetics, and molecular gas dynamics), sometimes principles of particle dynamics, and digital image processing. Computational flow imaging offers unique features that are impossible to obtain with optical methods alone because it is done completely in the computer. Although computational flow imaging is still in its early stages of development, its predictions for optical measurements are being used to plan experiments, to obtain understanding of real optical measurements and the accompanying errors, and to validate flow-field predictions of computational fluid dynamics and new optical diagnostic techniques. Computational flow imaging offers significant new benefits to the study of fluid mechanics, including the wherewithal to design, build, and validate data-processing algorithms; to optimize experiments and assess measurement expectations using pretest optical analyses; to assess the effects of errors and thereby establish bounds on uncertainty; and to perform code validation in the domain of nearly raw experimental data. But the wherewithal to actually see the visualization of a flow prediction in these new formats is the paramount benefit of computational flow imaging.

Visualization methods. Methods for flow visualization can be classified in three categories according to the bases of the phenomena being sensed: (1) changes in the refractive index, such as shadowgraph and schlieren photography; (2) electrooptically induced molecular transitions, such as planar laser-induced fluorescence (PLIF) and electron beam fluorescence (EBF); and (3) radiation scattered from tiny seed particles or molecules, such as particle image velocimetry (PIV) and Rayleigh scattering. Computational flow imaging establishes a fourth category that, unlike the others, is entirely theoretical. In place of experimental apparatus, computer models describing the interaction between optical waves and theoretical flow-field predictions are used to generate digital images that simulate the real observations. Particular formats are distinguished by appending the surname of the optical process to the acronym CFI, as in CFI-schlieren, CFI-PLIF, CFI-interferometry, or CFI-PSP (pressure-sensitive paint).

Practical issues in CFI. The concept of computational flow imaging is simple, but the practice is challenging because the goal is to generate images that are congruent with their optical counterparts. As such, accurate computational flow images must include the artifacts and noise of real optical data as well as the physics of the wave-flow interactions. Aberrations and imperfections common to optical data, treatment for the information lost by partitioning optical signals into matrixes of discrete values (pixelization), and coping with mismatched comparisons between instantaneous real images and synthetic images based on time-averaged or steady-flow predictions are typical issues of concern. Essentially, all the factors that separate truth from measurement and truth from prediction need to be understood in order to construct optical replications of nature.

Development of CFI. In the mid-1970s, the concepts of computational flow imaging were used to study high-speed compressible flows, to validate comparable flow-field predictions, to estimate numerical error in data-reduction algorithms, and to assess the effects of noise and flow anomalies in optical measurements. The computational flow imaging formats were line-drawn graphical plots of interferometric fringe centers or of flow-field isopycnals (density contours) because digital image processors did not exist (**Fig. 1**). In the early 1990s, high-resolution digital CFI shadowgraph, schlieren, and interferometric images strongly resembling the continuous-tone photographic flow visualizations of the corresponding flows were produced (**Fig. 2**). Optical formats from the other two categories, molecular excitation and particle seeding, have also

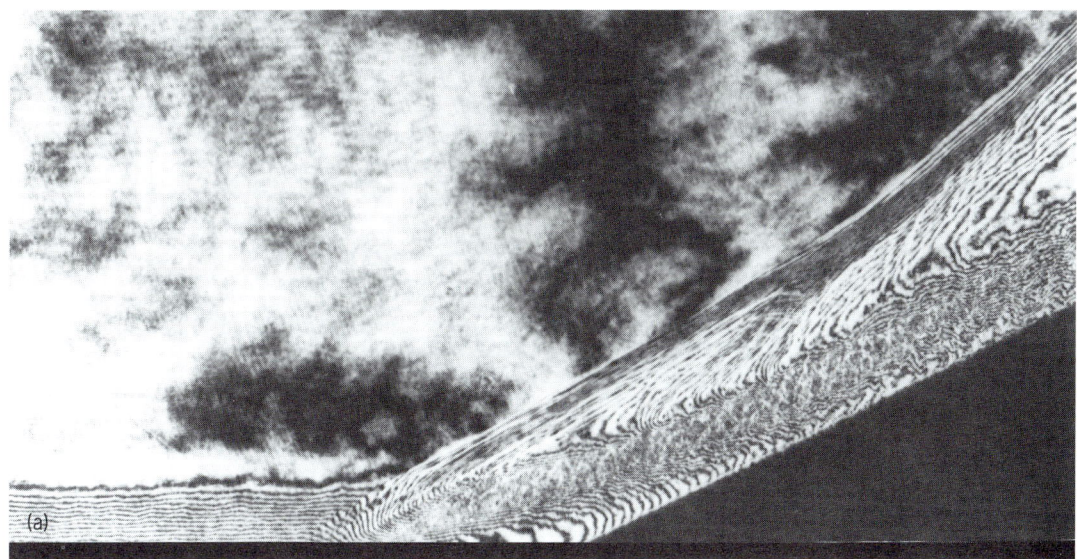

Fig. 1. Supersonic turbulent boundary-layer flow separation over a two-dimensional compression turn. (*a*) Holographic interferogram (*from A. G. Havener and R. J. Radley, Jr., Turbulent boundary-layer flow separation measurements using holographic interferometry, AIAA J., 12:1071–1075, 1974*). (*b*) Computational flow imaging line plot for the computed isopycnals (*from J. Shang and W. L. Hankley, Numerical solution for supersonic turbulent flow over a compression ramp, AIAA J., 13:1368–1374, 1975*).

been constructed. For instance, CFI-PLIF images aiding the development of computational fluid dynamics and PLIF applications to high-enthalpy flow fields have been made (**Fig. 3**), and CFI-PIV images are used regularly to assess camera resolutions and data-reduction algorithms. CFI-PSP and CFI-DFS (digital Fourier schlieren) formats are being developed to help implement and extend both diagnostics to a wide class of wind-tunnel applications.

Benefits of CFI. Image enhancement, pretest planning, and code validation are three important benefits of computational flow imaging.

Image enhancement. The value of image enhancement lies in the opportunity to gain knowledge from the computer capabilities unique to digital image processing. False color painting an image, for example, heightens visual awareness of subscale details and subtle boundaries between changing conditions. Computer zooming and perspective viewing allow enlargements, images of single planes or windowed regions of interest, and off-axis views of specific flow structures to be made. In the future, computational flow imaging may be used to create virtual-perspective views of theoretical flow fields, a feature akin to flow-field virtual reality. These computer-controlled capabilities provide new opportunities to obtain different insights on the fluid mechanics of some classic flow-field problems such as transition, separation, and shock–boundary layer interactions.

Pretest planning. Used in conjunction with flow-field predictions, computational flow imaging can help reduce costs and test requirements. From

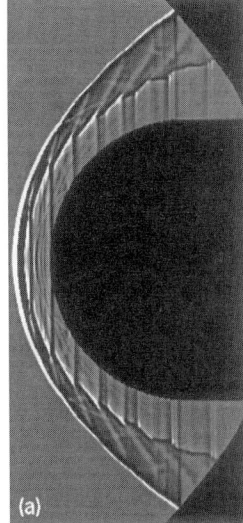 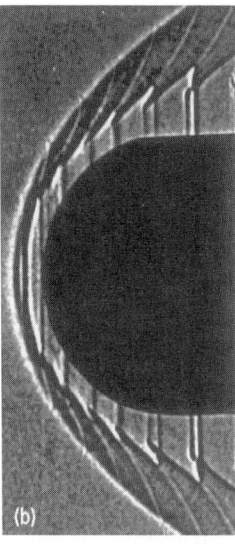

Fig. 2. Unsteady shock-induced hydrogen-air combustion for a hypersonic spherically blunt cone flow. (a) CFI shadowgraph. (b) Optical shadowgraph measurement. (*From L. A. Yates, Images constructed from computed flowfields, AIAA J., 31:1877–1884, 1993*)

pretest computational flow imaging analyses, enhancements of system designs and optimum test conditions are identified and thereafter used as guides for conducting the actual test, and the pretest computational flow imaging data also provide a basis upon which measurement uncertainty factors are identified. For example, a pretest CFI-interferometric analysis of high-frequency instabilities in hypersonic boundary layers revealed flow conditions for null measurements, test conditions that were subsequently omitted from the experimental study.

Code validation. Comparing experimental data to predictions for temperature, pressure, density, or velocity is the crux of code validation. The normal and preferred procedure involves reducing a collection of optical measurements to one or more flow properties. However, the mathematical correspondence between the optical measurements (for example, light intensity and phase) and the flow properties is usually nonlinear, so reducing the optical measurements is often difficult, and sometimes impossible. Computational flow imaging offers an alternative because, although the calculations needed to generate the images can be tedious and encumbered with complicated formulae and grid-mapping issues, the optical data and thus the images are always computable from the theories, a task reasonably well posed for a computer. Moreover, many of the questionable assumptions and shortcomings inherent in data reduction are avoided by using the computational flow imaging approach.

The previously mentioned CFI-PLIF application is a case in point. PLIF technology requires that a thin sheet of monochromatic laser radiation be

Fig. 3. Supersonic (Mach 2) nitrogen (N_2) gas flow over a 10° spherically blunt axisymmetric cone. Bow shock wave is clearly visible ahead of cone. (a) PLIF measurement of fluorescence from nitric oxide (NO) that was seeded in the nitrogen gas. (b) CFI-PLIF image. (*From W. Ruyten et al., Computational flow imaging for planar laser induced fluorescence applications (CFI-PLIF), AIAA Pap. 94-2621, 18th Aerospace Ground Testing Conference, Colorado Springs, Colorado, June 20–23, 1994*)

used to excite a particular molecular species in a flow. Through optical absorption, the excited molecules are redistributed into higher vibrational and rotational energy levels. As these molecules rapidly transition to their electronic ground state, they release a near-instantaneous radiant emission called fluorescence. The magnitude of the fluorescence is a function of the number density of the excited molecular species and the temperature that defines the electronic ground state. Temporarily putting aside other factors that affect the observed fluorescence (for example, uniformity of the input laser energy, molecular quenching, background irradiance, and detector responses), a single PLIF image cannot be reduced completely because there is only one measurement, optical intensity, for two unknowns, number density and temperature. Experimental methods are being developed to overcome this problem, but for now the approach is to estimate the temperature and then reduce the number densities from the PLIF measurement. However, the uncertainty in the estimated temperature makes this approach unsatisfactory, so the computational fluid imaging approach was used to generate CFI-PLIF images that compared well to the corresponding PLIF measurements. The application was for the spectrophysical response of nitric oxide in a high-temperature supersonic flow over a spherically-tipped axisymmetric cone (Fig. 3). Without the CFI-PLIF approach, meaningful data comparisons aiding the development of both the computational fluid dynamics prediction and the PLIF diagnostic process would have been impossible.

New formats. CFI-PSP and CFI-DFS are developing applications that offer new benefits to aerodynamic wind-tunnel testing.

CFI-PSP. In the PSP technique, ultraviolet light is used to excite specific molecules in a special paint affixed to a test surface positioned in a wind-tunnel flow. As in PLIF, the optically excited molecules emit radiation when they decay from their elevated energy states; however, the molecular transitions are different and the emission is a relatively long-duration visible glow called phosphorescence. Interestingly, oxygen molecules in direct contact with the paint optically quench the intensity of the phosphorescence. For wind-tunnel flows, the amount of oxygen in contact with the paint is a function of the aerodynamic static pressure, so a calibration of phosphorus glow and pressure allows acquisition of highly resolved spatial distributions of surface pressure across the sensitized paint. But as in PLIF, many factors influence successful use of PSP, among them temperature, curvature, look angle, illumination, and paint thickness, and CFI-PSP is being used to study these effects.

CFI-DFS. Fourier schlieren imaging relies on the principle that the distribution of optical intensity in the focus of an imaging optic (a positive lens or a concave mirror) is mathematically equivalent to the power spectrum of the Fourier transform of the imaged optical waves. In wind-tunnel schlieren systems, the second schlieren optic brings the light transmitted through the tunnel to a focus. A knife edge positioned in the focus is used to sense a reduction in illumination by blocking specific light rays that are refracted by the flow-field density gradients. Physically, the second schlieren optic produces the analog Fourier transform of the light, and the knife edge is the analog application of a half-plane or bandpass filter.

In CFI-DFS, the schlieren process is performed synthetically in the computer. The discrete two-dimensional Fourier transform of the optical waves transmitted theoretically through a predicted flow field is used to generate the optical simulation of the focused light, digital filtering is used for the knife edge, and an inverse discrete Fourier transform is used to obtain the schlieren image. However, unlike the single knife-edge filter signal processing theory and CFI-DFS allow a variety of digital filters to be designed and used to study optical effects related to specific spatial frequencies. Noise and unwanted background disturbances may be removed, whereas other filters may be designed to enhance particular flow-induced frequency effects. DFS can also be applied to the analyses of real shadowgraph images, but CFI-DFS is needed first to develop filters that do not destroy useful information convoluted in the shadowgraphs.

For background information SEE FLUORESCENCE; INTERFEROMETRY; PHOSPHORESCENCE; SCHLIEREN PHOTOGRAPHY; SHADOWGRAPH; SHOCK-WAVE DISPLAY; SIMULATION in the McGraw-Hill Encyclopedia of Science & Technology.

George Havener

Bibliography. American Institute of Aeronautics and Astronautics, *Proceedings of 18th Aerospace Ground Testing Conference,* Colorado Springs, Colorado, June 20–23, 1994; L. A. Yates, Images constructed from computed flowfields, *AIAA J.,* 31:1877–1884, 1993.

Food engineering

Recent research in commercial food processing involves the use of high hydrostatic pressure, ohmic heating for continuous sterilization of foods, and magnetic resonance imaging for the study of physical and chemical changes.

Processing at Ultra-High Pressures

The food industry is continually interested in new and innovative methods for the processing of food products. More efficient and economical processes are sought, as well as products that resemble their fresh counterparts in flavor, texture, appearance, and nutrient value while maintaining shelf life and safety. High hydrostatic pressure is a rediscovered food process used by the food industry in the preservation and processing of foods and beverages.

High hydrostatic pressure. Food processing at high hydrostatic pressure is analogous to the conditions found in the ocean depths. For example, in the deepest point in the Marianas Trench of the Pacific Ocean, pressures of approximately 1100 atmospheres (around 16,170 lb/in.^{2} or 110 megapascals) are obtained, but even in this hostile environment life exists. Pressures in the range of 1500–10,000 atm (around 150 MPa–1 gigapascal) have been examined for use in the treatment of foods. This technology has been evaluated for the preservation of foods, that is, for the inactivation of spoilage microorganisms and food-borne pathogens, as well as enzymes degradative to product freshness. Examples of other applications of hydrostatic pressure for foods are pressure-induced improvements in the functional properties of food components that lead to enhanced or unique sensory characteristics, improved digestibility of pressure-treated foods, and the tenderization or decolorization of meats from exposure to relatively low levels of pressure. Desirable changes in foods thawed under pressure are also being examined.

History. The concept of using hydrostatic pressure for the preservation of foods is not new. In the 1890s, B. Hite, a chemist at the Agricultural Experiment Station in Morgantown, West Virginia, first evaluated this idea. He built a machine capable of generating pressures in excess of 7000 atm (700 MPa). The foods were packaged in metal cans or tubes. He and his coworkers conducted storage studies with a wide range of foods and beverages to evaluate the effectiveness of pressure treatment as a means of preserving foods. High-acid products appeared to have the most promise for long-term preservation using hydrostatic pressure. The response of bacteria, fungi, and viruses to pressure treatment was also examined.

In the early 1980s, D. Farkas resurrected this technology, which was aggressively pursued in Japan where pressure-treated foods have been available since 1991. These products include fruit preserves and jams, and more recently, salad dressings and grapefruit juice. In North America and Europe, no commercial products have been released that incorporate a high hydrostatic pressure treatment.

Principles. Cooking involves energy transfer, that is, the formation of a gradient as a result of heat transfer through the product. This transfer takes time. The application of hydrostatic pressure is uniform and instantaneous throughout a food; the processing times for commercial batch treatments are usually in the range of 10–20 min. For some foods, hydrostatic pressure has the potential to replace or minimize the use of cooking temperatures or chemical preservatives to process a food or food ingredient.

Inactivation of microorganisms. To effectively preserve food, the pressure treatment must at least partially inactivate the microorganisms that are present. Hydrostatic pressure will kill microorganisms, but microbial sensitivities to pressure vary, and the intrinsic parameters of the food will greatly affect the response of microorganisms to the pressure treatment.

High hydrostatic pressures induce a number of changes to the morphology, biochemical reactions, genetic mechanisms, and cell membrane of microorganisms; however, it is generally believed that the immediate death and injury inflicted upon microorganisms at pressures above 1000 atm (100 MPa) involve disruption of the membrane accompanied by denaturation of critical proteins associated with it. Under pressure, a reduction in volume of the membrane bilayers occurs along with a reduction in the cross-sectional area per phospholipid molecule. Pressurized membranes normally show altered permeabilities. If the membrane is extensively permeabilized, cell death results.

Protein reactions. Protein denaturation has long been recognized as a result of the application of high hydrostatic pressure to foods. Raw red meats exposed to pressures of 3000–4000 atm (300–400 MPa) show surface protein denaturation similar to that resulting from light cooking. Egg whites will irreversibly denature into a solid gel (like in a fried egg) when exposed to 6000 atm (600 MPa) for 4 min. Pressure appears to affect biochemical reaction systems by reducing the available molecular space and by increasing interchain reactions. Although covalent bonds are not affected by pressure, hydrogen bonds and hydrophobic interactions are; therefore, it is the disruption of hydrogen bonds and hydrophobic interactions and their reassociation with new groups that occurs in the unfolding of proteins from their native state.

Inactivation of enzymes. Enzymatic reactions are inactivated by pressure because of denaturation, but some enzymes are very resistant to pressure-induced denaturation; under some pressure conditions enzyme-catalyzed reactions can be enhanced or modified because of conformational changes to the active site of the enzyme and surface changes to the substrate. For some pressure-resistant enzymes, such as pectinesterases and polyphenol oxidase, a blanching step may be necessary to ensure inactivation of these spoilage enzymes during the pressure processing of foods.

Practical considerations. To date in Japan, Europe, and North America, many if not most common foods and beverages have been examined for potential application of pressure treatment for commercialization. For the majority of these food items, an unresolved limitation of pressure treatment for food preservation has been the inability to inactivate bacterial endospores at neutral pH. A combination with other accepted methods of food processing is usually required. Extreme pressures alone will inactivate spores of *Bacillus* and *Clostridium* (well beyond 4000 atm or 400 MPa), but these high magnitudes appear commercially unfeasible and can result in detrimental changes to food quality. In products naturally possessing an acidic pH, or processed foods that can be acidulated to a pH less than 5.0, growth of sporeformers can be minimized or eliminated; however, as with canned foods, a pH of 4.6 is the critical boundary for food safety. It is generally accepted that *C. botulinum* cannot grow below a pH of 4.6, and foods in this category cannot cause botulism, a food-borne disease with lethal consequences. Therefore, foods with a pH less than 4.6 (more acidic) should be safe from outgrowth of spores of *C. botulinum* and other spore-forming bacteria; pressure treatment of such foods would be directed against acid-tolerant non-spore-forming bacteria and fungi principally involved in the spoilage of acid foods.

Besides pH, other factors that impact upon the effectiveness of the pressure treatment are the types and concentrations of the food microflora, water activity, redox potential and nutrient concentration of the food, presence of antagonistic (preservativelike) compounds, as well as the atmospheric composition and temperature during treatment and product storage. Given the multitude of factors in such a new approach to food processing, most of the work has been empirical.

A number of other important factors enter into the calculations. Cost factors, which include capital investment for new equipment, fees associated with securing possible regulatory approval, profit limits, and marketing costs to ensure consumer acceptance, are such that foods produced from the application of hydrostatic pressure technology will span a limited range of products, at least until the technology matures.

Acid foods are likely candidates for this technology since sporeformers cannot grow in these products. Another example of a commodity food group that appears to be a possible match with pressure preservation is raw seafood. Consumers of products such as oysters and sushi prefer the aroma, flavor, and mouthfeel of the raw, unfrozen product. Pressure treatment would preserve the sensory characteristics of the raw product but inactivate parasites and food-borne pathogens, such as *Vibrio parahaemolyticus*, that pose health risks. However, the question of whether the consumer of raw seafood would pay extra for this processing step to enhance product safety is unresolved.

The status of high hydrostatic pressure to food processing is now one of development and commercial exploration. New products and applications continue to be investigated. Any commercial food application in the United States will probably be combined with other established processing methods for shelf-life extension. For example, the combination of low pH, mild heat, and hydrostatic pressure appears to be a viable approach in the processing of some foods to allow extended shelf life and safety with organoleptic properties closer to fresh.
Dallas G. Hoover

Ohmic Heating

Ohmic heating involves passing an alternating electric current through a food product, which then heats up internally, and thus the food behaves as a heater element in an electric range or a space heater. The rate of heating depends on the voltage applied, the distance between the electrodes, and the properties of the food. For example, foods with ionic constituents such as salt would heat rapidly under ohmic treatment. Products containing high proportions of nonionic constituents, particularly fats and oils, generally do not heat well by this method. Depending on the heater design and the product, it is possible to heat foods at extremely rapid rates by ohmic heating. Further, if the product is properly formulated, the heating is remarkably uniform, with all parts of the food achieving nearly the same temperature. By contrast, conventional heating is considerably less uniform, because foods heat from the outside after contact with a hot medium. Microwave heating also suffers from problems of nonuniformity because of complicated standing wave patterns and loss of power as the waves enter the food.

History. Ohmic heating is not a new concept; however, successful applications are somewhat more recent. A number of patents showing that flowable materials may be heated by passage of an electrical current date back to the 1880s. In fact, electric pasteurization of milk was in practice in six states in the United States in the late 1920s and early 1930s. The technology appears not to have been successful apparently because of a number of factors, including the lack of suitably inert electrode materials and appropriate controls as well as the high cost. Also, at the time it was thought that bacteria were killed by electricity rather than by the heating effect. It is now known that extremely high electric field strengths (far greater than those used for ohmic heating) are required to inactivate microorganisms by electricity alone, and any bacterial destruction by ohmic heating is due purely to the heat generated.

Technologies for sterilization. One potentially promising application of ohmic heating is in continuous sterilization and aseptic packaging of foods. Its advantages need to be understood in relation to canning and conventional aseptic processing technology.

Canning technology. The concept of heating a packaged food to extend its shelf life was developed by N. Appert in the early 1800s. Numerous advances and refinements have been made but the basic approach has not changed. The technology involves filling a container with a food product, sealing under a vacuum, and heating for a sufficient time at sufficiently high temperature to ensure the absence of disease-causing microorganisms from even the coldest regions of the food. Although a sterile product is safe for human consumption, and can last for up to 1 year without refrigeration, the severe heat treatments required for sterilization can sometimes degrade quality. In addition, although the metal can is robust and an effective barrier against oxygen, it is heavy, expensive, and requires considerable warehouse floor space for storage of empty containers.

Continuous sterilization. This process involves aseptic processing and aseptic packaging. In continuous sterilization, foods and packages are sterilized separately, and filling and sealing is accomplished in a sterile environment. Products are typically sterilized in a continuous manner by using heat exchangers with steam or hot water as the heating medium, while packages may be sterilized by a variety of methods, the most common being exposure to hydrogen peroxide. The advantages include continuous operation, lower package costs, potentially improved product quality, and greater convenience. Recent developments in polymer/paperboard/metal foil laminates have resulted in the development of packages that possess the advantage of lower weight and cost, increased convenience in handling, as well as recyclability of most package constituents. Unlike traditional canning, where heating can be extremely slow for large can sizes, aseptic processing involves smaller heat transfer dimensions, and therefore potentially more rapid heating and higher end-product quality.

This technology has been successful for liquid food products, including milk, fruit juices, and viscous liquid products. However, continuous sterilization of products containing large solid pieces (such as stews or soups) poses a considerable challenge because, unlike canning, it is not presently possible to measure temperatures within solid particles moving continuously through process equipment. Thus temperatures must be predicted theoretically by using mathematical modeling, and verified by using biological indicators. Since consumer safety is the foremost consideration, conservative approaches are necessary to ensure that all parts of the food are sterile. The result is that in many cases the product may be overcooked to ensure product safety, making it difficult to realize the promise of high product quality associated with aseptic processing.

Perhaps the principal difficulty with this approach is that the solids are not heated directly but rather by first heating the liquid carrier medium, which then transfers heat to the solid. If the solid pieces are large, heat is transferred slowly to their centers by conduction, resulting in a product that may have an overprocessed liquid component, even though particle centers may not be sufficiently heated to be sterile. Relatively few products worldwide are processed in this manner.

Ohmic system. Unlike conventional heat exchangers, ohmic heating generates heat within the food; thus it is possible for the interior of solid particles to heat at the same rate or even faster than the surrounding fluid. Products of far greater quality than those obtainable by conventional processing can result. Studies show that foods in the form of solid-liquid mixtures can be heated with a remarkable degree of uniformity by using this method. Ohmic heating appears to work well with products having high concentrations of solids.

The critical parameters affecting the rate of ohmic heating are the applied voltage, the distance between the electrodes, and the electrical conductivity of the food. Since the electrical conductivity can be altered by varying food ingredients, it is possible to develop foods suited to ohmic heating. It is possible, for example, to adjust the salt content in a mixture to enable it to heat rapidly in an ohmic heater, and subsequently to reduce the salt concentration of the final product before packaging by mixing with a separately processed unsalted sauce.

Since the electrical conductivity of foods generally increases with temperature, heating actually accelerates as the material warms up. In contrast, in conventional processing heating is initially rapid but slows down as the product temperature approaches that of the heating medium. Conventional heating also involves an upper temperature limit (that of the medium), whereas ohmic heating has no such limit.

Design. Ohmic heaters may be made in a variety of different configurations; however, some popular configurations involve flow between parallel plates that serve as electrodes, so that current flows across the product flow path; in a variation, current flows through a tube with electrodes placed in line with the flow so that the current flows parallel to the product. A number of other designs are available or are in development.

Products. Currently, only one company worldwide has been producing sterile products by using ohmic technology because research has been conducted in only a few laboratories. However, industrial research in this area is intensifying and, if successful, may result in a variety of ohmically processed products becoming available.

Sudhir K. Sastry

Magnetic Resonance Imaging

The processing of foods often involves moisture removal, heating, mixing, or size reduction. The understanding of these processes is often limited because only the initial and final products can be studied. Magnetic resonance imaging (MRI), a noninvasive medical diagnostic technique, has recently been used to study the transfer of moisture in foods, the heating of foods, and size reduction as these processes are occurring. Thus, there is new information on the physical and chemical changes occurring during food processing. More importantly, this information assists in developing new types of engineered foods such as low-fat spreads and dressings.

Principles. Nuclear magnetic resonance spectroscopy is an experimental technique based on the phenomenon of magnetic resonance. Atomic particles, electrons, and the nucleus interact with an applied external magnetic field. This interaction between the particles and the magnetic field resembles the response of iron fillings placed in a magnetic field; a magnetic field is induced in the particles and the fillings become oriented. These systems display resonance, because they emit and absorb energy at specific frequencies. The frequency depends upon the atomic particle and the strength of the applied magnetic field. When the atomic particle is the nucleus, the phenomenon is known as nuclear magnetic resonance. Only nuclei with magnetic moments exhibit magnetic resonance. Common nuclei in foods with magnetic moments include hydrogen-1 (^{1}H), carbon-13 (^{13}C), sodium-23 (^{23}Na), and phosphorus-31 (^{31}P); nuclei without magnetic moments include carbon-12 (^{12}C) and oxygen-16 (^{16}O).

In applications of nuclear magnetic resonance spectroscopy and imaging in the medical and food industries, the term nuclear is often omitted so as to prevent confusion with experimental techniques that use radioactive materials.

Nuclear magnetic resonance is a particularly useful spectroscopy for the study of foods because it is sensitive to a number of material properties. These material properties include moisture and fat content, rheological properties, thermal properties, phase of the material, and motion of the material. Nuclear magnetic resonance provides a bulk value of material properties. Magnetic resonance imaging is an extension of nuclear magnetic resonance that enables spatially resolved measurements of a property. This spatial resolution can be in one, two, or three dimensions. A magnetic resonance image of an ear of corn freezing is shown in the **illustration**.

Magnetic resonance images are usually displayed in a gray scale format with bright regions indicating more signal. For instance, in the illustration the signal is primarily from water. As the ear of corn freezes, the ice that is formed does not give an observable signal in this experiment; hence, the ear of corn seems to shrink or disappear because of the advancing of the ice interface.

The core of a magnetic resonance imaging spectrometer is a magnet that acts to polarize the nuclei in the sample. The sample is contained within a probe that couples the spectrometer to the sample. The probe enables energy to be transferred to the sample and energy emitted by the sample to be recorded. The spectrometer also includes radio-frequency hardware, amplifiers, analog-to-digital converters, and a computer system.

Mass and energy transport. The data from magnetic resonance imaging can be obtained during processing by using custom-designed prototypes of processing equipment that are compatible with strong magnetic fields. The custom equipment allows for studies of changes occurring during mixing, heating, cooling, drying, or pumping.

Emulsions. In the manufacture of many foods, such as salad dressings, ice cream, mayonnaise, butter, and margarine, an emulsion is formed. An emulsion is the dispersion of one liquid as small droplets in another liquid, for example, oil droplets in water. The quality and processability of these foods depend upon the stability of the emulsion. Destabilization of the emulsion occurs when the

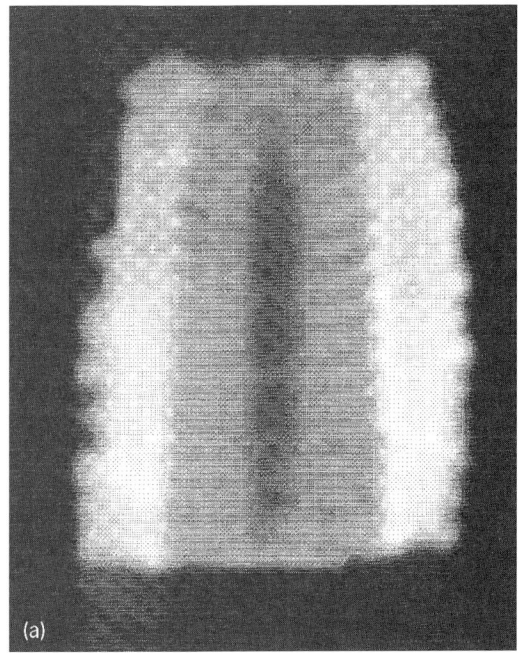

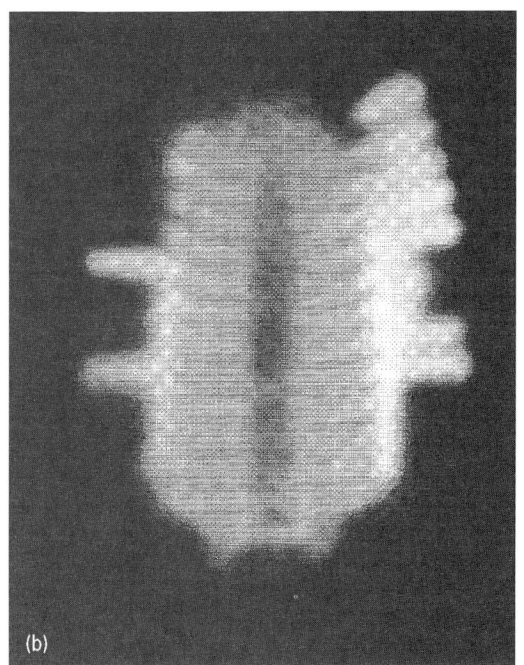

Magnetic resonance image of the unfrozen moisture in an ear of sweet corn during air blast freezing. (*a*) At the beginning of the experiment. (*b*) After the formation of ice, which does not give an observable signal.

dispersed phase separates to either the top or bottom of the container. Monitoring of the stability of emulsions has traditionally been performed by invasive techniques, which either destroy or disturb the sample. Magnetic resonance imaging provides a noninvasive method for studying emulsions and provides data on volume fractions and particle size of the dispersed phase as a function of height and time after emulsification. Current theoretical analysis of emulsion stability relies on the use of Stokes' law or its modifications, and analysis and testing of these relationships is just beginning. Data available from measurements made with magnetic resonance imaging, in combination with other methods, will provide important tests of these theories as well as data for proposing new models.

Dehydration. The removal of moisture from foods has been used for thousands of years to preserve wholesomeness and to make new foods. However, little is known about how the structure of the food changes and what role this has in controlling the drying process. Magnetic resonance imaging is being used to measure both the movement of moisture and the structure of foods as they dry. By improving the understanding of changes in foods during drying, it will be possible to improve the efficiency of drying operations and optimize dried food quality. This approach will also be important for baked goods, such as cookies and crackers, where significant amounts of moisture removal occur during processing.

Studies of the drying of corn kernels have revealed that the development of stress cracks is related to internal moisture gradients. The paths of moisture loss were observed to be through the glandular layer of the scutellum and from the surface through internal layers.

Temperature profiles in solid foods. Magnetic resonance imaging is used to measure temperature within solid particles during processing by calibrating the mobility of nuclei with temperature. In the case of a phase change from liquid water to ice, the mobility of the water molecule is greatly reduced. When the mobility of the molecule is reduced, the signal from this molecule in a magnetic resonance imaging experiment decays very rapidly and is not recorded; thus the loss of signal occurs in the image where ice is being formed. Most importantly, the corn image (see illus.) shows that the freezing process is not always characterized by a well-defined freezing interface. In an ear of corn, each kernel may freeze at a slightly different temperature as a result of a different composition. Additionally, this type of image can be used for process control of the food freezing operation.

Other magnetic resonance imaging methods for measuring temperatures in solids are actual calibrations of mobility with temperature. The calibrations are based on the change in either the signal decay rate in the experiment or the self-diffusion coefficient of the water molecule (or other probe molecule in the system). Methods based on these techniques can measure temperatures in solid foods during aseptic processing or ohmic heating with an accuracy of ±1°C (1.8°F). The current limitation is that the solid particle must be stationary during the experiment. By combining the temperature profiles in solids and measurements of residence time in the heating section of the process, food processors will be able to optimize the process to improve the quality of the product.

For background information SEE EMULSION; FOOD ENGINEERING; FOOD MANUFACTURING; MAG-NETIC RESONANCE; NUCLEAR MAGNETIC RESONANCE (NMR) in the McGraw-Hill Encyclopedia of Science & Technology.

Michael J. McCarthy

Bibliography. D. G. Hoover, Pressure effects on biological systems, *Food Technol.*, 47(6):150–155, 1993; D. G. Hoover et al., Biological effects of high hydrostatic pressure on food microorganisms, *Food Technol.*, 43(3):99–107, 1989; M. J. McCarthy, *Magnetic Resonance Imaging in Foods*, 1994; S. K. Sastry and S. Palaniappan, Ohmic heating of liquid-particle mixtures, *Food Technol.*, 46(12):64–67, 1992; G. W. Schrader, J. B. Litchfield, and S. J. Schmidt, Magnetic resonance imaging applications in the food industry, *Food Technol.*, 46(12):77–83, 1992; R. P. Singh and F. A. R. Oliveira (eds.), *Minimal Processing of Foods and Process Optimization: An Interface*, 1994.

Forest

Collecting, regularly updating, and distributing information on the extent and distribution of forest cover, deforestation, afforestation, biomass, productivity, and other related parameters are essential for forest planning, management, and public understanding and decision making. Monitoring is the mechanism that provides, on a continuous basis, the baseline and rate of change information for the forest and forest ecosystem. In forest inventories there are two categories of information: area information (such as the extent of forests and distribution) and area-related tree information (such as species and timber volume). Area information is normally assembled with the help of available topographic maps, aerial photos, and satellite image data, whereas area-related information such as timber volume estimation is completed on the basis of measurements taken on the ground by field survey parties.

Remote sensing. An essential tool for forest cover monitoring, remote sensing is the acquisition without physical contact of information about an object. Currently, remote sensing in forestry covers an array of techniques such as aerial photography, airborne videos, multispectral scanners, and RADAR (radio detection and ranging). The value of remote sensing in forestry was realized in the late nineteenth century, when a German forester took a single photograph of a forest region in Germany from a balloon. In 1920, for the first time, a Canadian company used vertical aerial photographs taken from an airplane for the preparation of forest stock maps. Subsequently, aerial photography was used to map tropical forests of the Irrawady delta in Burma (now Mynamar). Aerial photographs are now used routinely to prepare forest-type maps both at small and large scales, to stratify forest types, and to predict timber volume per unit area on the basis of photomeasurable parameters such as stand height and crown density.

The use of sequential aerial photography can be effective for monitoring changes in forest cover. However, the high cost of photoacquisition and the time required for interpretation and preparation of derived maps are determents to their use, particularly for large areas.

Satellite sensing. Satellite remotely sensed data, with its synoptic view and repetitive coverage, provides the best approach for monitoring changes in forest cover over large areas. The Earth Resources Technology Satellite (*Landsat 1*), the first satellite designed specifically to collect data of the Earth's resources, was launched in 1972. Since then, Landsat satellites have been obtaining multispectral imagery and transmitting raw data to ground stations in different parts of the world. *Landsat 1, 2,* and *3* carried multispectral scanner sensor systems that measured reflected solar radiance from the Earth's surface at a spatial resolution of 187 ft × 259 ft (57 m × 79 m) in four discrete bands in the visible and near-infrared regions of the electromagnetic spectrum. The data set is organized into a matrix of rows and columns. The individual elements of this matrix are referred to as pixels (picture elements), and consist of four values, one for each spectral band. *Landsat 4* and *5* carried nearly the same multispectral scanners as *Landsat 1, 2,* and *3,* but in addition they carried a scanner called the thematic mapper, which had an improved resolution of 98 ft × 98 ft (30 m × 30 m; thermal band 394 ft or 120 m) and operated in seven somewhat narrower spectral bands, including the middle and thermal infrared parts of the spectrum. The Landsat program is the longest running exercise in the collection of terrestrial data from space. More than 3×10^6 images have been acquired. In 1986, Landsat was joined by French *SPOT 1* (système pour l'observation de la terre), the first in a continuing series of SPOT satellites, and in subsequent years, a number of other satellite systems. The resolution of data ranges from high, 33 ft (10 m; SPOT) to coarse, 13,124 ft (4 km; NOAA AVHRR). Data from the optical sensors are complemented by meteorological and radar satellites. In addition, Russian (Priroda) satellites have been collecting space photography of 16-ft (5-m) resolution.

Data analysis and interpretation. Satellite data are available in standard photographic formats and also in digital form for further interpretation and analysis. Visual interpretation of satellite imagery is done on the basis of tone, texture, and context similar to photointerpretation of aerial photography. In computer-aided analysis, images are treated as two-dimensional arrays of numbers, and results of the computer processing are new arrays of numbers representing enhanced images or thematic classifications.

The basic principle of satellite sensing of Earth surface features is based on the characteristic spectral response of various cover types in different parts of the electromagnetic spectrum. The spectral response is recorded by the sensors and converted into appropriate form for interpretation. For example, vegetation absorbs visible light but reflects infrared light, bare soil reflects both visible and infrared light, and water absorbs both visible and infrared light. The repetitive coverage provided by satellites offers the possibility of obtaining data over large areas for different dates and seasons.

Forest assessment. Applications of satellite data to forestry could be broadly categorized into forest-type mapping, forest-cover monitoring, forest-fire-danger monitoring, damage assessment, forest seasonality study, vegetation production estimation, and wildlife habitat analysis. The results obtained through the use of satellite data in Brazil, India, Thailand, and elsewhere have attracted worldwide attention toward the adverse impacts of tropical deforestation (see **illus.**). Estimation of the rate of tropical deforestation with ground-based surveys is almost impossible because the regions where changes are occurring are very large, changes are occurring rapidly, and information is not readily accessible. Remote sensing has also been used for measuring gaseous emissions from biomass burning and reducing uncertainty in the estimate of terrestrial sources and sinks of carbon.

Microwave sensing systems. Because the bulk of remote sensing data is acquired by using passive systems that record solar radiation reflected from Earth's surface, their ability to provide complete coverage of the tropical regions of the world throughout the year is limited by the frequent cloud covering of these areas. Remote sensing satellites such as *ERS 1* (European Space Agency), *JERS 1* (Japan), *Almaz 1* (Russia), and the *RADARSAT* (Canada; scheduled to be launched in 1995), with a spatial resolution of 30–330 ft (10–100 m), carry synthetic aperture radar (SAR), a powerful microwave instrument that can penetrate clouds and darkness, obtaining detailed images of Earth. In the future, data from these microwave satellite systems will be increasingly utilized in global forest monitoring.

Limitations. Although remote sensing is a valuable tool for forestry studies, it is not without limitations. Data are not available for many areas of the world (particularly in the tropics); clouds and atmospheric conditions contaminate imaging; computer processing of data requires substantial investments in computer hardware and software and training; a direct estimation of forest biomass is not possible; and the commercialization of high resolution data in the mid-1980s has made its accessibility costly for researchers.

Future trends. At present, more than 12 satellites provide data about Earth and its environment. A number of missions are planned which will help to improve the understanding of the global environment. In 1996, the VEGETATION instrument (*SPOT 4* and *5*), specifically designed for the monitoring of the continental biosphere, will provide appropriate information for land-cover change. Toward the end of the century, the National Aeronautics Space Administration (NASA) Earth Ob-

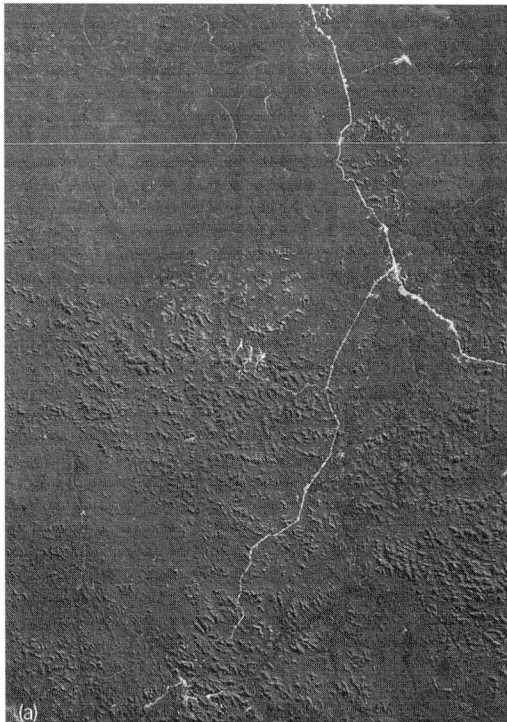

Deforestation monitoring in Rondonia, Brazil, using Landsat multispectral scanner images. (*a*) June 19, 1975. (*b*) August 1, 1986. The fishbone designs indicate the pattern of forest clearing due to settlements. (*EROS Data Center, South Dakota*)

serving System (EOS), consisting of a series of satellites, will provide high-resolution spectral data capable of making comprehensive observation of Earth from space. In addition to a number of government-sponsored satellite systems, some private ventures plan to build satellites that will carry up to 9.8-ft (3-m) resolution sensors with a revisit time of about 2 days. It is anticipated that the cost of data acquisition and transmission will be substantially lower in the near future.

The trends in technology indicate that remote sensing will play a significant role in environmental monitoring. Since the mid-1970s, tropical deforestation has emerged as a serious global environmental issue and is a key parameter in many areas of research and policy, including climate change, carbon modeling, ecosystem fragmentation, and loss of biodiversity. Hence, the programs which are being initiated to improve our understanding of these issues will rely on the satellite-derived data. The analysis of remote sensing data in conjunction with other geographically referenced data will provide valuable information for decision making.

For background information SEE FOREST MANAGEMENT; FOREST RESOURCES; FORESTRY, URBAN; REMOTE SENSING in the McGraw-Hill Encyclopedia of Science & Technology.

Ashbindu Singh

Bibliography. T. M. Lillesand and R. W. Kiefer, *Remote Sensing and Image Interpretation*, 2d ed., 1987.

Fractals

Fractal structures have caught the imagination of the engineering and scientific community because of their beautiful and intricate patterns and simplicity of formation. One of the first applications of fractals to electrical engineering was the realistic computer modeling of landscapes. Applications of fractals have progressed considerably since these first images. Fractals are now being used to model and predict the behavior of complex systems, to understand natural phenomena and noise, and to synthesize and design new devices and systems. Fractal mathematics has helped engineers and scientists to discover and describe the hidden order in many chaotic phenomena such as bistable circuits, turbulent flows, and volcanic eruptions, and has allowed engineers to understand, describe, and classify electrical signals from the heart and brain. In addition, researchers are now examining ways in which fractal mathematics can be applied to medical imaging, computational biology, the structure of music, and even the behavior of economic markets.

Construction and dimension. A fractal is a shape made up of parts similar to the whole in some manner. Thus, fractal structures contain repetitions of a given shape on all scales. Therefore, fractals are said to be self-similar or scale-invariant.

The concept of self-similarity gives rise to the generator-initiator method of making geometrical fractals through iteration. In this construction, an initiator is first chosen which determines the general shape of the fractal. Then a generator is chosen and repetitively applied at increasingly small scales to fill in the fine structure of the object. Examples of two common fractals are shown in **Fig. 1**. The Sierpinski gasket (Fig. 1*a*) is formed from an initiator equilateral triangle and a generator made of a smaller excising triangle. The excising operation is applied repeatedly with appropriate change in scale as the fractal is grown by iteration. The Cantor bar (Fig. 1*b*) becomes the Cantor dust in the case where the height of the bar approaches zero. In each case, magnification of a small portion of the figure yields the original figure.

For the Sierpinski gasket the resulting veillike structure does not fill all of two-dimensional space. Likewise it is clear that it takes up more space than a one-dimensional object. Analogous statements apply to the Cantor bar, suggesting that a generalization of the common or euclidean definition of dimension would be useful in characterizing fractal objects. This generalization is called the fractal dimension and is an indication of the space occupied by the fractal or its roughness. Such dimensions should take on values close to two for a smooth undulating surface and a value closer to three for a rough volume-filling curve.

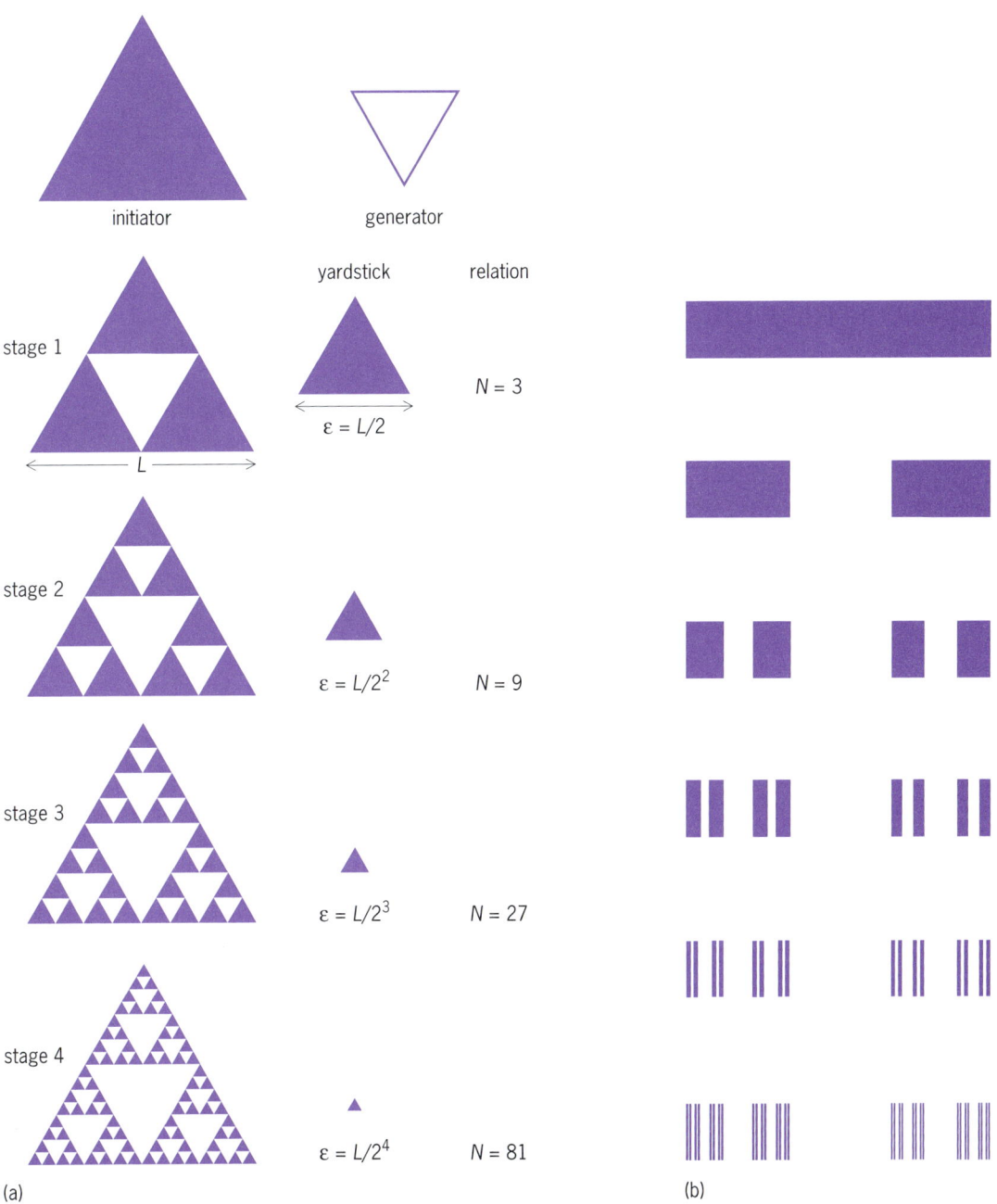

Fig. 1. Fractals constructed by the generator-initiator method. (*a*) Sierpinski gasket. The length of the yardstick, ε, that determines fractal dimensions and number, N, of yardsticks required to cover object are also shown. (*b*) Cantor bar. (*After H. N. Kritikos and D. L. Jaggard, eds., Recent Advances in Electromagnetic Theory, Springer-Verlag, 1990*)

The disk-covering method can be used to determine the fractal dimension by counting the number N of yardsticks of side ε that are needed to cover or fill the object of side L. The disk-covering fractal dimension D is defined by the equation below.

$$D = \lim_{\frac{\varepsilon}{L} \to 0} \frac{[\ln N(\varepsilon)]}{[\ln (L/\varepsilon)]}$$

The fractal dimension of the Sierpinski gasket (Fig. 1a) is given by $D = \ln(3)/\ln(2) \approx 1.58496$, using the above equation. Likewise, the fractal dimension for the Cantor dust is $D = \ln(2)/\ln(3) \approx 0.63093$.

Electrical engineering examples. Fractals in electrical engineering can be appreciated from three examples.

Scattering from fractal surfaces. The first example concerns the scattering of radar or optical waves from a rough fractal surface such as water or a fractured material, respectively. **Figure 2** shows the scattering that occurs from this interaction of incident wave (incoming arrow) as a polar plot of the relative scattered power. When the fractal dimension increases, the surface roughness increases and the scattered power is dispersed over an increasingly wide range of angles. This result is expected from simple physical principles in which each facet reflects the incident wave such that the local angle of reflection is equal to the local angle of incidence. This scattering pattern can be easily quantified so that the slope of the scattered power as a function of the sine of the normalized scattering angle (on a log-log plot) is linearly related to the fractal dimension. In short, there is a simple relation between the scattering parameters and the fractal dimension of such surfaces. Thus, a method is provided for the remote identification and classification of rough surfaces and has given rise to an economical and physically realistic model of many surfaces previously thought to be random.

Fractal image encoding. The second application is taken from the emerging area of efficient fractal image encoding. For such coding, images are modeled by basic shapes, and suitable mathematical transformations are used to encode image information by mapping the images into fractal patterns. Here the redundancy of image information is used to develop efficient fractal block codes. These methods are particularly intriguing for use in image

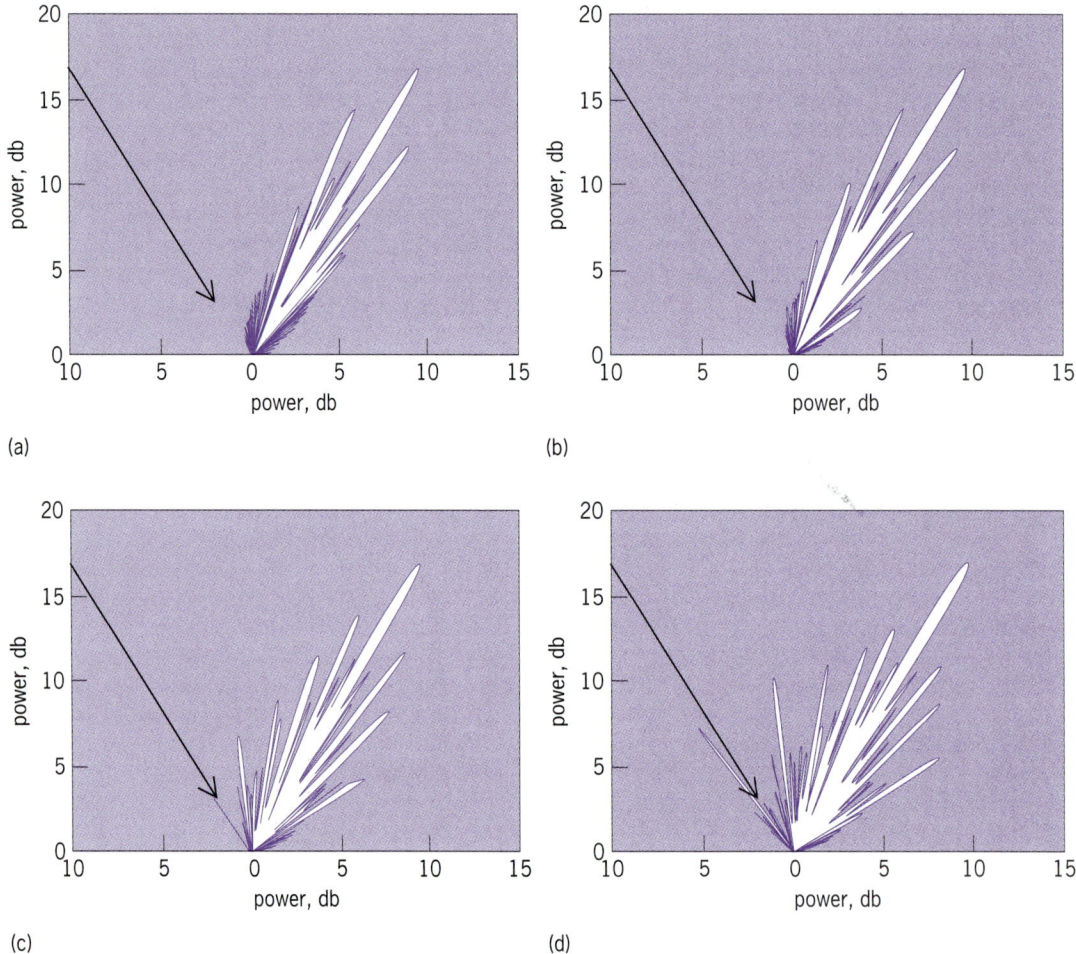

Fig. 2. Polar plots of angular scattered power distribution from fractally corrugated surfaces. Cuts through these surfaces have a fractal dimension D of (a) 1.05, (b) 1.30, (c) 1.50, and (d) 1.70. The arrows indicate the direction of the illuminating wave. db = decibel. *(After D. L. Jaggard and X. Sun, Scattering from fractally corrugated surfaces, J. Opt. Soc. Amer., A7:1131–1139, 1990)*

compression, where it is desirable to send image information over channels that would otherwise not have the necessary capacity. The first work in this area was concerned with the encoding of images of naturally occurring objects by using global transformations (single transformations that are used across the entire image to encode it). The first images encoded by using these methods tended to contain a number of self-similar objects, making them particularly suited for transformation. However, these methods required human intervention. Recently, fully automated iterative algorithms have been developed that depend on local transformations (wherein many different transformations are used on each piece of the image) rather than their global counterparts. Such techniques are under investigation for fractal image compression.

Fractal drumhead vibrations. Vibrations of drumheads with fractal boundaries have interested engineers and scientists because they model phenomena ranging from the characteristics of complex resonators, to the oscillation of water in lakes and river basins, to the acoustic response of musical instruments and auditoriums. Several modes of vibration are shown in **Fig. 3** for drumheads with fractal boundaries. These models demonstrate the rich behavior of the vibrations due to the rough edges, and the rapid damping of the drum displacement near the edge due to boundary roughness. Thus, the vibration amplitudes of these drumheads tend to be localized away from the boundary, in contrast with the vibrations of their smooth-edged counterparts. It has been conjectured that such structures may have applications to the nature of porous structures such as are used in the design of microwave anechoic chambers by electrical engineers and by coastal engineers for the design of breakwaters.

Other application areas. The self-similarity of fractals and the self-similarity of many geophysical processes led naturally to the use of fractals in both the generation of realistic landscapes and in the classification of natural scenes and textures. In more recent advances in fractal image processing, fractals are used both in sophisticated scene analysis and in automated image recognition systems.

Medical imaging. In one area of image analysis, fractals are being applied to medical images to automate diagnostic systems. In some cases, correlations have been found between fractal dimension and the presence of disease. For example, these techniques are being investigated for the identification and classification of liver disease, lung disease, and osteoporosis. Similar techniques can be used for the classification of tissue from ultrasound images or in the automated classification of a wide variety of other medical images.

Signal processing. Waveforms are investigated by using fractal techniques to examine signals for significant features. For example, in medicine, there is an interest in identifying pathological features through analysis of electrical signals from the heart and brain. Likewise, in utilizing radar waveforms, there is an interest in locating targets in the presence of clutter. There is considerable activity in both these areas, and both use the underlying structure of fractals. One rapidly developing and related area is the use of wavelets in signal and image processing. SEE WAVELETS.

Chaos. Certain nonlinear circuits with several stable states provide ideal test beds for investigating the order behind chaos. An analysis of the phase space constructed by plotting two voltages in such a circuit as the time is varied yields fractal pat-

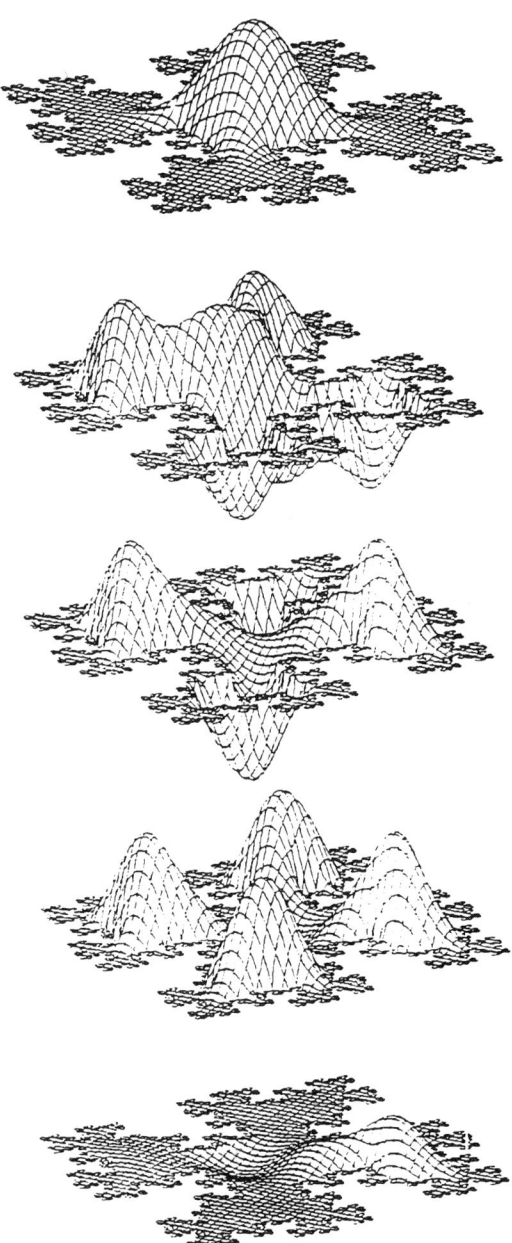

Fig. 3. Modal displacements of a vibrating drumhead with fractal boundaries. The decay of the displacement near the fractal boundary is clearly visible. *(After B. Sapoval, T. Gobron, and A. Margoline, Vibrations of fractal drums, Phys. Rev. Lett., 67:2974–2977, 1991)*

terns which are the footprints of chaos. Chaotic systems are also under study to examine irregular heartbeats and brain waves and the operation of electron and optical devices. Here fractals provide a key to the order underlying the seemingly random behavior of such systems.

Prospects. Fractal mathematics, originally inspired by the mathematicians of the nineteenth century and plainly observable physical phenomena, has had a successful history in the description and characterization of nature and natural processes. Additional physical, chemical, and biological systems formed by iteration or appropriate rules of growth will be brought under the fractal umbrella. In parallel, there are a number of areas in which fractal concepts are being used in electrical technology. These range from the signal and image processing and image compression noted above to other areas of interest. For example, fractal antenna arrays are under investigation for their properties of robustness under element failure. Fractals are being investigated for use in the design of efficient computer architectures and large networks. Finally, work has started on the use of fractals in computational biology and the long-range correlations of gene sequences. Fractals are rapidly moving from the language of description into the arena of engineering application.

For background information SEE CHAOS; FRACTALS; IMAGE PROCESSING; MEDICAL IMAGING; VIBRATION in the McGraw-Hill Encyclopedia of Science & Technology.

Dwight L. Jaggard

Bibliography. M. Barnsley, *Fractals Everywhere*, 2d ed., 1993; C. Baum and H. Kritikos (eds.), *Symmetry in Electromagnetics*, 1995; Fractals in electrical engineering, *Proc. IEEE*, 81:1423–1533, October 1993; B. B. Mandelbrot, *The Fractal Geometry of Nature*, 1983.

Fullerene

Fullerenes are a family of molecules that contain an even number of carbon atoms in a closed cage (**Fig. 1**). First discovered in connection with astrophysically based experiments in chemical physics, one member was discovered in minute quantities in soot (an amorphous, structureless form of elemental carbon). In recent years, a number of fullerene-type molecules have been synthesized, and this new area of research has proved to be significant in carbon chemistry. However, with the exception of the tiny amounts found in soot, until recently there had been no hints of the existence of fullerenes on Earth over geological time. Naturally occurring fullerenes have now been found in geologic formations that give evidence of their early existence. Discovery sites include a metaanthracite from Russia (reported in 1992), rocks that had been struck by lightning (reported in 1993), clays from the Cretaceous-Tertiary (K-T) boundary (re-

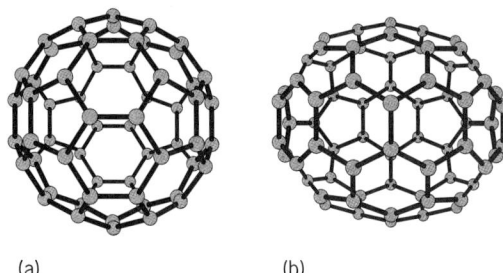

Fig. 1. Structures of two of the most commonly occurring forms of fullerene. (*a*) C_{60}. (*b*) C_{70}.

ported in 1993), and the Sudbury impact crater in Canada (reported in 1994).

K-T boundary sites. The Cretaceous and Tertiary are geologic periods, with the former ending and the latter beginning about 65 million years ago. The geological boundary between these periods (the K-T boundary) is marked worldwide by mass extinction of genera; hence the Tertiary rocks can be readily distinguished from the underlying Cretaceous rocks by their distinctly different fossil floras and faunas. The cause of the extinction is still a matter of controversy, with the impact hypothesis apparently most widely accepted. It is based on the assumption that a large meteorite or a comet, some 10–20 km (6–12 mi) across, may have collided with Earth and formed an impact crater of possibly 300 km (180 mi) in diameter. Many geologists believe that the remains of an ancient crater discovered recently at Chicxulub, Yucatan, Mexico, are actually the remains of this impact crater. According to the impact hypothesis most animals and plants, including trees, died soon after the event. It has been suggested that the transiently lowered atmospheric temperatures freeze-dried forests, which then caught fire, resulting in the injection of huge quantities of soot, and perhaps some fullerenes, into the stratosphere. There these particles attached themselves to iridium-rich rock dust and eventually settled to form the iridium- and soot-rich Tertiary clay seams on top of the uppermost Cretaceous rocks at K-T boundaries worldwide. SEE DINOSAUR.

Soot-rich clays. In 1993 it became known that C_{60} and C_{70} occur in small amounts in soot from the free burning in air of toluene and candle wax. This discovery rendered it more likely that the K-T wildfires might have produced fullerenes, which then "tagged along" with the soot and the iridium. This hypothesis triggered a search for fullerenes in the soot-rich clays from the K-T boundary sites at Woodside Creek and Flaxbourne River in New Zealand (**Fig. 2**). Here, a length of a few tens of meters of 1–2-cm-thick (0.4–0.8-in.) inclined clay beds are exposed at the surface and can be sampled. Control samples of limestones and shales from above and below the K-T boundary were also collected. The K-T clays contained fullerenes but the control samples did not. More recently,

Fig. 2. Cretaceous-Tertiary seam (position of the geologist's fingers) at a K-T boundary site in a limestone quarry on South Island, New Zealand. Tertiary limestone is to the left of the seam and Cretaceous limestone, to the right. (*Courtesy of R. R. Brooks*)

fullerenes were found in the K-T boundary clay from Caravaca, Spain, and Sumbar, Turkmenistan. Caravaca is almost at the antipodes of New Zealand; hence it appears that the K-T fullerenes were deposited over large areas of the globe.

Evidence for occurrence. The occurrence of fullerenes was substantiated by digesting the powdered clays with toluene, a solvent in which the fullerenes dissolved well. Fluids and solids were separated by centrifugation, and the analysis for fullerenes was done by high-pressure liquid chromatography. At Woodside Creek and Flaxbourne River, C_{60} occurs in amounts ranging from 0.06 to 5.4 nanogram per gram (parts per billion) of clay. C_{70} is also present. The C_{70}/C_{60} ratio is 0.30 ± 0.05. At Caravaca, C_{60} occurs in amounts of 11.9 ppb, and the C_{70}/C_{60} ratio is 0.22.

The average mass of C_{60} deduced to have settled per square centimeter at these sites is 12 ng, which corresponds to 6×10^{10} g in a uniform worldwide cover. At least this amount of C_{60} must have been synthesized at the K-T boundary. If wildfires were the cause of fullerene production, if the C_{60} content of the soot formed was 50 parts per million (a value found in the soot from the burning of toluene), and if other efficiency factors are taken into account, it can be estimated that the incineration of at least 8×10^{16} g of biomass was required, or 3.8% of the estimated total biomass available at Cretaceous-Tertiary time, which is consistent with the estimated mass burned to form the K-T boundary soot. The process that formed the fullerenes at that time must have been immense.

At the surface of Earth, fullerenes, once formed, are vulnerable to decomposition by photochemical reactions, especially to oxidation by oxygen and ozone. Sunlight was blocked for a considerable length of time after the Chicxulub impact by dust in the atmosphere; therefore, fullerenes may have become deposited in their marine environments without substantial exposure to sunlight. The lifetimes of fullerenes in contact with small concentrations of ozone, however, are much shorter than 65 m.y.; hence the clays must have provided some kind of protection against destruction by ozone.

Other occurrences. Fullerenes have also been discovered at Sudbury, Ontario, Canada. Although the K-T and Sudbury discoveries have in common that fullerenes were found, there is no evidence to suggest that the Sudbury event triggered wildfires of the magnitude of the Chicxulub impact. The process of formation at Sudbury is thought to have involved carbon chemistry in the explosion plume of the atmosphere, with most of the carbon provided by the impactor (an asteroid or comet). Thus the question is raised as to whether the fullerenes in the K-T clay could have formed likewise in the explosion plume of the Chicxulub event if it is assumed that the impactor was a 10-km (6-mi) carbonaceous chondritic body containing 1 weight % carbon, corresponding to a total of 2×10^{16} g of carbon. A cumulative yield of transformation to C_{60} and of fullerene deposition at K-T sites of only $1:10^6$ is required for the deduced C_{60} surface coverage of 2×10^{10} g, which seems to be possible. However, if the impactor was an ordinary chondrite, then the cumulative yield would have to be greater by one or even two orders of magnitude, and the impact hypothesis for the K-T fullerenes would become more difficult to support.

For background information SEE ATMOSPHERE, EVOLUTION OF; ATMOSPHERIC CHEMISTRY; CRETACEOUS; METEORITE; TERTIARY in the McGraw-Hill Encyclopedia of Science & Technology.

Dieter Heymann

Bibliography. L. Becker et al., Fullerenes in the 1.85 billion-year-old Sudbury impact crater, *Science*, 256:642–645, 1994; P. R. Buseck, S. J. Tsipursky, and R. Hettich, Fullerenes from the geological environment, *Science*, 257:215–217, 1992; D. Heymann et al., Fullerenes in the Cretaceous-Tertiary boundary layer, *Science*, 256:645–647, 1994; F. Radicati de Brozolo et al., Observation of fullerenes in an LDEF impact crater, *Nature*, 369:37–40, 1994.

Fundamental interactions

When atoms, accelerated to a fraction of the speed of light, collide, they can lose most or even all of their electrons. In this way bare uranium nuclei (uranium atoms with all 92 electrons removed) are made by passing a beam of uranium ions, accelerated to 0.4 GeV/nucleon (70% the speed of light), through thin metal foils. Lower velocities produce beams of uranium with predominantly one, two, or three electrons: the lower the velocity, the greater the number of electrons remaining on the atom. Bare and few-electron uranium and other heavy ions are used to study new effects that take place during ultrarelativistic atomic collisions, and to test quantum-electrodynamic (QED) calculations of few-electron atoms. Both these research areas involve the study of fundamental electromagnetic interactions in very strong electromagnetic fields.

Quantum-electrodynamic effects. In elements with large numbers of protons (high atomic number Z), such as gold ($Z = 79$), lead ($Z = 82$), or uranium ($Z = 92$), the large nuclear charge produces a strong electromagnetic field near the nucleus that greatly enhances both the relativistic and quantum-electrodynamic effects on the binding energy of the inner electrons. Although relativistic and quantum-electrodynamic effects are well understood in hydrogen and in one-electron ions, only recently have techniques been developed for calculating the spectra of atoms and ions with more than one electron. Theoretical progress in calculating these effects in few-electron ions has been stimulated by experiments that can measure quantum-electrodynamic effects in few-electron very high Z ions produced in tokamaks, electron-beam ion traps, and especially in heavy-ion accelerators.

The measurement of the 281-eV transition energy between the $2^2S_{1/2}$ ground state and $2^2P_{1/2}$ first excited state of lithiumlike uranium (U^{89+}, with the three electrons) is presently the most rigorous test of these calculations. The quantum-electrodynamic effects contribute about 43 eV to this transition energy, of which about 2 eV comes from quantum-electrodynamic effects involving more than one electron.

Lithiumlike uranium is excited to the $2^2P_{1/2}$ state by passing U^{89+} ions, traveling at $\beta = v/c = 0.42$ (where v is the velocity and c the speed of light in vacuum), through a thin aluminum foil. The energy of the 281-eV photon produced by its decay back to the ground state is measured by using a Doppler-tuned spectrometer (**Fig. 1**). Because the photons are emitted from a moving source, their energy is seen Doppler-shifted in the laboratory according to the equation below, where ω_{ion} is the energy of the

$$\omega_{ion} = \frac{\omega_{lab}(1 - \beta \cos \theta_{lab})}{(1 - \beta^2)^{1/2}}$$

emitted photon, ω_{lab} is the energy of the emitted photon as seen in the laboratory, θ_{lab} is the viewing angle, and $\beta = v/c$. When viewed through a column of argon gas, the photons pass through the gas as long as they cannot efficiently excite the argon atoms. However, at 244.39 ± 0.01 eV the photons excite the $2p \rightarrow 4s$ resonance transition. The photons are absorbed as the viewing angle is rotated to Doppler-shift the photon energy, ω_{lab}, through this energy. For $\beta = 0.42$, this shift occurs near $\theta_{lab} \approx 95°$. Measuring the angle θ_{lab} and the velocity determines the energy of the emitted photon.

The Doppler-tuned spectrometer (Fig. 1) uses position-sensitive x-ray detectors. In front of each such detector is a tapered Soller-slit collimator and an argon gas cell (pressure of 0.9 kilopascal or 7 torr) with very thin windows. The collimators are tapered so that each element of the x-ray detector views the same segment of the U^{89+} beam from a slightly different angle, transforming the position response of the detector into an angular response. The collimators are focused on a section of the beam downstream from the target where the $2^2P_{1/2}$ excited state is formed. The detectors are multiwire proportional counters, and the entire apparatus is under vacuum.

The measured value of 280.56± 0.1 eV agrees with present calculations, but a few small effects have not yet been included in the calculations. Although they are not expected to affect the agreement between experiment and theory, it is worth remembering that quantum electrodynamics was invented, in part, to explain a small effect in the spectrum of atomic hydrogen.

New types of atomic collisions. In atomic collisions, the nuclei pass close to each other but do not touch. They do, however, pass through the cloud of atomic electrons and may transfer enough energy

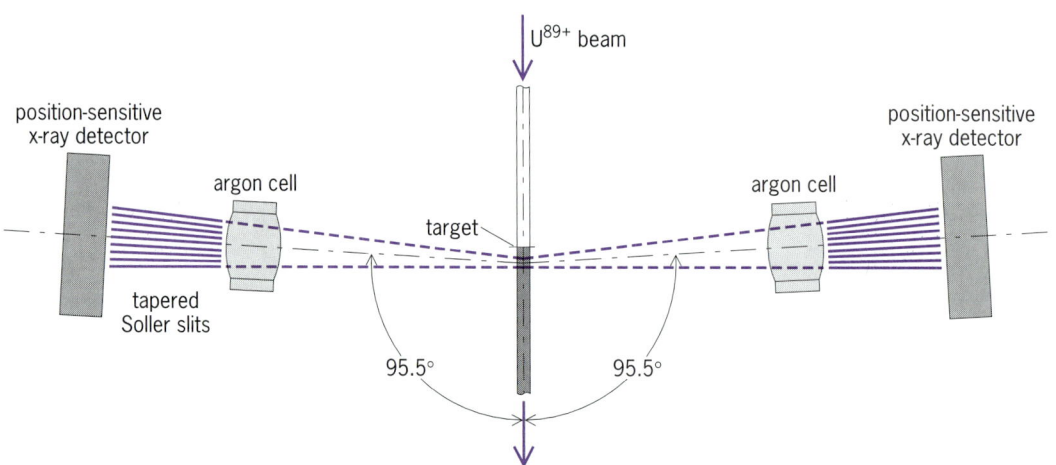

Fig. 1. Doppler-tuned spectrometer for measuring the energy of 281-eV photons emitted in the $2^2P_{1/2} \rightarrow 2^2S_{1/2}$ decay of lithiumlike uranium. *(After J. Schweppe et al., Measurement of the Lamb shift in lithiumlike uranium (U^{89+}), Phys. Rev. Lett., 66:1434–1437, 1991)*

to remove or add one or more of these electrons. Thus, a balance between electron capture and loss is possible. Until recently, it was thought that all the processes for electron capture had cross sections that decreased rapidly with increasing collision energy, and that electron capture would be insignificant at ultrarelativistic energies.

In relativistic, and especially ultrarelativistic collisions ($\beta > 0.999$), however, time dilation greatly increases the already large electromagnetic fields of the passing heavy nuclei. The strong electromagnetic fields can spawn electron-positron pairs, with the electron from the pair sometimes emerging from the collision bound to one of the ions (**Fig. 2**). The process is known as electron capture from pair production. For 0.96-GeV/nucleon ($\beta = 0.85$) bare uranium ions colliding with a gold target, the electron from the electron-positron pair emerges from the collision bound to the uranium ion about 40% of the time.

The first experiment to observe capture from pair production detected uranium ions that changed charge state from U^{92+} to U^{91+} in coincidence with the positron from the electron-positron pair (**Fig. 3**). The U^{91+} was separated from the U^{92+} in a magnet and was detected by a scintillator photomultiplier detector. The positron was detected by a thin scintillator photomultiplier, and was confirmed by its subsequent annihilation (with an electron in the detector) into two 511-keV photons, one of which was detected in a thick scintillator photomultiplier.

Other experiments and calculations show that for heavy ions above about 20-GeV/nucleon ($v/c = 0.999$) capture from pair production will be the dominant electron capture mechanism. In essence, the cross section for capture from pair production increases with energy because of the increased numbers of electron-positron pairs that are produced by stronger electromagnetic fields at higher energies. The cross section for capture from pair production also increases with increasing atomic number. The increase is due to the larger nuclear charge producing stronger fields and also a larger electron binding energy, thus increasing the probability that the electron will bind.

Since the electron that is captured is produced in the collision, it is possible for two bare nuclei to pass by each other and for one or both of them to emerge from the collision with an electron attached. Exactly this situation will exist at the Relativistic Heavy Ion Collider (RHIC) under construction at the Brookhaven National Laboratory in Upton, New York, and, when heavy ions are introduced, at the Large Hadron Collider (LHC) under construction at CERN in Geneva, Switzerland. RHIC will collide opposing beams of 100-GeV/nucleon bare gold and other nuclei to search for a quark-gluon plasma. Because RHIC collides opposing beams, its 100-GeV/nucleon energy per beam is equivalent to a 20,000-GeV/nucleon beam passing through a fixed target.

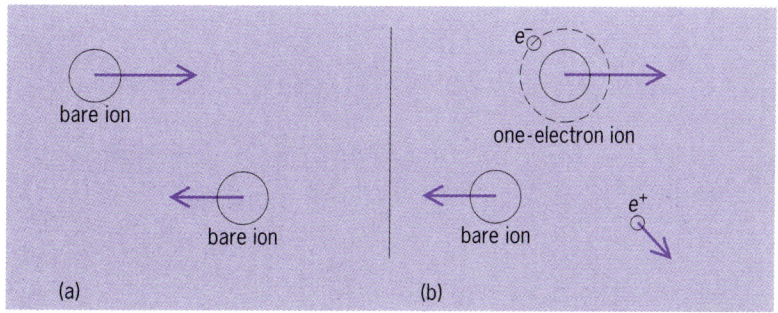

Fig. 2. Electron capture from pair production. An atomic collision of two (bare) ions spawns an electron positron pair with the electron emerging from the collision bound to one of the ions. (a) Initial state (no electron). (b) Final state: bound electron (e^-) + free positron (e^+).

Capture of an electron by one of the bare ions changes the ion's orbit in the confining magnetic field and the ion is lost from the collider. Calculations show that this effect will be important but should not limit the effectiveness of RHIC. However, until RHIC is operating the only accelerators that can test the calculations are fixed-target accelerators with collision energies hundreds to thousands of times lower.

Prospects. At the ultrarelativistic energies of RHIC and LHC, the electromagnetic fields produced by the atomic collision of two ions become strong enough to spawn particle-antiparticle pairs that are much heavier than electron-positron pairs. These pairs include taus (short-lived cousins of the electron with 3500 times the electron mass), mesons (quark-antiquark pairs) containing charm (c) quarks or bottom (b) quarks, and, possibly Higgs bosons. Some of the time, the negative particle of the pair will emerge bound to one of the ions (as an atom). Muons, pi mesons (an antiup and a down quark), K mesons (antiup and a strange quark), and of course electrons have been studied in atoms. Much of what is known about the physics of these particles comes from these studies. But taus and c- and b-containing mesons (D and B mesons) have not been studied in atoms. Since the

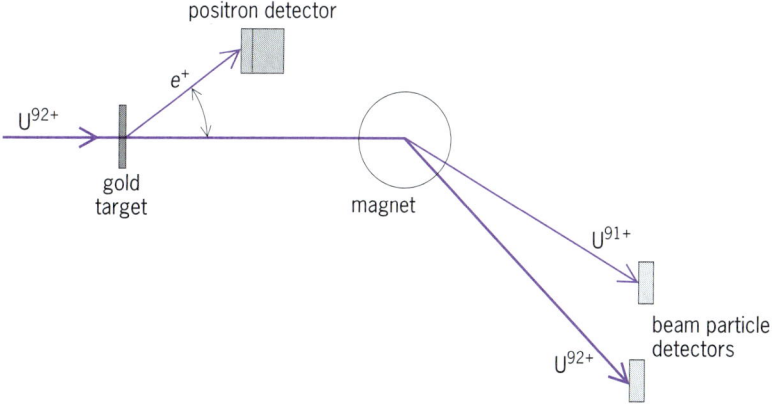

Fig. 3. Experiment to observe electron capture from pair production. The U^{92+} ion colliding with a gold atom creates an electron-positron pair with the electron bound to the uranium ion and the positron (e^+) escaping.

tau, D, and B lifetimes are typically 1 picosecond or less, capture from pair production is likely to be the only way to form and study them.

For background information SEE ATOMIC STRUCTURE AND SPECTRA; BEAM-FOIL SPECTROSCOPY; DOPPLER EFFECT; ELECTRON-POSITRON PAIR PRODUCTION; ELEMENTARY PARTICLE; HIGGS BOSON; LEPTON; MESON; PARTICLE ACCELERATOR; QUANTUM ELECTRODYNAMICS; SCINTILLATION COUNTER in the McGraw-Hill Encyclopedia of Science & Technology.

Harvey A. Gould

Bibliography. A. Belkacem et al., Measurement of electron capture from electron-positron pair production in relativistic heavy ion collisions, *Phys. Rev. Lett.*, 71:1514–1517, 1993; I. Lindgren, I. Martinson, and R. Schuch (eds.), *Heavy-Ion Spectroscopy and QED Effects in Atomic Systems*, Proceedings of Nobel Symposium 85, 1993; P. F. Schewe and B. P. Stein (eds.), *Physics News in 1992*, 1992; J. Schweppe et al., Measurement of the Lamb shift in lithiumlike uranium (U^{89+}), *Phys. Rev. Lett.*, 66:1434–1437, 1991.

Gamma-ray bursts

Gamma-ray bursts are brief sources of high-energy celestial radiation. Their exact origin is as yet unexplained, although many theories have been proposed. About once a day, a burst of gamma rays appears from a random location in the sky. For a short time it completely outshines all other sources of gamma rays in the sky. Afterward, no trace of an object remains at that location in gamma rays or in any other wavelength region. There is no analogous phenomenon in the astronomy of other wavelengths. Not only is the cause of the bursts completely unknown, but the distance to the burst sources is still being debated.

The study of gamma-ray bursts and development of theories of their origin is one of the most active research fields in high-energy astrophysics. It is quite likely that the problem of gamma-ray bursts will be considered as one of the greatest mysteries in astronomy of the twentieth century. The intriguing possibility exists that the gamma-ray bursts may represent an entirely new class of objects or an emission process that is unobservable by other means. Also, if the burst sources are near the edge of the observable universe, as many now propose, they are the most luminous objects known in the universe. As such, they may become an important probe of the conditions in the oldest and most distant parts of the universe.

Observation. Gamma-ray bursts were discovered by accident in the late 1960s by the Vela spacecraft. These spacecraft were designed to detect clandestine nuclear test detonations in space. From the spacecraft data, it was noticed that widely separated spacecraft were simultaneously detecting bursts of gamma rays that were not coming from the direction of Earth or any other body in the solar system. At first these observations were classified, but in 1972 they were published. Although gamma-ray bursts are studied by hundreds of scientists, observations have been limited to a relatively few satellite experiments and research groups. More than 100 theories have been proposed for the possible source of gamma-ray bursts.

Considerable observational progress has been made in the past few years as more sensitive spaceborne detectors have become available. Most of the recent observations of bursts have been made by the Burst and Transient Source Experiment (BATSE) on the Compton Gamma Ray Observatory. Other, smaller experiments are being developed in the United States to study gamma-ray bursts.

Time profiles. Perhaps the most striking features of the time profiles of gamma-ray bursts are their diversity and large range of duration. Coupled with this diversity is the inability to place many gamma-ray bursts into well-defined classifications. The durations of gamma-ray bursts range from about 10 ms to over 1000 s in the energy range in which most bursts are observed. However, recent observations by the EGRET experiment on the Compton Observatory show high-energy emission over 90 min after the initial part of the burst. Submillisecond structure has been detected in at least one burst. Some burst profiles are chaotic and spiky with large fluctuations on all time scales, whereas others show rather simple structures with few peaks. In some cases, both characteristics are present within the same burst. Weaker bursts show the same diversity as the stronger bursts, although the spiky features are of lower statistical significance. No periodic structures have been seen in gamma-ray bursts.

An important result from recent analysis of time profiles is the observation of a systematic widening or stretching of gamma-ray burst time profiles as bursts become weaker. The observed stretching of burst profiles is consistent with that expected from the effects of time dilation from bursts at cosmological distances. However, the time-dilation observation and its interpretation are not universally accepted. A time dilation of the wavelength of radiation from rapidly receding objects results in a redshift. Observation of such a redshift has also been claimed in the weaker, and presumably more distant, gamma-ray bursts.

Spectral characteristics. The unique feature of gamma-ray bursts is their high-energy emission: Almost all of the power is emitted above 50 keV. Some bursts show emission at energies as low as 1 keV, but the power emitted at these energies is less than 1–2% of the total power. Most bursts show rather simple continuous spectra which appear similar in shape when integrated over the entire burst and when sampled on various time scales within a burst. Although the spectral shapes of many bursts are similar, the energy at which peak power is emitted changes greatly from burst to burst and is seen

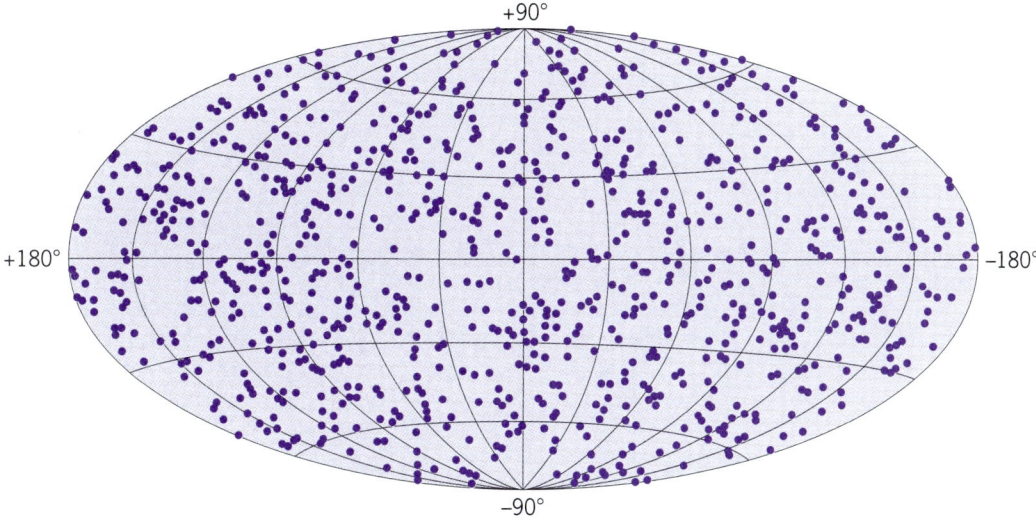

Distribution of 921 gamma-ray bursts observed by the BATSE experiment on the Compton Gamma Ray Observatory, 1991–1994. The map is plotted in galactic coordinates.

to rapidly change within a burst. Recently, the EGRET experiment has seen significant flux and power into the gigaelectronvolt energy range from several bursts. Many of these high-energy photons are delayed with respect to the bulk of the lower-energy emission. A search for gamma-ray line features with BATSE has thus far been unable to confirm earlier reports of spectral line features from gamma-ray bursts.

Counterparts. It is thought that a great advance in the understanding of gamma-ray bursts can be attained through successful correlated observations of bursts at other wavelengths. To that end, there have been major efforts to find a counterpart to a gamma-ray burst in other wavelength regions as evidenced by either simultaneous emission or afterglow emission. Comprehensive studies of archival astronomical photographic plates have been made. There have been several suggestions for counterparts, although the results are often debated. In view of the importance of the implied results, further observational evidence is needed before these results are accepted. Some of the world's most powerful ground-based facilities for radio and optical astronomy, observation of high-energy air showers and atmospheric Cerenkov radiation, and neutrino and gravitational-wave astronomy are involved in attempts at correlated burst observations. Space-borne correlated observations of well-located gamma-ray bursts have also been attempted in the ultraviolet, extreme-ultraviolet, and x-ray regions. There have been no confirmed observations of counterparts to gamma-ray bursts in other wavelength regions.

Distribution on the sky. The **illustration** shows the distribution of 921 gamma-ray bursts observed by BATSE, plotted in galactic coordinates. There appears to be no concentration of bursts near the galactic plane or near the galactic center. In fact, this distribution is consistent with an isotropic distribution of bursts across the sky, even when corrections are made to account for the uneven exposure of the experiment to different parts of the sky. The BATSE intensity distribution, derived from the same ensemble of bursts, shows a deficiency of weak gamma-ray bursts, which indicates that the sources are not homogeneously distributed in space in three dimensions. The isotropy of the BATSE gamma-ray burst distribution, coupled with its inhomogeneity (as measured by the deficiency of weak gamma-ray bursts), continues to be the most surprising observation, and the one that has eliminated the expected galactic distribution models.

Because it is difficult to construct reasonable models of gamma-ray bursters distributed in the Milky Way Galaxy according to these observations, most workers in the field now believe that the bursts are coming from cosmological distances such that the observed inhomogeneity is caused by a redshift of the faintest (and farthest) sources. There are no comprehensive models of what those sources might be, although highly beamed jets of radiation are thought to be involved.

For background information SEE ASTROPHYSICS, HIGH-ENERGY; GAMMA-RAY ASTRONOMY in the McGraw-Hill Encyclopedia of Science & Technology.

Gerald J. Fishman

Bibliography. G. J. Fishman, J. J. Brainerd, and K. Hurley (eds.), *Huntsville Gamma-Ray Burst Workshop*, Amer. Inst. Physics Conf. Proc. 307, 1994.

Gravitational collapse

One of the most exotic predictions of A. Einstein's general theory of relativity is the gravitational collapse of matter to form a black hole. This phenomenon occurs in the course of some dynamical processes when matter or energy becomes sufficiently concentrated, and the gravitational field becomes sufficiently strong, that a localized region forms from which no physical signal can escape. Such a region is known as a black hole. Astrophysically, gravitational collapse to black holes is significant since it is believed to occur at the end point of stellar evolution of stars too massive to become stable neutron stars or white dwarfs. However, gravitational collapse is also studied from a purely theoretical perspective because of the wealth of information it provides concerning the nature of the gravitational interaction as described by Einstein's theory. These theoretical studies increasingly rely on computational techniques implemented on powerful computers to allow researchers to directly simulate various models of gravitational collapse.

Dynamics of general relativity. General relativity describes gravitation in terms of the geometry of four-dimensional space-time: matter-energy tells space-time how to curve; space-time curvature tells matter-energy how to move. The geometry of space-time is encoded in the metric tensor, $g_{\mu\nu}$ (the indexes μ and ν range from 0 to 3), which can be viewed as a collection of 10 space-time fields that generalizes the single gravitational field of newtonian theory. The metric is governed by the Einstein field equations given by Eq. (1), where c is the speed

$$R_{\mu\nu} - \frac{1}{2} g_{\mu\nu} R = 8\pi \frac{G}{c^2} T_{\mu\nu} \qquad (1)$$

of light and G is Newton's gravitational constant. The Ricci curvature tensor, $R_{\mu\nu}$, and the Ricci curvature scalar, R, which appear on the left-hand side of Eq. (1), are very complicated, nonlinear functions of the metric and its first and second derivatives in space and time. The stress energy tensor, $T_{\mu\nu}$, which appears on the right-hand side, describes the total matter-energy content of the space-time. The gravitational field equations must be solved in concert with the equations of motion for each and every matter field that contributes to $T_{\mu\nu}$.

In the early 1960s researchers developed a formulation of the Einstein field equations based on a splitting of space-time into three-dimensional space plus one-dimensional time. This formalism has proven useful for simulating gravitational collapse and other dynamical (time-dependent) gravitational phenomena. In the space-plus-time approach, the geometry of space-time is viewed as the time history of the geometry of a three-dimensional space, known as a hypersurface, which represents a so-called instant of time of the universe. Just as the geometry of space-time is described by a four-dimensional metric tensor, $g_{\mu\nu}$, the geometry of a hypersurface is described by a three-dimensional metric tensor, g_{ij} (whose indexes i and j range from 1 to 3). The Einstein field equations can be written as a set of partial differential equations of evolution for the six independent components of g_{ij}. Thus, once values for the g_{ij} and their first time derivatives have been given at some initial time, the evolution equations permit the computation of g_{ij} at any future (or past) time. Initial values must also be specified for any matter fields which are present in a manner consistent with certain components of the field equations. In fact, even the initial values of g_{ij} and their first time derivatives must satisfy consistency relationships, called the constraint equations, which are a direct consequence of the coordinate independence (general covariance) of the field equations.

Computational techniques. In order to simulate dynamical gravitational phenomena on a computer, it is first necessary to reduce the infinite number of degrees of freedom represented by the field variables to a finite number. The most common technique is the method of finite differences. In this approach, the space-time continuum is approximated by a lattice (also known as a mesh or grid) of events, where the coordinate distance between neighboring lattice points is characterized by a discretization scale, h. Further, whereas the continuum three-dimensional hypersurface is spatially infinite, the corresponding finite-difference lattice is of finite extent. Thus, at any given time the lattice contains a finite number of points at which approximations to the field variables are computed. Field values are prescribed on some initial lattice and then advanced in discrete increments of time to generate an approximate space-time. The field updates are accomplished by using discrete versions of the equations of motion obtained by replacing partial derivatives by finite-difference quotients. Equation (2) defines a

$$\frac{\partial f(x^i,t)}{\partial t} \approx \frac{f(x^i,t+h) - f(x^i,t-h)}{2h} \qquad (2)$$

typical finite-difference quotient used to approximate the partial time derivative of a function f. In principle, in the limit that the discretization scale h tends to 0, the solution of the finite-difference equations should converge to the continuum solution of the field equations. In practice, for fixed initial conditions the change in the finite-difference solution as a function of h is of great importance since it provides an intrinsic measure of the level of error in the calculations.

An important improvement of the basic finite-differencing technique is provided by algorithms which allow the discretization scale h to vary locally and dynamically in response to the development of features in the solution. These adaptive methods are particularly vital for gravitational-collapse simulations because such calculations are often characterized by a large variation in the length and time scales on which phenomena must be resolved in order to produce accurate results.

Results of simulations. Many simulations of gravitational collapse have been performed by using a variety of matter-energy sources, including particles, idealized fluids, scalar fields, and even the gravitational field itself (in the collapse of gravitational waves). Much attention has been devoted to the question of singularity formation in these calculations. A singularity in space-time, typically characterized by infinite gravitational forces, signals a breakdown in the predictability provided by the dynamical formulation of physical laws. It has long been known that black holes contain singularities, but because what goes on inside a black hole cannot be seen by observers outside it, there is no loss of predictability for regions of space-time outside of black holes. A more worrisome possibility are collapse scenarios which produce singularities not hidden from external view by black holes. Interestingly, although many highly idealized models exhibit these so-called naked singularities, in virtually all cases the existence of the singularity can be shown to be intimately connected with some special feature of the model. Thus, there is considerable evidence for R. Penrose's cosmic censorship hypothesis, which conjectures that singularities that form from generic gravitational collapse will always be hidden in black holes.

Numerical calculations provide the means for easing some of the restrictions imposed by idealized models, and several computational studies provide evidence for the formation of naked singularities. However, these calculations have invariably involved idealizations of their own, and given the intrinsic difficulty in accurately simulating a phenomenon characterized by infinite field values, the relevance of these computations to the cosmic censorship hypothesis is still a subject of debate. Recently, a new class of simulations has been performed in which, under certain conditions, the appearance of a mild type of naked singularity seems well established.

Typical of this new class is a study based on the collapse of a massless scalar field in spherical symmetry. The initial conditions are set up so that each computation begins with the implosion of a pulse of scalar radiation. Each simulation is characterized by the setting of a single parameter, p, whose value controls the ultimate strength of the gravitational field as the field collapses toward the center of symmetry, $r = 0$. If p is sufficiently small, gravitational effects are negligible and the pulse of radiation implodes through $r = 0$, then disperses to infinity. If p is sufficiently large, gravitational effects dominate and black-hole formation occurs. The threshold of black-hole formation is then defined by some intermediate critical value, $p = p^*$, and a detailed study of the regime where $p \approx p^*$ reveals many interesting features.

In particular, as p approaches p^*, the evolution is increasingly dominated by an apparently unique solution which has an echoing property. This echoing is characterized by the recurrence of the same basic dynamics on ever-decreasing spatial and temporal scales. During one echoing period, most of the mass-energy of the scalar field is expelled from the central region but a small amount is left in a spatially scaled-down copy of the original strongly self-gravitating configuration. If conditions are right, another echo can occur, and in fact, by tuning p sufficiently close to p^*, an arbitrary number of echos can in principle be produced. Several conclusions follow immediately from the echoing property, including the theoretical possibility of generating arbitrarily small black holes and the existence of a naked singularity for the precisely critical case $p = p^*$. Similar features have been observed in calculations of the collapse of gravitational waves and idealized fluid models.

For background information SEE BLACK HOLE; GRAVITATIONAL COLLAPSE; RELATIVITY in the McGraw-Hill Encyclopedia of Science & Technology.

Matthew W. Choptuik

Bibliography. D. M. Eardley, Cosmic censorship, *Ann. N.Y. Acad. Sci.*, 688:408–417, 1993; C. R. Evans, L. S. Finn, and D. W. Hobill, *Frontiers in Numerical Relativity*, 1989; C. W. Misner, K. S. Thorne, and J. A. Wheeler, *Gravitation*, 1973; A. Zeyher, Adaptive mesh yields discoveries in black hole studies, *Comput. Phys.*, 7:122–124, 1993.

Hantavirus

Hantaviruses are a group of viruses that belong to a family called Bunyaviruses, but uniquely among Bunyaviruses, they lack any established insect vector. Recognized hantaviruses are zoonotic viruses that naturally infect wild rodents; the association between individual hantaviruses and their rodent hosts is species specific. The epidemiology of the hantaviruses and their associated illnesses in humans is intimately linked to the biology of the specific rodent hosts. Human infection with hantavirus is associated with two different acute clinical illnesses, hemorrhagic fever with renal syndrome and hantavirus pulmonary syndrome.

Characteristics. At least six well-characterized hantaviruses have been identified: Hantaan, Seoul, Pumaala, Dobrava, Prospect Hill, and Sin Nombre. In 1978, the prototype hantavirus, Hantaan, was isolated from the lungs of a striped field mouse and found to be a causative agent of Korean hemorrhagic fever.

Morphologically, hantaviruses are spherical, enveloped virions approximately 100 nanometers in diameter. The genetic material of hantaviruses consists of three single-stranded ribonucleic acid (RNA) segments, large (L RNA, about 6.5–8.5 kilobars), medium (M RNA, about 3.6 kb), and small (S RNA, about 1.7 kb). The three RNA segments are complexed with N protein, coded for by S RNA, to form three nucleocapsids. In addition, hantaviruses contain two envelope glycoproteins (G1 and G2), coded for by M RNA, and a viral RNA polymerase, coded for by L RNA.

Hantaviral infection of humans and rodents results in the production and circulation of hantavirus antibodies in the infected host. These antibodies cross-react with the different hantaviruses because of commonly shared antigens between the different hantaviruses. Diagnosis of infection is most easily made by measuring these antibodies with specific serological tests. SEE ANTIGEN.

Epidemiology. Hantaviral infection of rodents is asymptomatic and lifelong. The virus is transmitted among rodents primarily by exposure to nesting materials contaminated with infectious secretions, by grooming behaviors, and by intraspecies biting. Infected rodents shed the virus in their saliva, urine, and feces for many weeks following infection. Each of the known hantaviruses has a specific primary rodent reservoir and a geographic distribution that parallels the distribution of the primary rodent host (see **table**). Secondary infection of other small mammal species can occur but is not believed to play a significant role in the maintenance of hantaviral infection in nature or in the transmission of the virus.

Hantavirus infection is transmitted to humans primarily through exposure to aerosolized rodent excreta, and occasionally through rodent bites. The risk of human hantavirus disease is a function of the density of the local rodent reservoir populations, the prevalence of rodent infection, and the frequency of activities that result in contact between humans and aerosolized virus-containing rodent excreta. Human infection has occurred among laboratory personnel working with experimentally infected rodents. No human-to-human transmission has ever been documented.

Clinical illness. Hantaviral infection of humans is associated with premonitory symptoms presumed to be accompanied by the presence of the virus in the blood. However, the onset of significant clinical symptoms is temporally associated with the emergence of an antibody response, and as in many viral diseases the significant pathology appears to be mediated as much by the immune response as by the virus itself.

Hemorrhagic fever with renal syndrome. Hemorrhagic fever with renal syndrome refers to a spectrum of clinical illnesses that occur in association with the Hantaan, Dobrava, Seoul, and Pumaala hantaviruses. These illnesses have been known as epidemic hemorrhagic fever, Korean hemorrhagic fever, Manchurian fever, Songo fever, and nephropathia epidemica. The most severe form, associated with the Hantaan hantavirus, was first recognized in North America in the 1950s when some 3000 soldiers serving with the United Nations forces in Korea contracted the disease. It is characterized by acute high fever, shock, hemorrhage, and renal failure. Mortality from this disease ranges from 5 to 15%. Infection with the Seoul hantavirus results in a mild form of hemorrhagic fever with renal syndrome, with a mortality rate of about 1% (see table). Although Seoul-infected rats have been found worldwide, acute human illness has most frequently been seen in urban areas of east Asia. The Pumaala hantavirus is associated with a usually benign form of hemorrhagic fever with renal syndrome. A severe form, associated with the Dobrava hantavirus, has also been described in the Balkans.

Hantavirus-associated disease varies both seasonally and annually. Hemorrhagic fever with renal syndrome is estimated to account for as many as 100,000–150,000 annual hospitalizations in China. Natural population cycles of rodents and certain human behaviors result in distinctive seasonal patterns of disease. In a placebo-controlled trial in China, intravenous administration of the antiviral ribavirin improved the survival rate from hemorrhagic fever with renal syndrome.

Hantavirus pulmonary syndrome. In the summer of 1993, the investigation of a cluster of unexplained respiratory deaths in the southwestern United States led to the identification of a new syndrome associated with a previously unknown hantavirus, Sin Nombre. The outbreak in the Southwest appears to have resulted from an increase in the density of deer mice in the area, which was due to

Hantaviruses recognized worldwide

Virus	Rodent host	Geographic distribution	Mortality, %	Clinical syndrome
Hantaan	*Apodemus agrarius* (striped field mouse)	Far East, Russia, Northern Asia, Balkans	5–15	Severe hemorrhagic fever with renal syndrome
Dobrava	*A. flavicollis* (yellow-necked field mouse)	Balkans	5–35	Severe hemorrhagic fever with renal syndrome
Seoul	*Rattus norvegicus, R. rattus*	Worldwide	1	Moderate hemorrhagic fever with renal syndrome
Pumaala	*Clethrionomys glareolus* (bank vole)	Europe, Scandinavia, western Russia	—	Mild hemorrhagic fever with renal syndrome (nephropathia epidemica)
Prospect Hill	*Microtus pennsylvanicus* (meadow vole)	United States	—	Unknown
Sin Nombre	*Peromycus maniculatus* (deer mouse)	North America	40–50	Hantavirus pulmonary syndrome

SOURCE: L. Chapman and R. F. Khabbaz, Etiology and epidemiology of the Four Corners hantavirus outbreak, *Infect. Agents Dis.*, 3:234–244, 1994.

increased rain. Hantavirus pulmonary syndrome is characterized by fever and pain in the muscles, followed by an acute onset of shock and respiratory failure. The respiratory failure is due to noncardiogenic pulmonary edema resulting from a leakage of fluids from capillaries into the lungs. Associated clinical laboratory findings include a low platelet count (thrombocytopenia), hemoconcentration, and an elevated white blood cell count with immature granulocytes. Chest radiography shows evidence of interstitial fluid in the lung. Pathological examination of postmortem tissues shows that the hantavirus antigen is present in endothelial cells in several different organs, including lungs, spleen, liver, lymph nodes, and kidneys.

In the year following the identification of hantavirus pulmonary syndrome, about 100 affected individuals were confirmed in the United States, and a few cases were confirmed in Canada. Most infected individuals are from areas where the deer mouse is found, but several cases have occurred outside of the geographic range of the deer mouse. Black Creek Canal virus, also associated with hantavirus pulmonary syndrome, has been isolated from cotton rats in the southeastern United States, and at least three additional variant hantaviruses have been reported.

The mortality from hantavirus pulmonary syndrome (40–50%) is higher than from hemorrhagic fever with renal syndrome. Confirmation of a beneficial effect of ribavirin on individuals with hantavirus pulmonary syndrome is lacking, and supportive respiratory care in an intensive care unit with careful attention to the balance of fluids is its cornerstone of management.

Preventive measures. Prevention of human hantavirus infection and the resulting disease is dependent on education about ways to decrease human exposure to rodents. Recommendations for risk reduction include guidelines on safer approaches for eliminating rodent infestations within the home, for cleanup of rodent-contaminated areas, for reducing the availability of food sources and nesting places inside the home, for preventing rodents from entering the home, and for reducing rodent shelter and food sources within 100 feet of the home. Special precautions should also be taken for cleaning homes of infected individuals and buildings with heavy rodent infestation, and also by campers and hikers and workers in affected areas who are exposed to rodents. In addition, biosafety precautions are recommended for laboratory personnel working with these viruses or handling animals infected with a hantavirus.

For background information SEE ANTIBODY; ANTIGEN; RESPIRATORY SYSTEM DISORDERS; RIBONUCLEIC ACID (RNA); RODENTIA; VIRUS in the McGraw-Hill Encyclopedia of Science & Technology.

Rima F. Khabbaz

Bibliography. R. B. Belshe (ed.), *Textbook of Human Virology*, 2d ed., 1991; L. Chapman and R. F. Khabbaz, Etiology and epidemiology of the Four Corners hantavirus outbreak, *Infect. Agents Dis.*, 3:234–244, 1994.

Hazardous waste

The disposal of hazardous waste and toxic materials has advanced considerably from the use of uncontrolled landfill. The developments have come about as a result of increasingly restrictive legislation that requires many forms of waste to be pretreated, often by incineration. The problem has been exacerbated by the increasing domestic and commercial use of chlorinated plastics such as polychlorinated biphenyls (PCBs), which produce dioxins and other hazardous chemical compounds, industrial waste such as slags or arc-furnace dust containing heavy metals, chemical-warfare agents, and nuclear waste, all of which require extensive and costly pretreatment before being disposed of, normally by landfill.

Existing incineration methods. Existing methods of destroying wastes by incineration use the rotary kiln furnace because of its versatility in handling almost all types of solid, liquid, and even gas or vapor waste material. The flue gas from the conventional rotary kiln incinerator, however, often contains hazardous products, whereas the residual ash may contain heavy metals. The rotary kiln, essentially a refractory-lined drum with a fuel burner on its axis, forms only a small part of an incinerator for burning hazardous waste such as polychlorinated biphenyls. The running and capital cost of the plant is much higher than that for conventional waste incinerators because of the need for ancillary equipment for chemical analysis of the input material, scrubbing equipment for the exhaust, and the higher operating temperature to ensure complete destruction of the waste. Chlorinated plastics, for example, require a temperature of at least 1250°C (2282°F) for 2 s with adequate turbulence for effective mixing in the presence of 6% excess oxygen. This requirement is at the limit of temperature of a fuel and air flow and, together with the need for ancillary equipment, makes the process very expensive.

Electric discharges, which are already used for vitrification of hazardous materials and for reclaiming heavy metals from metallurgical flue dust as an alternative to disposal by landfill, provide several different solutions to problems of waste destruction. Thermal processes using electric discharges make it possible to obtain higher temperature and energy densities than can be achieved by fuel-fired combustion processes; thus, nonequilibrium processes and alternative chemical routes with greater conversion efficiencies are possible, and selective chemical transitions can be carried out.

Electric discharges. The synthesis or destruction of chemical compounds can be achieved by using the different energy transition processes that may be optimized by control of electric discharge

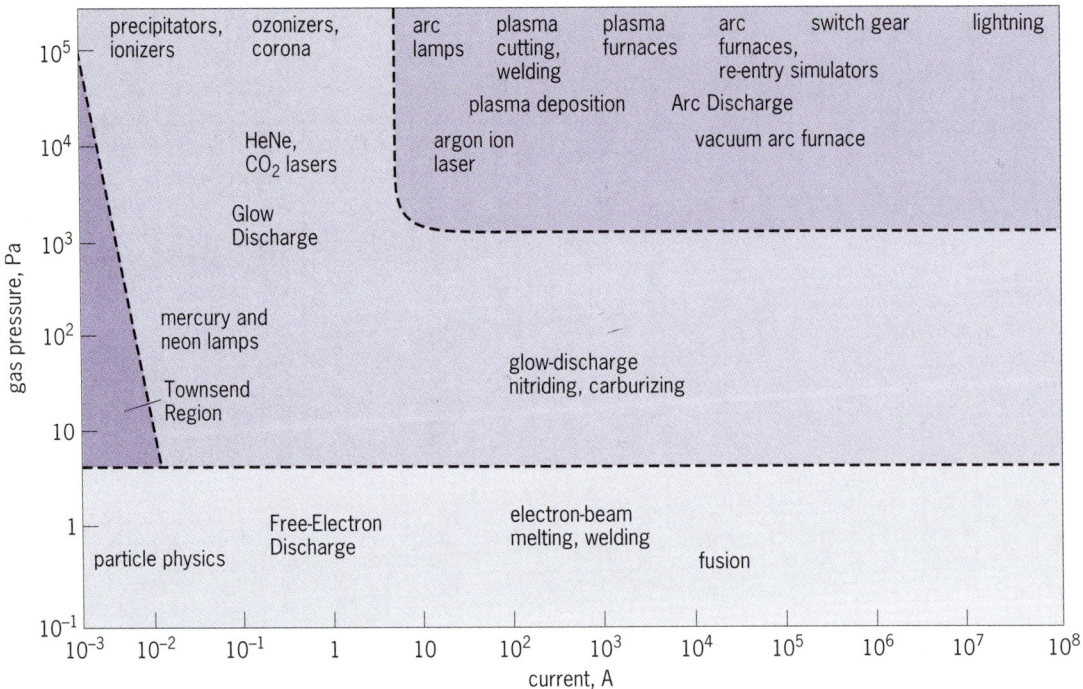

Fig. 1. Ranges of discharge current and gas pressure of various electric discharge regimes. 1 pascal = 10^{-5} bar.

parameters such as current and gas pressure (**Fig. 1**). In order of increasing energy, these processes are excitation, dissociation, and ionization; their relative magnitudes depend on the gas pressure and discharge current. The characteristic variation of the voltage with current of an electric discharge is shown in **Fig. 2**, which also shows the principal modes of the discharge.

Electric discharges can be used to provide a high temperature and energy density for thermal dissociation by using an arc (thermal equilibrium) in the same way as combustion, but at higher energy densities and temperatures and without the need for the air required for combustion. The process can be carried out in a controlled atmosphere, and the heat source (plasma torch) can be kept separate from the reaction. Alternatively, glow and corona discharges can be used to carry out nonthermal reaction processes, do not require the thermal energy needed to heat up the mixture, are potentially more efficient, and enable nonequilibrium processes to be carried out.

Glow or corona discharge. Such discharges (Fig. 2) can be used to carry out selective nonthermal chemical transitions in the gas or vapor phase which are more efficient than high-temperature equilibrium processes since in theory only the energy for the required chemical reaction is required. The normal low-current limitation of glow discharges is overcome by using large electrodes with a large surface area or corona discharges with rod or wire electrodes.

Corona discharges normally operate at atmospheric pressure and are characterized by an electrode region similar to the low-pressure glow discharge. An additional method of control of the energy transition is possible by pulsing the corona discharge. Corona discharges are widely used at atmospheric pressure for the large-scale manufacture of ozone (O_3) from atmospheric oxygen (O_2) for water purification, and for polymerizations.

Corona discharges are in use in a pilot plant for reducing the sulfur dioxide (SO_2) and nitrogen oxide (NO_x) content of flue gas from power stations by dissociation and recombination to less harmful compounds, and for the dissociation of hydrogen sulfide (H_2S). Compounds such as dioxins might also be destroyed in this way with only small energy inputs compared with the existing method of incineration.

Volatile organic compounds (produced during coating processes that use organic solvents, such as paint spraying, dry cleaning, and printing) con-

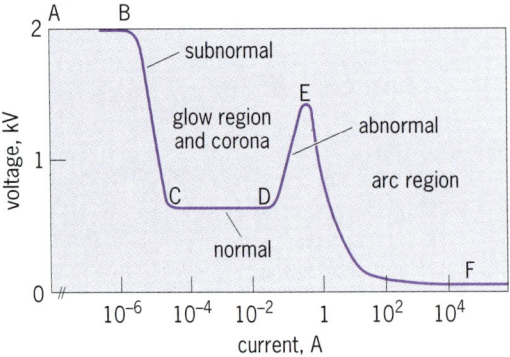

Fig. 2. Voltage-current characteristic of an electric discharge. A glow or corona discharge occupies portion B-E of the curve, and an arc discharge occupies portion E-F.

Fig. 3. Experimental installation using electric discharges for combustion of volatile organic compounds. *(Plasma Systems, SIP Analytical LTD., Sandwich, Kent, U. K.)*

tribute to the formation of ground-level ozone. A pilot plant has been developed for the depollution of air-xylene mixtures, which can be incinerated by using so-called cold discharges, which are inherently more efficient as they are not in thermal equilibrium (**Fig. 3**).

Arc discharge. These discharges (Fig. 2) can be used to produce high temperatures and molecular bonds in a way similar to combustion processes. The higher temperatures and energy densities enable a greater degree of dissociation without the dilution by the air or oxygen that is necessary for combustion, and also reduce the quantity of effluent produced.

Several different thermal plasma processes are possible for waste destruction. These processes include dc and three-phase arc furnaces similar to those used for steel manufacture and plasma torches. The plasma torch has the advantage of enabling the discharge to be separated from the material being heated by using a gas flow to transfer the energy from the arc. Plasma torches use either a rod cathode electrode with gas stabilization or a hollow cathode and magnetic rotation, and can operate in a nontransferred mode, in which the discharge is to an electrode within the torch, or a transferred mode, in which the discharge is to an external anode.

An incineration plant using hollow-cathode plasma torches has been developed and is in use for the destruction of highly hazardous materials such as polychlorinated biphenyls. As in the rotary kiln furnace, analysis of the waste prior to treatment and discharge in the flue gases, as well as postcombustion treatment, are part of the process, but the higher temperatures and the reduction of exhaust gases by eliminating the air required for a burner enable low levels of hazardous compounds to be obtained.

The very high temperatures possible by using arc discharges enable melting of refractory materials such as slags containing heavy metals or other refractory waste from municipal incinerators or industrial sources, and when cooled these materials form a glasslike solid. The process also results in a reduction in volume, and the vitrified product is impervious to leaching and can be used as landfill. A similar process is used to vitrify radioactive waste.

The disposal of automobile tires is a growing problem. Conventional pyrolysis using low-temperature combustion processes results in the production of fuel gas, liquid and solid hydrocarbons, and ash. With a high-temperature thermal plasma, carbon monoxide (CO) and hydrogen-rich combustible gases are produced and inorganic waste is solidified in a vitrified slag.

Thousands of tons of furnace dust arise from the production of stainless steel. This dust contains heavy metals and therefore cannot be used untreated for landfill, whereas the heavy metals have a residual value. The dust is in the form of a sticky brown fume and is reduced in a dc submerged-arc furnace in an argon atmosphere.

About 15% of hospital waste is contaminated with infectious agents. The waste is packaged in nonrecyclable materials, mainly plastics. The emissions of the plastics-rich waste have been analyzed to include ethane, ethylene, propylene, and hydrochloric acid, as well as furans and dioxins. Hospital incinerators have a relatively low flue stack, lack equipment for analysis of the flue gas, and operate at temperatures too low for complete destruction of refractory organic materials. A compact plasma reactor using hollow electrodes and magnetic rotation is intended for use by hospitals, industry, and research laboratories for the destruction of wastes close to their point of production. It provides a high temperature and energy density which together with the ionized reactive medium increases the rate of reaction and its degree of completion.

A three-phase ac arc plasma reactor rated at 200 kW with three plasma torches using water-cooled copper electrodes stabilized by low-current dc pilot arcs is being used to study the destruction of liquid waste in the United Kingdom. The ac supply is connected to the water-cooled electrodes and provides a high-power superimposed plasma which is stabilized by the pilot arcs. Liquid chemicals are injected around the arc region, and the exhaust gases are quench-cooled at the exit of the reactor and analyzed. Model chemicals used include trichloroethane, benzyl alcohol, dichlorobenzene, and transformer oil with up to 1.2% polychlorinated biphenyls. Destruction efficiencies of 99.999% are obtainable.

For background information SEE ARC DISCHARGE; ELECTRICAL CONDUCTION IN GASES; GAS DISCHARGE; GLOW DISCHARGE; HAZARDOUS WASTE; KILN in the McGraw-Hill Encyclopedia of Science & Technology.

John E. Harry

Bibliography. J. E. Harry, The destruction of waste and toxic materials using electric discharges, *IEE Eng. Sci. Educ. J.,* 2(4):171–176, 1993; J. E. Harry,

Electric Discharges for Heating, 1993; International Union for Electroheat, *Plasma Technology for a Better Environment,* 1992.

Headache

The age of customized drugs coupled with a radically new understanding of what constitutes a migraine has created an explosion of new drug treatments for migraine headaches.

Until recently, it was believed that migraine was caused by constriction and dilation of cerebral vasculature, but in the 1980s studies of blood-flow changes during migraine headaches disproved this theory. Thus, a reexamination of the entire phenomenon of migraine was stimulated, challenging scientists to devise new theories to explain this common problem. New drug treatments for migraine are being developed based on these new theories.

Symptoms. Some 10% of migraine headaches are preceded by a very special transient focal neurologic dysfunction called aura. Although the aura can be any focal neurologic dysfunction (for example, sensory loss in an arm, weakness in a leg, or difficulty speaking), it is most likely visual. The aura appears to spread across the surface of the brain (the cortex) at 0.08–0.16 in./min; (2–4 mm/min) and consists of three phases: an active leading edge (the scintillation), a loss of vision (the scotoma), and a restoration of vision. The phases follow the same pathway, so that vision returns at the point of origin while the scintillation and the scotoma continue to spread. The entire three-phase cycle takes about 15 min.

Spreading depression is a phenomenon that can be induced on the cortex of the brain of laboratory animals; it has many similarities to migrainous aura. This phenomenon spreads across the brain cortex in a similar three-phase wave: a leading edge of hyperactive neurons, a period of electrical silence, and a recovery of neuronal activity. The time it takes for this sequence to occur and its speed of propagation across the cortex are about the same as for migrainous aura. Spreading depression is more easily induced in that cortex with a high ratio of neurons to glial cells (cells that support and nourish the neurons), which in humans is the visual cortex.

If spreading depression exists in humans (a disputable point) it would be a reasonable explanation for the aura of migraine headaches. However, it would not explain the accompanying pain.

Causes. The pain of a migraine headache appears to be the result of chemical interactions between the endings of peripheral sensory nerves (normally responsible for feeling sensations) and of sympathetic nerves (normally responsible for controlling automatic functions such as blood flow and body temperature) which terminate in the arteries inside the skull. The chemicals released from each type of nerve activate the other, and the activation of sensory nerves produces the pain. In addition, the activation of each nerve involved causes an increased local release of that nerve's chemical contents. The result is a self-sustaining reciprocal activation, the process continuing until the chemicals involved are depleted.

Multiple chemicals are involved in these sensory-sympathetic nerve interactions, probably including substance P from the sensory nerves, leukotrienes and neuropeptide Y from the sympathetics, and nitric oxide from the artery itself. Significantly, each distinct chemical step represents a potential specific therapeutic intervention. For spreading depression and the chemical interactions to be sensible explanations for the aura and pain of migraine headaches, a link must exist between the two analogous to the link between the aura and the pain. The cause of the aura needs to precede the cause of the pain, but with variable timing and variable intensity. Volume conduction provides such a link. This term refers to the chemical changes that occur outside of the neurons when neuronal activity occurs. Probably the external chemical fluxes that occur in relation to spreading depression are large enough to activate the small sensory and sympathetic fibers directly over the involved cortex. However, the speed of the activation and its robustness would vary from episode to episode.

Treatment. Each step in the neurochemical reactions that occur represents another potentially useful intervention in controlling migraine. Such an approach is already being used. For example, it is likely that antiserotonin drugs such as methysergide and dietary restriction of glutamate both work by reducing spreading depression.

The most promising possibilities from a clinical perspective are those interventions that work on sensory and sympathetic nerves, which are located in the head but are not part of the central nervous system. These nerves could be treated with drugs that do not cross into the central nervous system, thus markedly reducing the potential of undesirable side effects.

The first truly customized drug for headache is sumatriptan. It was developed to bind to, and activate, serotonin 1_D receptors, and minimally binds to anything else. The serotonin 1_D receptor is found on small peripheral nerves terminating in the arteries inside (but not outside) the skull. The activation of the serotonin 1_D receptor stops an important component of the perpetuation of the pain associated with migraine. Activation likely occurs by suppressing the release of a sensory nerve chemical such as substance P, or a sympathetic nerve chemical such as leukotrienes or neuropeptide Y. This serotonin 1_D receptor is the same site of action as that used by several older, well-established antimigraine drugs such as ergotamine and dihydroergotamine (DHE). Differences between these three drugs involve other attributes. Sumatriptan's specificity minimizes its side-effect liability. Most important, the nausea that can be prominent with DHE and the vasoconstriction that can be prominent with ergotamine are minimal with sumatriptan.

Several large clinical trials have shown that sumatriptan rapidly and effectively aborts migraine headaches. In about 80% of the instances, the migraine is reduced to mild or no symptoms within 90 min with minimal side effects (primarily local irritation at the injection site).

DHE has similar efficacy, but produces more side effects. For example, in a study of 170 migraines treated with IV DHE, 81% achieved an average of 89% relief but nausea was a common side effect.

Sumatriptan's major liabilities are its relatively short duration of action and its expense. The short duration of action results in a large recurrence of otherwise responsive headaches. Studies indicate that 40% of the headaches successfully treated with sumatriptan recur within 12 h, and 60% within 24 h.

Although both oral and injection forms of sumatriptan exist, only the injection form has received U.S. Food and Drug Administration (FDA) approval. Thus, usefulness of sumatriptan in its present form is limited. Oral sumatriptan is easier to use, and should cost less than the injected version. Its availability will enlarge the suitable patient population for this drug.

By the year 2000, second-generation serotonin 1_D agonists will be available; they will have longer half-lives and hopefully cost less. In addition, other novel agents, such as peripheral antisubstance P agents, will give physicians and researchers more tools for controlling specific components of migraine headaches.

For background information SEE BRAIN; CENTRAL NERVOUS SYSTEM; SOMESTHESIS in the McGraw-Hill Encyclopedia of Science & Technology.

Marvin J. Hoffert

Bibliography. M. J. Hoffert, A new era in the treatment of migraine, *Amer. Fam. Phys.*, 49:633–638, 1994; M. A. Moskowitz, Neurogenic versus vascular mechanisms of sumatriptan and ergot alkaloids in migraine, *Trends Pharmacol. Sci.*, 13:307–311, 1992; J. Olesen and L. Edvinsson (eds.) *Basic Mechanisms of Headache*, 1988.

Horse

Two equine diseases have received recent attention from researchers: equine motor neuron disease (EMND) and hyperkalemic periodic paralysis.

Equine motor neuron disease. Equine motor neuron disease is a spontaneous disease of horses and ponies that was first described in 11 horses examined at Cornell University (Ithaca, New York) between 1985 and 1990. Postmortem investigations of these and subsequently affected animals reveal degenerative changes in brain stem and spinal motor neurons that are remarkably similar to those described in humans afflicted with a form of sporadic amyotrophic lateral sclerosis (ALS), also known as Lou Gehrig's disease. Studies on the cause of this recently identified disease with weakness and wasting in horses could have direct bearing on the cause of spontaneous amyotrophic lateral sclerosis.

Clinical signs. Equine motor neuron disease usually occurs sporadically, affecting only one of several or many horses at a stable. At large stables only occasionally have two or more animals been affected concurrently or consecutively. Typically, the horse progressively weakens and loses weight (100–300 lb or 45–135 kg) despite a good or voracious appetite. The weakened horse experiences generalized muscle wasting, and close inspection usually reveals localized muscle twitches called fasciculations. When standing, the horse trembles and attempts to compensate for its weakness by drawing both hind- and forelimbs under its body. It lacks the normal ability to lock its knees in extension, and as a result continually shifts its weight-bearing hindlimb. With so little residual strength, the animal must exert itself to remain standing, and this effort may be accompanied by profuse sweating. At the walk, an affected horse often appears stronger, but it is obviously short-strided and may scuff its hooves or stumble. Faster gaits cannot be sustained because of exercise intolerance.

As the disease progresses the horse spends increasing time lying down. Electrodiagnostic testing provides evidence of muscle denervation. Serum levels of muscle enzymes are mildly to moderately elevated. Although some recumbent horses die, and others experiencing severe weakness and wasting are destroyed at the request of their owners, many survive. In the latter case, the progression of weakness and muscle wasting remits or abates at some point, leaving the animals with varying degrees of permanent disability. Although the diagnosis of equine motor neuron disease can be confirmed only by postmortem demonstration of motor neuron degeneration and loss, the study of spinal accessory nerve biopsies has proved 95% reliable in establishing a clinical or antemortem diagnosis. Because of their residual weakness, horses surviving equine motor neuron disease pose a risk to riders.

Pathology. Although equine motor neuron disease has often proved more benign, the spinal cord and brain stem motor neuron changes in this disease closely resemble both in type and distribution those of progressive muscular atrophy, a form of amyotrophic lateral sclerosis, the degenerative changes involving the same cranial nerve and spinal neuron groups. Affected neurons are correspondingly pale and swollen and contain abnormal nuclei. As in amyotrophic lateral sclerosis, the pallor of the neuronal cytoplasm reflects extensive loss of ribosomes, and the swelling is caused by marked accumulations of neurofilaments. When present, the nuclei of these ghost cells are irregular, condensed, and encrusted by cytoplasmic densities. Eosinophilic cytoplasmic inclusions in some degenerating equine neurons resemble the inclusions found in nerve cells (Bunina bodies) in amyotrophic lateral sclerosis. Dead motor neurons are removed through the phagocytic activity of microglial cells. However, pigmented glial cell scars persist as long-term markers at sites of prior neuron loss. Associated changes in axonal processes include proximal swellings and distal fragmentation. Indicators of earlier axon loss are cords of proliferated Schwann cells in spinal motor roots and peripheral nerves.

Histochemical preparations suggest that there is a preponderant denervation of type I muscle fibers, the slow, oxidative fibers that are metabolically equipped to sustain their activity in antigravity muscles. Such preferential involvement suggests a greater vulnerability of the motor neurons to muscle fibers undergoing sustained contraction rather than those serving muscles undergoing phasic or burstlike contraction patterns. Such differential involvement may explain why affected horses display greater weakness while standing than when walking or running.

Epidemiological findings. Since affected horses in the initial study had originated in the northeastern states, it was suspected that equine motor neuron disease might be confined geographically. However, as knowledge of the disease spread, the number of confirmed cases increased and the size of the reporting area expanded. The more than 100 cases reported in the United States are distributed throughout the entire country. Affected animals have also been identified in Canada, Brazil, Belgium, England, Ireland, Holland, Switzerland, and Japan. In the United States, affected horses often come from boarding stables and riding academies. Although affected horses vary in age, breed, and sex, mature quarterhorses and retired thoroughbreds are overrepresented, and the majority are geldings. It is strongly suspected that these characteristics fit the general profile of horses at American boarding and riding stables, and that management procedures rather than the breed and sex are predisposing. In this regard, at most of these stables, the affected horses have little or no access to pasture or green grass and are fed hay of poor quality.

Causative factors. Elucidation of the cause of equine motor neuron disease may prove especially significant in view of the disease's resemblance to amyotrophic lateral sclerosis. As in the familial form of amyotrophic lateral sclerosis, the cause of equine motor neuron disease may be related to oxidative stress. However, in equine motor neuron disease the stress has been associated in part with an acquired deficiency of vitamin E rather than a gain-of-function genetic mutation in the enzyme superoxide dismutase as in familial amyotrophic lateral sclerosis. The vitamin E deficiency can be related to a lack of green pasture at the affected stables. Yet, it seems that vitamin E deficiency is only one factor in the genesis of this disease, because some unaffected stablemates of the affected horses with equine motor neuron disease have also had low plasma levels of this antioxidant, and therapeutic supplementation with vitamin E has not been reliably associated with arrest of the disease or with improvement in condition.

Preliminary findings currently indicate that more than one antioxidant is deficient. Since cases of equine motor neuron disease have not been retrospectively identified prior to the 1980s, efforts are focused on recently developed procedures or substances that could exacerbate the oxidative stress imposed by antioxidant deficiency and thereby trigger motor neuron degeneration. Elucidation of the causative factors in equine motor neuron disease may provide valuable clues to the cause of spontaneous amyotrophic lateral sclerosis.

John F. Cummings; Alexander de Lahunta; Hussni Mohammed

Hyperkalemic periodic paralysis. Hyperkalemic periodic paralysis is an inherited disease of the muscle tissue that is caused by a genetic defect in the alpha subunit of the adult skeletal muscle sodium channel. The disease in horses is analogous to the heritable disease in human beings known as hyperkalemic periodic paralysis, or adynamia episodica hereditaria; both are caused by point mutations of the sodium channel gene. Presence of the defective sodium channel makes the horse's muscle overly excitable (that is, it fires more readily), leaving the horse susceptible to sporadic episodes of muscle tremors or paralysis that last from minutes to hours. However, in between episodes affected horses appear clinically normal, many even being highly successful show horses or used for pleasure and trail riding.

Clinical signs. The episodes of muscle tremors or paralysis are usually associated with modest to marked hyperkalemia (excessive amounts of potassium in the blood), hemoconcentration (increase in the concentration of blood cells and total solids resulting from fluid shifts), and normal to modest increases in muscle-derived serum enzymes. Between episodes, physical examination and laboratory parameters are within normal limits. Electromyography reveals generalized abnormalities during and between clinical episodes of paralysis. These abnormalities include increased insertional activity and spontaneous activity consisting of fibrillation potentials, positive sharp waves, and complex repetitive discharges. These electromyographic abnormalities are present following administration of neuromuscular blocking agents, and are indicative of muscle membrane hyperexcitability.

There is a wide range of clinical severity of disease among different horses carrying the same mutation, some being asymptomatic and others requiring daily medication to control episodes. Diet and exercise are believed to be responsible for some of the variation in clinical severity. In addition, there is a gene dosage effect, the homozygote affected horse being more severely affected than the heterozygote affected horse. Episodes of muscle tremors and paralysis are often accompanied by an increased respiratory rate and noisy breathing. The homozygote affected horse will exhibit noisy breathing even between episodes of paralysis. Endoscopically, the homozygote affected horse will exhibit upper airway obstruction due to prolonged contractions of the laryngeal and pharyngeal muscles. Sometimes a horse will die following a severe episode of hyperkalemic periodic paralysis, the cause of death presumably cardiac or respiratory failure.

Cause. Hyperkalemic periodic paralysis is inherited as an autosomal dominant trait and occurs with equal frequency in both sexes (stallions and geldings and mares). The appearance of clinical signs of disease is unpredictable; however, all horses with the defect demonstrate muscle fasciculations and paralysis following oral administration of potassium chloride (KCl).

The mutation affects a highly conserved portion of the sodium channel protein near the cytoplasmic face of the membrane. The mutation can be identified in samples of whole blood, hair roots, or tissues using the polymerase chain reaction–based test which detects the presence of a phenylalanine to leucine amino acid substitution of the sodium channel. High conservation of the channel protein across species suggests that the exact sequence of amino acids is vital for proper function. Sodium channels are voltage-sensitive ion channels responsible for the initiation and propagation of the action potential. During an action potential, sodium channels open or activate, then synchronously inactivate. When the sodium channel is activated, an inward sodium current causes membrane depolarization. Upon inactivation, the channel pore is closed, and repolarization occurs by active transport of sodium and potassium by the sodium-potassium adenosinetriphosphatase. SEE CLINICAL PATHOLOGY.

Laboratory conducted patch-clamping experiments in horses and humans experiencing hyperkalemic periodic paralysis demonstrate that amino acid substitutions result in a failure of inactivation of the sodium channel. During an episode, the sustained influx of sodium results in either prolonged muscular contractions or paralysis, depending on the number of noninactivating sodium channels and muscle fibers involved. The parallel efflux of potassium from the muscle leads to dramatic increases in plasma potassium and can further exacerbate the weakness. Environmental factors that predispose the horse to episodes of weakness include fasting, ingestion of feed containing a high potassium content, general anesthesia, and rest following exercise.

Treatment. Emergency treatment for paralytic episodes includes intravenous fluid therapy for hyperkalemia. Intravenous calcium gluconate and dextrose solutions bring on a rapid resolution of signs of weakness or paralysis. For mild episodes of muscle tremors, exercise or oral glucose supplements often abort an attack. The disease is usually readily managed through diet by decreasing the potassium content of and avoiding drastic changes in feed, and in many cases through the use of diuretics such as acetazolamide or hydrochlorothiazide. In addition, owners and prospective buyers of affected horses should be informed about the heritable nature of the disease.

For background information SEE HORSE PRODUCTION; MUSCULAR SYSTEM DISORDERS; NERVOUS SYSTEM DISORDERS; VITAMIN E in the McGraw-Hill Encyclopedia of Science & Technology.

Sharon J. Spier; Eric P. Hoffman

Bibliography. J. F. Cummings et al., Equine motor neuron disease: A new neurologic disorder, *Equine Prac.*, 13:15–18, 1991; J. F. Cummings et al., Equine motor neuron disease: A preliminary report, *Cornell Vet.*, 80:357–379, 1990; T. J. Divers et al., Equine motor neuron disease: Findings in 28 horses and proposal of a pathophysiological mechanism for the disease, *Equine Vet. J.*, 26:409–415, 1994; H. O. Mohammed et al., Risk factors associated with equine motor neuron disease: A possible ALS model, *Neurology*, 43:966–971, 1993; B. A. Valentine et al., Acquired equine motor neuron disease, *Vet. Pathol.*, 31:130–138, 1994.

Ink

According to the U.S. Environmental Protection Agency (EPA), the lithographic printing industry in the United States annually emits more than 10^9 lb (4.5×10^8 kg) of volatile organic compounds into the air through the use of organic cleaning solvents. These solvents have been used for many years to clean oil-based inks from printing press components. Although highly effective, they have come under increased scrutiny because of the environmental, safety, and health concerns they present. For example, volatile organic compounds contribute to ozone formation in the lower atmosphere.

Faced with greater pressures from regulatory groups, the printing industry has struggled to find effective solvent replacements that minimize emissions of volatile organic compounds and hazardous air pollutants. Although conventional replacement approaches have been unsuccessful, a recently developed alternative approach completely eliminates ink removal solvents and their related volatile organic compounds from the lithographic printing process.

This new printing system, which became commercially available in 1994, incorporates selective solubility in oil-based ink formulations, resulting in an ink that performs lithographically during printing but requires a water-based solution during cleanup.

Lithography process. Discovered in 1798, lithography has become the world's most popular printing method. In the United States alone, lithography accounts for 50% of all printing and is used to print newspapers, magazines, books, and advertising items. Lithography is based on the principle that oil and water do not mix; thus, a lithographic printing press consists of an oil-based ink system and an aqueous fountain solution system that apply ink and water solution to chemically treated printing plates, which accept the ink and water solution in specified image and nonimage areas (**Fig. 1**). During printing, the ink and the aqueous fountain solution are mixed on the press roller train under significant shear. The mixture then travels to the plate and separates, the oil-based ink adhering to the image area and the aqueous solution wetting the nonimage area. The ink from the plate is then transferred (offset) to an intermediate blanket cylinder and finally to the paper substrate.

Periodically during and, especially, after a printing run, inks need to be cleaned from press components. Because these inks are oil-based and water-repelling, organic solvents are the most effective cleaners. These solvents are applied with shop towels or shop rags that are subsequently sent to industrial launderers for cleaning or directly to landfills for disposal. Automated press-washing systems are also used.

Recently, the EPA identified the cleaning solvents used by the lithographic industry as a significant source of emissions of volatile organic compounds and moved to curtail their use. Generally, cleaning solvents consist entirely of volatile organic compounds, and proposed EPA emission guidelines call for levels of these compounds below 30% by weight.

As the printing industry has sought alternatives to organic solvents, three approaches have been tried. The first reduces vapor pressure in solvents and thus evaporation of volatile organic compounds; the second dilutes solvents with water; and the third uses emulsification and detergency. These approaches have been unsuccessful because they either do not eliminate the use of solvents or are generally ineffective against the inherent ability of oil-based ink to resist water.

Each of these approaches operates within the orthodox view of modern lithography, which considers the components of ink, aqueous fountain solution, and solvents as independent elements. Lithographic inks, for example, have evolved from a perspective of perfecting print quality and on-press performance; no consideration was given to clean-up properties. Similarly, solvent replacement research has accepted lithographic ink as immutable, and has attempted to develop materials to replace traditional solvents.

The new printing system revises the orthodox view by considering the elements of lithography as a process, one that consists of interdependent, not independent, chemical compositions. In the new approach, press clean-up properties become equal to ink quality, press performance, and fountain solution in the overall lithographic printing process.

Solubility conversion. Once lithography is viewed as a process, the key to the new approach is a fundamental understanding: lithographic ink needs to remain water-insoluble only until it reaches the offset blanket. After the image has been transferred to the paper substrate, it is immaterial that the ink remain water insoluble. Thus, the application of the principles of solubility conversion—the concept that ink can be made selectively soluble depending on the aqueous solution with which it interacts—becomes a possibility.

However, to be accepted by the printing industry, where tradition and large investments in equipment present formidable barriers, the new approach would need to conform to prevailing quality standards in ink properties and cleaning performance. In addition, the solubility conversion of the ink would need to occur quickly and under ambient conditions. Moreover, the new system would need to work with existing printing technology and procedures to minimize impact on printing operations, and would need to use existing raw materials so as to be cost-effective. Lastly, the mechanism that effects the solubility conversion would need to be an easily isolated, easily controlled chemical key.

After many possible systems were considered, it was decided to use pH as the conversion mechanism capable of changing the solubility of oil-based inks. The selection of pH was made for a number of reasons: (1) solution equilibrium allows development of a defined isoelectric point (pI) for the ink with well-defined solubility behavior; (2) pH permits the design of amphoteric systems, which allow solubility conversion over a wide range of pH; (3) pH is unaffected by fountain solution additives, such as salts, surfactants, and cosolvents; (4) pH is easy to measure and generally remains constant during a printing run; and (5) pH is cost-effective, simple, and familiar to printers.

To accomplish a solubility conversion acceptable to the printing industry, the solubility of the ink needs to be a strong function of pH and to mimic a step function. The ideal behavior of solubility conversion using an acidic fountain solution and alkaline wash is a step function (**Fig. 2**). Increasing the pH triggers the conversion. Conversely, decreasing the pH after the conversion causes the ink to precipitate from the water-based solution, allowing its removal by filtration. The solubility conversion also has to be stable and predictable. On either side of

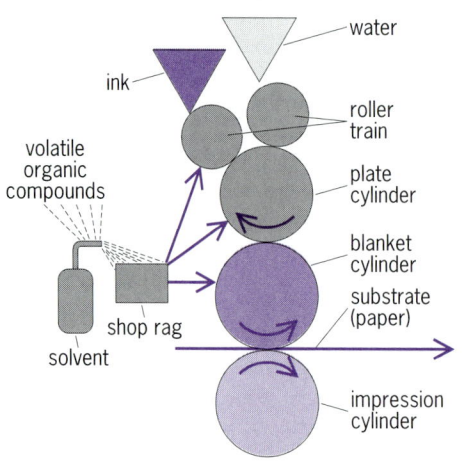

Fig. 1. Components of the offset lithography process.

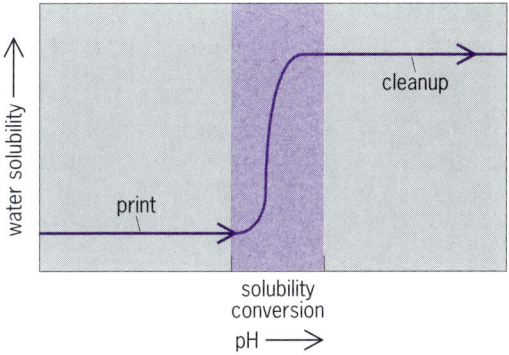

Fig. 2. Ideal behavior of solubility conversion. Water solubility is plotted as a function of pH; an increase in the pH corresponds to an increase in alkalinity.

the steplike conversion, the curves need to remain flat to prevent small pH changes that would cause swings in the hydrophilic properties of the ink.

Water-reducible resins. Lithographic inks are produced by heating resins such as rosin esters or cyclopentadienes in a vegetable oil or petroleum distillate to produce a varnish, which is the major component of the ink. Pigment is used as colorant, and is dispersed in the varnish by milling or other techniques. Additives such as waxes, driers, and dispersants are also commonly used to produce desired characteristics.

To make lithographic ink amenable to solubility conversion, the new system includes water-reducible resins as part of conventional ink formulations. Using resins with acid functional groups in unneutralized form produces an acid-print, alkaline-wash system. Using resins with amine functional groups produces an alkaline-print, acid-wash system. Although both systems are used in lithographic printing, the acid-print system is the dominant one. Resins that possess a carboxyl group (an acidic functional group) are ideal for this system. The pI of the resins is approximately 4–5, which is the most common pH of acid fountain solutions. At the pI of a resin, maintaining the acidic form is strongly favored, resulting in the resin remaining water insoluble.

As the shift to the neutralized form occurs, ionic sites are created, causing the ink to become water soluble. Meanwhile, the ionic behavior causes a shift in the oil-based varnish that destroys the integrity of varnish and enhances the ability to clean with an aqueous solution. During cleanup, two forces occur simultaneously to produce a rapid conversion. The first is varnish destabilization and the second is conversion to a water-soluble form.

Benefits. The new system offers several benefits. The system remains oil-based, uses the existing raw materials from which inks are produced, and offers quality and performance equal or superior to conventional inks. More important, the new system allows the move from petroleum distillates to completely vegetable oil–based inks, a renewable resource. The new system also works with conventional printing technology, including various combinations of printing plates and fountain solutions.

Apart from the performance issues, the new system offers significant benefits related to environmental, safety, and health issues. Besides eliminating the volatile organic compounds related to solvent use, the system eliminates the water pollution that was previously associated with conventional printing, mainly because the conversion can be reversed by changing the pH of the water-based solution. The ink then reverts to a water-insoluble form, precipitating from the solution as solid particles. Filtration removes the ink particles with 99% efficiency. The water-based cleaner helps mitigate the problems involving the safety and health of the printing plant employees. Like the ink, the wash is nontoxic. It is also noncorrosive and free of hazardous air pollutants, and its use eliminates the need for solvent storage.

For background information SEE INDUSTRIAL HEALTH AND SAFETY; INK; pH; PRINTING; RESIN in the McGraw-Hill Encyclopedia of Science & Technology.

Thomas J. Pennaz

Bibliography. R. H. Leach (ed.), *The Printing Ink Manual,* 4th ed., 1988.

Insects

Several recent discoveries have challenged many traditional views of the fossil record and the evolutionary history of insects. These new advances include a more intensive study of the fossil record of insects, particularly broad-scale analyses of their taxonomic, morphologic, and ecologic diversity in the fossil record; analyses of the relationships of hexapods (insects and relatives) and the related myriapods (centipedes, millipedes, and related forms) to other major arthropod groups; and insights regarding the phylogenetic relationships of a late Paleozoic group of insects, the Protorthoptera. In addition, there has been a renewed effort to analyze tissues, cells, and biomolecules of amber-entombed insects in order to reconstruct phylogeny. Finally, the traditional phylogenetic relationships of a major insect group, the Strepsiptera (twisted-wing parasites), may be altered because of new insights from the phylogeny of modern forms.

Fossil evidence. One of the most perplexing aspects of insect paleobiology has been the nature of their fossil record. Although the insect fossil record was traditionally considered inferior in quality to the fossil records of other major taxa, a recent, extensive study of family-level diversity of fossil insects has demonstrated that they compose a high (63%) representation of extant taxa as fossils.

Since the early Carboniferous (about 330 million years ago, or m.y.a.), insect diversity has significantly exceeded tetrapod diversity. Although insect diversity suffered a modest decline during the terminal Carboniferous, and a drastic reduction during the Permo-Triassic extinction, the Triassic rebound in diversification essentially has continued to the present, unaffected by the terminal Cretaceous event during which many shallow marine and terrestrial groups, including ammonites and dinosaurs, were destroyed (see **illus**.). During this Mesozoic and Cenozoic interval, the semilogarithmic rate of increase in insect diversity was achieved not by high origination rates but by low extinction rates. Even so, diversification rates actually slackened during the early Cretaceous, contemporaneous with the angiosperm radiation, indicating that at the family level, the ecological dominance of flowering plants had no or even a negative effect on the origin of major insect lineages.

This trend in taxonomic diversity is also supported by morphological disparity in the form of mouthpart classes, with 65–88% of all modern insect mouthpart classes occurring prior to the ecological dominance of flowering plants. Collectively, these findings indicate that the angiosperm radiation was not instrumental in producing either major taxonomic lineages or major morphological mouthpart novelties in insects; this pattern is explained as a consequence of insect diversity propelled by earlier seed plant radiations during the late Paleozoic and early Mesozoic.

Lepidopteran fossils. Another approach used to assess the potential role of insects during the angiosperm radiation is the evaluation of highly stereotyped, insect-mediated plant damage in early angiosperm deposits. Of all major insect orders, the most highly coevolved with angiospermous plants is the Lepidoptera, consisting of moths and butterflies. Three types of moth-induced leaf mining in well-preserved leaves from the mid-Cretaceous Dakota formation of Kansas and Nebraska were recently documented. One example showed highly diagnostic damage done by the leaf-mining moth, *Ectoedemia,* to an ancestor of modern *Platanus*, the sycamore, recording an association between *Ectoedemia* and its sycamore host for the past 97 m. y., an association that began during the latter stages of the angiosperm ecological expansion on land. This association is the most geologically ancient known for any herbivore interacting with a flowering plant. Early lepidopteran body-fossil evidence suggests that the basal radiation of major lineages of the Lepidoptera occurred during the Late Jurassic, amid a gymnospermous global flora.

Phylogenetic interpretation. A perennial issue in the study of fossil arthropods and insects in particular has been the interpretation of the array of appendages and body sclerites, and the use of these structures for reconstructing the phylogeny of major arthropod lineages. Until recently, functional interpretations of these structural features have concluded that the Arthropoda was not phylogenetically cohesive and was composed of at least three major lineages: Crustacea, Chelicerata (horseshoe crabs, scorpions, spiders, and relatives), and Uniramia. Each evolved arthropodlike features independently from a nonarthropod, presumably annelid, ancestor. According to this view, the Uniramia consists of the Onychophora (velvet worms), Myriapoda, and Hexapoda. Several separate studies of arthropod phylogeny have now shown that all arthropodlike taxa are phylogenetically related and are united by several, uniquely defined characteristics. The Uniramia excludes the Onychophora and consists of the Myriapoda and the Hexapoda, which share sister-group status with Crustacea. Thus, the concept of Uniramia has been discarded.

The heterogenous assemblage of generalized and primitive fossil insects, the Protorthoptera has become disassembled, as some species became allocated to modern lineages. Several features of protorthopteran wing veins, mouthparts, and mouthpart-related head structures have been identified as clues for assignment of particular taxa to the plecopteroids (stone flies), orthopteroids (grasshoppers, earwigs, and relatives), blattoids (roaches, termites, and mantids), hemipteroids (thrips and cicadas), and endopterygotes (beetles, true flies, moths, wasps, and relatives). This reassignment of protorthopterans is the first major attempt to address the polyphyletic nature of the Protorthoptera, in which independently derived lineages with distinct histories have been gathered together into a single taxonomic order. This reevaluation is considered controversial by some, in part, because it uses a morphological ground plan approach in which specific structural features are not explicitly suited to characterization for phylogenetic analysis.

Preservation in amber. Post-Paleozoic, three-dimensionally preserved amber has revealed exquisite detail of insects at the histological, cellular, and organellar levels. Soft-part preservation in amber-entombed insects has only been demonstrated for the past decade, and its exceptional nature dispels earlier notions that the interiors of amber insects consisted only of cavities devoid of tissue and surrounded by chitinous husks. The examination of Dominican amber (25–40 m. y. a.) by scanning electron microscopy has revealed detailed histological information such as the external ornamentation of the digestive tract proventriculus, the structure of respiratory tracheoles, and the fibrillar organization of flight muscle. At higher magnifications achieved with use of the transmission scanning electron microscope, cellular and organellar ultrastructure has been resolved, including nuclear membranes, endoplasmic reticula, and mitochondrial cristae. These discoveries expose a new role for insect amber fossils by providing histological, cellular, and especially biomolecular data in the understanding of insect evolutionary history. SEE *RADIOCARBON DATING*.

The most striking application of insect soft-part preservation in amber has been the use of biomolecules to determine phylogenetic relationships. For example, insect deoxyribonucleic acid (DNA)

Insects 157

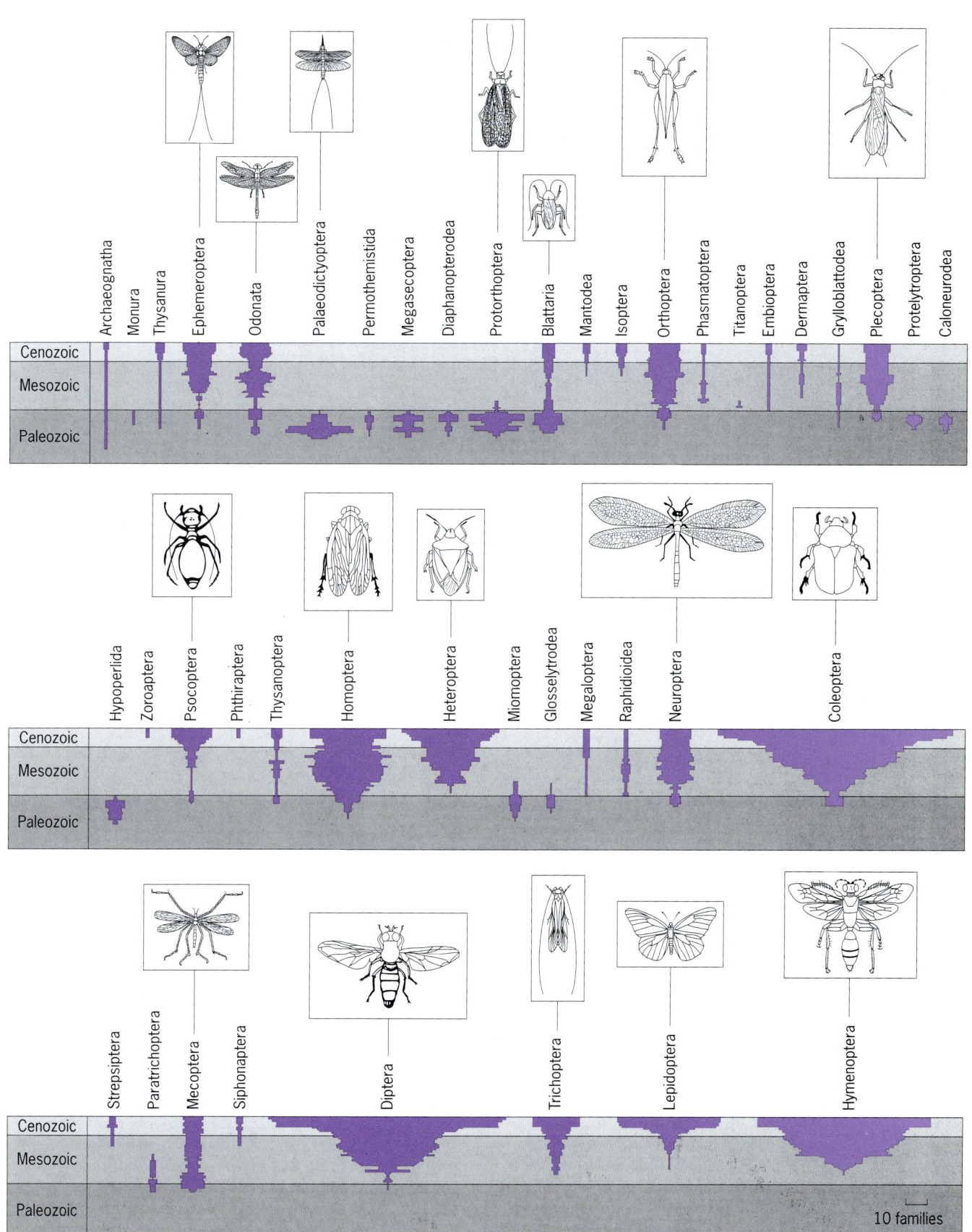

Spindle diagrams displaying the diversities of fossil insect families within orders in geologic stages of the Phanerozoic. Note the scale bar representing 10 families in the lower right.

from both Dominican and Lebanese (120–135 m. y. a.) ambers have been sequenced to reveal the gross relationships of mastotermitid termites and nemonychine weevils, respectively, to their modern insect relatives. The DNA chains from these two ancient taxa exhibited the greatest sequence similarity with modern congeneric and confamilial taxa, as expected. However, in the case of the termite *Mastotermes*, inclusion of the fossil DNA in the phylogenetic analysis resulted in a monophyletic group, the Isoptera (termites), or specifically one that incorporated all of its ancestors as a distinct group with respect to other related insect orders. However, exclusion of the *Mastotermes* DNA data resulted in a phylogeny whereby the Isoptera was a paraphyletic group (derived from more than one common ancestor). This indicates that incorporation of fossil DNA data can potentially transform the understanding of insect relationships and render invalid certain phylogenetic hypotheses based only on modern taxa.

Modern evidence. In instances where fossil representatives are absent or rare, evidence from modern insects alone can reveal phylogenetic events such as the origin of the rare, parasitic, and poorly fossilized order Strepsiptera. Phylogenetic analyses of both molecular and morphological evidence have indicated that the Strepsiptera may be the likely sister group to the Diptera (true flies) rather than the sister group to or even a member of the Coleoptera (beetles) as traditionally thought. A genetically based, developmental mechanism has been proposed for this novel sister group relationship, namely, a homeotic mutation of the strepsipteran thorax, which reversed the relative position of the paired vestigial wings (halteres), and the large, functional wings for each order. By proposing phenotypic expression of the ultrabithorax homeotic mutation in an ancestral larval strepsipteran, an opposite placement can result from the typical dipteran condition of halteres on the third thoracic segment and wings on the second thoracic segment. The plausible genetic basis for such a transformation partly explains the unusual morphology of this group.

For background SEE AMBER; ARTHROPODA; FOSSIL; INSECTA; MICROSCOPE in the McGraw-Hill Encyclopedia of Science & Technology.

Conrad Labandeira

Bibliography. R. DeSalle et al., DNA sequences from a fossil termite in Oligo-Miocene amber and their phylogenetic significance, *Science*, 257: 1933–1936, 1992; D. G. Grimaldi et al., Electron microscopic studies of mummified tissues in amber fossils, *Amer. Mus. Novitates*, 3097:1–31, 1994; A. A. Henwood, Soft-part preservation of beetles in Tertiary amber from the Dominican Republic, *Palaeontology*, 35:901–912, 1992; J. Kukalová-Peck, The "Uniramia" do not exist: The ground plan of the Pterygota as revealed by Permian Diaphanopterodea from Russia (Insecta: Paleodictyopteroidea), *Canad. J. Zool.*, 70:236–255, 1992; C. C. Labandeira and J. J. Sepkoski, Jr., Insect diversity in the fossil record, *Science*, 261:310–315, 1993; M. F. Whiting and W. C. Wheeler, Insect homeotic transformation, *Nature*, 368:696, 1994.

Intermetallic materials

Recent research involving structural materials suitable for high-temperature applications such as components for aircraft engines and aerospace vehicles has focused on intermetallic materials. The drive to improve performance and cost-efficiency has placed an unprecedented demand on materials requirements, including enhanced high-temperature capabilities and decreased weight. It is anticipated that vehicles being designed for the twenty-first century will fly significantly faster and at higher altitudes, mandating alternative engine materials. Although several of these needs are now being met by advanced nickel-base superalloys, the density of these superalloys (~ 7.5–8.0 g/cm^3) is a disadvantage; additionally, these alloys are being used to as high as nine-tenths of their melting point, leaving little room for further enhancement. Ceramic materials are capable of sustaining higher temperatures and are less dense, but they are brittle and lack the desired damage tolerance. In addition, ceramics are difficult to process, machine, and join by conventional techniques. Ordered intermetallics bridge the gap between metals and ceramics. They exhibit physical properties characteristic of metals and can be processed by conventional casting techniques, but their mechanical properties resemble those of ceramics. Some intermetallics also have poor resistance to oxidation and sulfidation attack at elevated temperatures.

The primary barrier to their use in load-carrying applications is their brittleness or low crack tolerance. The reasons for this behavior include complex crystal structure and deformation behavior, embrittlement due to defects (interstitials, vacancies) and to hydrogen, and notch sensitivity. The major effort in research and development on intermetallics is concerned with increasing the ductility and toughness while maintaining good high-temperature strength, stiffness, creep resistance, and resistance to oxidation and sulfidation.

Ordered structures. An ordered intermetallic is an alloy of two elements, A and B, in specific ratios (for example, AB, AB$_2$, AB$_3$, A$_2$B, A$_3$B). In the fully ordered condition, the two species are arranged on specific lattice sites. Thus, in **Fig. 1**a, a metal with a body-centered cubic (bcc) crystal structure such as iron (Fe) has two types of sites: a corner site X and a body-centered site Y, yielding two atoms per unit cell. When an alloying element is added to this metal, it can occupy both sites at random. However, if the material is fully ordered, the second species can occupy only one of the two sites. If all such sites are occupied, a compound of

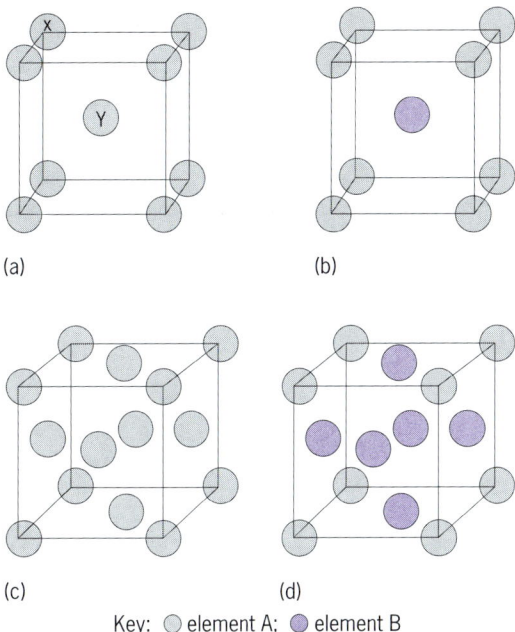

Fig. 1. Ball-and-stick models showing the unit cells for (*a*) body-centered cubic, with corner site X and body-centered site Y; (*b*) B2; (*c*) face-centered cubic; and (*d*) L1$_2$ structures.

the type AB results, with a crystal structure known as B2 (Fig. 1*b*). A face-centered cubic (fcc) structure (Fig. 1*c*) has an ordered equivalent called in standard crystallography terminology the L1$_2$ structure (Fig. 1*d*), where the face-centered sites are occupied by the major species and the corner site by the minor species, yielding the AB$_3$ stoichiometry. Similarly, the ordered version of a hexagonal close-packed (hcp) structure is called a D0$_{19}$ structure with an AB$_3$-type stoichiometry. The ordered structure bestows certain unique characteristics to these materials, such as a low diffusion coefficient that translates to improved creep resistance—a measure of the material's high-temperature capability. Similarly, an ordered structure can be expected to adversely affect low-temperature ductility by increasing the unit slip distance, making dislocation motion difficult.

Depending on their constitution, ordered intermetallics are divided into groups such as aluminides, silicides, and beryllides. The aluminides show the most promise for the near term, although the silicides, with higher melting points, offer higher temperature capabilities. The beryllides tend to exhibit the lowest densities; however, they are brittle, require excessive precaution in processing, and are technologically the least advanced. Of the aluminides, iron, nickel, and titanium aluminides are receiving the most attention. Pairs of relevant compounds in each of these systems are Fe$_3$Al and FeAl, Ni$_3$Al and NiAl, and Ti$_3$Al and TiAl. **Table 1** shows the physical properties of these aluminides.

Nickel aluminides. The two major nickel aluminides are Ni$_3$Al and NiAl.

In the crystal structure of Ni$_3$Al, the nickel atoms occupy the face-centered sites and the aluminum atoms occupy the corner sites of a face-centered cubic unit cell. Ni$_3$Al has excellent elevated-temperature strength and good oxidation resistance; there is an anomaly in the yield strength in that it increases, rather than decreases (as is the case with most other structural materials), with increasing temperature. Single crystals of Ni$_3$Al are highly ductile, whereas polycrystalline materials are brittle at ambient temperatures due to the brittleness of the grain boundaries. Polycrystalline Ni$_3$Al is also prone to environmental embrittlement at both ambient and elevated temperatures. The ambient temperature embrittlement is due to hydrogen from moisture in the air, and the elevated temperature embrittlement is caused by oxygen that penetrates along the grain boundaries, causing intergranular fracture. The worst embrittlement occurs in air at 600–800°C (1100–1500°F). The mechanical properties of Ni$_3$Al can be improved by alloying additions of boron, chromium, hafnium, molybdenum, and zirconium. Binary Ni$_3$Al alloys are being examined for alternate applications, such as turbocharger rotors in diesel engine trucks.

The alloy NiAl has two-thirds the density of nickel-base superalloys and, in addition, exhibits significantly higher (four to eight times depending on composition and temperature) thermal conductivity and excellent oxidation resistance. Although binary NiAl is brittle even in single-crystal form, minor alloying with iron, molybdenum, and gallium has recently been shown to significantly improve tensile ductility in the ⟨110⟩ crystallographic direction. Various forms of NiAl, including

Table 1. Physical properties of the nickel, titanium, and iron aluminides

Compound	Crystal structure	Melting point, °C (°F)	Critical ordering temperature, °C (°F)	Density, g/cm^3	Young's modulus, GPa
Ni$_3$Al	L1$_2$	1390 (2530)	1390 (2530)	7.5	178
NiAl	B2	1640 (2980)	1640 (2980)	5.9	230*
Ti$_3$Al-base	DO$_{19}$	~1650 (3000)	~1160 (2120)	4.1–4.7	110–145
TiAl-base	L1$_0$	~1450 (2640)	~1450 (2640)	3.7–3.9	160–180
Fe$_3$Al	DO$_3$	1540 (2800)	540 (1000)	6.7	141
	B2		760 (1400)	1540 (2800)	
FeAl	B2	1250 (2280)	1250 (2280)	5.6	260

*Per M. R. Harmouche and A. Wolfenden, Modulus measurements in ordered Co-Al, Fe-Al, Ni-Al alloys, *J. Testing Eval.*, 13 (6):424–428, 1985.

single crystals of single-phase and multiphase alloys, directionally solidified eutectics, and particulate- and fiber-reinforced composites, are being produced, characterized, and compared for their performance capabilities. Of these, single-crystal NiAl (single-phase and multiphase) appears to exhibit the most desirable combination of properties and is currently being evaluated for turbine blades and vanes. Such blades can be machined out of a single-crystal bar, or alternately an oversized net-shape single crystal blade can be produced and then be machined to final dimensions. A photograph of a single-crystal NiAl blade is shown in **Fig. 2a**.

Iron aluminides. The two major iron aluminides are Fe_3Al and FeAl. Both have excellent oxidation resistance and corrosion resistance because they form protective scales at elevated temperatures in hostile environments. They also are low in cost and have low density and good strength up to intermediate temperatures. Their main drawbacks are poor ductility and brittle fracture at ambient temperatures and poor strength and creep resistance above 650°C (1200°F). The poor ductility at ambient temperatures is caused by the environmental embrittlement due to hydrogen from the moisture in the air. Their inadequate high-temperature strength and creep resistance has discouraged research on these materials for gas-turbine engine applications in the aircraft industry. Instead, these compounds are being investigated as potential replacements for stainless steel in selected applications in the chemical, petroleum, and coal industries.

Titanium aluminides. The major titanium aluminides are Ti_3Al and TiAl. They are well suited for aerospace applications because of their low density compared to the standard engine materials, nickel-base superalloys. The density of titanium aluminides is about half that of superalloys. Titanium aluminides also have good high-temperature strength but poor ductility and toughness at ambient temperatures.

Alloys corresponding to Ti_3Al have been developed as two-phase alloys with compositions based on Ti with 23–25% Al and 10–30% niobium (Nb) [a phase is a constituent of an alloy that is physically distinct and is homogeneous in chemical composition]. The oxygen impurity level is a very important consideration, because it has a large effect on the ductility. Ti_3Al has limited ductility at ambient temperatures. The mechanical properties depend on the composition and the microstructure. Increasing the Nb content increases the ductility, but the toughness and creep resistance remain low. Increasing the Al content increases the strength. The microstructure can be altered significantly by thermal mechanical processing. Ti_3Al is susceptible to stress corrosion cracking and hydrogen embrittlement. The oxidation resistance of Ti_3Al is not very good, but adding Nb tends to increase it. Research focused on minimizing or eliminating these drawbacks by alloying and coating and by reinforcing Ti_3Al with particulates and fibers is in progress.

Promising materials corresponding to TiAl have the following compositions: Ti with 46–52% Al and 1–10% M, where M is at least one element from the group of metals chromium, molybdenum, niobium,

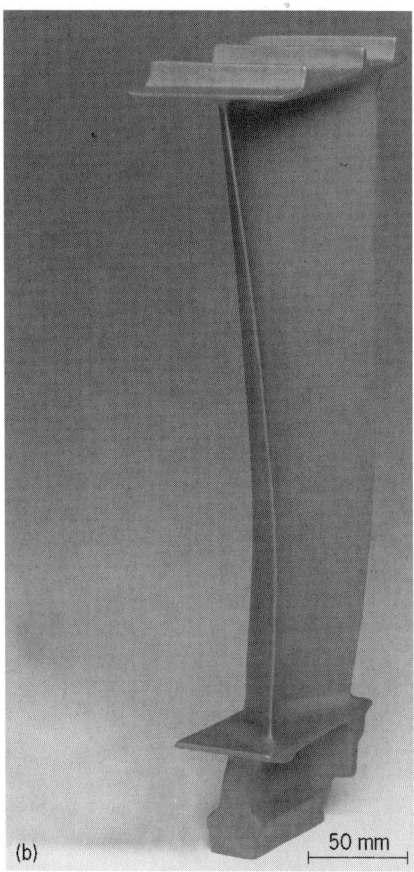

Fig. 2. Applications of aluminides. (*a*) Single-crystal NiAl high-pressure turbine blade (*courtesy of R. Darolia of General Electric Aircraft Engines*). (*b*) Polycrystalline, investment-cast TiAl-base, low-pressure turbine blade (*courtesy of D. Larsen of Howmet Corp.*).

Table 2. Properties of titanium alloys, titanium aluminides, and superalloys				
Property	Ti-base	Ti$_3$Al-base	TiAl-base	Superalloys
Structure	hcp/bcc	DO$_{19}$	L1$_0$	fcc/L1$_2$
Density, g/cm^3	4.5	4.1–4.7	3.7–3.9	7.9–8.5
Modulus, GPa	95–115	110–145	160–180	206
Yield strength, MPa	380–1150	700–990	350–600	800–1200
Tensile strength, MPa	480–1200	800–1140	440–700	1250–1450
25°C (77°F) ductility, %	10–25	2–10	1–4	3–25
25°C (77°F) fracture toughness, MPa$\sqrt{m}$	12–50	13–30	12–35	30–100
Creep limit, °C (°F)	600 (1100)	750 (1380)	750–950 (1380–1740)	800–1090 (1470–1990)
Oxidation, °C (°F)	600 (1100)	650 (1200)	800–950 (1470–1740)	870–1090 (1600–1990)

SOURCE: Y.-W. Kim, Ordered intermetallic alloys, pt. III: Gamma titanium aluminides, *J. Met.*, 46(7):30–39, 1994.

tantalum, vanadium, and tungsten. The materials are produced as either single-phase or two-phase materials. The two-phase materials contain both the Ti$_3$Al and the TiAl phases. Additions of niobium or tantalum increase the strength and oxidation resistance of single-phase materials. In the two-phase materials the morphology of the phases can be changed from lamellar to equiaxed by means of thermal mechanical processing.

The mechanical properties of these alloys are sensitive to these microstructural modifications. Unlike the NiAl alloys, TiAl alloys are produced in the polycrystalline form; at present, investment casting appears to be the preferred route to producing net-shape components (Fig. 2b). **Table 2** compares typical mechanical properties of TiAl-base alloys with other Ti-based systems and superalloys. A number of gas turbine engine components have been identified for TiAl-base alloy applications: rotational and stationary compressor components such as high-pressure compressor blades; turbine components such as low-pressure turbine blades; combustor components such as diffuser case and swirler; and nozzle components such as flaps and outer skins. A low-pressure turbine wheel with cast TiAl-base alloy blades has passed a rigorous simulated engine test.

Molybdenum disilicide. High-temperature silicides are a new class of materials with potential applications in the temperature range 1200–1600°C (2200–2900°F). However, there are several issues to be solved before they can be used as structural materials. Such issues include (1) improving the intermediate-temperature oxidation behavior to avoid pesting (the disintegration of a silicide material in an oxygen-bearing environment, generally occurring between 300 and 900°C or 570 and 1700°F); (2) increasing the ambient-temperature ductility and toughness; and (3) improving the high-temperature creep resistance. The silicide that seems most promising is molybdenum disilicide (MoSi$_2$) because of its high melting temperature (2200°C or 4000°F) and excellent oxidation resistance.

The material with the composition MoSi$_2$ has a tetragonal crystal structure. A major problem is its absence of ductility at temperatures up to 1000°C (1800°F). It also has poor high-temperature strength due to the presence of a grain boundary silicon-rich phase that may become viscous at very high temperatures. Recent research has shown that this problem may be solved by the addition of carbon to the material. Although MoSi$_2$ is not adequate for structural applications in either the single crystal or polycrystalline form, recent work has shown that by using MoSi$_2$ as a matrix material, composites can be produced that may be suitable for high-temperature structural use.

Production. The primary processing methods for production of intermetallics are melting, casting, ingot processing, and powder metallurgy. The secondary processes include forming, machining, and chemical milling.

Primary processing. The two major methods of producing intermetallics are by melting and casting or by the production of powders. In the melting of aluminides the large difference between the melting temperature of aluminum and that of iron, nickel, and titanium must be considered. Other pertinent factors are (1) the large amount of aluminum in the intermetallic; (2) the melting temperature of the intermetallic (it may be higher than that of either constituent); and (3) the general reactivity of the elements to be melted. Similar issues exist for MoSi$_2$, except that the difference in the melting temperatures of the elements is smaller.

The melting methods used are air induction melting (AIM), vacuum induction melting (VIM), vacuum arc remelting (VAR), vacuum arc double electrode remelting (VADER), electroslag remelting (ESR), plasma melting, and electron-beam melting. Ni$_3$Al and the iron aluminides can be melted in air, although the high amount of aluminum promotes the rapid formation of a continuous aluminum-oxide film on the top of the molten materials. For many applications it is best to use processes that do not involve air melting, such as VIM, VAR, VADER, and ESR. The extreme reactivity of molten titanium with moisture and with oxidizing and carburizing environments makes the melting of titanium aluminides in conventional melting crucibles impossible. Titanium aluminides are melted by a process known as the induction skull method (ISM), which combines features of consumable arc melting and conventional ceramic crucible induction melting. Also the VAR process and plasma arc cold hearth melting (PACHM) process have been used to melt titanium aluminides as well as $MoSi_2$. Many applications of intermetallics require cast components. Casting methods include sand, investment, centrifugal, directional solidification, and near net shape methods.

Fine-grain wrought products are required in other applications. The commercial use of intermetallics at competitive cost requires fabrication by conventional hot working operations. The primary processing of cast ingots is feasible, but the requirement for hot working is more stringent than for commercial metallic alloys. The secondary processing of intermetallics is very difficult and varies from intermetallic to intermetallic.

Powder metallurgy offers the most flexibility in producing intermetallics. The problem of using powder metallurgy for this purpose is that these production methods often result in surface contamination of the powders. Each of the powder consolidation methods for producing intermetallics from powders has processing difficulties. These methods are hot pressing, hot isostatic pressing, powder injection molding, extrusion, and explosive compaction. The reaction synthesis process has been uniquely applicable for intermetallics and has been used to produce many different materials.

Secondary processing. Secondary steps such as machining and joining are critical in using these advanced materials for various applications, and extensive efforts are under way to develop these technologies. Innovative joining techniques such as friction welding, capacitor discharge welding, flash welding, laser welding, welding using the combustion synthesis concept, welding using microwaves and infrared waves, electron-beam welding, brazing, and diffusion bonding are under evaluation. A variety of machining techniques, including electrodischarge machining, water-jet cutting, ultrasonic machining, and laser cutting, are available to precision-machine complex geometries and contours. Conventional grinding, diamond drilling, and boring techniques have seen limited applications in machining TiAl alloys.

Intermetallic matrix composites. Intermetallic matrix composites have recently received considerable attention, and a variety of matrices and reinforcements have been examined to date. Reinforcement type, volume fraction, size, shape, and distribution have been shown to affect microstructure and mechanical properties. Several innovative approaches ranging from conventional techniques, such as mechanical alloying, to more exotic techniques, such as reactive consolidation and magnetron sputtering, have been used to produce these composites. Process models are being formulated and coupled with state-of-the-art sensors technology to optimize parameters to produce composite materials of high integrity.

Issues related to reinforcement selection include availability, mismatch in thermal expansion coefficients, and chemical compatibility with the matrix. Significant advances in characterization have been made in continuously reinforced Ti_3Al-based alloys (SiC fibers in Ti_3Al + Nb alloys) and particulate-reinforced TiAl alloys (TiAl + TiB_2 particulates), in directionally solidified (DS) eutectics of NiAl, and in discontinuously reinforced $MoSi_2$. In most cases, the major emphasis has been on obtaining a desirable balance between creep resistance and low-temperature fracture toughness. In this endeavor, micromechanical modeling has been used to identify, understand, and quantify the critical material parameters that control these properties, thereby permitting microstructural design to obtain the desired combinations of these properties. Potential areas for future research include the need for viable machining and joining techniques and novel but reliable nondestructive evaluation methods.

For background information SEE ALLOY; CERMET; COMPOSITE MATERIAL; CRYSTAL STRUCTURE; HIGH-TEMPERATURE MATERIALS; INTERMETALLIC COMPOUNDS; WELDING AND CUTTING OF METALS in the McGraw-Hill Encyclopedia of Science & Technology.

K. Sharvan Kumar; Jeffrey Waldman

Bibliography. R. Darolia, NiAl alloys for high-temperature structural applications, *J. Met.*, 43(3):44–49, 1991; E. P. George et al., Ordered intermetallics, *Annu. Rev. Mater. Sci*, 24:409–451, 1994; Y.-W. Kim, Ordered intermetallic alloys, pt. III: Gamma titanium aluminides, *J. Met.*, 46(7):30–39, 1994; J. H. Westbrook (ed.), *Intermetallic Compounds*, 1967.

Irrigation (agriculture)

Agricultural erosion research has focused primarily on rainfall-induced soil loss, but erosion losses associated with surface irrigation practices can be equally severe. Of the estimated 2.5×10^8 hectares (6×10^8 acres) irrigated worldwide, at least 60% are surface irrigated. In the Pacific Northwest, approximately 1.5×10^6 ha (3.7×10^6 acres) of the most erosive soils in the United States are surface irrigated. Typically, 5.5–55 tons of soil per hectare per year (5–50 metric tons per acre per year) can be lost from furrow-irrigated fields, and three times that amount from near the furrow inlets at the upper end of fields.

Soil erosion. Soil erosion is highly detrimental, both on- and offsite. Furrow erosion can be reduced effectively by using various approaches, including settling ponds, minibasins with buried-pipe runoff control, furrow straw mulching, and sodded furrows. However, farmers have resisted using these alternatives for various reasons. In some cases, the techniques cannot be conveniently incorporated into existing farm plans; in others, the philosophical or economic inducements are not great enough to stimulate the additional effort. If erosion control is to be more uniformly implemented, farmers require a simple and economical erosion prevention method that permits them to use familiar tillage and crop cultural practices. New technology employing water-soluble polymers may provide such a method. The technique involves the application of 10 mg/liter (0.0013 oz/gal; 10 parts per million) of polyacrylamide polymer to the furrow irrigation stream. Optimal anionic-polyacrylamide applications have reduced mean soil loss from furrow-irrigated fields 94% (80–99%), while net water infiltration into soil increased 15%. Treatment with polyacrylamide also improves furrow tailwater (surface runoff) quality by decreasing levels of phosphorus, nitrate, and biochemical oxygen demand.

Water-soluble polyacrylamide. The most successful polymer to date is a water-soluble anionic organic compound known as polyacrylamide (PAM). Anionic polyacrylamide is composed of 10,000 or more three-carbon molecular units linked into a single linear chain. Of the repeating units, 82% have the structure below. The remaining 18%

$$\left[\begin{array}{cc} H & H \\ | & | \\ -C - & -C - \\ | & | \\ H & O=C-NH_2 \end{array} \right]_n$$

are similar, except the —NH_2 group is replaced by a —[ONa] ionic pair. This pair dissociates upon dissolution, giving the polymer a moderate negative charge. The properties that enable polyacrylamide to prevent soil erosion are its ionic charge type and density, water solubility, and very high molecular weight ($5–15 \times 10^6$ g mol^{-1}).

The ability of polymer manufacturers to synthesize polyacrylamides with increasingly more effective configurations and properties has improved significantly since their introduction to agriculture as soil-stabilizing amendments in the mid-1940s. Historically, polymers were spread in the field and incorporated into the plow layer with tillage. The resulting well-aggregated field surface was significantly more resistant to raindrop impact and concentrated flow, which tend to break down soil aggregates, detaching soil particles, reducing infiltration, and increasing runoff and erosion. Unfortunately the polymer application rates (250–500 kg/ha or 220–450 lb/acre) required to obtain acceptable results were not cost effective. New polymer technologies apply only 0.5–3 kg/ha (0.45–2.7 lb/acre) polyacrylamide per application. At projected prices of $6.50–11.00 per kg, and total seasonal applications of 1–7 kg/ha (0.9–6 lb/acre), polyacrylamide can be economical for all but the lowest-valued irrigated crops. Although new polyacrylamides are actually more effective agents, additional efficiency of the current polyacrylamide application technology results because polymer is applied in the irrigation water. Hence, only that part of the soil surface subject to erosive and seal-forming processes is treated (that is, the wetted furrow perimeter).

Environmental safety. At concentrations used in field application, anionic polyacrylamide is benign, having little or no toxic effect on humans and other mammals, aquatic vertebrates and invertebrates, or plant life. In the United States it has been listed by the Environmental Protection Agency as an acceptable drinking water additive and has gained a variety of approvals by the Food and Drug Administration for food additive applications. In soil, polyacrylamide acts like other naturally occurring, persistent forms of organic matter, degrading to water and carbon dioxide at a rate of approximately 10% per year. At higher rates of polyacrylamide application to soils (>25 mg/kg or 0.0033 oz/lb), research has shown some impacts on soil microorganism populations, although results have not been consistent or pronounced in the studies that have been completed. Much of the current research on polyacrylamide-amended irrigation water seeks to better understand its potential short- and long-term impacts in soil and aquatic systems.

In the United States, by law manufacturers must market only pure polyacrylamide products. The main concern is the manufacturing contaminant, acrylamide monomer, a known toxin. Its concentration in marketed anionic polyacrylamide is strictly regulated (<0.05%). At these levels, acrylamide monomer is of little concern in soils or surface waters because it rapidly biodegrades, decomposing in a matter of days.

Soil interactions. The basis of polyacrylamide's erosion control ability is its propensity to bind with both soil particles and other polyacrylamide molecules. In dilute solution, strands of anionic polyacrylamide polymers exist as random coils. When a strand collides with a soil particle, several segments of the coil become adsorbed to the particle surface. The polymer is held to the surface at numerous points along the contacting segments by any one of several forces, which include electrostatic attraction, hydrogen bonding, van der Waals forces, and chemical bonding. Cation bridging is an important form of electrostatic bonding associated with anionic polyacrylamide. Here, a divalent cation [such as the calcium ion (Ca^{2+})] acts as a positively charged bridge between the negatively charged site on a soil particle and a negatively charged site on the anionic polyacrylamide segment. Because bonding occurs at several locations along the particle/polymer-segment contact, the attraction between the two is very strong and essentially irreversible. A significant fraction of the polymer segments are not in contact with the particle, but extend into the solution. The nonadsorbed segments may then come into contact with, and become adsorbed to, other soil particles. They may also contact another polymer strand, and become either entangled or linked together by cation bridging or hydrogen bonding. These interactions facilitate polyacrylamide's soil stabilizing and flocculating activity during furrow irrigation. When soil colloids suspended in a polyacrylamide solution collide, polymer strands help bind the particles together in flocs. These flocs encounter and bind to other particles. The aggregating masses rapidly become so large that they settle out of suspension by a process known as flocculation.

Furrow irrigation processes. Furrow erosion is controlled by two main factors, furrow stream hydraulics and soil characteristics. Velocity of the flow determines the amount of shear or drag forces available to detach soil particles. Velocity also determines the flow's sediment transport capacity, which along with sediment/aggregate size and density characteristics determines the amount of detached soil that can be transported down the furrow. The soil characteristics aggregate stability and soil cohesion determine to what degree soils are susceptible to flow shear force, and they also control the characteristics of sediment and aggregate size distribution.

To understand how polyacrylamide acts to control erosion, it is necessary first to consider the effects of rapid wetting and flow shear during irrigation of an erodible soil. Prior to irrigation of a newly cultivated furrow, soils are typically very dry and cloddy, and the surface quite rough. Rapidly advancing water is quickly absorbed by soil, causing aggregates to slake and soil particles to disperse. Flow shear easily detaches dispersed soil particles and transports them down the furrow. The rough surface of the wetted furrow is smoothed as soil clods break down. Dislodged soil particles then fill surface cavities along the furrow perimeter. This smoothing increases the velocity and erosiveness of the furrow stream. Initially, infiltration is high and flow rate is low. Infiltrating water soon carries flow-suspended sediment into the soil, where it blocks soil pores and initiates formation of a slowly permeable depositional layer, or surface seal. The seal reduces infiltration, and as a consequence runoff and soil losses increase.

The introduction of polyacrylamide into irrigation water, even at low concentrations, has several impacts on furrow conditions. During initial wetting, polyacrylamide contacts and binds a 1–3-mm-thick (0.04–0.12-in.) layer of soil on the surface of furrow clods and along the wetted furrow perimeter. Treated soil is more cohesive and stable, that is, more resistant to slaking, dispersion, and shear forces. Any fine soil particles in the furrow stream are flocculated and settle as aggregates. Together these processes produce a well-aggregated system. Consequently, surface roughness in the furrow is better maintained, and the depositional layer formed along the wetted perimeter is more porous. Thus, infiltration rates remain high and runoff rates are lower, and soil detachment is inhibited. Sediment transport capacity of the flow is also reduced, because stream velocity is lower and the average aggregate in the system is larger and less easily transported. Polyacrylamide may also increase viscosity of flowing water, resulting in lower turbulence and smaller shear forces.

Application strategies. The mode of polyacrylamide application can be altered simply by (1) varying polyacrylamide concentration in the irrigation water; (2) changing the timing of application (at beginning of irrigation, only intermittently, or continuously); (3) adjusting the length of application period relative to the time required for water to initially traverse the dry furrow (furrow advance time); or (4) changing the form of polyacrylamide added to irrigation water (aqueous stock solution versus a directly introduced crystalline solid). For example, an initial high strategy applies a high dose of polyacrylamide only during early stages of an irrigation. An initial episodic strategy applies polyacrylamide intermittently during the entire irrigation, but dosage rates are lower and individual application periods are shorter compared to the initial high strategy. Another strategy is the continuous low, where polyacrylamide is applied throughout the irrigation but at very low concentration.

Initial high and initial episodic strategies are equally effective at controlling furrow soil loss in the initial treated irrigation, even though the total polyacrylamide applied for initial episodic was 50% of that applied for initial high. When initial high and continuous low strategies were compared, it was found that the continuous low treatment did not protect the furrow from the high loss of loose and easily detached soil particles that typically occurs early in an irrigation: as the irrigation proceeded, the more stable soils remaining in the furrow were more successfully protected. In contrast, the initial high treatment protected both loose and cohesive soil, and was clearly the more effective treatment for the given conditions. Compared to control furrows, it reduced soil loss by 93%, in contrast to a 51% reduction for the continuous low application. A continuous or intermittent application strategy may be more effective under circumstances in which flow shear is relatively high (for example, steeper slopes or high flow rates).

The initial high strategy is recommended, because it has been shown effective for a variety of soils and slopes. If polyacrylamide application is restricted to the furrow advance period, this method also minimizes the amount of polyacrylamide lost in tailwater. (During a 10 mg/liter or 0.0013 oz/gal polyacrylamide application, tailwater contains an average 5–7 mg/liter or 0.0007–0.0009 oz/gal of polyacrylamide.) Although the initial episodic strategy may possibly be as effective, it has not been as thoroughly tested, and may be more difficult to implement in actual farming situations. Preliminary studies comparing initial high, polyacrylamide solution and solid applications (10 mg/liter or 0.0013 oz/gal) indicate that they control soil loss equally well. The advantages and disadvantages of these methods are listed in the **table**.

Research has shown that anionic polyacrylamide technology successfully controls furrow irrigation–induced soil loss under a variety of circumstances; however, efficacy of polyacrylamide

Comparison of solution and solid polyacrylamide application methods					
		Total polyacrylamide employed		Accuracy of polyacrylamide application	
Method	Erosion control efficacy	Volume applied	Mass applied	Concentration	Timeliness*
Solution	Excellent	Large	Large	Excellent	Excellent
Solid	Excellent	Small	Small	Poor	Poor

*The applicator's ability to make timely adjustments to polyacrylamide concentration in irrigation water.

treatments has been shown to vary among irrigations. The effectiveness of polyacrylamide treatments is influenced by the properties of the polymer, the characteristics of polyacrylamide application and of the field under treatment, the nature of irrigation and irrigation water, and the chemical and physical soil characteristics.

Further research is needed to determine how factors such as furrow soil properties, slope length, and inflow water quality influence anionic polyacrylamide efficacy in irrigated furrows.

For background information SEE EROSION; IRRIGATION (AGRICULTURE); POLYMER; SOIL in the McGraw-Hill Encyclopedia of Science & Technology.

Rodrick D. Lentz

Bibliography. F. W. Barvenik et al., Polymers in irrigation water: Symposium, *Soil Sci.*, 158(4):233–300, 1994; D. F. Cook et al., Polymer soil conditioners, *Soil Sci.*, 141(5):311–397, 1986; R. D. Lentz et al., Preventing irrigation furrow erosion with small applications of polymers, *Soil Sci. Soc. Amer. J.*, 56:1926–1932, 1992; T. F. Tadros (ed.), *Solid-Liquid Dispersions*, 1987.

Laser

Ultrashort pulsed lasers, which generate light pulses between 1 femtosecond (10^{-15} s) and 1 picosecond (10^{-12} s) in duration, have found a variety of applications in engineering, chemistry, and physics. Light pulses from these sources can be manipulated to provide time-resolved measurements of ultrafast events, such as photochemical reactions, charge-carrier dynamics in semiconductors, or microcircuit response times; these events occur in a time regime well beyond the capabilities of conventional electronic instruments such as pulse generators and oscilloscopes. The technology of ultrashort laser pulse generation and manipulation will be examined and current applications discussed.

Ultrashort light pulses. Traditional measurements of fast dynamical events are performed by using high-speed electronic instrumentation such as oscilloscopes and transient digitizers. This instrumentation has evolved dramatically, yielding subnanosecond ($<10^{-9}$ s) performance. However, advances in laser technology have enabled the routine generation of optical pulses with durations on the order of 10 fs. The terminology ultrafast and ultrashort is used to refer to optical pulse durations or physical phenomena which fall in the femtosecond to picosecond time regime.

The generation of ultrashort laser pulses is made possible by mode locking. This technique is achieved by a device which modulates the loss (or gain) in a laser cavity in the time period required for a pulse to complete one round trip within a cavity of length L (**Fig. 1**). The effective generation of very short duration optical pulses requires that the laser cavity and gain medium be capable of supporting several thousand longitudinal cavity modes whose optical frequencies are separated by $\Delta f = c/(2L)$, where c is the speed of light. The combination of many oscillating laser modes and the synchronous modulation of cavity loss results in a locking together of the phases of the cavity modes such that discrete laser pulses are emitted at an interval equal to the cavity round-trip time of $2L/c$. The duration of these discrete pulses is inversely proportional to the number of longitudinal modes that are locked in constant relative phase. This mode-locking phenomenon provides a stable constructive and destructive interference between the otherwise randomly phased oscillating cavity modes, resulting in a discrete series of ultrashort optical pulses.

A dilemma arises concerning the measurement of ultrashort pulse durations. Typically, the temporal variation of laser output is monitored by means of electronic photodiodes connected to oscilloscopes, but conventional electronic instrumentation cannot measure a laser pulse of subpicosecond duration.

The technique of autocorrelation is utilized to infer information about such pulse widths. A short

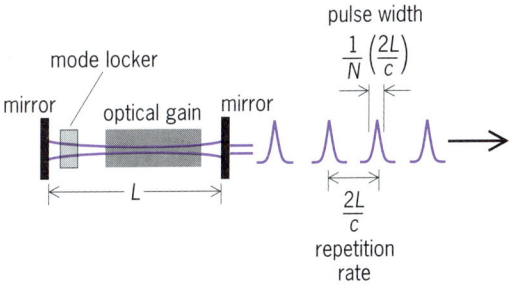

Fig. 1. Operation of mode-locked laser of cavity length, L, supporting N longitudinal cavity modes. c = speed of light.

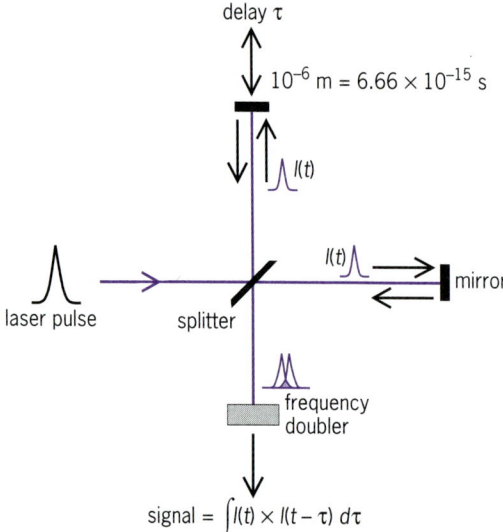

Fig. 2. Optical pulse autocorrelator. The input pulse is split into two equal-intensity pulses. Their recombination in the frequency doubler yields an optical signal proportional to the pulse overlap, which is systematically scanned by the optical path-length delay, τ. The output signal is described by the integral function.

optical pulse is divided into two equal-intensity pulses by a partially reflective mirror (splitter) (**Fig. 2**). The two pulses retain the same temporal profile $I(t)$ as the incoming laser pulse, and are rejoined after traversing independent optical paths. The pulses are combined in a frequency doubler, a nonlinear optical material which produces a signal field proportional to the product of the incoming optical fields. By purposely changing the path length that one pulse travels relative to the other, the pulse overlap at the doubler can be systematically varied to yield an output signal proportional to the autocorrelation of the two input-pulse intensity profiles. If, for simplicity, the input profiles are assumed to be symmetric, the output profile can be used to infer the input pulse shape and width.

The effective optical delay time is simply the path-length difference divided by the speed of light. Thus, the time resolution depends on the measurement of small spatial distances, on the order of 10^{-6} m, yielding a time resolution of about 3 fs.

Time-resolved measurements. The method of trading distance for time to achieve ultrafast time resolution provides the basis for performing time-resolved measurements of events in the subpicosecond time regime. Such measurements are initiated by exciting a material with one ultrashort light pulse, and probing the excitation-induced changes with another synchronized light pulse which is sequentially delayed relative to the excitation pulse (**Fig. 3**). This time-resolved optical probing usually takes the form of monitoring a change in the optical properties of the sample which results from absorption of some of the excitation pulse. This absorption can produce heating or thermal expansion, initiate chemical reactions, induce electronic transitions, excite charge carriers, or result in other changes. These phenomena in some way perturb the optical properties of the sample, and the relaxation of the perturbation is monitored as changes in optical reflection, transmission, absorption, or polarization of the sequentially delayed probing pulse. In some cases, changes in electrical properties can be monitored by using the probe pulse to trigger a semiconductor switch to permit voltage or current measurements at specific time delays.

Ultrashort pulsed lasers. Since the earliest demonstrations of mode locking in the 1960s, there has been a steady evolution in laser materials, techniques, and shorter pulse widths. Since Fourier analysis implies that the spectral bandwidth of a pulse is inversely proportional to the pulse duration, ultrashort optical pulses have a broad spectral bandwidth (range of color) compared to continuous-wave or long-pulsed lasers, which are relatively monochromatic. Thus, laser materials with exceptionally broad-gain bandwidths are favorable for ultrashort pulsed operation. Organic dyes in solution, and numerous solid-state media have been used to generate ultrashort laser pulses with central wavelengths spanning the ultraviolet to the near-infrared region.

The generation and manipulation of ultrashort optical pulses requires careful selection of optical elements for the laser cavity and the measurement apparatus. Because the refractive index of ordinary optical materials changes with wavelength (dispersion), pulse broadening and distortion can arise. For instance, as a pulse propagates through a material with positive wavelength dispersion, the short wavelength components of the pulse will be delayed relative to the longer wavelength components. This distortion can be compensated for by introducing an optical device which exhibits negative dispersion. Such devices can be constructed from grating

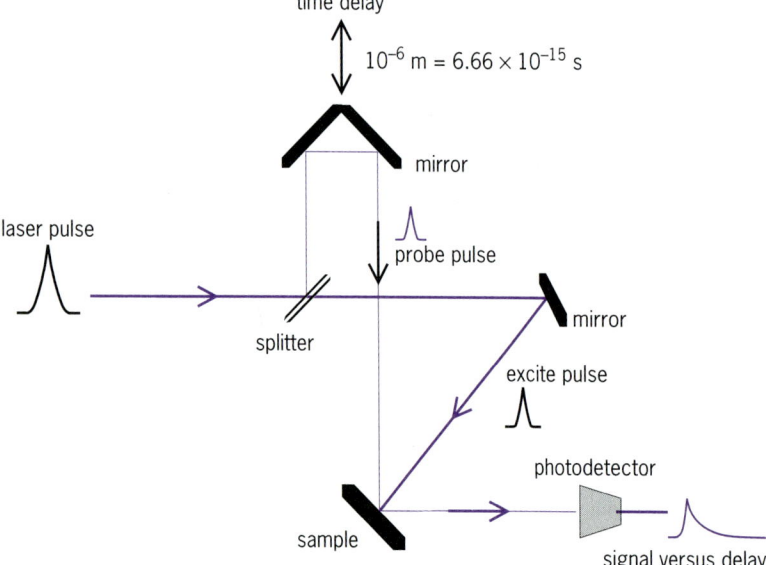

Fig. 3. Ultrafast measurement scheme. An incoming laser pulse is split into two unequal intensity pulses. The stronger pulse is used to perturb the sample, whereas the weaker one probes the resulting optical changes as a function of delay time.

or prism pairs, and effectively retard the red (longer) wavelengths relative to the blue (shorter), compensating for prior distortions in pulse shape.

These principles have also been used to compress the output pulses of lasers. The most widely used scheme relies on an optical fiber to broaden the spectral content of an input ultrashort pulse by means of a process called self-phase modulation. This nonlinear optical interaction results from the intensity-dependent refractive index of the fiber. Linear dispersion in the fiber also produces some temporal broadening, which is more than compensated for by passing the spectrally broadened (chirped) pulse through a negative-dispersion prism pair. Again, because of Fourier considerations, the net result is an output pulse shorter in duration than the original laser pulse. Of course, the concept of color is not well defined for visible femtosecond pulses, which contain only a few optical cycles. Thus, advances into the subfemtosecond domain will require laser sources that produce short ultraviolet or x-ray wavelengths.

Numerous types of lasers and techniques are used for ultrashort pulse generation. Among the most notable are lasers using titanium-doped sapphire as the lasing medium. This host medium has a gain-bandwidth range covering the 700–1000-nanometer wavelength region, and is capable of directly producing pulse durations below 100 fs. A remarkable feature of this laser is that it is self-mode locking, even though it is usually excited by a continuous-wave laser. If a temporary cavity loss perturbation is introduced (such as banging on the supporting table), the laser output spontaneously mode-locks. The general consensus is that the self-mode-locking mechanism relies on an optical Kerr-effect nonlinearity intrinsic to the host material.

The relative simplicity of this mode-locked laser has resulted in a number of commercially available systems offering a wide range of tunable wavelengths, pulse widths, and powers. It has been pivotal in expanding the accessibility of ultrafast optical measurements to a wide range of research and development efforts.

Applications. Three generic areas of research and development impacted by ultrafast optics are materials characterization, device characterization, and optical communications (see **table**).

Materials characterization. The pursuit of new semiconducting materials (in particular gallium arsenide) with promise for high-speed computing and communication applications was a driving force in the development of ultrafast optical characterization tools. The ability to monitor energy relaxation and recombination of electrical carriers is crucial to understanding the dynamics occurring in semiconducting materials that can be tailored chemically and structurally for specific applications. Ultrafast optical studies continue to confirm and extend the understanding of charge-carrier dynamics in metals and superconductors as well.

Advances in the technology of amplifying ultrashort optical pulses have led to the attainment of optical intensities in the 10^{12} W/cm^2 range. As a result, generation of hot-electron plasmas producing x-ray pulses is possible, and the range of studies regarding nonlinear optical interactions and material ablation has been extended.

Chemical reactions proceed by the transformation of chemical bonds in times on the order of femtoseconds. The ability to take snapshots of chemical reactions as they occur at surfaces, in biological cells, and in all phases of matter enhances the understanding of these phenomena. There has been rapid

Ultrafast optical studies and applications

General areas	Application areas	Specific applications
Materials characterization	Charge-carrier dynamics	Metals Superconductors Semiconductors
	Chemical-reaction dynamics	Surfaces Solid, liquid, gas Biological
	Ultrafast heating; ultrafast expansion	Thermal diffusion Acoustic pulse propagation
	Imaging through turbid media	Human tissue Fog
	High intensity: $I = 10^{12}$ W/cm^2	Hot electrons; x-ray pulses Material ablation Nonlinear optics
	$I = 10^{18}$ W/cm^2	Relativistic field strengths
Device characterization	Electrooptic sampling scope Picosecond electrical and electromagnetic pulses	High-speed circuit response Transmission-line response
Optical communications	High-speed data links	100 Gbit/s

growth in femtochemistry research, and advances in more complex photochemistry are expected.

An intriguing application of ultrafast optics involves imaging through turbid media. This time-resolved imaging is based on the understanding that photons (particles of light) that undergo multiple scattering events during transmission through translucent media such as fog are delayed relative to photons that undergo fewer scattering events. By illuminating an obscured object with an ultrashort optical pulse and temporally gating the collected image light, the object image and location can be reconstructed. Ultrafast optical schemes can image through several millimeters of animal tissue.

Device characterization. Ultrafast device characterization is a natural outgrowth of more fundamental semiconductor studies. Electrooptic schemes have been devised for triggering ultrafast electrical impulses in electronic devices, and subsequently monitoring their propagation. The generation of terahertz (10^{12} cycles/s) electromagnetic bursts by ultrafast optical excitation of semiconductors has proven useful for characterizing superconducting transmission lines as well as for providing a source of terahertz radiation for spectroscopic studies of materials.

Communications and instrumentation. The advent of high-repetition-rate ultrashort-pulsed semiconductor lasers has improved the feasibility of using mode-locked lasers in applications that were previously unreasonable or uneconomical. Because semiconductor lasers are easily interfaced and integrated into high-speed electronic packages, applications ranging from high-bit-rate communications to new types of instrumentation (such as an electrooptic sampling scope) are conceivable.

For background information *SEE LASER; LASER PHOTOCHEMISTRY; NONLINEAR OPTICS; OPTICAL PULSES; ULTRAFAST MOLECULAR PROCESSES* in the McGraw-Hill Encyclopedia of Science & Technology.

Gary L. Eesley

Bibliography. H. Messenger, Technology of ultrafast laser and electro-optics expands rapidly, *Laser Focus World,* pp. 69–83, September 1993; J. D. Simon, Ultrashort light pulses, *Rev. Sci. Inst.,* 60(12):3597–3624, December 1989; Ultrafast optics and optoelectronics, *Optics Photon. News,* vol. 3, no. 5, May 1992.

Lighting systems

Since the introduction of gas-discharge lamps, electrical circuits have been used to operate lamps. Typically, these circuits, generically referred to as gear, have consisted of simple laminated iron chokes and transformers. The pursuit of high-quality lighting and improved energy efficiency has led to the replacement of these circuits with fully electronic gear and the addition of electronic circuits to existing 50- and 60-Hz gear. The use of electronics has improved the performance and controllability of lamps, both gas-discharge and incandescent. These improvements have resulted not only in the use of lamps in new applications but also, more recently, in the development of high-performance light sources, which can be operated only by exploiting the controllability offered by an electronic ballast. These developments have transformed lighting technology since the mid-1980s.

Almost all lighting circuits are essentially power converter systems, taking in mains supply and converting it to a stabilized higher-frequency voltage source. Hence, the increased use of electronics has been made possible only by technological advances, in particular, advances in power-supply technology, semiconductor devices (both discrete devices and customized integrated circuits), and electronic assembly techniques. These advances have enabled equipment suppliers to produce products at a price which can be justified on an economic and performance-benefit basis.

Electromagnetic compatibility. Because lighting systems form a significant part of the electrical energy consumption of a building, typically 40% in offices, it is necessary to ensure that lighting circuits do not distort the mains supply or emit radio-frequency interference (RFI), which can interfere with the operation of other electronic equipment. Electromagnetic compatibility (EMC) standards have existed for a number of years, determining the levels of electrical pollution that lighting equipment is allowed to generate. These standards cover the three key areas of mains-current harmonic distortion and both radiated and conducted radio-frequency interference.

Electronic transformer. Probably the simplest and one of the most common circuits found in lighting applications is the electronic transformer, which is used to power compact tungsten halogen lamps. The lamp's small size and use of a dichroic reflector has provided designers with an ideal light source for high-technology, high-fashion luminaires. These lamps require a low-voltage source, typically 12 V. Conventional 50- and 60-Hz transformers are large and heavy. High-frequency transformers are significantly smaller, thereby complementing the compactness of the light source.

Electronic transformers are essentially full-bridge rectifiers followed by a high-frequency, self-oscillating, half-bridge inverter (**Fig. 1**). On startup, C_2 is charged via R. When the voltage across C_2 reaches the breakdown voltage of the diac, current flows through the diac and into the winding of the drive transformer connected to the transistor T_2. Thus T_2 is caused to turn on, allowing current to flow through the full-bridge rectifier, via C_3 and the output transformer, into the primary winding of the drive transformer. Current flowing in the primary winding in this direction causes the secondary windings to turn T_2 off and T_1 on. Current now flows through T_1, the primary winding of the drive transformer, the output transformer, and C_4. Thus,

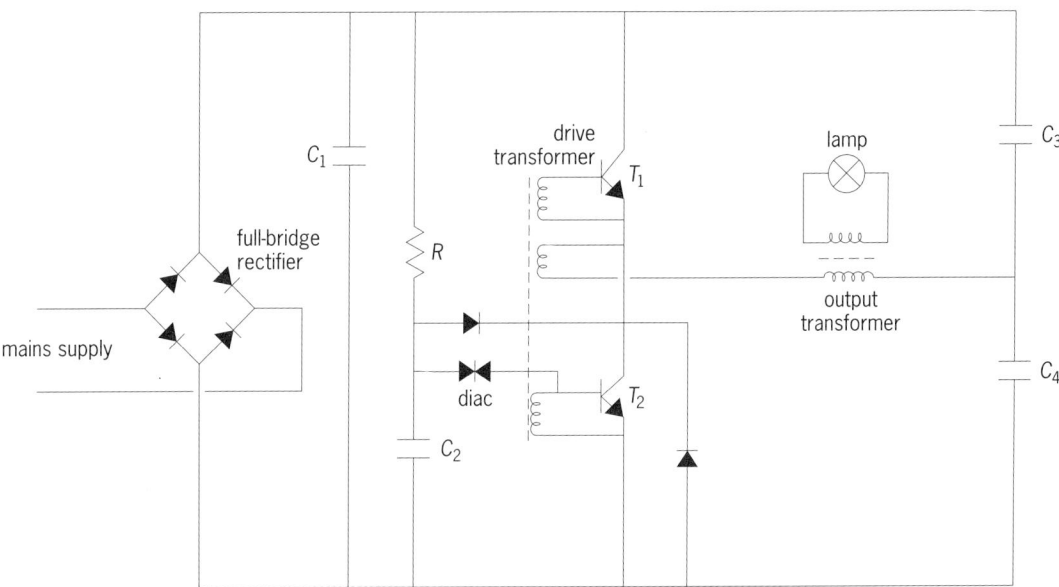

Fig. 1. Electronic transformer circuit.

the previous action is reversed, turning T_2 on and T_1 off. Hence, the circuit self-oscillates. As current alternates through the output transformer, current is coupled into the transformer's secondary winding and into the lamp. The output from this circuit is thus a chopped 50- or 60-Hz waveform and by the use of appropriate filters a sinusoidal current can be drawn from the mains supply.

Electronic ballasts. Electrically, gas-discharge lamps may be characterized by a negative impedance and, as a consequence, need a series current-limiting component, referred to as a ballast, to maintain stable operation. Another feature of discharge lamps is the almost constant voltage drop across the electrodes. The dynamic impedance of the lamp at low frequencies continuously changes over the voltage cycle. Above 1 kHz the lamp represents a substantially resistive load, resulting from the recombination time of the ionized gas being longer than the period of the lamp's current waveform. The main advantages of operating gas-discharge lamps on a high-frequency ballast are an unmodulated light source and improved energy utilization. Discharge lamps operated on 50- or 60-Hz gear generate light that fluctuates at twice the supply frequency. A number of medical ailments have been attributed to this effect.

An electronic ballast generally consists of three parts: a preregulator, an inverter, and a starter. The preregulator converts the mains supply to a high-voltage dc output, typically 400 V. The inverter then chops this dc supply into a high-frequency ac source to drive the lamp. Finally, a starter circuit is used to initiate the gas discharge.

Compact fluorescent lamps. Over the past few years, low-wattage compact fluorescent lamps have become popular as they provide a very energy-efficient light source compared to the conventional filament lamp. By using an electronic ballast, a compact, lightweight lamp can be produced. Compact fluorescent ballasts are very similar in design to electronic transformers, with a current-limiting inductor replacing the main output transformer. The main difference is the use of a large dc storage capacitor to maintain a near-constant dc voltage across the half-bridge inverter (**Fig. 2**). In addition, a starter circuit is incorporated into the inverter. Generally, the starter circuit takes the form of a capacitor C_3 in series with the output inductor L.

The circuit operates in a similar fashion to the electronic transformer. In order to start the lamp, a large voltage must be applied across the lamp. When the lamp is unlit, the inductor L and C_4 form a tuned resonant circuit. As current in the circuit oscillates, by virtue of the action of the tuned circuit L and C_4, voltage builds up across the lamp and C_4. Once this voltage is sufficiently large to break down the lamp, current flows through the lamp generating the desired gas discharge. As the impedance of the lamp is significantly lower than C_4, the lamp short-circuits C_4. Often a thermistor is included across the starting capacitor to dampen the tuned circuit. As the current through the thermistor increases, its resistance changes, allowing the voltage across the lamp to increase gradually with respect to time. However, because the power consumption of these lamps is very low, the mains supply will not experience significant distortion. Consequently, these lamps are often excluded from electromagnetic compatibility standards.

Fluorescent ballasts. For higher-wattage lamps, in order to comply with current harmonic distortion requirements the dc capacitor is replaced with a preregulator, which generates a high-voltage dc source while drawing a sinusoidal current from the supply. These circuits usually take the form of a switch-mode power-supply boost converter.

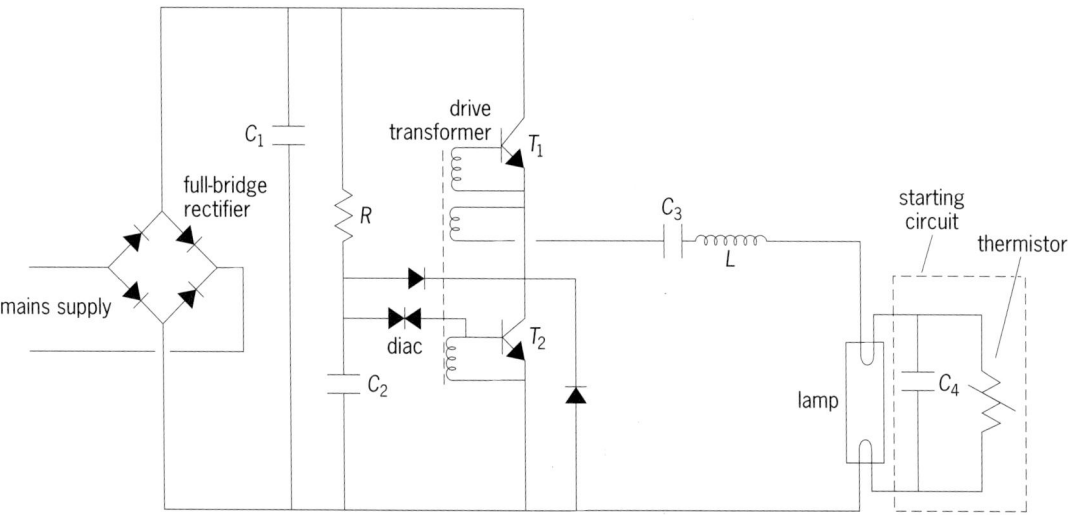

Fig. 2. Ballast circuit for compact fluorescent lamp.

As with compact fluorescent lamps, higher-wattage ballasts use a half-bridge inverter to drive the lamp. Operating fluorescent lamps on high-frequency ballasts increases the efficacy of the lamps, because of the higher efficiency of the ballast and the increased efficiency of the gas discharge. An added benefit of the system is that the lamps can be dimmed by changing the oscillation frequency of the inverter. As well as providing better energy utilization, dimming enables the lighting environment of offices to be varied to take into account daylight conditions, building usage, and individual preferences and task lighting needs.

Induction lamps. The most recent developments in fluorescent lamps have been in the field of inductively coupled lamps. Because no electrodes exist in these lamps, the key failure mechanism (electrode failure) in conventional fluorescent lamps has been eliminated. Lamp life in excess of 60,000 h has been projected (compared with 8000 h for compact fluorescent lamps). In these lamps energy is coupled into the gas discharge by means of an electromagnetic field. A transformer arrangement is used to couple the energy, the primary winding of the transformer being connected to the ballast circuit and the discharge lamp's gas plasma forming the secondary winding. Efficient electromagnetic field coupling can be achieved by operating the circuit at a very high frequency, typically 3 MHz. As with conventional fluorescent lamps, initiation of the discharge is achieved by application of a high voltage. Once the plasma has been established, magnetic-field coupling energizes the plasma. As with compact fluorescent lamps, a half-bridge inverter is used. However, in order to operate efficiently at these very high frequencies, complex transistor switching techniques are used to limit the power loss in the transistors. Relaxation of the radiated radio-frequency interference limits have been permitted in order to facilitate the introduction of the new technology.

High-intensity discharge lamps. Although electronic ballasts for fluorescent lamps have been available since the mid-1980s, ballasts for high-intensity discharge (HID) lamps have generally been limited to special applications. This limitation can be attributed to the instability of high-intensity discharge lamps when driven at high frequencies. The instabilities are associated with dynamic pressure waves, and the phenomenon is referred to as acoustic resonance. Fluctuations in the gas temperature, and therefore gas pressure, caused by the modulation in power derived from the oscillating supply voltage and current set up standing pressure waves in the arc tube. As a result the arc becomes unstable. Three techniques have been used to operate high-intensity discharge lamps at high frequency: operating the lamps at a frequency above the threshold limit of acoustic resonance (typically above 1 MHz); frequency shifting, ensuring that the voltage frequency changes before the acoustic resonance builds up; and square-wave operation (**Fig. 3**). Because the power component of a square wave is constant, no power modulation occurs in square-wave operation, and hence acoustic resonance is avoided.

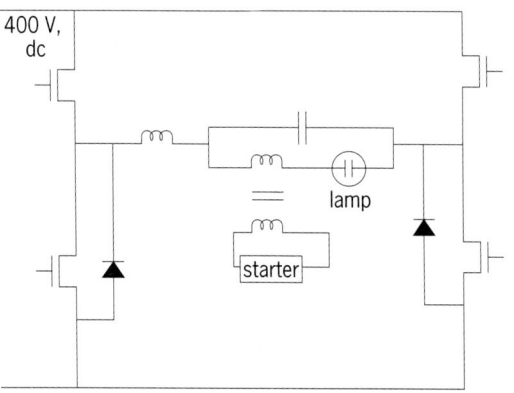

Fig. 3. Circuit for square-wave operation of a high-intensity discharge lamp.

Improving ballast performance. The 50- or 60-Hz operation of lamps can be greatly improved by the addition of control electronics, as is exemplified by the electronic starter. This device is able to control the starting sequence and the ignition voltage so that damage to the electrodes is limited and a cleaner start is achieved. As a result a longer lamp life is obtained. As well as improving the performance of lamps in terms of lamp life, uniformity of color, and efficiency, the additional control provided by electronics has made possible the development of new lamp technologies.

For background information SEE ELECTROMAGNETIC COMPATIBILITY; ELECTRONIC POWER SUPPLY; FLUORESCENT LAMP; RECTIFIER; VAPOR LAMP in the McGraw-Hill Encyclopedia of Science & Technology.

E. J. P. Mascarenhas

Bibliography. M. A. Cayless and A. M. Marsden (eds.), *Lamps and Lighting*, 3d ed., 1983; E. J. P. Mascarenhas, Applications of electronic circuits in lighting, *IEE Proc. A,* 140(6):435–442, 1993; D. O. Wharmby, Electrodeless lamps for lighting: A review, *IEE Proc. A,* 140(6):465–473, 1993; H. L. Witting, Acoustic resonance in cylindrical high pressure arc discharges, *J. Appl. Phys.,* 49(5):2680–2683, 1978.

Lyme disease

Lyme disease, which is caused by a tick-borne bacterium, is a multisystem disorder in humans with pronounced dermatologic, neurologic, and rheumatologic manifestations. It has become the most commonly reported vector-borne disease in the United States and has been reported from Europe, Asia, Africa, and Australia. The syndrome in dogs, most properly referred to as Lyme borreliosis, shares many clinical components and epidemiologic features with human Lyme disease. However, the incidence, clinical spectrum, and optimal treatment regimen of canine Lyme borreliosis is still ill defined compared with its human counterpart. The efficacy of a canine Lyme borreliosis vaccine has not yet been definitively assessed.

Etiologic agent. The etiologic agent of the disease is a recently identified bacterium named *Borrelia burgdorferi. Borrelia burgdorferi* is a motile, gram-negative, microaerophilic, slow-growing spirochete with a generation time in culture of about 12 h. The organism is fastidious and relatively difficult to cultivate in the laboratory. Strain variability exists for *B. burgdorferi,* and there is evidence that varying degrees of pathogenicity and infectivity between strains may be related to differing plasmid profiles and to heterogeneity of outer surface proteins. This antigenic variation of outer surface proteins could be significant in the search for an effective vaccine because vaccination with surface proteins from one strain of *B. burgdorferi* may not cross-protect against other strains.

Distribution. No formal reporting system exists for canine Lyme borreliosis; thus, the incidence of the disease for this species is unknown. After a dramatic rise in the number of reported Lyme disease cases in humans in the United States during the late 1980s, the annual case totals have stabilized somewhat, averaging approximately 8800 between 1989 and 1993. The disease in this country is largely a regionalized problem, with the vast majority of cases occurring in the Northeast, upper Midwest, and northern portions of California. These endemic areas coincide with the geographic range of the known tick vectors of *B. burgdorferi.*

Vectors. Hard ticks of the *Ixodes ricinus* complex are the only proven vectors of *B. burgdorferi.* In the United States, the vectors associated with Lyme borreliosis are the ticks *I. pacificus* on the West Coast and *I. scapularis* in the Northeast, upper Midwest, and Southeast. The northern populations of *I. scapularis* (commonly called the deer tick) were until recently considered a distinct species named *I. dammini.* Reports indicate that *B. burgdorferi* infection rates in the northern populations of *I. scapularis* vary from 12 to near 100%. Infection rates of generally less than 5% have been found for the southern variant of *I. scapularis* and for *I. pacificus,* possibly because these ticks commonly feed on lizards, which do not serve as reservoirs for the spirochete. Although no longer classified as a separate species, the northern population of *I. scapularis* is important because of the deer tick's unique ecology and epidemiologic association with the vast majority of Lyme disease cases in the United States.

Each of the three life stages of *I. scapularis* (larva, nymph, and adult) feeds on a separate host, taking one blood meal per life stage. Each stage remains continuously attached to its host for several days while feeding. When replete with blood, the tick detaches from its host and later molts to the next life stage, or, in the case of the adult female, lays eggs from which the larvae emerge. The juvenile stages (larva and nymph) of the deer tick feed primarily on small rodents, particularly on the white-footed mouse (*Peromyscus leucopus*). This mouse is the primary mammalian reservoir for *B. burgdorferi,* and is typically the source of tick infection. The adult stage of the tick has a strong predilection for white-tailed deer, which are the usual site for tick mating.

All three stages of the deer tick can be found on humans and domestic animals. However, because transovarial transmission of *B. burgdorferi* rarely occurs, unfed larvae are generally free of infection. Although both the adult and the nymphal deer tick are capable of transmitting *B. burgdorferi,* the majority of human infections are associated with bites sustained from nymphs, presumably because their small size (0.04–0.08 in. or 1–2 mm) permits them to go unnoticed during feeding. (The tick bite itself is painless.) Thus, most humans with Lyme disease are infected during late spring or early summer, coinciding with the time of peak nymphal activity. However, the seasonality of canine Lyme borreliosis is not well described. It is believed that dogs are likely to continue to acquire the infection through autumn (the time of peak adult tick activity) because even larger adult ticks can easily go unnoticed in a dog's fur.

Borrelia burgdorferi has been isolated from several species of blood-feeding arthropods, notably fleas, mosquitoes, American dog ticks, and tabanid flies. However, there is no epidemiologic evidence to suggest that these species play any role in transmission of the spirochete, and laboratory studies indicate that these arthropods are incapable of transmitting the organism.

Clinical manifestations. The spectrum of well-defined clinical signs and symptoms of canine Lyme borreliosis appears to be narrower than in human Lyme disease. Notably, the majority of data on naturally acquired Lyme borreliosis consist of isolated case reports, and some discrepancies exist between descriptions of the naturally occurring disease and experimentally induced infection.

The incubation period of the disease in nature is unclear; however, experimentally infected dogs began to experience symptoms 2–5 months after *B. burgdorferi*–infected ticks were permitted to feed on them. The predominant clinical signs of canine Lyme borreliosis are fever, lethargy, and an acute onset of lameness. The lameness is primarily the result of arthritis, with the large joints of the carpus, elbow, tarsus, and stifle most commonly affected, and with joint effusion frequently present. The arthritis typically is nonsymmetric, affecting only one or a few joints. Commonly, it is recurrent and may shift from one joint to another. In separate case series describing naturally acquired and experimentally induced Lyme borreliosis (using an infected-tick model), each episode of lameness lasted fewer than 5 days and the majority of dogs had only one or two episodes. In the series involving naturally infected dogs, the average period between episodes was several months, whereas experimentally infected dogs averaged only about 3 weeks between lameness episodes. The majority of dogs are feverish during lameness episodes, and lymph node enlargement is common.

Clinical manifestations that have been much less commonly described include myocardial dysfunction, complete heart block, and renal lesions suggestive of glomerulonephritis (a kidney disorder). The development of rheumatoid arthritis subsequent to Lyme borreliosis has been reported, as has the transplacental transmission of *B. burgdorferi* from an experimentally inoculated female to her pups. Although most humans with Lyme disease develop a characteristic skin rash during the early stages of the disease, skin erythema is rarely reported in canine Lyme borreliosis. Whether it does not occur or simply goes unnoticed because of the fur is not known. Similarly, neurologic abnormalities, a common feature of human Lyme disease, have never been definitively ascribed to canine Lyme borreliosis.

Diagnosis. Attempts to isolate *B. burgdorferi* from clinical specimens are impractical in most veterinary clinical situations because of the paucity of organisms present and the fastidious nature of the spirochete. Serologic assays that detect antibodies directed against *B. burgdorferi* are widely available. These tests may have low sensitivity (a high proportion of false negative results) in the early stages of canine Lyme borreliosis. Poor specificity (a high proportion of false positive results) also may be a problem in some patients because antibodies directed against certain proteins on other microorganisms can cross-react in these assays. Asymptomatic infection with *B. burgdorferi* can occur; asymptomatic dogs in endemic areas are commonly seropositive. Because of these limitations, the currently available serologic assays should not be used to screen clinically normal animals for Lyme borreliosis. Positive serologic tests performed on healthy dogs have been shown to have no predictive value for the subsequent development of lameness.

The diagnosis of Lyme borreliosis is made on the basis of clinical findings and a history of exposure (or potential exposure) to ticks in an endemic area. Serologic testing can be a useful adjunct to diagnosis, but cannot by itself establish a diagnosis of canine Lyme borreliosis. Before canine Lyme borreliosis is definitively diagnosed as the cause of a lameness, attempts should be made to rule out other possible causes, such as trauma, cruciate ligament rupture, intervertebral disk herniation, degenerative joint disease, septic arthritis from other agents, rheumatoid arthritis, and systemic lupus erythematosus.

Treatment. No controlled randomized studies of antibiotic treatment regimens have yet been conducted to assess either the most effective antimicrobial agent or the optimal duration of therapy for canine Lyme borreliosis. However, based on isolated case reports and extrapolation from human medicine, the tetracyclines (tetracycline or doxycycline), and the beta-lactam antibiotics (penicillin, amoxicillin, and certain third-generation cephalosporins) can be effective. Steroidal anti-inflammatory agents should probably be avoided because their use in humans has been associated with nonresponsiveness to antibiotic therapy.

Prevention. Minimizing a dog's exposure to potentially infected ticks through the use of tick collars, sprays, and dips can be of value in endemic areas. Tick repellents containing DEET (*N,N*-diethyl-*meta*-toluamide) or permethrin are marketed for veterinary use, and have proven effective in human studies. Dog owners are advised to consult with their veterinarians before using these repellents and acaricides. Animal studies have shown that *B. burgdorferi* transmission from the tick to its host is minimal during the first 48 h of attachment. Therefore, daily inspection and removal of ticks should be performed on dogs that have been in a tick habitat.

A vaccine for the prevention of canine Lyme borreliosis is currently available. This killed, whole-cell vaccine is meant to be administered prior to *B. burgdorferi* exposure. Reports regarding the efficacy of this vaccine are encouraging but not definitive; well-controlled evaluations of this vaccine using an infected-tick challenge model have not yet been conducted. There are no data to demonstrate that the use of the vaccine has any human public health significance.

Because of the highly regional nature of canine Lyme borreliosis, dog owners considering either the vaccine or chemical tick-control measures for their dogs are well advised to learn about the incidence of the disease in those areas that are frequented by the dogs. State health departments can provide information on the incidence of indigenously acquired Lyme disease and the presence of vector ticks in the geographic area of interest.

For background information SEE ANTIBIOTIC; LYME DISEASE in the McGraw-Hill Encyclopedia of Science & Technology.

James J. Kazmierczak

Bibliography. M. J. Appel et al., Experimental Lyme disease in dogs produces arthritis and persistent infection, *J. Infect. Dis.*, 167:651–664, 1993; H. J. Chu et al., Immunogenicity and efficacy study of a commercial *Borrelia burgdorferi* bacterin, *J. Amer. Vet. Med. Assoc.*, 201:403–411, 1992; J. J. Kazmierczak and F. E. Sorhage, Current understanding of *Borrelia burgdorferi* infection, with emphasis on its prevention in dogs, *J. Amer. Vet. Med. Assoc.*, 203:1524–1528, 1993; A. C. Steere, Lyme disease, *N. Engl. J. Med.*, 321:586–596, 1989.

Mad cow disease

Bovine spongiform encephalopathy (BSE), or so-called mad cow disease, was first recognized in Great Britain in 1986. The disease has killed more than 140,000 cattle in Britain and has spread to 10 other countries, thereby impacting on the export and import of livestock worldwide, and on the use of bovine tissues in drug manufacture.

BSE is caused by cows eating the infected protein of other cows and has occurred because of the long-standing practice of feeding animal protein to livestock. Dead cattle are rendered to produce meat and bone meal, much of which is fed back to other cattle. The rendering process cooks the carcasses to a temperature that inactivates most pathogenic microorganisms, but not the transmissible agent that causes BSE. Once ingested, the agent spreads slowly to the central nervous system where it gradually destroys nerve cells, resulting in a progressive neurologic disease that ends in complete debilitation and death.

The average incubation period for BSE is 4 years. During this period, there are no means of detecting infection in the animals, although brain tissue may contain large amounts of the transmissible, infectious agent. Thus, although the practice of feeding ruminant animal protein was banned in 1988, it has taken 6 years for a reduction in the occurrence of the disease in Britain.

Clinical signs. The first clinical signs of disease in BSE-affected cattle are usually behavioral, such as nervousness as manifested by apprehension and frenzy. In addition, some abnormalities of posture or movement may also be seen, including a lack of hind-limb coordination (ataxia) and falling. Many BSE-affected cattle become overly aggressive. For example, a previously docile cow may assume the behavior of a bull, laying its ears along its head and charging other cattle, humans, or inanimate objects.

Over a period of weeks, the affected animal becomes progressively more debilitated until it becomes totally recumbent. During the early stages of the BSE epidemic, these infected animals were sold and their tissues were rendered to be fed back to other cattle.

Diagnosis. BSE can be diagnosed by a histopathologic examination of affected brain tissue. Areas of the midbrain and brainstem contain foci of spongiform degeneration of the gray matter accompanied by an enlargement and thickening of the astrocytes, macroglial cells in brain tissue that are responsible for neuronal homeostasis. Their hypertrophic reaction in BSE is a likely response to nerve cell damage.

Etiology. The most interesting feature of BSE is that the disease appears to be caused by a protein, not by a virus. Evidence suggests that BSE and other transmissible spongiform encephalopathies (TSEs; diseases characterized by the brain becoming permeated with holes) are caused by the modification of a normal host protein into a form that cannot be degraded and that accumulates as an abnormal protein (amyloid) that kills nerve cells. This prion protein is a 33,000–35,000-dalton glycoprotein of unknown function that is coded for by a single gene on chromosome 20 in humans. Abnormalities in this gene can cause rare inherited TSEs in humans, but most incidents of these diseases in humans and animals are sporadic, with no familial pattern of occurrence.

The disease-specific form of prion protein can be distinguished only by its resistance to proteolytic digestion. The mechanism by which the protein acquires this resistance is presently unknown but is believed to reside in posttranslational modification of some type. This modification and the means by which the modified protein is able to convert its normal counterpart are of great interest. The presence of the modified form of prion protein in affected brain tissue confirms the diagnosis of BSE. Tissues that contain histopathologic evidence of BSE infection can be extracted with detergents that enrich for insoluble prion protein, which can be detected by Western blot analysis after protease treatment to remove the normal protein.

Other transmissible encephalopathies. BSE is the most recently recognized TSE, but a similar disease in sheep, scrapie, has been known for more than 300 years. In addition, naturally occurring TSEs in other species include transmissible mink encephalopathy in ranch-raised mink, chronic wasting disease in captive mule deer and elk, and kuru and Creutzfeldt-Jakob disease in humans.

Epidemiologic intrarelationships. Some epidemiologists believe that BSE began by feeding animal protein from scrapie-infected sheep to cattle; others discount a sheep-to-cattle cycle. The rapid evolution of the epidemic and its long duration indicates that the agent was well adapted to cattle before the disease was ever recognized. This seemingly trivial aspect of the origin of the disease has important implications for BSE-free countries. If BSE infection is already present at low levels in cattle populations, the major emphasis should be on changing current practices of feeding ruminant animal protein as well as on reducing the prevalence of sheep scrapie.

Species susceptibility. The mechanism of host restriction at the molecular level is not understood. Therefore, it cannot be predicted if the modified prion protein from one species will convert or modify the prion protein of another species. Experimental and natural transmissions usually occur most easily, but there have been some surprises in homologous species. BSE is naturally transmissible to domestic cats and to some captive wild felines that ingest contaminated bovine tissues. This would not have been expected based on previous experimental findings. These observations suggest that factors other than the primary amino acid homology of prion protein are necessary for the recognition and conversion of normal proteins. One of these factors is likely to be the posttranslational modification responsible for strain specificity of these unusual proteinaceous pathogens, known as prions. Prions produced in the same species can have distinct biological properties affecting incubation period, clinical signs, neuronal cell targeting, and host range.

Perhaps the best example of strain diversity can be seen in recent experiments testing the susceptibility of cattle to American sources of sheep scrapie. Intracerebrally inoculated animals developed progressive neurologic disease, but the clinical signs and histopathologic lesions were atypical of BSE. The affected animals were not aggressive, but gradually became so debilitated that they were unable to stand. More importantly, the brains of these affected animals showed no histologic evidence of spongiform degeneration. These results suggest that some strains of prions are capable of producing clinicopathologic disease in cattle that differs from BSE.

Implications. For decades the heat resistance of the sheep scrapie agent was known, but the implication of this physicochemical property for the origin of the BSE epidemic was not foreseen. It is possible to learn from this experience and prevent the occurrence of BSE-like diseases in other countries. The European Union presently has a ban on the feeding of ruminant animal protein. It is to be hoped that all countries will implement similar restrictions.

The disease has lasted longer and had a greater economic impact than expected. TSEs are not like conventional diseases that can be controlled or eradicated over a relatively short period of time by testing for infected individuals and then instituting quarantine and treatment. The long incubation period allowed years of exposure before BSE was recognized, and additional years passed while preventative measures followed their slow course.

The magnitude of the public concern over possible human health risks with BSE was unexpected. Despite assurances from the British government and epidemiologic studies showing no evidence of transmission of TSEs from animals to humans, the public perception is that it had not been definitively established that BSE cannot infect humans. This translated into a 25–30% reduction in beef consumption in Great Britain, and British beef was banned from school menus in London for a few weeks immediately following the revelation that BSE could be naturally transmitted to cats.

Richard F. Marsh

Bibliography. R. H. Kimberlin, Bovine spongiform encephalopathy: An appraisal of the current epidemic in the United Kingdom, *Intervirology*, 35:208–218, 1993; R. F. Marsh, Bovine spongiform encephalopathy: A new disease of cattle?, *Arch. Virol.*, 7:255–259, 1993; S. B. Prusiner, The prion diseases, *Sci. Amer.*, pp. 48–57, January 1995; J. W. Wilesmith and G. A. H. Wells, Bovine spongiform encephalopathy, *Curr. Top. Microbiol. Immunol.*, 172:21–38, 1991.

Magnetic switch

The basis of integrated-circuit technology is the electronic properties of electrons in semiconductors, specifically the charge of the electrons. In contrast, the magnetic properties, or spin, of the electrons have not been exploited in the design of devices because creating and controlling a current of a particular spin is more difficult than doing the same with a current of a particular charge. Spin may take on two values, up or down, which can be randomized by thermal forces or by forces due to other electrons, whereas the charge of an electron is a fundamental constant. Even with such challenges, a significant advance in controlling the spin of electrons has been made recently with the realization of a magnetic switch. The action of this switch is based on spin instead of charge populations and is controlled by a small magnetic field instead of an electric field. Unlike conventional electronic devices, such magnetic devices are predicted to perform better as their size is reduced, thus facilitating miniaturization. The work on the magnetic switch is one aspect of continued efforts to control spins in small magnetic systems, work which has important repercussions for the magnetic recording industry as storage densities are increased.

Bipolar spin device. The structure of the magnetic switch, or more accurately the bipolar spin switch, is in some ways parallel to the structure of the transistor, which could also be called a bipolar charge switch. Both devices are composed of three regions. The spin transistor consists of three metallic layers, a paramagnetic layer P between two ferromagnetic layers $F1$ and $F2$ (see **illus.**), whereas an ordinary transistor is composed of a base semiconducting layer of one charge type surrounded by emitter and collector semiconducting layers of the opposite charge type. In an ordinary transistor, the charge populations are controlled by doping the semiconductor layers with different impurities. In the spin transistor, the spin populations are controlled by using a particular ferromagnetic material such as iron, nickel, cobalt, or an alloy such as Permalloy, Ni_xFe_{1-x}.

A ferromagnet has a majority of electrons with a particular spin leading to a net macroscopic magnetic moment. The spins ordinarily point in one direction only within a given domain, not uniformly throughout a ferromagnet. In a small enough ferromagnet, there will be only a single domain, and if in addition the ferromagnet is made thin enough, the net moment will lie in the plane of the film. Typical dimensions of the ferromagnetic layers in the magnetic switch are 100 micrometers on a side and 70 nanometers thick. The paramagnetic layer P between the two ferromagnetic layers is composed of gold. In a paramagnet, the spins are randomly oriented, and therefore P can be biased either spin up or down by the adjacent ferromagnetic layer. To transmit the information of the spin population from $F1$ to $F2$, P must also be thin (100 nm).

The biasing is accomplished by connecting a battery between $F1$ and P. The charges from the battery are unpolarized; however, when they pass through $F1$, spin-up electrons preferentially carry the current since they lie at higher energy than the spin-down electrons if $F1$ has its net moment down. Polarized current then flows into P, increasing the population of up spins while decreasing the population of down spins in order to preserve charge neutrality. Thus, $F1$ acts similarly to a polarizer of light. However, the polarization is not maintained throughout P because magnetic scattering events act to randomize the spin of the polarized electrons. For this reason, the gold layer must be made thinner than the spin depth of the electrons, defined as the distance traveled by electrons before suffering a scattering event. The state of the switch is read by measuring the voltage between the second ferromagnetic layer $F2$ and a nonmagnetic counterelectrode N deposited on P. The highest voltage will develop if the magnetization of $F2$ points in the same direction as $F1$, similar to the case in optics where the maximum intensity of light is passed through two polarizers when they are aligned. The state of $F2$ is controlled by applying a small magnetic field on the order of 10 oersteds (800 A/m). (By comparison, Earth's magnetic field is on average 0.5 Oe or 40 A/m.) To bias the device, a current of 0.1–10 mA is used, and the voltages produced are on the order of 10 nV, which is read by a superconducting quantum interference device (SQUID). *See* SQUID.

Miniaturization and stability. One promising feature of the magnetic switch is that miniaturization should increase the size of the effect. If the current through $F1$ is kept constant while the size of P is decreased, the magnetization (the magnetic moment per unit volume) should increase in P, thus producing a larger voltage in $F2$. Another promising feature is that the fundamental switching time of the device is determined by the spin-flip time of the electrons in $F2$, which is on the order of a picosecond, although this limit has not yet been reached because of other coupling effects that slow the device. In contrast, the fastest modern transistors switch on a time scale of 100 picoseconds. Although the magnetic switch has been operated only from 4.2 to 65 K (−452 to −343°F), the effects should also be observable at room temperature.

The greatest challenge is to design a device that also has power gain in order to be able to drive other devices. In a transistor, a small input current controls a larger current from a power supply. If the input current does not saturate the transistor, the output current will be proportionately greater than the input current; hence, the transistor amplifies the input signal. However, the only current in the magnetic switch is the small bias current, which is only reduced in passing through the paramagnetic layer. In order to achieve power gain, the magnetic switch will have to control a current larger than the bias current. The stability of the device will ultimately be limited by quantum effects: with dimensions on the order of 10 nm or smaller, the magnetization may be able to tunnel through the anisotropy barrier and effectively randomize the polarization of the film.

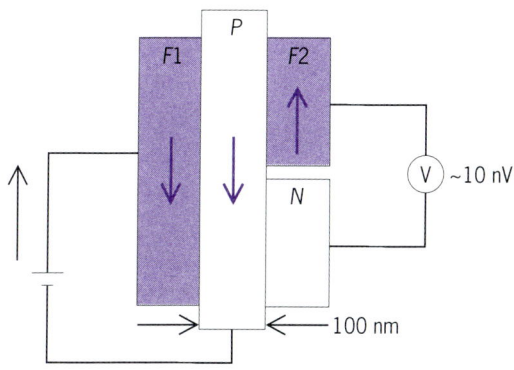

Schematic of bipolar spin device in off state. F_1, F_2 = ferromagnetic layers; P = paramagnetic layer; N = nonmagnetic counterelectrode; current I = 0.1–10 mA.

Prospects. Spin devices are one aspect of a range of efforts to understand magnetic switching in small particles. With current advances in microscopy, it is possible to probe the dynamics of switching in a single particle instead of having to study a distribution of particles and interpret the data based on assumptions about the distribution. A magnetic-force microscope can be used to study an isolated, single-domain γ-Fe_2O_3 particle of dimensions 300×60 nm. Such particles are commonly used in commercial magnetic storage media. Classical theory predicts that the switching of a small single-domain particle should be governed by thermal activation over a single potential barrier. The barrier is due to a preference of the magnetization to lie along a particular axis of the particle (the easy axis), and can result from either the shape of the particle or an inherent anisotropy in the crystal. However, recent experiments indicate that the switching is more complicated than activation over a single barrier. The time dependence does not fit a simple exponential law, suggesting that not all the magnetic iron ions in the particle experience the same anisotropy barrier. Rather than a single barrier or peak a complicated anisotropy landscape may exist. Similar effects were seen in the past in a system of particles, but were attributed to the distribution of particles in the system.

Efforts are also being directed to make magnetic particles even smaller by using an adaptation of another type of microscopy, namely, scanning tunneling microscopy. Iron particles as small as 10 nm have been fabricated by introducing a metallorganic precursor gas, iron pentacarbonyl, $Fe(CO)_5$, in a scanning tunneling microscope (STM) chamber and disassociating the gas with the STM to deposit iron atoms locally. When the tip of the STM is close to a substrate, a tunneling current with an associated high electric field is produced that can break the bonds between the iron and the carbonyl ligands. The iron is deposited below the tip, whereas the carbonyl ligands, now in the form of carbon monoxide, are removed by the vacuum system. The shape of the particles can also be controlled by maintaining the current through the STM tip as it is withdrawn from the substrate for deposition. Arrays of such particles (about 500) have been found to be magnetic at temperatures up to at least 100 K ($-280°F$). As in the work on γ-Fe_2O_3, it would be desirable to be able to study a single STM-grown iron particle. SEE MICROSCOPE; TRANSISTOR.

For background information SEE DOMAIN (ELECTRICITY AND MAGNETISM); ELECTRON SPIN; FERROMAGNETISM; INTEGRATED CIRCUITS; MAGNETISM; SCANNING TUNNELING MICROSCOPE; SQUID in the McGraw-Hill Encyclopedia of Science & Technology.

David Awschalom; Savas Gider

Bibliography. D. D. Awschalom, D. P. DiVincenzo, and J. F. Smyth, Macroscopic quantum effects in nanometer-scale magnets, *Science,* 258:414–421, 1992; M. Johnson, Bipolar spin switch, *Science,* 260:320–323, 1993; M. Lederman et al., Measurement of thermal switching of the magnetization of single domain particles, *J. Appl. Phys.,* 75:6217–6222, 1994.

Mammalia

The diet of ancient mammals, including early humans and their ancestors, can provide important clues for paleoenvironment and foodweb reconstruction. This knowledge leads to a better understanding of evolution and of early civilization. Recent research has pointed to the chemical signature in fossil bones and teeth as an important source for paleodietary information.

Trace element and isotope systematics. The crystalline phase of bones and teeth is composed of the mineral apatite, which is principally calcium (Ca), phosphate (PO_4), and a hydroxyl group (OH). Trace elements and anions substitute for calcium, phosphate, or hydroxyl without adversely affecting the organism. Among the most studied for paleodietary analyses are strontium (Sr) and carbonate (CO_3). The organic phase of teeth contains carbon (C) and nitrogen (N), which also preserve paleodietary information. The ratio of Sr to Ca and the isotopic composition of Sr, C, and N in inorganic and organic phases are analyzed for paleodiet analysis.

Because most elements have at least two isotopes, small differences in mass lead to the fractionation of one isotope relative to another in many biochemical reactions such as bone and tooth synthesis, and are reflected in the isotopic ratios of the resulting product. The isotopic ratio can be measured very precisely with modern mass spectrometers. By convention and for easier measurement, geochemists express the isotopic ratio in an unknown sample as a deviation in parts per mille relative to widely used standards, expressing the result in a delta notation as shown in the equation below, where R_{sample} and $R_{standard}$ are the

$$\delta = \left(\frac{R_{sample}}{R_{standard}} - 1 \right) \times 1000$$

measured isotopic ratios in the unknown sample and the standard, respectively. The isotopic ratios of paleodietary interest are $^{13}C/^{12}C$ ($\delta^{13}C$) and $^{15}N/^{14}N$ ($\delta^{15}N$). The standards to which samples are compared are air (for $\delta^{15}N$) or a carbonate known as the Pee Dee belemnite (PDB; for $\delta^{13}C$). The isotopic composition of strontium is usually expressed as the ratio of two isotopes ($^{87}Sr/^{86}Sr$) without reference to a standard.

Trace elements and isotopes in paleodietary analysis. The isotopic and trace element composition in the diet, along with physiological processes in the organism, determine the composition of bones and teeth. Sr is discriminated against relative to calcium in organisms; thus, the ratio of Sr to Ca in a predator is lower than the ratio in the prey. Carbon isotopes in the organic and inorganic phases of bones and teeth undergo similar, predictable patterns of enrichment. The nitrogen isotopic ratio has a more complex relationship, reflecting not only diet but also the effects of water stress in the animal. Isotopes of carbon and nitrogen are strongly fractionated by physiological processes because the difference in mass of the isotopes is large (6–8%). The ^{87}Sr to ^{86}Sr ratio directly reflects the diet without the influence of physiologic processes because the mass difference between these two isotopes (about 1%) is too small for the isotopes to be strongly fractionated by physiological processes. For herbivores, the ^{87}Sr to ^{86}Sr ratio of plant foods reflects the ^{87}Sr to ^{86}Sr ratio of local soil and water. However, the ^{87}Sr to ^{86}Sr ratio of seawater, and of marine organisms, is different from most rocks found on continents, allowing marine versus terrestrial components in the diet to be distinguished. Therefore, combination of several independent chemical systems provides the best evidence for paleodietary analysis.

Ancient hominids. Combined analyses of Sr to Ca ratios and δ^{13}C of fossil hominid bones has shed new light on dietary adaptations of human ancestors. The hominid *Australopithecus robustus* (approximately 1.8 million years old) and other taxa, including leopards, hyenas, and several herbivores, for example, baboons from the Swartkrans site in South Africa, have been very well studied. The chemical signatures of these bones indicate that *A. robustus* was probably an omnivore, having a Sr to Ca ratio intermediate between the leopard and hyenid, on the one hand, and hares, baboons, and hyraxes, on the other. Based on the δ^{13}C signature, the diet of *A. robustus* probably included nuts and fruits, and possibly rhizomes and leaves of C$_3$ plants (mostly trees and shrubs). Just as important as determining what ancient hominids may have eaten, is determining what may have eaten them. The fossil leopards from Swartkrans record a signature that excludes *A. robustus* from likely prey species. The Swartkrans leopards most likely preyed upon the baboons and hyraxes.

Cave bears. The cave bears *Ursus spelaeus* and *U. deningeri* are relatively common fossils from the late Pleistocene and Holocene of Europe. During periods of glacial oscillations in this time frame, western Europe underwent large shifts in climate and vegetation. Cave bears are suspected of being herbivorous based on the morphology of cheek teeth. This interpretation is supported by δ^{13}C and δ^{15}N analysis of collagen from fossil bones and teeth of cave bears and other herbivores recovered from the same localities. The δ^{13}C and δ^{15}N of cave bear bones and teeth overlap with those of other herbivores (horses, aurochs, hares, and goats), and are quite different from that of a panther found at the same locality.

Evidence of the physiology of ancient cave bears is also preserved. The chemical signature of deciduous teeth of the cave bears is different from that of permanent teeth, reflecting the influence of nursing. The deciduous teeth form before the cubs have been weaned, and the isotopic composition of milk is very different from the herbivorous diet of adult animals. Furthermore, the large difference between δ^{13}C of the bears and that of other herbivores may indicate oxidation of fat during hibernation.

Elephant ivory sources. The isotopic composition of elephant ivory and bones reflects the different ecological and geological influences at southern African wildlife refuges. As with humans, the ^{87}Sr to ^{86}Sr ratio of elephant bones reflects that of plants that are consumed. The ^{87}Sr to ^{86}Sr ratio of plants, in turn, reflects the ^{87}Sr to ^{86}Sr composition of the local soil and water. Lead (Pb) isotopes have a similar relationship in elephant bones. The combination of Sr and Pb isotopes (which monitor local geology) with δ^{13}C and δ^{15}N (which reflect local plants and level of water stress of the elephants) leads to unambiguous distinction between elephants from different parks.

These records have implications for controlling elephant poaching because the source regions can be identified, and ivory can be "fingerprinted" to monitor sources of products from endangered African elephants. Another possible application is the determination of migratory patterns of ancient elephants and other taxa as recorded in the fossil record. An animal with a bone chemistry signal that does not reflect the local environment is presumed to have migrated into the region. For example, the chemical signature within seasonal growth rings in elephant tusks may record seasonal patterns of migration between areas with different conditions of geology, vegetation, and water stress.

Future research. Future research may focus on predator-prey relationships among multiple components of a fauna to completely reconstruct an ancient terrestrial foodweb. Another application would be to delimit migratory patterns in fossil taxa. Because δ^{15}N and δ^{13}C are very sensitive to changes in flora and water stress, it may be possible to reconstruct details of ancient paleoenvironmental conditions in greater detail. For example, current archeological applications include determining the dietary behavior of early people and the introduction of agriculture and other food sources, such as marine fish or mammals. SEE PALEOCLIMATOLOGY

For background information SEE BONE; FOODWEBS, ISOTOPE; TOOTH; MAMMALIA; PALEOBIOCHEMISTRY in the McGraw-Hill Encyclopedia of Science & Technology.

James Daniel Bryant

Bibliography. B. J. MacFadden and J. D. Bryant (eds.), Stable isotope and trace-element geochemistry of vertebrate fossils: Interpreting ancient diets and climates, *Palaeogeog. Palaeoclimatol. Palaeoecol.*, 107:199–328, 1994; T. D. Price (ed.), *The Chemistry of Prehistoric Human Bone*, 1989; A. Sillen and G. Armelagos (eds.), Second advanced seminar in paleodietary research, *J. Archaeol. Sci.*, 18:225–416, 1991; J. C. Vogel, B. Eglington, and J. M. Auret, Isotope fingerprints in elephant bone and ivory, *Nature,* 346:747–749, 1990.

Marine geology

Recent research in marine geology has involved the exploration of the oceans with satellite altimeters and the study of the precipitation of hydrothermal minerals at the sea and subsea floors.

Satellite Altimetry

The surface of Earth consists of about 30% land and 70% ocean. Many individuals could sketch the outlines of the continents and perhaps label the major mountain ranges and river basins, yet most are unfamiliar with the large topographic features beneath the deep oceans. For example, the Pacific-Antarctic rise, which has an area about equal to South America, is a broad rise of the ocean floor caused by sea-floor spreading between two major tectonic plates: the ridge axis of spreading (sea-floor spreading ridge; **Fig. 1**), and to the west the Louisville seamount chain, a chain of large undersea volcanoes having a length equal to the distance between New York and Los Angeles. These features are unfamiliar because they were discovered fewer than 20 years ago. The Louisville seamount chain was first detected in 1972 by using depth soundings collected along random ship crossings of the South Pacific. Five years later the full extent of this chain was revealed by a radar altimeter aboard the *Seasat* spacecraft. Guided by these new satellite observations, the chain was charted in greater detail in 1986 by using a multibeam echo sounder aboard a Scripps Institution of Oceanography research vessel. In an age when the surfaces of Venus and Mars are being mapped in great detail, it is difficult to believe that so little is known about the planet Earth.

Ocean charting. The reason that the ocean floor, especially in the Southern Hemisphere, is so poorly charted is that electromagnetic waves cannot penetrate the deep ocean (3000–5000 m or 2–3 mi). Instead, depths are commonly measured by timing the two-way travel time of an acoustic pulse (velocity of 1500 m/s or 1 mi/s in seawater). An array of echo sounders, mounted on the hull of a large ship, can be used to map a 10–15-km-wide (6–9-mi) swath. However, because research vessels travel quite slowly (6 m/s or 12 knots) it would take approximately 125 years to chart the ocean basins using the latest swath-mapping tools. It might be possible to commission 12 ships, and the oceans could be mapped in just 10 ship-years; but in addition to the high cost of such a mapping effort, many countries restrict scientific exploration in their territorial waters. To date, only a small fraction of the sea floor has been charted by ships.

Fortunately, such a major mapping program is largely unnecessary because the ocean surface has broad bumps and dips that mimic the topography of the ocean floor. These bumps and dips can be mapped by using a very accurate radar altimeter mounted on a satellite.

Ocean surface bumps. To understand why the ocean surface is bumpy, it is useful to consider an ocean surface that is undisturbed by winds or currents (**Fig. 2**). Since water seeks its own level, the

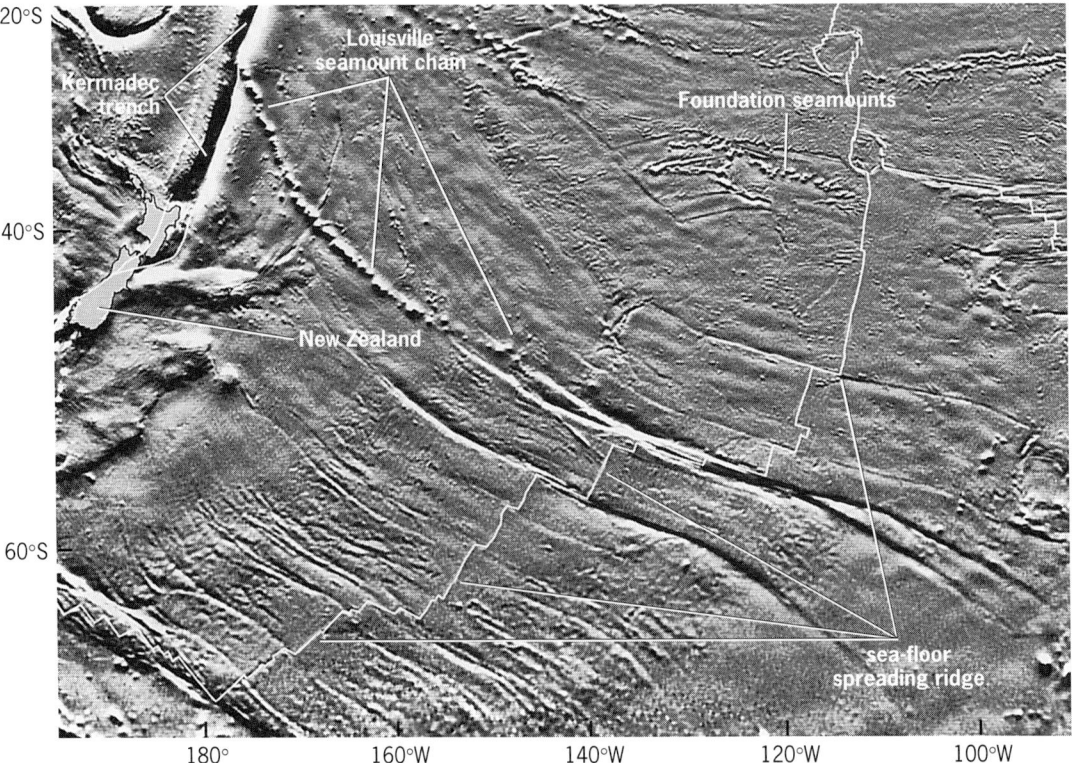

Fig. 1. Geoid height over the South Pacific (illuminated from the north) based on satellite altimeter profiles from the *Geosat, ERS 1* and *Topex/Poseidon* spacecraft.

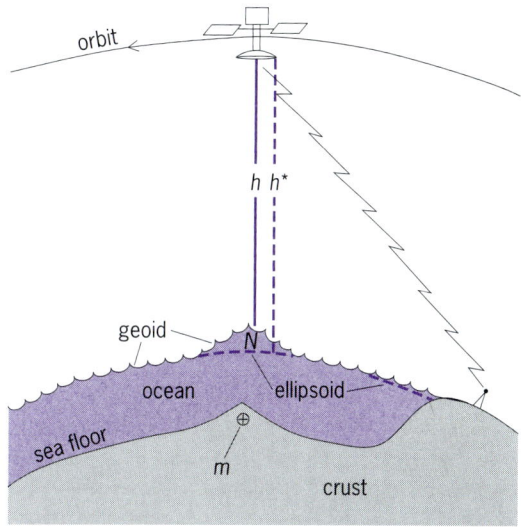

Fig. 2. Satellite altimeter measurement of geoid height (N). m = the local excess in mass. h = the height of the satellite above the closest ocean surface. h^* = the height of the satellite above the ellipsoid. The topography of the ocean surface is highly exaggerated.

ocean surface follows a surface of constant gravitational potential called the geoid. To first order, the surface of the Earth (mostly ocean) is a sphere of average radius R_o (6371 km or 3981 mi). However, because the Earth rotates, its shape is much better matched by an ellipsoid of revolution where its polar radius is 21 km (13 mi) less than the equatorial radius. Superimposed on this ideal ellipsoidal shape are small bumps and dips in the equipotential surface that are caused by local excesses and deficiencies in mass (Fig. 2). A typical bump may be 0.1 m high (4 in.) and 100 km (60 mi) in radius, so the slope of the ocean surface, relative to the ellipsoid, is very low (~10^{-6} radians).

Some simple physics is needed to understand why features on the ocean floor cause perturbations in the equipotential ocean surface. The constant value of the potential (U_o) on the ocean surface is given by Eq. (1), where G is the gravitational constant,

$$U_o = -\frac{G M}{R_o} \qquad (1)$$

6.67×10^{-11} m^3/(kg)(s^2), and M is the mass of Earth, 5.98×10^{24} kg. This potential represents the amount of energy released when an object (of unit mass) is lowered to the ocean surface from a point infinitely far from Earth. However, a disturbance in the potential can be caused by a volcano or any irregular topography on the ocean floor. Undersea volcanoes are typically 1–4 km (0.6–2.5 mi) tall and 10–40 km (6–25 mi) in diameter. They consist of basaltic rock having a density of about 2800 kg/m^3, and they displace seawater having a lower density of 1030 kg/m^3. This density difference, times the volume of the volcano, produces a local excess in mass (m) on the ocean floor. The change in potential (ΔU) caused by this additional mass is given by Eq. (2), where r is the distance between the ellipsoid

$$\Delta U = -\frac{G m}{r} \qquad (2)$$

and the mass. Near the volcano the total potential on the ellipsoid ($U_o + \Delta U$) is now lower than the value on the ocean surface. The equipotential ocean surface must bulge outward (that is, farther from the center of Earth) to remain at a constant potential U_o. The height of the ocean surface above the ideal ellipsoidal shape of Earth (N) is called the geoid height. This simple physics is used to relate the amplitude of the sea-surface topography to the amplitude of the sea-floor topography. The ratio of the topography of the sea surface to that of the sea floor is typically 1:1000, although it can be even smaller for large-scale features, because large volcanoes sink into the outer layer of Earth and form crustal roots with negative mass anomalies. If both the ocean-surface topography and the ocean-floor topography are measured, the mass of the crustal root can be estimated.

Measurements. The tiny bumps and dips in the geoid height can be measured by using a very accurate radar mounted on a satellite (Fig. 2). For example, the *Geosat* satellite was launched by the U.S. Navy in 1985 to map the geoid height at a horizontal resolution of 10–15 km (6–10 mi) and a vertical resolution of 0.03 m (1 in.). *Geosat* was placed in a nearly polar orbit to obtain high latitude coverage (±72°); its orbit is nearly circular and has a relatively low altitude (800 km or 500 mi) in order to optimize the performance of the radar. The *Geosat* altimeter orbits Earth 14.3 times per day, resulting in an ocean track speed of about 7 km/s (4 mi/s). Earth rotates beneath the fixed plane of the satellite orbit, so over a period of 1.5 years, the satellite maps the topography of the surface of Earth with a ground-track spacing of about 6 km (4 mi).

Two very precise distance measurements must be made in order to measure the topography of the ocean surface to an accuracy of 0.03 m (1 in.). First, the height of the satellite above the ellipsoid (h^*) is measured by tracking the satellite from a globally distributed network of lasers and Doppler stations. More recently, the space-based Global Positioning System (GPS) was used to track the *Topex/Poseidon* radar altimeter. The trajectory and height of the satellite are further refined by including the orbital perturbations due to noncentral gravitational attractions of the Earth, Moon, and Sun. Second, the height of the satellite above the closest ocean surface (h) is measured with a microwave radar operating in a pulse-limited mode on a carrier frequency of 13 GHz. (The ocean surface is a good reflector at this frequency.) The radar illuminates a rather large spot on the ocean surface about 45 km (28 mi) in diameter. SEE APPLICATION SATELLITES; SATELLITE NAVIGATION SYSTEM.

A smaller effective footprint (1–5 km or 0.6–3 mi in diameter) is achieved by forming a sharp radar pulse and accurately recording its two-way travel time. The footprint of the pulse must be large enough to average out the local surface irregularities due to ocean waves. The spherical wavefront of the pulse ensures that the altitude is measured to the closest ocean surface. A high repetition rate (1000 pulses/s) is used to improve the signal-to-noise ratio, especially when the ocean surface is rough. Corrections to the travel time of the pulse are made for ionospheric and atmospheric delays, and known tidal corrections are applied as well. The difference between the height above the ellipsoid and the altitude above the ocean surface is approximately equal to the geoid height ($N = h^* - h$).

Applications. As the spacecraft orbits Earth, it collects a continuous profile of geoid height across an ocean basin. Profiles from many satellites, collected over many years, are combined to make high-resolution images of the geoid height (Fig. 1, illuminated from the north). The geoid height image reflects many sea-floor structures associated with plate tectonic activity in the South Pacific. The sea-floor spreading ridge between the Pacific and Antarctic plates is offset in many places by transform faults; the extensions of these transform faults (that is, fracture zones) are antisymmetric across the ridge and record the fossil sea-floor spreading direction. The Louisville seamount chain was formed as the Pacific plate moved over an upwelling plume of hot mantle rock (hotspot). The Pacific plate is being subducted back into Earth at the Kermadec trench. Many of these features were unknown prior to their discovery by satellite altimeters. The Foundation seamounts still have not been surveyed by ships.

The best horizontal resolution of the geoid height derived from satellite altimetry is about 10–15 km (6–9 mi). The major limitation is the natural smoothing of the potential field (geoid height) due to the distance between the sea floor and the sea surface. Nevertheless, this resolution provides significant new information about the ocean-floor topography, especially in the remote southern oceans where the typical spacing between shipboard bathymetric profiles is more than 100 km (60 mi). The total swath width of a multibeam mapping system on a ship is 10–15 km (6–9 mi), so the geoid height maps provide the perfect reconnaissance tool for planning the more detailed shipboard surveys. Other applications include identification of fracture zones and transform faults for detailed plate-tectonic reconstruction models. The U.S. Navy uses the precise gravity information to improve the accuracy of submarine-launched ballistic missiles. The fishing industry sometimes uses this information to locate shallow areas on the crests of uncharted seamounts where fish and lobster are abundant. Petroleum exploration companies are particularly interested in using gravity anomaly data for locating uncharted sedimentary basins on the shallow continental shelves.

David T. Sandwell

Ocean-Floor Hydrothermal Deposits

Recent advances in understanding the physicochemical and geological controls that result in the precipitation of hydrothermal minerals (sulfides, sulfates, silicates) at the rock-water interface (sea floor) and subsea floor, stem from the combined efforts of geoscientists studying modern sea-floor systems and ancient fossil hydrothermal systems on land.

Hydrothermal sites and mineral precipitation. Hydrothermal activity on the modern sea floor is restricted to regions with high geothermal gradients. These areas are composed of volcanically active domains along oceanic ridge crests and of associated off-axis sedimented ridges and seamounts, back-arc extensional environments, and mantle hot spots. The presence of such gradients above thin oceanic crust or shallowly emplaced intrusions results in the down-draw of seawater through the overlying fractured ocean floor. Chemical reactions of the now heated seawater with cooling rocks results in the conversion of seawater into a buoyant acidic hydrothermal fluid capable of transporting high concentrations of sequestered base and precious metals back to the sea floor.

These processes result in the formation of hydrothermal mineral accumulations with concomitant alteration of the rock substrate within and on the sea floor. Minerals precipitate as a result of conductive cooling and quench mixing of hydrothermal fluid with cold seawater. Other important parameters affecting the mineralogy of precipitates include pH change; metal, sulfur, and oxygen contents of the fluids; water depth; and the nature of reservoir rocks where rock-water reactions occur. Deposits formed at Mid-Oceanic Ridge crests are copper-, iron-, zinc-, lead-, and barium-bearing due to the leaching of immiscible sulfides and ferromagnesium minerals from the basaltic source rocks (such as at the Axial Seamount, the East Pacific Rise, and the Galápagos Rift). The mineral assemblage is dominated by sphalerite, pyrite, chalcopyrite, pyrrhotite, and isocubanite.

Deposits formed from felsic volcanic rocks in an island-arc or back-arc environment are typically enriched in copper, iron, and zinc, with relatively more arsenic, lead, and barium (such as in the Lau Basin, the Okinawa Trough, and the Manus Basin). Other deposits are formed in sediment-covered rifts (such as in the Middle Valley, the Escanaba Trough, the Guaymas Basin, and the Red Sea), where the basalt-derived hydrothermal fluid composition is further modified by interaction with sedimentary strata. Chemical buffering by the sediment and contained organic matter causes the fluids to have lower metal contents, higher pH, and higher lead content and to be more reducing than bare-ridge crest deposits. The resulting mineralogy is dominated by pyrrhotite and pyrite. The main sites of mineral precipitation in hydrothermal fields on and near the sea floor include chimney and mound structures, sedimented plume particulates, and stockworks with a chambered texture.

Chimney and mound structures. Low-temperature discharge (50–250°C or 120–480°F) can be diffuse, especially during the early stages of the evolution of a hydrothermal system (as in North Gorda or the North Cleft); at the margins of high-temperature (>300°C or 570°F) vent fields [Trans-Atlantic Geotraverse (TAG), Axial Seamount] and during the waning stages of a hydrothermal field (Middle Valley). Diffuse discharge forms iron-manganese-silicon dioxide deposits and manganese oxide crusts, as well as isolated barite-marcasite-silica chimneys (as are found in the Galápagos, the Axial Seamount, and Middle Valley).

White smokers (low-temperature focused discharge; 50–250°C or 120–480°F) result from slowly effusing fluids that contain gray particles of silica, barite, and anhydrite. Venting results in the formation of sulfide-poor chimneys composed of barite, amorphous silica, sphalerite, pyrite, and clay minerals. These chimneys grow from continuous mineral precipitation from cooling-vent fluid mixing with seawater. Amorphous silica and iron oxyhydroxides may enclose tube worms and bacterial mats living at the vents and, together with barite and marcasite, form dispersed crusts around zones of low-temperature diffuse venting.

Black smokers, sulfide-rich chimneys, form from high-temperature (>300°C or 570°F) vent fluids due to the rapid precipitation of sulfide particulates from sulfur and metal-rich fluids upon mixing with seawater. These chimneys are characterized by a central core of chalcopyrite and würtzite. Black smoker chimney structures in the early growth stages have a chalcopyrite, pyrrhotite, and isocubanite core with an outer annular rim of sphalerite and anhydrite. As the chimney grows, the high-temperature chalcopyrite-rich core displaces the lower temperature anhydrite, sphalerite, and marcasite outward. Up to 90% of the sulfur and metals in the hydrothermal fluid may be dispersed in the plume smoke during these stages.

Hydrothermal mounds associated with active venting are produced by the collapse and coalescence of growing chimneys and by inflation as a result of efficient mineral precipitation within the pile of sulfide-sulfate. The original pore space on the flanks of the mound may be filled in by lower-temperature secondary minerals such as amorphous silica, clay minerals, barite, marcasite, and iron oxides, effectively sealing the mound margins. Internal dissolution, mobilization, and reprecipitation of metals are important processes (that is, zone refining) in the formation of large sulfide mounds (as are found in Middle Valley, TAG, and the Galápagos).

Hydrothermal plumes. The black smoke in buoyant plumes is composed of tiny particles of sulfides and sulfates produced by the rapid precipitation of these phases from the quench mixing of cold seawater and hot fluid. Dissolved forms of iron and manganese are retained in buoyant plumes because of their solubility. The formation of the iron oxyhydroxide hydrothermal particles involves scavenging phosphorus, vanadium, arsenic, chromium, aluminum, and thorium from seawater and their eventual deposition in sediments. These plume particles can be traced up to tens of kilometers from specific vent fields based on their compositions by utilizing specific tracers (certain elements that can be used to determine the signature of the plume, such as helium, maganese, and iron). Discovery of megaplumes, extraordinarily large pulsed fluxes of hydrothermal effluent, indicates that the heat and chemical fluxes from Mid-Oceanic Ridge systems are episodic, thereby reflecting the incremental nature of ridge volcanic accretion processes.

Ocean Drilling Program. In 1992, the Ocean Drilling Program (ODP) provided the first opportunity to drill an active hydrothermal system (Middle Valley, northern Juan de Fuca Ridge) and enabled geoscientists to observe the third dimension of a modern vent field. The source of the metals is the basaltic basement beneath a sediment rift fill. Zones of high permeability within the sedimentary strata were defined and may be the site of subsea-floor metal deposition. Subsea-floor replacement, recrystallization, precipitation of sulfides, and mineral-zone refining are evident in a subcropping sulfide ridge measuring 60 × 120 m (200 × 400 ft) that was drilled to 95 m (310 ft) below the sea floor. The primary assemblage of pyrrhotite, isocubanite, chalcopyrite, and sphalerite has been replaced by pyrite, magnetite, sphalerite, chalcopyrite, and massive pyrite. Chalcopyrite is concentrated in the core of the recrystallized body with an upper part rich in sphalerite (zinc). The outer low temperature silica-rich crusts contain lead-arsenic-antimony sulfosalts. Copper-zinc zone refining is due to continued hydrothermal fluid flushing through primary mineral assemblages. The present-day mineralogy of this massive sulfide deposit is the product of extensive reaction of the older high-temperature sulfide assemblage with cooler (<276°C or 529°F), sulfur-depleted fluids that have equilibrated with the sedimentary pile and are now precipitating anhydrite, pyrrhotite, and clay chimneys.

Metal zonation in large surficial mounds such as TAG reveal high copper-zinc ratios in the core and low copper-zinc ratios at the margins of the mound, analagous to the zonation in chimney structures. It was anticipated that Ocean Drilling Program deep drilling of the large (3–5 × 10^6 metric tons or 3.3–5.5 × 10^6 tons) active, high-temperature, TAG deposit in the fall of 1994 would define the hydrological pathways, chemical composition, and processes of formation of the sulfide deposits precipitated below the sea floor. Ancient copper-zinc deposits show evidence of such major redistribution of hydrothermal minerals and mobilization of metals within a large deposit (that is, zone refining). Zone refining is a process well documented in ancient deposits; however, the roots of a modern hydrothermal system are essentially inaccessible without drilling the active system.

Alteration of oceanic crust. Alteration of the oceanic crust is the result of down-welling seawater and discharge of hydrothermal fluid interacting with the host rocks. It is characterized by magnesium metasomatism and calcium depletion, and it is accompanied by depletion of seawater in magnesium and sulfate. The discharge zones, below the surficial chimney and mound structures of modern sedimented hydrothermal systems, have been observed only through shallow piston coring and Ocean Drilling Program drilling (Middle Valley) and by fortuitous exposures along fault block walls (Galápagos, 9°50′N East Pacific Rise). The roots of the east Galápagos hydrothermal vents are uniquely exposed in three dimensions along horst scarps, revealing 40×150 m (130×500 ft) of stockwork mineralization and alteration.

Volcanic-hosted massive sulfide deposits, the ancient analogs for black smoker fields and hydrothermal mounds, are well exposed on land through mining access. The geochemical characteristics and mineral paragenesis of discordant alteration pipes <100 m (330 ft) below ancient copper-zinc deposits have been intensively studied, although the focus of more recent research is the definition of broad semiconformable alteration zones that outline the fluid reservoirs of the hydrothermal convection system.

Research gains and future directions. Development of the capability for rapid response to volcanic and seismic events (T-phase detection), by using remotely operated vehicles and manned submersibles, is the most significant recent advance. Studies of the volcanic eruptions on the coaxial segment (Juan de Fuca Ridge) and the East Pacific Rise (10°N) in 1992–1993 permitted direct observations of the relations between volcanic, hydrothermal, and biological processes. These areas are now targets for the establishment of long-term deep observatories to monitor future events. Other significant discoveries include that of active shallow sea-floor venting associated with fore-arc alkaline volcanism in the Tabar-to-Feni island chain, Papua New Guinea. It is generating submarine epithermal-type gold mineralization.

For background information SEE ALTIMETER; APPLICATIONS SATELLITES; HYDROTHERMAL VENT; MID-OCEANIC RIDGE; ORE AND MINERAL DEPOSITS; PLATE TECTONICS; SATELLITE NAVIGATION SYSTEMS in the McGraw-Hill Encyclopedia of Science & Technology.

Doreen E. Ames

Bibliography. D. E. Ames, J. M. Franklin, and M. D. Hannington, Mineralogy and geochemistry of active and inactive chimneys and massive sulfide, Middle Valley, northern Juan de Fuca ridge: An evolving hydrothermal system, *Can. Min.*, 31:997–1024, 1993; R. W. Embley, R. A. Feely, and J. E. Lupton, Volcanic and hydrothermal processes on the southern Juan de Fuca Ridge, *J. Geophys. Res.*, 99(B3):4735–5017, 1994; M. D. Hannington et al., Hydrothermal activity and associated mineral deposits of the seafloor, *Geol. Surv. Can.*, 2915C, 1994; D. R. McConathy and C. C. Kilgus, The Navy *Geosat* mission: An overview, *Johns Hopkins APL Tech. Digest*, 8:170–175, 1987; W. I. Ridley et al., Hydrothermal alteration in oceanic ridge volcanics: A detailed study at the Galápagos fossil hydrothermal field, *Geochim. Cosmo. Acta*, 58(11):2477–2494, 1994; D. T. Sandwell, Geophysical applications of satellite altimetry, *Rev. Geophys., Supplement,* pp. 132–137, April 1991; R. H. Stewart, *Methods of Satellite Oceanography*, 1985.

Materials processing

One of the most promising, fast, low-cost methods for processing materials is known as self-propagating high-temperature synthesis (SHS). The **table** presents a comparison of this combustion synthesis process and the other synthesis processes.

Combustion synthesis. Nonoxide materials (such as carbides, nitrides, borides, chalcogenides, hydrides, intermetallic compounds, and sulfides) have been produced by combustion synthesis. This self-propagating high-temperature synthesis, originally developed in the late 1960s, usually involves exothermic reactions above 2500°C (4500°F). The combustion process itself can be stable or unstable and does not require a furnace. The high temperatures remove any volatile contaminants by vaporization.

Combustion synthesis can involve several different types of reactions. An oxidation-reduction, or thermite, can produce multiphase compositions such as cermets. For example, thermite occurs when a mixture of aluminum powder and iron oxide powder reacts, causing strong heating and yielding aluminum oxide plus a white-hot molten mass of metallic iron.

For example titanium (Ti) and boron (B), two starting metallic elements, can be combined to make titanium boride (TiB_2). There are also combinations of these two types of reactions, as well as reactions requiring chemical activators. Preheating is required when the heat of reaction is insufficient for liquid-phase diffusion. The exothermic reaction may occur simultaneously throughout the whole material or may be propagating where a synthesis wave passes through the material. For example, the reaction of titanium and carbon powders to form titanium carbide releases 737 cal/g, with a calculated adiabatic temperature (no heat moved in or out of the system) of 3240°C (5860°F). Such strongly exothermic reactions have been used as sources of energy in many kinds of practical applications. For example, so-called thermite reactions have been used to weld metals, as in the welding of cracks in railroad rails.

Synthesis of refractory materials. Since the mid-1980s the work of a group of Russian scientists has led to the development of a process for the synthesis of refractory materials. An example of this process is the reaction between transition metals and carbon powders. Cold-pressed compacts of these mixed powders are ignited at one end in an argon atmosphere or in a vacuum by means of hot metal wires, pulsed lasers, electric arcs, or high-intensity lamps.

Comparison of various synthesis processes

Process	Advantages	Disadvantages	Compositions
Carbothermal reduction	Possible automation; some control of chemistry	Can be somewhat energy and capital cost-intensive; usually requires milling, which can produce impurities; large scale-up may be difficult; expensive raw materials	Titanium boride, silicon carbide, other nonoxides
Solid-solid, solid-gas, combustion synthesis	Usually self-propagating (requiring no external heat source) and fast (within seconds)	Exothermic, volatile reactions; sometimes low density or low yields; densification may require high pressures; addition of dopants may be required	Titanium boride, titanium carbide, silicon nitride, other nonoxides, composites
Vapor-phase synthesis	High purity; no aggregation; ease of preparation; narrow size distribution; versatility; homogeneity	Limited chemistry, ternary compounds difficult; low yields; reactant gases expensive	Oxides, nonoxides (nitrides, carbides), metals, binary compounds
Laser synthesis	Wide range of composition; short reaction times; uniform heating rates; improved process control; uniform size distribution; minimum agglomeration	Volatile reactants; expensive equipment; powder yields can be low; contamination a problem for certain reactions	Refractory materials (nitrides, borides, silicides, carbides), transition metal compounds, oxides with other emission lives
Plasma synthesis	Highly efficient, simple, continuous; homogeneous mixtures; high surface areas; oxides with wide range of available starting materials; very fast quench rates	Requires high power (10^3 kW); large capital and operating costs; higher surface areas cause greater pyrophoricity; health hazards due to inhaling; carbides, nitrides are sensitive; low powder yields; some agglomeration; nonreproducibility	Oxides, carbides, nitrides, mixtures (silicon oxide + aluminum nitride, silicon carbide + silicon carbide, aluminum + aluminum nitride)

Upon ignition of the compact, a combustion wave rapidly (several seconds) propagates through the mass, converting the reactants to metal carbides (see **illus**.). Besides carbides, ß-sialon, and microcomposites, silicon carbide + aluminum oxide and titanium carbide + aluminum oxide have been formed.

As a class of refractory materials, the nitrides, such as aluminum nitride, titanium nitride, zirconium nitride, hafnium nitride, boron nitride, and silicon nitride, have all been formed by self-propagating high-temperature synthesis. The process for synthesizing nitrides proceeds in a fairly predictable manner. The metal powders are cold-pressed into porous compacts. Subsequently, these compacts are ignited in nitrogen gas or liquid nitrogen by a resistance-heated tungsten wire. The self-propagating combustion wave quickly transforms the metal powder into porous solid nitrides.

Reaction temperature. In combustion synthesis, both particle size and starting composition affect the reaction temperature. The heat of reaction is a function of the amount of liquid phase formed, and this fact ultimately determines the porosity of the product. Particle-size distribution controls initiation of and rate of reaction, and increasing the starting-element particle size can decrease the propagation rate for certain reactions. The final particle size, as well as phase percentages, are functions of the starting particle size of either reactant. The addition of dopants can reduce porosity and increase density, and the addition of inert materials, which have the intended final composition, can decrease the reaction rate, wave velocity, and combustion temperature.

Densification. Since the final product is often in a porous powder form, subsequent densification may be necessary. Such methods include pressureless sintering, uniaxial pressing, isostatic pressing, casting, and hot rolling. The reactor capacities available vary from 2.5 to 30 liters (0.088 to 1.1 ft^3) and a block of three 16-liter (0.57-ft^3) reactors can average 90 kg/h (200 lb/h) of product in continuous production. Production at the pilot facility in Russia for TiC recently reached a level of more than 1000 metric tons (900 tons) per year.

Control of the reaction. During the course of development of self-propagating high-temperature synthesis technology in Russia, various techniques evolved to control the synthesis reactions. Of four such innovations, one is used to slow down (kinetic braking) and three to intensify the reactions (thermal explosion, chemical furnace, and chemical activators), as follows:

1. Kinetic braking. This technique is used to reduce the reaction intensity of the self-propagating high-temperature synthesis by inclusion of the previously reacted product within the mixture of reactants (sometimes in portions as large as 65% of the total). This variation of self-propagating high-temperature synthesis has been used in the production of aluminum nitride and titanium boride.

2. Thermal explosion. This technique is used to ignite some self-propagating high-temperature synthesis reactions when a higher reaction temperature is desired. The mass of reactants is simply preheated (usually to 149–319°C or 300–606°F) until it self-ignites (and continues thereafter by self-propagation).

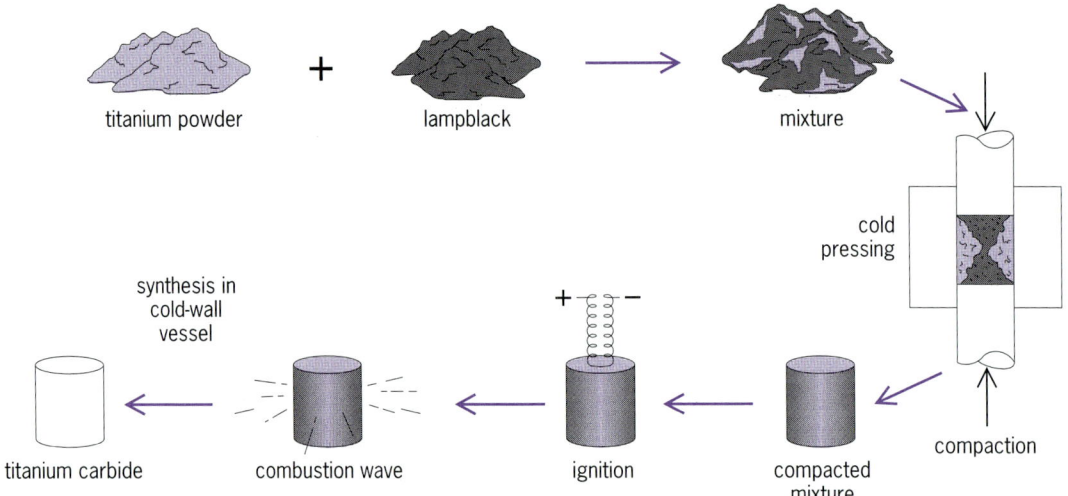

Self-propagating high-temperature synthesis of compounds without external energy.

3. Chemical furnace. This process provides a means of intensifying some self-propagating high-temperature synthesis reactions, especially when a high mass element (such as tungsten) is the metal reactant. A blanket of material that attains a more intense reaction than the desired product surrounds it and serves as a reaction booster.

4. Chemical activators. These activators are added to various reactant mixtures to intensify the reactions. One example is the use of an oxide and metallic reducing agent. This technique has been used in Russian operations involving self-propagating high-temperature synthesis to produce cast (and dense) ceramic shapes.

Reaction hot pressing. This process utilizes self-propagating high-temperature synthesis reactions in place in a uniaxial hot press to form densified ceramic products. Reaction hot pressing takes advantage of the favorable thermodynamics of self-propagating high-temperature synthesis reactions to rapidly form and densify product phases. This process offers the potential for the formation of unique phase assemblages and cost-effective conditions for the production of standard materials. Among the variety of reaction-hot-pressed materials produced are titanium boride/aluminum oxide, titanium boride/aluminum oxide/silicon nitride, and aluminum oxide/titanium carbide. A number of these materials have been prototyped and tested in various applications, including metal cutting inserts, pump seals, and bearings. One product of reaction hot pressing, titanium boride/aluminum, has been successfully developed for applications of aluminum electrorefining.

Reactive sintering. This type of self-propagating high-temperature synthesis is also known as gasless combustion. It has been used to produce intermetallic compounds such as nickel aluminide. In this process an exothermic reaction provides enough heat to produce a liquid phase that markedly reduces sintering time. If isostatic pressure is applied during sintering, the process is known as reactive hot isostatic pressing.

Shock compression. Chemical reactions can also be initiated by shock compression of powder mixtures. This type of dynamic processing technique, referred to as shock-induced reaction synthesis, utilizes the simultaneous application of very high pressure and temperature generated during the passage of shock waves through a powder mixture. Rapid increases in pressure and temperature induce extensive plastic deformation, turbulent flow and mixing of powders of different species, and rubbing and fracturing of powder particles. The resulting densification and generation of intimate contacts between cleansed surfaces of dissimilar powder particles can lead to reaction initiation at temperatures often lower than those required for self-propagated high-temperature synthesis reactions.

The fundamental processes that control shock-induced chemical reactions leading to the synthesis of compounds are not completely understood. Because of the exothermic nature of these reactions, the reaction temperature often exceeds the product melt temperature. Thus, once the reaction is initiated, the product exhibits a microstructure typical of melting and rapid solidification, not of the high-pressure shock state.

Unlike the process of self-propagated high-temperature synthesis, shock-induced chemical reactions occur in only a few microseconds, thus limiting time-resolved monitoring of the events occurring during the reaction process. The dynamic nature of shock-induced chemical reactions is, in some respects, analogous to the detonation phenomenon in explosives. However, detonation of explosives is accompanied by a large buildup in pressure (due to generation of gaseous reaction products), which can be monitored by instruments using various types of pressure-sensing probes. However, shock-induced chemical reactions are gasless reactions and result only in generation of high temperatures. Dynamic monitoring of temperature, although attempted to a limited degree, is a difficult task.

For background information SEE CERAMICS; CERMET; COMPOSITE MATERIAL; MATERIALS SCIENCE AND ENGINEERING; SINTERING in the McGraw-Hill Encyclopedia of Science & Technology.

Mel Schwartz

Medical mycology

In the 1980s and 1990s, the aggressive use of immunosuppressive therapy and broad-spectrum antibacterial antibiotics to treat or manage a variety of medical problems, and the advent of acquired immunodeficiency syndrome (AIDS), has changed the clinical picture of infectious diseases. As a consequence, fungal infections have become one of the leading causes of morbidity and mortality in AIDS patient populations and in individuals who have intrinsic immunological defects. Fungal infections in the immunocompromised and debilitated patient often appear clinically as an acute, life-threatening disease, which allows little time for a definitive diagnosis to be made either by culture of the organism or by serological procedures. In order to manage these diseases properly, treatment must be rapid and aggressive, and the patient's underlying condition of immunosuppression or debilitation corrected.

Virtually all fungi that have been implicated in diseases of humans are found in nature in the soil or on vegetation as free-living saprobes (organisms that live on decaying matter). The exceptions are various species of yeastlike fungi belonging to the genus *Candida*, such as *C. albicans*, and the lipophylic yeast *Malassezia furfur*, both commensals (organisms that obtain food or other benefits from another without damaging or benefiting it) that are part of the normal flora of most healthy individuals. Species of *Candida* are frequently isolated from the oropharyngeal, perianal, and vaginal mucosa of healthy individuals and *M. furfur* is found in the absence of clinical disease on areas of the skin that are rich in sebaceous glands. Only under certain conditions of immunosuppression or host debilitation (**Table 1**) do these organisms become clinically significant as the etiologic agents of disease. Parasitizing individuals is not required for perpetuation of fungal species, and fungi gain no obvious benefits by infecting susceptible hosts. In general, all fungal infections of humans can be described as being opportunistic in that fungi take advantage of favorable situations for growth. SEE YEAST INFECTION.

Fungi and their infectious particles (such as conidia, spores, and hyphal elements) are ubiquitous in the environment, but healthy immunologically competent individuals have a high degree of natural resistance to these organisms. At least two features appear to be common among the increasing number

Table 1. Predisposing factors for opportunistic infections

Factor	Examples
Mechanical	Trauma
	Burns
	Abrasions
	Occlusion (moisture and/or maceration)
	Dentures
	Obesity
	Occlusive dressings
Systemic	Endocrinopathies
	Diabetes mellitus
	Cushing's syndrome
	Hypoparathyroidism
	Hypothyroidism
	Hypoadrenalism
	Intrinsic immunodeficiency states
	DiGeorge's syndrome
	Nezelof's syndrome
	Severe combined immunodeficiency syndrome
	Myeloperoxidase deficiency
	Chediak-Higashi syndrome
	Chronic mucocutaneous candidiasis
	Chronic granulomatous disease
	Acquired immunodeficiency syndrome (AIDS)
	Hematological disorders
Iatrogenic	Barrier breaks
	In-dwelling catheters
	Intravenous drug abuse
	Surgical procedures
	Radiation therapy
	Corticosteroids and other immunosuppressive drug therapy
	Broad-spectrum antibacterial antibiotic therapy

of fungi that have been implicated in disease processes (**Table 2**): a form of trauma precedes appearance of the disease, or an immunosuppressed or physiologically debilitated state exists in the host. However, of the growing number of fungi that have been implicated in diseases of humans, only *Histoplasma capsulatum*, *Blastomyces dermatitidis*, *Paracoccidioides brasiliensis*, *Coccidioides immitis*, and *Cryptococcus neoformans* possess the inherent ability to infect healthy hosts in the absence of these conditions and are considered primary systemic fungal pathogens; the others are termed opportunistic mycotic agents. Notably, the primary systemic pathogens are the cause of most cases of life-threatening fungal infections in immunocompromised patients. Certain mycological, epidemiological, and clinical features of the diseases caused by these agents are important from a diagnostic and clinical viewpoint. Four of the five primary systemic pathogens (*H. capsulatum*, *B. dermatitidis*, *P. brasiliensis*, and *Coccidioides immitis*) are dimorphic (occurring in two distinct forms). In nature (and at 77°F or 25°C) these organisms grow in a filamentous mycelial form as molds, but when they infect a suitable host (or are grown at 98.6°F or 37°C), they change to a unicellular morphology. In contrast to these dimorphic pathogens, *Cryptococcus neoformans* is a unicellular budding yeast at 77°F (25°C), 98.6°F (37°C), and in tissues. The unique morphological feature of this yeast is that it is surrounded by an

Table 2. Fungal infections frequently seen in the immunocompromised host

Disease	Organism	Unique features
Candidiasis	*Candida* spp., especially *C. albicans*	Mucosal disease common in AIDS, but systemic involvement is rare; systemic disease in individuals who are deficient in neutrophilic leukocytes; fatal if not treated rapidly
Cryptococcosis	*Cryptococcus neoformans*	Most frequent cause of fungal central nervous system disease; important cause of mortality in individuals with AIDS, lymphoma, leukemia, sarcoidosis, and those receiving chronic corticosteroid therapy
Histoplasmosis	*Histoplasma capsulatum*	In the endemic area, an important complication of AIDS
Coccidioidomycosis	*Coccidioides immitis*	A complication often seen in individuals with AIDS who live in or have visited endemic areas
Aspergillosis	*Aspergillus* spp., especially *A. fumigatus*	An important complication in individuals who are neutropenic, receiving steroid or broad-spectrum antibacterial therapy, or who have coexisting hematological malignancies
Mucormycosis*	*Rhizopus* spp., especially *R. oryzae* *Absidia corymbifera* *Mucor* spp. *Sakseneae vasiformis*	Predisposing factors: uncontrolled diabetes mellitus (hyperglycemic and acidotic states), aggressive steroid therapy, and leukopenia
Blastomycosis	*Blastomyces dermatitidis*	No definitive patterns of immunosuppression identified yet
Sporotrichosis	*Sporothrix schenkii*	Lymphocutaneous disease most frequently seen; pulmonary disease with hematogenous dissemination to various organ systems reported in severely immunosuppressed states
Phaeohyphomycosis*	*Alternaria* spp. *Curvularia* spp. *Bipolaris* spp.	Subcutaneous, sinusitis, or rhinocerebral infections
Systemic pityriasis versicolor	*Mallasezzia furfur* *M. pachydermatis*	Hematogenous infections in infants and immunocompromised individuals receiving intravenous lipid therapy
Miscellaneous*	*Fusarium* spp. *Trichosporon beigelii* *Pseudallescheria boydii* (*Scedosporium apiospermum*) *Rhodotorula* spp., especially *R. rubra*	Increasing frequency of reports of these organisms as a complication in individuals being treated with cytotoxic drugs or otherwise immunocompromised; some of these organisms very refractory to antifungal therapy

* Partial listing of organisms implicated in these diseases.

acidic mucopolysaccharide capsule which enables it to resist phagocytosis by macrophages.

Primary systemic pathogens. With rare exceptions the initial site of infection of the primary systemic pathogens is the lung. Infectious fungal particles are inhaled, lodging on mucous membranes of the respiratory tree and the alveoli where they are rapidly phagocytosed by macrophages. In a majority of cases where the individual is healthy and immunocompetent, these exposures result in subclinical or asymptomatic disease, the individual responding with no aftereffects of disease. However, this absence of aftereffects depends on the degree of exposure and the virulence of the organism. In a small number of cases, the diseases become progressive and affect many organ systems. Left untreated, they result in a chronic debilitating disease condition or cause death. The manifestations of the primary systemic fungal infections in the immunocompromised host (Table 2) take on a different clinical picture, depending on the basis of the immune defect and the etiologic agent.

Opportunistic fungal infections. Only under certain unusual situations (Table 1) do these organisms gain access to body tissues, infect, and—depending on the degree of host debilitation—cause disease. At least one of two features are responsible for the increasing number of fungi that have been implicated in these disease processes: some form of trauma precedes appearance of the disease, or the existence in the host of an immunosuppressed or physiologically debilitated state. When such diseases occur in the immunocompromised or debilitated patient they are very difficult to treat. Fungi classified in this group have been termed emerging fungal pathogens, and the list of such fungi will continue to grow in number as long as there are immunocompromised patients. The importance of some of these fungi is their tolerance to some of the currently available antifungal agents.

Diagnosis and treatment. In clinical settings, it is important that fungi be considered as the possible etiology of the infectious process. Definitive diagnosis, relying on observing fungal elements in clinical material and culture, is important so that therapy can be rapidly instituted. The number of currently available antifungal agents used in the treatment of life-threatening infections is small and falls into three major classes: polyene, azole, and pyrimidine derivatives.

Amphotericin B, a polyene macrolide antifungal agent produced by various species of *Streptomyces*,

is the drug of choice in treating life-threatening fungal infections. This drug, administered intravenously, acts by binding to the sterol ergosterol in the membranes of fungi. However, since host cell membranes contain cholesterol, amphotericin B can also bind to and perturb host cells. As a result, although amphotericin B is very effective against fungi, it can also be toxic to host cells. Thus, the benefits of therapeutic effectiveness must be weighed against the possibility of host toxicity, and individuals treated with amphotericin B should be carefully monitored for toxicity.

An alternative group of compounds currently being used for the management of life-threatening fungal infections are the azole derivatives, such as fluconazole and itraconazole. These compounds are effective inhibitors of ergosterol biosynthesis and have been used in treating various fungal infections. These compounds have the advantage over amphotericin B in that they are less toxic and are given orally. Unfortunately, the azoles inhibit fungal growth in the concentrations used, and in some cases, therapy must be continued for long periods of time (or in the case of AIDS for the life of the individual). Another problem that has further complicated treatment of these fungal infections is the development of resistance to the azoles in some individuals. In this situation, the underlying immunological defect must be corrected for maximum therapeutic benefits.

Flucytosine or 5-fluorocytosine, a fluorinated pyrimidine derivative, is given orally and is effective against various species of *Candida, Cryptococcis neoformans*, and some dark-colored fungi. However, strains resistant to 5-fluorocytosine rapidly emerge, and for this reason this agent is not used alone but in combination with amphotericin B. In susceptible strains, 5-fluorocytosine is converted to 5-fluorouracil and then to 5-fluorouridylic acid, where it is incorporated into ribonucleic acid (RNA) and inhibits protein synthesis; or it is metabolized to 5-fluorodeoxyuradylic acid monophosphate, a potent inhibitor of thymidylate synthase resulting in inhibition of deoxyribonucleic acid (DNA) synthesis.

For background information SEE ACQUIRED IMMUNE DEFICIENCY SYNDROME (AIDS); FUNGI; IMMUNOSUPPRESSION in the McGraw-Hill Encyclopedia of Science & Technology.

George S. Kobayashi

Bibliography. D. J. Drutz, *Systemic Fungal Infections: Diagnosis and Treatment*, vols. 1 and 2, 1988, 1989; W. B. Powderly and J. W. Van't Wout, *Fluconazole: Anti-fungal Agents*, vol. 1, 1992; E. S. Vartivarian, E. J. Anaisse, and G. P. Bodey, Emerging fungal pathogens in immunocompromised patients: Classification, diagnosis, and management, *Clin. Infect. Dis.*, 17:(Suppl. 2):487–491, 1993; T. Walsh, Management of immunocompromised patients with evidence of an invasive mycosis, *Hematol. Oncol. Clin. N. Amer.* 7:1003–1026, 1993.

Medical testing

Advances in chemistry and immunology have led to the development of a number of easy-to-use medical diagnostic tests. These tests provide rapid results, and their performance is often comparable to state-of-the-art clinical analyzers. The simplicity and reliability of these tests have allowed users to receive important information on their health conditions in the privacy of their homes.

Glucose measurement. The first test to be developed, during World War II, was for the measurement of glucose. It involved the addition of a readily soluble reagent tablet into a diluted urine sample. Reaction of copper sulfate in the reagent tablet with glucose in the sample produced a colored solution that, when compared with a standard color chart, provided the estimate of glucose in the urine. Because of its complexity, this system was replaced in the early 1950s by a solid-phase dipstick test, involving a chemical-coated paper strip that when dipped in urine produces a color change related to the concentration of glucose. This system has become the basis of present-day dipstick tests for important clinical markers in urine, such as glucose and ketones (for diabetes), proteins (for proteinuria), and nitrites (for urinary tract infections).

Both chemical reagents and enzymes have been used on the dipstick pad. The glucose test, by far the most important test, uses glucose oxidase and horseradish peroxidase along with a substrate on the pad. Addition of the reagent pad into urine samples reconstitutes the reagents (by dissolving the dry materials) and initiates reactions (1) and (2), ultimately depositing a thin film of a

$$\text{Glucose} \xrightarrow{\text{glucose oxidase}} \text{hydrogen peroxide} \quad (1)$$

$$\text{Hydrogen peroxide} \xrightarrow[\text{substrate}]{\text{horseradish peroxidase}} \text{colored product} \quad (2)$$

colored product, the shade of which gives the estimated concentration of glucose. Since these dipstick tests involve visual color comparison with standard color charts, the results obtained are at best semiquantitative.

Diabetes is an incurable life-threatening disease caused by a deficiency of insulin, a hormone that functions as a chemical regulator to maintain a constant supply of glucose to energize billions of body cells. Although valuable for an indication of the disease, the urine-based semiquantitative glucose dipstick test is of little or no value to a diabetic who must tightly control blood glucose levels by insulin intake for effective management of the disease. The urine-based glucose tests have been further refined by fine-tuning of chemistry and incorporation of a red-blood-cell separation mechanism on a test strip. Filtration of blood provides plasma that reacts with

Fig. 1. Hand-held instrument for blood glucose analysis. (*Life Scan*)

reagents on the test strip pad to provide a color change. To quantify the exact concentration of glucose, portable spectrophotometers precisely register the color change. One successful commercially available system for blood glucose analysis employs a test strip pad in the form of a membrane that acts both as a red blood processor and as a converter of glucose into a readily quantifiable indamine dye. The reader that accompanies this system (**Fig. 1**) is a sophisticated hand-held instrument that acts both as a spectrophotometer and a minicomputer to store data on previous glucose tests. More recent systems incorporate electrochemical detectors for accurate measurement of glucose, and have decreased instrument size to that of a credit card.

Cholesterol levels. Although a small instrument has been very cost-effective in monitoring blood glucose levels for diabetics who must measure their glucose levels frequently (the insulin-dependent type I diabetic, for example, needs to measure blood glucose levels more than once a day), such a device may not be useful for other infrequently measured components of blood. One such example is cholesterol, which is implicated in arteriosclerosis, one of the leading causes of heart attack. In the United States the National Cholesterol Education Program of the National Institutes of Health has issued guidelines for lowering blood cholesterol to the desired 200 mg per deciliter level (1 dl = 100 cm^3).

A novel technology quantifies cholesterol from blood samples in minutes without the use of an elaborate instrument. This device is a disposable, hand-held plastic cassette (**Fig. 2**) that contains all the reagents necessary to run the test and a sophisticated blood separation and automated sample measuring device. The system also has built-in "test working" and "test complete" indicator windows. The user pushes a simple button fingerprick device to release a few drops of blood into the sample well. After a 2-min wait to ensure complete conversion of the entire blood sample into plasma, the user pulls the plastic tab. The window turns green to indicate that the test is complete, and the user reads the produced blue-colored band much like a thermometer. The height of the color bar is read in millimeters, then checked against the result chart for the corresponding cholesterol concentration in milligrams per deciliter. The test is accomplished in

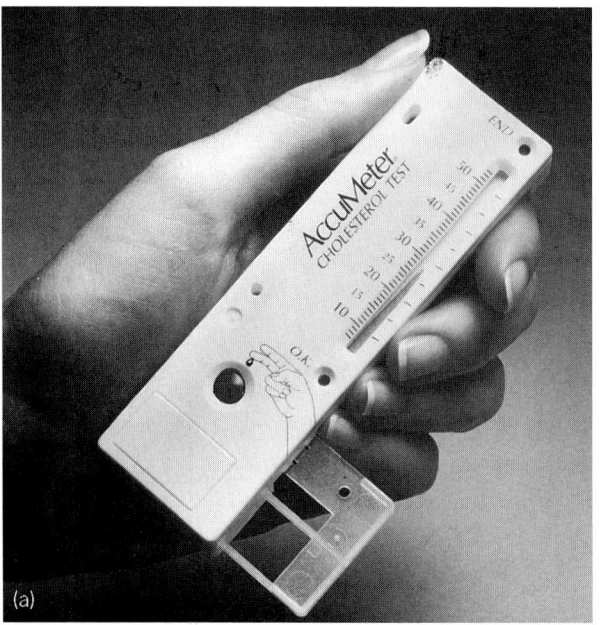
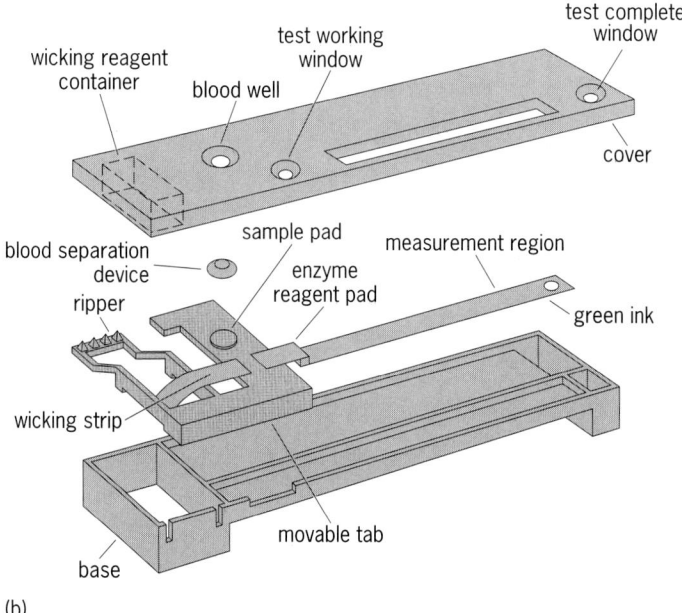

Fig. 2. System for a self-contained cholesterol test. (*a*) Hand-held instrument showing a test result from a whole blood sample; migration height of approximately 0.04 in. (3 mm) corresponds to a concentration of 210 mg/dl cholesterol (1 dl = 100 cm^3). (*b*) Instrument components. (*ChemTrak*)

about 15 min. The results obtained are highly accurate. In a number of studies, this cholesterol test device has been shown to be as accurate as the state-of-the-art clinical analyzers, and a precision of greater than 95% is routinely obtained when replicates of a sample are analyzed.

The cholesterol-testing cassette houses, along with the blood separation and metering device, a number of precisely aligned papers and all liquid and dry reagents required to run the chemistry of the test. The test uses enzyme chemistry very similar to that used in clinical analyzers. The added blood is filtered into plasma, precisely measured onto a sample pad, and reacted with the enzymes cholesterol esterase and cholesterol oxidase. These enzymes stoichiometrically transform cholesterol in the sample into hydrogen peroxide. The hydrogen peroxide produced along with the horseradish peroxidase is transferred by a wick onto a measurement strip that contains an immobilized analog of N,N-dimethyl aniline and 3-methyl-2-benzothiazoline hydrazine. The indimine dye thus produced is measured as a blue-colored band, which varies in length proportionately to the concentration of cholesterol. This technology is highly versatile, and it has also been used for the quantification of high-density lipoprotein (HDL) cholesterol (the so-called good cholesterol) and the measurement of the antiasthmatic drug theophylline, among others, in unprocessed blood samples.

Immunodiagnostics. With the advance in the preparation of monoclonal antibodies, a new class of solid-support immunodiagnostics has emerged, allowing consumers to track important changes in their bodies. The use of immunochemical reagents to detect minute quantities of specific markers in urine such as human chorionic gonadotropin (hCG) and luteinizing hormone (LH) has resulted in products that provide powerful information for early detection of pregnancy and the identification of the most fertile time in the menstruation cycle of a woman.

Pregnancy test. The pregnancy test, the first of the immunodiagnostic tests to be made available to consumers, has undergone major improvements since its introduction in 1977 as a 2-h multistep ring test, culminating in a fast one-step dipstick test introduced in the late 1980s. Intervening were a number of iterations of protocols involving solid-phase sandwich immunoassay and multiple steps. In the one-step pregnancy test, the user introduces the porous end of the dipstick into the urine stream and observes the formation of a colored band on the reaction window of the test device. This test, which essentially is complete in less than 5 min, uses the principle of sandwich immunoassays requiring two antibodies.

A monoclonal anti-hCG antibody labeled with colloidal gold particles is provided in the lower porous end of the dipstick and the second antibody, often polyclonal, is immobilized on the reaction window. Application of urine to the porous end of the dipstick starts capillary migration, carrying labeled monoclonal antibodies with the urine; the monoclonal antibodies are trapped by the immobilized antibodies on the reaction window. The formation of a blue-colored band indicates the presence of hCG in the urine and hence a pregnancy. A second region provides an internal control to indicate that the test has been properly performed. Pregnancy tests have an accuracy of up to 99%.

Luteinizing hormone. The wide acceptance of the pregnancy test has resulted in the extension of this technology to semiquantitative measurement of levels of LH in urine. Normally, during each menstruation cycle the level of this hormone in a woman's body will suddenly surge, accompanied by an increase of the level of the hormone in the urine. The increased presence of LH causes the release of an egg from the ovary 24–36 h later, making it available for fertilization by the sperm of the male partner. Accurate prediction of the LH surge in a woman thus increases the chances of pregnancy. The LH test design is essentially similar to the pregnancy test and uses immunocapillary migration sandwich immunoassay. Use of the specific antibodies for LH and appropriate amounts of reagents on the dipstick provides a test that shows a positive signal after certain amounts of LH have appeared in the urine. As with the pregnancy tests, a one-step version of the test provides results in minutes.

For background information SEE CHOLESTEROL; DIABETES; GLUCOSE; INSTRUMENTATION; INSULIN; PREGNANCY in the McGraw-Hill Encyclopedia of Science & Technology.

Prithipal Singh

Bibliography. M. P. Allen et al., A noninstrumented quantitative test system and its application for determining cholesterol concentration in whole blood, *Clin. Chem.*, 36:1591–1597, 1991; J. A. Hunt and N. C. Alojado, A new, improved test system for rapid measurement of blood glucose, *Diabetes Res. Clin. Prac.*, 7:51–55, 1989; H. K. Naito, Y. S. Kwak, and R. S. Galen, Hypercholesterolemia screening in one step, *Lab. Med.*, 25:444–448, 1994; P. Singh, B. P. Sharma, and P. Tyle (eds.), *Diagnostics in the Year 2000*, 1993; G. E. Valkirs and R. Barton, Immunoconcentration: A new format for solid-phase immunoassays, *Clin. Chem.*, 31:1427–1431, 1985.

Memory

A recent advance in memory research is the integration of cognitive psychology and neuropsychology to define systems of memory. In this approach, memory is divided into procedural and declarative memory, each containing several different forms of memory. When neurological damage selectively affects one of these two memory types, it provides support for cognitive theories that also distinguish between the two.

Procedural and declarative memories. Procedural memory includes memory for skills, such as playing a musical instrument or riding a bicycle. Declarative memory includes various forms of factual information, such as memory for a word, a picture, or other sensations and thoughts. Declarative memories may be described verbally, are open to introspection, and may be formed through a single experience. Procedural memories are easier to perform than describe, and usually take time to accumulate. For instance, memory for riding a bicycle takes time to form and is difficult to describe, but memory for a word is quickly learned and easily defined.

Amnesic individuals with damage to the hippocampus, diencephalon, and related brain structures were once believed to be unable to form any new long-lasting memories. Research now indicates that amnesic individuals are able to form some new memories, and it has been proposed that they possess intact procedural memory but have impaired declarative memory. For example, in the mirror tracing task, individuals are shown a figure in a mirror, and are asked to draw the figure in its correct orientation. This process is repeated in several different sessions. With practice, amnesic individuals show increased accuracy; but they do not remember the earlier sessions. They cannot describe the earlier events, or recognize the testing apparatus. Thus, amnesic individuals show intact procedural memory for the task despite impaired declarative memory for the previous study trials. More recent efforts propose that the amnesic individuals are impaired at explicit memory, or memory with conscious awareness. That is, they are able to show declarative memory when they are not asked to consciously recollect the information.

Semantic and episodic memory. Semantic and episodic memory are two types of declarative memory identified through cognitive and neuropsychological investigation. Semantic memory includes factual information, such as the predicted temperature for the day. Episodic memory includes contextual information about where, when, or from whom information was first learned. Both types can occur independently. For example, it is possible to remember the weather report and forget when it was learned, where it was learned, or who provided it.

Experiments involving normal individuals revealed that memory of a piece of information and its learning episode can be dissociated. In one test, individuals were asked to memorize a list of words and later to recall the words. They were then provided with all the words and asked to reconstruct the temporal order of the list. However, experimental factors can selectively affect item and temporal-order memory. For instance, when words are drawn from several categories (such as fruits or vehicles) individuals remember a greater number of words, but their recall of the temporal order of the words does not improve. Thus, item memory appears to be separable from the temporal memory about that information. Likewise, item memory and spatial memory, another form of episodic memory, can also be affected independently.

A similar dissociation has been found in individuals with damage to the frontal lobe of the brain. This area has been implicated in higher-order processes such as reasoning and problem solving. Individuals with frontal damage and control subjects were asked trivia questions, such as "What is the location of Angel Falls?" The individuals were taught the answers to the questions they did not know. In a subsequent test session, they were asked the same questions. Individuals with frontal lobe damage and control subjects showed similar memory for the answers they learned in the first session, but differed in their memory for where the facts were learned. The control subjects attributed the memory for the answers to the earlier session. In contrast, the individuals with frontal lobe damage remembered the session but did not identify it as the context or source of learning for the trivia facts. Thus, these individuals were described as possessing a source memory deficit.

Future classifications. One goal in memory research is the classification of memory by type of information retained and by the brain systems that process each type. As research progresses, dissociations between different forms of episodic memory, including temporal, spatial, and person memory, and the neurological basis of each may follow.

For background information SEE AMNESIA; BRAIN; MEMORY; PSYCHOLOGY, PHYSIOLOGICAL AND EXPERIMENTAL in the McGraw-Hill Encyclopedia of Science & Technology.

Paul Jurica

Bibliography. J. R. Janowsky, A. P. Shimamura, and L. R. Squire, Source memory impairment in patients with frontal lobe lesions, *Neuropsychologia,* 27(8):1043–1056, 1989; L. R. Squire, *Memory and Brain,* 1987; E. Tulving, What kind of a hypothesis is the distinction between episodic and semantic memory?, *J. Exp. Psychol.,* 12:307–311, 1986.

Metal cluster compound

Studies of metal clusters, that is, small metal particles containing a number (n) of atoms in the range $n = 10$–$10,000$, are of interest not only for fundamental science but also for potential applications, for example, in the fields of catalysis, microelectronics, and magnetic recording media. Fundamental scientific questions are mostly related to the so-called quantum-size effects. Basically, for such small particles the cluster size becomes comparable to characteristic physical length scales such as the de Broglie wavelength of an electron at the Fermi energy of the (bulk) metal, the superconducting coherence length, and the wavelengths of lattice waves (phonons) or magnetic waves (magnons). As a consequence, the familiar bulk behavior is lost, and the physical properties reflect the quantum mechanical phenomena, such as the wave nature of the electron. These size-dependent effects will probably be exploited to create materials with novel magnetic, optical, dielectric, or electronic transport properties.

Monodisperse particles. To explore the evolution of the physical properties with cluster size, the ideal procedure would be to dispose of assemblies of clusters of one and the same size (monodisperse), well separated from one another to prevent coalescing, and with the possibility of varying the cluster size in the assembly from the molecular range of a few atoms up to the sizes typical of large metal colloids. In addition, the samples available for investigation should preferably be macroscopically large, that is, in quantities exceeding 10 mg range, in order to permit the use of a wide scope of experimental solid-state techniques. Until recently, most experimental methods for producing metal clusters suffered from large distributions in size within the assembly, as for instance in catalysts, metal-nonmetal composites such as cermets (ceramic metals), or the usual metal colloids, colloidal metal particles that are stabilized with ligands similar to those present in the metal cluster molecules. Alternatively, mass-selected clusters may be produced by cluster-beam technology, but the yield of such methods is extremely small, allowing only optical spectroscopic tools to be applied.

Recently, it became apparent that the ideal of macroscopically large assemblies of monodisperse metal particles is closely approximated in metal cluster compounds. These materials are chemical compounds, of well-defined stoichiometry; they consist of (macro-)molecules composed of so-called cores of metal atoms surrounded (chemically coordinated) by so-called shells of ligands. The type of metal atom and the number of metal atoms in the core (cluster) varies from compound to compound, as does the type of ligand, which can be an atom or a molecule. The metal cluster macromolecule can be ionic, in which case it may form an ionic crystalline solid together with suitable counterions; or it can be neutral, forming a Van der Waals–type molecular solid that can be crystalline or amorphous. The cluster solids so formed must be homogeneous chemical phases, so that a given compound will contain only a single type of metal core. Thus, metal cores that are not too small (>10 atoms) may be viewed as nanosized, monodisperse metal clusters embedded in the dielectric matrix formed by the ligand shells and, if present, the counterions. The yield of these chemical compounds can be quite large, in the range of milligrams to grams; in some cases even sizable single crystals are available.

Large clusters. This intuitive conceptual view can be meaningful only for large enough metal cores. As a matter of fact, metal cluster compounds with (relatively) small metal cores have been known in chemistry for quite some time; by the mid-1980s several hundred such compounds had been synthesized, with core sizes ranging up to 10–20 atoms and with a great many different transition metal elements. Since then, research has been focused on creating ever larger metal cluster

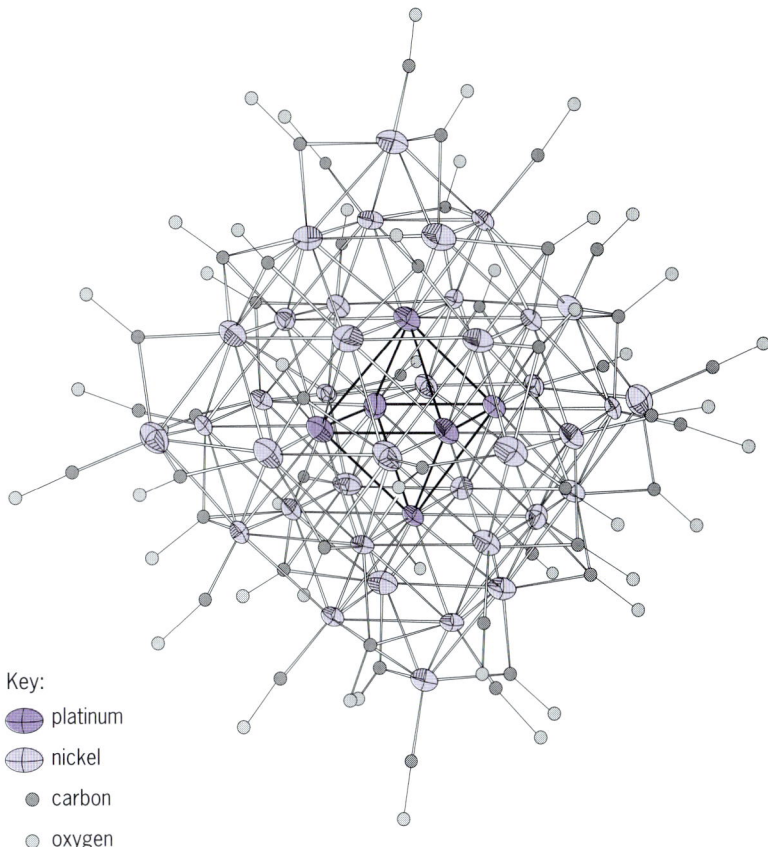

Fig. 1. Structure of the metal cluster molecule $[Ni_{38}Pt_6(CO)_{48}H]^{5-}$. Hydrogen is not shown.

molecules. For example, within the subgroup of metal carbonyl clusters [metal cores coordinated by carbonyl (CO) ligands], very large metal cores containing up to 34 nickel (Ni) atoms or up to 38 platinum (Pt) or palladium (Pd) atoms were synthesized, as well as bimetallic cores consisting of 6 Pt atoms surrounded by 38 Ni atoms (**Fig. 1**). By using phosphine (or related) groups as ligands, neutral cluster macromolecules with large groups of nickel and selenium (Se) atoms or copper (Cu) and Se atoms as cores have been obtained. Bulk Cu_2Se has metallic properties. These materials form molecular crystals, and in fact $Cu_{146}Se_{73}(PPh_3)_{30}$ is the largest cluster compound to date that has been fully characterized by x-ray analysis.

Giant magic-number clusters. Extremely large clusters have been found in a series of giant metal cluster molecules, the metal cores of which are members of the series of so-called magic-number (full-shell) clusters (**Fig. 2**). They are obtained by surrounding an atom progressively with additional shells of like atoms. The resulting 1-shell, 2-shell, and multishell clusters possess the magic numbers of atoms 13, 55, 147, 561, and so forth. These numbers are the same for icosahedral or cuboctahedral (face-centered cubic) packing of atoms, but experimental examples such as $Pt_{309}Phen^*_{36}O_{30}$ and $Pd_{561}Phen_{36}O_{200}$ have cuboctahedral structure.

So far, no single crystal samples of these giant cluster compounds have been obtained. The neu-

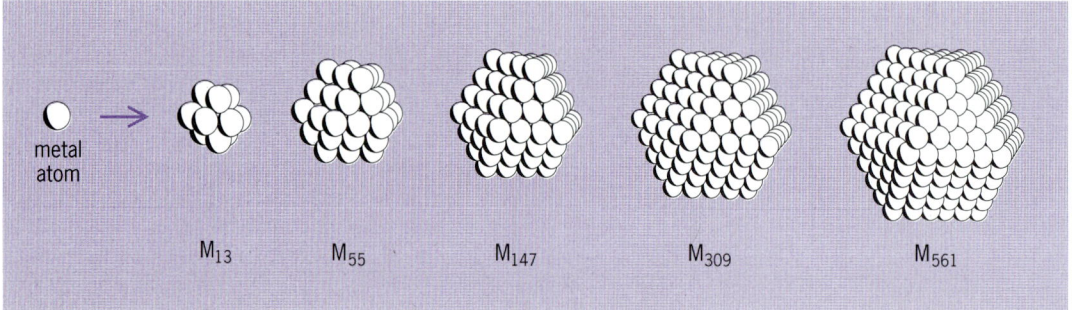

Fig. 2. Magic-number clusters (M_n), shown for cuboctahedral packing.

tral macromolecules form dense but randomly packed solids with only short-range order, as in a glass. In the absence of full x-ray analyses, the precise chemical stoichiometry is uncertain. Notwithstanding, a wealth of direct or indirect physical data are available that appear to agree consistently with the above given formulations, coming from such analytical techniques as high-resolution electron microscopy, extended x-ray absorption fine structure (EXAFS) spectroscopy, Mössbauer spectroscopy, nuclear magnetic resonance (NMR) spectroscopy, and calorimetry. From the 3-shell cluster onward, the metal cores in these macromolecules are large enough to be studied individually by x-ray scattering. The observed reflections confirm the cubic close packing, with metal-metal distances indistinguishable from the corresponding bulk values. In addition to these giant cluster compounds, it is possible to synthesize new types of Pd and Pt colloids by using the same ligands. These colloids are available for experiments in powder (solid) form, with a very high metal fraction and small size distributions (<10%), and thus complement the metal cluster molecules. Together they offer a scale of cluster sizes ranging from 10 to 100,000 atoms. Therefore, the most pertinent physical data, revealing the size evolution to metallic behavior, have been taken on these materials.

Evolution to metallic properties. Among the items to consider in the evolution to metallic properties are the effect of the ligand shell, magnetic moment, quasicontinuous energy bands, and electron charge density.

Ligand shell. The first item to consider in this regard is the effect of the ligand shell. The ligands that are so beneficial in stabilizing the metal cores also have a strong influence on the electronic structure, notably on the surface metal atoms of the cluster to which they are chemically bonded. Such effects should come in addition to the bare surface effects that occur in naked (unligated) metal clusters, or at the surfaces of bulk metals. For instance, it is well established that, because of a reduced number of neighboring atoms, the electronic structure of an atom at the surface of a metal will be different from the atoms within the bulk.

From the recent physical experiments on the large ligated metal clusters, it has been determined that the presence of the ligand shell indeed largely affects the surface atoms, but that at the same time its influence is to a very good approximation restricted to these atoms. Therefore, the inner-core atoms, being unaffected, can be viewed as forming a minute, embryonic piece of the bulk metal, with strong quantum-size effects evolving to bulk behavior with increasing size.

This conclusion is in keeping with experimental and quantum-theoretical studies of chemisorbed molecules on metal surfaces. Such theories have also been applied to ligated metal cluster molecules, yielding the electronic level structure and the way it is affected by the addition of the ligand shell. Lately, these calculations have reached a very high level of sophistication; the calculations are being extended to increasingly larger sizes, including the $Ni_{38}Pt_6$ cluster (bare and ligated) shown in Fig. 1. For this particular cluster the theory predicts a complete quenching of the magnetic atomic moments of the Ni atoms (which are all at the surface; Fig. 1) by the ligation with the CO ligands. Very careful and sensitive magnetic experiments have indeed evidenced the absence of an intrinsic magnetic moment on this cluster.

Magnetic moment. Another experimental example is the strong reduction of the magnetic moment with respect to the bulk, as was observed recently in the series of Pd clusters and colloids, namely the 5-shell Pd_{561} cluster, a sample containing a 50/50 mixture of the 7- and 8-shell clusters Pd_{1415} and Pd_{2057}, and a Pd colloid with average particle diameter of about 3 nanometers. The behavior of the magnetic susceptibility (χ) versus temperature is shown in **Fig. 3**; besides the strong reduction, a striking feature apparent from Fig. 3 is the very slow evolution to bulk behavior, thus implying that surface effects alone cannot explain these results. For instance, for the Pd colloid the inner-core fraction has already increased to about 0.8, as compared to 0.5 for the 5-shell Pd cluster; the low-temperature ratio of χ/χ_{bulk} is 0.6 and 0.3, respectively. Evidently, the inner-core Pd atoms also do not yet show bulk behavior, and to understand this it is necessary to consider in more detail the consequences of quantum-size effects on the electronic level structure of the metal.

Quasicontinuous energy bands. When an itinerant electron is confined to a volume comparable to

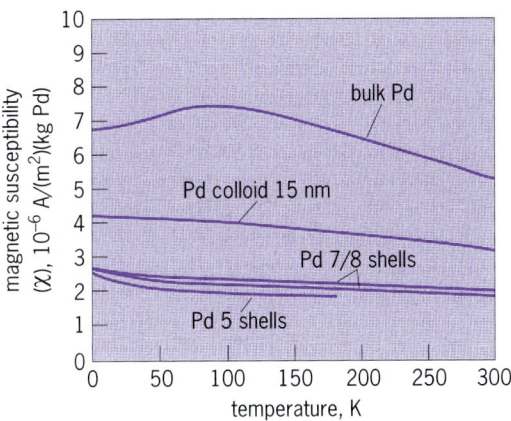

Fig. 3. Size evolution of magnetic susceptibility of large palladium (Pd) clusters and colloids toward the bulk behavior.

its de Broglie wavelength, its energy spectrum becomes discrete, as for a π electron delocalized over a benzene molecule. The average distance δ between levels will be inversely proportional to the particle dimension, and may be estimated as $\delta = E_f/N$, where E_f is the Fermi energy (the energy of the highest occupied orbital) and N the number of occupied orbitals. For example, if $E_f = 5$ eV as is typical for bulk metals, and a cluster of 200 atoms is considered, with 2 electrons per orbital, a value of $\delta = 0.05$ eV is obtained for an alkali metal with one valence electron per atom and $\delta = 0.005$ eV for the transition metals, in which both d- and s-valence electrons are involved. These δ values correspond to thermal energies of 500 K (440°F) and 50 K (−370°F), respectively. For the bulk metals, these quantum gaps evidently have disappeared, and are replaced by the quasicontinuous energy bands responsible for typical metallic properties such as the electronic contribution to the magnetic susceptibility and specific heat. In fact, several of these properties are directly related to the density of states, $D(E_f)$, which is the number of orbitals per unit of energy and per atom, evaluated at E_f. From studies of surface effects at bulk metal surfaces, it is well known that $D(E_f)$ may be much lower in a surface layer, increasing to the bulk value deeper inside the metal. This concept has been transposed to (bare) metal clusters, assuming, for example, a position-dependent $D(E_f)$ inside the cluster that decreases (exponentially) toward the value at the (bare) surface.

The same idea has been applied advantageously to explain the size-dependent susceptibility of the Pd clusters. It should be noted that for free, noninteracting electrons the magnetic susceptibility χ_e is proportional to $D(E_f)$. However, for transition metals such as Pd exchange-correlation interactions between the electrons strongly enhance the susceptibility above the χ_e value. Because this enhancement is also proportional to $D(E_f)$, it is reduced for a small cluster, as $D(E_f)$ diminishes with decreasing particle size.

Electron charge density. A similar division in contributions from surface and inner-core metal atoms was successfully applied to the analyses of NMR and Mössbauer spectroscopic data, notably on the 4-shell Pt_{309} cluster and on the 2-shell Pt_{55} and Au_{55} clusters. In the Mössbauer spectrum not only surface and inner-core sites but also different surface sites may be distinguished.

In particular, the Mössbauer parameter of isomer shift is a measure of the total (integrated) s-electron charge density, and it could be shown that for the inner-core atoms of the Pt_{309} cluster this quantity is already equal to the bulk-metal value. Because the isomer shift measures the total charge density, it is rather insensitive to the appearance of the size-induced quantum gaps in the energy spectrum. By contrast, whenever there appears a gap between highest-occupied and lowest-unoccupied levels around E_f, physical quantities such as the NMR line, the electronic specific heat and susceptibility will deviate from metallic behavior as soon as the thermal energy becomes smaller than this gap. Thus, the complexity of the size-induced transition to bulk-metal behavior is illustrated: it depends not only on the physical property considered but also on the temperature of interest. Indeed, a cluster of a given size can show metallic behavior above a certain temperature, and undergo a metal-to-nonmetal transition below it.

For background information SEE COORDINATION CHEMISTRY; CRYSTAL STRUCTURE; DELOCALIZATION; MAGNETIC SUSCEPTIBILITY; METAL CARBONYL; METAL CLUSTER COMPOUND; QUANTUM (PHYSICS) in the McGraw-Hill Encyclopedia of Science & Technology.

L. Jos de Jongh

Bibliography. W. P. Halperin, Quantum-size effects in metal particles, *Rev. Mod. Phys.*, 58:533–606, 1986; L. Jos de Jongh (ed.), *Physics and Chemistry of Metal Cluster Compounds: Model Systems for Small Metal Particles*, 1994; G. Schmid (ed.), *Clusters and Colloids: From Theory to Applications*, 1994.

Metamorphism

The study of metamorphic rocks provides an important avenue to understanding thermal and mechanical processes that occur in the Earth's crust. Plate tectonic activity creates changes in the structure and thickness of the lithosphere (which consists of the Earth's crust and lithospheric mantle). Plate interactions such as continent-continent collisions (Himalayas, Asia), the subduction of oceanic crust underneath continents (Andes, South America), or continental rifting (Basin and Range Province, western United States) result in deformation of the lithosphere and either thickening or thinning of the crust. Deformation is typically concentrated along discrete zones at plate margins (orogenic or mountain belts), but can be transmitted thousands of kilometers into the mountain belt foreland, as evidenced by fault movement interior of plate margins.

Interrelationship with deformation. During crustal thickening, sediments and rocks once at the surface become buried to depths of 3–15 mi (5–25 km) or more. The resulting growth of new minerals (porphyroblasts) and deformation of the rocks is manifested in the formation of planar (foliations) and linear (lineations) structures. These rocks are then brought back to the Earth's surface through a combination of tectonism and erosion. The different tectonic settings create variation in heat flow as well as the magnitude and timing of deformation with respect to changes in the thermal structure of the crust. One of the goals of metamorphic petrology is to decipher the spatial, temporal, and mechanistic relationships between rock deformation and metamorphism that occurs during the process of mountain formation. Understanding the spatial and temporal relationships between metamorphism and deformation narrows the range of interpretation of data on the tectonic evolution during metamorphism by providing information on the relative timing between heating of rocks, which results in the growth of metamorphic minerals, and deformation. Since both heat flow (metamorphism) and deformation involve the addition of energy to rocks, it is also important to evaluate the potential for these processes to behave synergistically where there may be a feedback mechanism between rock deformation and metamorphism.

Spatial and temporal relationships. An important advance in recognizing the relation between tectonic deformation and metamorphism was the realization that the pressure and temperature conditions responsible for high-grade regional metamorphism were not attainable with a normal geothermal gradient (**Fig. 1**). Studies using computer models showed that re-equilibration of geotherms following crustal thickening resulted in elevated temperatures consistent with observations of metamorphic rocks in the field. This modeling showed also that individual rocks in the crust followed unique pressure (P) and temperature (T) paths during the process of mountain formation (path A in Fig. 1). For collisional tectonic activity P-T paths are clockwise, with the highest pressure encountered prior to the highest temperature. Anticlockwise paths are marked by high temperatures prior to elevated pressures, and are typically found in terranes invaded by numerous plutons (large bodies of intrusive igneous rock), which supply the heat required to drive metamorphism (path B in Fig. 1).

Field mapping of metamorphic isograds and thin-section examination of rocks are used to decipher the temporal relationships between deformation and metamorphism, and therefore the tectonic environment in which these rocks formed. The distribution of metamorphic minerals, especially where index minerals such as garnet, biotite, or staurolite are first observed, are represented as lines on a map called metamorphic isograds. For clockwise P-T paths, isograds should postdate the regional deformation and crosscut observed structures such as folds and faults. These relationships are complicated if the rocks have undergone a complex thermal and deformational history. For counterclockwise paths, isograds developed from heating associated with regional plutonism or fluid flow may be folded, indicating that some deformation postdated the metamorphism. In map view, these isograds may be spatially related to exposed plutons or batholiths.

Petrography, the study of textures and mineralogy of rocks in thin section, is a powerful tool for deciphering the temporal relation between deformation and metamorphism. In general, three end-member possibilities show distinctive relations between mineral growth and fabric development. First, mineral growth may predate the deformation (**Fig. 2***a*). In this case, deformation fabrics that are defined by alignment of phyllosilicates (layer silicates) wrap around porphyroblasts, the relatively large crystals formed in metamorphic rocks; pressure shadows consisting of quartz typically form in spaces around the mineral crystal. Second, mineral growth may be simultaneous with deformation (Fig. 2*b*); porphyroblasts formed during deformation should show pronounced effects of deformation, such as flattening of mineral grains and continuity with external fabrics. Third, mineral growth may postdate deformation (Fig. 2*c*); post-deformation porphyroblasts appear to overgrow fabrics with aligned mineral grains passing undisturbed from the matrix through the porphyroblast.

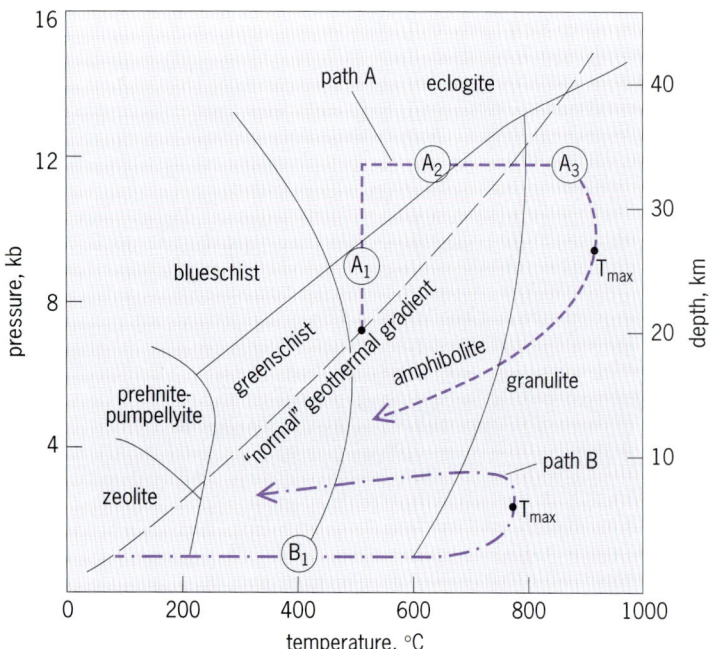

Fig. 1. Pressure (P; depth) versus temperature (T) plot with conditions for different metamorphic facies and so-called normal geothermal gradient. 1 kb = 10^8 Pa. 1 km = 0.6 mi. °F = (°C × 1.8) + 32.

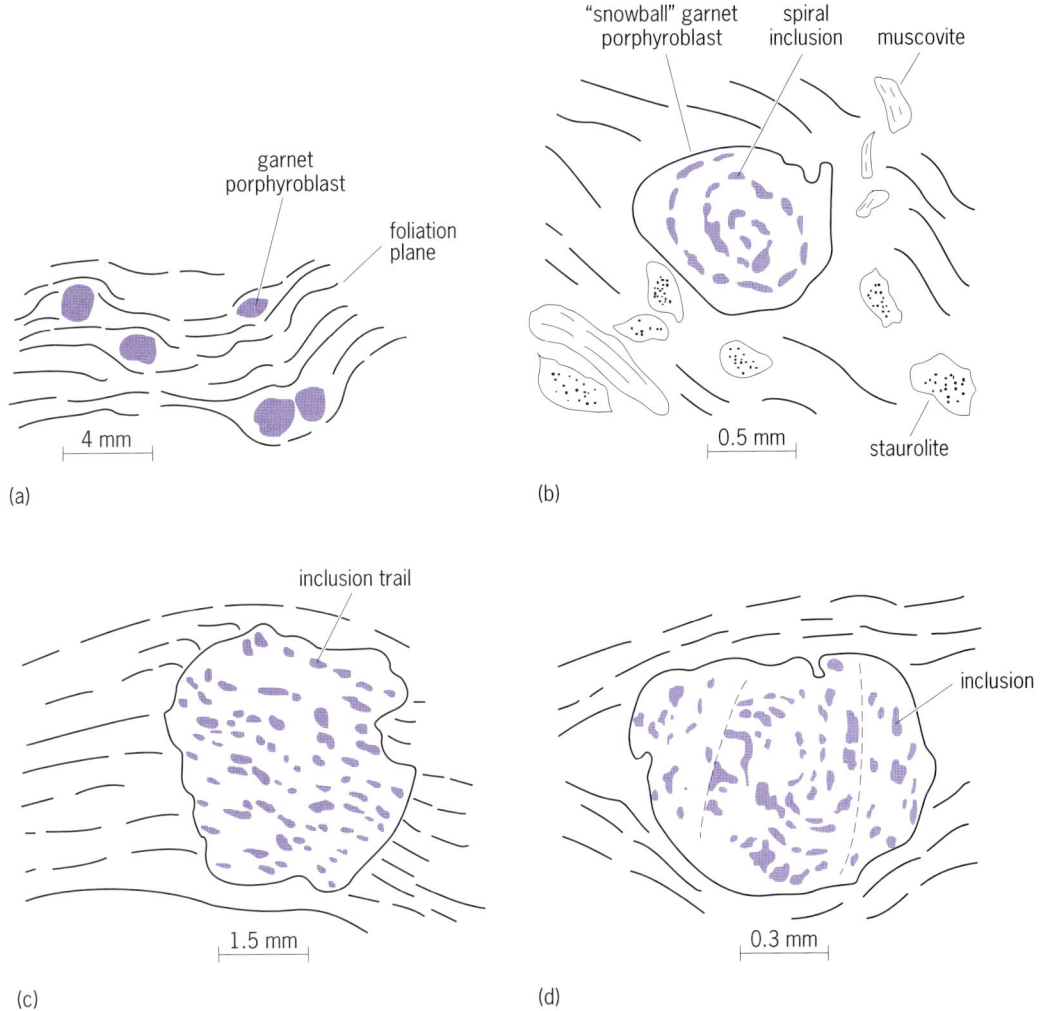

Fig. 2. Examples of timing between porphyroblast growth and deformational fabrics. (*a*) Garnet porphyroblasts wrapped by foliation planes, indicating that mineral growth occurred prior to foliation. (*b*) "Snowball" garnet porphyroblast with spiral inclusions, mostly quartz, with other minerals in the matrix; spiral inclusion trails indicate that garnet growth occurred during rotation of the matrix. (*c*) Garnet porphyroblast overgrowing a crenulated foliation. Inclusion trails within garnet crystal are continuous with the matrix, indicating that mineral growth occurred after deformation of the fabric. (*d*) Garnet porphyroblast with inclusions whose geometry indicates a complex deformational and metamorphic history.

In rocks that have undergone multiple heating or deformational events, the relationship between fabrics and mineral growth is more complicated. In the garnet porphyroblast shown in Fig. 2*d*, the geometry of the inclusions indicates a complex deformational and metamorphic history. The inner portion of the garnet may have grown during deformation, preserving a so-called snowball texture. Continued deformation produced a fabric that was subsequently overgrown by new garnet, with a final phase of deformation producing a foliation surrounding the garnet porphyroblast. Resolving these relationships provides the only avenue to fully document the deformational and metamorphic history of the rocks.

Mechanistic relationships. During dynamothermal metamorphism there are two end-member mechanistic interrelationships between deformation and metamorphism. On the one hand, deformation can enhance metamorphic processes by (1) grain-size reduction, which can promote fluid ingress or enhance intergranular diffusion rates, and can also increase grain boundary volume available for reaction; (2) increase in dislocation densities of minerals, which can in turn change the thermodynamic stability of phases; (3) enhancement of volume diffusion by providing fast diffusion pipes along dislocation cores and other defect microstructures; and (4) provision of nucleation sites along defect microstructures, thus lessening activation energy barriers for initial stages of grain growth. In addition, dislocation creep (intragranular movement of crystal defects) or cataclasis (grain-size reduction during brittle deformation without loss of rock cohesion) may significantly alter the physical properties of grain boundaries, thus dramatically affecting intergranular diffusion rates and hence the progress of chemical processes.

On the other hand, mineral reactions and change in heat or fluid flow can alter the physical properties of rocks and therefore change the amount of strain that can be accommodated as well as induce changes in deformation mechanisms. The ductility of rocks can be affected by several factors: (1) A decrease in the grain size of reaction products would facilitate mechanical processes such as grain boundary sliding, a process known as reaction-enhanced ductility. (2) Release of fluids during dehydration reactions would promote cataclasis or produce transient changes in the rate at which rocks deform. (3) Volume changes may enhance intracrystalline deformation of grains. (4) Variation of local chemical potential gradients can affect diffusive transfer deformation mechanisms that involve shape changes through dissolution along highly stressed grain boundaries and precipitation in low-stress areas of the rock.

Detailed chemical and structural study of fault zones in the middle (6–9 mi or 10–15 km) and lower (9–15 mi or 15–25 km) crust have provided important data on the mechanistic interaction between deformation and metamorphism. Middle- and lower-crust fault zones, typically called ductile shear or mylonite zones, are often continuous with shallow, brittle faults such as those exposed at the present surface. Shear zones range in size from inches to miles and contain rocks that show grain-size reduction and increases in fabric development as compared to the surrounding rocks. In addition, these zones may show distinct changes in rock and mineral chemistry with respect to the relatively undeformed host rocks. Because relatively undeformed host rock is immediately adjacent to strongly deformed rock within these zones, detailed chemical and structural study of shear zones provides a unique opportunity to investigate the interrelationship between metamorphism and deformation. Furthermore, rocks within shear zones may be more reactive than their undeformed counterparts, and therefore record portions of the metamorphic evolution not detected elsewhere.

Shear-zone formation in the middle and lower crust can occur either while rocks are heating (prograde shear zones) or after the peak metamorphic temperature during cooling (retrograde shear zones). In each case the mechanistic interactions between deformation and metamorphism are distinct. Prograde shear zones are relatively rare, because higher temperatures during and after shear-zone formation typically homogenize the effects of deformation. In general, prograde shear zones form without any significant changes in bulk chemistry of rocks, in part due to a lack of significant fluid flow through the shear zone. In prograde shear zones a feedback mechanism between metamorphic changes is likely in response to heating and the rates and mechanisms of deformation. Rocks deformed within a prograde shear zone are typically more reactive because of a decrease in grain size and increases in grain boundary areas available for reaction. Movement of dislocations through individual porphyroblasts may also enhance metamorphism by altering the thermodynamic stability of minerals, leading to changes in mineral reactivity. However, mineral reactions within the shear zone may yield finer grain products, leading to increases in the rate of deformation or changes from intracrystalline deformation to a grain boundary sliding mechanism. Also, prograde metamorphic reactions release fluid that could change the nature of the rock deformation, typically inducing fracture.

Retrograde shear zones are always marked by formation of hydrous minerals within zones of high deformation as compared to the relatively undeformed host rocks. Changes in isotopic and chemical compositions of shear-zone rocks as compared to the relatively undeformed host rock have been interpreted to indicate significant fluid flow through retrograde shear zones. Formation of hydrous minerals within retrograde shear zones typically weakens the rocks by a process known as reaction-enhanced ductility, where more easily deformable hydrous minerals in the shear zone result in the focusing of deformation. Deformation may also enhance the reactivity of the rock; fracturing of the rock can facilitate fluid movement into the shear zone, resulting in hydration.

For background information SEE EARTH CRUST; METAMORPHISM; OROGENY; PETROGRAPHY; PLATE TECTONICS in the McGraw-Hill Encyclopedia of Science & Technology.

Matthew W. Nyman

Bibliography. P. C. England and A. B. Thompson, Pressure-temperature-time paths of regional metamorphism, pt. I: Heat transfer during the evolution of regions of thickened continental crust, *J. Petrol.*, 25:894–928, 1984; A. B. Thompson and P. C. England, Pressure-temperature-time paths of regional metamorphism, pt. II: Their influence and interpretation using mineral assemblages, *J. Petrol.*, 25:929–955, 1984.

Microbial compound

A large number of compounds from fungi (including both cellular components and secondary metabolites) have been shown to affect the immune system and, therefore, have potential for the treatment of diseases or for prophylaxis as health food ingredients or dietary supplements. In practical medicine, compounds that stimulate the immune system could be useful as adjunct therapy to chemotherapy, radiotherapy, and surgery in the treatment of cancer, in the treatment of immunodeficiency disease and generalized immunosuppression caused by drug treatments, in combination therapy with antibiotics to increase host resistance against infection, and as adjuvants for vaccines (an adjuvant enhances antigenicity). In contrast, compounds that suppress the immune system could be utilized as immunosuppressants in organ transplantation, in the treatment of autoimmune diseases, and in the treatment of gastrointestinal tract disease.

Immunomodulators or immunoregulators, also known as biological response modifiers (BRMs), are substances that can either stimulate or suppress the natural immune system. Immunomodulators can also be classified as nonspecific or specific. Nonspecific immunostimulants are administered alone to elicit a generalized state of resistance to pathogens or tumors. Specific immunostimulants are administered along with an antigen, as in vaccines, to boost the immune response. Nonspecific immunosuppressants reduce the ability of the immune system to respond to antigens by interfering with the function of cells in the immune system. No specific immunosuppressants are known at this time.

Immunostimulants. Stimulation of the immune system offers an attractive alternative to conventional chemotherapy, especially when the host's defense mechanisms are impaired, in the prevention of the spread of opportunistic infections in individuals at risk, and in the therapy of malignant diseases.

High molecular weight polysaccharides. The high molecular weight polysaccharides, which are naturally occurring components of the fungal cell wall, have been found to stimulate both nonspecific host resistance and specific immunologic reactivity, and to exert inhibitory effects against tumors (see **table**). Most tumor studies have been conducted in animal models by using sarcoma 180. This tumor can be easily transplanted into female mice having cells with the molecular marker CD-1, where it grows rapidly over a period of 30 days and can be measured without killing the animal. The extracted fungal polysaccharides are usually dissolved in a saline medium and injected within the peritoneal cavity at 24-h intervals for 10 consecutive days. Typically, tumor growth is not influenced during the first 10 days. After this time the tumor regresses, indicating the indirect, noncytotoxic effect of the polysaccharide. Under optimal conditions, in 30 days the animal is free of tumor cells.

In most cases, effective fungal polysaccharides are glucans composed of glucose units with different types of linkages. Certain structural features are required for antitumor activity: a molecular weight of more than 300,000 in the primary structure, β-1,3 linkages in the main chain of the glucan with β-1,6 branch points, and a triple helix structure.

Four fungal polysaccharides are already in clinical use: lentinan, schizophyllan (SPG), krestin (PSK), and PSP. Lentinan extracted from the shiitake mushroom *Lentinula edodes* and schizophyllan from the bracket fungus *Schizophyllum commune* possess the basic structure of a β-1,3-glucan with β-1,6-glucopyranosidic branches and appear to be T-lymphocyte (T-cell) adjuvants. Krestin is a protein-bound polysaccharide isolated from the bracket fungus *Coriolus versicolor* that activates macrophages. It is a β-1,4-glucan with β-1,6-glucopyranosidic side chains and some peptide residues. Krestin's solubility in water and effectiveness when administered orally are probably due to its protein content. PSP, also from *C. versicolor*, is a peptide-bound polysaccharide with slightly different linkages.

The clinical usefulness of these polysaccharides has been demonstrated in individuals with head and neck, lung, gastric, and cervical cancers. These polymers have been used alone or in combination with chemotherapy and radiotherapy, but in most cases, they are used after the surgical removal of the primary tumor. As a result of treatment, the immune response of the individuals improved. The survival periods were considerably prolonged; side effects were rarely observed. In general, intramuscular injections of these polysaccharides were tolerated by cancer patients.

Antitumor activity among fungal polysaccharides differs. Pachyman from the sclerotium of the wood-decaying fungus *Poria cocos* has no effect, but when converted to the β-1,3-linear glucan, pachymaran, there is strong activity, thus providing evidence for the importance of higher molecular structure. DiLuzio's yeast glucan obtained from *Saccharomyces cerevisiae*, the active component of zymosan, stimulates the reticuloendothelial system. D-Mannan and D-glucan, also from *S. cerevisiae*, increase phagocytosis. Vesiculogen, a polysaccharide from the coprophilic cup fungus *Peziza vesiculosa*, mainly activates B-lymphocytes (B cells).

Antitumor polysaccharides from fungi

Polysaccharide	Fungal source	Chemical structure
Basidiomycetes		
Lentinan	*Lentinula edodes*	β-1,6; β-1,3-glucan
Schizophyllan (SPG)	*Schizophyllum commune*	β-1,6; β-1,3-glucan
—	*Ganoderma lucidum*	β-1,3-glucan
Pachymaran	*Poria cocos*	β-1,3-linear glucan
PSK	*Coriolus versicolor*	β-1,4; β-1,3; β-1,6-glucan-protein complex
PSP	*C. versicolor*	α-1,4; α-1,2; β-1,3; β-1,6-glucoside peptides
Ascomycetes		
Scleroglucan (SSG)	*Sclerotinia sclerotiorum*	β-1,6; β-1,3-glucan
Oomycetes		
Glucan A_1	*Phytophthora parasitica*	β-1,6; β-1,3-glucan
Yeasts		
DiLuzio's yeast glucan (DLG)	*Saccharomyces cerevisiae*	β-1,3-glucan
Mannozyme	*S. cerevisiae*	Glucomannan

Another particularly promising polysaccharide is called glucan A_1, a branched glucan extracted from the cell walls of the plant pathogen *Phytophthora parasitica*. It is extremely potent with an antitumor activity in animal models equal to or greater than that of schizophyllan. Tests have confirmed that it has no direct cytotoxic effect on cancer tissue but operates by stimulating the host's immune system.

Low molecular weight substances. Only a few low molecular weight immunostimulants have been identified. The protein Ling Zhi-8 (LZ-8), extracted from the mycelia of the bracket fungus *Ganoderma lucidum*, contains 110 amino acids. The molecular structure resembles the variable region of the immunoglobulin (antibody) heavy chain, and possibly accounts for its immunomodulating function. Another protein, named α-amanitin, a cyclooctapeptide isolated from the liver-toxic mushroom *Amanita muscaria*, has been shown to induce interferon, a glycoprotein that exerts antiviral activity. The antifungal antibiotic coriolins and a derivative, diketocoriolin B, from the bracket fungus *C. consors*, prevent the growth of microbial and animal cells by inhibiting the enzymes located on cell surfaces or membranes. This reaction may be the stimulus that induces antibody formation in cells that are able to do so. Diketocoriolin B inhibits the membrane enzyme Na^+-K^+-ATPase and acts directly on B cells, stimulating their proliferation or differentiation into antibody-producing cells.

Thus, although there are several suitable candidates for the development of useful immunostimulatory drugs in the class of high molecular weight compounds, especially those polysaccharides with a β-glucan structure, the low molecular weight compounds demonstrate no structure-activity relationships.

Immunosuppressants. Compounds capable of specifically stimulating thymus suppressor-cell populations are useful in organ transplantation and to treat immune diseases, such as rheumatoid arthritis, polyarthritis, chronic active hepatitis, myasthenia gravis (chronic progressive muscular weakness), multiple sclerosis, psoriasis, uveitis (eye inflammation from an autoimmune reaction), and non-insulin-dependent diabetes.

Cyclosporins. The most successful immunosuppressant is cyclosporin A, a weakly antifungal metabolite produced by the microscopic soil fungus *Tolypocladium inflatum*. Cyclosporin A is a cyclic peptide consisting of 11 amino acids. It blocks the function of the T cells that cause transplant rejection but does not kill the bone marrow cells. Its effect on T cells is reversible. Screening in culture and in animal models has shown that cyclosporin suppresses delayed-type sensitivity, inhibits antibody production, and lowers tissue graft rejection in rodents. Cyclosporin A is currently used in human heart transplants and has been associated with improving the effectiveness of kidney and liver transplants, but higher doses induce renal dysfunction and other side effects. It can also interfere with the production and release of interleukins 1 and 2 (IL-1 and IL-2), which are important inflammation mediators in rheumatoid arthritis and other autoimmune conditions. A total of 25 cyclosporins have been isolated from the fermentation broth of *T. inflatum*. Other than cyclosporin A, the majority are minor metabolites of the same structural type.

Besides *Tolypocladium*, the microscopic hyphomycetes *Cylindrocarpon lucidum*, *Fusarium solani*, *Beauveria bassina*, and *Stachybotrys chartarum* and the ascomycete *Neocosmospora vasinfecta* have been reported to produce metabolites belonging to the cyclosporin family. Cyclosporin A binds with high affinity to ubiquitous, cytoplasmic proteins, and a good correlation has been observed between binding affinity and immunosuppressive activities of cyclosporin analogs.

Other immunosuppressive substances. An immunosuppressive metabolite (ISP-I) [10–100 times more potent than cyclosporin A] has been found in the insect pathogen *Isaria sinclairii*. The structure of the lipophilic amino acid, which is simpler than that of cyclosporin A, is identical to the antifungal agent myriocin (thermozymocidin). ISP-I prevents proliferation of antigen-stimulated lymphocytes and suppresses the immune response in culture and in mice, making it a good candidate for use in organ transplantation and for treatment of autoimmune diseases. A derivative, 14-deoxomyriocin, has a potency 10 times greater than myriocin.

The mammalian organism responds to infection, tissue injury, and the intrusion of foreign materials by activating a number of effector systems. In an immune reaction, antibodies or lymphokines (molecules other than antibodies produced by lymphocytes) can act directly on the target or indirectly by activating complement. The complement system is a humoral system (as opposed to a cellular system) composed of at least 15 distinct serum proteins that exist in blood as inactive precursors, which must interact in a specific order to become biologically functional. The consequences of complement activation include viral neutralization, enhanced phagocytosis, leukocyte mobilization, a release of vasoactive amines and lysosomal enzymes, the generation of anaphylatoxins, and interactions with the cellular immune system. Occasionally an organism's immune system reacts against its own tissues (an autoimmune reaction). Persistent tissue damage causes the release of more autoantigen. In a limited number of cases, autoimmune reactions become chronic, and autoimmune diseases may ensue. In the autoimmune diseases—rheumatoid arthritis, lupus erythematosus, and glomerulonephritis—complement activation apparently contributes directly to cellular and tissue destruction. As a result, there is interest in any agents that are potent inhibitors of complement. K-76 monocarboxylic acid produced by the hyphomycete *Stachybotrys complementi* inhibits the formation of a factor in human complement serum. Since it has been shown to suppress nephrotoxic nephritis (an inflammatory kidney disease) in rats, it may be useful in immune complex diseases, allergic diseases, and inflammation. SEE ANTIGEN; CELLULAR IMMUNOLOGY.

Trichothecene mycotoxins are very toxic molecules that affect the organs essential for maintaining immunological competence, namely, the thymus, spleen, liver, and bone marrow. Fusarenon-x, a mycotoxin from the hyphomycete *F. nivale*, suppresses mitogen (a mitotic activator) responsiveness and antibody synthesis both in culture and in the living organism and may induce nonlymphocytic suppressor cells. It has been shown to prevent antibody synthesis to sheep red blood cells and delay allograft (a graft between genetically nonidentical species) rejection in mice.

A variety of other compounds with immunosuppressive activity have been identified. Cyclomunine, a hexacyclodepsipeptide extracted from *F. equisetti*, inhibits the proliferative response of human lymphocytes to mitogens and prolongs skin allograft survival. The pyrrothine antibiotics, which prevent aggregation, appear to have anti-inflammatory activity. Heruline, produced by the hyphomycete *Penicillium herquei*, is an example of an inhibitory alkaloid. Most natural alkaloids produced by species of the ergot fungus *Claviceps* (for example, ergocristin) are also efficient immunosuppressors, probably because of their structural similarity to important mediators, such as norepinephrine, serotonin, or dopamine. Metacytofilin (MCF), produced by a species of *Metarhizium* found in the soil, has an immunosuppressive effect with low toxicity on the mixed lymphocyte culture reaction and immune responses in mice. Cytochalasin isolated from a species of the plant pathogen *Pestalotia* also has immunosuppressive properties. A glycoprotein derived from the mycelia of the bracket fungus *G. lucidum* acts as an immunosuppressive agent in the treatment of allergic diseases and cell-mediated immune diseases. Immunosuppressive components have also been found in the mushroom *Lactarius flavidulus*.

Immunotherapy. The realization that a functional immunological defense system can successfully alter the course of a disease has generated a great deal of interest in immunotherapy with the subsequent isolation and identification of a diverse group of immunomodulators. From experimental work it is apparent that some immunomodulators may exert totally different effects depending on the dose, route of administration, and the clinical status of the host or patient at the time of exposure. It is also evident that similar types of effects can be elicited by structurally diverse molecules.

For background information SEE ANTIBODY; ANTIGEN; CANCER (MEDICINE); FUNGI; IMMUNOLOGICAL DEFICIENCY; IMMUNOSUPPRESSION; MEDICAL MYCOLOGY; TUMOR in the McGraw-Hill Encyclopedia of Science & Technology.

Jeannette M. Birmingham

Bibliography. G. Chihara, Y. Y. Maeda, and J. Hamuro, Current status and perspectives of immunomodulators of microbial origin, *Int. J. Tiss. Reac.*, 4:207–225, 1982; S. C. Jong, J. M. Birmingham, and S. H. Pai, Immunomodulatory substances of fungal origin, *EOS J. Immunol. Immunopharmacol.*, 11:115–122, 1991.

Microscope

Recent advances in the field of microscopy include the development of near-field optical microscopy and high-speed scanning tunneling microscopy. These two techniques are quite different from one another, but both make use of a movable probe that can be scanned over the object to achieve imaging, and they both employ optical methods.

Near-Field Optical Microscopy

A fundamental law of optics, the so-called diffraction limit, states that two objects can be imaged as separate entities only if their distance is larger than about one-half of the wavelength of visible light, which ranges from 400 to 700 nanometers. As a consequence, conventional optical microscopy is restricted to a resolution of about 200 nm, which is not enough for many important observations, for instance observations of the structure of viruses or of microelectronic elements.

In the past, therefore, scientists had to resort to nonoptical methods such as electron microscopy. Imaging by means of electrons indeed allows extremely high resolution, but also requires special sample preparation and operation in vacuum. Furthermore, electron microscopy is limited to conductive objects and provides information mainly on the topography of an object, whereas details of composition are difficult to obtain.

In contrast, optical microscopes readily provide information regarding such qualities as color, luster, transmissivity, and birefringence, which are sensitive indicators of material composition and status. Furthermore, optical microscopes operate at ambient conditions, a prerequisite for observations of living organisms. It therefore has long been a goal of microscopists to find an optical technique that would circumvent the diffraction limit. The development of scanning near-field optical microscopy has realized this goal. Although this imaging technique was first demonstrated in 1985, it received broad attention only recently when spectacular results were obtained, such as images of single-molecule fluorescence, magnetooptic writing and reading, photon tunneling through liquid-metal films, and local large-amplitude plasmon excitations.

Principles. Four requirements must be met to eliminate the diffraction limit: a tiny light source must be available, this light source must be brought into the close vicinity of the object, the light emitted by this optical probe must be detected, and the light source must be scanned across the object surface.

The ability to image an object is closely related to the ability to confine radiation to a narrow spot.

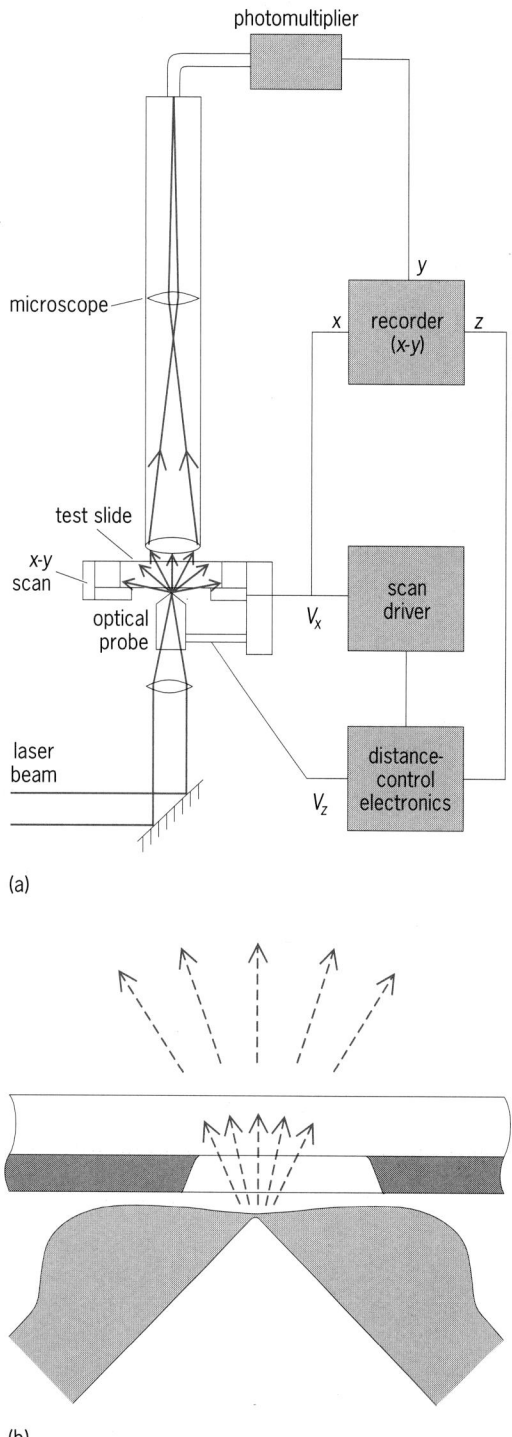

Fig. 1. Most widely used scanning near-field optical microscope (SNOM or NSOM). (*a*) Setup. (*b*) Optical probe next to a metallic film with small holes. (*After U. Dürig, D. W. Pohl, and F. Rohner, Near-field optical scanning microscopy, J. Appl. Phys., 59:3318–3327, 1986*)

a single atom or molecule. When excited to radiate, for instance by an electrical discharge, it behaves like a small dipole antenna.

Light emerging from such a small light source remains confined over only a distance comparable to its size, the so-called near-field zone. Beyond that, that is, in the far field, it diverges rapidly. The light source, (termed an optical probe after this) hence must be brought so close to the object to be investigated that its near field is disturbed by the object. This requirement can best be satisfied by mounting the optical probe at the apex of a sharply pointed tip. The most common optical probe is an aperture in an opaque metallic coating which covers the corner of a transparent crystal or the end of an optical fiber. This probe is prepared such that a sharp tip is formed.

Optical detectors such as photomultipliers are fairly large and therefore cannot be mounted in the near field. The disturbance of the near field by the object, however, also influences the far field. This property is illustrated by the case in which the optical probe is a small aperture and the object is a strongly absorbing metal film with small holes. If the probe is positioned directly above a hole, the light passing through the aperture can also pass the object without much attenuation; otherwise, the metal film will act like a lid, reducing the light transmission the closer the aperture is to the film surface.

An image of the object surface can be composed by scanning the probe line by line over the object surface and recording the intensity of radiation as a function of position. An image as it would appear to the eye can be easily created by associating the signal intensity with the gray scale on a computer display; color images can also be created in this way.

The scanning motion is conveniently generated by means of piezoelectric elements which can be made to bend sideways or expand in length when voltage is applied in an appropriate way. Such scanners were originally developed for scanning tunneling microscopy (discussed below). They can be used simultaneously for distance control.

Microscope design. Various types of scanning near-field optical microscopes (common acronyms are SNOM and NSOM) have been developed on the basis of the principles outlined above. The most widely used one (**Fig. 1**) utilizes the aperture at the apex of a metal-coated transparent tip as the optical probe.

Very small apertures, and hence high resolution, can be obtained with etched quartz crystals whose facets form three-sided pyramids with a large angle of apex and atomically sharp tips. However, optical fibers are now used more frequently. They are pulled apart while being heated in such a way that one end forms a tip with a radius of curvature of 30–100 nm. A metallic coating is evaporated from the side. The fibers are rotated along their axes during evaporation. When these axes are properly aligned, this rotation results in the shadowing of the very apex and hence the formation of a small aperture. These optical probes are very easy to prepare

Lenses and mirrors, the common focusing elements, are subject to the diffraction limit, and hence can be ruled out. When illuminated, however, holes in an opaque screen or small scattering particles can also act as confined sources of radiation, and they can be made much smaller than half a wavelength. The smallest possible light source indeed is

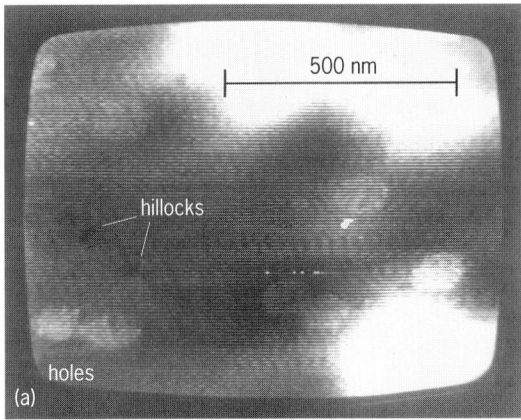 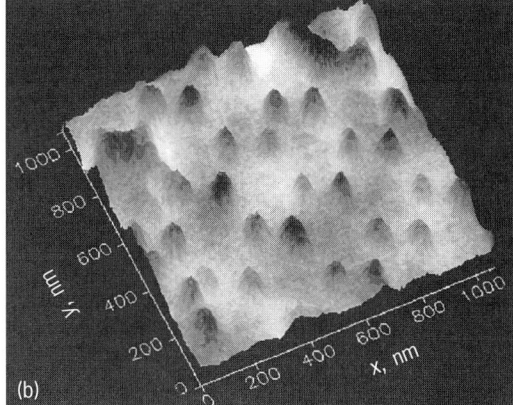

Fig. 2. Scanning near-field optical microscope images. (a) Metallic film with holes (*from U. Dürig, D. W. Pohl, and F. Rohner, Near-field optical scanning microscopy, J. Appl. Phys., 59:3318–3327, 1986*). (b) Tiny metal patches forming a hexagonal pattern on a glass substrate; superposition of optical with topographical image (*from H. Heinzelmann et al., Scanning near-field microscopy in Basel, Ruschlikon, and Zurich, vol. 34, no. 8, August 1995*).

and handle, but provide somewhat lower resolution than the quartz-crystal probes.

The optical probe is usually illuminated with a monochromatic laser beam; it is also possible to use a conventional light source. The forward-transmitted light is collected by a classical microscope focused on the optical probe. An aperture in the image plane allows undesired stray light to be blocked out. A photomultiplier or another sensitive light detector mounted behind the pinhole converts the transmitted radiation into an electrical signal, which is fed into a computer. There, the scan image is generated by combination with the lateral (x-y) position signal applied to the piezoactuator. The process of image generation is very similar to the one used in television but on a much slower time scale (typically, 3–60 s per frame).

The distance between the optical probe and the object surface has to be kept constant and very small during this time, requiring a control mechanism for electrical or mechanical contact. It is possible, for instance, to vibrate the tip slightly sideways by applying a small alternating-current voltage to the electrodes of the piezoactuator. At resonance, the vibration amplitude depends sensitively on the amount of damping. Hence, when the probe comes into the slightest contact with the object, the amplitude is considerably reduced. The change in amplitude can be readily measured by a small auxiliary optical arrangement which is sensitive to the variation in shadowing associated with the vibration. A feedback mechanism can then be used to keep the amplitude reduction at a constant, low level, ensuring the gentle, nondestructive sliding of the probe over the object surface.

Examples of imaging. One of the first images (1985) made with a scanning near-field optical microscope (**Fig. 2***a*) has bright spots corresponding to holes in a semitransparent metal film and dark ones corresponding to small protrusions called hillocks, which are frequently formed during vacuum evaporation. The resolution is about 20 nm. The diameter of the holes is 100 nm, meaning that the pair of holes in the lower left-hand corner cannot be resolved with conventional optical microscopy. The near-field image, however, clearly shows the contours of the rims of the holes; it is apparent that these holes are not perfectly circular but have subtle irregularities, such as the tiny peninsula at the right-hand rim of the hole pair. The width of the peninsula does not exceed 25 nm, and similarly small are the hillocks further up on the left-hand side of the image.

Computer imaging was still in its early stages of development in 1985; present-day pictures have a much higher degree of perfection (though not higher resolution). Figure 2*b* shows a hexagonally structured pattern of tiny metallic patches about 10 nm thick and 50 nm in lateral extension. The optical probe of the microscope produces a topographic image simultaneously with the near-field optical image. Superposition of the two images creates a vivid impression of what the object would look like if it were about 30,000 times larger. Moreover, the superimposed image is illuminated from below, so that the metallic "mountains" absorb light and therefore appear dark on the bright planes of the transparent glass substrate. With a conventional microscope, the object appears slightly gray and completely unstructured because the tiny metal patches cannot be resolved.

Near-field optics. The behavior of radiation passing a tiny, subwavelength-sized aperture cannot be predicted on the basis of classical optical theory, which is restricted essentially to structures larger than the wavelength. Many optical effects manifest themselves in a novel way if the relevant dimensions are small compared to the wavelength. Subwavelength-sized objects, for instance, can appear distorted and with unexpected contrast.

Some insight may be obtained from antenna theory because the radiation properties of so-called pointlike light sources are indeed very similar to those of broadcast antennas. The intensity and directional distribution, in particular, depends not only on the intensity of excitation and the shape of

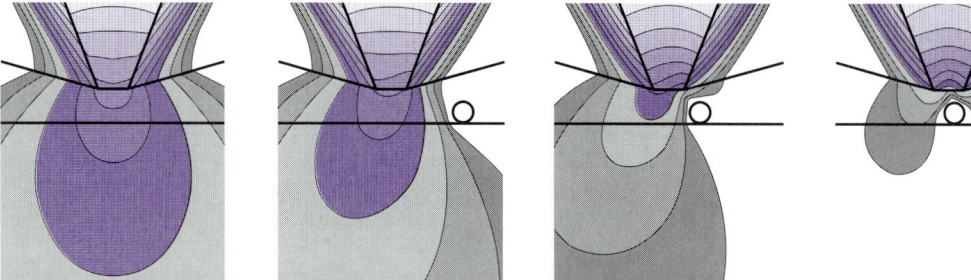

Fig. 3. Field patterns in and next to optical probe scanning a cylindrical particle. Contours indicate electrical energy density, with a factor of three between successive contours. Polarization is perpendicular to the plane of the image. (*After L. Novotny, D. W. Pohl, and P. Regli, Light propagation through nanometer-sized structures: The two-dimensional scanning near-field optical microscope, J. Opt. Soc. Amer. A, 11:1768–1779, 1994; courtesy of L. Novotny and O. Martin*)

the antenna but also on the dielectric, that is, the optical properties of its immediate environment, the near-field zone. Thus, the range of a walkie-talkie device, for instance, is different when used on dry land, in a boat at sea, or in a hang glider far above the ground. The record of the received radiowave intensity from the transmitter held by a person walking in an area of, say, different degrees of humidity is strictly analogous to the near-field optical microscope scan line of an optical probe that is moved along a piece of material with varying indexes of refraction.

Another useful analogy is that of the stethoscope: The medical doctor locates the position of the heart by moving the end piece over the patient's chest while listening to the heart beat. Again, imaging occurs beyond the diffraction limit, because the acoustic waves have a length of several meters.

Quantitative understanding requires the solution of Maxwell's equations, the basic set of relations that describe the formation and propagation of electromagnetic waves. The adaptation of these equations to the complicated shape of the scanning near-field optical microscope is possible only with extensive computations. **Figure 3** shows the result of a field calculation simplified to two dimensions: an illuminated narrow slit moving over a small cylindrical object. The contours indicate the local light intensity. It can clearly be seen that the radiation is confined next to the exit of the slit and that its propagation is strongly influenced by the presence of the object.   *D. W. Pohl*

High-Speed Scanning Tunneling Microscopy

The scanning tunneling microscope (STM), invented by G. Binnig and H. Rohrer in 1982, revolutionized the field of surface science, making it possible, for the first time, to image conducting surfaces with atomic resolution and obtain maps of the local density of electronic states belonging to the surface atoms. The scanning tunneling microscope also spawned a large number of related technologies, from atomic force microscopy, which can image atomic forces on conducting as well as nonconducting surfaces, to scanning force microscopes that can map electric and magnetic fields across the surface of technologically important samples. The scanning tunneling microscope system is an elegant combination of mechanical, electromechanical, and electronic subsystems that work together under the control of a computer. The computer processes the information yielded by the scanning tunneling microscope and produces atomically resolved images. Until recently, however, the speed at which the scanning tunneling microscope could operate was limited by the mechanical and electronic components of its system, which is on the order of a fraction of a millisecond. To overcome these speed limitations and enable it to respond on a nanosecond time scale, optical technology was incorporated into the scanning tunneling microscope system using several approaches.

The realization of a high-speed scanning tunneling microscope, operating at nanosecond or picosecond time scales can open the door to explorations of fast phenomena occurring at surfaces. Chemical processes, conformation of biological samples, and artificial nanostructures are among the areas that will benefit from this technology. Extensions of this technology to the areas of atomic force microscopy and near-field optical microscopy (discussed above) are in prospect.

Principles of STM. At the heart of the scanning tunneling microscope (**Fig. 4***a*) is a mechanically (or chemically) sharpened tip with a single atom at its apex, placed in close proximity to a conducting sample. The tip is mounted on a piezoelectric structure, usually with cylindrical geometry, that has several electrodes attached to it. A computer drives power supplies that control the voltages applied to these electrodes, making it possible to position the apex atom relative to the atoms belonging to the probed sample with subangstrom accuracy. The piezoelectric structure also scans the tip across the surface of the probed sample. A bias voltage, V, ranging from several millivolts to several volts, is applied to the gap s separating the tip and sample, and a sensitive amplifier is used to monitor the current as the gap is decreased. Once the electronic clouds belonging to the apex atom and the adjacent surface atom begin overlapping, a tunneling current is registered. This current is used to control the

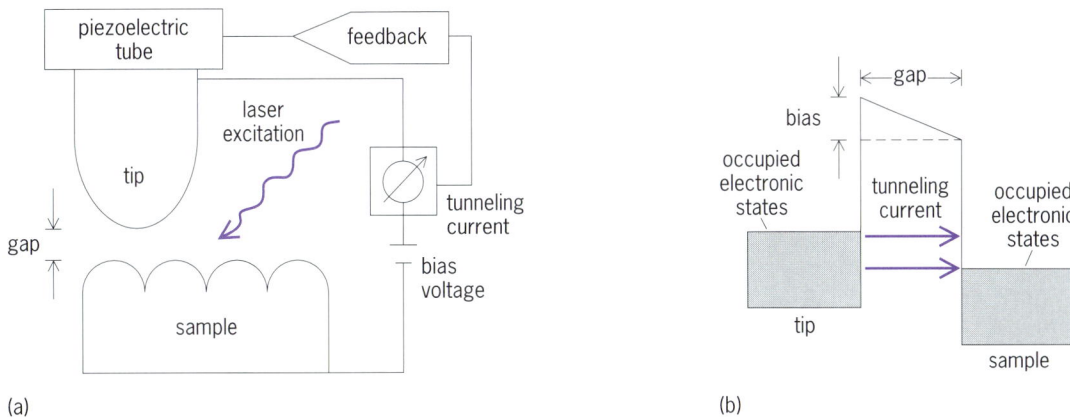

Fig. 4. Scanning tunneling microscope system. (*a*) Microscope with laser beam focused onto its junction. (*b*) Tunneling process.

gap, via a negative feedback mechanism, and measure the electronic structure across the probed surface.

The tunneling process takes place through a potential barrier between the apex atom and the atoms belonging to the probed surface (Fig. 4*b*). Electrons can flow from the occupied electronic states of the tip and sample. By reversing the bias voltage, either the occupied or empty states of the sample can be probed, and by changing the magnitude of this voltage, the local density of states, ρ_{sample}, can be probed as a function of energy. The tunneling current, I, can be expressed approximately by Eq. (1), where κ is proportional to $\sqrt{\Phi}$,

$$I \propto V \rho_{sample} e^{-2\kappa s} \qquad (1)$$

and Φ is the work function of the probed surface. Since the local density of states of a semiconductor is a function of energy (expressed here in terms of the applied tip-sample bias voltage), the current-voltage characteristic of the tunneling for semiconductors is usually nonlinear. The local density of states of the probed surface at an energy equal to the difference between the Fermi energy, E_F, and the applied bias voltage can be obtained by plotting the local conductivity, dI/dV, as a function of V, using Eq. (2).

$$\frac{dI}{dV} \propto \rho_{sample}(E_F - eV) \qquad (2)$$

The ability to image electronic structures of conducting surfaces has provided a major contribution to the understanding of a variety of surface-science phenomena that no other system but the scanning tunneling microscope can furnish. Decreasing the response time of the scanning tunneling microscope to a subnanosecond scale was made possible by employing optical methods.

Use of lasers. A variety of fixed- or variable-wavelength lasers are available, having continuous-wave optical beams as well as high-intensity, short-pulse beams. Such lasers have been used, in a variety of ways, to convert the otherwise slow operating scanning tunneling microscope into a fast-response system. This transformation is possible because the time involved in the tunneling process, t_{tunnel}, is very short. This time is given by Eq. (3),

$$t_{tunnel} = s \sqrt{\frac{m}{2\Delta E}} \qquad (3)$$

where ΔE is the difference between the barrier height and the energy of the electron, and m is the mass of the electron. For example, for $s = 0.2$ nm and $\Delta E = 1$ V, the value of t_{tunnel} is 0.34×10^{-15} s, which is comparable to one period of an optical wave. To exploit the fast time response of the tunneling junction, two strategies can be employed: modulating optically some known property of the probed sample or switching optically the electrodes of the scanning tunneling microscope that apply the bias and measure the tunneling current.

Optical modulation of samples. One important property of a semiconductor is its response to optical excitations. The absorption of photons having energies above the band gap results in the generation of electron-hole pairs. The existence of surface states in the band gap, because of dangling bonds or adsorbates, for example, will bend the energy bands in the vicinity of the surface (**Fig. 5**). The excited electron-hole pairs will migrate to their respective bands, generating a surface photovoltage, which will flatten the band bending. This surface photovoltage can be measured by the scanning tunneling microscope via tunneling or capacitive pickup. The following two sets of experiments, (Fig. 4*a*), were aimed at measuring the fast response time of the surface photovoltage by using laser excitation techniques. In designing these experiments, careful consideration had to be given to eliminate thermal effects associated with the absorption of optical radiation by the scanning tunneling microscope tip and semiconducting samples.

Pulsed laser excitation. The original experiments, which probed the temporal response of the surface photovoltage across silicon surfaces, were carried out using a mode-locked laser that produced picosecond pulses. In one implementation, the optical beam was split into two parts, and one part was

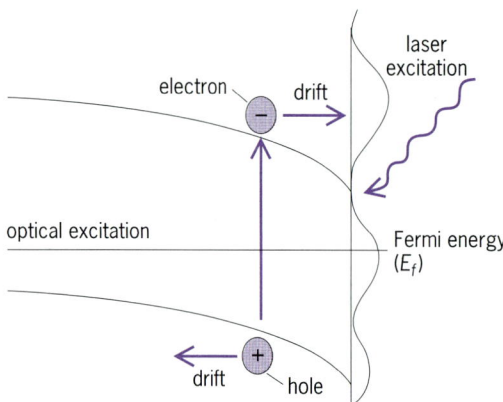

Fig. 5. Generation of surface photovoltage by photoexcited electron-hole pairs that flatten the band bending.

delayed relative to the other and then recombined with the other part. The combined beam, consisting of pairs of optical pulses with a controllable time delay between the constituents of the pair, was focused on the tunneling junction and generated a surface photovoltage. Monitoring the surface photovoltage as a function of the time delay and the intensity of the optical excitation yielded the lifetime of the photoexcited charge carriers.

Continuous-wave laser excitation. Another method for probing the temporal response of the surface photovoltage of a semiconductor employed a multimode continuous-wave laser whose beam was focused on the tunneling junction of an scanning tunneling microscope. The samples for this experiment were the n- and p-type semiconductors, molybdenum sulfide (MoS_2) and tungsten selenide (WSe_2), respectively, which could be easily imaged by a scanning tunneling microscope. The 380-MHz beat frequency of the longitudinal modes of the beam modulated the generation of the photoexcited electron-hole pairs, giving rise to a surface photovoltage and a tunneling current at this frequency. This high-frequency current was plotted as a function of the tunneling gap, yielding an exponential decay, as expected from a tunneling process. Also plotted was the dependence of this current on the applied bias voltage, revealing that the surface photovoltage was largest for a backward-biased tip-semiconductor configuration.

Optical switching of electrodes. The two methods described above show how a scanning tunneling microscope can yield local electronic properties of a semiconducting surface using optical excitations that produce a surface photovoltage. A different method for probing fast processes on any conducting surface employs a scanning tunneling microscope whose tip and sample electrodes can be switched in rapid succession (**Fig. 6**). In one implementation, subpicosecond pulses from a laser were focused on silicon-on-sapphire switches. By splitting and recombining the train of pulses, the electrodes leading to the tip or sample bias could be switched on and off under controlled conditions. The experimental results demonstrated a 5-nm spatial resolution and a 2-ps temporal resolution on a conducting sample. In another experiment, pulses from a mode-locked laser were focused on a transmission line, consisting of a pair of gallium arsenide (GaAs) photoconductive switches leading to and from the sample electrode of the scanning tunneling microscope. A 130-ps temporal resolution on a conducting sample was demonstrated.

For background information SEE ANTENNA (ELECTROMAGNETISM); MAXWELL'S EQUATIONS; OPTICAL MICROSCOPE; SCANNING TUNNELING MICROSCOPE; SEMICONDUCTOR; SURFACE PHYSICS in the McGraw-Hill Encyclopedia of Science & Technology.

Dror Sarid

Bibliography. U. Dürig, D. W. Pohl, and F. Rohner, Near-field optical scanning microscopy *J. Appl. Phys.*, 59:3318–3327, 1986; M. J. Gallagher et al., Nanosecond time-scale semiconductor photoexcitations probed by a scanning tunneling microscope, *Appl. Phys. Lett.*, 64:256–258, 1994; R. J. Hamers, Ultrafast time resolution in scanned probe microscopies: Surface photovoltage on Si(111)-7 × 7, *J. Vac. Sci. Technol. B*, 9:514–518, 1991; H. Heinzelman et al., Scanning near-field optical microscopy in Basel, Ruschlikon, and Zurich, vol. 34, no. 8, August 1995; L. Novotny, D. W. Pohl, and P. Regli, Light propagation through nanometer-sized structures: The two-dimensional-aperture scanning near-field optical microscope, *J. Opt. Soc. Amer. A*, 11:1768–1779, 1994; G. Nunes, Jr., and M. R. Freeman, Picosecond resolution in scanning tunneling microscopy, *Science*, 262:1029–1032, 1993; D. W. Pohl and D. Courjon (eds.), *Near Field Optics*, NATO ASI Series E 242, 1993; S. Weiss et al., Ultrafast scanning probe microscopy, *Appl. Phys. Lett.*, 63:2567–2569, 1993.

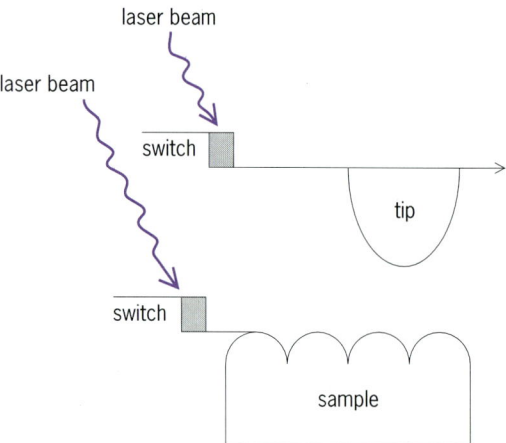

Fig. 6. High-speed scanning tunneling microscope system using optical switching of the tip and sample electrodes.

Mineralogy

In recent years, an array of sophisticated instrumental techniques have become available that are being applied to studies of chemical composition and reactions on minerals. Some of these techniques are now well established (for example, x-ray diffraction and infrared spectroscopy). Others (for example, scanning tunneling microscopy and laser-ablation microprobe inductively coupled plasma mass spectrometry) are under development.

Chemical composition. New methods that focus on in-place microbeam analysis of trace elements, light lithophile elements, and isotopes include secondary ion mass spectrometry, proton-induced methods, and microprobe sampling methods.

Secondary ion mass spectrometry (SIMS). This technique uses a focused ion beam to remove ions from a mineral surface, and analyzes their mass and charge with a mass spectrometer. Secondary ion mass spectrometry can analyze very low levels of rare earth elements, high field-strength elements (such as zirconium, tantalum, or niobium) and light lithophile elements, and is contributing significantly to work on geochemical processes in Earth's mantle. This type of spectrometry is also a very sensitive isotopic technique, and it is effective for high-resolution microprobe dating of zoned minerals using radiogenic isotopes. As an ablation technique, secondary ion mass spectrometry can continually analyze the surface of a mineral and slowly drill though the surface, providing a depth profile of chemical composition at a resolution of the order of an angstrom.

The fluorescent emission of characteristic x-rays by a mineral or rock has long been used for bulk rock analysis by x-ray fluorescence. The recent availability of high-intensity x-ray radiation from synchrotron sources (which accelerate electrons in closed orbits) has allowed the development of a synchrotron x-ray fluorescence (SXRF) microprobe. This device has an extremely low background signal, and is perhaps the ideal trace-element microprobe method; it is under active development and shows great promise.

Proton-induced methods. These methods use protons to excite characteristic x-rays or gamma rays. Because the mass of the proton is large, the background radiation is low in proton-induced emission, rendering the methods suitable for trace-element analysis. Proton-induced x-ray emission (PIXE) is the most common technique; it has been particularly important in the analysis of trace amounts of precious metals in sulfide minerals. Proton-induced gamma-ray emission uses the intensities of gamma rays emitted from nuclear transitions in the sample, and has considerable potential for in-place microbeam analysis of light lithophile elements. In addition, scanning techniques have recently been developed, and trace-element mapping of mineral grains has become possible.

Microprobe sampling methods. The availability of lasers has greatly expanded the versatility of many mass-spectrometer techniques, and important development work has been done on microprobe sampling methods. Of particular interest is the laser-ablation microprobe inductively coupled plasma mass spectrometer. In this in-place microprobe technique, a neodymium:yttrium-aluminum-garnet laser removes material from a mineral and transports it by a stream of argon into an inductively coupled plasma (**Fig. 1**), where it is dissociated and ionized; the mass spectrometer then measures the charged masses produced in the plasma. This method can analyze trace elements at the parts per million to parts per billion level; it also shows promise as an isotopic microprobe. Compositional zoning of major elements is common in

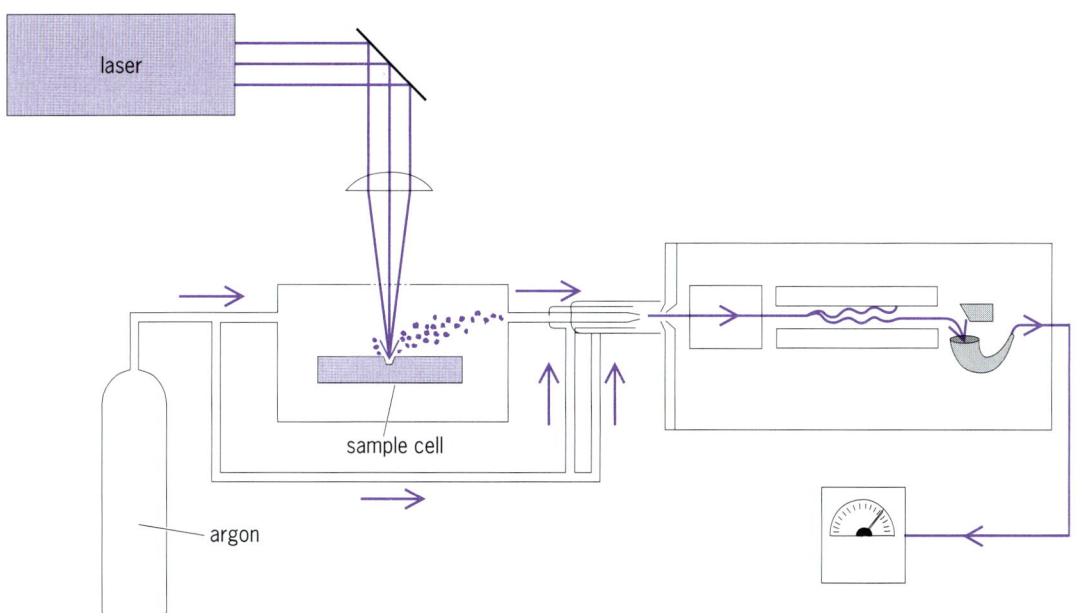

Fig. 1. Components of a laser-ablation microprobe inductively coupled plasma mass spectrometer. (*After B. J. Fryer, S. E. Jackson, and H. P. Longrich, The design, operation, and role of the laser ablation microprobe coupled with an inductively coupled plasma-mass spectrometer in the earth sciences, Can. Mineral., vol. 33, 1995*)

many minerals. Initial work with this method suggests that the same is true for trace elements and isotopes, and that zoning patterns of major elements in the same mineral grain are not necessarily coherent; this information has great potential as an indicator of petrologic and geochemical processes.

Atomic arrangements. Most information on mineral structures has resulted from x-ray and neutron diffraction data. Spectroscopic techniques have been important in providing very specific structural probes that augment the results of the well-established methods.

X-ray techniques. When x-rays are absorbed by an element, there is a gradual change in the amount of absorption with energy, interrupted by a dramatic increase in absorption at a specific energy that is characteristic of the absorbing element. This feature is known as an absorption edge; it is caused by the excitation of an electron from a deep core level to an empty bound state or a continuum state. When the photoelectron is ejected from the absorbing ion, it may be scattered by the local environment to produce modulations close to the absorption edge in the spectrum, known as the x-ray absorption near-edge structure (XANES), and weaker oscillations at higher energies in the spectrum, known as the extended x-ray absorption fine structure (EXAFS). Interatomic distances and coordination numbers of major and minor elements in both crystalline and noncrystalline materials can be derived. X-ray absorption spectroscopy has provided the first definitive information on the environment of petrologically important trace elements in complex rock-forming minerals and glasses.

Nuclear magnetic resonance techniques. Nuclear magnetic resonance (NMR) involves the resonant absorption and emission of radio-frequency electromagnetic radiation by an atomic nucleus via the interaction of its magnetic multipole moment with an applied magnetic field. Transition energies are sensitive to the local electric-field gradient around the nucleus, and hence nuclear magnetic resonance is a useful technique for the characterization of the degree of order in diamagnetic minerals. The complexity of the spectrum is greatly reduced if the sample is spun at the magic angle, giving rise to the technique known as magic-angle spinning, which has revolutionized the application of nuclear magnetic resonance to mineralogy. Prior to the development of magic-angle spinning in 1980, very little work was done on minerals by using nuclear magnetic resonance. With the application of magic-angle spinning nuclear magnetic resonance to silicates, a great deal of research was undertaken, first on nuclei of spin ½ (of which silicon-29 is the most important) and then on quadrupolar nuclei such as aluminum-29, sodium-23, and oxygen-17. The strength of the method is that it can be used on diamagnetic minerals and noncrystalline materials (glasses, melts, gels); it has been an important technique in the derivation of aluminum-silicon distributions and local structure in aluminosilicate minerals [feldspars, feldspathoids, zeolites (**Fig. 2**), and glasses].

Mineral surfaces. When minerals participate in chemical reactions or processes such as adsorption or dissolution, the important interactions often occur at the mineral surface. The major breakthrough on surface characterization of minerals came with the development of scanning tunneling microscopy and atomic force microscopy. Scanning tunneling microscopy works on electrical conductors such as native metals and platinum-group minerals and semiconductors such as sulfides. In this technique, an atomically sharp conducting voltage-biased tip is brought close to the surface of a mineral. Electrons tunnel from the mineral to the tip, or from the tip to the mineral. The spatial variation in current images the surface at atomic or near-atomic resolution; single adsorbed atoms can be imaged, and there is the potential for identifying the chemical species of the atom. Atomic force microscopy produces similar-scale images for insulators; it senses the Born-type repulsion between the atoms of the tip and the surface and produces an atomic or near-atomic scale topographic map of the surface via the displacement of the tip. These last two techniques have provided scientists with the capability of examining such processes as crystallization, dissolution, adsorption, and alteration at the atomic scale.
Frank C. Hawthorne

Reactions in the Earth. Minerals in the Earth can undergo structural and chemical reactions in response to changes in conditions (such as pressure, temperature, or presence of water) that occur as a result of a variety of geological processes. An understanding of these reactions provides insights into the mechanisms and conditions under which they occurred and narrows the range of interpretation on the nature and duration of geological and geochemical processes. In the Earth, mineral reactions are often extremely slow and the transformations do not proceed to completion. Known as stranded

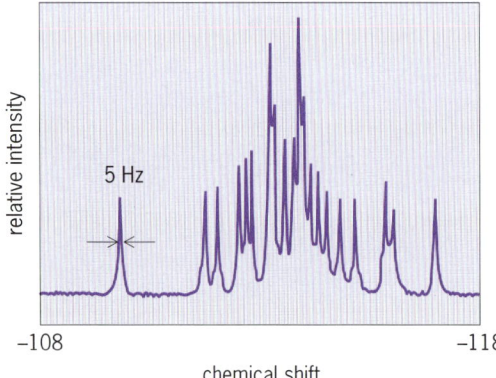

Fig. 2. Ultra-high-resolution silicon-29 magic-angle spinning nuclear magnetic resonance spectrum of the highly siliceous zeolite ZSM-5, showing resolved peaks corresponding to 20 of the 24 crystallographically distinct silicon sites. (*After C. A. Fyfe, J. H. O'Brien, and H. Strobel, Ultra-high resolution of ^{29}Si MAS NMR spectra of highly siliceous zeolites, Nature, 326:281–282, March 19, 1987*)

reactions, they allow the relationships between reactant and product phases to be studied in detail.

Significant new insights into the nature of mineral reactions have come from studies using the transmission electron microscope, an extremely powerful instrument with a unique combination of analytical capabilities. A modern instrument allows textural, crystallographic, and chemical information to be obtained from areas as small as a square nanometer, and features that are as little as 0.14 nm apart can be resolved. By comparison, most minerals have crystal structures whose basic structural units (unit cells) are greater than 0.4 nm^2. The high resolution offered by a transmission electron microscope allows the very earliest stages of mineral reactions to be studied before any evidence can be observed by lower resolution techniques, such as optical microscopy.

Low-temperature replacement reactions. Hydration reactions occur at the surface of Earth because of weathering processes or at moderate depths involving hydrothermal fluids, at temperatures up to about 300°C (570°F). These reactions play important roles in the formation of soils and sediments and in the mass transfer of elements in the geochemical cycle. Many examples involving important rock-forming minerals such as olivines, pyroxenes, amphiboles, micas, feldspars, and aluminosilicate minerals have been studied by transmission electron microscopy. The products of these reactions are hydrous minerals, such as amphiboles and sheet silicate minerals (for example, micas, serpentines, and clay minerals). The minerals that form are controlled by the composition of the reacting phase and the degree of hydration, as well as by other factors such as the chemistry of the reacting solutions. A common feature of hydration reactions is that the crystal structure of the reacting phase frequently has a strong influence on the mechanism of the reaction. Alteration often proceeds parallel to crystallographic planes that have atoms (especially oxygen atoms) packed closely together.

Pyroxene, amphibole, and sheet silicate alteration. Pyroxenes, amphiboles, and sheet silicates commonly show the effects of low-temperature hydration, although they react more slowly than olivine. Depending on the degree of hydration, pyroxene [$(Mg,Fe)SiO_3$ or $(Ca,Mg,Fe)Si_2O_6$] can alter to amphibole [for example, $(Mg,Fe)_7Si_8O_{22}(OH)_2$], pyriboles [for example, $Mg_{10}Si_{12}O_{32}(OH)_4$], or sheet silicates such as serpentine [$Mg_3Si_2O_5(OH)_4$], chlorite [$(Mg,Al,Fe)_{12}(Si,Al)_8O_{20}(OH)_{16}$], and talc [$Mg_6Si_8O_{20}(OH)_4$]. Studies made with transmission electron microscopy show that the reactions can occur by multiple reaction paths and by different mechanisms. For example, during hydration pyroxene can react directly to form sheet silicate minerals or by a stepwise path in which amphibole forms as an intermediate step in the reaction. Different minerals may form as a result of variations in the chemistry of the local fluid. High concentrations of magnesium (Mg) stabilize serpentine, whereas amphibole or talc formation is favored when the concentration of silicic acid is high. The replacement of pyroxene, as well as other minerals such as aluminosilicate minerals, can occur by at least two different reaction mechanisms. The first involves the nucleation and growth of oriented lamellae of the product phases within the pyroxene. The pyroxene crystal structure controls the orientation of the new phases, and some structural elements of the pyroxene are retained by the product phase. In contrast, in the second mechanism bulk replacement of pyroxene occurs on a broad reaction front, and the product phases show no orientation relationships with the altering pyroxene. In this case, complete dissolution of the pyroxene by the fluid phase probably occurs, and the new phases precipitate from solution.

Feldspar alteration. Feldspars [$CaAl_2Si_2O_8$; $(Na,K)AlSi_3O_8$] are abundant minerals in igneous rocks in Earth's crust and alter to clay minerals and micas. The breakdown of feldspars in granitic rocks can involve direct transformation to a new phase or dissolution followed by reprecipitation of new minerals. The earliest stages in feldspar alteration involve the formation of a disordered layer on the surface of the feldspar (<1 micrometer thick), which is depleted in calcium (Ca), sodium (Na), potassium (K), and silicon (Si) relative to the bulk feldspar composition but enriched in iron (Fe). This layer is an intermediate phase in the reaction and recrystallizes to form clay minerals (for example, smectite). Highly localized variations in the compositions of the product smectite show that distinct chemical environments exist on the microscale. Smectite is a transitional phase in the reaction, and it reacts progressively to form kaolinite [$Al_4Si_4O_{10}(OH)_8$]. Other phases may form depending on the conditions of alteration.

At higher temperatures (200–300°C or 390–570°F), for example, in fumarole systems associated with young volcanic systems, alteration of compositionally zoned plagioclase feldspar by highly acidic chloride and silica-rich fluids results in the preferential replacement of calcium-rich portions of the crystals by amorphous silica. Transitional aluminum-rich sheet silicate minerals are precipitated close to the reaction interface. At the reaction interface two compositionally distinct, leached surface layers, less than 0.1 μm thick, develop. The inner layer adjacent to the feldspar is enriched in chloride, which forms a complex at the surface of the altering feldspar, consistent with experimental laboratory studies of alteration of feldspar by highly acidic solutions (pH = 2–4).

High-temperature reactions. A number of high-temperature reactions involving the breakdown of sheet silicate minerals (muscovite, biotite, and chlorite) to very fine grained reaction products have also been studied by transmission electron microscopy techniques. These reactions occur in rocks adjacent to igneous intrusions at temperatures of 800–1100°C (1470–2000°F) when heating takes place rapidly. The reaction products nucleate within the sheet silicates and are crystallographically oriented with respect to the reacting mineral. For muscovite, the reaction products are K-feldspar ($KAlSi_3O_8$), mullite ($3Al_2O_3 \cdot 2SiO_2$), corundum (Al_2O_3), and biotite [$K(Mg,Fe)_3AlSi_3O_{10}(OH)_2$], whereas for biotite, spinel [$(Mg,Fe)Al_2O_4$], K-feldspar, and a new, more magnesium-rich biotite are the product phases. An understanding of these reactions provides insights into the behavior of minerals under disequilibrium conditions (conditions far from equilibrium) and the duration of heating events associated with the intrusion and eruption of magmas.

For background information SEE ELECTRON MICROSCOPE; NUCLEAR MAGNETIC RESONANCE (NMR); PROTON-INDUCED X-RAY EMISSION (PIXE); SCANNING TUNNELING MICROSCOPE; SECONDARY ION MASS SPECTROMETRY (SIMS); X-RAY FLUORESCENCE ANALYSIS in the McGraw-Hill Encyclopedia of Science & Technology.

Adrian J. Brearley

Bibliography. J. F. Banfield and R. A. Eggleton, Analytical transmission electron microscope studies of plagioclase, muscovite and K-feldspar weathering, *Clays Clay Min.*, 38:77–90, 1990; A. J. Brearley, An electron optical study of muscovite breakdown in pelitic xenoliths during pyrometamorphism, *Mineral. Mag.*, 50:385–397, 1986; P. R. Buseck (ed.), *Reviews in Mineralogy*, vol. 27, 1992; F. C. Hawthorne (ed.), *Spectroscopic Methods in Mineralogy and Geology: Reviews in Mineralogy*, vol. 18, 1988; D. L. Perry (ed.), *Instrumental Surface Analysis of Geologic Materials*, 1990; D. R. Veblen and P. R. Buseck, Hydrous pyriboles and sheet silicates in pyroxenes and uralites: Intergrowth microstructures and reaction mechanisms, *Amer. Mineral.*, 66:1107–1134, 1981.

Molecular magnet

Molecular magnetism emerged as a new field of research in the mid-1980s. This field deals with the synthesis and the study of the physical properties of molecular assemblies possessing units with unpaired electrons. It is essentially interdisciplinary, involving organic, organometallic, and inorganic chemists, as well as theoreticians, physicists, and materials scientists. Molecular magnetism is closely related to the field of molecular electronics.

The heart of molecular magnetism concerns the design and the synthesis of new molecular assemblies exhibiting bulk properties, in particular long-range magnetic ordering with a spontaneous magnetization below a critical temperature T_c. These compounds are usually called molecular-based magnets.

Design strategy. The strategy for designing molecular magnets is based on positive and negative spin densities. Since the 1960s, synthetic chemists began to speculate about the design of molecular-based magnets. However, the first compounds of this type were not produced until 1986.

The basic strategy used for obtaining molecular-based magnets consists first in synthesizing molecular units possessing a large positive spin density in one region and a small negative spin density in another region. The units are then assembled in such a way that the large spin density of one unit preferably interacts with the small negative spin density of the nearest neighbor unit. Most often, the interaction between two local spin densities is of the up-down type (that is, antiferromagnetic), which in the present case leads to an overall parallel alignment of the molecular spins. **Figure 1** shows such a molecular entity with both large positive and small negative spin-density regions. This entity contains two metal ions, manganese(II) [Mn(II)] and copper(II) [Cu(II)], linked by a conjugated bridge (an oxamide) transmitting very efficiently the electronic effects between these metal ions. Mn(II) carries a local spin $S_{Mn} = 5/2$, and Cu(II) a local spin $S_{Cu} = 1/2$. Owing to the presence of the bridge, the two ions strongly interact, so that the local spins align in an antiparallel fashion in the ground state, affording a molecular spin $S = S_{Mn} - S_{Cu} = 2$. Therefore, in the ground state, there is a large positive spin density around Mn(II) and a weak negative spin density around Cu(II). A polarized neutron diffraction study, which leads to spin-density maps, confirms this situation.

Ferromagnetic ordering of one-dimensional ferrimagnets. The following step requires the assembly of the Mn(II)Cu(II) units so that the Cu(II) ion of one unit interacts with the Mn(II) ion of the adjacent unit. First the bimetallic chain compounds are synthesized with a perfect alternation of Mn(II) and Cu(II) ions. In the ground state, the S_{Mn} spins tend to align along one direction, and the S_{Cu} spins along the opposite direction, which confers a large magnetism to each chain. Such chains are said to be ferrimagnetic. However, the magnetic ordering is most generally a three-dimensional phenomenon. In no case can it occur in one dimension. Therefore, it is necessary to create some interchain interactions and to control their nature. In 1986 the first compound exhibiting a spontaneous magnetization below $T_c = 4.6$ K ($-451°$F) was synthesized. Making the chains closer to each other through a mild thermal treatment may result in an increase of T_c up to 30 K ($-406°$F).

Interlocking of two-dimensional networks. A more recent synthesis produced two-dimensional compounds of formula $(cat)_2Mn_2[Cu(opba)]_3 \cdot L$ where cat$^+$ stands for a monovalent cation such as tetraalkylammonium, L for solvent molecules, and [Cu(opba)]$^{2-}$, with opba = orthophenylenebis(oxamato) for the copper(II) precursor, structure (I).

(I)

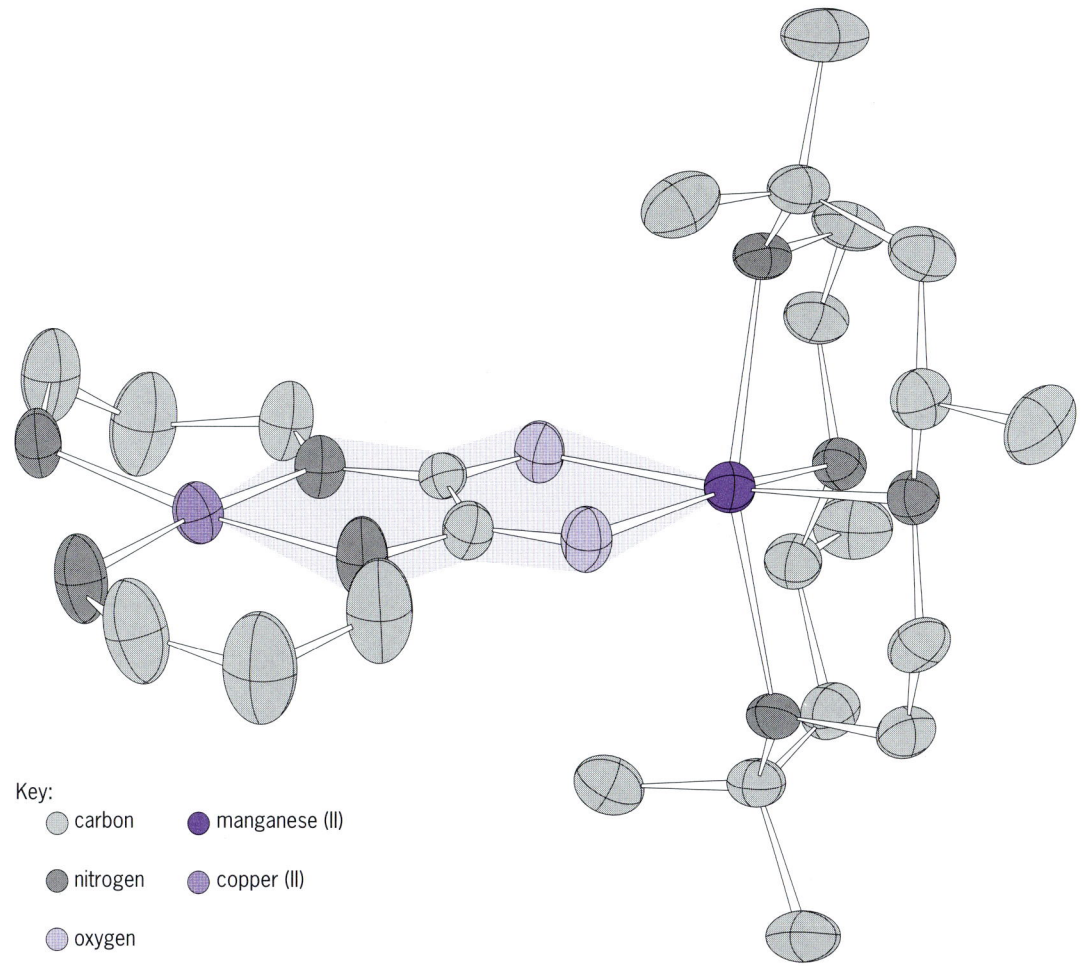

Fig. 1. Structure of the binuclear cation [Mn(Me$_6$-[14]ane-N$_4$)Cu(oxpn)]$^{2+}$, in which the Mn(II) and Cu(II) are antiferromagnetically coupled, so that there is a large positive spin density around Mn(II) and a small negative spin density around Cu(II). Shaded area represents the conjugated bridge (oxamide).

These compounds exhibit long-range magnetic orderings with T_c ranging from 12 to 22 K (−438 to −420°F).

Increasing the dimensionality further requires linking together metal ions occupying different layers. To accomplish this the radical cation [abbreviated rad$^+$; structure (II)] was used. This cation carries both a positive charge and a spin S_{rad} = 1/2 equally shared between the two nitrogen-oxygen (N-O) groups. A compound was produced with the formula (rad)$_2$Mn$_2$[Cu(opba)]$_3$(DMSO)$_2$ · 2H$_2$O (DMSO is dimethylsulfoxide, which was used as solvent); this compound possesses an unprecedented crystal structure. This unusual structure consists of two almost perpendicular two-dimensional networks. Each network is made up of layers stacked on each other, as in graphite, with an interplane separation of 1.48 nm. A layer consists of edge-sharing hexagons (**Fig. 2**) with a Mn(II) ion at each corner and a Cu(II) ion at the middle of each edge. The mean length of an edge is 1.09 nm. The Mn(II) ions present a perfect alternation of Λ and Δ chiral sites [Λ and Δ define the absolute configuration for the two possible arrangements of the Cu(opba) groups around the Mn atom]. Nearest neighbor Mn(II) and Cu(II) ions are bridged by oxamato groups, favoring the antiparallel alignment of the S_{Mn} and S_{Cu} spins. The two graphitelike networks interpenetrate with a full interlocking of the hexagons, forming a kind of three-dimensional wire netting as shown in **Fig. 3**. In the center of each hexagon of the tinted network (A) is a Cu(II) ion of a hexagon of the untinted network (B), and vice versa. The two networks are further connected by the rad$^+$ cations that bridge two Cu(II) ions, one from each network, resulting in Cu$_A$-rad-Cu$_B$-rad chains.

Interlocking of supramolecular rings is an area of fundamental importance in chemical topology. Of

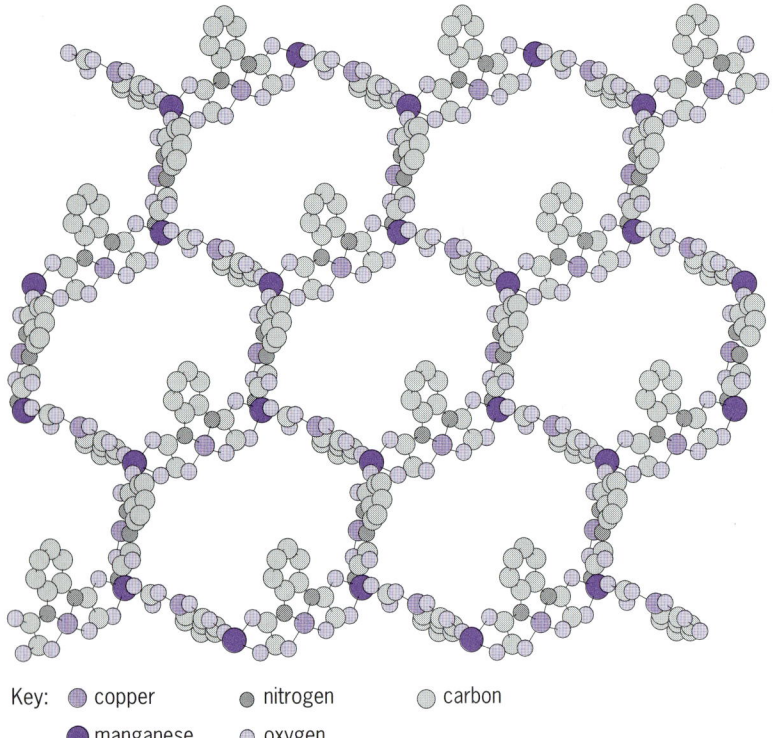

Key: copper, nitrogen, carbon, manganese, oxygen

Fig. 2. Layer in $(rad)_2Mn_2[Cu(opba)]_3(DMSO)_2 \cdot 2H_2O$, showing the honeycomb topology with edge-sharing Mn_6Cu_6 hexagons.

the many two- or three-dimensional molecular structures, very few present a complete interlocking of independent networks. The structure of $(rad)_2Mn_2[Cu(opba)]_3(DMSO)_2 \cdot 2H_2O$ has some additional subtleties, such as the presence of three kinds of spin carriers or the chirality or the Mn(II) sites.

The study of the magnetic properties of this interlocked molecular-based magnet reveals a critical temperature $T_c = 22.5$ K ($-419°$F). Below T_c, within a very small magnetic field, the S_{Mn} spins are aligned along the field direction, and the S_{Cu} and S_{rad} spins, along the opposite direction. When increasing the field, the S_{rad} spins tend to align along the field direction, which progressively increases the magnetization. For a classical magnet, the magnetization reaches a saturation value when applying a very small field (which may be the magnetic field of the Earth), and then does not increase further as the field is increased.

Large coercivity. A molecular-based magnet possesses a large coercivity. One of the main applications of magnetic substances is information storage (for instance, with magnetic tapes). The memory effect is conferred by the hysteresis loop in the field dependence of the magnetization. To store information, the coercive field (that is, the field neces-

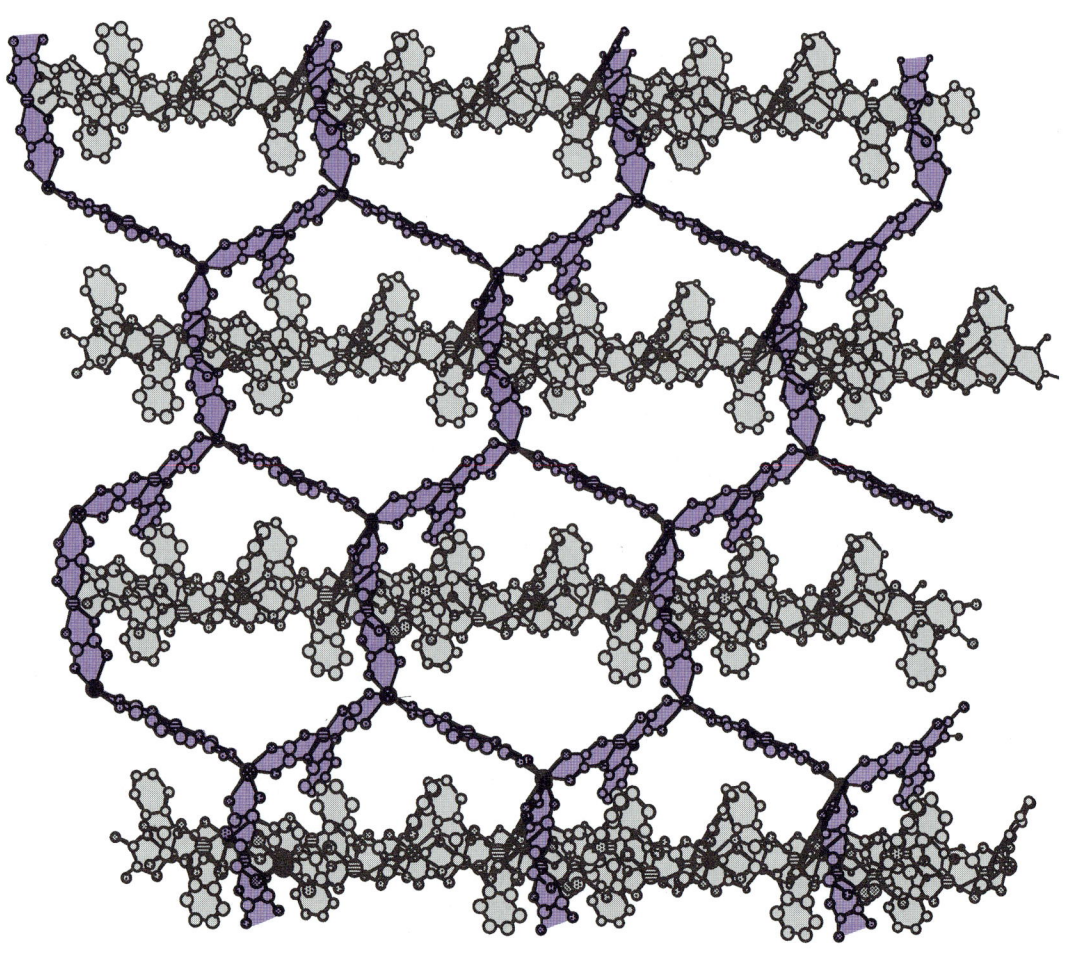

Fig. 3. Interpenetration of the two networks, A and B, in $(rad)_2Mn_2[Cu(opba)]_3(DMSO)_2 \cdot 2H_2O$.

sary to cancel the magnetization) must be sufficiently large, at least a few hundred oersteds. The compound $(rad)_2Mn_2[Cu(opba)]_3(DMSO)_2 \cdot H_2O$ as well as the other Mn(II)Cu(II) magnets have very weak coercive fields, of the order of a few oersteds, and until recently there was speculation as to whether molecular-based magnets could display large coercivities. The value of the coercive field at a given temperature depends on both the chemical nature of the substance and structural factors such as the size and the shape of the grains within the sample. Concerning the chemical nature, the key role is played by the magnetic anisotropy of the spin carriers, which prevents the domains from rotating freely when a magnetic field is applied. Replacing the magnetically isotropic Mn(II) ion by the strongly anisotropic Co(II) ion results in a dramatic enhancement of the coercive field. The compound $(rad)_2Co_2[Cu(opba)]_3 \cdot 3H_2O$ is a magnet below $T_c = 34$ K ($-398°F$). At 4.2 K ($-452°F$), the coercive field is as large as 3.1 kOe; that is, it is much larger than the coercive field of chromium dioxide (CrO_2) in the presently used magnetic tapes.

Molecular magnetism is still in its infancy, and the properties of molecular-based magnets are far from being completely known or understood. These compounds are usually weakly colored, and the synergy between magnetic and optical properties seems to be one of the important issues. In any case, more and more research groups are involved in the design of novel molecular assemblies behaving as magnets. This field of research is one of the most original components of supramolecular chemistry.

For background information SEE BIOELECTRONICS; ELECTRON SPIN; MAGNETIC FIELD; MAGNETIC MATERIALS; NEUTRON DIFFRACTION in the McGraw-Hill Encyclopedia of Science & Technology.

Olivier Kahn; Yu Pei

Bibliography. O. Kahn, *Molecular Magnetism*, 1993; J. S. Miller and A. J. Epstein, Organic and organometallic molecular magnetic materials: Designer magnets, *Angew. Chem. Int. Ed. Engl.*, 33:385–415, 1994; H. O. Stumpf et al., A molecular-based magnet with a fully interlocked three-dimensional structure, *Science*, 261:447–449, 1993; H. O. Stumpf et al., Chemistry and physics of a molecular-based magnet containing three spin carriers, with a fully interlocked structure, *J. Amer. Chem. Soc.*, 116:3866–3874, 1994.

Molecular materials

Matter of nanoscale dimensions often displays properties unlike those of isolated gas-phase molecules or the bulk liquids or solids which they make up. These unusual properties are often attributable to the unique bonding, structure, and morphology which small systems of finite dimensions adopt, whereas in other cases they result because the small sizes bring about new phenomena arising from what is termed quantum confinement, the restriction of space available to electrons. Because of the special properties that these cluster systems (aggregates of atoms or molecules) display, they have become an active subject of basic investigation in recent years. Other impetus for such research has arisen from the promise that new advanced materials might be produced by assembling together nanoscale systems.

In 1992, investigations into the reactions of small hydrocarbons with transition-metal ions, atoms, and clusters led to the discovery of an unexpected dehydrogenation reaction leading to the assembly of a new molecular cluster family termed metallocarbohedrenes (met-cars). The first member of the class discovered was composed of 8 titanium atoms and 12 carbon atoms (Ti_8C_{12}). Through extensive studies of their chemical and physical properties, met-cars have been established as having a cagelike structure. Subsequent experimental work has revealed the unusually large stability of the cation (positively charged) Ti_8C_{12}, the anion analog (negatively charged species), and, most importantly, the neutral molecular cluster itself.

Met-car composition and structure. The surprising finding that a heretofore unknown compound composed of transition-metal and carbon atoms was very stable, and displayed unique bonding far different than that of the well-known cubic structure of bulk titanium carbide, prompted a search for other clusters with similar compositions. Ensuing research on a wide variety of early and late transition metals interacting with hydrocarbon vapors in a plasma reactor ultimately revealed the existence of a comparatively wide class of met-cars with the composition M_8C_{12}, where M is titanium (Ti), zirconium (Zr), hafnium (Hf), vanadium (V), niobium (Nb), chromium (Cr), or molybdenum (Mo); analogous species also have been detected for iron (Fe). By contrast, scandium (Sc), yttrium (Y), tantalum (Ta), tungsten (W), and copper (Cu), for example, have failed to reveal similar structures and composition. Most of these other elements are observed to assemble into the more common bulk carbide structures. Tantalum-carbon clusters are a good example, undergoing a cubic buildup of microcrystals in gas-phase experiments rather than development of the met-car cage.

All experimental findings are compatible with the met-cars having a structure corresponding to a pentagonal dodecahedron. But it was recognized from the beginning that this representation is idealized for the geometry of these clusters, and that not all bonds between atoms have identical lengths. Since their discovery, deducing the exact geometry of met-cars has been the subject of extensive theoretical and experimental research. A number of alternative structures have been proposed, including cubes of metals with C_2 units embedded in the faces, various capped structures with carbons bonded at the waist, metal cubes decorated with carbons, and carbon spheres decorated with metal atoms. All of these suggestions have failed to survive experimental and theoretical tests, and there are now essentially two caged structures which are viable alternatives: the originally proposed pentagonal dodecahedron and a distorted version in which the C_2 units are twisted between adjacent metal atoms. Theoretical computations favor the distorted structure, whereas most experiments are compatible with the more regular pentagonal dodecahedron. Work to establish the exact geometry of this unique and broad class of materials continues to be actively pursued.

Mechanisms of formation. The original discovery was made during the course of studies being conducted with a laser plasma reactor. The plasma source was composed of a laser beam impinging on a rotating metal target. Appropriately timed gas pulses of hydrocarbons flowed over this target, expanded out of a small reaction channel into a vacuum chamber, and were sampled with a mass spectrometer. The observation of complete loss of hydrogen atoms from the hydrocarbons, an ensuing reaction of the carbon atoms with the metal atoms, and the ultimate assembly of these into the met-car structure was surprising, and studies to deduce the mechanisms responsible continue to be pursued. However, other formation methods have since been found.

The first successful alternative method involved the laser vaporization of well-mixed powders of selected metals and graphite, followed by the entrainment of the formed product species in pulses of helium gas. Met-cars were readily produced with appropriate conditions of laser power and metal-to-carbon ratio mixtures. Thereafter, it was found that met-cars could be produced from a mixture of metal carbides along with carbon and metal powders by vaporizing them in a small crucible suspended in vacuum, without any need for a gas to transport the plasma or assist in its cooling.

An important advance toward exploring the potential use of these molecular clusters as new materials came from successful attempts to synthesize bulk quantities of met-cars. Synthesizing was first accomplished with an arc-discharge technique similar to ones which have been used to produce so-called buckyballs (fullerenes). In this technique, rods of appropriate composition composed of metal and graphite mixtures are assembled, and an arc is struck between two rods which are a small distance apart. This procedure led to the formation of a soot mixture containing small amounts of met-cars. Building on the aforementioned observations that laser vaporization also could be used to effect the production of met-cars has led to the development of another successful method for synthesizing bulk amounts of these materials, namely, vaporizing appropriate carbide samples into a stagnant noble gas. Work is actively being pursued to find suitable techniques for extracting the met-cars from the unreacted soot and determining their exact geometry through conventional structural characterization techniques.

Other structures. Studies of zirconium and niobium met-cars have turned out to be among the most revealing. Investigations into the growth patterns of zirconium-carbon complexes revealed another surprising and totally unexpected building pattern for species with sizes beyond that of Zr_8C_{12}. In contrast to the typical structures found for other molecular clusters, met-cars grow into multicaged species composed of a network of pentagons (where the pentagonal dodecahedron represents, in terms of an idealized structure, the first member of the family), with a similar growth pattern persisting at larger sizes. Upon further growth, a common face is shared such that double cages ($Zr_{13}C_{22}$ and $Zr_{14}C_{21}$), a triple cage (at $Zr_{18}C_{29}$), and a quadruple cage (at $Zr_{22}C_{35}$) develop. This unique growth pattern distinguishes the class of metallocarbohedrenes from the regular metal-doped fullerenes.

Work with niobium has been particularly revealing for two reasons. First, niobium has also been found to develop into multicage networks analogous to the double cage seen for the zirconium system. Also, through a judicious selection of experimental conditions, it is a system whose structural development can be altered between cubic growth patterns characteristic of bulk carbides and the met-car structure. These findings suggest that an inherent aspect of met-car formation is associated with the kinetics of the building process. In fact, met-cars are formed when small units such as MC_2 and M_2C_3 are present, but cubic structures develop when the dominant gas species is composed largely of MC species. These findings provide significant clues to methods for producing met-cars by alternate routes.

Binary metal met-cars. Perhaps one of the most significant findings since the original discovery of met-cars is that met-cars composed of more than one metal type can be formed. Through an appropriate selection of experimental conditions, it has been found that various atoms can be incorporated into the 8-12 cage, even including some metals that do not form the pure single-component met-cars. In particular, it was first observed that the titanium met-car could be induced to undergo extensive substitutions with zirconium, leading to a full complement of structures composed of $Ti_xZr_yC_{12}$, where $x + y = 8$. Subsequent work revealed that substitutions of hafnium and niobium were also possible, which is perhaps not surprising in view of the successful formation of the mixed zirconium-titanium system. Particularly unexpected was the observation that other metals which do not form pure met-cars, such as yttrium, tantalum, and even tungsten, thorium (Th), and silicon (Si), could be introduced into the titanium met-car structure. Surprisingly, silicon, which commonly behaves more like carbon than a metal, substitutes for a metal atom rather than a carbon unit. This result is undoubtedly due to the strong bonding of the species SiC_2, providing additional insight into the mechanisms of met-car formation and their rather appreciable stabilities.

These new findings shed some additional light on the structure of met-cars. For example, the distorted structures mentioned above are predicted to have two different metal-atom types present in the cage lattice. Thus, a preference for a stable 4-4 mixed metal system compared to ones of single-metal composition would be expected. However, clusters formed via the substitutions of zirconium for titanium failed to reveal any preference for the $Ti_4Zr_4C_{12}$ structure over that of Ti_8C_{12}. Indeed, the degree of substitution appears to be statistical in nature, an observation in keeping with expectations for, though not proving, the dodecahedral structure. Other predictions for the distorted cage structure suggest that yttrium should facilely substitute, again a prediction not in accord with the experimental observations. In contrast, predictions for the case of the more symmetrical dodecahedral structure suggest the ease of substituting niobium and tantalum for titanium, which is consistent with recent observations.

Status and prospects. Met-cars have been established as a new class of molecular cluster materials which are likely to have unusual properties and many potential applications. It has been observed that these molecular clusters readily ionize even at very low intensities of light of widely varying wavelengths, suggesting that they have a substantially delocalized electronic character and that they may display unique optical and perhaps electrical properties. Moreover, studies focusing on their chemistry reveal that they can display various classes of reactions, including ones involving bond breaking, as well as association (weak attachment of interacting molecules). All the evidence points to their structure being one in which the metal and carbon atoms are bound in a well-defined lattice with one type of metal site. The formation of these interesting materials appears to be dominated by kinetic effects involving MC_2 building units. The clusters are found to grow by a unique mechanism leading to the development of a multicaged extended network structure.

Work is actively being pursued to isolate large amounts of bulk material and study the exact geometry and properties of met-cars using conventional chemical and physical techniques. Of particular interest is elucidating their optical, electronic, and reactive properties, subjects which are currently being studied through the use of molecular beams, lasers, and flow reactors.

For background information SEE ARC DISCHARGE; ATOM CLUSTER; CHEMICAL DYNAMICS; DELOCALIZATION; PLASMA PHYSICS in the McGraw-Hill Encyclopedia of Science & Technology.

A. Welford Castleman, Jr.

Bibliography. S. F. Cartier et al., The production of metallocarbohedrenes by the direct laser vaporization of the carbides of titanium and zirconium, *Chem. Phys. Lett.*, 220:23–28, 1994; S. F. Cartier, B. D. May, and A. W. Castleman, Jr., Binary metal metallocarbohedrenes of titanium and group IIIA, VA, and VIA metals, *J. Amer. Chem. Soc.*, 116:5295–5297, 1994; S. F. Cartier, B. D. May, and A. W. Castleman, Jr., $Ti_xZr_yC_{12}$ and $Ti_xHf_yC_{12}$ ($x + y = 8$): Mixed metal metallocarbohedrenes, *J. Chem. Phys.*, 100:5384–5386, 1994; H. T. Deng et al., Formation and stability of metallocarbohedrenes: $Ti_xM_yC_{12}$ ($x + y = 8$, M = Nb, Ta, Y, and Si), *Int. J. Mass Spectr. Ion Proc.*, 138:275–281, 1994. S. Wei et al., Formation of met-cars and face-centered cubic structures: Thermodynamically or kinetically controlled?, *J. Amer. Chem. Soc.*, 116:4475–4476, 1994.

Molecular transport

The concept of molecular transportation, from the perspective of biotechnology, pharmaceuticals, and bioengineering, entails the delivery, presentation, or immobilization of biochemically active molecules. Applications under consideration for molecular transportation include drug delivery for conventional pharmaceuticals, gene delivery for gene therapy, antigen presentation for vaccines, and even oxygen transport for use in blood substitute preparations. There have been many technological approaches to molecular transportation, and all are complicated by a series of problems that arise directly from fundamental constraints of biophysics.

The molecular transport assembly (MTA) that is under development combines material science, chemistry, and biology to successfully stabilize, target, and deliver a broad range of biochemically active molecules in their active form. The molecular transport assembly is a three-layered compound consisting of a ceramic solid-phase core, a glassy polyhydroxyl oligomeric (carbohydrate) intermediate coating, and an outer layer composed of noncovalently bound biochemically active agents for which transport is desired. Together, these three components yield a fundamental platform with the potential for stabilizing, targeting, and delivering such agents as synthetic blood, pharmaceuticals, vaccines, and even genetic material.

The destinations for delivery of these targeted molecules are not limited to points within the human anatomy. Traditionally, volatile substances such as explosives may become more manageable after being surface-immobilized with the molecular transport assembly. Cosmetics and personal care products may also be improved with the molecular transport assembly technology, which promises to stabilize color and moisturizing properties as well as eliminate the need for alcohol or oil-based suspension media.

Self-assembly. The principles of self-assembly are key for understanding how complex molecules and compositions such as the molecular transport assembly are formed. Three types of interactions are the catalysts for self-assembly in an aqueous environment: (1) interactions between charged groups of molecules that have adsorbed ions, causing them to assume specific prescribed structural orientations in three-dimensional space; (2) hydrogen bonding that brings molecules together and imparts order to the water surrounding the assemblies; (3) interactions caused by van der Waals forces, which as a result of the intrinsic weak polarity of molecules cause the noncovalently bonded structural assembly of macromolecules to be hard or soft relative to both surface and external physical forces. Although the strength of the dipole-induced attractions may not be solely responsible for self-assembly, this factor can determine the probability of adhesion or separation of the individual compounds.

Self-assembly is a simple yet elegant process by which macromolecules form subsequent to inherent forces, drawing resources from their biological surroundings. When harnessed, as they have been for the molecular transport assembly, these forces yield a broad delivery platform. When unconstrained, these forces become the basis for surface-induced denaturation and molecular inactivation.

Surface interactions. The surface energy of most core molecules is so great that when a biochemically active compound adsorbs to the surface, it is irreparably damaged. The damage renders the biologically active material inactive, or at least unrecognizable when it arrives at its destination. For years there have been attempts to bind materials to cores of various construction for the purpose of molecular delivery. In most instances, the binding has involved covalent forces that are far greater than van der Waals forces and that tend to induce significant denaturation. In other instances, noncovalent forces have been used for surface immobilization, but these forces have also largely altered the shape and the function of the molecule. In the molecular transport assembly, surface science plays a major role in ensuring that the biological agent that is to be adsorbed to the core for delivery can be recognized and active when it arrives at its destination.

Stabilizing film. The use of a film only 1 nanometer thick of polyhydroxyl oligomers (carbohydrates) is a remedy for the excessive surface energy that causes molecular surface denaturation. In particular, a group of disaccharide-related molecules have proven effective as surface energy reducers and molecular stabilizers. The carbohydrate is a cushion that can buffer the surface energy while remaining able to be adsorbed to by the third and final layer desired for delivery. The process of adsorption of the disaccharide to the core is easily accomplished in an aqueous environment. The removal of water enhances binding of the nanometer-thick layer of sugar to the core because of the hydrophilic properties of hydroxyl-rich disaccharides. The result is a quasiaqueous surface that has a reduced energy, and thereby is not prone to denature the biochemically active agent when the latter is adsorbed to the outer layer. Further stability to the biochemically active agent is reinforced by the glassy physical state assumed by the carbohydrate film and the ability of the numerous hydroxyl groups within the film to engage in both hydrogen donation and acceptance roles with the macromolecules of interest.

Materials. Two commonly used ceramic cores in these experiments have been carbon (diamond) and calcium-phosphate dihydrate (brushite). A typical process for producing clean diamond cores and coating them with cellobiose begins with diamond powder which is sonicated at 39.2°F (4°C) in $12N$ hydrochloric acid (HCl) for 16 h and washed with ultrafiltered water until the pH is near 7. Activated diamond dispersions in ultrafiltered deionized water are then introduced to the polyhydroxyl oligomeric solution (cellobiose) and lyophilized (rapid freezing at a very low temperature followed by rapid dehydration) for 24 h. Calcium phosphate (brushite) is prepared by streaming equal volumes of calcium chloride ($CaCl_2$) and sodium phosphate (Na_2PO_4) against each other. The immediately formed precipitate is sonicated and washed with distilled water. To inactivate the highly reactive ceramic surface, arrest further particulate growth, and provide an hydroxy-rich surface, pyridoxal-5-phosphate (PLP) is added for certain drug delivery applications. The mixture is then lyophilized overnight, washed, and sonicated again. A layer of biochemically active molecules is allowed to adsorb to the surface-modified ceramic cores, and the molecular transport assembly is thus formed over a 25-h period.

Present and future applications. For every biochemically reactive pair of compounds there seems to be an application for molecular transport assembly. This technology facilitates stable delivery while preventing compound deactivation, and may be able to deliver biologically active agents in a viable state, more efficiently, to more places in the body than any other delivery method. Moreover, specific cellular targeting arising from the preservation of surface antigens has been observed.

Blood. The ability to produce a universal blood substitute would be invaluable in times of war, disaster, and medical emergency. By serving as a carrier for hemoglobin, the molecular transport assembly could offer a type-free and disease-free artificial blood with long-term shelf life. Laboratory prototypes have been fabricated, and commercial production is under development.

This technology preserves the shape and function of biochemically active molecules such as hemoglobin so that they can be delivered into the body safely and effectively. As in all molecular transport assembly applications, the core is made of inert carbon ceramic nanoparticles or calcium-phosphate dihydrate particles. The surrounding layer is a sticky nanometer-thick cellobiose coating. This proprietary coating glues the biochemically active molecules onto the assembly and prevents them from changing shape. Finally, the biochemically active hemoglobin molecules form a third, outer layer over the core. The molecules retain their natural size and shape, and, consequently, their biochemical function.

Drug delivery. The ability to preserve the integrity of the therapeutic agent means efficient delivery of a wide range of pharmacological products with decreased and reduced side effects. Currently in the research stage is the application of molecular transport assembly to enable the prescription of formerly toxic pharmaceuticals such as chemotherapeutic agents derived from taxines. Insulin is an example of a drug that has a relatively short half-life and poor shelf stability, two conditions that can be improved with the molecular transport assembly. When insulin was bound to the molecular transport assembly it showed comparable serum glucose reduction activity but for a prolonged period of time. Moreover, the construct with bound insulin was stable for 7 months, while unbound insulin lost activity rapidly after 2 months.

Vaccines. The molecular transport assembly can be used as an effective viral decoy, delivering noninfectious viral material into the body to elicit an immune response against deadly viruses. The process involves removing the infective genetic core of a virus, replacing it with ceramic nanoparticles, and allowing the outer viral proteins to self-assemble. The result is a vaccine that has the same shape and size as a virus but that cannot reproduce or cause disease (see **illus.**). The simplicity of the chemistry means that this technique is applicable to virtually any virus for which the body can raise antibodies. Indeed, preliminary data indicate that this vehicle system will work even with human immunodeficiency virus (HIV) and simian immunodeficiency virus (SIV).

With the advent of genetic mapping, more and more deoxyribonucleic acid (DNA) sequences are becoming known. Molecular transport assembly may be able to deliver specific genetic material to a cell; when the cell reproduces, the genetic material will be propagated. Through the body's own cell division systems, the new genetic material has the power to repair at an exponential rate. Thus, genetic material delivery could cure both congeni-

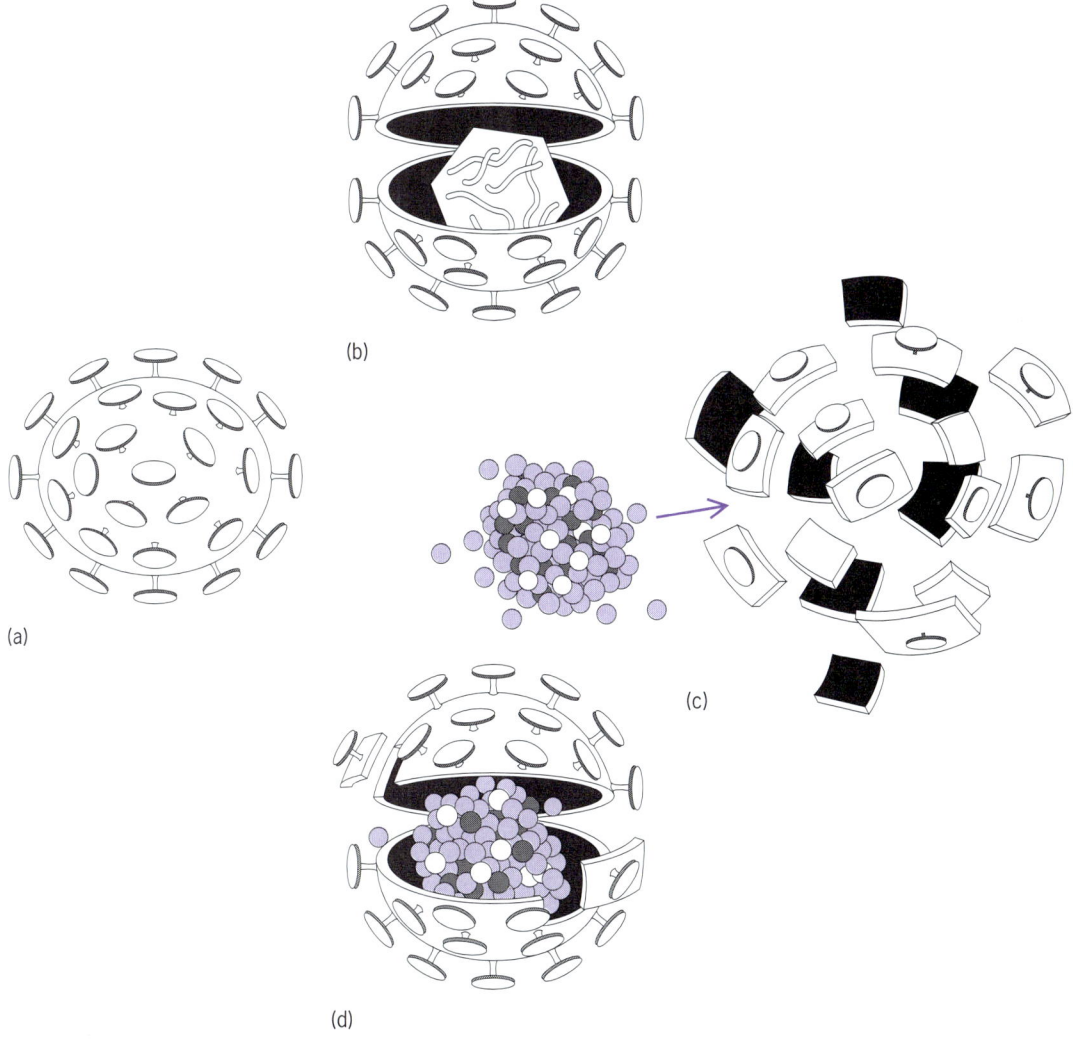

Production cycle of a decoy virus (vaccine) utilizing the molecular transport assembly technology. (*a*) A virion is isolated. (*b*) The virus is opened with mild detergents and the genetic material is removed. (*c*) The carbon particulate core (diamond), modified with a carbohydrate coating, is introduced to replace the virion core. (*d*) The virus's outer shell reassembles around a new core, resulting in a viral decoy.

tal and acquired diseases such as cancer, cystic fibrosis, diabetes, and high blood pressure.

For background information SEE ACQUIRED IMMUNE DEFICIENCY SYNDROME (AIDS); BIOLOGICAL SPECIFICITY; BIOTECHNOLOGY; GENETIC ENGINEERING; MOLECULAR BIOLOGY in the McGraw-Hill Encyclopedia of Science & Technology.

Nir Kossovsky; Andrew Gelman; Edward Sponsler

Bibliography. J. N. Israeliachvilli, *Intermolecular and Surface Forces*, 1985; N. Kossovsky et al., Conformationally stabilizing self-assembling nanostructured delivery vehicles for biochemically reactive pairs, *J. Nanostr. Mat.*, 5(2):233–247, 1995; N. Kossovsky et al., Self-assembling nanostructures, *Bio/Technology*, 11:1534–1536, 1993; G. M. Whitesides, J. P. Mathias, and C. T. Seto, Molecular self-assembly and nanochemistry, *Science*, 254:1312, 1991.

Moon

The United States returned to the Moon in 1994 for the first time since the end of the Apollo lunar missions. The *Clementine I* satellite, a joint mission of the Department of Defense (DoD) and the National Aeronautics and Space Administration (NASA), was launched on January 25, entered lunar orbit on February 19, and spent 71 days mapping the Moon's 38×10^6 km^2 (15×10^6 mi^2) in 11 colors in the visible and near-infrared spectrum. The spacecraft also collected thousands of high-resolution and thermal images, charted the lunar topography with a laser-ranging experiment, improved understanding of the surface gravity field through radio tracking, and carried a charged-particle telescope to characterize the solar and magnetospheric particle environment. Scientists were able for the first time to acquire global, multispectral image data for an entire planetary body. The satellite collected, in total, more than 1.8×10^6 images of the Moon, making it the most completely mapped of any celestial body (including Earth). The collected data set will prove very important in understanding the Moon's evolution over time.

Satellite. *Clementine I* was designed, operated, and built by the U.S. Naval Research Laboratory. It carried sensors, attitude control systems, and software supplied by Lawrence Livermore National Laboratory (**Fig. 1**). The spacecraft's mission was sponsored by DoD's Ballistic Missile Defense Organization as part of a program to test lighter-weight, more power- and cost-efficient space components and systems. It was built in only 22 months.

Seven primary sensors were carried by the *Clementine* spacecraft to cover the spectrum from the visible to infrared regions: ultraviolet-visible, near-infrared, long-wavelength, and high-resolution cameras; two wide-field star-tracker cameras; and a laser ranging system capable of making range measurements up to 640 km (400 mi) with a resolution of 40 m (130 ft).

Fig. 1. *Clementine* spacecraft being prepared for testing at Naval Research Laboratory, Washington, D.C.

Lunar shape and internal structure. Data acquired by *Clementine* have produced a topographic lunar model that gives the first reliable global characterization of surface heights. Combining data on lunar topography and gravity, the model shows that the Moon exhibits a dynamic range of about 16 km (10 mi). This range is approximately 30% greater than earlier estimates obtained from Apollo measurements. The difference in range results from more complete coverage of the lunar surface by *Clementine*. The greatest variations in lunar topography are on the lunar far side over a broad latitude band. Earlier measurements were limited to equatorial swaths taken with Apollo laser altimeters.

The most pronounced topographic feature on the Moon is the South Pole–Aitken Basin (**Fig. 2**). The basin is approximately 2500 km (1500 mi) in diameter (with a central depression of approximately 2000 km or 1200 mi in diameter) and extends up to 10 km (6 mi) in depth. The initial cavity of excavation for such a basin must have been at least 1000 km (600 mi) in diameter and would have quarried material from depths as great as 120 km (75 mi). Thus, the rocks that make up the basin's floor likely contain materials from the lunar mantle. South Pole–Aitken is the largest and deepest impact basin yet discovered in the solar system.

The south pole of the Moon lies in perpetual shadow and could function as a cold trap, rising possibly only 40°C (72°F) above absolute zero. Under these conditions, the pole might serve as a collection point for water ice, the residue from many millions of years of comet impacts. Use of bistatic radar, an adaptation of *Clementine*'s com-

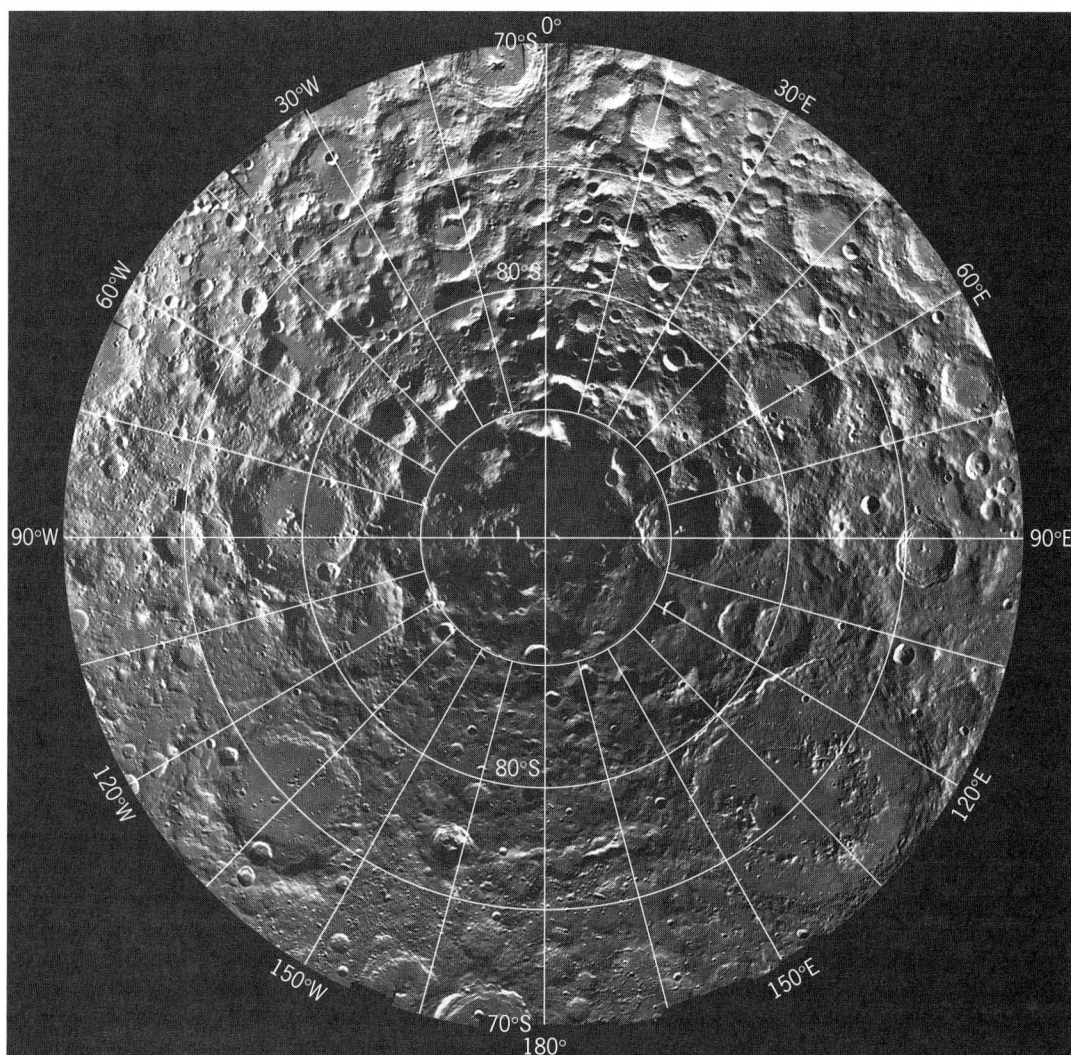

Fig. 2. Mosaic of 1500 *Clementine* images, taken through a red filter, of south polar region. (*Naval Research Laboratory*)

munications antenna made while the mission was in process, indicates that there is some evidence for enhanced radar reflection from the region, a possible indication of the presence of water ice. However, the data are still inconclusive.

Gravitational data on the Moon were determined from velocity perturbations of *Clementine* as measured through the Doppler shift of its radio tracking signal. Mathematical manipulation and correction of the data allowed gravity mapping at the lunar surface. Gravity highs, or mascons, that correlate with the major near-side basins are clearly evident. The highland terrains appear, as in earlier determinations, to be in a state of isostatic compensation. However, the lunar basins display a broad range of compensation states. Thus, the strength of the lunar lithosphere has likely been marked by significant spatial variability since the major basins formed. A complex lunar thermal history is implied.

The simplest interpretation of topographic and gravity observations is that subsurface mass differences are solely a consequence of crustal thickness variations. Under this interpretation, the mapping shows a minimum crustal thickness of 4 km (2.5 mi) beneath the Orientale Basin. Other areas of thin crust include the Crisium (10 km or 6 mi), Smythii (15 km or 9 mi), and South Pole–Aitken (20 km or 12 mi) basins. Lunar basins as a whole show varying degrees of crustal thinning, most likely the results of excavation and mantle rebound from the impact process. The thickest crust (106 km or 66 mi) is on the far side in the vicinity of the Korolev Basin, also the area of highest topography.

Lunar topography, as shown by the *Clementine* studies, appears to be supported by multiple compensation mechanisms. Highland crust produced by a lunar magma ocean may be supported by deep mantle density variations, some large basins may be compensated by shallower crustal thickness variations, and other basins may be uncompensated. Topographical and gravimetric studies, therefore, have demonstrated a much more complex lunar structure and history than hitherto realized.

Surface features. Data from *Clementine* have greatly extended knowledge of the Moon's surface

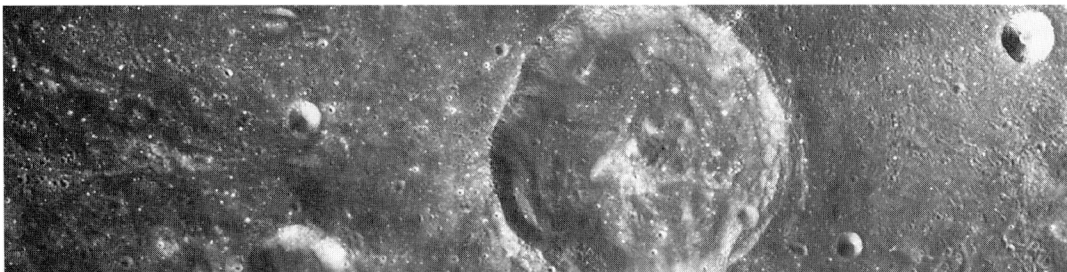

Fig. 3. Rydberg crater from an altitude of 485 km (300 mi). (*Naval Research Laboratory*)

compositional features. Laser altimetric observations have detected a population of ancient, multiring basins in a variety of sizes, morphologies, and states of preservation. These rings make up the basic geological and structural feature of the lunar crust. The presence and form of newer basins, such as that of Orientale, have long been known. However, the new data have provided extensive information on lesser-known structures, such as the Mendel-Rydberg (**Fig. 3**) and Coulomb-Sarton basins. The laser profile of Mendel-Rydberg clearly shows the basin depression, averaging about 5 km (3 mi) in depth. The rings are marked as inflections in the relief profile. The basin appears to be partially filled by mare lavas.

Additionally, *Clementine* observations have confirmed the existence of a number of previously undetected ancient and degraded multiring features. Two probable such cases are depressions northeast of Mare Moscoviense, approximately 330 km (205 mi) and 450 km (280 mi) in diameter. Confirmation that a large population of such basins exist in the lunar highlands strengthens arguments that the highlands are filled with craters of very large diameter.

Clementine sensors also provided important new information on lunar impact craters. Impact cratering is a fundamental geological process throughout the solar system. Its influence is especially clear on the Moon where other environmental processes are not present to obscure its effect. Multispectral images of the Copernicus, Tycho, and Giordano Bruno craters, among others, have presented new information about target stratigraphy and how preimpact materials were excavated, melted, mixed, and deposited.

In Copernicus, impact melt is readily detected in large sections of the floor and in smaller sections along the walls, indicating a nonuniform distribution. Lithic materials in the wall of the crater maintained much or most of their distinctive properties despite the huge energies involved in crater formation. Evidence indicates an enhanced concentration of iron-bearing materials along the crater's southern wall. Basaltic materials (magnesium- and iron-rich rocks), along with more common feldspathic (aluminum- and calcium-rich) crustal materials excavated from the crater, appear to have been incorporated into the wall. The impact excavated a thin layer of basalt overlying breccias, which in turn overlay more ancient crustal materials. Some of the materials were melted and recognizable mixtures were produced. Excavated and uplifted materials in the crater remained within approximately 30 km (19 mi) of their origin.

New insights were also gained into the composition and geological history of the Aristarchus crater and plateau. Altimetry shows the plateau dipping about 1° to the north-northwest and rising about 2 km (1.25 mi) above the surrounding lavas of Oceanus Procellarum to the south. Dark reddish pyroclastic (volcanic) glass is seen to cover most of the plateau to depths of 10–30 m (30–100 ft). The latter is evidenced from the 100–500-m (300–1500-ft) diameter craters that have exposed materials below the pyroclastics. These craters also show that mare lavas are much more common on the plateau than previously thought. The eastern walls of Aristarchus crater reveal olivine-rich lavas, and its central peak includes 2-km (1.25-mi) outcrops of anorthosite. The latter could be either a result of local intrusive magmatism or a remnant of the ferroan anorthosite crust that formed over the global magma ocean of earliest lunar history.

For background information *SEE* ISOSTASY; MOON; SPACE PROBE in the McGraw-Hill Encyclopedia of Science & Technology.

David K. van Keuren

Bibliography. Special section on the *Clementine* mission, *Science*, 266:1835–1862, December 16, 1994.

Muscular system

The skeletal muscles of vertebrates contain active force-generating cells (muscle fibers or myofibers), motor nerves that activate from one to several hundred myofibers, and elastic elements, which are able to store and return mechanical energy with little or no loss, in the manner of a spring. Some of the elastic elements are within the muscle fibers. Indeed, in active myofibers, the contractile myofilaments themselves behave elastically. Other elastic elements are external to the myofibers, and consist of the muscle-associated connective tissues. These connective tissues compose the tendons and aponeuroses by which the muscles are connected to bones: the epimysium, which surrounds an entire muscle; the perimysium, which surrounds the bundles (fascicles) of myofibers; and the endomysium, which surrounds individual myofibers. Collagen, the most abundant structural protein in animals, is the principal organic constituent of these connective tissues. Recent studies have pointed to some unexpected roles of collagen and passive muscle fibers in normally functioning vertebrate skeletal muscles, and have raised some complex questions concerning the neural regulation of certain muscles.

Eccentric contractions. The lengthening of active skeletal muscle fibers occurs when the external load exceeds the contractile force. Such so-called eccentric contractions are used to decelerate bodies in motion. For example, the eccentric contractions of the human quadriceps muscles at the front of the thighs brake during every step during a walk downhill or downstairs. In a braking contraction, the motor nerves activate a sufficient number of myofibers to slow the motion. In this case, the muscles serve to reduce the potential or kinetic energy of the body without using the elastic capacity of the myofibers. In other activities (for example, jumping) enough myofibers may be activated to prevent more than a very small muscle stretch, and the kinetic energy taken from the body as the foot is set down may be stored as so-called strain energy and subsequently returned as an elastic recoil. This stored energy can be added to the energy produced by a shortening contraction to increase the total energy available to accelerate the body.

Strain energy. Stretching a tendon by a few percent of its unstretched length also stores strain energy that can be recovered to increase the acceleration of the body. It has been calculated that the relative contribution of a muscle and its tendons to the total strain energy is directly proportional to the relative lengths of the muscle and tendons. A striking example of the use of energy stored in this way is the hop of the kangaroo. An important consequence (for the kangaroo) is that the use of strain energy, which has very low metabolic cost, greatly decreases the cost of locomotion at high speeds. Similar cost reductions have been estimated for other animals as well.

Aponeuroses. Most myofibers do not attach directly to a tendon but rather to intramuscular or surface expansions of tendons called aponeuroses. Aponeuroses are broad sheets of collagenous tissue to which muscle fibers attach at some angle. The number of aponeurotic sheets determines the overall architecture of the muscle, including the number of angular directions taken by myofibers. Unipennate muscles have two sheets, (one on each surface) connected by myofibers, which have approximately the same angular orientation to the aponeuroses. Bipennate muscles have three sheets, and multipennate muscles have more than three aponeurotic sheets, and correspondingly more layers of muscle fibers with at least two different angular orientations.

Geometric models. Recently developed mathematical models account for the curved surfaces of skeletal muscles as well as the curvilinear trajectories of individual myofibers by requiring that the models have no unbalanced torques or pressures. These new models have been able to predict with fair accuracy the overall shape of real muscles, including the human gastrocnemius (a large muscle of the leg), as well as the pressure distributions within them. Some architectures of muscle tissue are predisposed to produce regional intramuscular pressures that can greatly exceed capillary pressure, thus potentially leading to local transient obstruction of the blood supply. The relationships between these model predictions and muscle cramps and pains have yet to be firmly established.

The new geometric models of muscle shape have explicitly incorporated the generally accepted assumptions that, at physiologic working lengths, the perimysium and endomysium of muscles contribute nothing to their force outputs, and that the myofibers have a constant cross-sectional area along their lengths, are the same length as the muscle fascicle, and are free to slide with respect to their nearest neighbors. The assumption concerning the perimysium is probably correct. It has been shown that the perimysium consists of wavy collagen fibers (a fiber being a bundle of collagen fibrils) arranged in a crossed helical pattern. That is, the fibers form right-handed and left-handed helices. At rest length, these fibers are positioned approximately 55° to the long axis of the myofibers. Only when the muscle has been extended to the extreme of its physiological length do the collagen fibers become straightened and sufficiently aligned with the myofiber axis to contribute a significant resistance to stretch. It is possible that the perimysium functions to prevent overstretching of the muscle.

For many skeletal muscles, however, the other three assumptions are invalid. Numerous skeletal muscles in vertebrate animals consist of muscle fibers that are shorter than the fascicles of which they are a part and that end by tapering within the endomysium. For example, the pectoralis muscle of the quail is mostly composed of fibers that are much shorter than the fascicles to which they belong, have single neuromuscular junctions located at approximately each fiber center, and taper at both ends (**Fig. 1**). Muscle fibers that end by tapering within the endomysium transmit force across their cell surfaces, through the endomysium, to adjacent muscle fibers. The cell surface and the endomysium therefore transmit shearing loads between muscle fibers. The tapered shape of the muscle fibers eliminates stress concentrations at the fiber surface, whereas the curvilinear weave of collagen fibrils within the endomysium is suited to the transmission of shearing loads.

It is possible that the rearrangement of endomysial collagen fibrils that occurs as muscles change length does not cause significant changes in the shear properties of the endomysium. The endomysium thus serves as an adhesive between adjacent

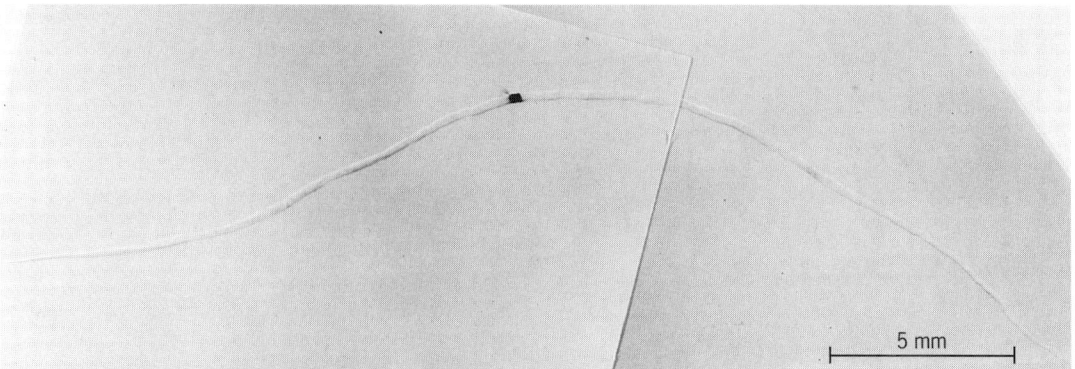

Fig. 1. Light micrograph of a muscle fiber isolated from the pectoralis muscle of a quail. The muscle fiber is about 1 in. (3 cm) long, and the dark spot in the middle represents the position of the neuromuscular junction. (*From J. A. Trotter et al., The composite structure of quail pectoralis muscle, J. Morphol., 212:27–35, 1992*)

muscle fibers at all muscle lengths. This inference can be made by comparing scanning electron micrographs of the endomysium with similar images of the endoneurium, the collagenous layer surrounding individual nerve fibers (**Fig. 2**). Whereas each nerve fiber has its own separate collagenous layer, adjacent muscle fibers share the collagenous layer between them.

Shear properties. Two further complications in the structure of skeletal muscles composed of fibers arranged in series arise from the findings that the muscle fibers that are innervated by the same motor nerve (that is, they are within a single motor unit) are not necessarily adjacent to one another in any cross-sectional plane through the muscle and that they cover varying cross-sectional areas at different planes along the muscle. The lateral nonadjacency implies that the force detected at muscle ends when a single motor nerve is stimulated must have been transmitted in part through shearing loads on passive muscle fibers. The longitudinal inequality implies that the force detected under similar conditions must have been transmitted in part through tensile loads on passive muscle fibers, the connective tissues associated with them, or both.

Recent studies on skeletal muscle have thus shown that the shear properties of the endomysium and passive muscle fibers are important in the normal performance of in-series muscle fibers and that the tensile properties of passive muscle may also be important. No studies have yet been published on the shear properties of either endomysium or skeletal muscle fibers. The complex lateral and longitudinal arrangements of relatively short muscle fibers in series-fibered muscles (muscles in which individual muscle fibers terminate at one or both ends within the endomysium) raise equally complex questions about the nervous control in such muscles. The outcomes of the types of studies suggested will be significant in the design of artificial control systems for pathologically denervated muscles.

For background *SEE MUSCLE; MUSCULAR SYSTEM* in the McGraw-Hill Encyclopedia of Science & Technology.

John A. Trotter

Bibliography. R. McN. Alexander, *Elastic Mechanisms in Animal Movement*, 1988; D. W. L. Hukins (ed.), *Connective Tissue Matrix*, vol. 2, 1990; T. A. McMahon, *Muscles, Reflexes and Locomotion*, 1984; J. A. Trotter, Functional morphology of force

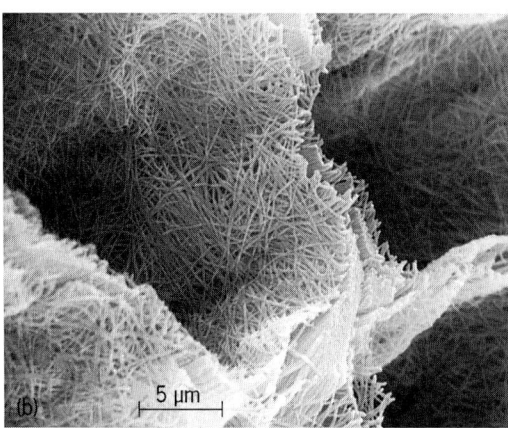

Fig. 2. Scanning electron micrographs of muscle and nerve fibers. (*a*) Endomysium in a bovine sternomandibularis muscle (from which the muscle fibers had been chemically removed, producing the hollow tubes of endomysium), showing the curvilinear weave of endomysial collagen fibrils and the absence of two separate endomysial layers at the cut edge toward the right (*after J. A. Trotter, F. J. R. Richmond, and P. P. Purslow, Functional morphology and motor control of series fibered muscles, Exer. Sport Sci. Rev., 23:167–213, 1995*). (*b*) Nerve from which nerve fibers have been chemically removed, producing a hollow and revealing the separate endoneurial layers of collagen fibrils associated with each nerve fiber.

transmission in skeletal muscle, *Acta Anat.*, 146:205–222, 1993; J. L. Van Leeuwen and C. W. Spoor, Modelling mechanically stable muscle architectures, *Phil. Trans. Roy. Soc. Lond. B*, 336:275–292, 1992; J. L. Van Leeuwen and C. W. Spoor, Modelling the pressure and force equilibrium in unipennate muscles with in-line tendons, *Phil. Trans. Roy. Soc. Lond. B*, 342:321–333, 1993.

Neandertals

Neandertal refers to a specific group of fossil humans who were first recognized in the midnineteenth century, when a partial skeleton was unearthed from a cave in the Neander Valley, near Düsseldorf, Germany. Since their initial discovery, the fossil record of Neandertals has expanded dramatically and currently includes thousands of bones representing almost 300 individuals, as well as a rich archeological record. This substantial pool of data is used by scientists to reconstruct patterns of Neandertal anatomy and behavior, and to examine their evolutionary relationship to our own species, *Homo sapiens*.

Geographic and geological associations. Neandertals have a restricted geographic distribution which encompasses all of the European continent below the 55th parallel and portions of western Asia and the Middle East (**Fig. 1**). It has long been known that Neandertals date to the Pleistocene Epoch, the geological period punctuated by the recurrent cycles of climatological change known as ice ages. Paleoenvironmental studies have shown that Neandertal sites are associated with the last major glacial cycle, near the end of the Pleistocene Epoch. Because they lived during an ice age, Neandertal habitats in much of Europe and western Asia were characterized by cold temperatures, great aridity, and reduced vegetation. In southern Europe and the Middle East, environmental conditions were more temperate, and it is probably no accident that the majority of sites are concentrated in the southern portions of their geographic range. Establishing a precise temporal framework for Neandertals has proven to be an important but challenging endeavor. Technologies have only recently been developed which permit an exact determination in years of the geological ages of these fossils. There is general agreement that most sites fall into a time range extending from 70,000 to 40,000 years ago, with the latest occurrence about 35,000 years ago. Dates for the earliest fossils are more tenuous, but recent studies suggest that fully developed Neandertals were living at the Middle Eastern site of Tabun 110,000 years ago, and perhaps even earlier in Europe.

Anatomical characteristics. Because their anatomy demonstrates an upright locomotor adaptation, Neandertals are placed in the biological family Hominidae, along with living humans and other extinct species of bipedal primates. The evolutionary history of hominids extends well over 4×10^6 years, and includes many primitive species only distantly related to humans. However, the overwhelming anatomical similarities shared by Neandertals and *H. sapiens* indicate the presence of a close genetic relationship. For example, both groups are characterized by similar patterns of brain asymmetry and large absolute brain size, as measured by cranial volume. Moreover, recent studies show that when brain size is standardized to body size, encephalization (or relative brain expansion), is comparable. Although information on gross brain morphology is suggestive, questions relating to specific cognitive and intellectual abilities, such as language development, need to be addressed in a larger context. Studies based on both the anatomical and cultural record suggest that Neandertal dependence on language, as well as the ability to produce speech, may have been limited.

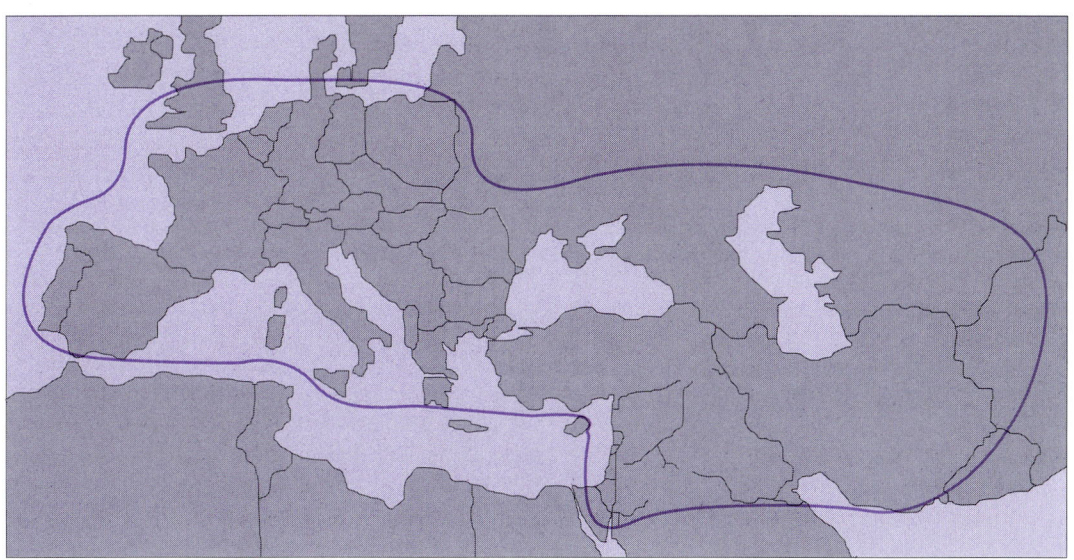

Fig. 1. Geographic range of Neandertals. (*After R. G. Klein, The Human Career: Human Biological and Cultural Origins, University of Chicago Press, 1989*)

Specialized adaptations. Neandertals also display many unique morphological traits which indicate the development of specialized adaptations. For example, their body proportions, which are most similar to modern populations living in arctic regions, probably represent an adaptation to the cold environments of the late Pleistocene. Overall, Neandertals were short and stocky people. Average height was about 5 ft 3 in. (160 cm) for females and 5 ft 6 in. (169 cm) for males, while body weight estimates range between 110 and 175 lb (50 and 80 kg). The short stature of Neandertals was achieved, in part, by reduction of the length of the lower leg and forearm. The shortness of the limbs helped to reduce heat loss from the extremities.

Neandertal skulls are quite distinctive, with a long barrel-shaped braincase, low sloping forehead, well-developed brow ridge, and a bulging protrusion at the back of the head, the occipital bun (**Fig. 2**). The face is long and heavily built, with a forwardly placed jaw, large front teeth, and an inflated, projecting nasal region (midfacial prognathism). Wear patterns on the front teeth indicate they were used habitually as gripping or clamping tools. It has been suggested that the morphology of the jaw and cranium may represent a structural adaptation to the biomechanical forces produced by heavy usage of the anterior dentition. Whether these forces also account for the peculiar projecting midfacial anatomy is more controversial. One alternative proposal for the function of the large nose and adjacent sinus cavities is that these structures helped dissipate heat, generated during periods of intense physical activity, from areas surrounding the brain. The suggestion that Neandertals were very active individuals, who may have required special adaptations to prevent overheating of the brain, is supported by evidence from the postcranial bones, which are heavily built and show large muscle attachment areas. Limb bones also display a high incidence of degenerative joint disease, which in modern young adult humans is often produced by habitual physical stress.

Life-span. Life-history studies suggest that Neandertals did not enjoy a long life-span, with less than 10% of the population living beyond the age of 35. Additionally, analysis of tooth formation and dental eruption patterns indicates that growth rates may have been accelerated, producing early maturation. Because most cognitive development in modern humans occurs during the long juvenile period, this finding, if confirmed, has important implications for the interpretation of Neandertal behavioral complexity.

Sites. While they are famous for inhabiting caves and rock shelters, the archeological record shows that Neandertals also built camps in open locations. Patterns of cultural accumulation indicate that sites were occupied on a temporary, sometimes recurrent, basis, suggesting that Neandertals were probably seminomadic hunter-gatherers. At some open sites ephemeral structures may have been constructed, but there is no evidence for the intensive manipulation of the environment or for the building of the substantial structures which characterize sites associated with modern humans. The presence of simple hearths attests to the use of fire, a virtual necessity of survival for populations living in the glacial habitats of Europe.

Subsistence. From studies of the food remains preserved at sites, it has been suggested that Neandertals lacked the cognitive skills necessary to carry on the complex planning and well-organized subsistence activities associated with modern humans. Plants, although an important resource, would have been available only on a limited basis in the northern latitudes. Large quantities of nonhuman bone, including the remains of deer, cattle, horse, sheep and goats, are present at all sites, suggesting a diet heavily oriented toward meat. However, analysis of faunal accumulations has prompted some researchers to conclude that Neandertal hunting practices were less sophisticated and much less efficient than those of modern humans. It has even been proposed that much of their meat was obtained not from hunting but by scavenging carcasses. Unlike modern humans, Neandertals did not make tools for fishing and fowling activities, and the lack of bones from these animals supports the conclusion that Neandertals were not able to exploit these resources. The overall picture indicates that Neandertals were opportunistic foragers who, of necessity, devoted most of

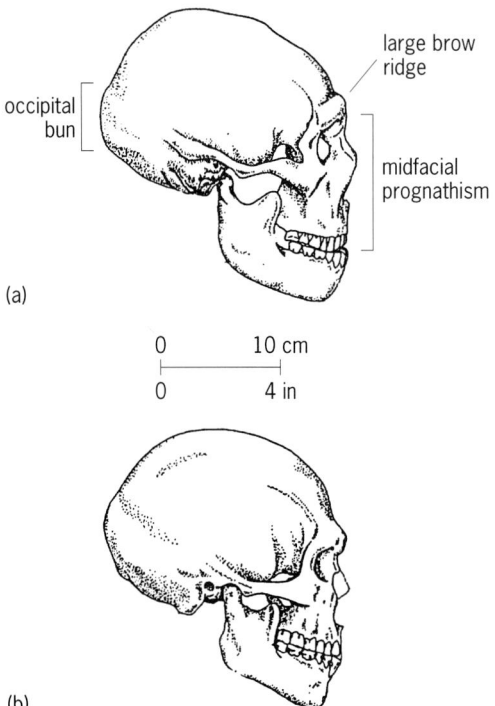

Fig. 2. Reconstructions of a Neandertal skull. (*a*) Skull from Shanidar Cave, Iraq. (*b*) An early modern human skull from Predmosti Cave, central Europe. (*After R. G. Klein, The Human Career: Human Biological and Cultural Origins, University of Chicago Press, 1989*)

their time and energy to filling their subsistence needs.

Artifact assemblages. The artifact assemblages commonly associated with Neandertals are collectively called middle Paleolithic (Mousterian, in Europe). In general, middle Paleolithic industries contain large quantities of flaked stone tools; indeed, one of the major technological achievements of this period involved the development of methods which provided greater control and efficiency in the production of flake tools. Although early modern humans commonly manufactured tools from a wide variety of raw materials, the presence of nonstone tools in middle Paleolithic assemblages is limited to a few wooden spear shafts and pieces of worked bone. Moreover, there is scant evidence for the production of the nonutilitarian objects, such as jewelry, beadwork, animal carvings, and paintings, which distinguish modern sites. Many archeologists associate the appearance of artistic expression and elaborate personal adornment with the development of human consciousness and symbolic belief systems.

Social behavior. While artifactual evidence for complex perception is lacking, there are other indications in the fossil record of humanlike social behavior and self-awareness. Studies of Neandertal pathologies show that some physically disabled individuals, whose survival would have depended on substantial amounts of care from kin or comrades, lived through long periods of incapacitation. Thus, it would appear that Neandertals did not measure human value solely in economic terms, and that strong emotional bonds existed between individuals.

In addition, studies show that, like modern humans, Neandertals practiced deliberate internment of the dead; many archeologists have cited evidence of ritual activity from these graves. Some researchers suggest that evidence of ritual activity is tenuous, and have proposed that Neandertal burials may represent simple, pragmatic acts of household cleanliness.

Origin. The fossil record shows that the initial stages of human evolution occurred in Africa. Between 1 and 2×10^6 years ago, one African species, *H. erectus*, underwent a population expansion and dispersed into broad portions of Europe and Asia. Subsequent events in hominid evolution during the middle Pleistocene (700,000–130,000 years ago) are not well understood, because the hominid fossil record throughout most of the Old World is quite spotty for this period. However, recent discoveries, including the recovery of more than 700 specimens representing the remains of at least 24 individuals from the cave site of Atapuerca in Spain, have significantly improved our understanding of pre-Neandertal human evolution in Europe. Although these fossils, which date to around 300,000 years ago, display many anatomical features typical of *H. erectus*, they also exhibit the development of some distinctive Neandertal traits, such as midfacial prognathism. In combination, the entire sample of European fossil hominids suggests that the Neandertal lineage represents a long sequence of local evolution which began early in the middle Pleistocene, with the arrival of hominids in this region.

Relationship to H. sapiens. Although the geographic origins and early evolution of the Neander-

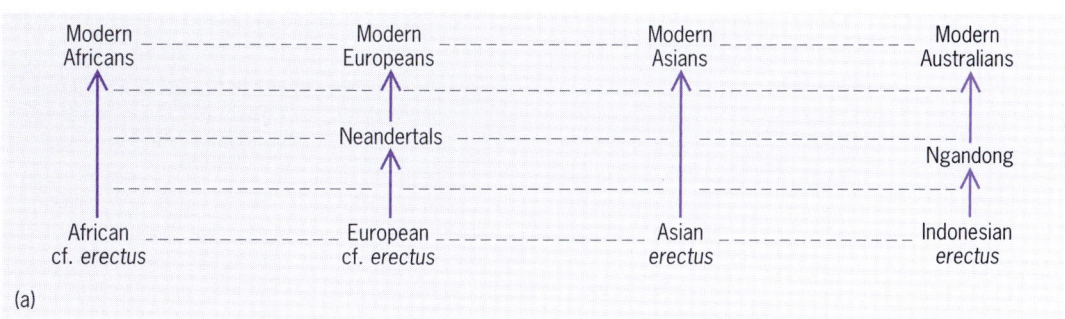

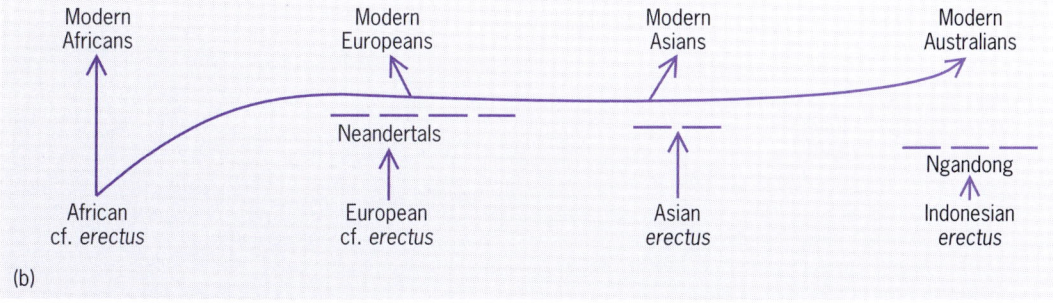

Fig. 3. Contrasting views on the relationship of Neandertals and *Homo sapiens*. (*a*) Multiregional model, postulating that Neandertals are directly ancestral to modern Europeans. (*b*) Single origin model, proposing a replacement of Neandertals by modern populations dispersing from Africa. Vertical arrows show regional evolution through time; broken lines indicate gene flow between geographic populations. (*After C. Stringer and C. Gamble, In Search of the Neandertals, Thames & Hudson, 1993*)

tals appear to have been clarified by the discovery of new fossils, one of the most contentious questions in paleoanthropology, the origin of *H. sapiens*, remains. The range of hypotheses concerning the taxonomic relationship between Neandertals and *H. sapiens* provides a good example of how widely interpretations of the fossil record can vary (**Fig. 3**). Some studies suggest that *H. sapiens* evolution began with the early Pleistocene dispersal event of *H. erectus* and proceeded through a series of gradual stages involving hominid populations on all three continents. Proponents of this multiregional hypothesis argue that Neandertal evolution culminated in the development of modern European and Middle Eastern human populations between 30,000 and 50,000 years ago. Because they feel that Neandertals are ancestral to some modern human peoples, these researchers favor classification of Neandertals as an archaic subspecies, or race, of *H. sapiens*. Other studies indicate that our species originated long after the initial dispersion of *H. erectus*, and evolved from a single geographically restricted hominid group, perhaps a population living in Africa. Proponents of this single origin hypothesis suggest that Neandertals were replaced by expanding populations of modern humans during a second dispersal event that began late in the middle Pleistocene. These researchers believe that Neandertals represent a specialized group of hominids whose genes are not preserved in any living populations. Consequently, they favor classification as a distinct but closely related species, *H. neandertalensis*.

For background information SEE FOSSIL HUMAN; NEANDERTALS in the McGraw-Hill Encyclopedia of Science & Technology.

Suellen C. Gauld

Bibliography. P. Mellors and C. Stringer, *The Human Revolution: Behavioral and Biological Perspectives on the Origins of Modern Humans*, 1989; C. Stringer and C. Gamble, *In Search of the Neandertals*, 1993; E. Trinkaus, *The Shanidar Neandertals*, 1983; E. Trinkaus and P. Shipman, *The Neandertals*, 1993.

Neural network

Many structures, such as bridges and buildings, can develop dangerous defects after being subjected to stresses such as earthquakes. It would be desirable to have an assessment of the safety of these structures without requiring a human inspector to carry out an inch by inch, time-consuming examination. Acoustic monitoring is aimed at accomplishing precisely this task, by sensing the vibrations of the structure and detecting any changes that occur.

Artificial neural networks. Earlier work with acoustic monitoring used statistical pattern-recognition software to make decisions. The work discussed in this article explores the usefulness of a global assessment method using artificial neural networks (ANNs). These networks consist of relatively simple, connected computational units. A weight is associated with each connection. The inputs to a unit are multiplied by the weights and summed, and a number called the bias is added to this total. The result is multiplied by a sigmoidal function, which produces an output between zero and one. In a common artificial neural network configuration the units are arranged in layers, with one input layer, one or more hidden layers, and one output layer. The units in a layer are connected only to the units in the adjacent layers, and there are no connections between units that are not in adjacent layers. This type of network is referred to as a multilayered feedforward artificial neural network.

Artificial neural networks can be trained by adjusting the connection weights until the desired output is generated for a given input pattern. An error function can be defined in terms of the deviations between the calculated and desired outputs. The so-called back-propagation algorithm is often used for this purpose.

The merit of artificial neural networks is that they can be made to approximate any desired input-output relationship. For example, a network might be trained to predict future values of a given acoustical response, using preceding values. If such a network is presented with data generated from a different system, poor prediction of the time series will most likely result. Hence, a monitoring system can be constructed by first training a network to predict the acoustical response of an intact object. Then the learning is stopped while continuing to take data from the same object. If, after some time, the trained network fails to predict the ongoing time series, the conclusion is drawn that the object has undergone a change.

Another way to use the artificial neural network depends on its connection weights. The network is first trained as described above for an intact system. Then, instead of halting the learning, the weights of the network are saved and the training continues. The weights will not be changed very much, provided that the structure being tested remains the same. However, observation of a rela-

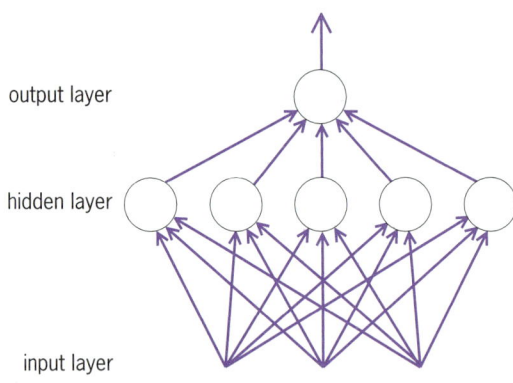

Fig. 1. Architecture of time-delay neural network (TDNN).

tively large change in the weights indicates that a defect has occurred in the system being monitored.

One of the most straightforward approaches to the use of artificial neural networks for predictive analysis is the time-delay neural network (TDNN). This feedforward network uses delayed samples of the time series as inputs. The output of the time-delay neural network is a sample of the time series more recent than the inputs. Thus, the network can be said to predict a future sample of the time series using past and present samples. This approach was used in the application discussed below (**Fig. 1**).

Data collection. Steel beams of rectangular cross section with dimensions of 36 × 2 × 0.75 in. (914.4 × 51.8 × 19.05 mm) were mounted in a test fixture. Vibrations were induced by the impact of a hammer 30 in. (762 mm) from the end of the beam and measured by accelerometers attached 10 in. (254 mm) from the same end of the beam. The hammer was equipped with a transducer to determine the impact level imparted to the structure, this level being used for normalization and calibration purposes. The outputs from the beam and hammer accelerometers were amplified, low-pass filtered to 10 kHz, and sampled at 20 kHz.

Measurements were first made on an intact beam. Then the beam was dismounted, and a 0.1-in (2.5-mm) slot was cut in the cross section at the center of the beam. The beam was then remounted and the measurements repeated. Next the beam was taken out, and the depth of the slot was increased by 0.1 in. (2.5 mm), after which the beam was remounted, and so on. A total of 62 measurements were taken, with slots up to 0.7 in. (17.5 mm) deep. Specifically, four repeat measurements were obtained for each of the slot depths (including 0.0 in., that is, the intact beam). Thirty additional measurements were available for the beam with a 0.4-in. (10-mm) slot, for a total of 34 recordings, which were obtained by mounting and dismounting the beam four times.

The weights were updated after each pass of all the data; this updating is referred to as epoch training. Training was halted after 30,000 epochs.

Fault detection. The trained neural nets were used to determine the similarity of the training and the testing signals (1) by comparing the prediction error of the trained artificial neural network for the training and testing sequences, and (2) by retraining the network on the testing signal, starting with the weights computed during the initial training, and then comparing the weights of the retrained network with the weights saved after the initial training.

The first approach involves freezing the network weights. The signal to be tested is then presented to the network to obtain a predicted signal. The predicted values are compared to the target signal, and an error function called NE is computed. A small NE indicates that the test signal was produced by the same system as originally used to train the network. A large NE indicates that a change in the system has occurred.

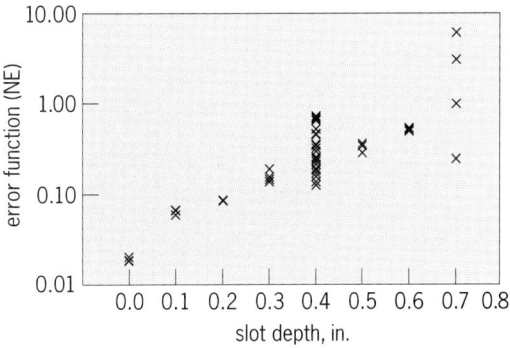

Fig. 2. Values of error function, NE, obtained by comparison of signals predicted from vibration testing of slotted steel beams with target signal from training on intact beams.

In the second approach, a copy of the artificial neural network is made, which forms the starting point for training on a test signal. Once this second training phase has been completed, the difference in weights between the new network and the stored one is computed and used as an indicator of similarity between the signals. The measure used for this comparison is called WDI.

Use of prediction errors. A time-delay neural network with a 10-8-1 configuration was trained on a single acoustic response from the intact beam for 30,000 epochs. The trained networks were presented with data from beams with slot depths cut in 0.1-in. (2.5-mm) increments to 0.7 in. (17.5 mm). The resulting values of NE are plotted in **Fig. 2**. There are more samples for the beam with the 0.4-in. (10-mm) slot than for the other slot depths. These data have a larger variance than the measurements obtained for other slot depths, because the data from the beam with the 0.4-in. (10-mm) slot included data from remounts.

Use of weights. The WDI measure as a function of slot depth is presented in **Fig. 3** for a 10-8-1 time-delay neural network. The networks were first trained on a signal from an intact beam and then retrained on the signal to be tested. Each point in the graphs corresponds to one retrained network. The weights before retraining were the same for all cases.

Figure 3 shows that the WDI values were larger for all the slotted beams than for the intact beam.

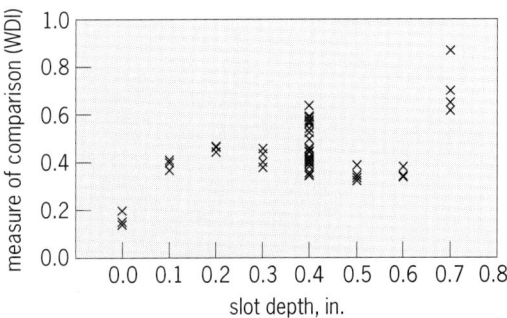

Fig. 3. Values of measure of comparison, WDI, between weights of artificial neural network trained by vibration testing on slotted steel beams and weights of network trained on intact beams.

This suggests that the WDI measure is sensitive to the presence of a defect. However, the trend of WDI is not monotonically increasing with slot depth, and the WDI is consequently not useful to determine the slot depth.

The WDI measure was tested for its discriminating power by performing statistical tests comparing the data from the intact beam to the data from the defective beams. It was found that WDI can discriminate between all the defective and the intact beams at a 99.5% level of confidence.

Results. The NE measure was shown to be a linear function of slot depth. Also, NE for the intact beams was significantly different from all the slotted ones at a 97.5% level of significance. In this respect, the WDI measure performed even better in that it could detect all defects at a 99.5% level of confidence. However, WDI did not monotonically increase with slot depth. Hence, this measure was not specific, that is, it could not be used to estimate the slot depth.

Prospects. An advantage of the approach is that no prior assumptions are made regarding the data. For any given architecture of the artificial neural network there are limitations on which data sets can be successfully modeled, but the general approach is likely to work for a very wide variety of applications. This generality would be especially useful for nonlinear, difficult-to-model systems.

The weight-based detection method is computationally intensive, which may cause problems for real-time applications. However, testing by means of NE can be done rapidly and is suitable for real-time applications.

The network-based approach has shown some merit for the relatively simple system of a single beam. It would be very interesting to apply the approach to a more complex structure, where the built-in nonlinearity of artificial neural networks could prove very useful.

For background information SEE NEURAL NETWORK in the McGraw-Hill Encyclopedia of Science & Technology.

Peeter M. Akerberg; Ben H. Jansen; Robert D. Finch

Bibliography. P. Y. Robin et al., Vibration monitoring of steel beams by evaluation of resonance frequency decay rates, *J. Acoust. Soc. Amer.*, 95: 2029–2037, 1994; D. E. Rummelhart and J. M. McClelland, *Parallel Distributed Processing*, 1990.

Nitric oxide

Although nitrates were used in the middle of the nineteenth century to treat individuals with constricting chest pains, not until a century later was the mechanism of nitrate action understood. Nitrates break down to release nitric oxide (NO), a small messenger molecule that causes blood vessels to dilate. Nitric oxide is also endogenously produced by blood vessels to cause vasodilatation.

Nitric oxide was discovered simultaneously by workers in vascular biology, immunology, and neuroscience. In 1980, it was found that the endothelial cells lining blood vessels released a substance that caused vessels to dilate, and in 1987 this endothelial-derived relaxation factor was identified as nitric oxide. At the same time, it was shown that macrophages killed tumor cells by producing nitric oxide; neurons also were found to produce nitric oxide.

Nitric oxide is an unusual endogenous messenger molecule: it is is small; it is uncharged, so it can diffuse freely across membranes; and it has an unpaired electron, making it a radical species. Nitric oxide is highly reactive with a half-life in biological fluids of about 5 s.

Nitric oxide synthases. Nitric oxide is made by NO synthase (NOS), which converts arginine and molecular oxygen into citrulline and NO. The purification and cloning of NOS reveals a large and complex protein, homologous to cytochrome P450 reductase, with discrete domains that bind cofactors involved in electron transfer. The carboxy terminal of NOS contains binding domains for reduced nicotinamide adenine dinucleotide phosphate (NADPH), flavin adenine dinucleotide (FAD), and flavin mononucleotide (FMN). The center of the molecule contains a binding site for calmodulin. The binding sites for the other cofactors, tetrahydrobiopterin and heme, and for the substrate arginine are not known, but perhaps are located in the amino-terminal portion of NOS.

Three highly homologous isoforms of NOS have been isolated and named after the cells in which they originally were found, neuronal NOS, endothelial NOS, and inducible NOS. The isoforms of NOS can be grouped into two classes, the constitutive isoforms (endothelial and neuronal) that transiently produce small amounts of NO, and an inducible isoform that continuously produces larger amounts of NO. The constitutive isoforms are always expressed in certain cells and are regulated by calcium calmodulin. The inducible isoform is normally absent but can be induced in macrophages and other cells given the correct stimuli; because it binds calmodulin with a high affinity, its activity is independent of calcium. A transient, low-output production from the constitutive isoforms is ideal for signaling between cells. In contrast, a continuous, high output of NO from the inducible isoform might be more suited for killing pathogens.

Physiological actions. The molecular targets of NO are diverse. The best-characterized action of NO is its ability to bind to iron, either in heme or in iron-sulfur clusters. For example, NO activates guanylate cyclase by binding to the heme group of guanylate cyclase, stimulating the production of cyclic guanylate monophosphate (cGMP). It can also induce the adenosine diphosphate (ADP) ribosylation of target proteins by unknown mechanisms. It is possible that NO can also bind to nitrosylate tyrosine and cystine residues of proteins or nucleoside residues of DNA. It has been proposed that NO can combine with the superoxide radical form to form peroxynitrate, which can decompose, yielding the toxic hydroxyl radical form. This result could occur in particular when an excess of NO is produced. Thus, NO can potentially affect a wide range of targets. Small amounts of NO are beneficial to the host, whereas large amounts can be harmful.

Vasodilatation and sepsis. Nitric oxide is the major endogenous vasodilator of mammals. Endothelial cells produce NO at a basal level, causing a continuous and active dilation. When these cells are stimulated by reactive substances, they can produce additional NO, mediating further vasodilatation. Reactive compounds such as bradykinin, acetylcholine, and thrombin raise the intracellular calcium concentration, causing the calcium-calmodulin complex to bind to endothelial NOS and thus activating the synthesis of NO. The NO synthesized by endothelial NOS diffuses out of the endothelial cells into adjacent smooth muscle cells, and then binds to the heme of guanylate cyclase, causing an increase in cyclic guanylate monophosphate, which causes smooth muscle relaxation, and leads to vasodilatation.

The expression of endothelial NOS can be increased by several stimuli, including estrogen and shear stress, and can be repressed by cytokines such as TGF-β. Although reduced NO production occurs in dysfunctional endothelial cells associated with diseases such as atherosclerosis, diabetes, and hypertension, the exact cause is unknown. During sepsis, when bacterial infection overwhelms the host, the NOS isoform is induced in many cells, including the smooth muscle and endothelial cells of the vasculature. The large amount of NO produced causes an excess of vasodilatation, leading to hypotension.

Neurotransmission and strokes. Neuronal NOS, first found in the central nervous system, produces NO which serves as a neurotransmitter. Excitatory amino acids such as glutamate are released from neurons, and bind to glutamate receptors of other neurons, causing an increase in intracellular calcium. This increase activates neuronal NOS to produce NO, which diffuses to other neurons and activates guanylate cyclase. Since NOS is particularly abundant in the cerebellum, NO might be expected to regulate motor coordination. However, mice that lack neuronal NOS (because of the laboratory technique termed genetic knockout) show normal behavior. NO has also been proposed to be a mediator of a form of learning, long-term potentiation (LTP), but the mice homozygously deficient for neuronal NOS also have normal LTP. Thus, the exact role of NO in the brain is unclear.

Although the studies involving neuronal NOS knockout mice have not revealed the normal role of NO, they have been useful in illustrating the harmful effects of overproduction of NO in the brain. During a stroke, an excess of excitatory amino acids such as glutamate are released from neurons, and are believed to stimulate the overproduction of NO from neurons, thus damaging or killing adjacent neurons. But NOS knockout mice are incapable of producing this extra NO, and vascular (ischemic) damage to the brain is reduced in knockout mice compared to normal mice. Thus, NO contributes to the neurotoxicity of a stroke.

NOS-containing neurons are found in certain peripheral nerves, defining a novel nitrergic nervous system. NO mediates the relaxation phase of peristalsis in the gut. The intestine is enervated by nerves containing NOS, and the mice deficient in NOS have large, hypertrophied stomachs, presumably because the smooth muscle relaxation of the gut and its sphincters is impaired. NO is also a mediator of penile erection. The nerves of the pelvic plexus that supply the penis contain NOS, and during erection they release NO which relaxes the smooth muscle of the paired cylindrical vessels that run the length of the penis, causing them to dilate, fill with blood, and produce an erection.

Immune killing. Inducible NOS was discovered in macrophages stimulated by infection, and subsequently it was shown that many cells are capable of expressing inducible NOS. NO can kill a variety of pathogens, including bacteria, fungi, viruses, and parasites. It has been proposed that the main function of inducible NOS is to kill intracellular pathogens, especially viruses and atypical bacteria such as *Mycobacterium tuberculosis*. The precise mechanisms by which NO kills pathogens is not known. It is possible that NO binds to iron-sulfur clusters of glycolytic, citric acid cycle, or respiratory enzymes, and inactivates them.

The transcriptional control of inducible NOS must be tightly regulated: inadequate NO production could lead to death by infection but excess NO could damage the host. Because the inducible NOS is always active once translated, its regulation is primarily transcriptional. Various stimuli, including bacterial products such as lipopolysaccharide, induce intracellular activation of nuclear factor κ B, a transcription factor which binds to a site in the regulatory region upstream of the inducible NOS gene, inducing its transcription. This induction can be amplified by interferon γ binding to the cell, inducing the phosphorylation of the interferon response factor 1, which binds to a deoxyribonucleic acid (DNA) site further upstream from the inducible NOS gene. The mechanism by which other cytokines such as interleukin 1 and tumor necrosis factor α induce NOS expression is unclear. Transforming growth factor β can reduce inducible NOS expression in several ways, by decreasing its transcription, the stability of the message, and the half-life of the protein.

In the short time since NO was discovered, the complex molecule that produces it, its diverse effects, and its physiological roles have been revealed. The unexpected discovery that the body synthesizes a radical molecule that acts as a messenger not only has opened a new field of radical biochemistry but has stimulated the exploration of other, novel messenger molecules such as carbon monoxide and ethylene gas.

For background information SEE ARTERIOSCLEROSIS; CYTOCHROME; CYTOKINE; MUSCULAR SYSTEM; NERVOUS SYSTEM (VERTEBRATE); NEUROBIOLOGY in the McGraw-Hill Encyclopedia of Science & Technology.

Charles J. Lowenstein

Bibliography. C. J. Lowenstein and S. H. Snyder, Nitric oxide: A novel biologic messenger, *Cell*, 70:705–707, 1992; C. Nathan and Q. W. Xie, Nitric oxide synthases: Roles, tolls, and controls, *Cell*, 78:915–918, 1994; M. A. Marletta, Nitric oxide synthase: Aspects concerning structure and catalysis, *Cell*, 78:927–930, 1994; R. M. Palmer, A. G. Ferrige, and S. Moncada, Nitric oxide accounts for the biologic activity of endothelium-derived relaxation factor, *Nature*, 327:524–526, 1987.

Nobel prizes

The Nobel prizes for 1994 included the following awards for scientific disciplines.

Physics. The prize was awarded to Clifford G. Shull of the Massachusetts Institute of Technology and Bertram N. Brockhouse of McMaster University in Hamilton, Ontario, both retired. Shull developed neutron diffraction and Brockhouse developed slow neutron spectroscopy (or, simply, neutron spectroscopy), methods that use neutron beams to study the atomic structure and dynamics of solids and liquids.

In 1946, Shull began working with neutrons from the research reactor in Oak Ridge, Tennessee, in a group headed by E. O. Wollan. According to quantum theory, neutrons have wavelike properties and therefore, like x-rays, undergo diffraction when they interact with matter, scattering in particular directions in patterns that depend on the arrangement of the atoms. However, neutron diffraction studies require monoenergetic beams of neutrons, whereas nuclear reactors generate neutrons with a broad range of energies. Shull and his colleagues obtained monoenergetic neutron beams by passing reactor neutrons through crystals, which acted as diffraction gratings. They then deduced atomic structures from the diffraction patterns formed when the neutrons interacted with matter. They showed that neutron diffraction could be used to distinguish elements (such as hydrogen), and even isotopes of the same element (such as deuterium), that are indistinguishable to x-rays. Based on the neutron's magnetic moment, they also developed a technique to study the arrangement of magnetic atoms in a sample, allowing the study of magnets on an atomic level.

While Shull and his colleagues studied the elastic scattering of neutrons, Brockhouse, who began working at the Chalk River research reactor in Canada in 1950, studied inelastic scattering, in which the neutrons either gain energy from or give up energy to the motions of the atoms in the sample. Brockhouse invented the triple-axis spectrometer, which used crystal diffraction to measure the energies of the scattered neutrons. Thus, he was able to study the motions of atoms in matter, that is, their dynamics, just as neutron diffraction studies atomic structure. In particular, he measured energy spectra of phonons, the quanta of atomic vibrations in a crystal lattice, which help determine a solid's electronic and thermal conductivities and play a central role in superconductivity.

Chemistry. George A. Olah, Director of the Loker Hydrocarbon Research Institute and Professor of Chemistry at the University of Southern California, Los Angeles, was awarded the prize for key chemical research that led to significant advances in the field of hydrocarbon chemistry.

Carbocations are positively charged fragments of hydrocarbon molecules. Prior to the work of Olah and his colleagues, carbocations had been postulated but not characterized, because they were extremely unstable and highly reactive. Olah and his group discovered that stable carbocations could be prepared by treating them with very powerful acidic media known as superacids. The first stable carbocation that they achieved was a trivalent alkyl cation, produced by dissolving alkyl halides in a superacid medium consisting of a mixture of hydrogen fluoride and antimony pentafluoride. As the research evolved, Olah and his colleagues found that they could use similar techniques to produce pentacoordinated carbocations. These reactive fragments, which previously had been too short lived to be investigated in detail, could now be stabilized and characterized with standard spectrometric techniques.

Reactions that in the past were too rapid to study could now be elucidated as a result of Olah's work, leading to the development of new varieties of hydrocarbon synthesis that could be carried out under relatively mild reaction conditions. Among these syntheses were new methods for isomerizing hydrocarbons and synthesizing higher hydrocarbons from methane, leading to the technology for producing gasolines with higher octane ratings. Based on his key work, hydrocarbon chemistry could be carried out with unprecedented precision. Advances derived from Olah's prize-winning work include the development of new fuels and new classes of compounds, for example, high-strength plastics.

Physiology or medicine. Alfred G. Gilman, Professor of Pharmacology at the University of Texas Southwestern Medical Center and biochemist Martin Rodbell, retired since June from the National Institute of Environmental Health Sciences, near Durham, North Carolina, shared the prize for the discovery of G proteins and their role in cellular signaling.

G proteins play a crucial role as intermediaries in the conversion of extracellular signals, such as hormones, drugs, neurotransmitters, and other chemical signals, to a form to which cells can respond and then generate a specific reaction, such as the activation of nerve cells in the retina to relay stimuli to the brain or the feelings of smell and taste.

In the late 1960s and early 1970s, Rodbell and his colleagues at the National Institutes of Health in Bethesda, Maryland, showed that signal transduction through the cell membrane involved not only a receiver, which initially responds to the signal (termed the first messenger), and an amplifier within the cell, which makes a signaling substance (termed the second messenger), but also a third element, mediating between the two, termed a transducer. This molecule in the cell membrane binds to an activator such as the enzyme adenylyl cyclase, thus triggering the synthesis of cyclic adenosine monophosphate (AMP), which reacts to the signal. The function of the transducer was thought to be dependent on the compound guanosine triphosphate (GTP).

In the 1970s, Gilman began working to isolate membrane proteins that synthesize cyclic AMP, and discovered that cell extracts lacking adenylyl cyclase could still produce cyclic AMP, thus proving the transducer's existence. Because it relies on GTP to activate synthesis, it was named G protein (guanine nucleotide–binding protein). G proteins are activated when the receptor binds to it, causing the G protein to break away from guanine diphosphate (GDP) and bind to GTP. This activation stimulates production of the second messenger. After 4–5 s, the G protein deactivates.

Several hundred receptors have been identified, and several types of G proteins, activated by specific receptors, are known. It is possible that several hundred different G proteins exist.

Understanding the role of G proteins is crucial not only for their importance in normal life processes but also because they play a part in disease. For example, mutated and overactive G proteins can cause some types of cancer. The diarrhea that is a symptom of cholera is the result of a toxin released by the cholera bacterium binding to a G protein, thus preventing the intestine from activating to absorb water and salt. G proteins are also implicated in the symptoms of diabetes and alcoholism.

For background information SEE CELL MEMBRANES; NEUTRON DIFFRACTION; ORGANIC CHEMICAL SYNTHESIS; PROTEIN; REACTIVE INTERMEDIATES; REFORMING PROCESSES; SLOW NEUTRON SPECTROSCOPY; SUPERACID in the McGraw-Hill Encyclopedia of Science & Technology.

Nondestructive testing

By some estimates, about 70% of the hundreds of tons of cocaine and other illicit drugs smuggled annually into the United States are transported across land and sea borders in cars, trucks, and containers. Another problem is the possible smuggling of nuclear materials or even entire nuclear weapons across borders. Prevention of the flow of these items is a daunting problem. A novel technique based on nonintrusive inspection employing fast neutrons addresses this hitherto intractable problem, namely, rapid and effective inspections of all containers at ports of entry (border crossings and seaports). This new tool is called pulsed fast neutron analysis (PFNA).

Fast neutron interrogation. Automatic nonintrusive inspection of large objects, such as trucks, requires the full use of all the attributes of nuclear techniques, namely, high sensitivity, high specificity, high speed, and high penetrability. Although thermal neutrons possess most of these attributes, they are not sufficiently penetrating to probe the entire volume of a truck or a shipping container. Fast neutrons, in the energy range of several million electronvolts, are far more penetrating in most materials, and their interaction with each chemical element generally stimulates unique characteristic gamma rays. The detection of these rays identifies the presence of the specific material and determines its quantity. Although high-energy x-rays are also sufficiently penetrating, they can only determine the surface density distribution of the entire object. This information is useful but generally not sufficient to establish the presence of contraband. Fast neutrons, especially those at 14 MeV, which are abundantly produced by electronic neutron generators, have been used for decades for oil logging and exploration. Interrogation with fast neutrons has found unique applications in many other industrial processes.

Fast neutrons are especially important in the detection of organic materials owing to their ability to interact with nuclei of oxygen, carbon, and nitrogen. The first is not detectable at all with thermal neutrons, and the second can only marginally be detected. The nuclear interaction that takes place is called fast neutron inelastic scattering. It generates gamma-ray energies of 6.13 and 4.43 MeV for oxygen and carbon, respectively, and 1.6, 2.3, and 5.1 MeV for nitrogen. Other unique gamma rays are stimulated in iron, silicon, chlorine, and so forth. Since all materials are made of either pure elements or chemical compounds of these elements, the determination of the gamma-ray energy identifies the material. Examples of gamma-ray spectra resulting from fast neutron interactions in an organic solvent (acetone), food (rice, coffee), plastic explosive (C-4), drug (cocaine), glass, and iron are shown in **Fig. 1**. Each material exhibits distinctive features, which enable the PFNA system to detect the presence of drugs, explosives, or other materials, including large household appliances and food, inside other materials.

PFNA system. The PFNA system combines the excellent interrogation capabilities of fast neutrons, that is, sensitivity, specificity, and penetration, with the time-of-flight technique to determine where the neutrons interact. The point or rather the volume where the neutrons interact reveals the presence of

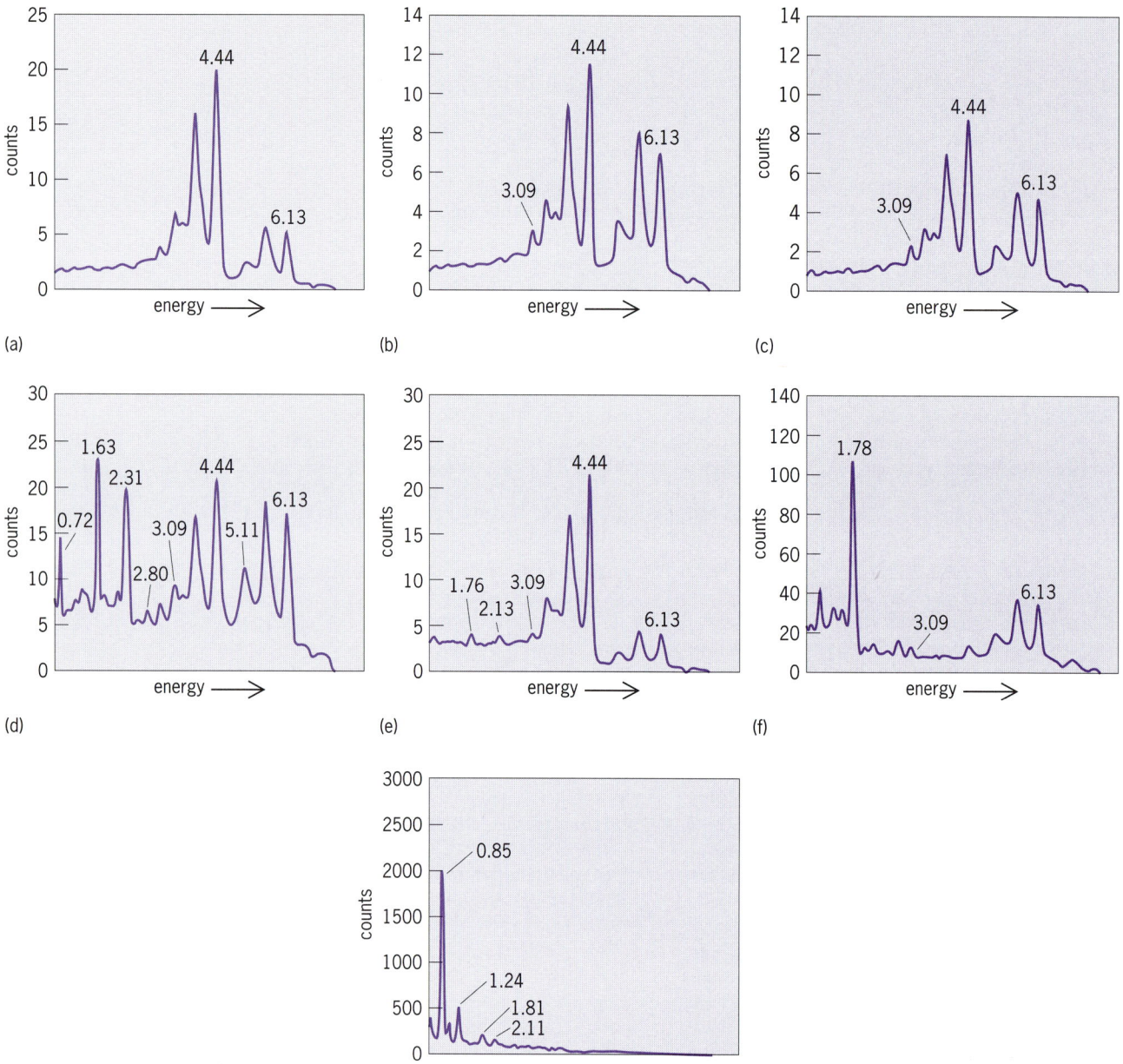

Fig. 1. Gamma-ray spectra of (*a*) acetone, (*b*) rice, (*c*) coffee, (*d*) C-4 (a plastic explosive), (*e*) cocaine, (*f*) silicon dioxide (SiO_2), and (*g*) iron resulting from fast neutrons (8 MeV) interacting with different materials and providing specific signatures. Most of the gamma-ray energy peaks are indicated on each spectrum by their energies in MeV.

the specific material. By scanning the entire object, a complete three-dimensional map of its material content is obtained. This map can be contrasted with high-energy x-ray radiography, which provides a two-dimensional projection (silhouette) not of the materials but of their combined density. Although the latter is generally useful, especially because of its high resolution, typically 0.2×0.2 cm² (0.08×0.08 in.²), it is not sufficient to determine the presence or absence of the specific materials of interest (such as contraband, dutiable items, and hazardous chemicals). In PFNA, the spatial resolution (voxel size) may be as small as $5 \times 5 \times 5$ cm³ ($2 \times 2 \times 2$ in.³). Usually, it is selected to be consistent with the size of the specific target: smallest for explosives in luggage and much larger for smuggled taxable goods.

In the PFNA system (**Fig. 2**), a pulsed beam of monoenergetic high-energy neutrons, typically in the energy range of 7–9 MeV, is generated. The neutrons are generated by the nuclear reaction of accelerated deuterons (ionized heavy-hydrogen atoms) with a target containing deuterium (heavy-hydrogen molecules). The deuterons, as well as the neutrons, are produced in narrow pulses, typically 1-nanosecond long and occurring $2–10 \times 10^6$/s. The neutrons produced tend to concentrate in the forward (that is, the deuteron-beam) direction. They are further collimated to provide as narrow a fan or pencil beam as needed. To inspect an object such as a shipping container, the deuteron and the resulting neutron beam is scanned up and down in a raster pattern (a series of parallel lines) to cover the entire

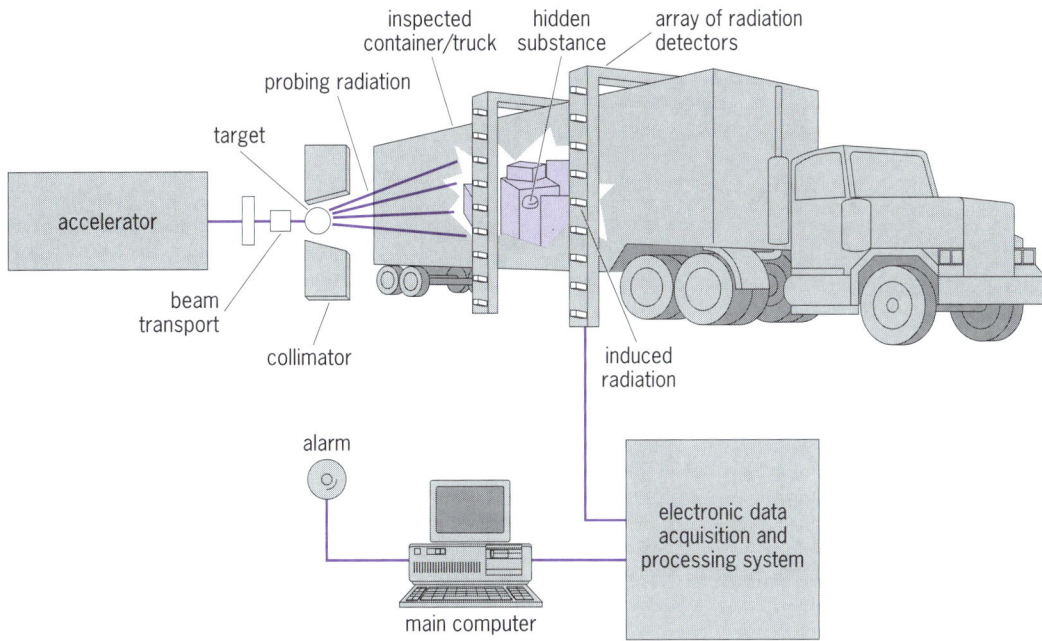

Fig. 2. Pulsed fast neutron analysis (PFNA) container inspection system. A computer carries out diagnostics and makes a decision automatically. (*Science Applications International Corp.*)

height of the object. The object is conveyed through the beam to be fully scanned along its length.

The materials in the path of the neutron beam are interrogated and emit their characteristic gamma rays, which are detected by detectors placed around the container on both sides of the beam line. Since gamma rays travel much faster than neutrons (the speed of light is about seven times higher than that of the neutrons), if a neutron creates a gamma ray along the beam line the ray will arrive almost instantly at the detector and will disclose where the neutron interaction occurs. For example, interactions of neutrons entering the container will show up much earlier than interactions of exiting neutrons identifying materials at the other side of the container. Without the unique time-of-flight imaging feature, the PFNA system could not deliver its promise of detecting the presence of concealed substances or goods inside the large volume of legal goods. Most of the legal goods are generally made of the same elements (though with different chemical compositions) as the illicit substances. Thus, if averaging over large parts of the container were done, the illicit goods would generally be missed. It is the different relative elemental compositions detectable by PFNA in a small volume of elements that make the detection possible. Furthermore, the PFNA system provides direct imaging, since it does not need a mathematical reconstruction process, which is required of x-ray computer-aided tomography (CAT), and is thus simpler.

PFNA operation. The operation of the PFNA system is straightforward. A truck or shipping container is automatically conveyed through a shielded area past the beam of neutrons. As the truck is being scanned, gamma-ray detector data are acquired and processed. By the time the truck exits the interrogation zone or within a short time afterward, three-dimensional images of the detected elements and detection features, calculated from the element data, are displayed for the operator. Locations in the truck which are likely to contain the illicit material are clearly identified through a color code. The operator can note the materials and their location for further action, which may include, for example, removal of the contents of the truck and a manual detailed search.

Depending on the specifics of the installation and the utilization of complementary techniques, such as a high-energy x-ray system, the PFNA inspection can take only a few minutes. The ability of PFNA to detect explosives, drugs, and many goods has been demonstrated in extensive laboratory tests.

For background information SEE ACTIVATION ANALYSIS; GAMMA-RAY DETECTORS; GAMMA RAYS; NONDESTRUCTIVE TESTING; NUCLEAR SPECTRA in the McGraw-Hill Encyclopedia of Science & Technology.

Tsahi Gozani

Bibliography. D. R. Brown et al., Application of pulsed fast neutrons analysis to cargo inspection, *Nucl. Inst. Meth.*, 353:684–688, 1994.

Nuclear fuels

The end of the Cold War and the subsequent collapse of the former Soviet Union at the end of 1991 have resulted in large stockpiles of excess nuclear weapon fissile materials from tens of thousands of missile delivery systems. These stockpiles contain highly enriched uranium and weapons-grade plutonium. The challenge is to devise a strategy for the safe ultimate disposition of the excess fissile materials. Safe disposal of excess plutonium is now under study, and the ultimate disposition strategy for the excess high-enriched uranium is that it will be converted to low-enriched uranium for use as a fuel for civilian nuclear power reactors. Because of its great fuel, and thus financial value, the excess military arsenals of highly enriched uranium are expected to become civilian power reactor nuclear fuel over the next several decades.

Weapons-grade uranium. The primary isotopic constituents of the natural uranium produced as a result of mining uranium ore deposits are ^{238}U and ^{235}U, which are present in 99.283 and 0.711 weight percentages, respectively. Trace amounts of the isotope ^{234}U are also present. In order to be useful as a fuel for United States–type civilian nuclear reactors, the weight percentage of ^{235}U must be increased to between 2 and 5. Weapons-grade uranium requires that the ^{235}U composition be increased to the order of 90 wt% or higher. The process of increasing the ^{235}U content is known as uranium enrichment, and the process of enriching is referred to as performing separative work and is usually measured in kilograms of uranium separative work units (kgU SWU or just SWU). The industrial enriching processes now in use, based on gaseous diffusion of uranium hexafluoride (UF$_6$) and the use of UF$_6$ gas centrifuges, result in a UF$_6$-enriched uranium product and in uranium tails output. The tails may be thought of as depleted uranium (less than 0.7 wt% ^{235}U).

The production of a kilogram of weapons-grade uranium, usually referred to as highly enriched uranium (HEU), requires the expenditure of about 45 times more enrichment separative work than does a kilogram of civilian reactor low-enriched uranium (LEU) fuel. The production of a pound of HEU also requires about 28 times more natural uranium enrichment feed material than does the production of an equal amount of typical LEU. These quantity differences are only approximate since the precise values will depend on the specific ^{235}U assays of the weapons and fuels being compared, as well as the efficiency of the specific enrichment process.

If HEU is converted to LEU, with the exception of nominal conversion process losses all of the original natural uranium is conserved. However, only the equivalent enrichment effort normally needed to produce such LEU is obtained as a result of the weapons material conversion process. Because the United States and several of the republics of the former Soviet Union together are believed to currently have stockpiles of HEU totaling approximately 2000 metric tons (2200 short tons), this material represents the equivalent of a source of natural uranium supply in the amount of approximately 10^9 lb (5×10^8 kg) of U$_3$O$_8$. This amount would be enough to fuel all the nuclear power reactors in the United States for about 25 years or those of the world for about 7 years.

Although the number of years of enrichment separative work represented by the HEU-derived LEU may be generally similar to that for other uranium, the actual worth depends on the ^{235}U concentration of the LEU under consideration.

HEU conversion to LEU. Highly enriched uranium can be converted to LEU by mixing it with either depleted or natural uranium, or with enriched uranium with a lower ^{235}U assay. From the commercial reactor fuel standpoint, the product is a substitute for conventionally produced LEU.

Before actual reduction of the fissile material can be performed, the warheads must be separated from their delivery systems, for example, missiles. The next step is separation of the so-called plutonium pits from the HEU and the secure storage of both of these fissile components. If the fissile HEU metal is to be reduced to LEU, then it must first be melted into shapeless buttons and its alloy composition altered in order to eliminate all classified

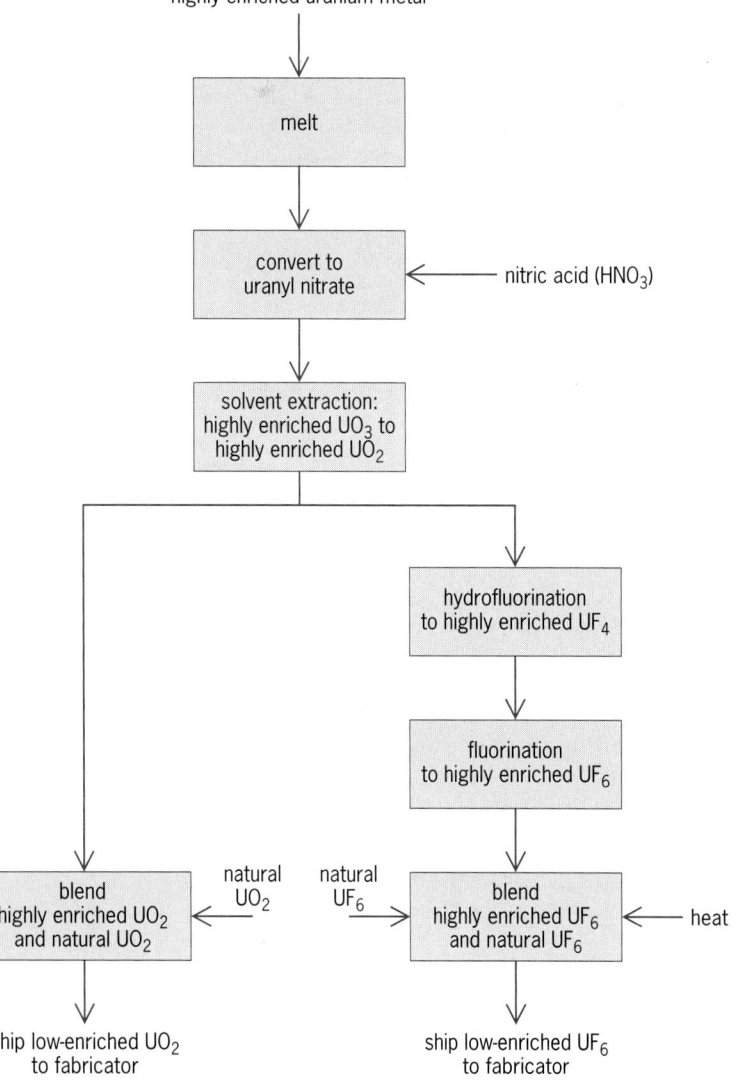

Highly enriched uranium (HEU) metal reduction processes.

information associated with weapons use. Safeguards verification would almost certainly be required at each step in the process so long as weapons-grade materials are involved.

The unique steps of the HEU-metal reduction processes (see **illus.**) consist of the reduction of uranium in metallic form, and purification and the removal of any alloying agents that may be present, followed by blending of the HEU with other uranium to produce LEU. Although not widely performed at this time, the processes are well known to the nuclear industry. Limited facilities may already exist in Russia for performing this task, as they do in the United States. However, in both the United States and Russia, existing weapons materials handling facilities will need to be modified or new facilities built to operate at a larger scale.

As mentioned above, the conversion of HEU to LEU can be achieved by blending it with a diluent material consisting of depleted, natural, or low-enriched uranium. This blending can be done either as solid UO_2 powder or as liquid UF_6. HEU metal can be dissolved in nitric acid and purified through solvent extraction, converted to UO_3, and subsequently reduced to UO_2. The HEU in pure UO_2 powder form may be acceptable for commercial fuel fabricators to blend with conventionally produced UO_2 powder just prior to the fabrication of fuel pellets. However, the fabricator would have to perform an added processing step of powder blending.

The alternative of blending UF_6 to achieve the isotopic content desired has been a standard practice at enrichment plants for decades. An important feature with regard to using HEU as one of the blending components is the wide difference in isotopic content of the ^{235}U and ^{238}U in the two components being blended. Special controls and geometrically safe process vessels are necessary to ensure that the HEU does not segregate in the low-enriched product vessel and result in a critical-mass configuration and associated criticality accident situation.

The choice of the diluent material to use for blending and the chemical form for blending will be made on the basis of the relative cost and availability of the diluent uranium, the cost of the specific processing function, UO_2 fuel-fabrication specifications, and the political logistics involved. The presence of contaminants in the HEU, such as ^{232}U, ^{234}U, ^{236}U, and plutonium, may dictate consideration of higher dilution (higher weight percentage of ^{235}U in the blend) than would otherwise be the case. Since both United States and Russian enrichment facilities were fed reprocessed uranium during the era of HEU weapons production, such contaminants are believed to be present. The ^{232}U and ^{234}U isotopes are sources of high-energy gamma rays and alpha particles respectively, which can present radiological protection issues during material handling. The ^{232}U and ^{236}U isotopes are so-called neutronic poisons, which can affect power reactor performance.

For background information SEE ISOTOPE (STABLE) SEPARATION; NUCLEAR FUEL CYCLE; NUCLEAR FUELS in the McGraw-Hill Encyclopedia of Science & Technology.

Julian J. Steyn

Bibliography. National Academy of Sciences, *Management and Disposition of Excess Weapons Plutonium*, 1994; U.S. Office of Technology Assessment, *Dismantling the Bomb and Managing the Nuclear Materials*, OTA-O-572, 1993.

Nuclear fusion

Fusion is the energy source of the universe and serves to power the stars including the Sun. If the fusion process could be replicated on Earth, it would provide a virtually unlimited supply of safe and environmentally attractive energy.

Fusion process. The basic fusion process consists of combining (fusing) two light nuclei to form a heavier nucleus with less mass, the missing mass being converted to energy. Typical fusion reactions produced in the laboratory are reactions (1)–(3),

$$D + D \longrightarrow {}^3He\ (0.82\ \text{MeV}) + n\ (2.45\ \text{MeV}) \quad (1)$$
$$D + D \longrightarrow T\ (1.01\ \text{MeV}) + H\ (3.02\ \text{MeV}) \quad (2)$$
$$D + T \longrightarrow {}^4He\ (3.6\ \text{MeV}) + n\ (14.1\ \text{MeV}) \quad (3)$$

where H, D, and T represent hydrogen and its isotopes deuterium (D) and tritium (T), n represents the neutron, and energies are expressed in millions of electronvolts (MeV).

The fusion reaction requires that the fusing nuclei, each positively charged, come into close contact. The nuclei can overcome their mutual electrostatic repulsion only if they are moving at each other at high speeds or energies, that is, at high temperatures. Typical fusion fuel temperatures (T) are 10^8 °C. At this temperature, the fusion fuel is a fully ionized gas called a plasma.

In order to produce net power from the fusion process, there must be a sufficient number of fusing particles, expressed as n (fuel ions per cubic meter), and the fusing particles must be confined for a time, τ, that is sufficiently long to produce more fusion energy than the energy required to sustain the plasma at high temperature. For D-T reaction (3), the requirements for a self-sustained fusion reaction are given by notation (4).

$$\begin{aligned} n\tau &\geq 6 \times 10^{20}\ \text{m}^{-3}\ \text{s} \\ T &\geq 10^8\ \text{°C} \end{aligned} \quad (4)$$

Since the Sun is so large, the value of $n\tau$ is correspondingly large, and the temperature required for fusion to occur in the core of the Sun is only 1.5×10^7 °C. Since the 1950s, fusion researchers have tried to develop techniques for achieving the laboratory conditions for fusion specified in notation (4).

Tokamak system. The fusion fuel can be confined by a magnetic field (a magnetic bottle) or by its own inertia. The most successful magnetic bottle is

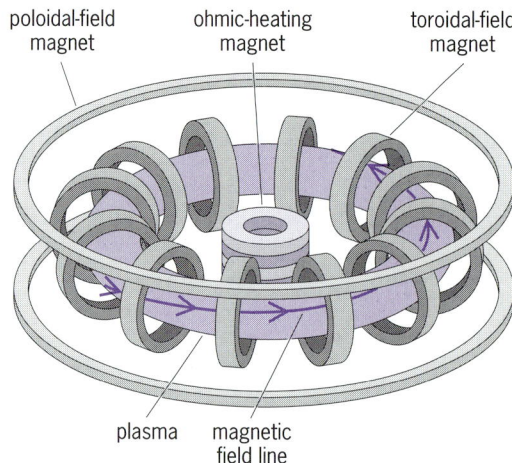

Fig. 1. Tokamak magnetic fusion confinement system.

called a tokamak, a Russian acronym for toroidal magnetic chamber (**Fig. 1**). In this system, the main magnetic field is produced by electric currents in a set of external toroidal field coils. This field, in conjunction with the field produced by a large current (typically 2×10^6 amperes) flowing in the plasma, forms an invisible toroidal or donut-shaped confinement system with spiraling field lines, which provides thermal insulation between the ultrahot fusion fuel and the surroundings. The magnetic field is typically limited by the strength of materials to about 5 tesla, which is 10^5 times stronger than the Earth's magnetic field. The magnetic field produced by the poloidal coils holds the plasma-filled donut in position against the natural forces that try to increase the major radius of the donut. Since the plasma is a very good electrical conductor, varying currents in the ohmic-heating transformer can induce an electrical current of several million amperes in the plasma, which heats the plasma to the required temperature. In magnetic confinement fusion, the fuel density is about 10^{20} particles per cubic meter, which is 10^{-5} the density of ordinary air. Therefore, to satisfy the fusion condition, notation (4), the hot plasma must be confined for about 6 s in order to make net fusion power.

Tokamak development. During the mid-1970s oil embargo, increased concerns about future energy supplies led to the start of construction of several large tokamaks, including the Tokamak Fusion Test Reactor (TFTR) at Princeton University in the United States, the Joint European Torus (JET) near Oxford, England, the JT-60 tokamak near Tokyo, Japan, and the T-15 tokamak in Moscow, Russia. At that time it was possible to produce only 0.1 watts of fusion power in the laboratory by using the fusion process. The goal of the TFTR was to demonstrate fusion energy production from the burning, on a pulsed basis, of deuterium and tritium. In particular, the TFTR was to produce between 10^6 and 10^7 joules of fusion energy per pulse, or, because the pulse length of the TFTR is about 1 s, 10^6–10^7 W of fusion power. This is an extrapolation of 10^8 in power over the output of the tokamaks that were used to design the TFTR.

The TFTR began operation in 1982, and experiments using deuterium fusion fuel, in 1984. From 1984 to 1993, the TFTR systematically increased the fusion plasma performance using deuterium as the fuel and first achieved reactor-fuel temperatures of 2×10^8 °C in 1986. The increase in fusion power of the TFTR and other large tokamaks over two decades has been faster than the growth of computer memory chip density (**Fig. 2**).

The interior of the 2-m-high (6.5-ft) TFTR reaction chamber (**Fig. 3**) is normally evacuated to a very low pressure. During operation, a magnetic field of 5.2 T is imposed around the chamber and a toroidal plasma carrying a current of 2×10^6 A surrounds the central column. The highest fuel temperature exists in the center of the plasma and gradually decreases toward the edge. The tiles on the central column are fabricated from three-dimensional carbon fiber composite material and serve to absorb the energy escaping across the magnetic field.

Deuterium-tritium experiments. In 1994, the first extensive experiments (approximately 250) using equal concentrations of deuterium and tritium were carried out on the TFTR. In 1991, the JET carried out two experiments using 10% tritium fuel concentrations. In the TFTR, the intense beams of deuterium and tritium neutral atoms are injected across the magnetic field and are absorbed in the plasma, providing deuterium-tritium (D/T) fuel for the fusion reaction as well as heating the plasma. The TFTR D/T plasmas have been heated to a record temperature of 5.1×10^8 °C. An important result is that the D/T fuel mix had a 20% better

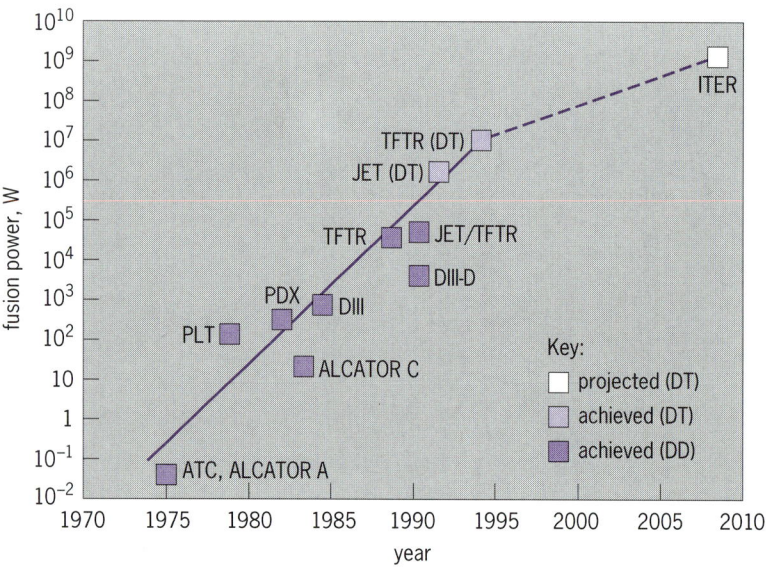

Fig. 2. Fusion power produced in the laboratory as a function of year since the construction of large tokamaks was started in the mid-1970s. PLT = Princeton Large Torus; PDX = Princeton Divertor Experiment; TFTR = Tokamak Fusion Test Reactor; JET = Joint European Torus; ITER = International Thermonuclear Experimental Reactor. ATC and TFTR are at the Princeton Plasma Physics Laboratory; Alcator A and C are at the Massachusetts Institute of Technology; and DIII and DIII-D are General Atomics tokamak experiments.

Fig. 3. Interior of the reaction chamber of the Tokamak Fusion Test Reactor (TFTR).

confinement than the standard D/D plasmas that were used as the test fuel previously. During the D/T experiments, a maximum of 1.07×10^7 W of fusion power was produced by injecting 3.95×10^7 W of neutral beam power. The fusion power density in the center of the TFTR is about 2 MW/m^3, which is near the value expected for fusion reactors of the future. The Sun is the best example of a working fusion reactor, but the fusion power density there is only about 1 W/m^3. Increasing the power density by a factor of 10^6 illustrates the tremendous technical challenge of producing a miniature star on Earth.

The greatest technical interest is directed at the helium nucleus (alpha particle) produced in the fusion reaction (3). The alpha particle is electrically charged and is, therefore, confined by the strong magnetic field. As the energetic alpha particle moves through the plasma, it gradually transfers its energy to the cooler plasma. If there are enough alpha particles, they can provide the energy to sustain the high temperature of the fuel without heat being applied from outside the plasma.

In the TFTR the presence of the alpha particles inside a tokamak has been measured for the first time. The D/T experiments on the TFTR confirm that the alpha particles are confined by the magnetic field as expected. In addition, the electron temperature of the plasma increased by approximately 15%, roughly as expected, from heating by the alpha particles.

After the alpha particles heat the plasma, they should be removed from the plasma or they could accumulate as cold alpha ash and quench the fusion reactions by excluding the fusing D/T fuel from the plasma. One of the TFTR's most important technical results has been to show that natural processes in the plasma tend to remove the cold alpha ash from the plasma core before it has a chance to accumulate. These data are extremely important for the design of future tokamak devices.

Prospects. Data gathered from the TFTR and other tokamaks will provide the information needed to design the next steps in the fusion development program. Because the TFTR device and other contemporary tokamaks use copper coils to produce the magnetic field and the electrical power demands are large, the field can be sustained for only a few seconds. In order to develop the physics and technology for a commercially attractive fusion reactor that will operate continuously, the United States is proposing to build the Tokamak Physics Experiment (TPX), which will use superconducting coils and be able to operate continuously. The TPX will be about the same size as the TFTR and is expected to be operational in the year 2001.

An engineering test reactor is needed to demonstrate the scientific and technological feasibility of fusion at reactor scale. The International Thermonuclear Experimental Reactor (ITER) is being designed as a collaboration among the United States, Europe, Japan, and the Russia Commonwealth of Independent States. If approved for construction, the tokamak device, scheduled to begin operation in 2007, would be capable of generating 1500 MW of power for 15 min. The 150-fold increase in power in the final step from the TFTR to the ITER is much smaller than the increase of

10^8, from 0.1 to 10^7 W, already achieved by the TFTR (Fig. 2).

For background information SEE NUCLEAR FUSION; PLASMA PHYSICS in the McGraw-Hill Encyclopedia of Science & Technology.

Dale M. Meade

Optical computing

Various schemes have been proposed using photons, or light particles, instead of electrons as the carriers of information in computer systems because of the speed at which these photons can travel through space or a material with a low index of refraction. Recently, the world's first fully functional digital optical computer has been demonstrated, fabricated from off-the-shelf telecommunications parts such as semiconductor lasers, fiber optics, and optical waveguide switches called directional couplers. This computer holds its program and data in memory in the form of optical pulses stored in a long loop of optical fiber and manipulates these data using an arithmetic logic unit, an accumulator, an instruction and control register, a program counter, a memory counter, and an address comparator and state controller. All these functions are found also in conventional electronic digital computers.

Analog optical computers. Optical computing for special purposes has been available for a long time. The first optical computing project, initiated in 1953, was an analog device. It was based on the fact that an optical lens will produce the Fourier transform of a spatial filter (an *x-y* pattern of opaque and transparent regions in the aperture plane of **Fig. 1**) when placed in the parallel coherent light input of a lens. Such optical computers are used, for example, to identify objects detected by pulsed radar signals by calculating the correlation coefficients of the radar signal with known patterns from various objects.

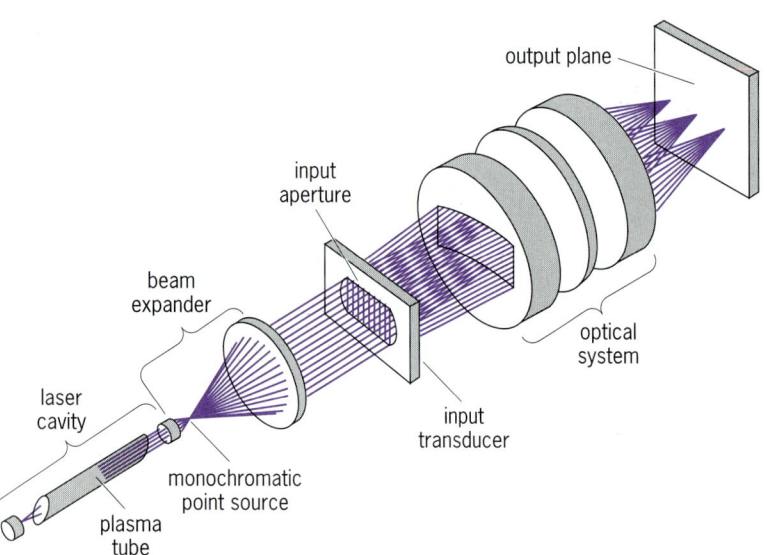

Fig. 1. Analog optical computer. (*After K. Preston, Jr., Coherent Optical Computers, McGraw-Hill, 1972*)

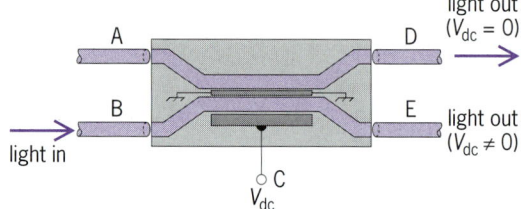

Fig. 2. Integrated optical-fiber electrooptic directional coupler. (*a*) Diagram. (*b*) Interchange of modal optical power between the two guides as a function of propagation distance z under phase-matched conditions. $P_A(z)$ and $P_B(z)$ are the powers in the two guides. (*After D. L. Lee, Electromagnetic Principles of Integrated Optics, John Wiley and Sons, 1986*)

Methods have been proposed to add logic to such a computer by using varying spatial patterns in the aperture plane. These patterns would be generated by a spatial light modulator, a device which can perform this variation by addressing pixels in the aperture plane electrically. In addition, some scheme for storing the logic patterns would be required to produce a digital optical computer. Storage or memory might be in the form of holographic images which can be written to and read from electrooptic crystals, whose optical properties can be modified by the strong electric fields generated by laser light. Such an optical memory has recently been demonstrated, but no completely functional digital optical computer had been demonstrated until the one cited above.

Directional couplers. Digital computers, optical or semiconductor, require switching devices to perform essential boolean logic functions. The minimum required logic functions are AND, OR, and NOT; NAND and NOR can be derived from these three. In addition, some sort of data multiplexing and demultiplexing is required and can be constructed from the three basic logic devices. The digital optical computer uses only one optical switching device, a directional coupler, to represent not only the three minimum logic functions but several others as well. This directional coupler is a commercially available optoelectronic device fabricated in lithium niobate (LiNbO$_3$) single-crystal wafers. Two closely coupled optical waveguides are formed in such a wafer by diffusing titanium atoms to increase the index of refraction in the diffused waveguide region. Optical waves, coupled to the waveguides by attached optical fibers, interact in the region where the two waveguides are physically parallel to each other and are separated by a distance of the order of several wavelengths of the incident light (**Fig. 2***a*). Because of this close coupling, part of the light in the first waveguide overlaps the second waveguide and vice versa. Thus, the optical power incident on the device can be cycled back and forth between the two guides, A and B (Fig. 2b).

The lengths of the waveguides can be selected so that a light pulse entering waveguide A emerges from waveguide B. By applying an electric field across the second waveguide (B) by means of an applied voltage, the index of refraction of waveguide B is changed, thereby changing the phase relationship of the two waves so that the light emerges from the other waveguide. The directional couplers used in the digital optical computer are in a crossed state (A → E and B → D) when the control-terminal voltage is zero and will switch to the so-called bar state (A → D and B → E) at a threshold voltage applied to the control terminal C. Such a switching device is not entirely optical because an input electrical voltage is needed to cause the optical signals to switch paths through the device. The directional coupler can be made to simulate an all-optical device by using an optical detector to convert optical signals into voltages at input C. Gallium arsenide (GaAs) high-speed electronic amplifiers and pulse-stretching circuitry amplify the electrical pulse from the optical detector at terminal C and provide a means of communication (input and output) between the optical computer and an external semiconductor computer from which data and programs are loaded and output is analyzed.

Delay-line architecture. Because of the very high cost of the lithium niobate switches, the architecture of this first digital optical computer is quite different from that of conventional electronic computers. There are no traditional latches (pairs of switches arranged to store information in one of two possible states). By using latch architecture in conventional electronic circuitry, memory is composed of extremely large arrays of pairs of transistors which can be latched into the 0 or 1 state. The architecture of the first digital optical computer, because of the cost of each switch, clearly could not be based on this scheme. A very old architecture was therefore employed in which data are stored serially in a long delay line. This architecture was devised early in the development of electronic computers, when cost per

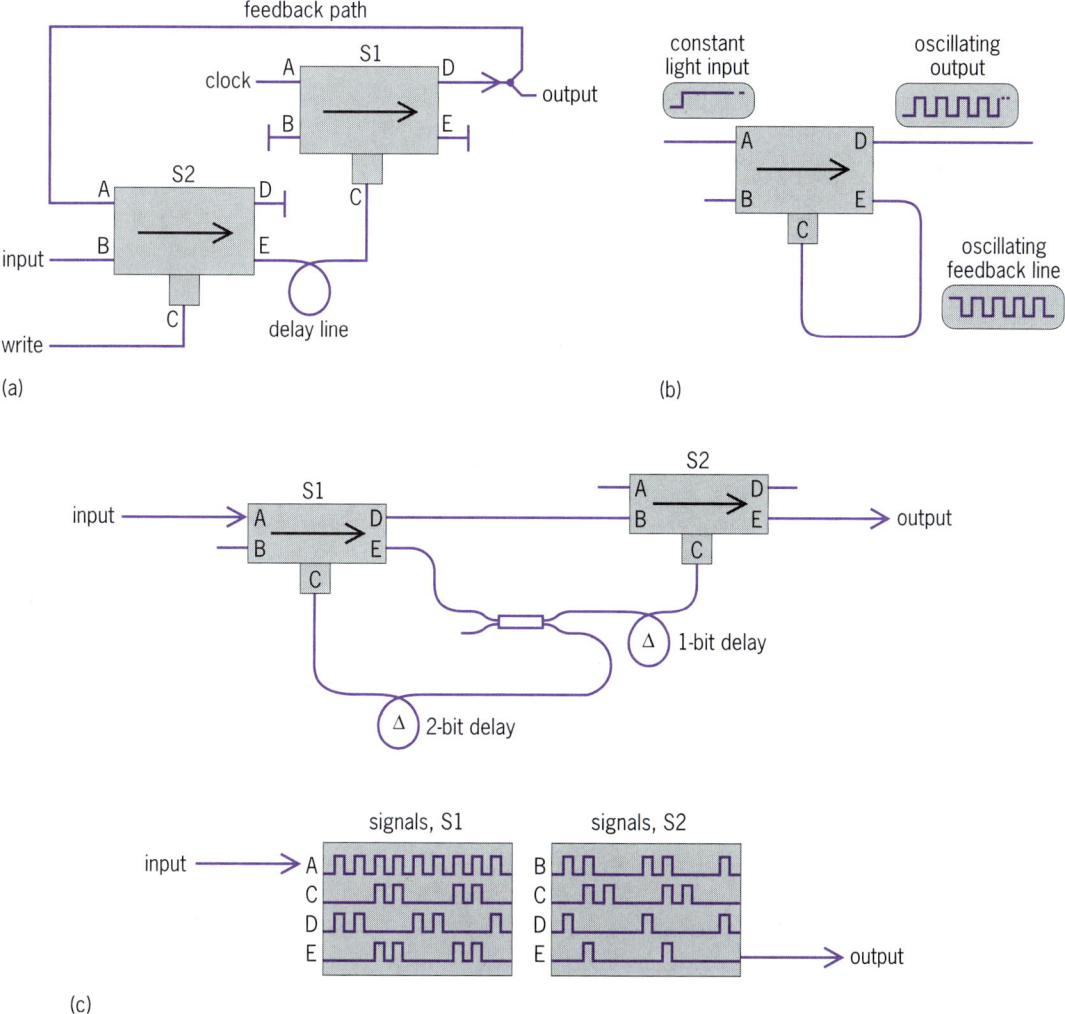

Fig. 3. Optical delay-line circuits. (a) Memory circuit (after D. B. Sarrazin, H. F. Jordon, and V. Heuring, Fiber optic delay line memory, Appl. Optics, 29:627–637, 1990). (b) Feedback clock circuit. (c) Divide-by-2 circuit (parts b and c after A. F. Brenner, H. F. Jordan, and V. P. Heuring, Digital optical computing with optically switched directional couplers, Opt. Eng., 30(12):1936–1941, 1991).

switch was also an important factor. This bit-serial architecture has led to the name bit-serial optical computer (BSOC). In this scheme, data and instructions in bit form are stored sequentially in a fiber-optic delay line. All of the light pulses (ones) and dark spaces (zeros) are entered into the delay line and exit the delay line in a serial fashion. In fact, this computer is analogous to a railroad switching yard in which railroad cars (bits) are constantly on the move and are switched to the appropriate tracks (fiber-optic delay lines) at the appropriate times. Memory consists of a single large loop of optical fiber. The light pulses, representing instructions and data, are constantly recirculated around the same memory loop and picked off by appropriate switching devices which are set up to wait until the correct word of information comes by. Registers and accumulators use this same memory scheme.

Delay-line circuits. It is instructive to investigate this serial memory loop since all of the BSOC's architecture is developed by variations of this basic circuit (**Fig. 3**a). Only two directional couplers (switches S1 and S2) are required to maintain this memory loop, to regenerate the data on each pass of the data around the loop, and to provide a means of entering and retrieving information. During operation, the write signal is normally set to 0 so that memory bits flow from the feedback path into the delay line (A → E of S2) and are "remembered." Bits coming from the delay-line memory are fed into input C of S1. On each bit pulse to terminal C, S1 is switched to the bar state and a clock signal pulse travels from A to D to replace the pulse at C. In this way the data bit in the loop is refreshed and regenerated; the loop would otherwise lose intensity and pulse shape in many passes around the memory loop. The memory can be written to at the appropriate time by changing the write signal to 1. In this event, S2 is switched to the bar state and the input line has access to enter bits into the delay-line memory.

An optical feedback oscillator can be used to generate a master clock signal (Fig. 3b). The constant light input from a dc laser at A is first switched to the delay line, which has an optical delay of one clock pulse. When this light arrives at terminal C, the light from A is switched to output D in the form of a pulse whose time length is the time of travel around the delay line. During this time, input C receives no light from E. Once the pulse is output, the circuit returns to its previous state and a second pulse is generated. The clock period is determined entirely by the length of the delay line.

A divide-by-2 circuit can be used to generate a word clock pulse at the beginning of each word (in this case two bits), for pulse bookkeeping (Fig. 3c). Pulses from the system clock of 1-bit length are input into S1. A 2-bit-delay feedback loop generates the pattern 11001100... and the complementary signal 00110011.... One output signal goes directly into the B input of S2 while the other is delayed by 1-bit period relative to the first before entering the C input of S2. The second switch operates as a logical AND, giving E = B · C. There is one bit of overlap, so that the circuit provides one output pulse every two clock cycles. To construct a word pulse for each 16-bit word, the length of the feedback loop is simply increased with a longer fiber delay.

Demonstration computer. The basic demonstration computer has a 50-MHz clock speed and a 64-word memory of 16-bit words, and is composed of 62 active components. Each word stored in memory either is an assembly-language instruction or consists of data. The computer instruction set is composed of a minimum number of instructions. Each instruction is a 16-bit word with the leading 6 bits representing the instruction and the remaining 10 bits being an address on which the instruction operates. The memory can therefore hold only 64 words (1024 bits) as any combination of instructions and data.

The size of the memory is basically limited by thermal expansion of the main memory delay loop, which must be 4 km (2.5 mi) long to hold 1024 bits at the clock speed of 50 MHz. The memory capacity could be increased by limiting the changes in fiber temperature or increasing the clock speed. There are plans to develop optoelectronic integrated circuits to reduce the size of the computer and greatly increase its speed and memory capacity. This new digital optical computer is primarily a demonstration of the principle that a complete optical computer can be built on optical-pulse, time-of-flight principles.

For background information SEE DELAY LINE; DIGITAL COMPUTER; DIRECTIONAL COUPLER; LOGIC CIRCUITS; OPTICAL INFORMATION SYSTEMS in the McGraw-Hill Encyclopedia of Science & Technology.

Jon M. Meese

Bibliography. A. F. Benner, H. F. Jordan, and V. P. Heuring, Digital optical computing with optically switched directional couplers, *Opt. Eng.*, 30(12):1936–1941, 1991; V. P. Heuring, H. F. Jordon, and J. P. Pratt, Bit-serial architecture for optical computing, *Appl. Optics*, 31(17):3213–3224, 1992; T. Main et al., Implementation of a general-purpose stored-program digital optical computer, *Appl. Optics*, 33(8):1619–1628, 1994; K. Preston, Jr., *Coherent Optical Computers*, 1972.

Optical fiber amplifiers

Light amplification by stimulated emission of radiation from atoms is, in essence, the concept underlying the laser acronym. However, in the laser oscillator, the amplifying medium is placed inside a reflecting cavity. Light initially generated by spontaneous emission from atoms is then trapped and undergoes multiple passes through the medium, which stimulates further emission and eventually yields the highly coherent laser light. Optical amplifiers can also be used to boost the energy of optical signals in a one-pass or traveling-wave fashion. In this case, spontaneous emission acts as a source of noise, adding to the amplified output signal as a parasitic background.

Using optical fibers as amplifying devices presents several advantages. First, optical fibers can guide light with very high power confinement. In typical single-mode fibers, an input power of only 1 mW corresponds to a guided intensity of 1 kW/cm² within the core. Thus, relatively low powers of light, such as are generated by miniature laser diodes, are able to efficiently pump the active medium forming the core. A signal input to such a fiber remains confined in the core over the whole length, which provides efficient single-pass amplification. Finally, relatively long lengths of fiber, from 1 m (3 ft) to 10 km (6 mi), can be used to achieve high signal gains, no matter how weak the net amplification over 1 cm. To give an idea of fiber amplifier performance, gains of 30 dB can be achieved at a signal wavelength of $\lambda = 1.5$ micrometers by using only 3-mW pump power ($\lambda = 0.98$ μm) launched into a 35-m-long (115-ft) erbium-doped fiber. (Amplifier gains of 10, 20, and 30 dB correspond to amplification factors 10, 10^2, and 10^3, respectively.)

Types. Optical fiber amplifiers are usually low-loss single-mode fibers made of basic silica glass (SiO_2), similar in design to those used for transmission of optical signals in telecommunications. The gain is generated along the fiber length by coupling pump light at either or both fiber ends, or at periodic locations in between. Amplifiers exist in two main categories. In nonlinear fiber amplifiers, nonlinear interactions between pump and signal cause positive power transfer during propagation, resulting in fiber gain. In rare-earth-doped fiber amplifiers, the fiber core is lightly doped with trivalent rare-earth ions, which have the property of absorbing light at certain pump wavelengths and emitting it at some signal wavelength through stimulated emission.

Nonlinear amplifiers. The nonlinear effects that cause signal amplification are stimulated Raman and Brillouin scattering and four-wave mixing. Relative to the signal-propagation direction, four-wave mixing and stimulated Brillouin scattering require copropagating and counterpropagating pumps, respectively, whereas stimulated Raman scattering can be pumped either or both ways. In scattering amplifiers, the optical signal frequency is shifted from the pump frequency by an amount fixed by silica glass (12 THz for stimulated Raman scattering, and 10 GHz for stimulated Brillouin scattering). Thus, scattering amplifiers can be tuned as a function of the pump wavelength. Whereas stimulated Raman scattering and four-wave mixing represent relatively weak nonlinearities, characterized by relatively high pump power thresholds, stimulated Brillouin is quite effective at milliwatt pump powers. However, the usable gain bandwidth in stimulated Brillouin scattering is relatively narrow (10–50 MHz), whereas it is rather broad in stimulated Raman scattering (7 THz). In the case of four-wave-mixing amplifiers, a phase-matching condition relating the pump and signal (and probe) velocities is required. The amplification effect in four-wave mixing is therefore most effective near the zero-dispersion wavelength of the fiber, which is located at $\lambda = 1.32$ and 1.55 μm in standard and so-called dispersion-shifted fibers, respectively.

Rare-earth-doped amplifiers. The trivalent rare-earth ions that can provide gain in glass fibers are Nd^{3+} (neodymium), Pr^{3+} (praseodymium), Tm^{3+} (thulium), and Er^{3+} (erbium). The corresponding signal wavelengths are given in the **table**. Rare-earth-doped fiber amplifiers exhibit broad gain bandwidths (2–4 THz), resulting from the effects of inhomogeneous line broadening and energy-level Stark splitting. For each rare-earth ion there exist several possible pump wavelengths, as determined by the positions of atomic energy levels. Pump wavelengths falling into the near-infrared region ($\lambda = 0.8$–1.5 μm) are the most suitable, because of the possibility of pumping with efficient and compact semiconductor laser diodes. The effect of pump excited-state absorption, which occurs in certain pump bands, reduces the pumping efficiency and further restricts the choice of possible pump wavelengths. In the case of Er^{3+}, the two infrared pump wavelengths free from adverse excited-state absorption are $\lambda = 0.98$ and 1.48 μm. The laser-diode sources amenable to generate high output power at these two wavelengths are based upon strained-layer indium gallium arsenide–aluminum gallium arsenide (InGaAs-AlGaAs) and indium gallium arsenide phosphide (InGaAsP) semiconductors, respectively. It is also possible to incorporate Yb^{3+} (ytterbium) ions as a codopant or sensitizer in erbium-doped fibers, which makes possible pumping with compact Nd^{3+}:YLF (yttrium lanthanum fluoride) or Nd^{3+}:YAG (yttrium-aluminum-garnet) laser sources ($\lambda = 1.05$–1.06 μm).

Comparison. Various required pump powers and corresponding gains can be achieved near 1.3-μm and 1.5-μm wavelengths in various types of fiber amplifiers (Er^{3+}-doped, Pr^{3+}-doped, and nonlinear). Overlooking stimulated Brillouin scattering, erbium-doped fiber amplifiers (EDFAs) pumped at $\lambda = 0.98$ μm are the most efficient, with gains of 20–40 dB in the 1–10-mW pump-power range. Considering that pump laser diodes operating at 0.98 and 1.48 μm can now generate powers in excess of 250 mW, efficient EDFA boosters having output signal powers in excess of 100 mW can be realized.

Development. The first rare-earth-doped fiber amplifiers were demonstrated in 1964, using Nd^{3+}, and laser-diode-pumped, Nd^{3+}-doped fiber lasers

Signal wavelengths amplified in glass fibers, in micrometers

Fiber	Rare-earth ion			
	Nd^{3+}	Tm^{3+}	Pr^{3+}	Er^{3+}
Silica fibers	1.06			0.85
	1.34			1.53–1.56
Fluoride fibers		0.81	1.28–1.33	2.7
		1.47		

were investigated as potential sources for communications in the mid-1970s. In 1985, the field of rare-earth-doped fibers was reactivated, and in 1987 gain measurements in erbium-doped fibers at the 1.5-μm communications wavelength were first reported.

EDFAs. The early investigations rapidly showed the superior performance of EDFAs as compared to other types of optical amplifiers. EDFAs are intrinsically polarization-insensitive. First, this requirement is fundamental in communication systems, because fiber trunks do not maintain polarization. Second, EDFAs can be operated in the regime of minimum spontaneous noise fixed by quantum physics. Thus, high-gain amplification can be achieved with minimal signal-to-noise ratio degradation. Third, EDFAs are immune to crosstalk under saturation because of their slow gain dynamics. (They require 100 microseconds to 1 ms to saturate or recover to steady state under transient excitation by a strong signal.) Thus, simultaneous amplification of several optical channels with no interchannel interference is possible. Another consequence is linearity: even under high saturation, there is minimal harmonic distortion, in contrast to microwave amplifier counterparts. Finally, EDFAs can be fusion-spliced to optical fibers with negligible coupling loss.

However, all these advantages do not alone account for the tremendous breakthrough generated by EDFAs in optical communications. For practical implementation, such as in transoceanic systems, an optical amplifier must be pumped by compact, efficient, reliable, and cost-effective laser diodes. The first demonstration of laser-diode-pumped EDFAs came in 1988 and 1989, and subsequent progress in manufacturing high-power pump laser diodes at 0.98 and 1.48 μm made this breakthrough possible.

PDFAs. Parallel developments concerned the 1.3-μm transmission window. The most suitable fiber amplifier for this window is based on Pr^{3+}-doping. However, praseodymium-doped fiber amplifiers (PDFAs) work only in fluoride-glass bases, which are rather delicate to manufacture. Additionally, they are far less efficient (20 times) under pumping than EDFAs, and novel pump sources ($\lambda = 1.017$ μm) must be developed. Yet the interest in 1.3-μm amplifiers is very high, because terrestrial fiber systems use a standard fiber type whose zero dispersion is at that wavelength. These systems can use $\lambda = 1.5$ μm for communications, with EDFAs as regenerators, but dispersion compensation must then be implemented.

Impact of EDFAs. The impact of EDFAs in optical communications represents a true revolution. Optical data can now be transmitted over transoceanic distances (9000 km or 5500 mi) at rates up to 10 gigabits per second, without in-line electronic regeneration. This performance represents a 100-fold increase in the product of transmission distance and capacity in 10 years. With digital coding techniques, such a huge capacity could provide 600,000 simultaneous voice channels in a single transmission fiber. The deployment of 5-Gbit/s transatlantic and transpacific EDFA-based links is under way. Several techniques are under study to increase capacity even further, by either increasing the line bit rate (time-division multiplexing) or transmitting several optical channels simultaneously (wavelength-division multiplexing).

In either case, fiber nonlinearities and dispersion represent fundamental limits. The elegant solution to this problem is to transmit data in the form of soliton pulses. Solitons exploit the combined effects of Kerr nonlinearity and fiber dispersion to retain the particlelike integrity of transmitted pulses. Soliton propagation in fibers was predicted in 1973 and demonstrated in 1980. Very intensive research is being pursued to explore theoretically and experimentally the ultimate capacity limits of amplifier-based soliton systems. Soliton transmission control techniques are being investigated, and controlled soliton transmission within a recirculating loop has been demonstrated over distances of 10^6 km (6×10^5 mi), with no measurable data errors. A variety of novel laser-diode-controlled fiber devices exploiting nonlinearities (made possible with EDFAs) have also shown potential for very high-speed (100-Gbit/s) all-optical signal processing.

For background information SEE LASER; NONLINEAR OPTICS; OPTICAL COMMUNICATIONS; OPTICAL FIBERS; SOLITON in the McGraw-Hill Encyclopedia of Science & Technology.

Emmanuel Desurvire

Bibliography. E. Desurvire, *Erbium-doped Fiber Amplifiers: Principles and Applications*, 1994; E. Desurvire, The golden age of fiber amplifiers, *Phys. Today*, 47(1):20–27, January 1994; A. Glass, Fiber optics, *Phys. Today*, 46(10):34–38, October 1993; H. A. Haus, Molding light into solitons, *IEEE Spectrum*, 30(3):48–53, March 1993.

Paleoclimatology

Paleoclimatology is the study and reconstruction of ancient climates. Knowledge of the ancient Earth's climate system and climatic responses to shifts in factors such as ocean currents, continental plate positions, and concentrations of trace atmospheric (greenhouse) gases, provides a means to test computer climate model simulations, the basis for predictions of future climates. Recent applications of biogeochemical methods to fossils has given new insight into ancient climate.

The nature of continental climate change is largely unknown; our knowledge of past climates is based principally on marine climate records. There is reason to expect that the history of continental climate differs from marine climate records; the effects of climate change in marine climate records are manifest not only by temperature change but also by interrelated changes in ocean circulation, isolation of major water bodies, distribution of continental land masses, glaciation, and other processes that affect ocean chemistry in complex ways.

The chemistry of marine microfossils cannot be interpreted simply as a climate signal. The chemical signal in fossil bones and teeth from deposits in continental interiors, however, responds principally to climate change. The chemistry of fossil bones and teeth provides a quantitative climate record, and a new tool for paleoclimate reconstruction.

Most chemical elements have two or more isotopes, forms of an element which differ in the number of neutrons in the nucleus. The number of neutrons affects the mass of an atom, leading to mass fractionation of one isotope relative to another in many biotic and abiotic chemical reactions, such as bone synthesis, which is reflected in the isotopic ratios of the products. Such ratios can be measured very precisely and form the basis of many quantitative climate records. Earth scientists express the isotopic ratio in an unknown sample as a deviation in parts per mille (‰) relative to standards and express the result in a notation as shown in Eq. (1), where

$$\delta = \left(\frac{R_{sample}}{R_{standard}} - 1\right) \times 1000 \quad (1)$$

R_{sample} and $R_{standard}$ are the measured isotopic ratio in the sample and standard. The measured isotopic ratios of interest for paleoclimatic studies are $^{18}O/^{16}O$ (expressed as $\delta^{18}O$), $^{13}C/^{12}C$ ($\delta^{13}C$), and $^{2}H/^{1}H$ (δD). The standard by which samples are compared is standard mean ocean water (SMOW; for $\delta^{18}O$ and δD in the hydrologic cycle), or a carbonate known as the Pee Dee belemnite (PDB; for $\delta^{13}C$ and $\delta^{18}O$ in carbonates).

Isotopes in continental precipitation. Oxygen and hydrogen isotopes in precipitation exhibit a strong relationship with temperature because of fractionation processes in the hydrologic cycle (**Fig. 1**). The temperature dependence of this relationship is reflected in the $\delta^{18}O$ and δD of surficial water bodies, such as lakes and rivers. In addition, the same ratios are preserved in the minerals which form in the water and the bones and teeth of vertebrates that drink and swim in the water. Thus, aspects of ancient climate can be inferred from fossil teeth and bone, assuming that chemical changes in the geologic environment have not altered the climate signature in the fossils through a process called diagenesis.

Oxygen isotopes in fish otoliths. Fish develop bonelike accretionary structures called otoliths in their ears. Otoliths are composed of aragonite, a calcium carbonate mineral for which $\delta^{18}O$ can be measured, and often have bands representing daily, monthly, and yearly growth cycles. Measurement of $\delta^{18}O$ from samples within bands can provide a proxy of ancient lake conditions, and hence paleoclimate, on a subannual scale. The $\delta^{18}O$ of otolith carbonate is related to temperature and $\delta^{18}O$ of the lake water in which the fish lived according to Eq. (2), where

$$\delta^{18}O_{CaCO_3} \approx \delta^{18}O_{H_2O} + \frac{1.856 \times 10^4}{T} - 33.49 \quad (2)$$

$\delta^{18}O_{CaCO_3}$ and $\delta^{18}O_{H_2O}$ are the isotopic compositions of otolith aragonite and lake water, respectively, and T is absolute water temperature.

Only one variable of the three in this equation, $\delta^{18}O_{CaCO_3}$, can be measured in the laboratory, and hence the equation cannot be solved directly. The two remaining variables ($\delta^{18}O_{H_2O}$ and T) are the unknown characteristics of the ancient lake system. One approach to this problem is to measure the $\delta^{18}O$ of otoliths from different fish that lived in the same lake: one from very deep, cold water, and one from shallow water. Deep, temperate lakes have a bottom temperature constant near 4°C (39°F), so water temperature can be fixed and the paleotemperature equation solved for $\delta^{18}O$ of the water for an otolith from a bottom-dwelling fish. In turn, this $\delta^{18}O$ estimate of water can be substituted into the paleotemperature equation, with analysis of $\delta^{18}O$ for an otolith from a surface-dwelling fish, to determine surface temperature and seasonal ranges. Such an analysis of fossil fish from the Pliocene (2–4.5 million years ago) of Idaho showed that the climate at that time was much more equable than today, with cooler summers and milder winters. Seasonal temperature range was probably from as low as near 32°F (0°C) in winter to as high as 70°F (21°C) in summer, compared with about 21°F (−6°C) to 72°F (22°C) today.

Oxygen and hydrogen isotopes in bones and teeth. The phosphate mineral component of teeth and bones (apatite), is deposited in isotopic equilibrium with body water and obeys an equation similar to Eq. (2). Because body temperature in large mammals is constant near 99°F (37°C), the $\delta^{18}O$ of biogenic apatite reflects the isotopic composition of body water. The $\delta^{18}O$ of body water is linked to the $\delta^{18}O$ of ingested water both from drinking and from eating plants and, therefore, climate. If the fractionation processes linking $\delta^{18}O$ of body water to $\delta^{18}O$ of ingested water are understood, then the $\delta^{18}O$ of

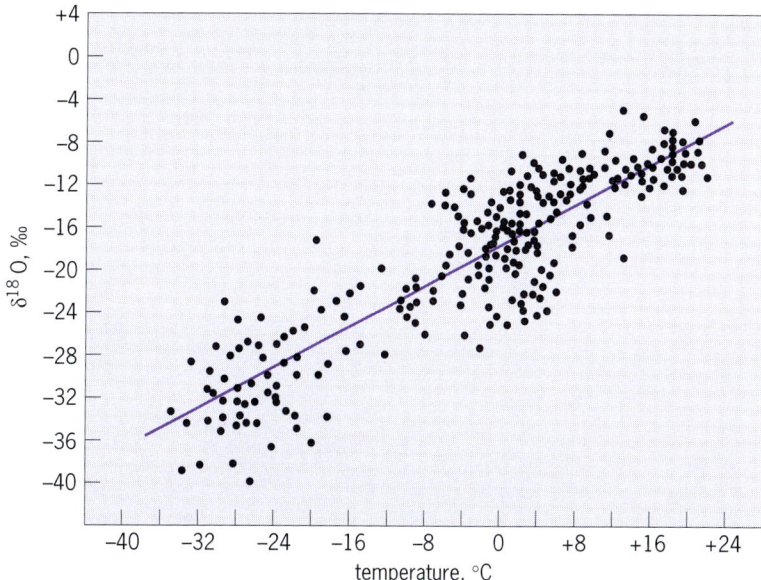

Fig. 1. Relationship between oxygen isotope composition of monthly mean precipitation and monthly mean temperature. °F = (°C × 1.8) + 32. (*After J. R. Gat and R. Gonfiantini, eds., Stable Isotope Hydrology: Deuterium and Oxygen-18 in the Water Cycle, International Atomic Agency, Tech. Rep. Ser. 210, Vienna, 1981*)

teeth can be used to reconstruct the $\delta^{18}O$ of ingested water and derive an estimate of temperature. Empirically determined relationships between $\delta^{18}O$ of biogenic phosphate and $\delta^{18}O$ of ingested water are roughly linear.

This relationship is also applied to the study of continental paleoclimates. An unknown facet of paleoclimate is how closely continental climate mirrors marine climate. One of the major events in Cenozoic climatic history was the initial onset of continental glaciation in Antarctica 33 m.y.a. This cooling event is reflected in an increase in the $\delta^{18}O$ composition of the shells of marine microfossils (**Fig. 2**). A similar shift is preserved in the $\delta^{18}O$ of fossil horse teeth from North America, indicating a major continental cooling event accompanied by marine climate change. However, the magnitude of the shift observed in the continental record seems too large to reflect only a temperature change. There are two possible explanations for the apparent discrepancy. The path followed by water vapor sources for precipitation may have changed; thus the relationship between temperature and the $\delta^{18}O$ of precipitation is not applicable. Alternately, the understanding of oxygen isotope fractionation processes in mammals may be flawed, in which case the relationship between $\delta^{18}O$ of body water and $\delta^{18}O$ of ingested water is not applicable. Thus, the biogenic phosphate system is not a perfect archive, but future research may refine understanding of the system.

Humidity affects the isotopic composition of water in plants and may overprint any temperature-related change. A solution to this problem is to combine analyses of $\delta^{18}O$ from phosphate and δD from collagen of the same bone. The δD of bone collagen is also related to δD of ingested water. Both undergo fractionation in animals and plants, but in response to different processes, and hence can be used to estimate humidity more reliably than either system alone. This combined system is very new and has not yet been applied to paleoclimate reconstruction, but it shows great promise for fossils with preserved collagen.

Ancient carbon dioxide. Climate change affects the terrestrial environment in many ways. The chemical composition of an animal's food is recorded in the bones and teeth of the animal, thus allowing reconstruction of other facets of ancient ecosystems. Two major types of grasses exist today, C3 and C4 grasses, distinguished by different physiologic approaches to photosynthesis. C4 grasses have adaptive advantages in environments with low carbon dioxide (CO_2), warm temperature, and little rainfall (for example, the modern Serengeti Plain), whereas C3 grasses are mostly restricted to regions with cooler, generally moister growing seasons. C3 and C4 grasses have distinct signatures in the $\delta^{13}C$ of their tissues which are transferred to the $\delta^{13}C$ of bones of animals that eat the grasses.

About 15 m.y.a., fossil horses underwent a transition from forest-adapted taxa with low-crowned teeth for browsing to savanna-adapted taxa with high-crowned teeth for grazing. A study of the $\delta^{13}C$ of fossil horse teeth (**Fig. 3**) shows that C4 grass expansion in the Great Plains of North America began 7 m.y.a., well after the transition from forest-adapted to savanna-adapted horse taxa, 15 m.y.a. Thus, it appears that reduced atmospheric CO_2 levels in the Miocene gave an adaptive advantage to C4 grasses, which displaced C3 grasses in savanna ecosystems.

Fossil bone chemistry is in its infancy compared to other areas of paleoclimatology. However, the available information adds a new dimension to climate reconstruction and has spurred new research.

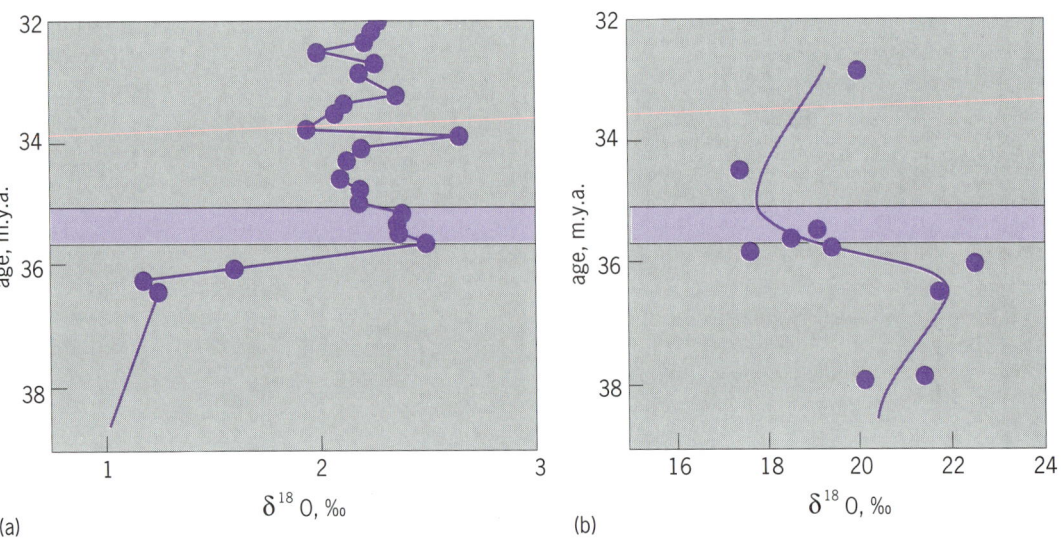

Fig. 2. Climate shifts recorded by the oxygen isotopic composition of marine and terrestrial fossil animals. (*a*) Marine climate record for the period 32–38 m.y.a., based on the oxygen isotopic composition of fossil marine microfossils (*after D. R. Prothero and W. A. Berggren, eds., Eocene-Oligocene Climate and Biotic Evolution, Princeton University Press, 1992*). (*b*) Terrestrial climate record for the same interval, based on the oxygen isotopic composition of fossil horse teeth from North America. Both records indicate global cooling corresponding to the growth of Antarctic ice sheets.

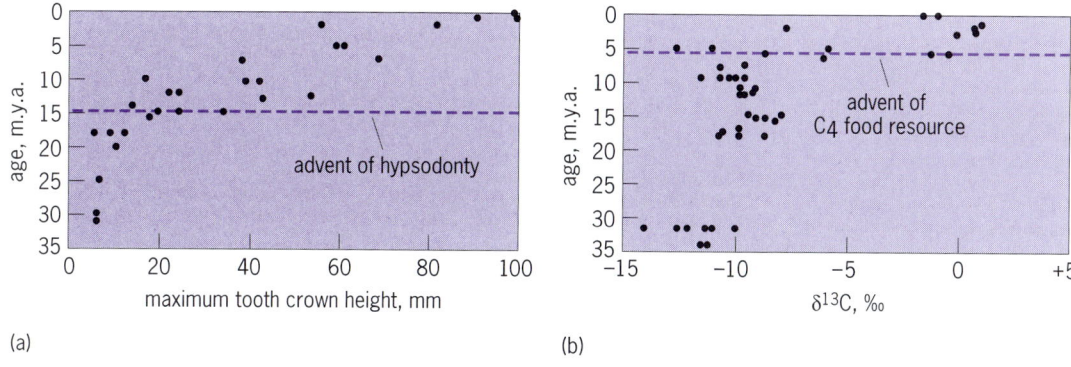

Fig. 3. Relationship between fossil horse evolution and the carbon isotope record of horse teeth. (*a*) Hypsodonty (tooth crown height) of fossil horse teeth. (*b*) Carbon isotope composition of fossil horse teeth. (*After B. J. MacFadden and J. D. Bryant, eds., Stable isotope and trace-element geochemistry of vertebrate fossils: Interpreting ancient diets and climates, Palaeogeog. Palaeoclimat. Palaeoecol., 107:269–279, 1994*)

Areas that are receiving considerable attention include studies that constrain biologic fractionation within an animal, such as the effects of rates of metabolism, bone and tooth growth, and timing of tooth formation. As the understanding of isotope fractionation in terrestrial animals improves, so will the number of possible applications.

For background information SEE CARBON; GEOLOGIC THERMOMETRY; ISOTOPE; ISOTOPE SHIFT; OXYGEN; PALEOECOLOGY in the McGraw-Hill Encyclopedia of Science & Technology.

J. Daniel Bryant

Bibliography. A. B. Cormie et al., Relationship between the hydrogen and oxygen isotopes of deer bone and their use in the estimation of relative humidity, *Geochim. Cosmochim. Acta*, 58:3439–3450, 1994; B. J. MacFadden and J. D. Bryant (eds.), Stable isotope and trace-element geochemistry of vertebrate fossils: Interpreting ancient diets and climates, *Palaeogeog. Palaeoclimatol. Palaeoecol.*, 107:199–328, 1994; P. K. Swart et al., *Climate Change in Continental Isotopic Records*, 1993.

Paleogeography

Recent investigations have revealed the occurrence of continent-continent collisions during the Paleozoic that were not recognized on the basis of previous work. The hypothesis of continental drift was first put forward on the basis of matching the shapes of the present-day continents (particularly of Africa and South America across the South Atlantic Ocean). Supporting this hypothesis is the apparent truncation of some ancient mountain ranges at continental margins (for example, the Appalachians in Newfoundland, Canada, and the Caledonian Mountains of the British Isles) as well as the distribution of fossils and indicators of ancient climates (notably the occurrence throughout the southern continents and India of remains of a distinctive fossil flora in Paleozoic strata overlying glacial deposits). The hypothesis was strengthened during the 1950s with the advent of paleomagnetic analysis by which the former position of a continent can be determined relative to the North and South Poles. For example, the shift of Europe from North America during the opening of the North Atlantic Ocean Basin is reflected in the rotation of the two continents relative to the magnetic field. However, the reality of continental drift was proved only when drilling of the deep ocean floors in the 1960s revealed that they are geologically very young and get older toward the edges of the continents. Therefore, the continents had formed by the process of sea-floor spreading.

Pangea. The supercontinent of Pangea, which can be reconstructed from evidence from the ocean floors and computer matching of the geometric outlines of the continents, came into existence only about 300 million years ago (m.y.a.), toward the end of the Paleozoic Era. Paleomagnetic results from rocks older than that indicate that Pangea was formed by the amalgamation of several continents as a result of continental drift during Paleozoic times. Some of these ancient continents, for example, North America (including Greenland), were essentially the same as the continents of today, because Pangea broke up along the weak zones joining their strong ancient nuclei. Several lines of evidence indicate that the Paleozoic continents were themselves fragments of an even more ancient supercontinent that existed from about 1000 to 750 m.y.a. First, the rock record shows that many continental margins originated between 750 and 550 m.y.a., near the end of the Precambrian Era (for example, all the margins of Laurentia, ancestral North America, and the Pacific margins of South America, Antarctica, and Australia). Second, transgression of the seas across all the continents at the start of the Cambrian Period about 540 m.y.a. indicates a period of rapid sea-floor spreading such as that accompanying the fragmentation of Pangea during the Mesozoic Era. Third, paleomagnetic results, although less extensive and more difficult to interpret than those from younger rocks, seem to support a clustering of continental masses on Earth during the interval 1000–750 m.y.a. The Russian name Rodinia has been proposed for this supercontinental entity.

Rodinia. The difficulty in reconstructing the paleogeography of Rodinia lies not only in its

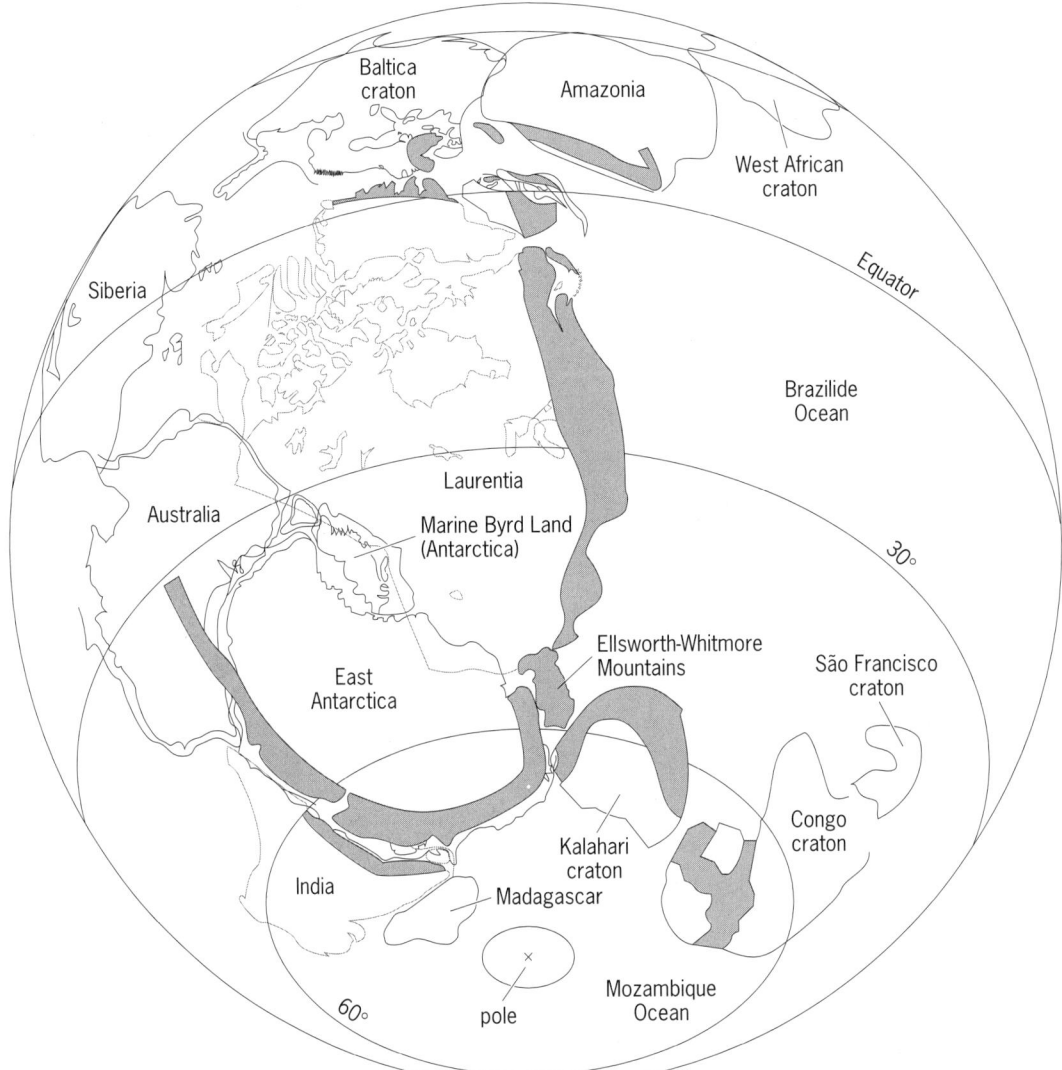

Fig. 1. Reconstruction of the hypothetical supercontinent Rodinia at 750 million years ago. The continents are juxtaposed by using geologic arguments, and the position of the pole is based on paleomagnetic data from Laurentia. Irregular areas in color are Grenvillian mountain belts. Whether the pole is North or South is uncertain.

antiquity. The main problem is that none of the rocks below today's oceans are older than about 200 m.y. The ocean basins formed during the fragmentation of Pangea are all younger, and although the Pacific Ocean Basin is older (it was even larger than it is today when Pangea existed), the parts of its floor that were older than 200 m.y. have all been thrust under the margins of the bordering continents during the more recent continental drift that formed the Atlantic, Indian, and Southern ocean basins. Hence, there is no road map for earlier periods of continental drift such as that leading to the assembly of Pangea.

Reconstruction. The problem of reconstructing the Rodinian supercontinent is of much greater interest and importance than if it were merely an ancient jigsaw puzzle. The changes that accompanied the fragmentation of that supercontinent included the waxing and waning of several ice ages, some of the most abrupt changes in the chemistry of ocean waters recorded (in the sedimentary deposits of the time) throughout the history of the planet, and the virtual explosion of multicellular life throughout the world ocean after nearly 3 billion years of single-celled life. It seems likely that the tectonic changes that led to and accompanied the fragmentation of Rodinia played a role in these environmental changes. This interval of rapid global change may have implications for the present-day Earth and its future.

Evidence from rocks. In the absence of data from the ocean floors to serve as a guide in determining the paleogeography of Rodinia, it is necessary to use the evidence from the rocks of the continents, including paleomagnetic results, just as geologists had to do in attempting to reconstruct Pangea before the exploration of the ocean floors. Any suggested configuration of Rodinia is therefore going to be hypothetical and in need of extensive testing. The hypothesis shown in **Fig. 1** has its origin in

Paleogeography 245

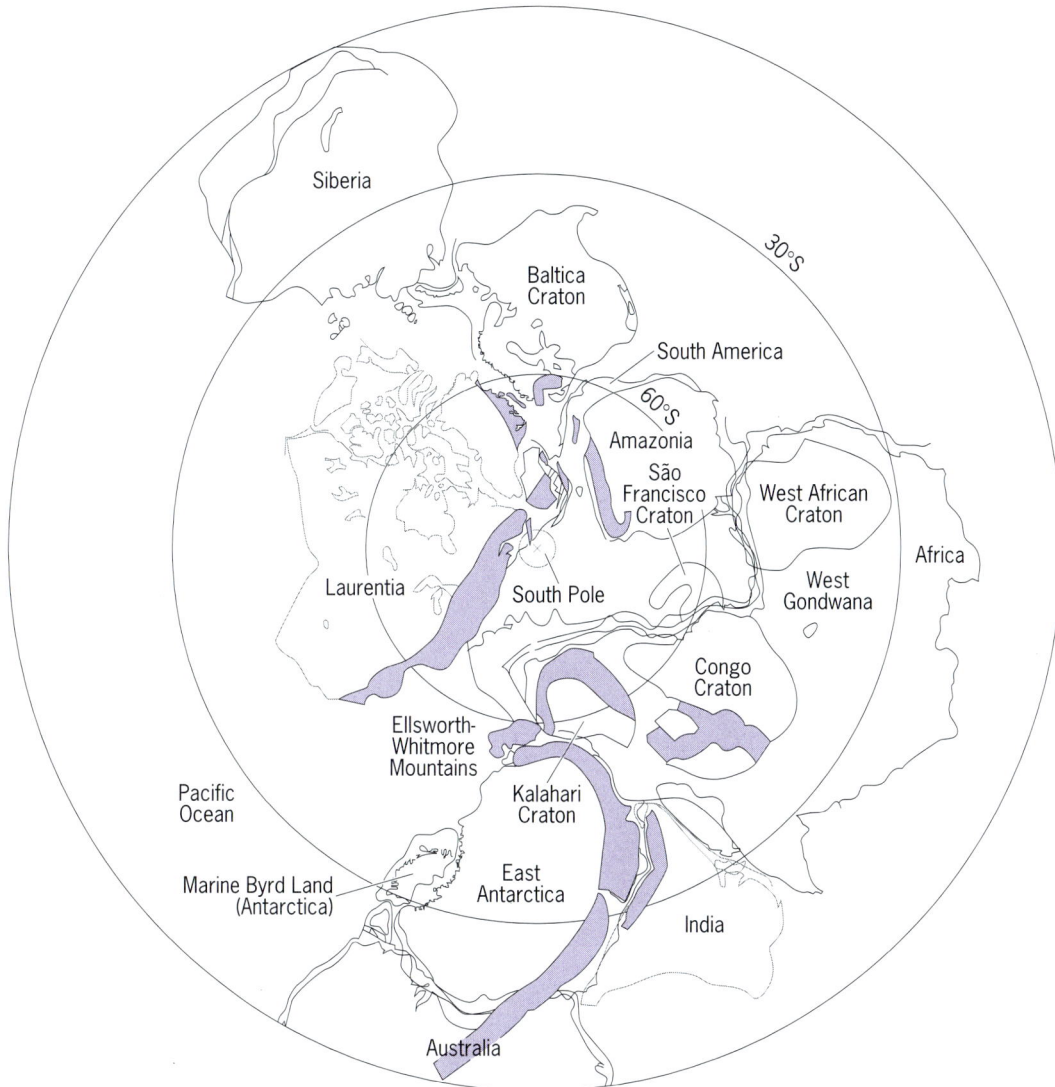

Fig. 2. Reconstruction of an unnamed supercontinent that may have existed in the latest Precambrian after opening of the Pacific Ocean between Laurentia and East Antarctica–Australia. The continents are juxtaposed by using geologic data, and the South Pole is based on paleomagnetic data from Laurentia. Irregular shaded areas are Grenvillian mountain belts.

studies of upper Precambrian rocks in western Canada and southeastern Australia. For many years, geologists from those countries had noted the similarity in these strata on either side of the Pacific Ocean Basin. However, close similarity in sedimentary successions is not necessarily evidence of juxtaposition at the time of deposition but could be the result of close similarity between environments in widely separated locations. Hence, these observations did not attract widespread attention.

In 1991, geologists in the United States started to compare the still older Precambrian rocks of the continental interiors of North America, Australia, and Antarctica, to apply some of the methods used in the early reconstructions of Pangea, and to attempt computer-aided reconstructions of Rodinia. Use of a computer has alleviated problems arising from the different scales and projections on geologic maps from parts of Earth that may once have been juxtaposed but are now widely separated, and has allowed accurate comparison of paleomagnetic data. The length of the late Precambrian rifted margin of Pacific North America (approximately 4500 km or 2800 mi) almost exactly matches those of East Antarctica and Australia combined. There is a general similarity in the distribution of earlier Precambrian rock units that compose the cores of the continents when they are juxtaposed as shown in Fig. 1. Most importantly, the well-studied roots of a mountain belt that extended from Labrador through eastern Canada and the United States, to southwestern Texas about 1 billion years ago, the Grenville Belt, has a possible continuation in East Antarctica. Like the Appalachian Mountains of North America and the Caledonian Mountains of Europe that were truncated during the breakup of Pangea during the Mesozoic Era, the Grenvillian mountain chain may have been truncated by the opening of the Pacific Ocean Basin late in Precambrian times.

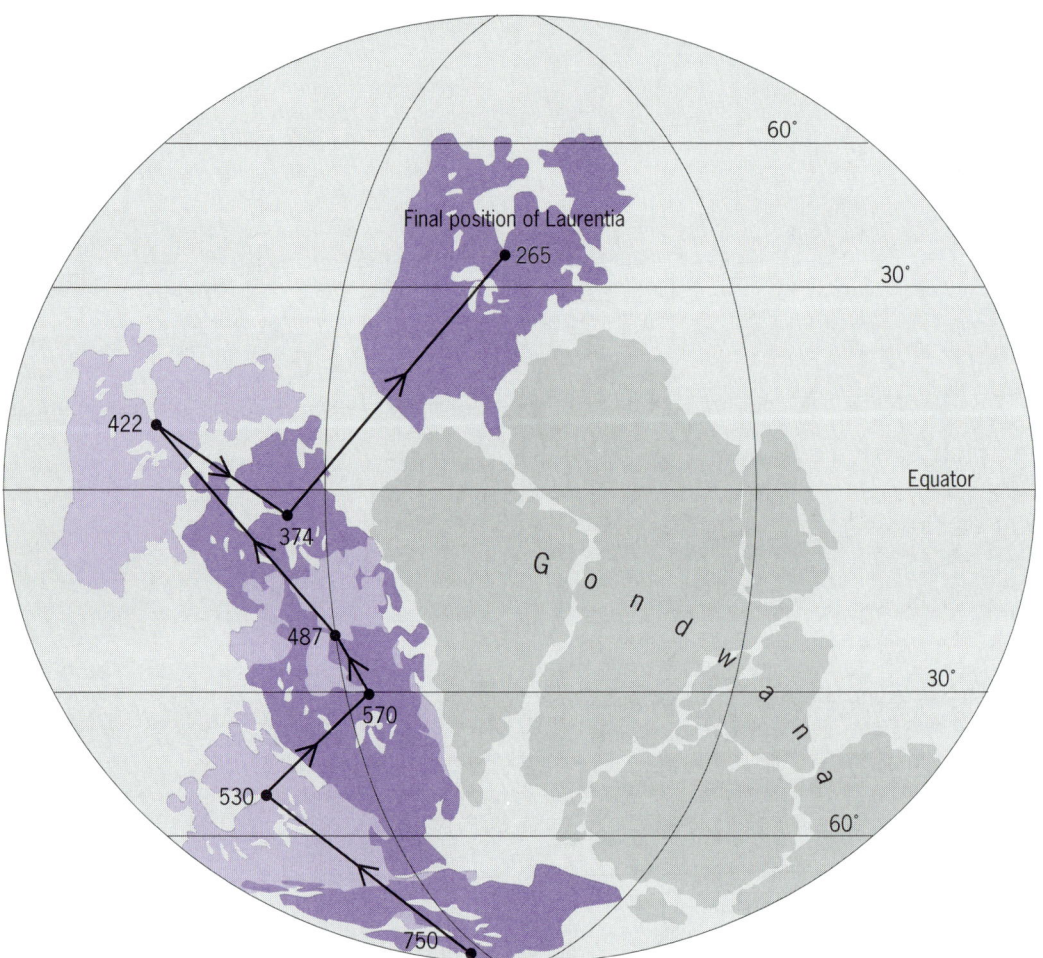

Fig. 3. Hypothetical so-called end run of Laurentia around Gondwana during the Paleozoic Era. South America is fixed in present-day coordinates. Numbers show times (in millions of years ago) of suggested positions of Laurentia with respect to Gondwana.

During the 1993–1994 Antarctic summer an American research group visited Coats Land in Antarctica and studied the possible continuation of the Grenville Belt there. They collected oriented rock samples, and determined the position of the North and South Poles relative to East Antarctica when these rocks cooled 1100 m.y.a. When these poles are rotated with East Antarctica to juxtaposition with North America, as suggested by the geologic evidence, they coincide with the position of the poles from rocks of the Grenville Belt from that continent. This result does not constitute an absolute proof of the hypothesis because of uncertainties inherent in the paleomagnetic method, but it is a positive result from the most quantitative test that can be applied and may provide valuable data for reconstructing Rodinia.

North America. Another line of evidence supporting the idea that North America may have been adjacent to what are now regarded as southern continents in the late Precambrian comes from South America. Whatever configuration is proposed for Rodinia 1000–750 m.y.a. in the late Precambrian must be compatible with the assembly of Pangea at about 300 m.y.a., late in the Paleozoic Era. Evidence from rocks from southern Peru seems to confirm that the proto-Appalachian margin of North America may have been rifted from the Pacific margin of South America in the position shown in Fig. 1. Grenvillian rocks are also present in Peru, as predicted in this reconstruction. The rifting along the Appalachian margin apparently took place at the very end of the Precambrian Era, approximately 540 m.y.a. Thus, the Pacific Ocean Basin may have opened in the time interval between approximately 750 and 540 m.y.a. Comparison of Fig. 1 and **Fig. 2** seems to indicate that the space occupied by the newly born Pacific Ocean Basin was made available (on an Earth of constant size) by the closure of late Precambrian oceans between east Antarctica, India, and Australia (East Gondwana), and various smaller continents that now make up Africa and South America (West Gondwana). This process could have happened in the same manner as today's Pacific Ocean Basin is closing, namely by thrusting beneath the surrounding continents. The Grenvillian mountain belts (around 1 billion years in age), shown in the reconstructions (Figs. 1 and 2) as long, dark irregular shapes, represent the sutures along which the supercontinent was assembled; the boundaries of the supercontinent are a sum of the outlines of all the entities that can be recognized by state-of-the-art techniques.

Following rifting of the proto-Appalachian margin of North America from South America at the end of the Precambrian Era, North America may have made a so-called end run around the proto-Andean margin of South America before colliding with northwest Africa in the final assembly of Pangea (**Fig. 3**). The presence of fragments of distinctly South American rocks in southern Mexico, and of distinctly North American strata in northwest Argentina, suggests that the two continents may have collided, possibly twice, during the Paleozoic Era. The Appalachian Mountain Belt as well as the ancestral Andes may thus have their origin in these encounters, which have modern counterparts in the interaction between Africa and Europe that generated the European Alps and the mountains of North Africa.

Implications. This scenario of previously unrecognized continent-continent collisions during the Paleozoic Era needs to be examined for its possible paleoenvironmental significance. Global changes in sea level during Paleozoic times as invertebrate and vertebrate life evolved and life moved on to the continents, as well as an ice age in South America and Africa during the Ordovician Period, approximately 440 m.y.a., may be related to such large-scale tectonic events.

For background information SEE CONTINENTAL DRIFT; EARTH, CONVECTION IN; PALEOGEOGRAPHY; PALEOMAGNETISM; PLATE TECTONICS in the McGraw-Hill Encyclopedia of Science & Technology.

Ian W. D. Dalziel

Bibliography. I. W. D. Dalziel, Pacific margins of Laurentia and east Antarctica–Australia as a conjugate rift pair: Evidence and implications for an Eocambrian supercontinent, *Geology,* 19:598–601, 1991; I. W. D. Dalziel, On the organization of American plates in the Neoproterozoic and the breakout of Laurentia, *GSA Today,* 2:237–241, 1992; P. F. Hoffman, Did the breakout of Laurentia turn Gondwanaland inside out?, *Science,* 252:1409–1412, 1991; E. M. Moores, Southwest U.S.–East Antarctic (SWEAT) connection: A hypothesis, *Geology,* 19:425–428, 1991.

Paleomagnetism

Paleomagnetism is the study of the ancient geomagnetic field of Earth, as recorded in almost all rock types of Earth's crust, and the implications of these ancient magnetizations recorded for geologic processes, ranging from core-mantle interactions to lithosphere plate motions, to deformation of segmented parts of the crust, to sedimentation, to variations in past climate. Magnetizations acquired at the time of a rock's formation or later in its history may be of thermal, chemical, or detrital origin. Paleomagnetic pole positions are derived from magnetizations that provide a sufficient average of short-period geomagnetic field behavior and assume an axial geocentric dipole for the long-term field. Paths of paleomagnetic poles for stable continental interiors track the past motion of lithosphere plates. Shorter-term, high-fidelity records of directional and intensity variations of the geomagnetic field over the past provide insight into the global behavior and origin of the field.

Basic principles. Study of the ancient geomagnetic field of Earth, recorded as permanent magnetizations in geologic materials—rocks and unconsolidated sediments—has played a major role in the geological sciences. A subdiscipline of paleomagnetism, rock magnetism, is concerned with mechanisms by which Earth materials acquire a permanent magnetization and has fostered an improved understanding of magnetic materials. Because direct observations of Earth's geomagnetic field span only a few centuries, paleomagnetic measurements provide the essential means for investigating the long-term behavior of the field over different time scales. The generation and behavior of the geomagnetic field are closely tied to other Earth as well as planetary processes. A general understanding of the behavior of the geomagnetic field is essential in interpreting the paleomagnetic record.

Magnetism and magnetic fields associated with magnetized bodies were studied as long ago as the early centuries A.D. By the sixteenth and seventeenth centuries, sufficient information had been acquired to permit speculation that the magnetic field of Earth originated from permanently magnetized material that contained two magnetic poles, essentially coinciding with the spin axis of the planet. It is now known that Earth's magnetic field originates in the liquid outer core, at a depth of about 1790–3190 mi (2880–5150 km). The field is maintained through a dynamo action, where a weak magnetic field is amplified in the core through the motions of an electrical conducting field.

The main elements of the geomagnetic field are declination, total field, and inclination. The deviation of a compass needle from true (geographic) north is the declination, measured positive eastward. A compass needle lies in the magnetic meridian containing the total field, which with the horizontal forms an angle termed the inclination. The inclination is taken as positive downward, as it currently is in most of the Northern Hemisphere in what is referred to as a normal polarity field, and negative upward, as in the Southern Hemisphere. The horizontal and vertical components of the total field may be resolved given the angle of inclination.

Both the total intensity and direction of the geomagnetic field at any given locality vary over geologically short periods of time. The directional variation (for example, a few degrees of direction per century on average) is referred to as secular variation; when it is recorded as a fossil magnetism, for example, in rapidly deposited lake sediments,

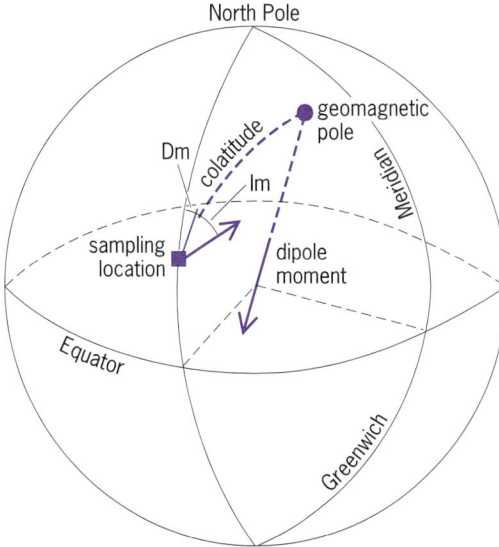

Fig. 1. Relationship between the magnetization recorded at a particular locality and the virtual geomagnetic pole position and associated axial geocentric dipole. Dm represents the declination and Im represents the inclination of the measured permanent magnetization of the sampled rocks.

earliest observations that rocks were magnetized parallel to Earth's magnetic field were made in the mid-1800s. In the early 1900s, work on lava flows and the material that they baked yielded convincing evidence that the natural remanent magnetization was a faithful record of the geomagnetic field. Unexpectedly, this work led to the first discovery of natural remanent magnetization in some lavas that was essentially opposite to that of the present field. Such behavior can be explained in essentially two ways. One requires that the geomagnetic field is capable of reversing itself at least once if not several times in the geologic past. The other implies that at least some rock types are capable of self-reversal, that is, magnetization is acquired that is opposite to the field applied. The first possibility, a self-reversing dynamo, has consequences for field-generating mechanisms and should, as pointed out by several workers in the 1920s, be able to be tested. Reversely magnetized rocks of similar age should be found the world over. Thorough research provided convincing evidence that the geomagnetic field has reversed its polarity several times in the past. Recently, detailed records from sequences of lava flows and recent sediments have actually revealed the morphology of the so-called transitional phase of a reversal. Rather than an actual flipping of the main dipole field, the short (approximately a few thousands of years in duration) transitional period is characterized by decay of the main dipole and domination by weaker nondipole fields that appear to persist from one reversal to another, and therefore may be controlled by relatively stationary core-mantle interactions (**Fig. 2**). With time, a main dipole of opposite polarity increases in its contribution to the field.

the term paleomagnetic secular variation is applied. Studies of recent geomagnetic secular variation and the directional and intensity variations of the field as recorded in rocks indicate that well over 80% of the field can be represented as a simple dipole, where the inclination increases in value from the Equator to the poles. At any instant in time, magnetic poles represent where the best-fit dipole intersects the surface of Earth (**Fig. 1**). Averaged over a geologically sufficient amount of time, possibly as long as several hundreds of thousands of years, the field appears to be reasonably represented by an axial geocentric dipole.

Records in rocks. Rocks provide records of ancient fields and geomagnetic field reversals. The

Paleomagnetic reconstructions. The earliest systematic studies in paleomagnetism focused on acquisition of paleomagnetic secular variation records (for example, in young lava flows and

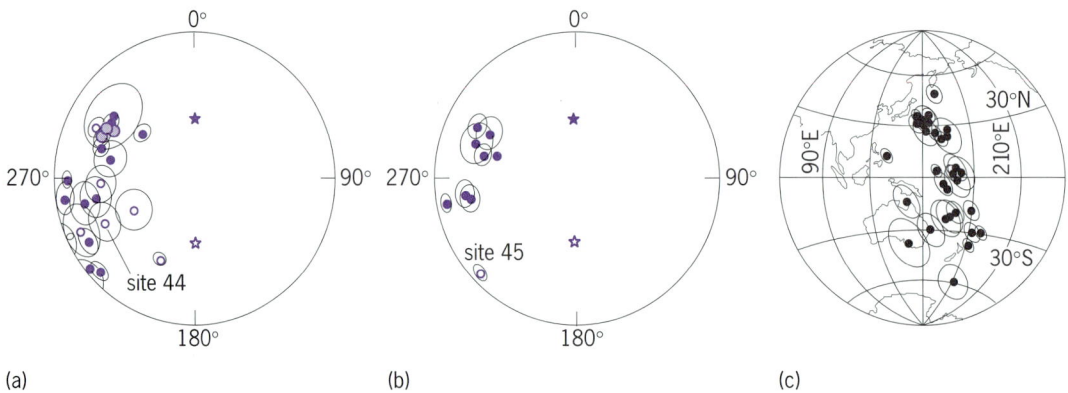

Fig. 2. Example of paleomagnetic data recording a geomagnetic polarity reversal in late Miocene time (about 8.5 million years ago). (*a,b*) Equal area projections of site mean directions (geographic coordinates), with circles of 95% confidence about the site mean direction, from mafic rocks exposed at Paiute Ridge, Halfpint Range, southwest Nevada. Open symbols refer to upper hemisphere projections; filled symbols refer to lower hemisphere projections. Stars represent the late Miocene expected axial geocentric dipole direction. Light-colored circles in *a* are site means from three lava flows; all other data are from either intrusions or host contact rocks. (*c*) Orthographic global projection of virtual geomagnetic poles derived from the site data. (*After C. D. Ratcliff et al., Paleomagnetic record of a geomagnetic field reversal from the late Miocene mafic intrusions, southern Nevada, Science, 266:412–416, 1994*)

glacial lake sediments). The development of a sensitive astatic rock magnetometer in the 1940s paved the way for a broader scope of research, including investigation of rocks dating back to the Precambrian period. In the 1950s workers from Great Britain began to acquire paleomagnetic data from Phanerozoic rocks of western Europe and North America to test the proposed continental drift versus the magnetic polar wandering hypotheses. By 1956, data sets indicated systematic discrepancies of more than 25° between Europe and North America, thus supporting the continental drift hypothesis.

The concept of the paleomagnetic pole was a significant product of early paleomagnetic research. Workers needed to compare magnetizations of identical age from widespread localities on a continent, because simple comparison of paleomagnetic directions was inappropriate. Assuming an axial geocentric model for the time-averaged field and given the location of the site, investigators could transform directions into poles, giving rise to the field at the locality studied. A paleomagnetic pole signified that the geomagnetic field has been well averaged through a set of independent measurements of the field. The term virtual geomagnetic pole was used for a representation of a short part of field time, for example, a single lava flow or sedimentary bed. Symmetry arguments show that the location of a paleomagnetic pole relative to a continent does not define the continent's paleolongitude. According to the dipole formula, $\tan(I) = 2 \cot(\phi)$, where I is inclination, and ϕ is colatitude, inclination is a function of the distance from pole to the sampled area, and thus it provides an estimate of paleolatitude.

A major goal of paleomagnetic research has been to provide a set of well-dated paleomagnetic poles from rocks which have remained part of the stable part of a continent since their formation. Accurate apparent polar wander paths are defined by paleomagnetic poles, each meeting a series of minimum reliability criteria, and provide a most powerful base by which the past motions of continents can be determined and compared with other continents or with lithosphere fragments that eventually are added to continents. This information provides the principal means of reconstructing continents prior to the early Mesozoic breakup of Pangaea, the postulated supercontinent composed of all the continental crust of Earth. Inferences based on path analysis of apparent polar wander are only as sound as the database, in terms of the quality of pole determinations and in turn of the accuracy of the age assigned to the magnetization. One approach to compiling apparent polar wander paths uses so-called running means with a particular time window to define the average paleomagnetic pole for some time. A more recent suggestion assumes that all lithosphere motion can be described by rotation about an Euler pole, a pole of rotation on Earth's surface that can be used to describe the history of a lithosphere plate for a particular period of geologic time, with rotation rates and positions of Euler poles varying with time. If true, age progressions of paleomagnetic poles should define a small circle and, thus, the location of the Euler pole. *See* PALEOGEOGRAPHY.

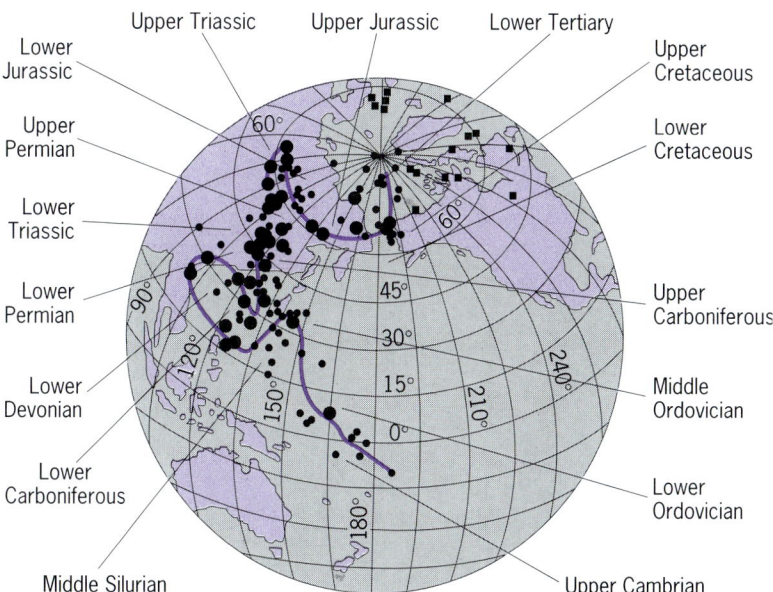

Fig. 3. Apparent polar wander path (colored line) constructed for the North American craton for most of the Phanerozoic (late Cambrian to present) on the basis of paleomagnetic poles (circles) that meet a number of reliability criteria. Poles that meet a greater number of criteria are indicated with larger circles. For comparison, selected paleomagnetic poles from rocks of the western borderland of North America are indicated by squares; these poles lie consistently clockwise of and farther from poles of similar age for stable North America. (*After R. Van der Voo, Paleomagnetism of the Atlantic, Tethys, and Iapetus Oceans, 1993*)

Figure 3 shows an apparent polar wander path that has been constructed for the North American craton for most of the Phanerozoic (Late Cambrian to present) on the basis of paleomagnetic poles that meet a number of reliability criteria, including but not limited to a sufficient quantity of samples and sites studied, adequate progressive demagnetization, coherence of the magnetization signal, no evidence of remagnetization, accurate estimate of age of magnetization, and tectonic coherence of the sampling area. Of all the continents, this apparent polar wander path for North America is probably the best determined, but controversy over both details and even the geometry of the path during some time periods still continues, fostering current efforts at better paleomagnetic determinations. The latest Precambrian through Phanerozoic North American apparent polar wander path records several global tectonic events. At and near the Precambrian-Cambrian boundary (around 545 million years ago) data suggest that North America was at high southern latitudes, possibly in close proximity to one of the Gondwana continents. From Cambro-Ordovician time (around 500 m.y.a.) to the early Mesozoic, poles are approximately 90° from the interior of North America, suggesting a near-equatorial position for the continent that is consistent with the kinds of sedimentary rocks deposited. The general similarity of poles for the late Paleozoic to earliest Triassic probably reflects the final closing of Iapetus and the stabilization of the supercontinent Pangaea. From Middle Triassic to earliest Jurassic time, North America experienced counterclockwise rotation. After earliest Jurassic time, however, a major change in plate motion occurred, and in Cretaceous time clockwise rotation combined with northward translation became dominant. Over this time interval, the continents bordering the North Atlantic separated.

The Mesozoic North American apparent polar wander path is an excellent example of current debate in the paleomagnetic literature. Part of the debate stems from the difference in approach to defining the path, by running means or paleomagnetic Euler poles. The controversies are exacerbated by the fact that most of the data have come from either well-exposed stratified rocks on the Colorado plateau of the southwestern United States or Mesozoic rift basins of the eastern seaboard. The Colorado plateau may have experienced up to about 13° of clockwise rotation since the late Mesozoic; the structural history of rift basins is also controversial. Although apparent polar wander paths have been established for the Phanerozoic for most continents, the database is not necessarily as comprehensive as that for North America. Data of sufficient quality exist to reveal many features of the past tectonic history of the continents. Paleomagnetic data from India document its rapid northward drift since the breakup of Gondwana beginning in Early Jurassic time. In comparison to North America, the Mesozoic and Cenozoic parts of the apparent polar wander path for South America indicate little motion of this continent. The few good Paleozoic data from South America, Africa, and Australia, however, indicate a complex motion history of these continents as part of Gondwana.

Growth and deformation of the continents. On smaller scales, paleomagnetism is a powerful tool for studying the construction of continents and their brittle deformation. Rocks of the western margin of North America, the subject of numerous paleomagnetic studies since the 1960s, have yielded aberrant data compared with the North American apparent polar wander path, in that their poles often fall to the right of or on the far side of the apparent polar wander path (Fig. 3). The results imply, in an absolute sense with respect to North America, clockwise rotation (eastward declinations) and northward translation (shallower inclinations). Rocks from Baja California to southern Alaska have revealed these paleomagnetic discordances. In conjunction with structural, stratigraphic, and paleontologic observations, the data demonstrate that much of the western borderland of the North America Cordilleran mobile belt consists of tectonically transported terranes of different sizes and origins, which eventually collided with North America. For some terranes, the timing of collision and suturing with North America has been estimated by direct comparison of paleomagnetic data.

Paleomagnetic studies may serve as impressive complements to other approaches in evaluation of crustal deformation, in particular when conventional field-based studies have limited predictive abilities. Studies include rotation (vertical axis) of crustal blocks in response to shear along one or sets of strike-slip faults or differential transport-contraction of thrust sheets, and tilting (horizontal axis) of crust during extension or compression. The magnitude of deformation may be determined in either absolute or relative reference frames. In the former, a sufficient averaging of the field must be obtained from the rocks studied because the data from the deformed region are quantitatively compared with expected directions from paleomagnetic pole data derived from the stable part of the continent. A large number of independent records of the field must be obtained from the area that is thought to be deformed. In the latter, a single geologic material with a coherent paleomagnetic signature is traced across a zone of deformation, and relative deformation is deduced by comparing data from undeformed and deformed areas.

Remagnetization. Virtually all geologic materials are subject to one process or another that could lead to partial or complete remagnetization of a pre-existing remanence. Paleomagnetists generally take the approach that most rocks are capable of being remagnetized; hence there must be careful progressive demagnetization work and field-based tests on the antiquity of a magnetization. Nonetheless, the recognition of pervasive secondary, yet geologically ancient, magnetizations has resulted in important avenues of paleomagnetic and rock magnetic research. One subjective method of assessing the antiquity of a magnetization involves comparing its paleomagnetic pole with available data. If the pole is dissimilar to those from rocks of younger age, then there is a strong likelihood that it is not a secondary magnetization. In the 1970s, data from several continents led to speculation about a widespread remagnetization episode, in late Paleozoic time, affecting older Paleozoic strata. A late Paleozoic age of remagnetization was supported by a similarity between paleomagnetic poles from older Paleozoic rocks and those derived from primary magnetizations in late Paleozoic rocks, as well as the uniform, reverse polarity of suspect magnetizations in older Paleozoic rocks. These magnetizations were arguably acquired during the Kiaman reverse polarity superchron, at least 60 m.y. in duration, including much of the Pennsylvanian and Permian periods. During this span of time Earth's magnetic field was essentially in a constant, reverse polarity state. For North America, carbonate rocks, hematitic sandstones and siltstones, and even underlying crystalline rocks have been remagnetized. The fact, among others, that the immobile portion of the continent (craton) is spanned by a distribution of rocks remagnetized in late Paleozoic time has prompted an association among remagnetization, plate tectonic processes (for example, closure of the Iapetus Ocean and Alleghenian deformation), and regional fluid migration. In at least some examples, remagnetization has been directly linked to growth of secondary magnetic phases.

Magnetostratigraphy and isotopic age. The significance of the geomagnetic polarity time scale was first recognized in the early 1960s. First-order estimates of age durations of polarity magnetozones (parts of the rock record having constant or normal polarity) are provided by marine magnetic anomaly data for the past 155 m.y. or so; still, the actual age of magnetozones remains open to question. Recently, the combination of high-precision argon-40/argon-39 isotopic age spectrum and magnetic polarity data for igneous rocks has led to improved estimates of ages of parts of the polarity time scale. Further work in this field promises to significantly improve the chronological framework provided by the geomagnetic polarity time scale and thus of estimates of ages of sequences of rocks and geologic processes based on it.

Magnetism and climate records. Remanence and associated magnetic property data from rocks of recent sedimentary origin have played an increasingly important role in diverse areas of environmental global change. High-fidelity secular variation records from recent well-dated strata have provided an important chronological framework for numerous related studies. The effects or causes of these short-term variations in the field are not completely known. Changes in atmospheric concentration of carbon-14 over periods of thousands of years have been attributed to fluctuations in the dipole moment and resulting effects on the cosmic ray spectrum. Periodicities on the order of 2400–3000 and 7000–12000 years recognized in records of paleomagnetic secular variation have been explained in terms of so-called standing oscillations or wave forms in the convecting liquid core. Although the frequency of geomagnetic field excursions is relatively poorly known, well-defined instabilities in the field appear to have occurred every 105,000 years or so, at least during the last 780,000 years that Earth's magnetic field has been in a normal polarity state. The relationship of such phenomena to orbital forcing dynamics is and will continue to be of interest.

Systematic study of variations in magnetic properties of recent sediments has revealed how the content and mineralogy of magnetic phases are controlled by factors such as climate, tectonism, depositional environments, and source regions. The field of biomagnetism arose through the recognition that magnetic precipitates can form through biochemically controlled processes, and research has shown the strong dependence by several organisms on Earth's magnetic field for orientation and navigation.

For background information SEE GEOMAGNETIC VARIATIONS; GEOMAGNETISM; GLACIAL EPOCH; PALEOMAGNETISM; PLATE TECTONICS; ROCK MAGNETISM in the McGraw-Hill Encyclopedia of Science & Technology.

John W. Geissman

Bibliography. R. F. Butler, *Paleomagnetism: Magnetic Domains to Geologic Terranes*, 1992; C. Kissell and C. Laj (eds.), *Paleomagnetic Rotations and Continental Deformation*, 1988; R. T. Merrill and M. W. McElhinny, *The Earth's Magnetic Field: Its History, Origin, and Planetary Perspective*, 1983; R. Van der Voo, *Paleomagnetism of the Atlantic, Tethys, and Iapetus Oceans*, 1993.

Photoselective chemistry

The highly developed scientific discipline of photochemistry is based on the promotion of chemical reactions by light. During a chemical reaction, one or several bonds of the reactants' configuration are broken, while new bonds are formed that are characteristic of the products. Energy is required to make this happen. Traditionally, chemical reactions are made to go faster by heating, where the control of the energy content of the reactants is via their temperature. Molecules have many modes of motion: they can move faster, they can rotate, their atoms can vibrate, and their electrons can acquire higher energy. When a sample of molecules is heated, each mode of motion gains energy. One way of depositing energy in a selective fashion is by the absorption of visible or ultraviolet (UV) light, which typically causes the excitation of the electrons and promotes alternative ways of reaction.

Selective excitation. The invention of lasers has opened up several novel possibilities as well as markedly improved known techniques. These advances all involve the fact that each molecule absorbs light of different wavelengths in its own characteristic fashion (it is customary to think of the light absorption as the so-called fingerprint of the molecule). Much work has been based on the well-defined wavelength of laser light. An exciting new venue is the control of the reaction as the atoms are moving. Since this motion is very rapid, guiding the atoms requires the use of the new femtosecond (1 fs = 10^{-15} s) lasers. These lasers access a time domain smaller than a typical molecular vibration (10–100 fs) and, just as in high-speed photography, can thereby provide a time-domain picture of the breaking and making of the bond during the course of the reaction. Thus, chemists can take snapshots of the nuclear motion.

To control chemical reactivity, energy must be pumped so as to lead preferentially to a desired chemical end. Photoselective chemistry seeks to use laser light sources to go beyond what can be achieved by heating or by conventional light sources. Lasers play a fundamental role in developing understanding of the acquisition, storage, and disposal of energy during the course of a chemical reaction. They provide a wide variety of ways of selectively exciting reactants as well as of probing the subsequent evolution of the activated molecule from energetic, spatial, and temporal points of view.

Working against the selective excitation are the myriad of dissipation pathways available to the excitation energy. In photoselective excitation, the energy is deposited in a localized site of the molecule, thereby creating a hot spot. This molecular hot spot will, in time, cool down by means of collisions with other molecules and by intramolecular energy redistribution. For a reaction to take place, the initial excitation needs to migrate to the site where it is needed (or to stay in its original location if that is where reaction is to take place) rather than undergo a random dissipation. Therefore, selective excitation is useful in achieving specificity in the products' formation if, simultaneously, the pathways followed by the excitation energy from the reactants' to the products' configurations can be controlled. This control depends crucially on the different time scales of the processes occurring in the activated molecule compared to the rate of the reaction of interest. Control can be passive or active. Passive control implies that once the reactants have been excited, they evolve on their own, without any further external influence. However, in the active control approach laser pulses are used during the course of the reaction to guide the evolution of the system toward the products' configuration.

Laser chemistry. The useful properties of the laser source (most notably, its frequency, the power output, a pulsed or continuous operating mode, and the polarization) as well as the properties of the absorbing molecule itself determine the nature and the degree of selectivity of the excitation. (In laser control an attempt is made to modify the properties of the molecule toward a desired goal.)

A molecule can absorb a photon if the laser frequency matches the energy difference between two of its molecular energy levels. The excited state can be a rotational, vibrational, or electronic state, depending on the wavelength of the laser source. The spectrum is a succession of such transitions in order of their increasing energy. At low internal energies, the molecular energy levels are well separated in energy, and the spectrum is discrete. As the amount of energy provided to the molecule increases, the spectrum becomes more congested because of the higher density of levels, turning first into a quasicontinuum and ultimately, for energies high enough for bond breaking, into a real continuum. The quasicontinuum as well as the continuum itself can be quite bumpy, reflecting a clustering of levels.

Laser excitation is species-selective, because it is in most cases possible to find a frequency range where, in a mixture of several types of molecules, only one component absorbs. This property is true of both single and multiphoton absorption, and it is widely used in applications. One important example is the selective removal of trace impurities when high-grade materials are required, as for microelectronics. Another application is the separation of isotopes, when only the molecule containing one isotope absorbs. In most cases, isotopic separation is achieved by the selective dissociation of one isotopic molecule in the mixture, induced by the sequential absorption by the same molecule of many infrared photons. Its applicability spans a wide range of the elements, including the heavy elements such as uranium-235 (^{235}U) and uranium-238 (^{238}U).

Lasers are also used to initialize or trigger chain reactions by breaking a bond in a particular component. At high pressures, high-power multiphoton activation induces chemistry similar to that obtained by thermal heating, with the advantage that it is a homogeneous heating without the side effects of degradation of material at the hot walls. Laser-induced chemical vapor deposition prepares high-quality thin films for photonics and electronics. Photoselective probing of molecules is used in many analytical applications, and is particularly important in environmental studies because of the long-range sensing made possible by the well-collimated nature of laser light.

Excited electronic states. The chemistry of the excited electronic states is often very different and much richer than that of the ground state. One intuitive way to describe the dynamics of electronically excited states, known as the Franck-Condon principle, is that during the very rapid electronic transition (less than 1 fs), the heavy nuclei are essentially at rest. The electronic excitation creates an excited state, but with the atomic arrangement of the ground state. Such a configuration is nonstationary, and its temporal evolution, in particular its delocalization, can be probed directly with femtosecond-pulsed lasers. Simultaneously, the development of time-propagation techniques allows theorists to compute this temporal evolution as well as to design active control schemes involving pulsed external fields. A schematic technical representation of the essence of photoselectivity, that is, the evolution of a nonstationary state created by electronic excitation, is shown in **Fig. 1**. The solid curves represent the two potentials for the motion of the atoms in a diatomic molecule AB versus the interatomic distance of the molecule. The excited molecule [AB]* prepared on the upper potential can dissociate into its constituent atoms in two different ways, giving either A* + B or A + B. The much faster motion of the lighter electrons permits an assumption that the heavy nuclei do not move, while the electrons adjust themselves and thereby determine the potential for the nuclei to move in. The potentials are different for the ground and the excited electronic states, and it is for this reason that upon electronic excitation, the chemistry, which reflects the motion of the nuclei, can be different. Light (typically in the visible or the ultraviolet range) can induce transitions between the two electronic states, and the nuclei hardly move during this fast transition. If before the

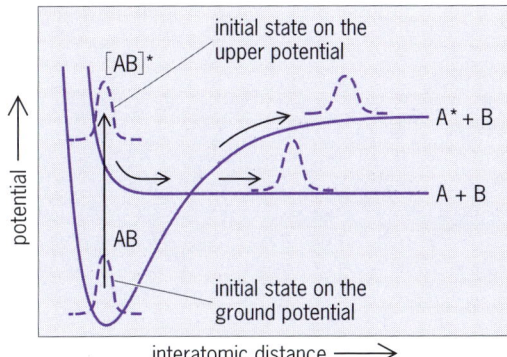

Fig. 1. Evolution of a nonstationary state of a diatomic molecule AB. The histograms (broken lines) indicate the various places where the atoms can initially be on the upper surface and their location after some time has elapsed.

excitation the molecule was fairly cold, the atoms are at or near their equilibrium position (bottom of the potential well) in the ground state. After the excitation, they are no longer in equilibrium, so that a force acts on them. As a result of the force the atoms will move. The histograms in Fig. 1 (broken lines) show various places the atoms can be during the course of the dissociation.

Specificity and intramolecular vibrational energy redistribution. Considerable insight in the interplay among selective excitation, intramolecular vibrational energy redistribution, and specific reactions is provided by laser-pump/laser-probe experiments, which if the time between the two is short enough avoid collisional relaxation. Mode-selective vibrational excitation on the ground electronic state can be achieved by the excitation of the vibrational levels of a specific mode, such as the C-H bond in acetylene (C_2H_2), and is called direct overtone excitation. Steric requirements are studied with polarized light, both in the pump and in the probe lasers. A useful technique to probe product states' distribution is laser-induced fluorescence. The products are first electronically excited, and their distribution is inferred from their light emission back to the ground state. Such experiments show that chemical reactions are not only selective in their energy requirements but also manifest specificity of the energy released. An illustration of the specificity of the energy release is shown in **Fig. 2** for the reaction below between a fluorine atom

$$F + H_2 \longrightarrow H + HF_v$$

(F) and a hydrogen molecule (H_2), where v enumerates the different final vibrational states of the product molecule hydrogen fluoride (HF). The potential energy profile as the reaction proceeds is shown in Fig. 2a. The channeling of the energy into the different vibrational states of HF is indicated by the arrows to Fig. 2b. In Fig. 2b, the observed nascent population of the final vibrational states of the HF molecule P_v is compared to the distribution expected for a fully nonspecific energy release, P^0_v. The inversion of population (which is required for chemical lasing activity) and the specificity of the energy disposal are evident. The specificity of the energy release is usually measured by the quantity $-\ln(P_v/P^0_v)$, which is called the surprisal (Fig. 2c). For the $F + H_2$ reaction (as well as for many other bimolecular reactions), the surprisal is linear, and a quantitative measure of the population inversion is provided by the slope λ_v of this plot.

Theoretical considerations show that intramolecular vibrational energy redistribution in polyatomic molecules is essentially an ordered process, proceeding in successive stages on different time scales. Experimental evidence is provided by stimulated emission pumping, which uses a two-step excitation scheme to access the high vibrational states of the ground state.

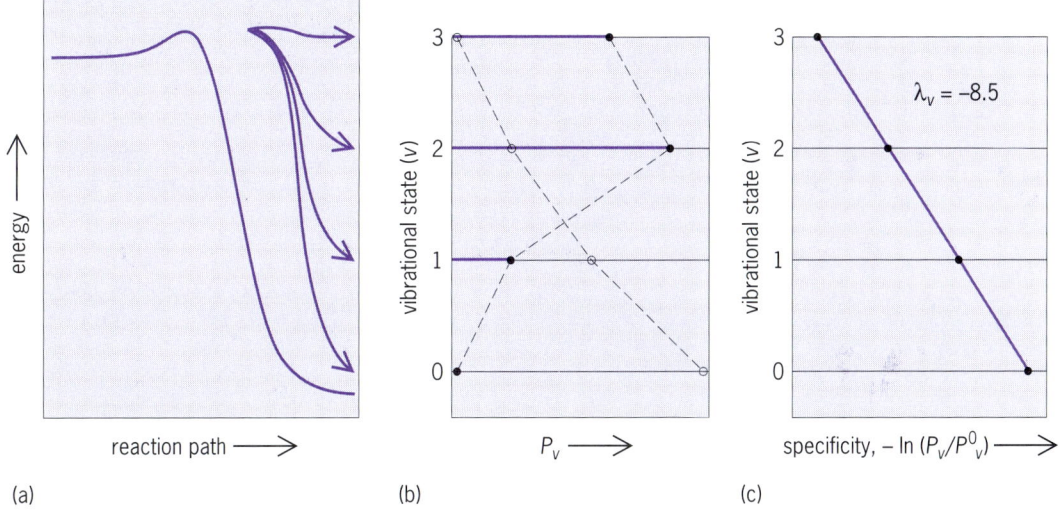

Fig. 2. Description of the specific energy release for the reaction $F + H_2$. (a) Potential energy profile for the reaction $F + H_2 \rightarrow H + HF_v$. (b) Observed nascent population P_v (solid circles) of the vibrational states of the HF molecule and the population P^0_v (open circles) which would result from a fully nonspecific energy release. (c) Measure of the specificity $-\ln(P_v/P^0_v)$ versus vibrational state.

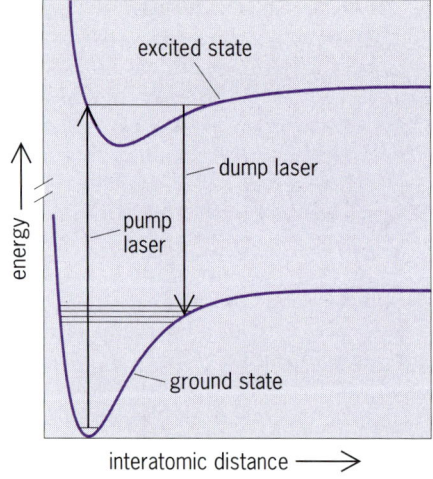

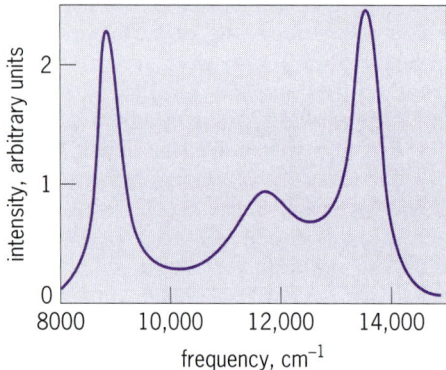

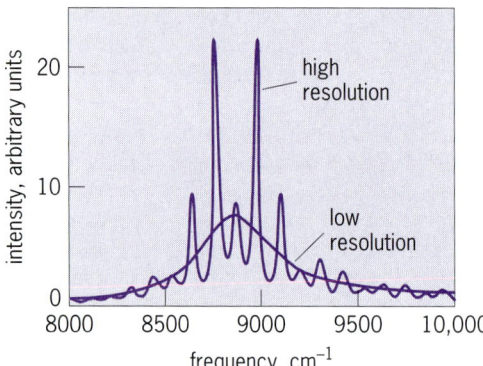

Fig. 3. Stimulated emission pumping. (a) Two-laser arrangement. (b) Low-resolution absorption spectrum. (c) Portion of curve in b (8000–10,000 cm^{-1} frequency range), showing high-resolution absorption spectrum superimposed on low-resolution spectrum.

Figure 3 shows absorption spectra resulting from the selective vibrational excitation of a polyatomic molecule via stimulated emission pumping. A pump laser is used to selectively access a rovibrational state of an excited electronic state. The transition down to the high vibrational levels of the ground state is then stimulated by the dump laser. The atoms hardly move during a change in the electronic state, so that it is possible to specifically access high vibrational levels of the ground state. The typical absorption spectra (Fig. 3b and c) exhibit a nested structure where an unresolved clump at the lower level of energy resolution specifically corresponds to a set of distinct lines at the higher resolution. Seen from the point of view of time evolution, the different inherent spectra observed at different levels of resolution imply a separation of time scales in the dynamics, thereby providing evidence for the sequential nature of the vibrational energy redistribution.

Specificity is thus a question of competition between intramolecular vibrational energy redistribution and the rate of the reaction of interest. Fast direct bimolecular reactions proceed via a concerted motion of the nuclei on a time scale of a vibrational period, so that there is essentially no time for intramolecular energy redistribution. In this simple case, it has been found possible to enhance or to decrease the rate constant compared to that observed when the same amount of energy is provided as heat. The opposite limit is that of unimolecular reactions for which the stage of the formation of the energy-rich intermediate is distinct and well separated in time from the reaction itself. Mode selectivity can be achieved by pumping the energy-poor molecule above the threshold for reaction. While direct overtone excitation of small polyatomic molecules indeed induces specific dissociation or isomerization, infrared multiphoton excitation, which occurs on a longer time scale, typically leads to complete randomization of the excitation energy prior to the bond rearrangement.

Perspectives. Laser light is expensive to generate, so photoselective industrial chemistry using lasers is likely to be restricted to cases where either a little light can induce a lot of chemical change or the value added is high, as in isotope separation. However, the exploration of selectivity has revealed much about how the atoms move when a chemical reaction takes place and has provided sharper tools for the promotion and analysis of the chemical act.

For background information SEE CHEMICAL DYNAMICS; LASER; LASER PHOTOCHEMISTRY; PHOTOCHEMISTRY; ULTRAFAST MOLECULAR PROCESSES in the McGraw-Hill Encyclopedia of Science & Technology.

R. D. Levine; F. Remacle

Bibliography. A. Ben-Shaul et al., *Lasers and Chemical Change*, 1981; F. F. Crim, State- and bond-selected unimolecular reactions, *Science*, 249:1387–1392, 1990; J. Jortner, R. D. Levine, and B. Pullman (eds.), *Mode Selective Chemistry*, 1991; V. S. Letokhov, *Nonlinear Laser Chemistry*, 1983; R. D. Levine and R. B. Bernstein, *Molecular Reaction Dynamics and Chemical Reactivity*, 1987; S. A. Rice, New ideas for guiding the evolution of a quantum system, *Science*, 258: 412–413, 1992; A. H.

Zewail and R. B. Bernstein, Real-time laser femtochemistry: Viewing the transition from reagents to products, *Chem. Eng. News,* 66(45):24–42, 1988.

Pipeline

Recent advances in pipeline technology permit the installation of utilities underground without opening a trench. Trenchless pipe installation involves placing a pipe below the ground surface without open-cut excavation. This trend toward less disruptive methods has resulted from a number of factors, including the changing needs of the utilities who install the pipes; growth in vehicle ownership and traffic density; greater concern for the environment; and development of technology, equipment, and materials to answer the challenge. Pipe and cable utilities have to replace or upgrade undersized, worn-out, and neglected pipes and cables to meet changing and increased demands. The traditional approach of digging a trench in the road to remove and replace the old pipe can be very disruptive to the community and damaging to adjacent structures and existing utilities. This kind of work is slow and expensive for the owner of the pipeline. It has been estimated that for a typical 6-in. (150-mm) diameter gas main replacement installation in an urban center, some 95% of the cost is accounted for by removing and replacing the road surface together with excavating and refilling the trench. Only 5% of the cost is accounted for by the permanent part of the works, that is, providing and laying the gas pipe. Trenchless technology offers two alternative approaches to the problem of replacing an underground pipe: on-line replacement, where the new pipe is simultaneously installed as the old pipe is removed, and off-line replacement, where the old line is abandoned and a new installation is made.

On-line replacement. This type of installation uses techniques that have been developed to deal with a wide range of pipe types and sizes. The simultaneous removal and replacement of direct buried cables is much less well developed. Where cables have been laid in ducts, their replacement is simple. On-line pipe replacement techniques can be divided into four main categories: bursting, extraction, splitting, and excavation.

Bursting. In this technique a device is inserted in the old pipe to apply radial forces that shatter it and force the fragments outward into the surrounding soil. Simultaneously, the new pipe is pulled or pushed in behind the burster device as it is drawn through. A somewhat larger hole than the old pipe size can be created where a larger new pipe is needed. The bursting device takes two main forms. One form is an air-powered impact hammer with an oversized head. The tool hammers its way along the old pipe, breaking it and pushing the fragments outward. The alternative approach is to use a hydraulic-powered device that in its contracted form will pass through the pipe, but that has powerful leaves that can be expanded to shatter the pipe and force the fragments outward.

Bursting is designed to break up brittle pipe materials such as unreinforced concrete, clay, brick, asbestos cement, and cast iron. It has found extensive use in the gas and water industries for replacing old cast iron lines. British Gas, which pioneered the method, has replaced several hundreds of miles of pipe in this way. The other main application is for replacement of old sewers. It is generally possible to undertake the sewer-bursting operation from within an existing access hole, thus greatly minimizing costs and disruption. Sewers with diameters from 4 to 24 in. (100 to 600 mm) have been replaced.

Extraction. These techniques are based on passing a wire line, chain, or steel rod through the old pipe and applying large tensile forces to the cap on the end, pulling out the old and simultaneously pulling in a new pipe. A variation used for the removal of lead water services (water connections made with lead pipe) involves passing a wire through the service pipe, filling the annular space with setting resins, and then pulling out the composite pipe that has formed.

Splitting. This method has been developed to deal with the problem of steel pipes, which are ductile and cannot be burst. The gas industry in the United States is particularly concerned with this problem. Large tensile forces are used to draw a cutting blade followed by an expander through the old steel pipe. The steel pipe is cut and opened outward into a C shape, allowing the new pipe to be pulled into the space created. This technique has been used so far for the smaller pipes of 6 in. (150 mm) or less in diameter.

Pipe excavation. This technique uses adaptations of the small-diameter remote-controlled tunneling machines called microtunnelers developed for laying sewer pipes 36 in. (900 mm) or smaller in diameter. By modifying the cutting head the microtunneler can break up and excavate an old sewer pipe while the new pipe is pushed behind the advancing machine. Initially the method was limited to pipes of brittle material such as unreinforced concrete and clay. Japanese manufacturers have now incorporated a clipping device to cut the steel rods in reinforced concrete pipes.

Advantages. On-line replacement of pipelines, when used correctly, offers the following advantages: reduction of surface disruption; use of the existing line, avoiding the need to find a new route; the ability to replace with a new pipe of choice, having upsizing possibilities; ease of replacement of old pipes that are badly damaged or deformed; and substantial cost savings compared to trenched replacement.

Disadvantages. Like all techniques certain limitations need to be taken into account: (1) The line has to be put out of operation, and service has to be suspended. (2) Replacement not only breaks up the old pipe but all the connections to that pipe not previously disconnected. (3) Disconnecting and reconnecting services is usually done from an open excavation; a large number of connections requires numerous such holes, which may be as disruptive as an open trench. (4) The expansion action results in movements in the surrounding soil, which can disturb and damage adjacent pipes and structures.

Off-line replacement. This method is based on setting a new line. Off-line replacement gives greater design freedom in terms of location, size, and depth of burial. Techniques are broadly divided by the type of pipe to be installed. Sewers, designed to operate by gravity, need to be laid to a constant gradient on straight lines, with close tolerances. Sewers are often laid much deeper than the other utilities and and present problems of installation. Gas and water pressure pipe and cables need not be laid to these close tolerances of either line or level. Distribution networks for pressure pipe and cables have traditionally been laid within 48 in. (1200 mm) from the surface, with transmission lines somewhat deeper. For both sewers and pressure pipes the final operational pipeline is installed. For cables either a duct is installed or the cable directly pulled in. Until relatively recently, for a variety of reasons it was necessary with these types of methods to lay a duct or carrier pipe and then insert the operational pipeline inside this duct, forming a pipe in a pipe. With developments in pipe and installation technology, this use of a duct has been eliminated; the operational pipeline (gravity and pressure) can be directly installed by the trenchless method. In the case of cables it is still common to use a carrier duct in which the cables are pulled through. However, it is also possible to pull cables directly into the ground in the same way that a plastic pipe would be pulled in.

Sewer installation techniques. These techniques are based on the principle of pipe jacking, where a cutting shield is pushed from a working shaft and pipe sections are used to line the bore that is formed by the cutting shield. Pipes from 6 in. (150 mm) to 180 in. (4600 mm) in diameter have been installed in this way. Methods fall broadly into two categories; for diameters up to 36 in. (900 mm) where entry and working by personnel is not possible or safe, remote control miniaturized tunneling machines are used. Excavation, spoil handling, monitoring, and steering all have to be done from a remote control panel.

For diameters of 42 in. (1070 mm) and greater, the choice of shield depends on the stability of the ground. Where soil and water table do not create instabilities, an open-face shield is possible, with personnel at the face working in a manner similar to traditional tunneling. However, where ground conditions are difficult, or for long lengths, remote-control closed-face tunneling machines are used. Equipment of this kind is designed or adapted to deal with particular ground conditions, which can range from solid rock to unstable silts and sands below the water table.

The standard pipes used for sewer construction have been adapted for use with jacking methods. Pipes can be installed within tolerances of ±2 in. (50 mm) for line and level over lengths of several hundred feet.

Pressure pipe and cable installation. Techniques for such installations are based on drilling methods, and they were developed in the United States in the early to mid-1980s and have found rapid worldwide acceptance. A guided drilling rig located at the surface rotates and drives drill pipe with a cutting head to form a pilot bore, which can be on a horizontal or vertical curve. The cutting head is fitted with a sonde which emits a fixed frequency signal. This signal is picked up by an operator with a detection instrument who can determine the boring head position for both line and depth. If a steering change is required, the head, which has a chisel profile and high-pressure fluid cutting jets, is set to the required course and pushed forward without rotation to move the head in the required direction. When the pilot bore, which is usually about 2 in. (50 mm) in diameter is completed to the required line and depth, a reamer or enlarger is attached to the drill rods and forms the final bore diameter by rotating and pulling back to the drill rig. The final stage is to attach the pipe to be laid behind the enlarging head and pull that back through the formed bore. Drill mud, based on bentonite, is normally used as the cutting fluid, which helps stabilize the borehole as it is formed and also lubricates and protects the pipe as it is pulled in. Gas and water distribution pipes with diameters ranging typically from 2–12 in. (50–300 mm) can be installed. Larger diameters and long drive lengths are possible by using more powerful machines.

The most widely used pipe for these installations is continuous length polyethylene, although other flexible pipes such as steel can be used. Replacement of failing power cables has been a major application.

Guided boring offers considerable advantages over trenching for urban and suburban locations because it is much less disruptive, quicker, and less costly. There are limitations to the type of soil conditions where this technique can be used. Normally, ground containing cobbles, heavy gravels, and rock is not suitable.

It is anticipated that in the coming years much greater use of trenchless methods for the installation and replacement of utilities will be seen worldwide.

For background information SEE PIPELINE; SEWAGE COLLECTION SYSTEMS in the McGraw-Hill Encyclopedia of Science & Technology.

James Thomson

Bibliography. S. Kramer, W. McDonald, and J. Thomson, *An Introduction to Trenchless Technology,* 1992; J. Thomson, *Pipejacking and Microtunneling,* 1993.

Plague

Plague, the cause of the infamous Black Death of the Middle Ages in which millions died, is maintained by rodents and their fleas. However, pet cats and dogs can also become infected and serve as a source of human illnesses and deaths.

Epidemiology. Although several great plague pandemics have occurred in Europe and Asia, the disease continues to be widespread throughout the Americas, Asia, and Africa. Plague was brought to the United States in the early part of the twentieth century by rats on ships arriving at the ports of San Francisco and New Orleans.

In the United States the disease primarily occurs in the western one-third of the country, from the Pacific coast to Kansas, Oklahoma, and Texas. Although cases can occur throughout the country, infection in pet dogs and cats and in humans has been reported primarily from New Mexico, Colorado, Arizona, and California. Between 1970 and 1993, the annual number of cases of human plague in the United States has varied from 1 to 40.

Infectious agent. Plague is caused by infection with the gram-negative bacteria *Yersinia pestis*. Rodents are the usual carrier for the bacteria, and rodent fleas are the most common vector for transmission. In the United States infected animals include wild rodents such as squirrels, chipmunks, prairie dogs, mice, wood rats, cottontail rabbits, and jackrabbits. Numerous other animals may show serologic evidence of infection, including coyotes, wildcats, badgers, dogs, and cats.

Clinical symptoms. Infections are typically separated into three types: bubonic, septicemic, and pneumonic. The term bubonic plague is used when an enlarged tender lymph node (a bubo) is found draining the area of bacterial inoculation. The term septicemic plague is used in the absence of bubos, when the infection is disseminated throughout the bloodstream. The classification pneumonic plague is used when pneumonia occurs. Laboratory-confirmed cases of plague without bubos or pneumonia and without bacterial isolation from the bloodstream are technically unclassified but presumed to be septicemic. Of 119 cats in New Mexico with plague, 53% were bubonic, 37% were septicemic or unclassified, and 10% were pneumonic.

In septicemic plague, the often flulike symptoms include nausea, vomiting, and diarrhea, sometimes progressing to full-blown sepsis, with shock and disseminated intravascular coagulation. Pneumonic plague can be either primary (from inhalation of organisms coughed up by a pneumonic plague case) or secondary (from the spread of the organisms to the lungs in an animal or individual already infected with bubonic or septicemic plague).

In all three types of plague, a fever is usually present, although it may be less frequently documented in animals than in humans. Other predominant symptoms include lethargy and anorexia (loss of appetite). Bubos can become abscessed, and if located in the mouth or neck area drainage from the mouth or nose can lead to coughing or sneezing and difficulty in breathing. In cats, bubos occur most frequently in the lower jaw (submandibular) region, probably resulting from bacterial inoculation through the mouth while eating infected rodents. Bilateral enlargement of the lymph nodes (lymphadenopathy) is more likely with submandibular infection.

The incubation period, or time between exposure and the occurrence of clinical symptoms, is typically 2–7 days, although it may be as short as 1 day for primary pneumonic plague. The survival rate varies from 47% for feline bubonic plague cases to 29% for septicemic or unclassified cases, and only 4% for pneumonic cases. However, these rates increase with use of appropriate antibiotic therapy.

The incidence of asymptomatic infections in cats is unknown, but as in humans they probably occur rarely. Serological studies indicate that dogs may become infected with plague even more often than cats, but dogs infrequently develop clinical symptoms. When symptoms do occur in canine species, they are similar to those found in feline species and in humans.

Although cases of plague can occur at any time during the year, the highest incidence in cats, dogs, and humans is during the summer months when the activity and interaction between rodents, dogs and cats, and humans is highest.

Pathogenesis. Plague is frequently transmitted from rodents to humans or dogs and cats through bites of rodent fleas. In addition, plague bacteria can be transmitted directly from clinically ill animals or humans through coughing infectious organisms into the air or inoculation of the organisms from draining abscesses into open wounds. Transmission from cat to human has occurred in at least 20 cases, either through inhalation of bacteria coughed into the area by pneumonic feline cases or by handling of feline plague abscesses with bare hands that have open wounds.

Animals and humans also have become infected through hunting. The plague bacteria are probably transmitted through flea bites when handling the hunted animal, direct inoculation into open wounds on the hands of a human hunter, or direct inoculation into the mouth of a hunting dog or cat. The predominance of submandibular lymphadenopathy in plague-infected cats indicates that they probably become infected with plague most often because of their hunting activities. Hunting brings dogs or cats into close contact with infected fleas living in rodent burrows, even without direct rodent contact.

Diagnosis. Analysis of bubos may show characteristic bipolar-stained gram-negative rods. Fluorescent antibody examination or culture of a bubo aspirate is diagnostic. Cultures of aspirated material or blood may take several days to form. A fourfold rise in antibody titer is also diagnostic of recent infection. Autopsy or necropsy specimens can also be analyzed.

In both animals and humans, one of these tests may indicate positive results for plague and others may show negative results; thus, submission of specimens for all types of diagnostic tests is recommended. Many local or private laboratories do not have the capability of identifying *Y. pestis*, and may identify the organism as *Y. pseudotuberculosis*, leading to a delay in diagnosis and appropriate treatment. When plague is suspected, specimens should be sent to a state diagnostic laboratory with the capability of identifying plague infections, and appropriate antibiotic therapy should be administered pending the laboratory results.

If the animal is not available for a convalescent serum specimen, a single titer of 32 or greater is presumptive. However, differentiating acute from past infections is difficult with only a single titer. Clinically ill cats and dogs (but more frequently dogs) with a single positive plague titer may be diagnosed with plague when in fact the recent infection is due to tularemia, another bacterial disease that produces similar symptoms. Tularemia titers take longer to rise, and thus the acute serum specimen may indicate a positive plague titer from a previous asymptomatic plague infection and a negative tularemia titer. Convalescent serum, preferably taken 3 weeks to a month after the acute serum and showing a fourfold rise in tularemia antibodies and a stable plague titer, will confirm the recent infection as tularemia and not plague.

Prevention and control. Critical prevention measures include flea control, avoidance of rodents, elimination of rodent harborage, wearing gloves when skinning animals, and wearing gloves and masks when handling ill cats, especially those with fever and bubos in plague-endemic areas.

Flea control can be accomplished by dusting or spraying animals weekly, or by using a flea collar. However, neck irritation can occur with flea collars, and it may be more difficult for the flea repellant to reach the base of the tail. Pyrethrins are effective chemicals for killing fleas, and are safer for animals and humans than other chemicals such as organophosphates or carbamates. Although treating animals for fleas throughout the year is ideal, in colder climates (particularly for nonhunting animals) owners may find flea control in the warm months sufficient. A thorough cleaning and vacuuming of the home is necessary to eliminate fleas brought into the house by dogs and cats.

Since dogs develop clinical symptoms less frequently, they are less likely to spread plague to humans directly, but nonetheless they have served as the source of human infections by bringing home infected rodent fleas. Individuals living in plague-endemic areas may wish to exercise caution about allowing pet dogs and cats, particularly those that are not treated for fleas, to sleep with them.

Prompt diagnosis and treatment will reduce mortality. The bacteria is susceptible to tetracycline, aminoglycosides, long-acting sulfonamides, and chloramphenicol. The tetracyclines can be also useful as preventive measures following exposure to plague bacteria in order to prevent development of symptoms.

For background information SEE PLAGUE; SIPHONAPTERA; TULAREMIA; YERSINIA in the McGraw-Hill Encyclopedia of Science & Technology.

Millicent Eidson

Bibliography. R. B. Craven et al., Plague and tularemia, *Infect. Dis. Clinics N. Amer.*, 5:165–175, 1991; M. Eidson et al., Clinical, clinicopathologic, and pathologic features of plague in cats: 119 cases (1977–1988), *J. Amer. Vet. Med. Assoc.*, 199:1191–1197, 1991; M. Eidson et al., Feline plague in New Mexico: Risk factors and transmission to humans, *Amer. J. Pub. Health*, 78:1333–1335, 1988; P. D. Hoeprich (ed.), *Infectious Diseases*, 1989.

Plant movements

In 1881, Charles Darwin reported that plants have the ability to orient their leaf laminae (surface) in relation to movement of the Sun. This ability, termed heliotropic leaf movement, enables plants to regulate the solar radiation heat load striking the leaf surface. Plants can orient their leaf surfaces parallel, obliquely, or perpendicular to the Sun's rays. Leaf movement has been reported in 16 different families, and is independent of the carbon-3 (C_3) or carbon-4 (C_4) photosynthetic pathways. Since leaf movements are readily reversible, they are not considered growth responses. All leaf movements can be classified into one of three categories: nyctinastic (sleep movements), seismonastic (movements in response to physical contact or shaking), and heliotropic (movements in response to the Sun).

Heliotropic movements. The two types of heliotropic leaf movement are paraheliotropic and diaheliotropic movement. Paraheliotropic leaf movement enables the plant to avoid direct solar radiation by orienting its leaves parallel or oblique to the Sun's rays; diaheliotropic leaf movement enables the plant to track the Sun by orienting its leaves perpendicular to the Sun's rays.

Paraheliotropic leaf movement is an adaptive mechanism used by plants to regulate leaf temperature to prevent heat injury and to maximize net photosynthesis. Since the enzymes that control photosynthesis function more efficiently in a narrow temperature range, plants have evolved mechanisms to maintain leaf temperature within that range. By altering the leaf angle relative to the angle of incident solar radiation, plants can regulate the amount of solar heat load striking the leaf surface, and therefore prevent leaf temperatures from becoming too hot. For example, the common bean (*Phaseolus vulgaris*) can maintain its optimum temperature for photosynthesis (84 and 90°C or 29 and 32°C) by orienting its leaf position relative to incoming solar radiation. In addition, leaf movement to avoid high leaf temperatures reduces the rate of transpiration (evaporative water loss through the leaf), and consequent plant water loss.

Diaheliotropic leaf movement enables the plant to orient leaves perpendicular to solar rays to maximize incident solar radiation. Common beans show diaheliotropic leaf movement by orienting their leaves perpendicular to incident solar radiation in the morning and late afternoon, when air temperatures are at or below the optimum for photosynthesis. Hence, the common bean is an example of species that are capable of both paraheliotropic and diaheliotropic leaf movement, depending upon the air temperature.

Pulvinus. The site of leaf movement is located in the leaf lamina in families such as Malavaceae and in the pulvinus in families such as Fabaceae. The pulvinus is a swollen jointlike region at the base of the petiole and at the base of each leaflet in the case of compound leaves. In cross-section, the pulvinus is composed of vascular tissue located in the central region, which is surrounded by a large region of loosely arranged parenchyma cells which are the agents of photosynthesis. Leaf movement is accomplished by changes in cell turgor of the parenchyma cells in the pulvinus and their concomitant contractions or expansion. Expansion of cells on one side of the pulvinus and contraction on the opposite side enables the leaf to move in the direction of the contracted cells.

The common bean, a member of Fabaceae, possesses four pulvini. One, located at the base of the petiole, regulates the angle of the leaf relative to the stem; three, located at the base of each of its leaflets, regulate the position of the leaflets relative to incident solar radiation. Movement of the leaflets is accomplished by changes in both the rotation and vertical inclination of the leaflet relative to the petiole. Movement of the two lateral leaflets has been shown to be correlated, but that of the terminal leaflet is usually not associated with that of the lateral leaflets.

Response to environmental factors. Environmental factors such as temperature and light (but not atmospheric humidity or carbon dioxide concentration) influence heliotropic leaf movements in the common bean. Under both laboratory and field conditions, leaf orientation is influenced by air temperature when light intensity is high (photon flux density above 50 μmol m^{-2} s^{-1}). When the air temperature is high, bean leaves orient more obliquely to the light source so as to reduce radiation heat load. Under laboratory conditions, leaves that are exposed to high levels of incident radiation and changing air temperature from 68 to 90°F (20 to 32°C), maintain their leaf temperature near the optimum for photosynthesis (~84°F or 29°C) by orienting at angles oblique to the light source. Under light conditions below 50 μmol m^{-2} s^{-1}, leaf response to temperature is negligible.

The sites of light perception that influence leaf movement in the common bean are the pulvini, which must receive direct light to activate leaf movement. The leaf can orient to light when the entire leaf except the pulvini is shaded. Conversely, if the pulvini are shaded and the leaf laminae and petiole are exposed to light, the leaf will not orient to sunlight. Furthermore, the leaf will orient only if the pulvini receive direct light, as opposed to diffuse light. This mechanism prevents the plant from expending unnecessary energy for shaded leaf orientation, since shaded leaves are not exposed to the heat load associated with direct solar radiation.

Genotype variability. A great deal of variation exists among and within species for the ability to express paraheliotropic leaf movement. Leaf movement can be measured by measuring the fraction of the direct solar radiation beam that strikes the leaf surface. This measurement takes into account the relative geometric positions of the leaf surface relative to the Sun's rays, and it is expressed as cosine (i), which can vary between 0 and 1. A cos (i) value of 0 is obtained when the leaf surface is parallel to the incident light rays, and a cos (i) value of 1 is obtained when the leaf surface is perpendicular to the light rays. Cosine (i) can be expressed as in the equation below, where P_C is the amount of

$$\cos(i) = \frac{P_C}{P_T}$$

photosynthetically active light (photosynthetic photon flux density) striking the leaf surface, and P_T is the total amount of photosynthetically active light striking a surface perpendicular to the direct solar beam.

Common bean genotypes differ in their ability to orient to midday solar rays. In general, beans that evolved in Central America have smaller, lighter leaves and a stronger heliotropic response to midday Sun than beans that evolved from the Andean region of South America. Both groups had cos (i) values less than 1 when fully expanded leaves exposed to the Sun were measured at ±2 h of solar noon, indicating that both groups can orient their leaves obliquely to the Sun's direct rays. Mean cos (i) values among common bean lines from the South American source were significantly higher than among those from Central America, 0.46 and 0.32, respectively. These results indicate that the Central American common bean lines oriented their leaves more obliquely to the Sun's rays than the South American common bean lines under the same field conditions. Leaf area and weight were also higher in the South American lines, and are possible factors in hindering leaf movement.

Leaf movement is also influenced by the amount of soil moisture available to the plant. Mean cos (i) measured within ±2 h of solar noon was 0.51 when common bean lines were grown under well-watered conditions, compared to 0.29 under water-stressed conditions. These results suggest that leaf orientation is influenced by the availability of soil water and by plant water status. The association between plant water status and leaf movement is not unexpected. Since water is used to cool the leaf via transpiration, leaves that have limited transpiration potential because of a shortage of available water need to reduce the solar radiation heat load more than plants that have adequate water for transpiration; therefore they orient their leaves more obliquely to the Sun's rays.

Relation to productivity. Leaf movement is an adaptive mechanism that enables plants to survive under warm arid conditions. It has been hypothesized that leaf movement and water use efficiency might be related, since leaf movement to avoid solar radiation can reduce the solar heat load striking the leaf surface and concurrently reduce transpiration. Further, the ability to regulate leaf temperature, which maximizes net photosynthesis, should also contribute to productivity by increasing net carbon gain via photosynthesis.

Studies have not found a consistent statistical association between leaf movement and productivity, as measured by cos (i). Significant associations between cos (i) and productivity were found among progeny derived from a group of diverse common bean parents; however, an association between productivity and cos (i) was not found among the parents of the progeny. It may be desirable to develop common bean genotypes that have a stronger heliotropic movement in order to improve productivity in arid environments, but the association between leaf movement and productivity is weak.

To better understand the genetic variation for control of leaf movement, common bean lines that differed in cos (i) were grown in several environments. There was significant genetic variation for leaf movement based on cos (i), and a heritability estimate indicated that the genetic source of variation was high relative to the environment. The results indicated that approximately 83% of the variation in leaf movement was caused by genetic factors, whereas only 12% was caused by the environment or interactions between the environment and the genotype. Thus, it should be possible to breed common bean lines for a stronger paraheliotropic leaf response. Such lines would be useful to further study the association between heliotropic leaf movement and productivity.

For background information SEE BREEDING (PLANT); LEAF; PHOTOSYNTHESIS; PLANT MOVEMENTS in the McGraw-Hill Encyclopedia of Science & Technology.

Mark A. Brick

Bibliography. C. R. Darwin, *The Power of Movement in Plants*, 1881; J. R. Ehleringer and I. Forseth, Solar tracking by plants, *Science*, 210:1094–1098, 1980; Q. A. Fu and J. R. Ehleringer, Heliotropic leaf movement in common beans controlled by air temperature, *Plant Physiol.*, 91:1162–1167, 1989.

Plant pathogens

Although world food production has increased substantially with the advent of modern or intensive agriculture, plant diseases frequently occur, in some cases causing extensive damage and loss of potential crop production. Plants are prone to many different diseases regardless of the soil used for their cultivation or the climatic conditions, and soil is a major source of plant pathogens. Effective new control measures are needed against these pathogens because of the high costs and detrimental environmental effects of chemical pesticides and the development of resistance to these chemicals by plant pathogens. In certain soils disease incidence is low, or on the decline, even though the pathogen and a susceptible host are present. This phenomenon has been termed disease suppression, and the soil in question called a suppressive soil. Soils where disease occurs are called conducive soils. Soils may be suppressive to many pathogens or specific to only a few related pathogens.

Suppressive soils. The first published reports on the occurrence of suppressive soils was in the late nineteenth century. Differences in soils were noticed in the incidence of wilt in cotton, caused by the fungal pathogen *Fusarium*. The disease was more prevalent in a sandy soil than in a clay soil. When cowpea was planted after cotton, the disease severity increased. In contrast, severity of Fusarium wilt of potatoes was reduced when the preceding crop was alfalfa. Crop alternation, commonly called crop rotation, is recommended as a means to increase yields, to reduce disease incidence, and to improve the soil quality. However, not all continuous crop culture practices, also called monoculture, were found to increase disease. Continuous wheat and barley cultivation is an example. A devastating fungal disease of wheat and barley called take-all, caused by *Gaeumannomyces graminis,* was reduced in Australia and also in the United States with years of monoculture. Controlling disease with years of monoculturing is an interesting proposition, but probably is not practical for modern farming.

Disease suppression is mainly attributed to soil microbial activity, as well as to physical and chemical characteristics of the soil and to crop and cultural practices. The reduced severity of damping off of peppers caused by the fungal pathogens in natural soil compared to heat-treated soil was one of the first indications that disease suppression could be biological. Microorganisms antagonistic to the pathogen were apparently destroyed in the heat-treated soil.

Mechanisms of disease suppression. Both plants and microorganisms are affected in many ways by physical factors such as soil temperature, moisture, aeration, and structure and also by chemical factors such as soil acidity or alkalinity (pH), organic matter, and mineral content. These factors affect the growth, movement, and activity of the microorganisms that are either beneficial or pathogenic to plants, and also the growth of the plant, especially the roots. Plant roots are not only the site of infection for many pathogens but also the site where saprophytes or nonpathogenic microorganisms colonize and obtain their nutrients without harming the host plant. Natural suppressiveness exists in soil because of characteristics that impede the establishment of the pathogen or reduce the ability of established pathogens to infect plants, although numbers of the pathogenic propagules may be high.

Soil characteristics. There are numerous examples documenting the importance of soil characteristics such as content of clay, sand, and silt on the movement and eventual infection of plant roots by the pathogenic fungi. For example, infection with fungal pathogens such as *Fusarium oxysporum* of banana plantations in South America and *Fomes annosus* of pine plantations in Virginia has been related to the soil textural components. The downward movement of the infecting pathogenic fungal spores in the soil was much faster in the porous light sandy soils than in heavier clay soils. Better soil structure can be achieved by using traditional farming methods or the so-called low-input sustainable agriculture used worldwide. This type of farming has caused minimal damage to the soil structure by minimum tillage and incorporation into the soil of organic matter from plant residues. The improved soil structure leads to less disease.

Mulches. The importance of organic mulches or green manure incorporation into soils has received considerable attention. In many ways it is a rediscovery of what traditional farmers have practiced for centuries. Two examples of the ancient agricultural systems, which are still in use today, were the chinampa system, practiced by the Aztecs in Central Mexico, and the popal system, practiced by the Mayans in the Yucatan area of Mexico. In both these systems, the crops (corn, pepper, and beans) were sustained by adding organic matter to the soil. The organic matter came from aquatic weeds, such as water hyacinth, from previous crops, and from animal manure. The soils from the above systems were highly suppressive to a range of soil-borne fungal pathogens. It is yet to be proven whether the suppression in this system is due to higher microbial activity, pH, or calcium ion content.

Organic farming. Organic farming can be used to reduce expensive agricultural inputs such as pesticides and fertilizers. High organic matter in the soil has been shown to improve not only the soil structure, fertility, water-holding capacity, aeration, and reduced water runoff but its biological activity. For example, cruciferous (cabbage, broccoli, mustard) residues that are decomposing reduce root rots in various crops by releasing toxic metabolites that kill or inhibit plant pathogens. Better soil structure also improves the rooting capacity of the plants. Plant roots, which are able to penetrate deeper into the soil profile in well-structured soils than into compact soils, can avoid the fungal pathogens usually present in the upper layers of soil.

Induction of suppression. Numerous new examples have been reported since the discovery early this century that *G. graminis* can be suppressed by monocropping wheat or barley. Monocropping can also induce suppression of damping-off diseases of radish caused by *Rhizoctonia solani* and potato scab caused by *Streptomyces scabies*. The suppressiveness did not increase in the presence of either the pathogen or the host plant alone; both had to be present.

The second important characteristic of biological suppression is that it is transferable to a conducive soil by adding small amounts of the suppressive soil. The effectiveness is directly proportional to the amount of suppressive soil added. For example, 600 g/m^2 (2 oz/ft^2) of a suppressive soil was sufficient to render a conducive soil suppressive to *Fusarium oxysporum*. Although it is difficult to differentiate biological suppression from physiochemically mediated suppression, the former can be manipulated to suppress the pathogen. With the rapid progress in biotechnology, in some cases the biological factor or the antagonist responsible for the suppression has been isolated, identified, and formulated for direct application to soil or as seed coating. Advances in molecular biology have further facilitated studies on the mode of action of antagonists, and also on the characteristics responsible for the biological suppression at the genetic level.

Biological agents. Analysis of microbial populations from suppressive soils has revealed that a vast number of soil bacteria or fungi belonging to diverse genera or species are antagonistic to a range of plant pathogens, including nematodes, pathogenic bacteria, and fungi. Suppression of pathogens occurs by either competition for nutrients at the site of infection on the hosts or the release of metabolites toxic to the pathogen. One of the first biologically based agents used in a large scale to control crown gall caused by *Agrobacterium tumefaciens* in peaches was an antibiotic-producing bacterium called *A. radiobacter* strain 84. Larger populations of the antagonist were found associated with the healthy plants than with the diseased ones. A similar example from France shows that the pathogenic wilt fungus *Fusarium oxysporum* is suppressed by a nonpathogenic strain of the same species by competition for the infection site on plant roots. The take-all decline in wheat has been attributed to a soil fungus (*Phialophora* spp.) closely related to the pathogen, and also to the presence of *Pseudomonas fluorescens*, a common soil bacterium. The suppression of *R. solani* by monoculturing radish has been related to increased numbers of a common soil fungus, *Trichoderma* spp. The soil bacterium *Burkholderia cepacia*, which has antifungal activity against a range of fungal pathogens, is often found associated with corn roots; monoculturing corn increases its populations. When used as a seed coating it reduces pre- and postemergence damping off, and improves plant growth.

Implications. In nature, disease suppression is at work on a larger scale and needs to be exploited. At least 10 biologically based pesticides are registered for use in the United States, and many more are in the process of being approved by the Environmental Protection Agency (EPA). This is the result of years of observation and study of biological suppression and also of the ability to exploit microorganisms for their biological control ability.

For background information SEE AGRICULTURAL SOIL AND CROP PRACTICES; PLANT PATHOLOGY in the McGraw-Hill Encyclopedia of Science & Technology.

K. Prakash Hebbar; R. D. Lumsden

Bibliography. R. J. Cook and K. F. Baker, *The Nature and Practice of Biological Control*, 1983; W. Lockeretz (ed.), *Environmentally Sound Agriculture*, 1983; R. D. Lumsden and J. L. Vaughn (eds.), *Pest Management and Biologically Based Technologies*, Proc. Beltsville Symp. XVIII, 1993; R. W. Schneider (ed.), *Suppressive Soils and Plant Disease*, 1982; D. Thurston et al. (eds.), *Slash Mulch: How Farmers Use It and What Researchers Know about It*, 1994.

Plant virus

The identification of a plant virus based on visual examination of the diseased plant is often equivocal because different viruses as well as other biotic or abiotic stresses cause similar symptoms. The definitive assay to determine the presence or absence of a plant pathogen is to transmit the pathogen to another plant and to reproduce the disease. This method is often laborious with viruses that may be transmitted only by a vector (such as an insect, fungus, or nematode) or in situations where the symptoms of virus infection may take weeks to months to appear (for example, virus diseases of fruit trees). Because the small size of viruses restricts the utility of light microscopy for viewing virus particles and the costs and time involved in electron microscopy are sometimes prohibitive, indirect methods for detecting viruses have been sought. Serology using polyclonal and monoclonal antibodies is one such method.

Monoclonal and polyclonal antibodies. Plant viruses are nucleoproteins with deoxynbonucleic acid (DNA) or ribonucleic acid (RNA) enclosed within a shell (capsid) that generally consists of a repeating protein subunit. Most protein subunits that form the capsid have a molecular weight of 17,000–40,000. Although both nucleic acids and proteins produce immune responses, the variability in the protein(s) found in different viruses facilitates the production of antibodies that react to a specific virus.

When a foreign protein is administered to an animal, a series of events occurs that results in stimulation of B lymphocytes to produce immunoglobulins specific in reaction to specific antigenic determinants (epitopes) of the virus. The major immunoglobulin produced in the animal is immunoglobulin G (IgG). A different lineage of B lymphocytes is stimulated for each exposed epitope on a virus. Thus, the serum fraction obtained from clotted blood collected from an animal after the administration of a plant virus as an antigen yields a collection of antibodies produced by different lineages of B lymphocytes that is referred to as a polyclonal antiserum, or as polyclonal antibodies.

The B lymphocyte cells have been found to be difficult to culture in the laboratory, but this problem was circumvented in 1975 by the development of techniques that permitted fusing a single B lymphocyte with a cell originating from a malignant B-cell tumor (myeloma). The hybrid cell (hybridoma) formed between the two cells can be grown in culture medium and produces immunoglobulin. Antibody produced from the cultured progeny of the hybridoma is referred to as monoclonal antibody because it reacts to only one epitope of the protein initially administered to the animal.

Thus, polyclonal antiserum consists of a collection of different antibodies that react to different epitopes of the viral protein, whereas a monoclonal antibody reacts to only one epitope. A collection of monoclonal antibodies that react to different epitopes of a plant virus is essentially a polyclonal antiserum; it lacks the proteins that are found in serum.

Uses. Polyclonal antibodies have been used to detect plant viruses since the 1920s, and since the 1980s monoclonal antibodies have also been used. There are advantages and disadvantages in the use of either polyclonal antibodies or monoclonal antibodies to detect plant viruses. The choice between polyclonal and monoclonal antibodies depends on the features needed for the serological test to detect a specific virus.

Preparation. The first step in preparing either form requires preparation of the antigen. Generally, isolated virus is used. To purify a plant virus, tissue from a plant infected by the virus is ground in a buffer in which the virus is stable. The virus is separated from the plant proteins by repeated differential and gradient centrifugations. If the virus preparation obtained for use as an antigen is not free of plant material, the animal used for antibody production will produce antibodies to both the virus and the plant material. Antiserum made to a preparation that is contaminated with plant proteins would react to virus-infected plants as well as to healthy plants. For a virus preparation not free of plant host material monoclonal antibodies can be developed because each cell line produces an antibody with specificity to a single epitope (such as an epitope either on the viral protein or to a host protein). Less virus generally is required to produce monoclonal antibodies compared to polyclonal antibodies. Techniques to clone plant virus genes and to express the gene in another host (for example, bacteria) may also be used to circumvent the problem of isolating a virus from plant host material for use as an antigen. This approach is also useful when low concentrations of a virus exist in the host plant. The cost of monoclonal antibody production must be weighed against the ease of preparation and the availability of antigen.

When a large quantity of uniform antibody is required, the use of monoclonal antibodies is advantageous. The titer of polyclonal antibodies varies from each blood sample taken from an animal, and the titer also varies between animals. Once a stable hybridoma is produced, the potential exists to produce an unlimited amount of the antibody. Furthermore, the hybridoma cell line can be stored. The cell line can again be cultured when additional antibody is needed.

Serological assays. The utility of monoclonal antibodies in serological assays needs to be thoroughly considered for each specific application. For example, the specificity of a monoclonal antibody may be useful if a particular strain or serotype of a virus is to be detected. Monoclonal antibodies are not useful when a general diagnostic assay is needed for a group of related viruses that do not share a common epitope. The specificity of a monoclonal antibody is an important consideration in the process of selecting hybridomas from the fusion of myeloma cells and B lymphocytes. Since the antigenicity of a virus, like other proteins, is determined by its three-dimensional structure, the presentation of the virus in a serological assay can modify the nature and number of epitopes that are exposed.

Epitopes have been categorized into cryptotope, neotope, and metatope. Cryptotopes are considered hidden unless the virus particle is disrupted by physical or chemical agents. Neotopes are antigenic sites that result from the association of subunits that compose the virus capsid. Metatopes are epitopes found on single subunits of the viral coat protein even when they are not associated in the capsid. Whether a monoclonal antibody binds to a cryptotope, neotope, or metatope depends on how the virus is presented during the initial screening process of the hybridomas. Different epitopes may be exposed if a reagent or process in the serological assay may distort the antigen, such as by binding the virus directly to an enzyme-linked immunosorbent assay (ELISA) plate, using western blot analysis, or if a detergent is used in the reagents. The type of assay used initially to select the hybridomas should also be the one used to detect the virus. The type of selection assay is generally not a consideration in production of polyclonal antibodies since the antiserum consists of a collection of immunoglobulins that react to different epitopes.

The presence of two antigen-combining sites on an antibody and the constant region of the heavy chain on the antibody are important features affecting the utility of either polyclonal or monoclonal antibodies to detect plant viruses. Thus, polyclonal antibodies and monoclonal antibodies can be used in the same types of assays, except for serological assays in which a positive reaction is evident only when enough of a lattice forms from the binding of antibody and antigen to cause a visible precipitate. This process is facilitated by the reaction of a mixture of antibodies to different epitopes with the result that many times a monoclonal antibody is not useful in these types of serological assays.

Since the 1960s, assays have become widely used in which antigen and antibody binding is indirectly determined. The constant region of the heavy chains of an antibody is not involved in the antigen-antibody binding. The approach has been to tag this area of the antibody with an enzyme or radioactive label. If antigen-antibody binding has occurred, then the label can be detected. Another approach has been to use an antibody that will bind to the antibody of another species. For example, murine IgG can be injected into a goat to produce goat antimouse antibody, which can be isolated and tagged. To detect the binding of a mouse monoclonal antibody to the target antigen, the tagged goat antimouse antibody is used.

Assays have also been developed where both polyclonal and monoclonal antibodies to the virus are used in the same assay. In an enzyme-linked immunosorbent assay, one antibody may be used to coat the ELISA plate to capture the antigen on the assay plate and the second antibody that is either tagged with an enzyme or detected by another antibody tagged to detect the second antibody. Many other combinations and techniques can be used.

For background information SEE ANTIBODY; ANTIGEN; ANTIGEN-ANTIBODY REACTION; CELLULAR IMMUNOLOGY; IMMUNOGLOBULIN; RADIOIMMUNOASSAY; VIRUS in the McGraw-Hill Encyclopedia of Science & Technology.

John L. Sherwood

Bibliography. R. A. C. Jones and L. Torrance (eds.), *Developments and Applications in Virus Testing*, 1986; R. E. F. Matthews, *Plant Virology*, 3d ed., 1991; M. H. V. Van Regenmortel, *Serology and Immunochemistry of Plant Viruses*, 1982; R. G. Webster and A. Granoff, *Encyclopedia of Virology*, 1994.

Plants, saline environments of

All soils and irrigation waters contain soluble salts, many of which are required for normal plant growth and development. However, many soils and waters, particularly in semiarid irrigated areas, contain excessive amounts of salts that are harmful to plants. The range of salt concentrations tolerated by crops varies greatly from species to species. Obvious effects of salt stress on plants are growth suppression and various symptoms of leaf injury. Growth suppression is related to the total concentration of soluble salts or osmotic potential of the soil solution in the root zone. The time and duration that plants are exposed also influence the degree of injury. Although salt-stressed plants are stunted, they usually appear normal in other respects. Leaves are often a darker green and, in some species, are thicker and more succulent than in nonstressed plants. Leaf injury results from the accumulation of toxic ions, for example, chloride and sodium, or from salt-induced nutritional deficiencies. Visual symptoms such as leaf-margin and tip burn, necrosis, and defoliation occur in some species, particularly in woody species, but are rare in herbaceous plants unless they are severely salt-stressed.

Salt tolerance. Salt tolerance refers to the ability of plants to cope with salinity. It can be quantified by expressing growth or yield as a function of the concentration of salts in the root medium. All plants tolerate salts up to some threshold level without adverse effects. As salt concentrations exceed the threshold, vegetative growth and fruit or seed yields are decreased. However, both the threshold level and the rate that yields decrease differ considerably among plant species.

The ability of plants to tolerate salinity depends on other environmental conditions. Climate and various soil, water, and cultural conditions affect the amount of salt that plants can tolerate. For example, plants are more sensitive to salinity during hot, dry weather than during cool, humid weather. The combined effects of salinity and high evaporative demand, whether caused by high temperature, low humidity, wind, or drought, are more stressful than salinity alone. Thus, absolute salt tolerance values cannot be assigned to plants. However, salt tolerance can be determined on a relative basis by comparing plant growth on saline soils to that observed on non-saline soils with all other conditions the same. Many agricultural and horticultural crop species have been tested in this way by growing them at several levels of salinity in small field plots. Growth or yield of the salt-treated plants is expressed as a percentage of that obtained under nonsaline conditions. When plotted as a function of increasing soil salinity, relative yield usually follows a sigmoidal relationship. However, within the range of soil salinities that permit acceptable economic yields, a single linear function adequately describes yield responses to salinity concentrations that exceed the threshold.

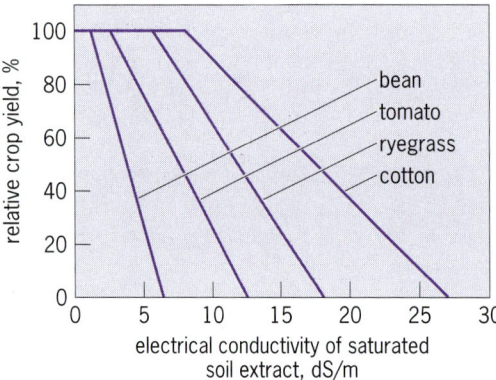

Fig. 1. Relative yields of bean, tomato, ryegrass, and cotton crops as functions of electrical conductivity of water extracted from a saturated-soil sample.

Variability. Reliable salt tolerance data require measurements that reflect the salt concentration that the plant experiences. Since plant roots encounter salt concentrations that vary with soil depth and time, an average value is obtained by measuring salinity at several depths in the root zone and at different times during the growing season. The concentration of salt in the soil can be measured or assessed in various ways. A common method involves measuring the electrical conductivity of water extracted from a saturated soil sample, which is directly proportional to its salt concentration. For example, the electrical conductivity of a 100 millimole NaCl solution is about 10.7 decisiemens per meter. The salt concentration of the soil solution bathing the roots is about twice that measured in the saturated-soil extract. Salt-tolerance data (**Fig. 1**) have traditionally been expressed in terms of this parameter. Researchers at the U.S. Department of Agriculture Salinity Laboratory in Riverside, California have compiled similar data for more than 90 different crop species.

Salt tolerance differs not only among plant species but among varieties of some species. The genetic variation for salt tolerance is usually less in cultivated varieties than in wild progenitors. This variability can be used to advantage by geneticists to increase salt tolerance through introgression of genes from tolerant lines into cultivated crops. In tree crops, tolerance is related in part to the ability of rootstocks to restrict salt uptake. Consequently, tolerance of citrus and various stone-fruit trees depends upon the particular rootstock used.

Implications for irrigation. Despite the many factors that affect plant response to salinity, salt-tolerance data are useful to farmers who must select crops that can be grown profitably on salt-affected lands. Salinity measurements made on field soil samples can be compared to crop-tolerance data to estimate potential yield losses. The data are intended for use where salinity is relatively uniform throughout the growing season. However, plants are more sensitive to salinity at some stages of growth than at others. Seedling emergence and early reproductive stages are two of the most sensitive stages. Computer simulation models are being developed that will predict plant growth responses to varying levels of salinity. If the tolerance of crops is known at different stages of growth, farmers can choose irrigation management practices that minimize salinity stress during the sensitive stages.

Eugene V. Maas

Sulfate and chloride tolerance. Soil salinity threatens agricultural sustainability in many arid and semiarid regions of the world. The major salts contributing to soil salinity are sulfates (sodium sulfate, magnesium sulfate, and calcium sulfate) and chlorides (sodium chloride, calcium chloride, and magnesium chloride). One of the more cost-effective strategies for coping with salinity is to grow crops and varieties that are tolerant of salts. Although sulfate salinity is the predominant form

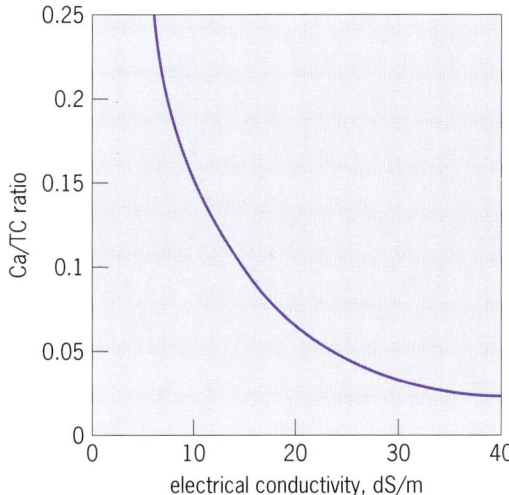

Fig. 2. Ratio of calcium to total cations in soil solution (Ca/TC) decreases as severity of sulfate salinity increases.

of salinity in most agricultural soils, the salt tolerance of crops is routinely evaluated by determining their response to chloride salts. Chloride salts are preferred because they are readily soluble and are easier to work with than sulfate salts. However, scientists are uncertain as to whether results obtained when plants are exposed to chloride salts are truly indicative of how they will perform in soils where sulfate salinity predominates.

Salt effects. Adverse effects of salinity on plants are due to (1) decreased osmotic potential of the root medium (making it more difficult for plant roots to absorb water), and (2) nutritional disorders or imbalances. The electrical conductivity of the soil solution, which is closely correlated with osmotic potential and is easily measured, is the standard index of soil salinity. In the past, the prevailing view was that crop damage from salinity was due primarily to osmotic effects, but it is now recognized that the ionic composition of salt medium can exert an important influence on plant responses.

Plant responses. Differential responses by plants to sulfate and chloride salts may arise either from differences in cation nutrition [especially from the cations calcium (Ca^{2+}) and potassium (K^+)] or from the effects of anions [chloride (Cl^-) versus sulfate (SO_4^{2-})]. These responses are collectively referred to as specific ion effects. Calcium plays a vital role in plant nutrition and physiology. It is essential for the efficient functioning of cell membranes; it stabilizes cell wall structure and plays a role in regulating the salt economy of plants. There is evidence that plants are more tolerant of salts when they are well supplied with calcium. One of the major differences between chloride and sulfate solutions is that the concentration of calcium can be low in sulfate systems because of the poor solubility of calcium sulfate ($CaSO_4 \cdot 2H_2O$). As the concentration of sulfate in the soil solution increases, a point is reached at which $CaSO_4 \cdot 2H_2O$ precipitates because its solubility is exceeded. As a result, the ratio of calcium to total cations (Ca/TC), which is a good indicator of calcium availability to plants, declines as sulfate salinity increases. At an electrical conductivity of about 13–15 dS/m in a sulfate salt system, the Ca/TC ratio can drop below 0.1 (**Fig. 2**), the value at which the supply of calcium may be inadequate for some plants.

Research has confirmed that certain plants (for example, barley) can suffer from severe calcium deficiency when exposed to moderately high concentrations of sulfate salts (**Fig. 3**). However, calcium deficiency is by no means universal in plants subjected to sulfate salinity. Plants that are adapted to saline environments (halophytes) are especially efficient at absorbing calcium. Kochia (*Kochia scoparia*), a plant that colonizes saline soils on the Canadian prairies, has been shown to maintain adequate levels of calcium in tissues even when the Ca/TC ratio in the soil solution is as low as 0.03. In the case of plants that are sensitive to salinity, the osmotic threshold may be exceeded before the Ca/TC ratio decreases below the critical value, in which case a shortage of calcium should not be a major contributor to salt stress.

Uptake of potassium, the metallic element needed by plants in largest amounts, tends to decrease as salinity and the ratio of sodium to calcium (Na/Ca) of the root medium is increased. Plants exposed to sulfate salinity, which characteristically has a high soluble Na/Ca ratio, can be low in potassium compared to plants salinized with chloride salts. As in the case of calcium, some plants, particularly halophytes, show a remarkable ability to selectively absorb potassium from root media dominated by sulfate salinity.

Chloride and sulfate contribute to salt stress, but in general these anions are themselves not toxic to annual crops. However, they can have indirect effects on growth by interfering with the uptake of nutrient anions such as nitrate. Chloride, which is taken up by plants more readily than is sulfate, can greatly inhibit absorption of nitrate from the soil

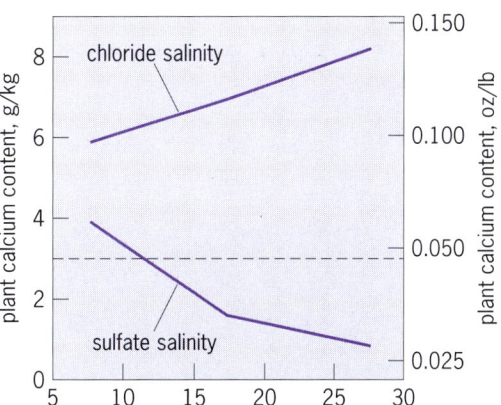

Fig. 3. Calcium levels in salinized barley seedlings; area below broken line represents calcium deficiency. (*After D. Curtin, H. Steppuhn, and F. Selles, Plant responses to sulfate and chloride salinity: Growth and ionic relations, Soil Sci. Soc. Amer. J., 57:1304–1310, 1993*)

solution. Chloride-induced nitrogen deficiency has been implicated as a factor causing growth reduction in wheat yields. In contrast to annual crops, long-lived woody plants (trees, vines, and shrubs) often show symptoms of severe leaf injury from chronic, toxic accumulations of chloride.

Because of the specific ion effects outlined above, sulfate and chloride salts may elicit somewhat different growth responses from plants. The magnitude of salt-type effects may vary from crop to crop and with the nutrient status of the soil. For many crop species, salt-type effects are sufficiently small that response functions generated with chloride salts will give a good indication of how well they will cope with sulfate salinity.

For background information SEE PLANT MINERAL NUTRITION; PLANTS, SALINE ENVIRONMENTS OF in the McGraw-Hill Encyclopedia of Science & Technology.

Denis Curtin; Harold Steppuhn

Bibliography. D. Curtin, H. Steppuhn, and F. Selles, Plant responses to sulfate and chloride salinity: Growth and ionic relations, *Soil Sci. Soc. Amer. J.,* 57:1304–1310, 1993; M. Pessarakli (ed.), *Handbook of Plant and Crop Stress,* 1994; C. G. Suhayda et al., Comparative responses of cultivated and wild barley species to salinity stress and calcium supply, *Crop Sci.,* 32:154–163, 1992; K. K. Tanji (ed.), *Agricultural Salinity Assessment and Management,* ASCE Manuals Rep. Eng. Prac. 71, 1990.

Plastics

Producing certain plastics generates huge amounts of hazardous waste in the form of organic solvents, such as toluene and chlorinated hydrocarbons, and contaminated water. The generated wastes create a great environmental threat. In an effort to minimize this environmental impact, the plastics industry is developing new processes to replace organic solvents and water with supercritical carbon dioxide.

Polymerizations involving supercritical fluids have been practiced for decades. Systems such as the polymerization of ethylene and the copolymerization of ethylene and vinyl monomers are good examples. However, studies of polymerization reactions utilizing inert supercritical fluids did not command attention until recently. Environmental concerns are rapidly shrinking the list of chemicals that are practical for use as solvents in reacting systems. Nontoxic supercritical fluids exhibit solvent properties that are pressure- and temperature-dependent; thus, they are an attractive replacement for the conventional solvents.

Among the inert supercritical fluids that have been studied, supercritical carbon dioxide, which is nontoxic, inexpensive, and nonflammable, is considered to have potential as an environmentally friendly reaction medium for polymerization reactions. Chemical processing based on carbon dioxide could save much of the waste associated with processes that use fluorocarbons or other organic solvents.

Supercritical carbon dioxide. The properties of supercritical carbon dioxide are very different from those of conventional hydrophilic and hydrophobic phases. A substance has three classical states: solid, liquid, and vapor. A critical point is defined as the point where the two fluid phases, liquid and vapor, become indistinguishable. A fluid is supercritical above the critical temperature (T_c) and the critical pressure (P_c). For carbon dioxide T_c is 31.06°C (87.91°F), P_c is 73.83 bar (73.83 × 10^5 pascals), and density at the critical point (D_c) is 0.467 g/cm^3. Supercritical carbon dioxide has weak van der Waals forces compared with hexane; in this respect it resembles the forces exhibited by fluorocarbons. In terms of polarity, supercritical carbon dioxide is roughly comparable to carbon tetrachloride.

Solubility characteristics of supercritical carbon dioxide could provide reaction parameters that could be easily controlled by changing temperature or pressure. As carbon dioxide approaches its critical point, certain physical properties such as surface tension, refractive index, viscosity, dielectric constant, and solvent strength become discontinuous and exhibit a strong dependence on temperature and pressure. Supercritical carbon dioxide can exhibit densities approaching that of liquids, with viscosities that are an order of magnitude lower, and diffusivities of up to an order of magnitude greater.

Homogeneous free-radical polymerizations. A primary advantage of performing free-radical polymerization, that is, polymerization by free-radical initiators, in carbon dioxide is that it serves as a relatively inert solvent toward free-radical centers. In conventional solvents, polymerization reaction rates are often limited by the increasing viscosity of the solution when a high percentage of monomers has been consumed and converted into polymers, and thus the rate of monomer diffusion to the reaction site is lower. However, supercritical carbon dioxide, with its lower viscosity and increased diffusivity, can overcome this limitation and allow for higher reaction rates and greater extents of monomer conversion for solution polymerization. Finally, carbon dioxide can be easily separated from the products simply by reducing the system pressure.

Homogeneous free-radical syntheses of high-molecular-weight fluoropolymers and copolymers of fluoromonomers, methyl methacrylate, butyl acrylate, styrene, and ethylene in supercritical carbon dioxide have been done. In some cases, the secondary monomer makes up as much as 53% of the total monomer feed. The advantages of performing these polymerizations in supercritical carbon dioxide is the elimination of chlorofluorocarbon solvents; in addition, because of the low viscosity of the reaction medium, the autoacceleration effects (the reaction rate increases with conversion) are minimized, leading to polymers with narrower molecular weight distributions (the ratio of the weight average molecular weight to the number average molecular weight).

Heterogeneous free radical polymerizations. Unfortunately, except for the fluoropolymers most polymers do not dissolve in carbon dioxide. Thus, the use of heterogeneous polymerizations becomes necessary to produce other types of polymer materials. There are four heterogeneous polymerization methods: precipitation, suspension, emulsion, and dispersion. The strong density dependence of the solubility parameters of supercritical carbon dioxide enables its use as a medium for achieving such heterogeneous polymerization reactions.

One of the distinctions of these four different heterogeneous polymerization techniques is the initial phase behavior of the polymerization mixture. The precipitation and dispersion polymerizations start as one-phase, homogeneous systems such that both the monomer and the polymerization initiator are soluble in the polymerization medium but the resulting polymer is not. As for the emulsion and suspension polymerizations, the monomers are not soluble in the continuous phase, and the polymerizations occur as two-phase reactions. Most of the monomers used are soluble in carbon dioxide; this solubility characteristic constitutes the criterion of the precipitation and dispersion polymerization. The only example utilizing a CO_2-insoluble monomer has been the inverse emulsion polymerization of acrylamide in supercritical CO_2. The study of emulsion and suspension polymerization reactions has been hindered by the high solubility of most vinyl monomers in CO_2.

Precipitation method. Typically, precipitation polymerizations are carried out by using liquid solvents in which it is difficult to accurately adjust the solubility threshold of precipitation of polymers from solution. In contrast, supercritical carbon dioxide exhibits solubility parameters as a function of both temperature and pressure. The solubility parameters can be finely tuned, so that polymer chains will precipitate from the solution after reaching a certain threshold molecular weight. Thus, much greater control of molecular weight and molecular weight distribution of the polymers can be achieved.

Polyethylene has been synthesized by polymerization of ethylene in inert carbon dioxide under supercritical conditions. The ethylene–carbon dioxide mixtures can exist as a single homogeneous phase, and polymers are found to precipitate from the solution as slightly swollen particles.

Precipitation polymerization in supercritical carbon dioxide has also led to the production of a variety of polymer materials, such as poly(vinyl acetate), poly(methyl methacrylate), poly(styrene), and poly(acrylic acid).

Dispersion method. Another heterogeneous polymerization technique that has been successfully practiced in supercritical carbon dioxide is dispersion polymerization. The conventional dispersion polymerizations of unsaturated monomers involve the addition of stabilizers known as interfacially active agents to stabilize the colloidal dispersion that forms during the polymerization. Interfacially active molecules contain chemical structures that can interact with two different phases at the phase boundary, for example, soap with oil and water. In supercritical carbon dioxide, the interfacially active agents are designed to have two different chemical moieties, the so-called carbon-dioxide-philic and the lipophilic groups (see **illus.**). The fluorine-containing structure serves as a carbon-dioxide-philic stabilizing segment that can dissolve readily in the carbon dioxide. The other end is lipophilic and can physically adsorb onto the surface of the growing particles. The polymer colloids produced during the dispersion polymerization can be stabilized by these interfacially active agents through a steric mechanism, which arises from the steric interaction between the stabilizer and the growing polymer particles on which it is adsorbed.

Chemical structure of an interfacially active, polymeric stabilizer, [poly(1,1-dihydroperfluorooctyl acrylate); poly(FOA)], depicting the proposed site of anchorage for the carbon-dioxide-philic steric-stabilizing moiety.

An important phenomenon associated with this procedure is the swelling effect (plasticization) induced by supercritical carbon dioxide. Poly(methyl methacrylate) can be plasticized with pressurized carbon dioxide. The glass transition temperature of poly(methyl methacrylate) decreases dramatically in the presence of high-pressure carbon dioxide. Under the conditions at which the polymerization of methyl methacrylate is carried out, poly(methyl methacrylate) is highly plasticized (that is, its effective glass transition temperature has been lowered), thus facilitating efficient monomer diffusion into the growing polymer particles.

Dispersion polymerizations carried out in supercritical carbon dioxide provide polymers with controlled morphologies. For example, poly(methyl methacrylate) can be formed as particles with diameters in the range of 0.8–2.8 micrometers. The ability to make plastic particles in supercritical carbon dioxide with controlled size is an important breakthrough that demonstrates the possibility of making various types of polymers with potential applications such as powder coatings, impact modifiers, and drug delivery vehicles.

Other polymerizations. Polymerizations in liquid carbon dioxide have also been investigated. A process such as the cationic polymerization of isobutyl vinyl ether and fluorinated oxetane is carried out in liquid carbon dioxide as a heterogeneous reaction. Polymerizations catalyzed by transition metals have also been achieved. For example, a ring-opening metathesis polymerization of norbornene, using a ruthenium complex as a catalyst, produces poly(norbornene). These reactions demonstrate the utility of carbon dioxide as an acceptable reaction medium for the transition metal-based or cationic polymerization reactions.

Prospects. The advantages of using supercritical carbon dioxide as a solvent can certainly be extended to reactions of small molecules. It is possible that the unique pressure-dependent physical properties of supercritical carbon dioxide can be used to increase the selectivity of a chemical reaction while still maintaining a high rate of conversion. The unusual changes in intra- and intermolecular interactions such as electron donor-acceptor forces could alter chemical structures in the transition state and lead to high chemoselectivity.

Reactions using gas-phase reactants often are impeded by the low solubility of the gas-phase reactant in the organic solvent. Hydrogen gas, for example, has a much higher solubility in carbon dioxide than in hexane, leading to more efficient reaction rates. The miscibility of carbon dioxide and other gases forms the basis for many other reactions, including a range of organometallic reactions such as hydrogenation and hydroformylation.

Another aspect of supercritical carbon dioxide is that it can be used as a thinner for industrial spray painting. As a result, a large reduction (as much as 80%) of the hazardous emissions associated with the spray painting process can be achieved.

Given all the advantages in phase behavior, the ability to tune density to control the outcome, and the benefits of waste minimization with environmentally acceptable solvents, scientists have incentives to find further uses for supercritical fluids.

For background information SEE CRITICAL PHENOMENA; FREE RADICAL; POLYMERIZATION; STEREOSPECIFIC CATALYST in the McGraw-Hill Encyclopedia of Science & Technology.

Yu-Ling Hsiao; Joseph M. DeSimone

Bibliography. J. M. DeSimone et al., Dispersion polymerizations in supercritical carbon dioxide, *Science,* 265:356–359, 1994; J. M. DeSimone, Z. Guan, and C. S. Elsbernd, Synthesis of fluoropolymers in supercritical carbon dioxide, *Science,* 259:945–947, 1992; P. G. Jessop, T. Ikarlya, and R. Noyori, Homogeneous catalytic hydrogenation of supercritical carbon dioxide, *Nature,* 368: 231–233, 1994; K. M. Scholsky, Polymerization reactions at high pressure and supercritical conditions, *J. Supercrit. Fluids,* 6:103–127, 1993.

Pneumocystis

Pneumocystis carinii is an organism of undetermined taxonomy found in the lungs of humans and lower mammals. Although it causes no discernible disease in the healthy host, *P. carinii* produces an extensive and usually fatal pneumonia in the severely immunocompromised host, such as an individual with cancer, organ transplantation, or acquired immune deficiency syndrome (AIDS). The disease can be effectively treated or prevented with antimicrobial drugs.

The organism. The two basic forms of *P. carinii* are the trophozoite and the cyst. The simplest form is the pleomorphic, thin-walled trophozoite ranging from 1 to 4 micrometers in diameter. The cyst form is thick-walled, oval to round, often cup-shaped and measures 5–8 μm in diameter. The mature cyst contains up to eight intracystic daughter cells referred to as sporozoites. Several major antigens to *P. carinii* have been identified. Karyotypic and antigenic differences have been observed between human and murine *P. carinii*. Studies of the ribosomal ribonucleic acid (rRNA) of *P. carinii* have shown greater homology to rRNA sequences of certain fungi than to those of protozoa. Other attributes of both fungi and protozoa have been described for this organism, leaving its exact classification unclear.

Epidemiology. Some 70–90% of healthy people worldwide acquire an antibody to *P. carinii* early in life, before approximately 4 years of age. In most cases, no discernible disease is associated with infection in the immunocompetent host. The organism is believed to lie dormant in the lung in a latent state for long periods. If the host's immune system, especially the CD_4^+ T lymphocyte, becomes impaired, pneumonitis may occur because of the activation of latent *P. carinii* or the acquisition of new organisms from the environment.

Pneumocystis carinii is also found in the lungs of rats, mice, ferrets, cats, dogs, monkeys, horses, cows, rabbits, and many other mammals. Animal-to-animal transmission of *P. carinii* has been demonstrated in experimental rats. Human-to-human and animal-to-human transmission have not been described.

Research on *P. carinii* is made possible through the use of experimental animal models. When conventional laboratory rats are immunosuppressed with corticosteroid drugs for 6 weeks or longer, 90% or more will develop *P. carinii* pneumonitis. Inoculation with the organism is not necessary because of the high prevalence of the organism in the lungs and in the environment.

Even within the broad category of immunosuppressed hosts, several conditions have been recognized as high risk. About 70% of untreated adults with AIDS will develop *P. carinii* pneumonia. Furthermore, *P. carinii* pneumonitis occurs in about 40% of children with severe combined immunodeficiency syndrome, 12% of individuals with leukemia, and about 6% of organ transplant recipients.

Outbreaks of *P. carinii* pneumonitis occurred in debilitated infants in Europe around the time of World War II. In these infants, the disease was an interstitial plasma cell pneumonitis, and recovery was spontaneous in 50% of the cases. The underlying predisposing entity was unknown, but malnutrition was suspected. Severe protein calorie malnutrition, as occurs with kwashiorkor, can provoke *P. carinii* pneumonitis.

Pathology. With rare exception *P. carinii* and its associated disease remain localized in the lung even in the most severely immunocompromised individual. Both cyst and trophozoite forms are abundant in the alveolar lumens and attach to the alveolar walls. Reactive alveolar macrophages add to the inflammatory response, resulting in an extensive, bilateral diffuse inflammation of the alveoli. Impairment of oxygen (O_2) and carbon dioxide (CO_2) exchange within the lung result, causing a reduction of arterial oxygen tension and an increase in arterial pH. The arterial carbon dioxide tension remains normal or slightly elevated with the typical pattern of respiratory alkalosis (high blood alkalinity caused by a loss of blood CO_2). The marked reduction of oxygen in the blood causes an increase in respiratory rate and labored breathing. A persistent cough and fever are also prominent clinical features. The pneumonitis becomes visible by chest radiograph as a bilateral diffuse infiltrate. In the infantile epidemic form, *P. carinii* pneumonitis involves the alveolar septum with interstitial plasma cell infiltration, but fever usually does not occur.

Although the diagnosis may be formulated by the clinical signs, symptoms, and chest radiograph, confirmation requires the demonstration of *P. carinii* in lung tissue or secretions from the lower respiratory tract. Such specimens for histopathology require a biopsy of the lung, induced sputum, or bronchoalveolar lavage (so-called lung washing). The lavage procedure is most frequently used for diagnostic purposes but biopsy is the most sensitive method.

Rare cases of extrapulmonary lesions due to *P. carinii*, involving the eye, bone marrow, heart, spleen, ear, and other sites, have been described .

Treatment. Untreated *P. carinii* is fatal in approximately 100% of affected immunosuppressed individuals. However, several effective drugs are available for treatment. Trimethoprim-sulfamethoxazole is the drug of choice and can be administered orally or intravenously. A course of treatment is 2–3 weeks. Individuals who are unable to take trimethoprim-sulfamethoxazole because of adverse reactions or who fail to respond to the drug combination are treated with one of the alternative drugs. Recovery occurs in 70–90% of cases treated. However, if the immunocompromised state persists, a recurrence of the pneumonitis will occur in at least one-third of all infected individuals who discontinue the antimicrobial drug therapy.

Supportive oxygen therapy is especially important in the treatment of moderate and severe cases. Efforts are made to maintain the arterial oxygen tension above 70 mmHg. However, if this requires a high fraction of inspired oxygen for prolonged periods, oxygen toxicity of the lung, often referred to as adult respiratory distress syndrome (ARDS), may occur. This condition mimics many of the features of *P. carinii* pneumonitis and may be irreversible when advanced. Thus, oxygen therapy must be carefully monitored to enhance survival and minimize toxicity.

Prevention. Disease caused by *P. carinii* can be prevented in most high-risk individuals through the use of chemoprophylaxis. Any of several drugs administered throughout the period of immunosuppression will prevent the pneumonitis, but none effectively eradicates the organism. Trimethoprim-sulfamethoxazole is the drug of choice. It is administered orally in one-fourth the dose used for treatment. Given daily or even only 3 days a week this drug is effective in the prevention of *P. carinii* pneumonitis in about 95% of high-risk individuals. However, it is protective only for the time administered, and individuals must stay on the drug until the immunocompromised state has been resolved, or indefinitely in the case of AIDS. For individuals who cannot tolerate the drug combination because of adverse reactions, other drug regimens with proven efficacy are available.

Prophylaxis is not necessary for all immunosuppressed individuals; certain categories at sufficient risk for prophylaxis can be identified within the compromised population. Risk for types of individuals with cancer is based more on the intensity of chemotherapy than the type of malignancy. For example individuals who are being treated for leukemia are at much higher risk than those treated for Wilm's tumor (a malignant tumor of the kidney). Individuals with impaired cell-mediated immunity are also at high risk because this arm of the immune system plays a much greater role than humoral immunity with regard to *P. carinii* pneumonitis, such as in individuals with AIDS for whom the number of CD_4^+ T lymphocytes in the peripheral blood serves as an excellent indicator of degree of susceptibility. Once the CD_4^+ T lymphocyte count decreases to values below 200 cells per cubic millimeter for adults, the individual enters a high-risk zone and a chemical prophylaxis is administered. For reasons that are not apparent, adverse reactions to trimethoprim-sulfamethoxazole occur much more frequently in AIDS (approximately 40%) than in non-AIDS patients (about 5%). These reactions include rash, neutropenia, and fever. When such reactions are not tolerable, other drugs may be used. SEE ACQUIRED IMMUNE DEFICIENCY SYNDROME (AIDS).

No vaccine is available or under development. However, any reduction of the immunosuppressive state, especially cell-mediated compromise, will likely enhance the individual's resistance to *P. carinii* pneumonia.

For background information SEE ACQUIRED IMMUNE DEFICIENCY SYNDROME (AIDS); IMMUNOLOGICAL DEFICIENCY; IMMUNOSUPPRESSION; MEDICAL PARASITOLOGY; OPPORTUNISTIC INFECTIONS; PNEUMONIA in the McGraw-Hill Encyclopedia of Science & Technology.

Walter T. Hughes

Bibliography. W. T. Hughes, *Pneumocystis carinii Pneumonia*, 1987; P. Pizzo and C. Wilfert, *Pediatric*

AIDS: The Challenge of HIV Infection in Infants, Children and Adolescents, 2d ed., 1994; A. G. Smulin et al., The biology of *Pneumocystis carinii, Crit. Rev. Microbiol.,* 18:191–216, 1992.

Polychaeta

Oxygen is a key parameter in the metabolic processes and the distribution of polychaetes. Although oxygen concentration in the sea can be much lower than air saturation [at 20 psu (practical salinity units), 25°C (57°F), and 1 atmosphere (10^5 pascals), 100% of air saturation is 5.3 cm^3 O_2/liter], respiration in most polychaetes does not appear to be significantly affected until extremely low concentrations are reached, about 2 cm^3 O_2/liter or less. Hypoxia is defined as starting at 2 cm^3 O_2/liter and anoxia, as the absence of measurable oxygen.

Tolerance to hypoxia and anoxia. Most species of the marine macroinvertebrates belong to four taxonomic groups: polychaetes, bivalves, echinoderms, and crustaceans. It is apparent that these groups exhibit different levels of tolerance to hypoxia. Studies in coastal areas in Scandinavia, the Adriatic, Japan, the United States, and the Black Sea and in offshore upwelling areas off Peru and California show that, in general, polychaetes are the most tolerant taxa, followed by some bivalves, with echinoderms and crustaceans being the least tolerant. Some species, such as the polychaetes *Heteromastus filiformis, Streblospio benedicti,* and *Paraprionospio pinnata* and the bivalves *Corbula gibba* and *Arctica islandica,* are found repeatedly in hypoxic-stressed areas, but echinoderms and crustaceans are rarely described from such areas.

Much of what is known about behavioral and physiological compensation of polychaetes to hypoxia and anoxia is derived from intertidal species. Intertidal organisms are regularly subjected to hypoxia during low tide when they are isolated from the water column. Compensation mechanisms that have evolved are all directed toward short-term survival of hypoxia but also allow organisms to utilize habitats that experience periodic hypoxia or are permanently marginal relative to dissolved oxygen concentration. Most species tested show some compensation for surviving in reduced oxygen conditions; however, there does not appear to be any long-term physiological mechanism for surviving chronic stable hypoxia. In polychaetes, and metazoan animals in general, the shift to anaerobic pathways during severe hypoxia and anoxia is energetically less efficient than the aerobic pathways utilized during normal oxygen conditions, and consequently is too costly to allow the completion of a life cycle.

Adaptations. Fluctuating and short-term hypoxia or anoxia is survivable through a combination of behavioral and physiological adaptations. The first response to declining oxygen is an increase in respiration rate. Mobile species first attempt to avoid hypoxia through migration. Sessile species capable of limited mobility may also attempt to escape. Sessile fauna that are unable to leave or escape initiate a series of sublethal responses related to the severity of hypoxia. A cessation of feeding occurs, and activities not related to respiration decrease. Increased periods of rest further depress metabolism. As oxygen continues to decline, tube dwelling and burrowing species move closer to the sediment surface or actually emerge from the sediment in search of higher oxygen concentrations. Under anoxia, tube irrigation decreases.

Emergence from sediment. The process of emerging from the sediment into hypoxic bottom water appears to be regulated by the need to find higher oxygen concentrations in the microgradients that may occur within centimeters of the bottom and to to escape the accumulation of toxic-reduced compounds, primarily hydrogen sulfide, that can accumulate in the sediments during hypoxia. Once on the sediment surface, different postures are assumed depending on the oxygen concentration. However, behavior appropriate for surviving or avoiding hypoxia is often inappropriate for avoiding predation. The moribund individuals lying on the surface of the sediment are potentially easy prey if bottom predators are able to enter the hypoxic water column to feed. Selective mortality occurs first among the more sensitive species. If hypoxia persists or intensifies, extensive mortality is seen in all but the most tolerant species. But, if hypoxia and anoxia have not resulted in the mass mortality of fish and crabs, the latter can quickly repopulate areas after hypoxic events in search of stressed benthic (bottom-dwelling) organisms that were unable to avoid the hypoxia. For example, the successful exploitation of oxygen-stressed benthic organisms by fish and crabs in the York River, Virginia, is related to the level of hypoxia (0.2–0.8 cm^3 O_2/liter), its short duration (6–14 days), and the relatively small area affected (only a few square kilometers). In contrast, the larger-scale severe hypoxia and anoxia that develops in the upper Chesapeake Bay, Maryland (<0.2 cm^3 O_2/liter, which occurs for months and over hundreds of square kilometers), does not allow for such a positive energy transfer. In this area benthic communities are drastically affected, and utilization of the stressed areas by fish and crabs does not occur again until the autumnal turnover breaks up the hypoxia and anoxia.

Trawling near the bottom of the ocean in hypoxic areas has shown that large numbers of sediment-dwelling species, including polychaetes, leave their protected positions in the sediment and lie exposed on the bottom. At a bottom water oxygen concentration of about 1 cm^3 O_2/liter (15% saturation) in the Kattegat, off the western coast of Sweden, almost no fish were caught, but 200–400 kg/hectare (180–360 lb/acre) of benthic invertebrates (such as the echinoderms *Brissopsis lyrifera, Echinocardium cordatum, Cucumaria elongata, Ophiura* spp., and *Amphiura filiformis* and the polychaetes *Polyphysia crassa,*

Aphrodite aculeata, and *Nephtys* spp.) were collected. Similar mass migration of species to the sediment surface has been recorded during trawling in the North Sea, off the northeast coast of New Zealand, and on the middle Atlantic continental slope off New Jersey. Many of the species involved in the trawls were deep dwelling and rarely taken in grab samples under normal dissolved oxygen conditions.

The spionid polychaete, *Malacoceros fuliginosus,* was observed in experiments to change its behavior in declining oxygen concentrations. Under fully aerobic conditions this species lives below the sediment surface in burrows. When oxygen concentrations drop below 2.4 cm^3 O$_2$/liter the animals emerge from their burrows and rise partly up in the water column. When oxygen concentrations fall below 0.5 cm^3 O$_2$/liter, the *M. fuliginosus* begin undulatory body movements and ultimately form rapidly weaving clumps of animals. This behavior likely causes water from higher up in the microgradient layer (presumably with higher oxygen content) to come in contact with the worms.

Metabolic switching. With the intensification or persistence of hypoxia there is a metabolic switching in favor of near anaerobic or anaerobic pathways. During anoxia most polychaetes have some ability for facultative anaerobiosis. As anaerobic conditions approach the sediment surface, sulfur and anaerobic microbes experience population explosions. Thus, the quality of the habitat for polychaetes is further reduced. With continued hypoxia, anoxic conditions and hydrogen sulfide eventually reach the sediment surface and extend into the water column, and mass mortality results.

Population dynamics. When bottom waters are reoxygenated, recolonization processes begin to restore polychaete populations. Larval tolerance of hypoxia is critical to the early recolonization of hypoxia-stressed habitats, particularly in organically enriched habitats which are prone to develop hypoxia and anoxia due to high chemical and biological oxygen demand. There is great advantage to being the first species to reenter a defaunated habitat, and polychaetes generally are the first. Larvae of spionid polychaete *S. benedicti,* for example, were found to be unaffected when exposed to short-term hypoxia (92 h; 14% saturation). This tolerance of hypoxia gives *S. benedicti* the opportunity to be the first to utilize food resources left behind during hypoxic and anoxic events.

The ecological consequences of periodic hypoxia on benthic organisms are varied, but it is clear that hypoxia functions as a mechanism for regulating benthic population dynamics. In affected areas, temporal and spatial patterns in benthic organism abundance and species composition are related to the occurrence of hypoxia. Because of their broad physiological tolerance of hypoxia polychaete species tend to dominate hypoxia-stressed estuarine and marine communities around the world. Two particularly tolerant species are the terebellid *Loimia medusa* and the spionid *Paraprionospio pinnata.* Other polychaetes (the spionid *S. benedicti* and the capitellid *Mediomastus ambiseta*) that are known to thrive in organic enrichment areas are more sensitive to hypoxia and are often eliminated during severe hypoxia and anoxia; however, these same species are the first to recolonize an area with the return of normal oxygen conditions. The timing of hypoxia relative to recruitment is also critical to survivorship of newly settled individuals. For example, the summer recruitment peaks for *Podarkeopsis levifuscina* and *Pseudeurythoe paucibranchiata* both declined with the onset of hypoxia.

Implications. At what point permanent damage will result is difficult to assess; to date there is no large system that has recovered after development of persistent hypoxia or anoxia. Exceptions may be small systems where point effluents have ceased and recovery can be initiated from surrounding nonaffected areas. The expanding occurrence of hypoxia and anoxia continues to bring about significant structural changes in benthic communities and to affect the cycling of energy within and between ecosystems. No other environmental parameter of such ecological importance to coastal marine ecosystems has changed so drastically in such a short period as dissolved oxygen. Areas experiencing hypoxia are spreading into shallower waters with increasing frequency, and have the potential to deleteriously affect benthic organisms and complete ecosystems.

For background material SEE ANNELIDA, FRESH-WATER ECOSYSTEM; HYPOXIA; POLYCHAETA in the McGraw-Hill Encyclopedia of Science & Technology.

Robert J. Diaz; Rutger Rosenberg

Bibliography. C. F. Herreid, Hypoxia in invertebrates, *Comp. Biochem. Histol.,* 67A:311–320, 1980; C. Mangum and W. van Winkle, Responses of aquatic invertebrates to declining oxygen tensions, *Amer. Zool.,* 13:529–541, 1973; R. V. Tyson and T. H. Pearson (eds.), *Modern and Ancient Continental Shelf Anoxia: An Overview,* Geol. Soc. Spec. Publ. 58, 1991.

Precipitation (meteorology)

Precipitation sustains nearly all terrestrial plants and animals; both rain-fed and irrigated agriculture depend on an appropriate amount and distribution of annual rain and snowfall. Precipitation also initiates important hydrological processes, such as infiltration, recharge, runoff, and erosion. Although these processes support natural plant communities and make agriculture possible, they may facilitate development of serious environmental problems, such as the contamination of surface waters with excess sediment, nutrients, and natural and manufactured chemicals and the adulteration of ground water via soil transport of surface-applied compounds. Understanding how precipitation may vary in the landscape permits identification of areas of high impact, better focusing of management efforts, maximization of agricultural productivity, and minimization of potentially adverse environmental hazards.

Wind-terrain interactions. Interactions of wind with Earth surface relief influence rainfall distribution in the landscape.

Wind-terrain interactions that alter rainfall uniformity across the landscape can occur at several spatial scales. Large-scale effects on rainfall distribution can occur over distances of 400 km (250 mi). Here, winds influence the trajectory of moisture-ladened air masses as they travel across hill or mountain ranges with relief of 300+ m (1000+ ft). A similar interaction occurs for small-scale landforms, where the extent of horizontal influence is less than 80 km (50 mi) and relief is less than 100 m (330 ft). Furthermore, microscale impacts can occur even over horizontal distances less than 30 m (100 ft), such as a farmer's field. In this case, winds traverse low hills less than 10 m (30 ft) high and influence only the trajectory of raindrops.

Wind-terrain interactions influence both cyclonic and convectional precipitation. Cyclonic storms produce precipitation when warm moist air is forced aloft by horizontal collision of airstreams in an area of low pressure, for example, the depression (low pressure) associated with a cold-front wave. These depressions are extensive, commonly encompassing an area in excess of 100,000 km^2 (38,600 mi^2). They produce generally continuous precipitation of low to moderate intensity, and raindrops are of relatively small or medium size. Convective storms are created when moist air that is warmed at the Earth's surface rises, or is forced to rise, into a cooler unstable atmosphere. The rising air mass cools, forming clouds and precipitation. Because the upper atmosphere is unstable, the parcel will continue to rise and produce further rainfall. Resulting thunderstorms usually produce high-intensity rainfall and more large-sized raindrops but are limited in duration (½–1 h per storm) and extent (20–50 km^2 or 7–20 mi^2).

Large-scale impacts. When a moist airflow intercepts a mountain front, it is deflected upward. This deflection may increase precipitation (1) by causing horizontal convergence and uplift as the airstream is channeled upward through mountain valleys or forced over steep windward slopes and (2) by promoting convective storm development by imparting an initial upward movement to surface air.

Upslope and cyclonic rain. In the first case, upslope rain is produced when a warm moist air mass is lifted into a stable atmosphere. Forced uplift brings the air mass to the saturation point and induces further condensation and raindrop formation. Upslope rain will also occur in conjunction with passing cyclonic storms. The effect that the topographic component has on precipitation becomes greater as the air mass is lifted to higher elevations, as is evidenced on the U.S. Pacific Coast mountain ranges at middle and higher latitudes. At lower latitudes, the topographic component reaches its maximum effect at middle elevations and decreases at higher altitudes because tropical moisture is concentrated near the surface. Cyclonic precipitation decreases significantly to the leeward of these topographic barriers owing to drying associated with descending air. This rain shadow produces a rainfall minimum on leeward mountain slopes and adjacent valley floors.

Lifting of airflow by its interception with a mountain front increases the frequency, duration, and intensity of precipitation events associated with cyclonic storm systems. Magnitude of forced uplift and precipitation values depends on (1) the character of each site (elevation, slope, aspect), (2) the local landscape milieu, such as upwind or downwind relief, and (3) the regional landform configuration, for example, distant valley or mountain features that, by virtue of their orientation, influence large-scale air movement across the site. Annual precipitation originating from cyclonic systems becomes greater as site location changes from leeward to more windward aspects, gentle inclines to steeper hill and mountain slopes, lower to higher elevations, and positions that feature gently rising leeward relief adjacent to steeply rising leeward relief.

Convective rain. Convective storms triggered by large-scale wind-terrain interactions tend to develop above the mountain summit and drift downwind. Hence, maximum precipitation occurs on leeward slopes and relatively lower amounts on windward aspects. The effect of mountainous terrain on convectional precipitation is less pronounced in rugged, mountainous landscapes. Complex relief is thought to reduce the efficiency with which developing storms assimilate latent energy from the atmosphere. The so-called drawing power of such storms is inhibited because the air mass lacks continuity. Thus, convectional storm activity in complex physiography consists of numerous, small rain cells that produce less intense rainfall. By comparison, storms that develop over hills surrounded by a smooth plain are able to access a larger pool of energy. Rain cells can grow larger and can produce more intense precipitation of longer duration and better-defined rainfall patterns.

Small-scale effects. Upslope rainfall is not produced when winds propel moist air over small hills. Air masses traverse the small landform too rapidly for raindrops to form. However, forced uplift over the summit does produce local low-level condensation and clouds. Raindrops produced by passing cyclonic or convective systems fall through the low summit clouds, intercept the cloud droplets, and grow larger. The resulting rainfall distribution has a maximum that is centered over the hill summit.

Researchers have observed strong topographic effects on convectional precipitation even when relief is low or moderate. For example, two zones of maximum summer rainfall observed in the Minneapolis–St. Paul region may be caused by the upward deflection of surface winds over a low, steeply sloping physiographic barrier. Winds from the E-SSE quadrant, which predominate during summer precipitation events, become channeled along similarly oriented 18-km (11-mi) segments of the Mississippi and Minnesota river valleys; the airstream is deflected upward at two sharp bends in the river valleys where the wind collides with steep 75–90-m (250–300-ft) valley sideslopes. Paths of severe thunderstorms and tornados coincide to a large degree with these zones of increased precipitation. In Missouri, convective storms that developed in, or passed over, hilly landscapes produced as much as 70% more rainfall than rain cells that were not associated with this type of terrain. Forced uplift and convergence of wind above the hills in response to topography causes intensification of hill-centered storms. In such cases, the resulting rainfall pattern may be more like that occurring over large-scale features, that is, with a leeward maximum.

Unique physiography. In areas smaller than 5 km^2 (2 mi^2) with relief less than 100 m (330 ft) research indicates that rainfall distribution is influenced by unique wind-flow patterns induced by local topographic features. In a level plain surrounded by low hills, rainfall was observed to increase in the direction of airflow along the plain and into the hills. Other researchers have measured greater precipitation in areas subject to concentrated wind flow. In one case, channellike topographic features gathered and conducted airflow toward the higher rainfall locations.

Windblown raindrops. For small-scale landforms, those hills and ridges with narrow summits can experience an additional local pertubation in rain distribution. Studies have shown that a consistent pattern of rainfall variability occurs along a path defined by the flow of wind over such ridges or hills. Frictional drag forces of moving air act on raindrops, causing them to drift in the direction of airflow. In the summit region, changes in windspeed and vertical direction alter the inclination angle of falling raindrops. This change in turn modifies amounts of rainfall received at the Earth surface, because rainfall intensity is a function of surface slope and orientation with respect to the direction of falling rain.

At larger landscape scales, rainfall is commonly measured by using a gage with a horizontal opening (meteorological rainfall). When comparing rainfall at small scales it is more appropriate to record the depth of rain actually received on the sloping Earth surface (hydrological rainfall). These measurements more accurately portray subtle rainfall differences that occur when relief varies over short distances. Hydrological rainfall is measured by using a gage with an opening oriented parallel to the Earth surface. For purposes of areal comparisons, hydrological rainfall from a sloping landform facet may be projected onto an equivalent horizontal area (map), and termed the hydrological projected rainfall.

Research has shown that, on a ridge with 40-m (130-ft) relief, hydrological projected rainfall (given as a percentage of the spatial mean) is at a maximum on the windward lower slope and decreases to 100% just leeward of the summit. From that point rainfall continues to decrease, but at a reduced rate, toward the lower leeward slope. Meteorological rainfall behaves in an opposite manner, starting at a minimum and increasing to the leeward.

Microscale impacts. Spatial rainfall variation at the microscale is produced by wind-terrain effects on raindrop trajectory. Field studies have shown that the velocity of surface wind flowing across a hill decreases near the windward foot slope, then increases to a maximum near the summit. Wind speed then decreases to a value that may be slightly below incident velocity at the leeward foot slope. The original speed is recovered at a distance of about 1½ hill widths leeward of the summit. A drift effect results when patterned wind flow alters local raindrop incident angles and therefore measured meteorological rainfall and hydrological rainfall intensities.

An instrumented full-scale hill model was recently developed, permitting researchers to conveniently measure microscale wind-terrain impacts. The spatial pattern produced by these interactions varies depending on the overall rainfall intensity, incident wind speed, and hill summit elevation. The locations of maximum and minimum hyrological rainfall across the instrumented hill apparatus can be generalized for cyclonic storms (low-intensity rainfall) and for convective storms (high-intensity rainfall) from data representing several incident wind speeds and summit elevations. On average, locations with maximum hyrological rainfall receive 1½ times more rainfall than locations with minimum rainfall. The magnitude of this variation is significant for hills with such slight relief. These rainfall differences, especially if they occur when crop or soil sensitivity to water inputs is high, can influence spatial patterns of seed germination or seed or fruit development and affect yields; or they can impact on spatial infiltration or soil detachment and transport patterns, and hence on soil erosion.

For rain-fed agricultural lands, it may be economically beneficial to manage soils in high-rainfall areas of the landscape differently from those in low-rainfall areas. Further research defining how topographically induced rainfall patterns influence crop growth and other soil and landscape processes is warranted.

For background information SEE AIR MASS; MOUNTAIN METEOROLOGY; PRECIPITATION (METEOROLOGY); RAIN SHADOW; WIND in the McGraw-Hill Encyclopedia of Science & Technology.

Rodrick D. Lentz

Bibliography. R. G. Barry and R. J. Chorley, *Atmosphere, Weather, and Climate,* 1993; D. Sharon, The distribution of hydrologically effective rainfall incident on sloping ground, *J. Hydrol.*, 46:165–188, 1980; R. B. Smith, The influence of mountains on the atmosphere, *Adv. Geophys.*, 21:87–230, 1979.

Prehistoric domestication of animals

The domestication of plants and animals made possible the development of civilizations by permitting human groups to directly control the availability of food resources and to produce food surpluses sufficient to support craft specialists and urban populations not directly engaged in subsistence activities. Archeological research over the past half century has demonstrated that domestication occurred independently several times in different parts of the world, in each case involving locally available plant and animal species. Archeological evidence points to the Near East as the oldest center of domestication, the process beginning there as early as about 10,000 years ago.

Near Eastern domesticates. The cultures inhabiting the Near East at the end of the Pleistocene exploited a wide array of wild plant foods, including the seeds of grasses, pulses (legumes), and nuts. They supplemented these plant foods by hunting a variety of game animals. From approximately 10,000 to 8000 years before present (B.P.) several plant and animal species were domesticated in the context of a transition from economic systems based on hunting and gathering to those based on food production. Of those species first domesticated for use as foods, the most important plants were cereal grasses of the genera *Triticum* (wheat) and *Hordeum* (barley), as well as pulses of the genera *Pisum* (field pea) and *Lens* (lentil). The most important of the animal domesticates were sheep (*Ovis*), goats (*Capra*), cattle (*Bos*), and pigs (*Sus*). Not all these species were domesticated concurrently; nor were all first domesticated within the same parts of the Near East. Among the earliest domesticated species were the cereal grasses and the ovicaprids.

The archeological evidence indicates that the cereal grasses were first domesticated approximately 10,000 B.P. by inhabitants of the relatively well-watered eastern Mediterranean seaboard (the Levant). From this core area, domesticated cereals, along with agricultural methods, appear to have rapidly spread to the surrounding highlands of eastern Anatolia, northern Iraq, and western Iran—the Taurus-Zagros arc. The evidence for the domestication of sheep and goats is sketchier, and points to a reverse of this pattern; the domestication of these species seems to have first taken place along the Taurus-Zagros arc, with domesticated ovicaprids subsequently spreading south into the Levant by 8000 B.P. Unfortunately, the relative paucity of excavated sites in the highlands has made it difficult to determine when exactly the domestication of sheep and goats took place. However, the evidence did suggest that they were the earliest animals domesticated for food in the Near East, with pigs following.

Sheep and goats. The greater economic importance of sheep and goats as compared to pigs at later prehistoric and historic sites in the Near East is well documented. It is believed that the preference for herding of sheep and goats over herding of pigs stems primarily from the fact that they can be more easily controlled. Thus, fewer people are needed to herd comparably sized groups of sheep and goats, enabling the maintenance of larger herds. Further, compared to pigs even large herds of sheep and goats can be more easily kept from foraging in stands of wild or cultivated cereals intended for human consumption.

Until recently, the data from the few excavated early sites along the flanks of the Taurus-Zagros arc indicated that, as in the Levant, the preagricultural inhabitants of this area also relied extensively on wild cereals for their subsistence. Thus, the conclusion that sheep and goats were the earliest economically important animal domesticates made theoretical sense and was generally consistent with the available data.

According to that view, along the Taurus-Zagros arc, there was a shift from mobile hunting and gathering to an increasingly settled mode of life, based on either the exploitation of wild cereals (earlier than 10,000 B.P.) or on the introduction of cereal domesticates and agriculture from the Levant (after 10,000 B.P.). Where possible, such increased sedentism resulted in attempts to compensate for local overexploitation of wild animals in the vicinity of now-permanent villages by attempts at animal husbandry. In the Levant, where the most economically important wild animal resource was the gazelle (*Gazella*), animal husbandry may not have been feasible or, if it was, did not lead anywhere. Along the Taurus-Zagros arc (the natural habitat zone for wild sheep and goats) it led to the domestication of these species and the subsequent spread of domesticated sheep and goats south into the Levant.

Controversial evidence from the site of Zawi Chemi in northern Iraq, excavated in the 1950s, was originally interpreted to suggest that the process of ovicaprid domestication was already under way as early as 11,000 B.P. However, subsequent excavations in this region at the sites of Çayönü and Ganj Dareh suggest that the domestication of sheep and goats may not have occurred until after 10,000 B.P. Of equal interest, but until now largely ignored for lack of corroborating evidence from other sites, was the discovery at Çayönü in eastern Anatolia that domesticated pigs were being exploited at least as early as domesticated ovicaprids.

Pigs. Recent discoveries at the site of Hallan Çemi in the Taurus foothills of eastern Anatolia now indicate that the domestication of pigs actually preceded that of ovicaprids along the Taurus-Zagros arc. The site of Hallan Çemi is the remains of a small, fully settled village community occupied for several hundred years toward the end of the eleventh millennium B.P. It is the earliest currently known fully settled village site along the Taurus-Zagros arc. The economy of the site's inhabitants revolved around the exploitation of the nuts and pulses that were abundant in the site's vicinity, as well as a variety of other plant foods. Evidence of exploitation of wild or domesticated cereals, however, was conspicuously absent.

Of the various animal species exploited by the site's prehistoric inhabitants, ovicaprids were the most economically important. However, whereas the ovicaprid remains are those of a wild population that was hunted, the pig remains appear that of a population in the process of domestication.

There are four primary lines of evidence. (1) Morphologically, the molars of pigs undergo reduction as a consequence of domestication, and the pig molars from Hallan Çemi are consistent with a population undergoing this process. (2) The survivorship curves support the morphological evidence. Approximately 75% of ovicaprids recovered at Hallan Çemi were more than 36 months of age at the time they were consumed, in contrast to only 36% of pigs of that age; approximately 29% were less than 12 months of age at death. This pattern of consuming disproportionally large numbers of relatively young animals is characteristic of animal husbandry and the culling of herds. (3) The sex ratio of pig remains indicates a disproportionate consumption of males that is consistent with husbandry. (4) The ratio of non-meat-bearing to meat-bearing bones for each species indicates that pigs were being killed and butchered at the site in much higher numbers than other animal species. Again, this pattern is typical of animal husbandry as opposed to hunting, wherein animals were typically butchered at the kill site and only meat-bearing portions of the carcass were selectively brought back to the camp.

Although ovicaprids can be more efficiently herded than pigs, pigs have certain characteristics that make them uniquely suitable for early attempts at domestication. The fecundity and growth rate of pigs makes them superior to all the other animal domesticates in the Near East as a renewable source of animal protein. In addition, because female pigs, like dogs, create nests and leave their young in them to go out on feeding forays, young pigs can be readily obtained. Also like dogs, young pigs imprint on humans, and are therefore easily tamed and do not necessarily require herding. In situations where pigs do not represent a threat to human resources, they can be left to forage (and return home) on their own.

The key point is that although pigs are more difficult to herd than ovicaprids, a situation could have existed where this trait was unimportant. The recent discoveries at Hallan Çemi show that, contrary to expectations, the economy of the earliest settled village communities along the Taurus-Zagros arc did not rely on the exploitation of cereals but instead on nuts and legumes. Thus, potential domesticates such as pigs were not potential competitors, and would not need to be herded away from stands of wild grains or agricultural fields. There was nothing to prevent experimentation with pig domestication; there was much to recommend it. Apparently, only with the shift to cereal exploitation sometime subsequent to 10,000 B.P. were the lessons learned on pigs applied to the domestication of ovicaprids, a more practical animal mainstay in the context of the changed economies.

Michael Rosenberg

Bibliography. J. Clutton-Brock (ed.), *The Walking Larder: Patterns of Domestication, Pastoralism, and Predation*, 1989; D. R. Harris and G. C. Hillman (eds.), *Foraging and Farming: The Evolution of Plant Domestication*, 1989; M. Rosenberg, Hallan Çemi Tepesi: Some further observations concerning stratigraphy and material culture, *Anatolica*, 20:121–140, 1994; W. Van Zeist, Some aspects of early neolithic plant husbandry in the Near East, *Anatolica*, 15:49–67, 1988.

Prospecting

New research and new ideas on the early evolution of the continental crust in the southwestern United States have important implications for mineral exploration in the region. Some of this continental crust formed late in the Archean Eon (3.8–2.5 billion years ago), but most of it formed in the early to middle part of the Proterozoic Eon (2.5–0.5 b. y. a.). Many ore deposits of the region formed contemporaneously with the development of this early continental crust, particularly in the period 1.8–1.7 b. y. a. However, other ore deposits in the region formed much later, during processes that modified the crust. Some studies are concentrating on the influence of the Proterozoic crust on the location and composition of ore deposits formed late in the history of the region, especially in the time interval 80–10 million years ago. A new hypothesis that Australia and Antarctica were once connected to western North America also has implications for mineral exploration in the region.

Current research includes three broad topics of considerable interest to economic geologists and to the mineral exploration industry in the southwestern United States: metallogeny of the Proterozoic provinces in the Southwest, controls of the Proterozoic crust on Phanerozoic ore deposits, and exploration implications of the proposed Australia-Antarctic connections.

Metallogeny. The study of broad crustal controls of ore deposition is termed metallogeny. As the Proterozoic provinces (**Fig. 1**) vary in age and character, so do their ore deposits (see **table**). Understanding the diverse metallogenic controls that characterize each province is critical to successful mineral exploration.

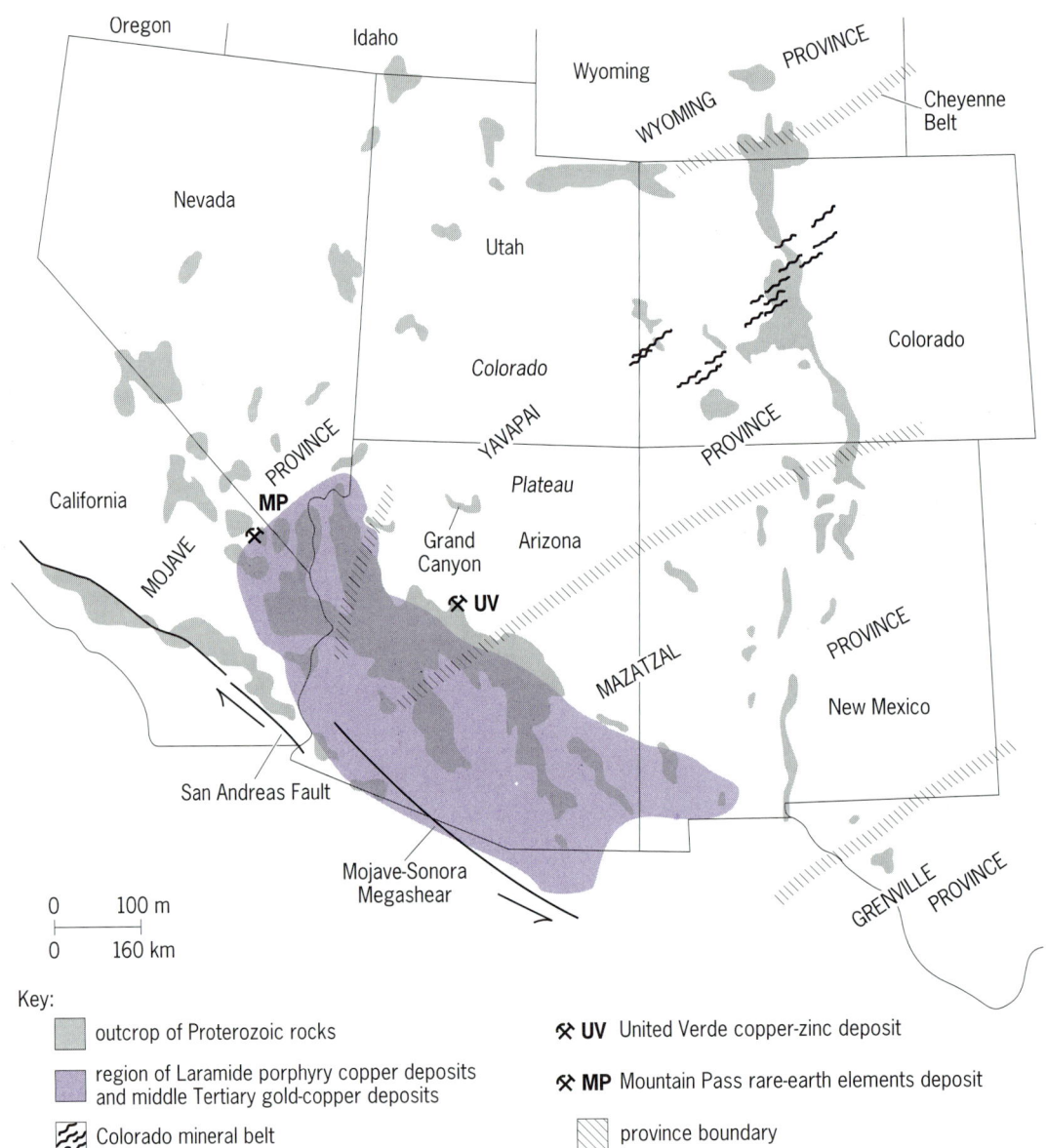

Fig. 1. Provinces and outcrop of rocks of the Proterozoic Eon in the southwestern United States. There is no outcrop over large areas because the thick Proterozoic crust is capped by a very thin layer of rocks of the Phanerozoic Eon.

In the Wyoming province near the Cheyenne Belt, minor uranium-thorium-gold deposits in sedimentary rocks are similar in origin to rich gold deposits of Archean age in Africa.

In the Mojave province few ore deposits are known that were formed during crustal growth in the early part of the Proterozoic Eon, although recent work indicates that some small copper-zinc deposits may be genetically related to earliest Proterozoic volcanism there. The early history is obscured because of metamorphism at very high temperatures and pressures.

The Yavapai province contains by far the richest and most abundant Proterozoic ore deposits in the Southwest: more than 100 known deposits of copper and zinc with secondary gold, silver, and lead, the most important being the giant United Verde deposit in Arizona (Fig. 1). This metal endowment is a result of the evolution of the Yavapai province as a series of largely submarine volcanic arcs, analogous to the modern volcanic island chains of the western Pacific Ocean or the western Aleutian island chain. The copper-zinc deposits formed in volcanic areas where metal-laden hot brines erupted onto the sea floor.

The numerous volcanic chains of the Yavapai province were driven northward by plate tectonic processes, colliding with the Archean Wyoming province (and possibly the Mojave province) along a complex structural zone called the Cheyenne Belt (Fig. 1). Strong deformation and complex structures throughout the province make modern structural geology techniques indispensable for mineral exploration.

Volcanic and sedimentary rocks of the Mazatzal province were deposited on a continental margin at

Proterozoic provinces and ore deposits of the southwestern United States

Province	Age, billions of years	Lithologies and tectonic associations	Contemporaneous ore deposits
Wyoming	3.0–2.5 (Archean) 2.4–1.9	Gneiss—high metamorphic grade Continental margin sedimentary rocks near Cheyenne Belt	Various Uranium-thorium-gold sedimentary; copper-zinc-lead volcanogenic
Mojave	2.5–1.7	Gneissic igneous and sedimentary rocks—high metamorphic grade	Copper-zinc volcanogenic(?)
Yavapai	1.8–1.7	Igneous and sedimentary rocks related to subduction	Copper-zinc-lead-gold-silver volcanogenic (includes world-class United Verde deposit, Arizona); variable metal-bearing vein
Mazatzal	1.7–1.65	Continental margin sedimentary rocks; locally extensive volcanic rocks related to melting of continental crust	Tin and other rare metals related to granite; iron and manganese related to quartz-rich sedimentary and volcanic rocks and locally with tin, tunsten, tantalum, and niobium
Grenville	1.35–1.1	Continental margin and volcanic arc rocks juxtaposed by thrusting	Various and minor
Mojave-Yavapai-Mazatzal	1.45–1.4	Granite of major transcontinental igneous event Carbonatite (Mountain Pass, Southern California) Anorthosite (Southeast Wyoming)	Tungsten; tin; uranium; rare metals Rare-earth elements Iron; titanium; aluminum
Mojave-Yavapai-Mazatzal-Grenville	1.1	Basalt dikes and sills Granite (Colorado, New Mexico, Texas) Layered intrusive (subsurface, Texas)	Talc and asbestos adjacent to limestone Tin and other rare metals Platinum, chromium

the southern edge of the Yavapai province and offshore in a deep sea to the south. No significant ore deposits were formed in the deep-sea deposits in what is now southern New Mexico and southeastern Arizona (Fig. 1). However, in northern parts of the Mazatzal province, including a broad overlap zone with the Yavapai province, granitic magmatic activity produced widespread, low-grade tin-beryllium deposits and left hints of potential for deposits such as Olympic Dam, a huge repository of copper, gold, iron, and uranium that was recently discovered in Australia.

Vast granite bodies were emplaced throughout the Mojave, Yavapai, and Mazatzal provinces (and across North America to eastern Canada) 1.45–1.4 b. y. a. during one of the greatest magmatic events in the history of Earth. Associated with these granites are numerous, mostly small, mineral deposits containing combinations of tungsten, uranium, tin, beryllium, and other metals. An important exploration method for locating these deposits is airborne gamma-ray spectrometry, which detects the high radioactivity characteristic of these granites. At Mountain Pass in the Mojave province (Fig. 1), the second-largest rare-earth-element deposit in the world is genetically related to small, rare carbonate-rich igneous bodies that were emplaced during the 1.45–1.4-b. y. a. magmatic event. Further exploration for rare-earth element deposits centers on the search for these carbonate-igneous rocks in the Mojave province.

Another singular, though not so widespread, igneous event occurred 1.1 b. y. a. Talc and asbestos deposits occur where basaltic magma intrudes limestone in western Texas, central Arizona, and Death Valley, California. A great layered intrusion of this age has been recognized from drilling and from geophysical anomalies in the subsurface of western Texas. This intrusion has potential for nickel, chromium, and platinum-group elements.

Controls on ore deposits. Many ore deposits in the southwestern United States were formed in the Phanerozoic Eon (0.5 b. y. a. to present). Most were deposited within a few miles of Earth's surface directly or indirectly from magmas produced at the base of the crust or in the uppermost mantle. The compositional character of the Phanerozoic magmas and their associated ore deposits and the positions of emplacement of the ore deposits are controlled to some extent by the thick Proterozoic crust through which the magmas rose.

It has recently been recognized that gold and silver contents in late Phanerozoic ore deposits in the tinted portion of Fig. 1 are controlled in some way by the Proterozoic crust. Ore deposits of Proterozoic, Laramide (80–60 m.y.a.), and middle Tertiary (30–10 m.y.a.) ages in the Yavapai and Mojave provinces have silver-gold ratios lower than 17.5:1 (the crustal average). Ore deposits of the same ages in the Mazatzal province have silver-gold ratios higher than 17.5:1. Regardless of the time of

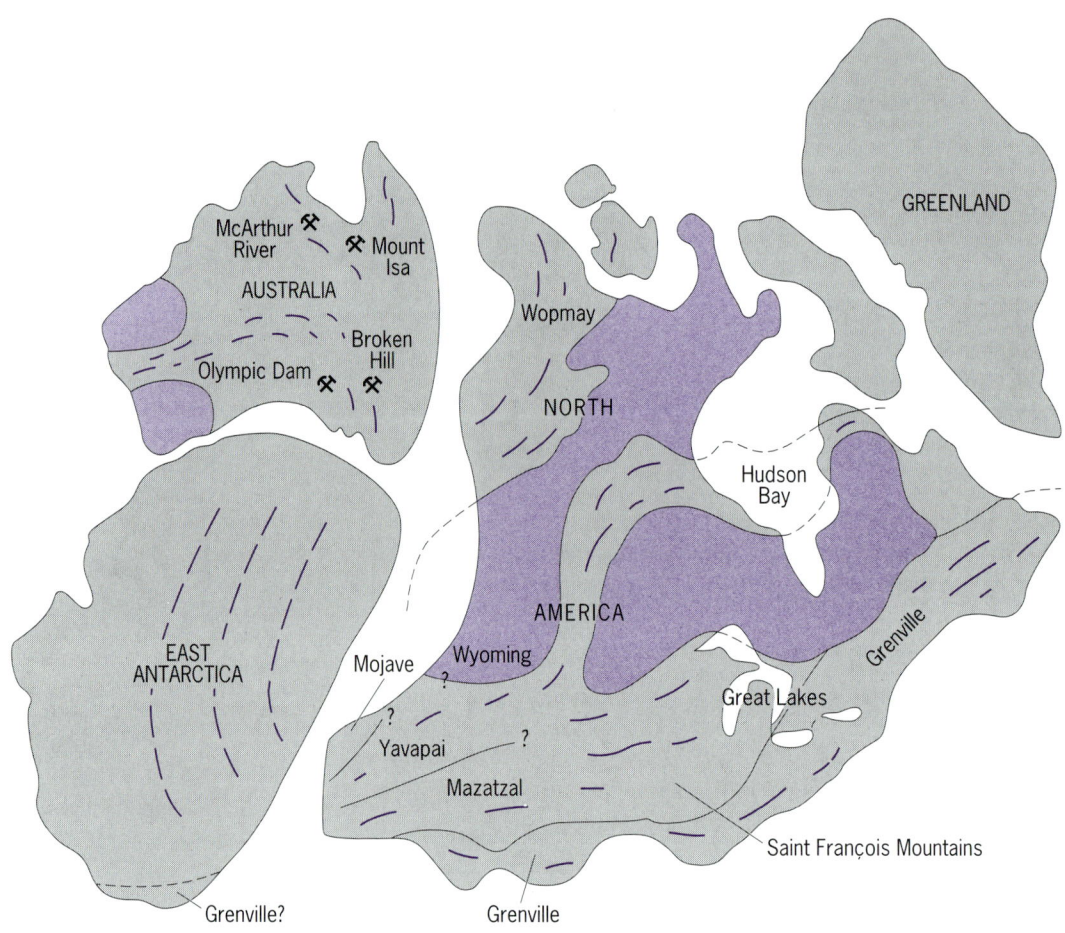

Key: ⚒ ore deposit

Fig. 2. Hypothetical relative positions of North America, east Antarctica, and Australia from approximately 1.8 to 0.7 billion years ago. Areas shown in color are Archean blocks; areas with broken colored lines, indicating trends of belts of deformed rocks, are Proterozoic belts, locally shown with province names.

ore deposition, the Yavapai and Mojave provinces clearly tend to produce relatively gold-rich, and the Mazatzal province relatively silver-rich, ores. Further research is necessary to explain this distribution, but simply recognizing the relationship is important in guiding mineral exploration.

There are other instances of Proterozoic crustal control on Phanerozoic deposits. It was noted many years ago that locations and trends of Tertiary ore deposits in the Colorado mineral belt (Fig. 1) were apparently controlled by northeast-trending Proterozoic structures. Recent discovery of hydrocarbons in late Proterozoic sedimentary rocks of the Grand Canyon suggests that some petroleum deposits in early Phanerozoic sedimentary rocks of the Colorado Plateau (Fig. 1) may have migrated upward from the Proterozoic rocks.

Proposed Australia-Antarctic connections. The southwestern-trending Proterozoic provinces of the Southwest have abrupt terminations near the Pacific Coast (**Fig. 2**), suggesting that these belts may be continuous into another continent long since rifted away from North America. Recent studies have suggested a prerift configuration for Australia, east Antarctica, and North America as shown in Fig. 2.

This configuration is based primarily on the possible extension of the Grenville province into east Antarctica, an independantly justified fit between Antarctica and southern Australia, and on paleomagnetic evidence that suggests the three continents were all in about the same position relative to Earth's magnetic pole during middle to late Proterozoic time. Also, there are similarities between the Wopmay province of northwestern Canada (Fig. 2) and the Proterozoic tectonic belts in central Australia.

In this hypothesis, the Proterozoic belts of southwestern North America are considered to extend westward then northward through east Antarctica and into Australia. Configurations other than that shown in Fig. 2 are possible. Australia could, for example, have been situated much closer to southwestern North America, whose truncated provinces have some striking similarities in details of geologic history to those in Australia.

In any case, continuation of the Proterozoic provinces of the southwestern United States into Australia has implications for mineral exploration in the Southwest. Volcanic-sedimentary sequences of the northern Mazatzal province are the same age (1.70–1.67 b. y. a.) as volcanic-sedimentary sequences that host three major lead-zinc-silver deposits in Australia (Broken Hill, Mount Isa, and McArthur River; Fig. 2). The volcanic rocks of the northern Mazatzal province are somewhat dissimilar in type to those that host the lead-zinc-silver deposits, but they are identical in character to those that host the Olympic Dam deposit (Fig. 2) of Australia. The Mazatzal province volcanic rocks are also identical to Proterozoic rocks that host ore deposits of the Olympic Dam type in the St. François Mountains of Missouri (Fig. 2).

Although there are differences, the parallels between the Proterozoic provinces of Australia and those of southwestern United States should stimulate exploration for deposits of both Broken Hill and Olympic Dam types in the Southwest.

For background information SEE CONTINENTAL DRIFT; EARTH CRUST; GEOLOGICAL TIME SCALE; MAGMA; ORE AND MINERAL DEPOSITS; PLATE TECTONICS; PROSPECTING in the McGraw-Hill Encyclopedia of Science & Technology.

Clay M. Conway

Bibliography. C. M. Conway, *Tectonomagmatic Settings of Proterozoic Metallogenic Provinces in the Southwestern United States,* U.S. Geol. Sur. Circ. 1062, 1991; E. M. Moores, Southwest U.S.–east Antarctic (SWEAT) connection: A hypothesis, *Geology,* 19:425–428, 1991; T. C. Pharaoh et al. (eds.), *Geochemistry and Mineralization of Proterozoic Volcanic Suites,* Geol. Soc. Spec. Pub. 33, 1987; S. R. Titley, The crustal heritage of silver and gold ratios in Arizona ores, *Geol. Soc. Amer. Bull.,* 99:814–826, 1987.

Psychoacoustics

In the past, auditory research focused on the perception of simple sounds, but in recent years the processing of complex sounds has been investigated, with some surprising results.

Simple sounds. The understanding of how simple sounds such as sinusoidal tones are heard is reasonably complete. In the early 1930s, G. von Békésy's research on cochlear mechanics established that different sinusoidal frequencies stimulate different areas along the basilar membrane, so that a frequency-to-place code exists. This finding confirmed the nineteenth-century speculation of G. Ohm and H. Helmholtz that the ear acts as a resonant system performing a rough Fourier analysis of an incoming signal. Later research has confirmed and refined these frequency-to-place maps. The basilar membrane of all mammals has a place of maximum motion that is stimulated by a sinusoidal signal, and this place changes linearly with the logarithm of the signal frequency. The region near the entrance of the cochlea responds most vigorously to high-frequency sounds, and lower frequencies stimulate more apical regions. The constants of this linear equation change for different mammals, so that size of the animal determines the range of frequencies to which it is most sensitive: larger animals are more sensitive to lower frequencies and less sensitive to the higher frequencies.

Auditory filters. After Békésy's findings, neurophysiologists used microelectrodes to record the information carried by first-order auditory neurons. Their experiments revealed that each auditory nerve fiber is tuned to a limited range of frequencies, and therefore acts as a bandpass filter. The bandwidth of these auditory filters is approximately one-sixth the center frequency. Thus, at 1000 Hz, the bandwidth of the auditory filter is about 160 Hz.

Masking. Researchers then turned to investigating how well changes in different aspects of the simple sinusoid could be discriminated. Extensive work was devoted to determining the minimal change in the frequency or intensity of a single sinusoid that was barely audible. A closely related topic is that of masking, that is, the process by which one sound (the masker) makes a second sound (the signal) difficult or impossible to hear. Two general principles emerged from these masking experiments, both consistent with the idea that the auditory system acts as a set of bandpass filters: The amount of masking or interference increases (1) as the frequency content of the masker and signal becomes more similar and (2) as the intensity or loudness of the masker increases, a result that is not surprising.

Complex sounds. Although complex sounds, containing a number of sinusoidal signals, characterize almost all the signals heard in everyday life, little research has been focused on this topic. The bulk of auditory research has concentrated almost exclusively on the perception of the simple sinusoid, because the sinusoid represents the elemental signal of auditory perception in the analytic approach that characterizes all science.

Profile analysis. Recent research on complex auditory signals has had several unexpected results. In a typical discrimination experiment, a listener hears one of two possible sounds. One sound, the standard, is a complex sound, consisting of a set of sinusoidal components all equal in amplitude. The components are equally spaced on a logarithmic frequency scale. Thus, the different sinusoidal components of the complex stimulate nearly equal distances along the basilar membrane. The second sound, the signal, is created by increasing the intensity of a single component of the standard complex; this component is called the signal component. In effect, the listener is asked to discriminate a change in the shape of the power spectrum of the stimulus, and hence this listening task is called profile analysis.

When the spacing between successive components is more than one-third of an octave, each component falls into a different auditory filter, and there is little interference or masking between components. The first experimental finding was that the signal (an increase in the central component) was easier to hear as the frequency spacing between the components was increased. It is easier to detect a change in the amplitude of a 1000-Hz component if the two other components are 200 and 5000 Hz, rather than, say, 900 and 1111 Hz. Second, for components widely spaced in frequency, increasing the number of components in the complex made the signal easier to hear. Detecting a change in the amplitude of a single component is easier if there are 20 other components rather than only 2. As the number of other nonsignal components increases, the total sound energy or loudness increases. Despite this increase in loudness, the ability to detect a change in the amplitude of a single component improves, thus disproving the dictum that masking always increases when the intensity of the masker is increased. Finally, for components such as these that are widely spaced in frequency, the phase of the components is unimportant. It is the power spectrum of the complex, not the time waveform, that is critical.

Simultaneous level comparisons. The listener is believed to make independent estimates of the level of each component of the complex. By comparing the average level of the nonsignal components with the level of the signal component, the listener detects the relative level of the signal component, and can thereby detect whether the spectrum has been altered. Such a process explains why it is better to have many nonsignal components: more nonsignal components provide a better estimate of the average nonsignal level, and hence allow smaller changes in the signal component to be detected. This process also explains that wide spacing of the components is preferable because the different level estimates are more independent for wider frequency spacings. The same general ideas can be applied to how other changes in the spectrum of a complex sound are heard, for example, tilts, ripples, or steps in the plot of the power spectrum at some frequencies. Depending on the nature of the change, the sound levels in all the auditory filters are simultaneously estimated and compared with each other to better detect the particular spectral change. These general ideas are often called the channel model of hearing, because the levels in all auditory filters are simultaneously compared in order to evaluate the nature of the sound spectrum.

Comodulation masking release. Another example of these simultaneous comparisons of several auditory filters involves listening to sounds that vary in time, such as when the listener is trying to hear a sinusoidal signal of fixed amplitude and frequency in a background of noise that slowly varies in level. Obviously, the signal-to-noise ratio of the auditory filter centered at the signal frequency will change over time. The signal-to-noise ratio will be high when the noise is low and low when the noise is high. If the noise fluctuations are the same over the entire auditory spectrum, then the same slow noise fluctuations will occur not only at the auditory filter containing the signal frequency but also at other auditory filters centered at frequencies remote from the signal frequency. These remote filters contain practically no signal energy, but do contain information about the relative level of the noise. The presence of these slow noise fluctuations, which are correlated between the nonsignal and signal bands, actually improves the listener's ability to detect the fixed sinusoidal signal. Because these slow, correlated noise fluctuations are often produced by amplitude modulation of the noise, the phenomenon, discovered in the early 1980s, is known as comodulation masking release (CMR). Prior to these experiments on profile analysis and CMR, it was not understood that a comparison of the information in many auditory filter channels could be used by the auditory system to detect and recognize different sounds. In fact, most theories ignore the listener's ability to compare the outputs of different auditory filters.

Source identification. In retrospect, it should not be surprising that the auditory system is simultaneously sensitive to the spectral shape of a complex sound and its temporal fluctuations. Different sound sources produce different complex spectra containing a number of components. The dynamics of these sounds, such as the onset or offset of these components and their fluctuations in level, often occur simultaneously over a wide frequency range. Thus, it has been argued that these new results simply confirm that the main goal of auditory perception is not to hear sound but to identify and classify the different sound sources in the listener's environment. Different sources have different spectral shapes or profiles. Correlations among the levels of these different components of the spectrum also indicate a single sound source.

The parameters of these interesting auditory processes remain to be determined; namely, what changes in spectral shape are easy to hear, and what range of dynamics the correlation detector can appreciate.

For background information SEE ELECTRIC FILTER; HEARING (HUMAN); MASKING OF SOUND; PHYSIOLOGICAL ACOUSTICS; PSYCHOACOUSTICS in the McGraw-Hill Encyclopedia of Science & Technology.

David M. Green

Bibliography. D. M. Green, *Profile Analysis,* 1988; J. Hall, M. P. Haggard, and M. A. Fernandes, Detection in noise by spectro-temporal pattern analysis, *J. Acous. Soc. Amer.,* 76:50–56, 1984; W. A. Yost, Auditory image perception and analysis: The basis for hearing, *Hearing Res.,* 56:8–18, 1991; W. A. Yost and C. S. Watson, *Auditory Processing of Complex Sounds,* 1987.

Quantum wells

The electrical properties of semiconductors are well understood and have been utilized by the electronics industry to create semiconductor devices which have become smaller, faster, and more efficient through continued development. However, magnetic effects in traditional semiconductors are too weak to be exploited in easily attainable magnetic fields. A relatively unexplored class of semiconductors, the diluted magnetic semiconductors (DMS), have interesting magnetic and magnetooptical properties at relatively modest magnetic fields. DMS materials are attractive for research and device applications because they can be grown in extremely thin magnetic layers. These thin magnetic layers can form quantum wells, which trap electrons and cause the semiconductor to exhibit magnetically controlled electrical effects. Further progress in understanding small DMS systems might make possible semiconductor magnetic storage, which would further integrate the magnetic recording industry with integrated-circuit technology.

Diluted magnetic semiconductors. In these semiconducting alloys some atoms of the host crystal are replaced by magnetic ions, usually manganese (Mn^{2+}) or iron (Fe^{2+}). Some examples of these are $Cd_{1-x}Mn_xTe$, $Zn_{1-x}Fe_xSe$, and $Hg_{1-x}Mn_xTe$. The magnetic ions mimic the valence electron configuration of the cations, cadmium (Cd), zinc (Zn), and mercury (Hg), and can therefore reside on the cation site. The interesting magnetic effects are the result of the large number of uncompensated electron spins in the magnetic dopant. Most semiconductors have an equal number of spin-up and spin-down electrons, resulting in a small magnetic moment. Each magnetic dopant in DMS materials can have upward of five uncompensated electron spins. These spins, depending on the concentration of magnetic dopant, can cause a DMS material to be a paramagnet or antiferromagnet, or to have a random frozen moment (that is, to be a spin glass). Additionally, when the semiconducting electrons of the host crystal interact with the electrons of the magnetic dopant, it is energetically favorable for the spin of the magnetic dopant to be transferred to a large number of semiconducting electrons. The result is that optical effects which normally occur in magnetic fields of several hundred tesla in standard nonmagnetic semiconductors occur in fields of only a few tesla in DMS materials. Examples of these effects are giant Zeeman splittings, large Faraday rotations, and magnetic polarons.

Well structure. These interesting magnetic properties are all the more attractive because DMS materials can be integrated with traditional semiconductors by the technique of molecular beam epitaxy (MBE). This technique allows for the growth of a semiconductor one atomic layer at a time. In this manner, ultrathin materials, which exhibit quantum-mechanical properties, can be grown. Studies on structures with ultrathin layers have shown that it is possible to trap electrons in a thin low-energy material by confining them between layers of another material with a high-energy barrier (**Fig. 1**). When these layers are thin enough to be a small multiple of an electron's wavelength, a quantum well is formed. Quantum wells have the property that free motion of the electron occurs only in the two directions perpendicular to the growth direction, the motion in the third direction being restricted by quantum mechanics to well-defined quantized values of energy. This ability to confine the electron has led to the development of quantum devices, such as the quantum well diode laser found in compact-disk players and the ultralow-noise transistors found in satellite amplifiers. All these devices are controlled by using the charge of the electrons, not their spins.

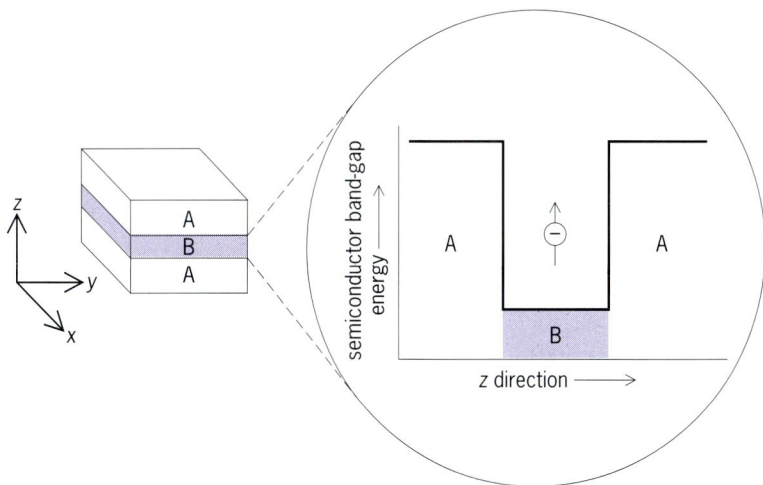

Fig. 1. Semiconductor quantum well. Electrons are confined in the low-energy material, labeled B, between layers of another material, labeled A.

Magnetic wells. Control over the electrons' spin in DMS materials is possible because of the strong energy exchange between the charge carriers and the magnetic ions. This energy exchange can be enhanced and more easily observed when there is a large amount of electron and magnetic ion overlap, which occurs most readily in magnetic quantum wells. In a magnetic quantum well, either the well or the barrier (or both) are made of a DMS material. The large electron population confined in a magnetic quantum well is forced to interact with the magnetic ions, since they consitute the magnetic well material. As a result, interesting magnetic effects occur.

Faraday rotation. One such effect is Faraday rotation. When some materials are subjected to a magnetic field, they rotate the plane of polarization of light traveling through them. This effect is unusually large in some DMS materials, in which the plane of polarization of the light can rotate upward of 1000° while transversing only 1 mm of material in a magnetic field of 0.01 tesla (100 gauss; approximately the field of a common refrigerator magnet).

This effect is employed in the construction of optical Faraday isolators (**Fig. 2**). A Faraday isolator

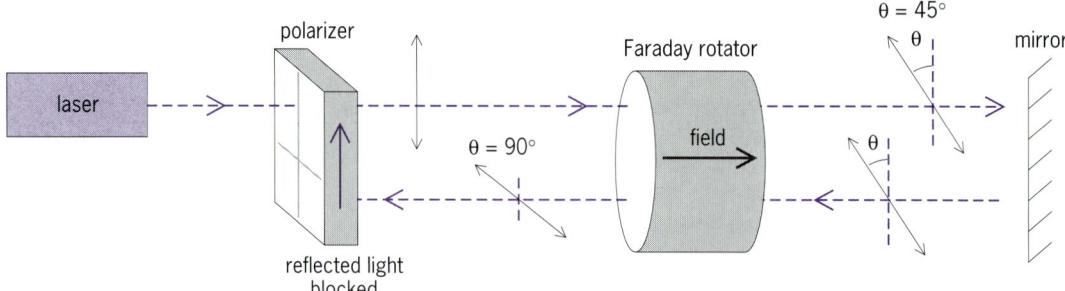

Fig. 2. Faraday isolator in operation.

is a device consisting of a linear polarizer and a Faraday rotator which is set to rotate the plane of polarization of light by 45°. Light which passes though the polarizer is oriented parallel to the axis of the polarizer. The Faraday rotator then rotates the plane of polarization of this light by 45°. Should the light reflect back through the rotator, it will acquire a second 45° rotation in the same direction as it passes back through the Faraday rotator. The orientation of the light is now such that it cannot pass though the linear polarizer, and as a result no reflections can get back to the light source.

Faraday isolators are used in the telecommunications industry to ensure that the quantum-well diode lasers used for high-speed fiber-optic communication are protected from light reflecting back into the laser, which would cause instability in the light output and compromise data integrity. Although other materials are currently in use, DMS materials can potentially be integrated into the same package as diode lasers by epitaxial growth directly on the diode lasers. This integration of laser and isolator will enable a greater number of optical devices to be contained in the same area, as is required by present technology, thus increasing the data-communication capacity of the telecommunications industry.

Bound magnetic polarons. A second interesting consequence of the large exchange energy interaction is the creation of bound magnetic polarons at temperatures below 10 K (−260°C or −440°F). A bound magnetic polaron consists of an electron attracted to a positive donor atom. The electron orbiting about this donor influences many magnetic ions, tending to align their magnetization with the direction of the electron's spin (**Fig. 3**). Thus, an oriented ball of roughly several hundred ions is produced, creating a very large magnetic moment.

In magnetic quantum wells, there are two types of mechanisms that produce polarons: (1) A polaron is formed by a carrier trapped at a quantum-well interface by local point defects or impurities. (2) A two-dimensional polaron is formed if the quantum-well width is thinner than the orbit of a trapped electron. This electron is then trapped by the well and is forced to orbit in its two free dimensions.

Carriers can be excited into the quantum well with light of the appropriate energy. The result is that the increased number of carriers in the quantum well align the magnetic ions, increasing the magnetization of the sample. These externally created polarons have two stable states at low temperatures: magnetization parallel or antiparallel to the direction of the light. This externally controlled bistable state could be the basis for a magnetooptical read-write memory. Problems presently exist in ensuring that a magnetic polaron memory bit does not decay due to the limited lifetime of the polaron state or because of temperature fluctuations.

Quantum dots. The properties of magnetic quantum wells are well established, but the major difficulty in utilizing their unique magnetic attributes is that all the interesting magnetic phenomena occur only at temperatures of 60 K (−210°C or −350°F) and below. One promising method of preventing these interesting effects from disappearing at elevated temperatures is to reduce the dimension of the magnetic system as a whole. When a potential is created that confines the carriers in all three directions, an artificial atom or quantum dot is produced. Quantum dots, like real atoms, are predicted to have discrete, widely separated energy levels. This increased separation of the energy levels can better resist transitions between levels caused by external perturbations, such as an applied electric field, magnetic field, or, more importantly, thermal fluctuations. By increasing the separation of the

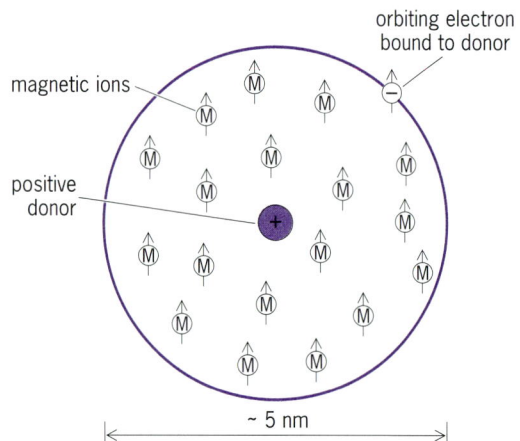

Fig. 3. Bound magnetic polaron formed by an electron orbiting a positive donor. The electron's interaction with the magnetic ions polarizes the magnetic moments (M) within its orbit.

energy levels, large thermal fluctuations at elevated temperature can be tolerated without affecting the magnetic properties of the quantum well. In addition, increasing the overlap of the semiconductor's electrons with the magnetic ions by confinement in a small area should increase the magnitude of many of the interesting magnetic effects. Advances in this technology might then be used to make more sensitive magnetic sensors to increase storage density of magnetic tapes and disks.

For background information SEE CRYSTAL GROWTH; FARADAY EFFECT; OPTICAL COMMUNICATIONS; OPTICAL ISOLATOR; POLARON; SEMICONDUCTOR; SEMICONDUCTOR HETEROSTRUCTURES in the McGraw-Hill Encyclopedia of Science & Technology.

David D. Awschalom; David Tulchinsky

Bibliography. J. K. Furdyna and J. Kossut (eds.), *Semiconductors and Semimetals,* vol. 25: *Diluted Magnetic Semiconductors,* 1988; F. Henneberger, S. Schmitt-Rink, and E. Göbel (eds.), *Optics of Semiconductor Nanostructures,* 1993; C. Weisbuch and B. Vinter, *Quantum Semiconductor Structures: Fundamentals and Applications,* 1991.

Quarks

Since 1960, experimental and theoretical elementary particle physicists have developed a picture of the elementary constituents of matter and how they interact. In this so-called standard model, matter is composed of quarks and leptons (see **table**). Protons, neutrons, and all nuclei in atoms are composed of quarks. The leptons are either electrically neutral (neutrinos) or charged. The charged leptons are the electron, which orbits around the nucleus in an atom, and two heavier particles similar to it (the muon and tau). In the standard model, the quarks and leptons are organized in pairs, and there are equal numbers of quark and lepton pairs. The first set (generation) of pairs consists of the up-and-down quark pair and the electron and electron neutrino lepton pair. The second generation has the charm and strange quarks and the muon along with the muon neutrino. The third generation contains the tau lepton and its neutrino, the bottom (b) quark, and a partner for the bottom quark, the top (t) quark. If the standard model is correct, the top quark must exist.

The search for this particle began shortly after the bottom quark was discovered in 1977. Since then, improved experimental sensitivity has allowed exploration over an increasing range of possible top quark mass. By 1992, the top quark was ruled out if it had a mass below 90 GeV. In April 1994, the first direct experimental evidence for the existence of the top quark was presented.

Producing the top quark. In the Fermilab Tevatron collider in Batavia, Illinois, top quarks can be created by the energy released in very high energy proton-antiproton collisions. In this reaction, a top

Three families of quarks and leptons

Constituents	Families	Particle name	Mass, MeV
Quarks	$\begin{pmatrix} u \\ d \end{pmatrix}$	up down	~5 ~10
	$\begin{pmatrix} c \\ s \end{pmatrix}$	charm strange	1500 ~200
	$\begin{pmatrix} t \\ b \end{pmatrix}$	top bottom	174,000 4500
Leptons	$\begin{pmatrix} \nu_e \\ e \end{pmatrix}$	e neutrino electron	<0.00001 0.5
	$\begin{pmatrix} \nu_\mu \\ \mu \end{pmatrix}$	µ neutrino muon	<0.3 106
	$\begin{pmatrix} \nu_\tau \\ \tau \end{pmatrix}$	τ neutrino tau	<30 1780

quark and its antimatter partner, the top antiquark, are produced together (see **illus.**). The top quark decays almost immediately into its quark-pair partner, a bottom quark, and a W boson, the carrier of the weak nuclear force responsible for the decay of the top quark. The W in turn also rapidly decays into any of the quark or lepton pairs, for example, an electron and an electron neutrino. Each final quark manifests itself as a collimated jet of particles in the detector.

Experimental apparatus. At Fermilab, protons, the nuclei of hydrogen atoms, are accelerated to nearly the speed of light. When they reach an energy of 150 GeV, they strike a metallic target and produce many subatomic particles, including antiprotons. When more than 10^{11} antiprotons have been produced and stored, they are injected into Fermilab's accelerators along with a larger number of protons. The protons and antiprotons rotate in opposite directions in the circular accelerator because they have opposite electric charge. The energy of the proton and antiproton beams are increased to 900 GeV. The beams then collide head-on in the center of each of two large detectors, CDF

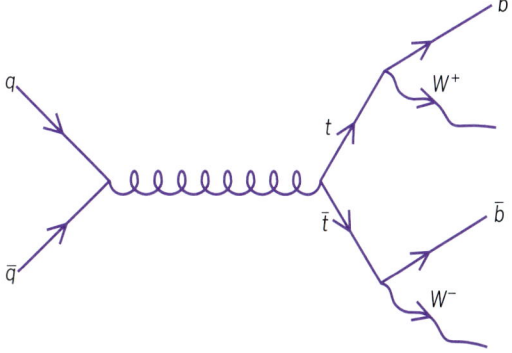

Top-quark production. A light quark (q) from a proton and a light antiquark ($\bar{q}$) from an antiproton annihilate to produce a top quark (t) and a top antiquark ($\bar{t}$), each rapidly decaying into a W boson and a bottom quark (b) or a bottom antiquark ($\bar{b}$).

and D0, that were built to study these proton-antiproton collisions.

The CDF apparatus is approximately 35 ft (11 m) wide, 35 ft (11 m) high, and 85 ft (26 m) long. It contains almost 100,000 individual particle detectors. Passing through the center of the apparatus is the Tevatron vacuum beam pipe which carries the counterrotating proton and antiproton beams into collision. Surrounding the beam pipe is a precision tracking chamber consisting of four layers of highly segmented solid-state detectors which can measure the location of a particle with a precision of approximately 0.01 mm. With this detector, the first one to be successfully operated at a proton collider, CDF is able to observe the few-millimeter flight path of b quarks before they decay. Surrounding that chamber is a tracking chamber that measures the location of the proton-antiproton collision point. Beyond that is a large charged-particle tracking chamber, embedded in a 1.4-tesla magnetic field, which precisely measures the momenta of the charged particles produced in a collision.

Outside of the tracking chambers are the electromagnetic and hadronic calorimeters. These highly segmented devices absorb all the energy of most incident particles, providing an electrical signal proportional to the particle's energy. Muons pass through the calorimeters without being absorbed, and chambers outside of the calorimeters measure their trajectories.

Signals from many of the detector systems are sent directly to the trigger electronics which analyze each proton-antiproton collision, selecting the most interesting collisions for further analysis. All the information from the apparatus for the selected collisions is read out into high-speed computers that perform additional analysis and store the most interesting events on magnetic tape. During the 1992–1993 data run, approximately 10^{12} proton-antiproton collisions occurred in the CDF detector. Of these, 1.6×10^7 were recorded and later fully analyzed.

Top-quark search. The CDF experimenters searched for the top quark in three ways. In the first, events were sought in which the W bosons from both the top quark and the top antiquark decayed into lepton pairs, either electron and electron neutrino or muon and muon neutrino. Two such events were observed, compared to approximately one-half of an event expected from sources other than the top quark (background). The other two methods searched for events in which one of the two W bosons decayed into an electron and electron neutrino or a muon and muon neutrino, with the other W decaying into a quark pair. The expected rate of such top events is considerably larger than in the first search mode, but there are also more copious sources of background. In order to reduce this background, both the second and third searches identified one or more of the b quarks in the event. The second search looked for the small flight distance a b quark travels before it decays. Six such events were observed, while the expected background was approximately two events. The third search looked for an electron or a muon from the decay of a b quark. Seven such events were seen, compared to a three-event expected background. Three of the events seen in the third search were also selected by the second method.

In order to test whether the data required the presence of a top quark, the experimenters calculated the probability that the entire observation was due to a statistical fluctuation of the background. This probability is 1 in 400. Although very small, scientists generally require an even smaller probability of a background fluctuation before a new phenomenon is firmly established. Thus, more data were needed to confirm the observation.

Interpretation of events. It is natural to interpret the excess events as the production of the top quark. The number of events is compatible with the theoretical expectation for the production rate. The particle energies observed in these events are consistent with the calculated mixture of background and top-quark signal. In order to exploit the different kinematic features expected in top-quark and background events, that is, the particle energies and angles, an additional analysis was performed looking for events in which the features are as expected in top-quark events. An excess of these events over background was found, including many of the events with an identified b quark.

Seven of the events have enough information to allow the top-quark mass to be calculated. The distribution of these masses is not that expected for pure background. Rather, six of the events cluster between 160 and 190 GeV, a range consistent with the mass resolution of the detector. The ensemble of events gives a top-quark mass of 174 GeV, with an uncertainty of ±16 GeV. This mass is close to the center of the top mass range predicted by the standard model, based on data from Z and W boson decay and neutrino interactions.

Background estimation. Much of the CDF effort on the top-quark search was focused on background processes that can mimic a top-quark event. It is the complexity of top-quark events that makes background estimation very difficult. For many of the elementary particles discovered in the past, an event contained two or three particle tracks. In contrast, for top-quark events, approximately 100 particles typically enter the CDF detector.

The most important backgrounds can be estimated from the CDF data. For example, the identification of b quarks plays an important role in the top-quark search. It is thus critical to know how often an identified b quark is actually a u or d quark that was misidentified in the detector. Therefore, the experimenters analyzed samples of events which were expected to be largely free of b quarks. Any of these events which appeared to contain a b quark were ascribed to errors in b-quark identification, and thus used to calculate the expected background in the top-quark search.

D0 detector. The D0 detector also searched for the top quark. Many of the search techniques were similar to those used by the CDF group, while others were somewhat different. Researchers observed seven candidate events, with three expected from background. They do not consider this number to be a significant excess, and thus do not feel that evidence exists for top-quark production.

Prospects. A second data run at the Fermilab collider, scheduled to conclude in mid-1995, was designed to produce approximately four times as much data at both CDF and D0 as was accumulated in the 1992–1993 data run. Thus, confirmation of the CDF top-quark result and improvement in the mass measurement by a factor of two should ensue.

Fermilab is building a main injector, an accelerator improvement that should increase the intensity of the collider's beams by a factor of 10. Experiments following its completion should collect 1000 top-quark events, a number sufficient to carry out a very sensitive test of the validity of the standard model.

For background information SEE ELEMENTARY PARTICLE; PARTICLE ACCELERATOR; PARTICLE DETECTOR; QUARKS; STANDARD MODEL in the McGraw-Hill Encyclopedia of Science & Technology.

Melvyn J. Shochet

Bibliography. F. Abe et al., Evidence for top quark production in $p\bar{p}$ collisions at $\sqrt{s} = 1.8$ TeV, *Phys. Rev. Lett.*, 73:225–231, 1994.

Radio communications

Satellite-based radio transmission provides the only practical technical means for airlines to offer live in-flight radio broadcasts. The physics of terrestrial radio-signal propagation have virtually precluded airlines from offering live broadcast services: an airliner flies out of range of a terrestrial FM broadcast station less than 30 min after signal acquisition. However, until recently the cost and size of satellite receiving antennas have prevented their installation on commercial aircraft. Recently developed equipment has Federal Aviation Administration (FAA) certified proprietary technology that allows aircraft to receive satellite signals with radio programming through a small, relatively inexpensive antenna. The airborne equipment is designed for overnight installation and couples directly into the existing aircraft audio system for distribution to passengers. Satellite-based radio is commercially available throughout the United States and can be heard live on more than 400 domestic commercial aircraft.

The radio program material is transmitted to equipped aircraft via an existing in-orbit domestic Ku-band communications satellite. On board the aircraft, the airborne equipment receives program material and inserts it into the passenger audiovisual system for distribution. Satellite-based radio currently provides three live radio broadcast channels containing news, financial reports, live play-by-play sports, national weather, special features, and talk programming. The program material is interspersed with regular advertising intervals. The service is customized for each airline, and includes periodic, brief announcements provided by the airlines to give the service a radio station–like identity.

Technically, the satellite receive hardware is optimized for the airline environment, imposing no operational penalty on aircraft performance. The system operates in all flight regimens throughout the United States. It weighs less than 35 lb (16 kg) without cabling, and the antenna has negligible drag impact on the aircraft. Without any operational intervention by the flight crew, the system automatically provides service from gate to gate throughout the normal taxi, takeoff, cruise, and landing flight envelope of commercial aircraft in domestic service.

The airborne equipment requires extensive testing and FAA certification prior to installation on commercial aircraft. Such certification is by aircraft type, and the equipment has already been certified for 10 types of aircraft.

System operation. Three channels of live audio are transmitted from a programming center in Arlington, Virginia, to airline passengers anywhere within the continental United States, its near offshore waters, and parts of southern Canada. The system uses standard Ku-band satellite technology to distribute its programming directly to the passenger. The system has been fully operational since 1992 and has clearly demonstrated a high level of reliability.

The satellite radio system functions much like a conventional private satellite data network. At the broadcast studios, the program material is prepared by on-air staff in real time and converted to digital audio. Each audio channel is compressed into a 32-kilobit-per-second data stream using psychoacoustic audio-signal compression processing. The three digital channels are then multiplexed into a single digital bit stream. This bit stream is routed to two redundant, geographically separated satellite uplink sites over redundant paths. At the operational uplink site, the bit stream is error protected, modulated onto a carrier, and transmitted to a leased Ku-band transponder on the *GSTAR III* domestic communications satellite in geosynchronous orbit. Equipped aircraft operating within the transmit footprint of the satellite receive the satellite radio signal. The aircraft receiver converts the received bit stream back into three channels of analog audio. These audio signals are then routed to passenger headsets via the aircraft audio system.

The satellite link is maintained through the normal flight envelope of the aircraft and provides uninterrupted audio from gate to gate. The system is designed with sufficient margin to operate flawlessly anywhere in the service area through aircraft roll, pitch, and yaw maneuvers.

Aircraft equipment. The satellite radio aircraft equipment is a small, lightweight, low-power, receive-only, Ku-band radio terminal tailored for installation on commercial aircraft. An engineering breakthrough in the design of this equipment allows economical installation of satellite terminals on commercial aircraft. Installation is performed during an overnight layover and does not interrupt aircraft service. The equipment consists of two assemblies: an externally mounted satellite antenna and a receiver mounted in the aircraft electronics bay.

The antenna design is based on proven aerospace military technology, and the antenna is built under stringent quality-control standards. Physically, the antenna is a ruggedized unit that is easy to install with minimum impact to the airframe. The antenna is a steered, medium-gain, Ku-band antenna contained in a top-mounted radome shaped to optimize aerodynamic performance. The antenna electronics locates the satellite signal and locks onto it. The antenna then tracks the satellite signal through the normal flight-profile movement of the aircraft around the roll, pitch, and yaw axes as well as through changes in aircraft speed. The antenna is steered in both azimuth and elevation to provide sufficient gain to receive the satellite signal over the normal flight envelope of a commercial aircraft.

The antenna provides full 360° azimuth coverage and 10–75° elevation coverage without so-called keyholes (dead spots in the coverage pattern). The low-profile antenna radome has been optimized to reduce drag turbulence and eliminate icing. The antenna assembly is mounted on the top center of the aircraft hull just aft of the wings. The antenna assembly has been installed on more than 400 B737, B757, B767, and A320 aircraft and is compatible with other commercial aircraft.

The satellite radio receiver is a very small, customized, high-performance, Ku-band satellite data broadcast receiver packaged for aircraft installation. The receiver processes the satellite signal and provides three independent demultiplexed audio signals to the aircraft audio system for distribution. The receiver uses advanced satellite signal processing, rate ½ Viterbi forward error correcting code, spread spectrum, and quadrature phase-shift keying (QPSK) modulation to enhance link performance and reduce the effect of noise. Extensive use is made of very large scale integrated circuit technology to improve reliability while reducing weight and power consumption. The receiver has a full set of built-in diagnostics to aid in testing. The audio output interface is compatible with currently installed aircraft audio systems. The audio bandwidth of each current satellite radio channel is 6 kHz, more than adequate for talk programming and wider than the bandwidth of current aircraft audio systems.

Each receiver is uniquely addressable, one at a time, by airline or by subflect, and each receiver has provision for storing promotional audio material. This feature is used extensively in the system to customize the channels for each airline and operates under control of the programming center.

System reliability. The system is designed to provide the highest availability of service in the demanding environment of a commercial airliner. The system was built with extensive multitiered redundancy to minimize signal loss due to random equipment failure. The transmission path to the satellite is fully redundant to ensure continuous broadcast to the fleet of equipped aircraft. Programming is routed through redundant signal processors, modems, and terrestrial circuits to the two satellite uplink Earth stations. These stations are placed at geographically different locations outside the Washington, D.C., area to provide protection against a catastrophic loss of one facility. The result is a very high measured level of reliability. Recent statistics show that a passenger has less than a 0.8% probability of not being able to hear the satellite broadcast radio programs.

For background information SEE ANTENNA (ELECTROMAGNETISM); COMMUNICATIONS SATELLITES; MODULATION in the McGraw-Hill Encyclopedia of Science & Technology.

Richard S. Cooperman

Radioactive beams

Recent technical developments permit the production of accelerated beams of radioactive nuclei (nuclei of isotopes which do not occur terrestrially) of sufficient intensity to be utilized for a variety of scientific enterprises. These endeavors range from studies of the structure of exotic unstable nuclei, to investigations of astrophysical processes occurring in the most violent cosmic environments, to a variety of material science topics.

Production. Two techniques are currently used to produce beams of radioactive ions: projectile fragmentation and isotope separation on line (ISOL).

In the projectile fragmentation technique, radioactive fragments of a specific nuclear species are collected from the collision of heavy nuclei by using a magnetic analysis system. To provide an efficient collection of the radioactive particles of interest, targets, usually of low atomic mass, are bombarded with high-energy heavy ions, producing high-energy radioactive fragments. This technique is well suited to yield a large variety of very high energy radioactive ions. Several high-energy heavy-ion accelerators devote more than half their scientific program to radioactive ion beam studies based on projectile fragmentation.

The radioactive ions produced by projectile fragmentation facilities are too energetic for many nuclear structure, nuclear astrophysics, and materials science applications. Another method of producing beams of radioactive ions, isotope separation on line, utilizes intense beams of light ions to produce radioactive ions. After production, the radioactive ions are allowed to diffuse from the target, are ionized and isotope-selected, and then are accelerated in another accelerator. In this technique, the secondary accelerator is chosen to provide the beam energy and quality best suited for the desired application. For studies of the decay properties of nuclei and materials science, where the ions are implanted in a solid, a very simple method of acceleration from a high-voltage platform is often sufficient.

Only recently have ISOL facilities with higher-energy secondary accelerators been considered. Such facilities, which are applicable to a variety of nuclear structure and nuclear astrophysics studies, greatly increase the scope of radioactive ion beam studies.

Nuclear structure studies. Atomic nuclei are composed of from a few to a few hundred positively charged protons and uncharged neutrons. These nucleons (a term used to describe protons and neutrons collectively) interact with each other via nature's strongest force, the nuclear force. The electrical repulsion between the positively charged protons also must be considered. The primary thrust of studies of the structure of the atomic nucleus traditionally has been the investigation of how the constituent nucleons interact and the effect of these interactions on the nuclear properties as a whole.

Nearly 290 different nuclear species occur naturally in terrestrial matter. The permanence of such stable nuclei indicates a special balance between the nucleonic forces acting among the constituent nucleons. If a specific configuration of nucleons does not represent a minimum in the energy, the nucleus will decay to a more stable configuration by emitting some type of radiation. Indeed, the detection of such radiation (for example, gamma rays, electrons, and positrons) provides much of the experimental information on nuclear structure. For even more extreme ratios of neutrons to protons, at the so-called proton or neutron drip line, the nucleus emits protons or neutrons directly. Neutron emission occurs so quickly (in an interval of $\sim 10^{-21}$ s) that no information can be derived for nuclei beyond the neutron drip line. Because of the charge of the proton, nuclei just beyond the proton drip line survive sufficiently long for experimental study; however, the lifetimes of very proton-rich nuclei, well beyond the proton drip line, also become vanishingly short.

Until recently, nuclear structure studies were confined to those species produced by the reactions of nuclei occurring in nature. Of course, unstable nuclei are produced and studied in reactions of stable nuclei. However, with such stable beams and targets the limits of nuclear stability as defined by the proton and neutron drip lines have been reached for only a few of the lightest nuclei. By utilizing accelerated beams of radioactive ions, which do not occur naturally, even more exotic nuclear species can be synthesized and their structure studied.

Among the interesting new nuclear species that have been investigated with radioactive ion beams are light neutron-rich nuclei, such as ^{11}Li (lithium-11; composed of three protons and eight neutrons). Such nuclei, lying at the neutron drip line, are only marginally stable to neutron emission. Studies indicate that for such light, weakly bound, neutron-rich nuclei a preponderance of neutrons lie near the

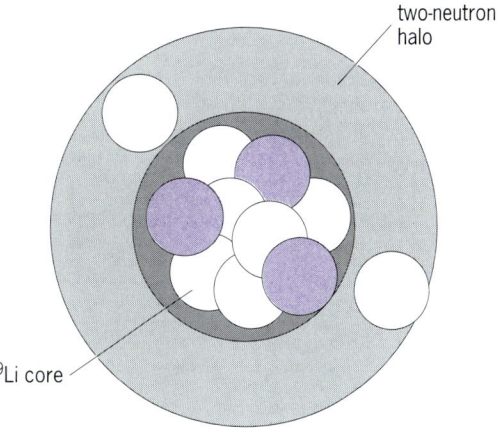

Fig. 1. Core and neutron-halo structures of lithium-11 (^{11}Li).

nuclear surface, forming a neutron halo (**Fig. 1**). Halos afford the unique possibility of studying relatively pure neutron matter at a smaller density than normal nuclear matter. Likewise, beams of halo nuclei may produce an enhanced transfer of neutrons leading to still more exotic nuclei.

Recent nuclear structure calculations predict dramatic changes for many familiar nuclear structure properties, such as proton and neutron magic numbers (associated with an increased nuclear stability similar to the increased atomic stability of the inert gases), nuclear pair correlations, and exotic nuclear shapes, near the drip lines. Some of these predictions for light nuclei are highlighted in **Fig. 2**. The tests of such predictions await a new generation of ISOL radioactive ion beam facilities, whose beams will be optimized for nuclear structure studies.

Nuclear astrophysics studies. Within 3 min after the big bang, the universe had cooled sufficiently for the deuteron (the simplest nucleus composed of one proton and one neutron) to be bound. Hence, since that time the particle-physics chapter of cosmology has been essentially complete, except for a few isolated sites where extremely high temperatures and densities occur. At lower temperatures (that is, for the last 1.6×10^{10} years) cosmology is dominated by nuclear and plasma physics.

At certain cosmological sites, such as supernova shock waves and the surfaces of white dwarf and neutron stars where there is an accretion of matter, the temperatures and densities become sufficiently large to produce proton- or neutron-capture reactions with subsecond reaction times. Any nucleus produced by such reactions with a lifetime comparable to or longer than the capture reaction will become a target for a subsequent reaction. Indeed, such reaction networks, which are the mechanism for synthesizing all but the lightest elements, do not follow stable nuclei.

Many of the properties of the rapid neutron capture process (*r* process) nuclei can be studied for the first time with radioactive ion beams. Likewise, a rapid proton capture process (*rp* process) also is predicted to occur near the proton drip line for light- and medium-mass proton-rich nuclei. Beams

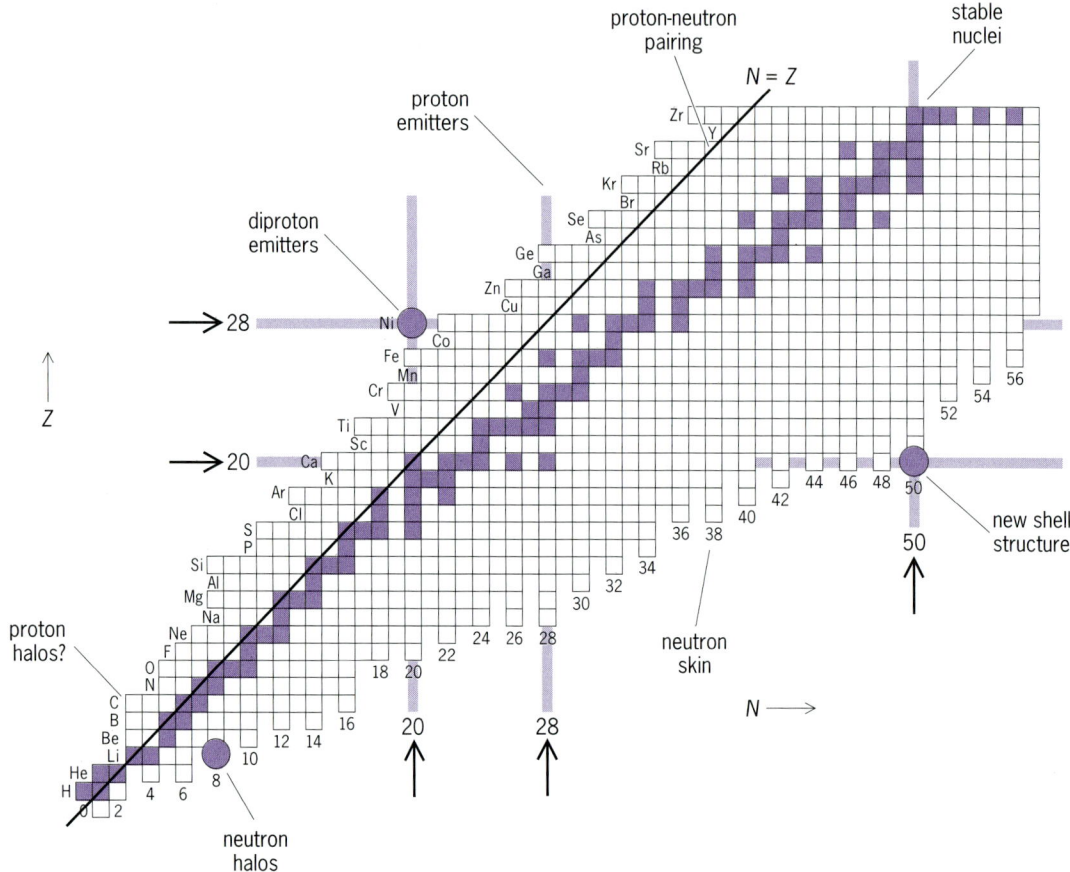

Fig. 2. Chart of light nuclides, based on proton number (Z) and neutron number (N). Some of the new opportunities for studies of nuclear structure of light nuclei afforded by accelerated beams of radioactive ions are shown. For reference, stable nuclei (which occur naturally) and proton and neutron closed shells (magic numbers $Z = 20, 28$ and $N = 20, 28, 50$) are also given. (After W. Nazarewicz, J. Dobaczewski and T. Werner, Physics of exotic nuclear states, Phys. Scripta, 1995)

of radioactive ions not only provide a means of studying the properties of rp-process nuclei but also allow direct measurements of the reaction rates in the rp-process network. For example, the probability of the capture of protons on ^{17}F (fluorine-17), producing ^{18}Ne (neon-18), can be measured directly by bombarding a hydrogen target (protons are the nuclei of the most common isotope of hydrogen, ^{1}H or hydrogen-1) with accelerated radioactive ions of ^{17}F. A program of measuring a sizable number of such reaction rates as well as the properties of nuclei associated with the rp-process is planned.

Materials science studies. Significant materials science research programs exist at the very low energy radioactive ion beam facilities. The higher-energy beams of radioactive ions becoming available at the newer ISOL facilities will allow radioactive ions to be implanted even more deeply into materials. For example, if a boron substrate is needed in a diamond crystal lattice to produce a diamond transistor, radioactive ^{11}C (carbon-11) can be implanted at the appropriate depth. The damage induced by the ^{11}C bombardment can be minimized by quickly annealing the crystal; ^{11}C will anneal as carbon. But after a few hours the ^{11}C (with a half-life of 20.4 min) will have decayed to ^{11}B (boron-11).

Longer-lived radioactive isotopes can be implanted in materials at various depths by varying the energy and the radioactive ion species. Such implantations are applicable to a variety of studies of the properties of advanced materials, such as diffusion times and wear reduction.

For background information SEE EXOTIC NUCLEI; ION IMPLANTATION; NUCLEAR STRUCTURE; NUCLEOSYNTHESIS; PARTICLE ACCELERATOR; RADIOACTIVITY in the McGraw-Hill Encyclopedia of Science & Technology.

J. D. Garrett

Bibliography. J. F. Bruandet (ed.), *Proceedings of the Workshop on the Techniques of Secondary Nuclear Beams,* Dourdan, France, 1992; F. Käppler and K. Wisshak (eds.), *Nuclei in the Cosmos,* 1992; D. J. Morrissey (ed.), *Proceedings of the 3d International Conference on Radioactive Nuclear Beams,* 1993.

Radiocarbon dating

The radiocarbon dating method has undergone major changes since the mid-1980s because of accelerator mass spectrometry (AMS), which allows high-precision dating of milligram amounts of carbon. This technological advance has greatly revised the understanding of natural and human prehistory for the past 40,000 years, the period that is accessed by the radiocarbon method. Recent research has involved using accelerator mass spectrometry carbon-14 (^{14}C) dating to study global climatic change, through the dating of insects incorporated into glacial deposits.

Glacial chronologies. Radiocarbon dating has provided the backbone for efforts in correlating glacial chronologies with other records of how global climate has changed, for example, cores taken from ice sheets, marine sediments, or lake sediments. A serious problem has been the lack of material suitable for ^{14}C dating in places where glaciers did not intersect the tree line. Where glaciers and trees coexist, such as in Alaska, large organic remains are often incorporated into till matrix for radiocarbon dating. However, in mid- and low-latitude settings, glaciers are at altitudes above the tree line. The result is that very few radiocarbon ages are available for glacial chronologies in these latitudes. Yet, there are abundant arthropods that are deposited onto glaciers, and then incorporated into till which can in turn be radiocarbon-dated by means of accelerator mass spectrometry. For example, grasshoppers occur in modern glacial ice in the Rocky Mountains of the western United States, and the ^{14}C analyses from accelerator mass spectrometry reveal that they were incorporated a few hundred years ago.

Fossil arthropods. The eolian (wind) biome, where life depends on nutrients carried by wind, is divided into the nival (snow), aquatic, and terrestrial subsystems. Life in the nival system is important here, and it is abundant and ubiquitous. Many people who visit snowfields in the summer are conscious of a red coloration caused by algae. Yet, much more than algae are found.

The diversity of arthropods in the nival system ranges from the carabid and staphylinid beetles of the glacial margin to the many types of insects that are deposited from the air and live on ice. Common in hard ice, moving along crystalline boundaries, are snow worms, which were also present in Pleistocene glaciers. Life is common in pools and ponds on glaciers. Himalayan glacier ponds, for example, have fairy shrimp, collembola, midge larvae, and a wingless midge. Rates of arthropod deposition are high in the nival eolian ecosystem; for example, 12,500 insects were counted on a snowfield in the Sierra Nevada of California, and 1500 arthropods/m^2 (140/ft^2) were found in the Pyrenees of Spain.

Fossil arthropods can be preserved for long periods of time under favorable conditions. The most stable material in insects is probably chitin, an amino-sugar polysaccharide that is second only to cellulose as a biosynthesized product in terms of global biomass. Chitin is well preserved in Quaternary deposits, and fossil arthropods have been used extensively to reconstruct Quaternary environments. Since lateral moraines (accumulation of glacial drift) include material that falls on a glacier, it should not be surprising that arthropod chitin is present within the matrix of glacial moraines well below the current zone of bioturbation.

In the most recent moraines (hundreds of years old), complete arthropods with wings can be preserved in silty matrix material. However, only pieces are found in moraines thousands of years old. When a hand sample of till matrix is examined, insects are not readily apparent. But when silty matrix material is subjected to wet-sieving procedures, arthropods are readily extracted.

Fossil arthropods have provided the first radiocarbon chronology for a sequence of glacial moraines in the Sierra Nevada of California. The **illustration** shows a comparison of arthropod radiocarbon ages for glacial moraines with tree-line altitudes in the Holocene and iceberg releases from the Laurentide Ice Sheet in eastern Canada. Holocene glacial advances correspond with times of tree-line lowering, but not all periods of tree-line lowering resulted in a dated glacial advance at Bishop Creek. Pleistocene glacial advances correspond with iceberg releases; one of the releases is missing, perhaps because it was overrun by a subsequent glacial advance that was more extensive.

Although reconstruction of the glacial history of a mountain system is an important part of natural history, one of the greatest uncertainties in global climate research rests in the assumption that moraines in different areas are truly penecontemporaneous. In reality, little numerical age control actually exists. Fossil arthropods in the till matrix, along with new advances in dating glacial moraines with cosmogenic nuclides, are providing a source of carbon to test this assumption.

Climatic change. Results of radiocarbon dating fossil arthropods relate to the current debate concerning the mechanisms of change of the global climate system. One argument is that climate changes slowly and gently, based on ratios of stable oxygen isotopes (^{18}O, ^{16}O) in marine foraminifera (protozoans having a shell enclosing an ameboid body) deposited in ocean sediment. When the world is in an ice age, more of the lighter oxygen (^{16}O) is locked up in the ice, and the oxygen in oceans (and in foraminifera shells) is enriched in ^{18}O. Foraminifera in cores taken from marine sediments, therefore, record changes in global ice volume. The general pattern for the last million years is that global ice slowly builds up over a period of about 100,000 years (with oscillations), and then the glacial world decays over a period of a few thousand years. The last glacial cycle ended about 10,000 years ago, and Earth has been in an interglacial world since then. These glacial cycles are thought to be brought about by rhythmic changes of the orbit of Earth, first quantified by M. Milankovitch. When less solar radiation was received at the latitudes where the great ice sheets built up (~65°N), a glacial world occurred. When more radiation was received, the

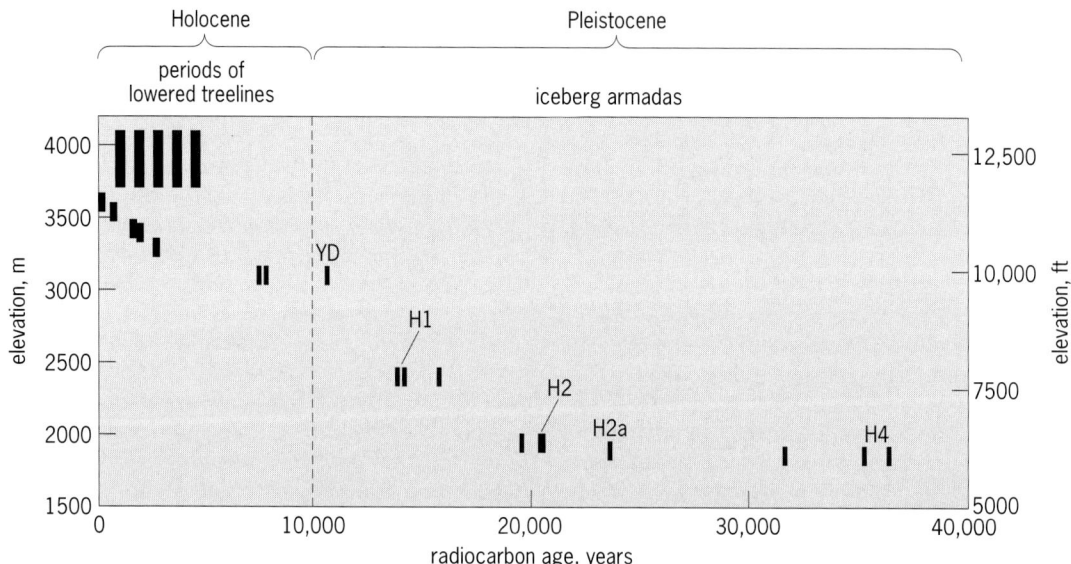

Accelerator mass spectrometry radiocarbon ages on arthropods extracted from the matrix of glacial moraines, compared with periods of lowered tree lines and iceberg releases in the North Atlantic. The Holocene glacial advances were at higher altitudes and the glaciers were quite small compared to Pleistocene glaciers, which extended to much lower elevations. The iceberg armadas are identified as Younger Dryas (YD) and Heinrich events (H1, H2, H2a, and H4).

ice sheets melted. This current paradigm is being challenged by results obtained through accelerator mass spectrometry radiocarbon dating.

An alternative theory is that the world's climate does not change gradually, but instead swings wildly and suddenly from one extreme to the other, in a time frame of a few decades to centuries. The cornerstone of this theory is the periodic release of armadas of icebergs from the great ice sheet over Canada into the northern Atlantic. Each of these Heinrich events (named for H. Heinrich, the oceanographer who discovered them) produced layers in marine cores rich in ice-rafted debris derived from Canada. The iceberg armadas, radiocarbon-dated by accelerator mass spectrometry, broke loose about 10,500 (Younger Dryas or YD event, another sudden climatic change, first recognized in northern Europe but now recognized as a global event), 14,000 (Heinrich event 1; H1), 21,000 (H2), 24,000 (H2a), 27,000 (H3), and 36,000 (H4) ^{14}C years ago (see illus.). There were also iceberg armadas released before the period that is accessible by radiocarbon dating.

A key uncertainty is whether glaciers advanced to the rhythm of astronomical forcings or Heinrich events. Until recently, radiocarbon ages have not been available to resolve this question. Evidence from fossil arthropods in glacial till indicate that there were multiple pulses of glaciations rather than a single event that could be predicted based on oxygen isotopes in marine cores.

The radiocarbon ages also suggest that there may be a linkage between iceberg armadas in the North Atlantic and glacial advances in the Sierra Nevada of California. One likely explanation for such a linkage that has been suggested is the ocean's conveyor belt. Iceberg releases changed the thermohaline circulation of the oceans, which in turn changed the global circulation of the atmosphere.

This debate is more than idle scientific curiosity about the nature of past climatic changes. If sudden changes in the oceanic conveyor belt bring about sudden changes in climate, a sudden change in climate could happen in the near future by this mechanism, although the change would not be as intense as those brought about by iceberg releases.

For background information SEE ACCELERATOR MASS SPECTROMETRY; BIOME; CLIMATIC CHANGE; GLACIAL EPOCH; MARINE SEDIMENTS; MORAINE; RADIOCARBON DATING in the McGraw-Hill Encyclopedia of Science & Technology.

Ronald I. Dorn

Bibliography. G. Bond et al., Correlations between climate records from North Atlantic sediments and Greenland ice, *Nature*, 365:143–147, 1993; J. S. Edwards, Arthropods of alpine aeolian ecosystems, *Annu. Rev. Entomol.*, 32: 163–179, 1987; S. A. Elias, *Quaternary Insects and Their Environments*, 1994; P. A. Mayewskiy et al., Changes in atmospheric circulation and ocean ice cover over the North Atlantic during the last 41,000 years, *Science*, 263: 1747–1751; L. A. Scuderi, Solar influences on Holocene treeline altitude variability in the Sierra Nevada, *Phys. Geog.*, 15:146–165, 1994; L. W. Swan, The aeolian biome, *Bioscience*, 42:262–270, 1992.

Reengineering

Responsiveness is a critical need for organizations. It involves providing enhanced products and services of demonstrable value to customers, and thereby to those individuals who have a stake in the success of the organization. Such actions must be accomplished by efficiently and effectively employing leadership and empowered people so that systems-management strategies, organizational processes, human resources, and appropriate technologies are brought to bear on the production of high-quality and trustworthy goods and services. Responsiveness also involves supplying appropriate technologies and systems to meet these objectives. It often requires examination and modification of existing ways of doing business and, on the basis of this examination, altering or reengineering the organization, the organizational processes, or the products that support an organization's activities.

The entity to be reengineered can be systems management, process, product, or some combination. In each case, reengineering involves a basic three-phase systems-engineering life cycle comprising definition, development, and deployment of the entity to be reengineered.

Systems-management reengineering. At the level of systems management, reengineering is directed at potential change in all business or organizational processes, including the systems-acquisition process life cycle itself. M. Hammer's definition of reengineering as "the fundamental rethinking and radical redesign of business processes to achieve dramatic improvements in critical, contemporary measures of performance, such as cost, quality, service, and speed," describes what is here called reengineering at the level of systems management. One major catalyst for reengineering is the creative use of information technology. However, reengineering is not just automation but the ambitious and rule-breaking study of everything about the organization to enable more effective and efficient organizational processes to be designed.

Systems-management reengineering, more commonly called organizational reengineering, may be defined as the examination, study, capture, and modification of the internal mechanisms or functionality of existing system-management processes and practices in an organization in order to reconstitute them in a new form and with new features, often to take advantage of newly emerged organizational competitiveness requirements but without changing the inherent purpose of the organization itself.

Life-cycle process reengineering occurs as a natural by-product of reengineering at the level of systems management. Although such process reengineering does not necessarily result in the reengineering of already existing products, generally it will do so, because new products and competitive strategies are a major underlying objective of reengineering at the level of systems management.

Process reengineering. Reengineering can also be considered at the levels of an organizational process. At the level of processes alone, the effort would be almost totally internal. It would consist of modifications to whatever standard life-cycle processes are in use in a given organization in order to better accommodate new and emerging technologies or new customer requirements for a system. For example, an explicit risk-management capability might be incorporated at several different phases of a given life cycle and accommodated by a revised configuration of management processes. It could be implemented into the processes for research, development, testing, and evaluation (RDT&E); acquisition; and systems planning and marketing. Basically, reengineering at the level of processes would consist of the determination, or synthesis, of an efficacious process for ultimately fielding a product on the basis of a knowledge of generic customer requirements, and with appropriate consideration for the objectives and critical capabilities of the systems-engineering organization.

Process reengineering may be defined as the examination, study, capture, and modification of the internal mechanisms or functionality of an existing process, or systems-engineering life cycle in order to reconstitute it in a new form and with new functional and nonfunctional features, often to take advantage of newly emerged or desired organizational or technological capabilities without changing the inherent purpose of the process that is being reengineered. It is possible to reengineer one or more of the processes for RDT&E, system acquisition or production, or systems planning and marketing.

Redesign of processes alone, without attention to reengineering at a higher level, may in many instances represent an incomplete and not fully satisfactory way to improve organizational capabilities. Thus, the candidates for reengineering should be high-level managerial processes as well as operational processes. Information technology is a major enabling catalyst for process reengineering. The benefits of using process reengineering include shorter development cycles, fewer engineering change orders, products that fulfill customer expectations, and reduced program and product development costs throughout the life cycle. Thus, the process improvement results ultimately in an increase in effectiveness of product for the same cost, a reduction in cost for the same effectiveness, or some blend of these two.

Product reengineering. The term reengineering could mean some sort of reworking or retrofit of an already engineered product, and could be interpreted as maintenance or refurbishment. Reengineering could also be interpreted as reverse engineering, in which the characteristics of an already engineered product are identified, such that the product can perhaps be modified or reused. Inherent in these notions are two major facets of reengineering: it improves the product or system delivered to the user for enhanced reliability, maintainability, or to meet a newly evolving need of the system users; and it increases understanding of the system or product itself. Thus, this interpretation of reengineering is almost totally product focused.

Thus, product reengineering may be defined as the examination, study, capture, and modification of the internal mechanisms or functionality of an existing system or product in order to reconstitute it in a new form and with new features, often to take advantage of newly emerged technologies without major change to the inherent functionality and purpose of the system. This definition indicates that product reengineering is basically structural reengineering with at most minor changes in purpose and functionality of the product that is reengineered. This reengineered product could be integrated with other products having rather different functionality than was the case in the initial deployment. Thus, reengineered products could be used, together with this augmentation, to provide new functionality and serve new purposes. There are a number of synonyms for product reengineering, including renewal, refurbishing, rework, repair, maintenance, modernization, reuse, redevelopment, and retrofit.

Example. A specific example of a product reengineering effort might be that of taking a legacy system written in Cobol or Fortran, reverse-engineering it to determine the system definition, and then reengineering it in C^{++} or Ada. Depending upon whether any modified user requirements are to be incorporated into the reengineered product, the product would be reengineered, either through a forward engineering effort, just after reverse engineering had determined either the initial development (technical) system specifications or user requirements and specifications, and then updating these. This reverse-engineering concept, in which salient aspects of user requirements or technological specifications are recovered from examination of characteristics of the product, predates product reengineering.

Reverse and forward engineering. Much of product reengineering is very closely associated with reverse engineering to recover either design specifications or user requirements. Then follows refinement of these requirements or specifications and forward engineering to achieve an improved product. Forward engineering is the original process of defining, developing, and deployment of a product, or realizing a system concept as a product; whereas reverse engineering, sometimes called inverse engineering, is the process through which a given system or product is examined in order to identify or specify the definition of the product at the level of either technological design specifications or system or user requirements. Reverse engineering may be broken down into redocumentation and design recovery.

Redocumentation is a subset of reverse engineering in which a representation of the subject system or product is re-created for the purpose of generating functional explanations of original system behavior and, perhaps more importantly, to aid the reverse engineering team in better understanding the system at both a functional and structural level. A number of redocumentation tools are available for software. One of the major purposes of redocumentation is to produce new documentation for an existing product where the existing documentation is faulty, or perhaps virtually absent.

Design recovery is a subset of reverse engineering in which the redocumentation knowledge is combined with other efforts, often involving others' experiences with and knowledge of the system, that lead to functional abstractions and enhanced product or system understanding at the level of function, structure, and even purpose.

Restructuring. This process involves transformation of the reverse engineering information concerning the original system structure into another representation form. Restructuring generally preserves the initial functionality of the original system, or modifies it slightly in a purposeful manner that is in accord with the user requirements for the reengineered system and the ways in which they differ from the requirements for the initial system.

Needs. Several needs must be considered if a product reengineering process is to yield appropriate and useful results.

1. Organizational and technological issues must be examined to develop useful product reengineering strategy.

2. Human, leadership, and cultural issues must be considered, including how they will be affected by the development and deployment of a reengineered product, as a part of the definition of the specifications for the reengineered product.

3. It must be possible to demonstrate that the reengineering process and product are, or will be, cost-effective and of high quality, and that they support continued evolution of future capabilities.

4. Reengineered products must be considered within a larger framework that also considers the potential need for reengineering at the levels of systems management and organizational processes, as it will generally be a mistake to assume that technological fixes will resolve organizational difficulties only at these levels.

5. Product reengineering for improved postdeployment maintainability must consider maintainability at the level of process rather than at the level of only product, such as would result in the case of software through rewriting source-code statements. Use of model-based management systems or code generators should yield much greater productivity, in this connection, than rewriting code at the level of source code.

6. Product reengineering must consider the need for reintegration of the reengineered product with existing legacy systems that have not been reengineered.

7. Increased conformance to standards must be a result of the product reengineering process.

8. Product reengineering must consider legal issues associated with reverse engineering.

For background information SEE INFORMATION SYSTEMS ENGINEERING; PROGRAMMING LANGUAGES; SOFTWARE ENGINEERING; SYSTEMS ENGINEERING in the McGraw-Hill Encyclopedia of Science & Technology.

Andrew P. Sage

Bibliography. R. S. Arnold (ed.), *Software Reengineering,* 1993; T. H. Davenport, *Process Innovation: Reengineering Work through Information Technology,* 1993; M. Hammer and J. Champy, *Reengineering the Corporation: A Manifesto for Business Revolution,* 1993; H. J. Harrington, *Business Process Improvement: The Breakthrough Strategy for Total Quality, Productivity, and Competitiveness,* 1991; A. P. Sage, *Systems Management for Information Technology and Software Engineering,* 1995.

Reproduction (plant)

Flowering plants (angiosperms) produce intricate, often beautiful structures, the flowers, within which their reproductive processes occur. In flowering plants, as in the less-advanced plants, a diploid spore-producing generation (the sporophyte) alternates with a haploid gamete-producing generation (the gametophyte). Both the male and female gametophytes of flowering plants are microscopic structures enclosed within the tissues of the sporophyte. The female gametophyte (embryo sac) is enclosed within an ovule in the pistil of a flower, and the initial development of the male gametophyte (pollen) also occurs in the flower, within an anther. The male gametophyte produces two sperm cells that function in the process of double fertilization unique to angiosperms.

Morphological events of pollen development. The sporogenous cells in the young anther are surrounded by cells of the tapetum which play a significant role in pollen development. The microsporocytes (pollen mother cells) are produced by mitosis in the sporogenous tissue. Following meiosis, each microsporocyte produces a tetrad of haploid cells, the microspores. On release from the tetrad, the microspores undergo a rapid increase in size and shape. This rapid growth is followed by a period of slower growth until the maximum volume of the pollen grain is reached some time before anthesis or anther dehiscence (the rupture of the anther and release of pollen). Mature pollen grains have thick walls composed of several layers that display elaborate and beautiful surface patterns. Depending on the plant species, pollen grains possess one or more pores or apertures from which a pollen tube emerges during germination of the pollen grain on the stigma of a pistil.

Following meiosis, there is an extended interphase period that terminates with a very unequal mitotic division (microspore mitosis) into a large vegetative cell and a small generative cell; both cells are included within the confines of the cell wall of the original microspore. This structure is termed a pollen grain. The generative cell that inherits a small amount of cytoplasm lies within the vegetative cell. The immature pollen grain advances through a number of developmental changes prior to release from the anther. At the time of anther dehiscence, the pollen grains of about 70% of flowering plant species are bicellular; that is, they contain a vegetative cell and a generative cell. In the remaining plant species, the generative cell undergoes a mitotic division within the pollen grain, forming two sperm cells. In pollens released when they are bicellular, the generative cell completes its division during the growth of the pollen tube in the style. At maturity, the male gametophyte consists of three cells: two sperm cells lying within the cytoplasm of a vegetative cell.

For a short period after anthesis, the mature pollen grain exists as a free organism. Then it is transported by insects, wind, or other agents to the stigma of the pistil in a flower, where it begins another phase of development. If the pistil is compatible, the pollen grain germinates by the outgrowth of a pollen tube, which grows into the style. The sperm cells are transported through the pollen tube to the embryo sac.

The pollen tube is a part of the vegetative cell; many of the biochemical and molecular events that occur in the vegetative cell of the pollen grain prior to anther dehiscence are related to the preparedness of the male gametophyte for the demands of germination on the stigma and growth of the pollen tube in the style and ovary.

Germination and pollen tube growth are relatively rapid events in most plants, the period from pollination to entry into the embryo sac ranging from about 1 to 48 h or longer. In some plants, rates of pollen tube growth as high as 35 mm per h have been reported. The pollen tube grows within the transmitting tissue of the style, enters the micropyle of the ovule, and contacts the embryo sac. It penetrates one of the synergids of the embryo sac, the tube tip ruptures, and the sperm cells along with some of the contents of the pollen tube are released into the synergid. The two sperm cells move across the synergid by unknown mechanisms. One sperm cell fuses with the egg cell to form the zygote; the other sperm cell fuses with the normally diploid central cell, giving rise to the primary endosperm cell. Thus, the process of double fertilization and the life of the male gametophyte are completed.

Gene expression during pollen development. Various lines of evidence suggest that the messenger ribonucleic acids (mRNAs), and many of the proteins required for the processes of germination and early pollen tube growth, are already present in the mature pollen grain and have been synthesized earlier during pollen development. Moreover, the ribosomes and transfer RNAs (tRNAs) required for protein synthesis in the germinating grain are synthesized in the microspore and stored for later use. The activity of a large number of genes is required for the development of the pollen grain and pollen tube. For example, the products of about 20,000 different genes are present in the mature pollen grain at the time of anthesis. Most of these genes are activated only after the stage of microspore mitosis, and have been termed late genes. Other genes are active in the microspores soon after meiosis; these are early genes. Many of the genes active in pollen development have been cloned, isolated, and sequenced, and have had their biochemical functions analyzed. Many of the late class of pollen-expressed genes appear to have primary functions in germination and growth of the pollen tube.

Pollen tube structure and growth. The function of the pollen tube is to transport the sperm cells through the tissues of the style and deposit them in the embryo sac. In many plants the tube must grow a long distance to traverse the style; for example, in maize the pollen tube has to grow through as much as 20 in. (50 cm) of style. The major activity of the vegetative cell of the pollen grain is the synthesis and assembly of the pollen tube wall. In most plant cells growth takes place over the entire surface of the cell. However, pollen tube growth is restricted to the region of the tube tip. The growth zone in lily pollen, for example, is between 3 and 5 micrometers long.

Pollen tubes contain distinct cytological zones. In the nongrowing region of the tube, there is an abundance of mitochondria, Golgi bodies, endoplasmic reticulum, and vesicles of various sizes. The growing tip region, however, contains a large number of vesicles and an absence of other organelles. In the pollen tube the cytoplasmic structures are aligned parallel to the long axis of the tube except in the tip region where they are more randomly distributed. The arrangement of mitochondria, vesicles, and other organelles appears to follow the cytoplasmic streaming patterns in the tube. The membrane-enclosed vesicles at the tip contain cell wall precursor materials. These vesicles fuse with the cell membrane at the tip, allowing the vesicle membranes to contribute to the plasma membrane; the contents of the vesicles are discharged to the outside, where they contribute to the growing pollen tube wall. The vesicles are produced by the Golgi apparatus some distance behind the tip, and they are transported by cytoplasmic streaming to the tip region. The microfilament cytoskeletal, contractile system appears to be responsible for cytoplasmic streaming. Both actin and myosin are involved in microfilament function, and these cytoskeletal molecules are present in pollen tubes.

Interactions in self-incompatibility. Self-incompatibility (SI) is the inability of a normal plant to accomplish fertilization when it is self-pollinated, such as when pollen from a flower is used to pollinate the stigma of the same flower or of another flower on the same plant. Self-incompatibility is a mechanism that has evolved to ensure outbreeding, and is considered to be largely responsible for the evolutionary success of flowering plants. The phenomenon of self-incompatibility is widespread among flowering plants. Self-incompatibility is in most cases controlled by a single gene locus, the S locus, which has a large number of different alleles. Self-incompatibility occurs as two main types: gametophytic and sporophytic. In gametophytic self-incompatibility, the self-incompatibility phenotype of the pollen is determined by the S genotype of the individual haploid pollen grain. In sporophytic self-incompatibility, the response of the pollen grain on the stigma is determined by the diploid S genotype of the plant (sporophyte) that produces the pollen. The pistil that is diploid contains two S alleles, whereas the haploid pollen grain contains just one S allele. The tapetum and the other parts of the anther contain two alleles because they are part of the sporophyte.

In plants with gametophytic self-incompatibility, pollen grains containing an S allele that is identical to at least one allele in the sporophyte are incompatible. Pollen grains germinate on the stigma and begin growing through the transmitting tissue of the style, but pollen tube growth in the style is arrested some distance before it reaches the ovary. Thus, fertilization cannot occur. For a compatible pollination to occur, the S allele present in the deoxyribonucleic acid (DNA) of the pollen grain must be different from both of the S alleles that are present in the pistil.

In plants that exhibit sporophytic self-incompatibility, whether a pollination is compatible or not is not dependent on the haploid S genotype of the pollen grain but is determined sporophytically by the diploid S genotype of the parent plant. When one or both of the alleles in the plant that produced the pollen are matched by alleles in the pistil, an incompatible response is elicited and germination of the pollen grain is arrested on the stigma surface. The two S alleles in the plant producing the pollen must be different from the two alleles present in the pistil for compatible pollination to occur.

The gene products of the individual S alleles in pollen and stigma or style must be responsible for the incompatibility reaction. The protein products of the S locus are known. In gametophytic self-incompatibility plants, the product is a protein with ribonuclease activity. In sporophytic systems, there appear to be two S locus–linked genes: a glycoprotein with a presently unknown function and a receptor kinase gene that is possibly a membrane-associated protein. Exactly how the products of the S genes bring about the specificity of the incompatibility response and how pollen germination or tube growth are inhibited are not yet known.

For background information SEE REPRODUCTION (PLANT) in the McGraw-Hill Encyclopedia of Science & Technology.

Joseph P. Mascarenhas

Bibliography. J. P. Mascarenhas, Molecular mechanisms of pollen tube growth and differentiation, *Plant Cell,* 5:1303–1314, 1993; J. B. Nasrallah and M. E. Nasrallah, Pollen-stigma signaling in the sporophytic self-incompatibility response, *Plant Cell,* 5:1325–1335, 1993; E. Newbegin, M. A. Anderson, and A. E. Clarke, Gametophytic self-incompatibility systems, *Plant Cell,* 5:1315–1324, 1993.

Respiratory system

The sustained activity that characterizes modern birds and mammals depends on the ability of these vertebrates to integrate locomotor and respiratory function and thereby avoid conflicts between the mechanisms of body propulsion and lung ventilation. Until recently, the significance of locomotor-respiratory interactions in vertebrates was generally unrecognized, but this situation is changing rapidly. New experimental approaches allow a comparison of the biomechanical dynamics of breathing in naturally exercising vertebrates. These studies reveal the importance of locomotor-respiratory interactions and offer new perspectives on the design and function of vertebrate respiratory systems, particularly that of mammals.

Locomotor-respiratory conflict in reptiles. Reptiles have relatively low capacities for sustained aerobic activity, perhaps because of an inability to run and breathe simultaneously. At speeds above a fast walk, lizards cannot maintain adequate levels of lung ventilation to meet the demand of their working muscles for oxygen. Contrary to what would be expected, the depth of breathing (tidal volume) declines dramatically as speed increases because of the conflicting mechanical demands that locomotion and respiration make on the same musculoskeletal elements of the body.

Both breathing and body propulsion require coordinated actions of the rib cage and associated trunk muscles. In reptiles, running involves pronounced lateral bending of the trunk, and consequently alternate contraction and relaxation of the trunk musculature on opposing sides of the body. However, effective breathing demands synchronous, bilateral activation of the same muscles. Furthermore, the lateral flexion of the trunk which reptiles experience while running compresses one lung while expanding the other. Such asynchronous inflation-deflation cycles of the lungs yield little net flow of air into and out of the respiratory tract. Therefore, during running the locomotor obligations of the trunk musculature assume priority, and effective breathing is not possible. Perhaps for this reason lizards tend to engage in activities that require short bursts of energy rather than endurance, and they also possess muscles with exceptional anaerobic capacities.

Locomotor-respiratory coupling. An inability to run and breathe simultaneously is probably a primitive characteristic of tetrapod vertebrates, and it seriously limits the physiological and behavioral potential of amphibians and reptiles, as well as their potential to inhabit ecological niches. Birds and mammals have independently achieved a level of locomotor-respiratory integration that makes stamina possible. There can be little doubt that this evolutionary achievement has contributed to the dominance of birds and mammals in most terrestrial vertebrate communities.

Birds. The association between locomotion and respiration in birds is reflected in the widespread synchronization or coupling of their wing-beat and breathing cycles. The ratio of locomotor (wing-beat) to respiratory cycles (or coupling ratios) ranges from a low of 1:1 (pigeon, crow) to as high as 5:1 (starling). Many species exhibit multiple coupling ratios with individuals shifting among these as flight and metabolic conditions demand.

The mechanical basis of avian locomotor-respiratory integration is not well understood. Coordination is such that during 1:1 coupling inspiration and expiration accompany (respectively) the up and down strokes of the wing cycle. It is believed that flight-induced loading of the thorax helps to drive air through the respiratory system by periodic compression and expansion of the numerous air sacs that constitute the bellows of the complex avian respiratory system. (The lungs are small compact structures positioned so as to be relatively immune to loading of the thorax.) High-speed x-ray motion pictures of flying starlings suggest that volumetric expansion and reduction of the air sacs is indeed linked to motions of skeletal structures associated with the wing-beat cycle. The most important of these is a dorsoventral rocking motion imparted to the large sternal plate by the contraction and relaxation of the primary wing depressor, the pectoralis. The exact fraction of each breath in a flying bird solely attributable to the pumping action of the locomotor musculature has not been determined; some preliminary estimates suggest that it may be quite small. Although the most extensively documented examples of locomotor-respiratory coupling in exercising birds involve the wing-beat cycle, breathing may also be synchronized with leg movements during walking and running.

Mammals. Running mammals routinely synchronize body motion and respiration. Several coupling ratios occur in the trot (for example, 1:1, 3:2, 2:1, 5:2, 3:1). However, a strict 1:1 coupling seems intrinsic to all galloping mammals regardless of body size and shape (for example, tiny gerbils and white rhinoceroses). The phase relations of the limb and respiratory cycles in the gallop assure that expiration occurs while the forelimbs are on the ground and that the chest walls are subjected to compressive loading. Inspiration takes place when the forelimbs are free of the ground and are being drawn forward in preparation for the next support interval.

The rigid 1:1 synchronization of body motion and respiration in galloping mammals suggests that perhaps forces arising from locomotor effort may help power the breathing cycle. Several mechanisms have been proposed that could conceivably provide for such a linkage: (1) pressure and volume changes within the thorax and abdomen associated with sagittal extension and flexion of the vertebral column and pelvis; (2) periodic compression and expansion of the chest walls arising from the loading and unloading of the forelimbs as they strike and leave the ground; and (3) rhythmic displacements of a visceral piston within the body cavity.

The visceral piston model offers a novel mechanism for driving lung ventilation during locomotor exercise. It proposes that the alternating acceleration and deceleration of the trunk of a running mammal will cause inertial, pistonlike displacements (for example, anterior-posterior sliding) of the heavy visceral mass within the abdomen. Since the liver, a principal component of the piston, attaches directly to the respiratory diaphragm, its movements could drive the diaphragm and hence respiratory flow. The model is based on anatomical reality, and its predictions are in keeping with the observed relationships between the locomotor kinematics and associated breathing patterns of galloping mammals.

The visceral piston model can be visualized through analogy with the vibration mechanics of a simple mass-spring system. In theory, a visceral piston would operate most efficiently as a respiratory driver if it behaved as a tuned oscillator vibrating at its resonant frequency. (Indeed, the respiratory rate of panting dogs is very close to the natural frequency of their visceral masses.) In practice, the stride frequency of a galloping mammal would have to match the resonant frequency of its guts. This constraint might help to explain why stride frequency remains nearly constant in a galloping mammal regardless of its speed (speed is changed by adjusting stride length). Nonetheless, it has been determined that a visceral piston is not utilized by galloping horses because the apparent phase relationship between visceral displacement and respiratory flow is inconsistent with such a mechanism. Instead, horses seem to rely on flexion and extension of the back and pelvis to couple gait and breathing. Other data suggest that a visceral piston is probably the primary driver of lung ventilation in hopping kangaroos.

High-speed cineradiography has recently provided considerable detail on locomotor-respiratory interactions in trotting dogs (see **illus.**). This approach permits direct visualization of the motions of the internal structures (such as the visceral mass, lungs, and diaphragm) and, for the first time, a determination of regional gas flow within the lungs of a running mammal. These observations show that trotting dogs employ a very sophisticated pattern of high-frequency (4–6 Hz) lung ventilation which is tightly integrated with locomotor mechanics. Among the major findings are that breathing is powered chiefly by a visceral piston; the diaphragm appears to be inactive much of the time; the various lobes of the lungs are ventilated asynchronously (inspired gas is shuttled between the diaphragmatic and apical lobes and is recycled extensively prior to expiration); and filling and emptying of the right and left apical lobes is out of phase and determined by the loading profiles of the two forelimbs. For example, during inspiration (illus. *a* and *c*), air flows preferentially into the diaphragmatic lobe contralateral to the supporting forelimb, bypassing the apical lobe on the same side. Air received by the ipsilateral diaphragmatic lobe is primarily that forced out of the corresponding apical lobe by chest wall compression. During expiration (illus. *b* and *d*), gas exiting from the diaphragmatic lobe contralateral to the stance limb is diverted into the apical lobe underlying the swing-phase forelimb; from the ipsilateral diphragmatic lobe it is forced out the trachea.

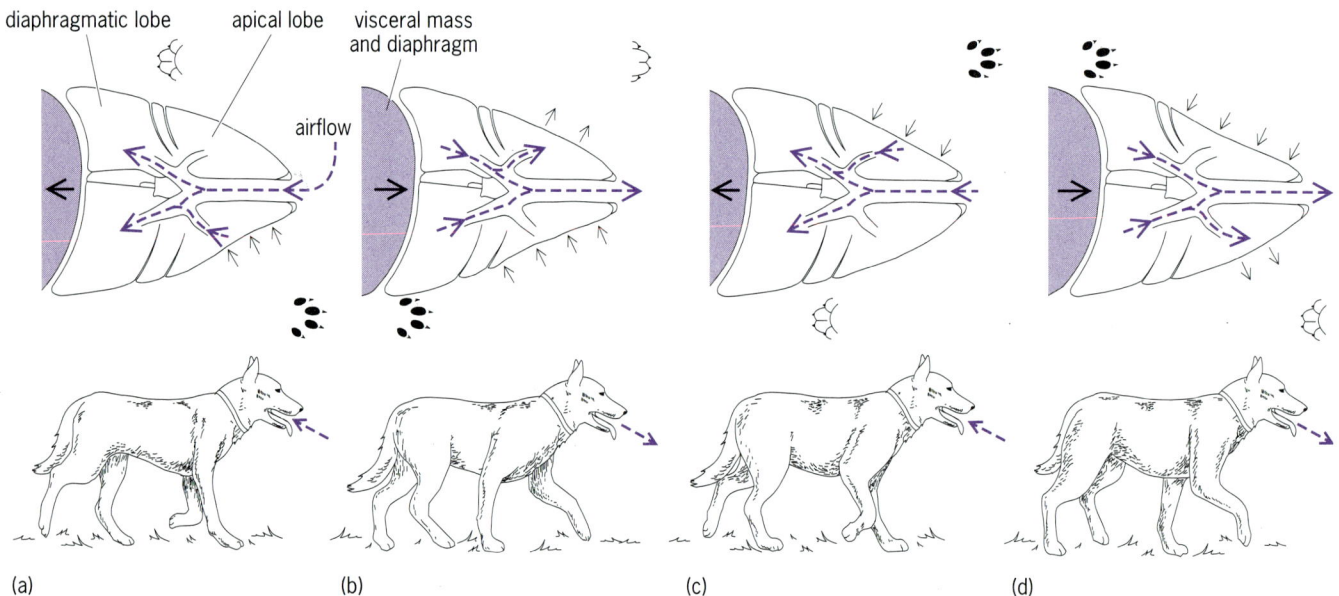

Model of locomotor-respiratory interactions in a trotting dog synchronized with recordings of respiratory flow. (*a,c*) Inspiration. (*b,d*) Expiration. The sequence depicts one complete internal ventilation cycle of the lungs, which corresponds to two breath cycles. The duration of the sequence is 0.45 s. The arrows inside the lungs indicate flow patterns. (*After D. M. Bramble and F. A. Jenkins, Jr., Mammalian locomotor-respiratory integration: Implications for diaphragmatic and pulmonary design, Science, 262:235–240, 1993*)

Asynchronous ventilation in running dogs is unanticipated by orthodox models of mammalian respiratory mechanics that do not consider the possible influence of locomotor dynamics. This form of lung ventilation may be widespread in exercising mammals, and may serve to increase the effectiveness of gas exchange by promoting a more thorough mixing of inspired air and reducing dead-space ventilation, which could be achieved by pumping the high-quality, unrespired air remaining within the airways at the end of inspiration into the apical lobes during expiration. The prospect of asynchronous ventilation provides a functional rationale for a unique design feature of the mammalian respiratory system, lobate lungs.

Implications. Partitioning of the lungs allows for regional independence of the respiratory phase, and thereby permits some portions to fill while others empty. One of the many unexamined issues concerns the directing of flow within the lungs during asynchronous breathing. In the absence of structural valves, it seems likely that mammals, like birds, utilize some form of aerodynamic valving mechanism. In fact, there are several possible but unexpected parallels between mammalian and avian respiratory systems. It remains to be determined how many of these mechanisms might be related to common issues surrounding the successful coupling of locomotor and respiratory function.

For background information SEE LUNG; RESPIRATORY SYSTEM (VERTEBRATE) in the McGraw-Hill Encyclopedia of Science & Technology.

Dennis M. Bramble

Bibliography. D. M. Bramble and F. A. Jenkins, Jr., Mammalian locomotor-respiratory integration: Implications for diaphragmatic and pulmonary design, *Science*, 262:235–240, 1993; P. J. Butler and A. J. Woakes, Heart rate, respiratory frequency and wing beat frequency of free flying barnacle geese, *Branta leucopsis*, *J. Exp. Biol.*, 85:213–226, 1980; D. R. Carrier, The evolution of locomotor stamina in tetrapods: Circumventing a mechanical constraint, *Paleobiol.*, 13:326–341, 1987; I. S. Young et al., The synchronization of ventilation and locomotion in horses (*Equus caballus*), *J. Exp. Biol.*, 166:19–31, 1992.

Risk analysis

The process of regulating nuclear power plants in the United States is becoming increasingly based on risk insights, resulting in risk-based regulation. This more modern form of regulation has the potential to increase safety levels while decreasing operating costs. This transition is occurring within the nuclear industry and at the U.S. Nuclear Regulatory Commission, and is developed, supported, and demonstrated by the major institutes and professional societies in the field.

Deterministic regulation. Nuclear power plants in the United States and in many other countries have been regulated on the basis of demonstrating that they could meet many conservative, deterministically derived safety criteria. Formerly, potential challenges to nuclear power plants, called design-basis accidents, were postulated (or determined) by senior government safety experts. Prior to full power operation, analyses or tests had to show that a plant's design was capable of placing the plant into a safe condition for each of these postulated accidents. In these formative years, only limited quantitative consideration could be given to the likelihood of these design-basis accidents occurring because of a lack of extensive operating experience. Although this traditional approach has been quite effective (that is, no member of the United States public has ever been exposed to significant doses of radiation from any nuclear power plant), the process is not a very efficient one and does not fully capture safety issues that may arise out of plant operation.

Risk-based regulation. Regulatory practices, much like the technologies they control, evolve over time. Increased operating experience, safety experiments, extensive reviews of actual accidents and incidents, and improved analytical capabilities mean that as the knowledge of a regulated technology matures, its regulatory processes can become increasingly precise and quantitative. Thus, the natural evolution of regulatory processes moves from the early qualitative deterministic phase to a more mature quantitative risk-based phase. The term risk, as used here, is defined as the product of the likelihood of an accident multiplied by the consequences (individual and collective health impairment or fatality, and property damage) that result from that accident.

Probabilistic safety assessment. The technical underpinning of risk-based regulation is probabilistic safety assessment (PSA), a relatively modern technology that is being used by an increasing number of industries and by various federal regulatory agencies. PSA, in simple terms, uses both national and plant-specific equipment operational data, descriptions of plant systems and their interactions, plant procedures, and analyses of human responses to various conditions. This information is placed into a large, often complex, logic. The analytical process accounts for the frequency of certain safety challenges occurring at a given power plant and for a great number of possible outcomes (sequences) for each challenge. A sequence is evaluated according to its probability of occurring and the consequences that may result. The results of many such sequences are then aggregated to give overall figures of merit, such as the likelihood of damaging the plant's radioactive fuel rods, that is, the core-damage frequency. Estimates are also made of the likelihood of core-damage events causing the failure of the heavy steel reactor vessel that surrounds the reactor fuel, and even the failure of the massive containment building that encloses the reactor vessel and its associated piping systems. Finally, analyses are made of the types and amounts of radioactive material that might enter the environment in the very unlikely event of core damage eventually leading to containment failure and the subsequent public health and property consequences.

Such PSA analyses not only can provide estimates of the likelihood of core-damage accidents and the risks to the public but can identify the strengths and weaknesses of a plant's design and operation, point out the dominant accident sequences, and be used to rank a plant's systems, structures, and components (SSCs) according to their importance to public risk. Virtually every nuclear power plant in the United States has its own specific PSA. The results of such PSAs could provide insights useful both to plant operational procedures and to evaluating the importance of equipment upgrades, thus reducing potential accident probability.

PSAs can also be used to judge if nuclear power represents a major risk to the public. Some years ago the U.S. Nuclear Regulatory Commission established health and safety goals for nuclear power. The basic intent was to limit the risks from nuclear power to such a low level that it would not be considered a major societal risk. This concept, when converted into numerical form, limited nuclear risks to one part in a thousand of the normal background health risks people are routinely exposed to. Risk estimates now available from completed PSAs show nuclear power plant risk levels below these strict safety goals by factors ranging from 25 to several thousand.

Distribution of resources. The fundamental principle of risk-based regulation is to distribute the resources of the regulator and the power-plant operator according to risk significance. Those items or events that are calculated to have high risk significance would receive the most attention, whereas those with little risk significance would command fewer resources. PSAs are the key to such allocations of resources because, as stated before, they can be used to rank a plant's SSCs according to risk. PSAs can also risk-rank the various safety challenges to a plant (also known as initiating events), and they can risk-rank human responses (operator actions) to the initiating events. Such risk rankings can result in a deemphasis of regulatory processes that burden plant operators and the public with little or no benefit, while simultaneously highlighting those areas that are most critical to public protection.

The concept of risk-based regulation was formally introduced at the Nuclear Regulatory Commission in 1992, but is already associated with a large and growing number of applications. Among the applications of risk-based regulation now being developed or implemented are risk-based maintenance, training, inspections, quality assurance, technical specifications, avoidance of high-risk configurations, testing, and cost minimization.

Application to surveillance testing. An example of how risk-based regulation may be used to both increase safety and decrease operating costs is given below for surveillance testing. Various pieces of equipment within a nuclear power plant are subjected to periodic surveillance tests to determine if they are functional. Historically, the determination of what equipment should be subject to surveillance tests and the frequency of such tests was based on so-called engineering judgment. However, by using PSA techniques it is possible to quantify the risk significance of each of these surveillance tests and then to optimize testing frequency.

Just such an analysis was performed on the surveillance tests of an actual nuclear power plant. Each surveillance test was examined as to its possible impact on core-damage frequency. The results of these analyses were placed into five categories (see **illus.**). Normalizing category 1 to 1.0, each successive category was a factor of 10 less important in affecting the core-damage frequency. Thus, the tests in category 4 were approximately $\frac{1}{1000}$ as important as the tests in category 1. About 71% of all the tests among this plant's surveillance tests had very little risk significance because they fell into categories 3, 4, or 5.

The reason that this distribution contains so many risk-insignificant tests is that prior to having PSAs there was no quantitative way of determining the risk significance of a nuclear plant's SSCs. It is now widely accepted that the number of SSCs that are risk significant is but a small percentage of all the SSCs in a nuclear power plant. In many cases the total elimination of particular SSCs would not increase the plant's risk level. Some surveillance test requirements placed on SSCs were themselves risk insignificant. If an SSC itself is not important to public protection, its surveillance test requirements could be relaxed. Similarly, many other regulatory requirements, in quality assurance, inspections, training, and so forth, were placed on risk-insignificant SSCs. Adjustments to the regulatory burdens in these areas are therefore under review.

Win-win situations. Studies such as that discussed above confer another benefit by pointing the way to so-called win-win situations. Win-win situations both increase plant safety and reduce costs. For example, the surveillance tests in categories 1 and 2 may be performed more frequently and those in categories 3, 4, and 5, less frequently or not at all. The risk impacts of increasing or decreasing the frequency of surveillance tests can be quantified with PSAs. PSA techniques can then be used to adjust testing frequencies so that there is an overall (or net) reduction in core-damage frequency, thereby improving plant safety. Since the tests in categories 1 and 2 are far more risk significant than those in categories 3, 4, and 5 the additional burden for making these tests somewhat more strict is not large. Most of the operating burden is carried by categories 3, 4, and 5. The modest increase in burden by having stricter tests in categories 1 and 2 is more than offset by relaxing testing requirements

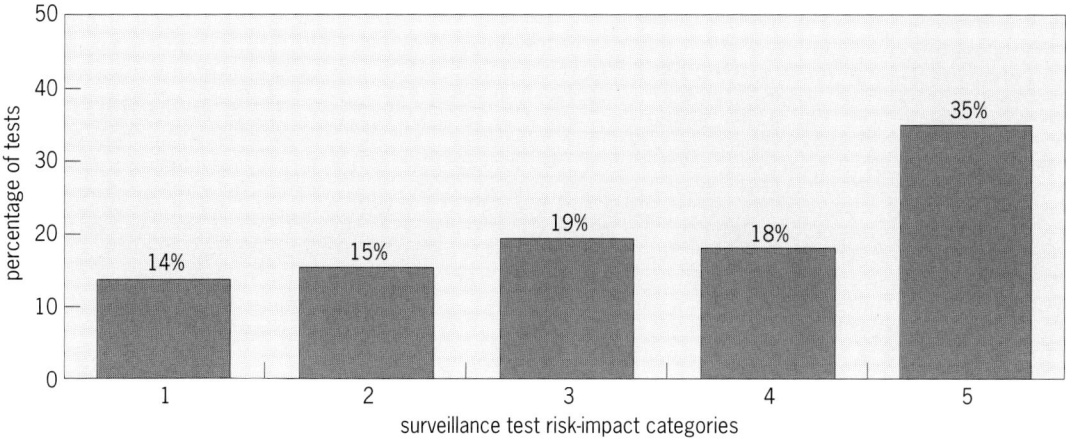

Risk impact of surveillance test requirements. Tests were placed in five categories by relative importance: category 1 = 1.0; category 2 = 0.1; category 3 = 0.01; category 4 = 0.001; category 5 = 0.0001.

in categories 3, 4, and 5. The total cost and time used in carrying out safety testing are reduced. Thus, both safety improvements and cost reductions are achieved.

The economic benefits of risk-based regulation are not yet precisely known. However, initial estimates indicate major cost savings for the entire nuclear industry. The use of risk-based regulation is not limited to nuclear power, and the potential savings for other regulated industries is very large.

For background information SEE NUCLEAR POWER; RISK ANALYSIS in the McGraw-Hill Encyclopedia of Science & Technology.

Herschel Specter

Bibliography. American Nuclear Society, *Risk-Based Regulation,* PPS-31, 1993; H. Specter, Risk-based regulation, *Trans. Amer. Nucl. Soc.,* 65:568, 1992.

Rocket propulsion

After three decades of space-flight growth, the momentum of the United States space program has slowed because of excessive space-launch costs, and future growth depends on reducing payloads. The more capable spacecraft required for future lunar and Martian flights could be launched less expensively if their propulsion efficiency were improved, because propellant dominates their size and mass. Nuclear thermal propulsion (NTP) uses less than half the propellant of the best chemical rockets. Nuclear thermal rocket engines (NTRE) accelerate reactor-heated hydrogen through a nozzle to higher velocities than are achievable with the heavier combustion products of chemical engines. Nuclear thermal propulsion with long-lived reactors can safely be developed because of the improved component and system technologies for second-generation reactors and engines.

Space-flight cost. Spacecraft have lightweight construction, and launch vehicles lift them against the Earth's strong gravity and air-drag forces within large, heavy, aerodynamic fairings. These launch vehicles are more than 10 times as massive as their payloads, and therefore account for the greatest portion of space-mission costs.

Specific impulse. Space flight requires velocity changes (ΔV) that are accomplished by accelerating spacecraft for a period of time with rocket engines. Specific impulse (I_{sp}) is engine impulse (thrust-time product) per pound (or kilogram) of propellant consumed. Spacecraft with higher specific-impulse engines use less propellant, thus saving money by requiring smaller propellant tanks. Vehicles with higher specific impulse and less propellant accelerate faster with the same engine thrust, and less impulse is needed for the same change in velocity. Therefore, rocket mass is exponentially dependent on the ratio of its change in velocity to its specific impulse ($\Delta V/I_{sp}$). Thus, specific impulse is more important on higher missions that require higher changes in spacecraft velocity.

Because planets are far apart, reasonable transit times require high velocities. Changes to high spacecraft velocity are necessary, especially for crewed missions, to protect against long exposures to interstellar radiation. It is not feasible to provide much shielding against these high-energy rays, but nuclear thermal propulsion with higher specific impulse shortens exposure times.

Rocket nozzle exhaust velocity determines specific impulse and is controlled by the square root of the ratio of the engine's exhaust-gas temperature to molecular weight ($\sqrt{T/m}$). The molecular weight of the exhaust from nuclear thermal propulsion is one-fifth that of the best chemical engines, so the specific impulse of nuclear thermal propulsion is twice as great in spite of its lower gas temperatures. Nuclear thermal propulsion will deliver at least 80% more crewed mission payload to the Moon than will chemical propulsion with equal launch weight.

Nuclear thermal rocket engines. The first integrated nuclear thermal rocket engine was tested about 1970 in Nevada during the nuclear engine for

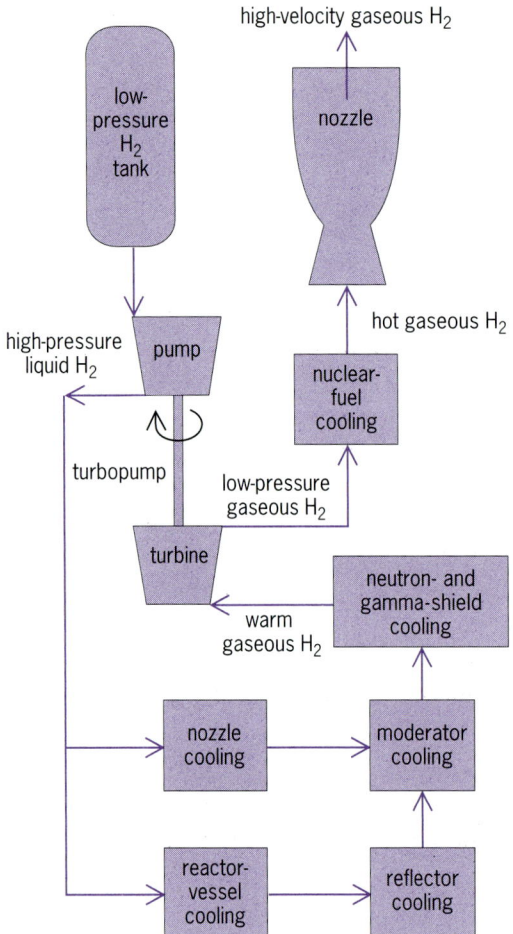

Fig. 1. Topping-cycle flow scheme for nuclear thermal rocket engine.

neutron moderator and reflector, and neutron and gamma shields. Next, warm turbine exhaust gas flows to nuclear fuel assemblies, which heat it to 3000 K (5000°F) to achieve nozzle velocities as large as 10 km/s (6.2 mi/s). This so-called topping cycle turbopump power scheme maximizes specific impulse, because all propellant flows through the rocket nozzle. The NERVA engine tapped hydrogen from its nozzle for pump power.

Heat recuperation. Spacecraft size and mass are significantly affected by their nuclear engines. Most of the length and mass of a nuclear thermal rocket engine resides in its reactor, shielding, and rocket nozzle. Engine operation at higher pressures reduces nozzle size and weight and has negligible impacts on other components. However, higher pressures require more turbopump power than is available with topping cycles. This dilemma is resolved with modern fabrication technology by converting the engine's gamma shield into a large heat exchanger, which can recycle turbine exhaust heat into the turbine drive gas (**Fig. 2**). Turbopump power and engine pressure can be tripled with this concept without loss of specific impulse. It enables nozzle length to be cut by 40% and nozzle mass by more than half.

rocket vehicle application (NERVA) program. Liquid hydrogen was pumped through engine components to protect them from nuclear fission heating, including the nuclear fuel elements of a solid-core reactor. Shielding, with mass greater than that of the reactor, protected nonnuclear components from neutron and gamma-ray damage and heating. Testing demonstrated the highest specific impulse ever achieved at high thrust for long durations. However, reactor degradation showed that nuclear fuel and reactor technologies needed improvement for practical space flight. Moreover, startup and shutdown were shown to be the most difficult phases of engine operation.

Topping cycle. A proper balance between reactor power and propellant flow is necessary to control reactor temperatures to avoid reduced specific impulse or reactor life. Propellant flow should depend on reactor power for good control, and can be achieved with second-generation nuclear thermal rocket engines by using a version of the coolant flow scheme shown in **Fig. 1**. A low-temperature loop drives hydrogen turbopumps with the nearly 5% of reactor power that deposits directly into engine components by nuclear radiation. These components include the reactor vessel and nozzle,

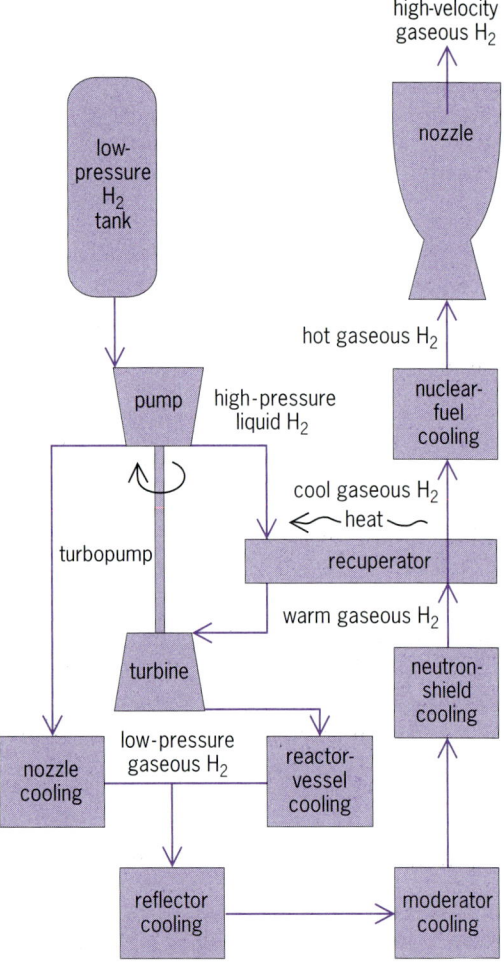

Fig. 2. Recuperated topping-cycle flow scheme for high-pressure nuclear thermal rocket engine.

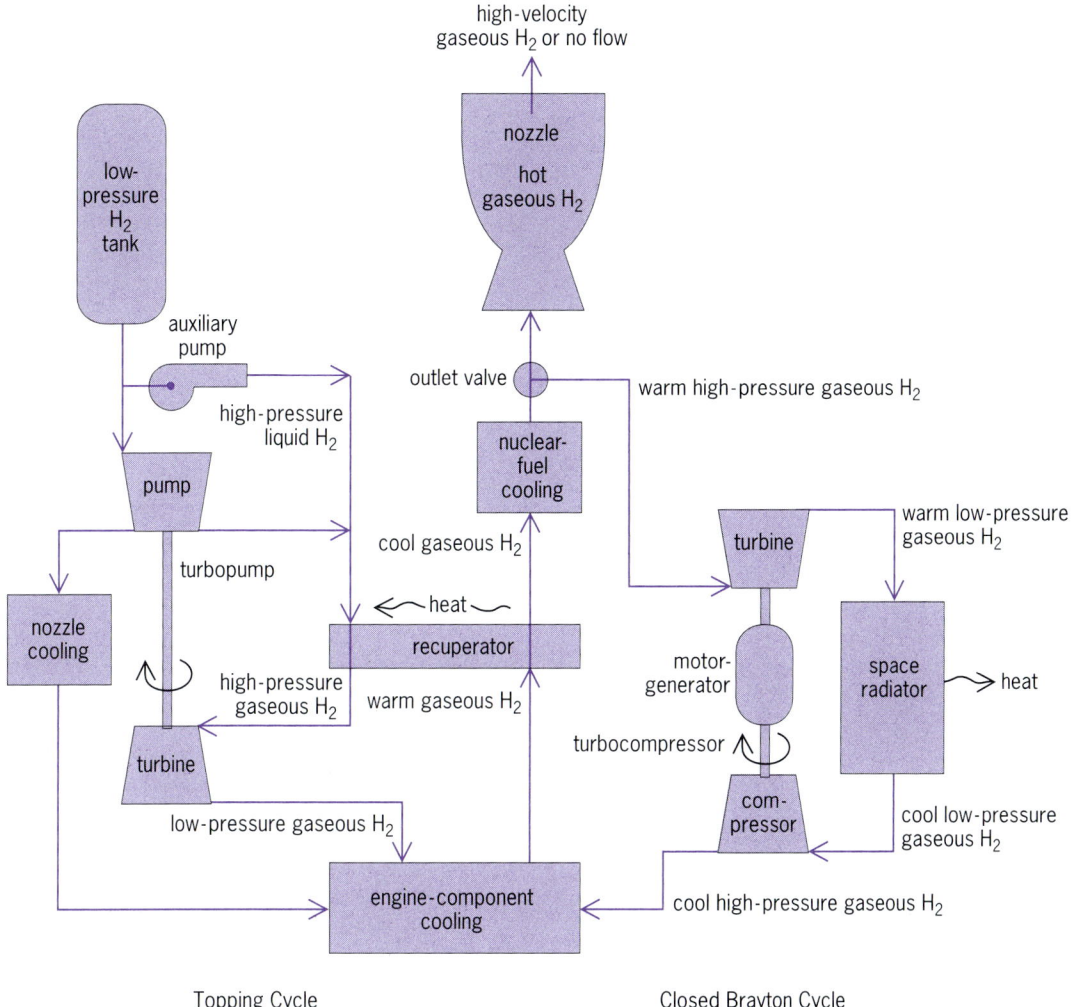

Fig. 3. Integrated topping and closed Brayton cycles for efficient, long-life nuclear thermal propulsion.

The power of the NERVA reactor was controlled with difficulty during startup, because liquid hydrogen increased its fission efficiency. In second-generation engines, all liquid hydrogen will be gasified in the recuperator before cooling the reactor. Thus, hydrogen density and mass will be reduced in the reactor by more than 90%.

Closed Brayton cycle. NERVA reactor shutdowns were complicated, because fission daughter products had accumulated in its nuclear fuel during operation. The daughters decayed, releasing energy, and engine cooling was required for a day after shutdown. Decay power reduces with time, rapidly at first, and then more slowly, and active cooling is needed until the reactor can radiate its heat to space. Turbopumped cooling wastes propellant, because it cannot reduce smoothly with time to control reactor temperature throughout the necessary power and flow range of approximately 100,000 to 1. Interrupting the flow at lower power levels saves propellant but creates thermal shocks throughout the engine that damage the reactor's brittle, high-temperature fuel.

Most residual energy is removed from the reactor at low power levels, and this fact enabled a new idea. A few minutes after reactor shutdown, decay power can be released through a reasonably sized space radiator designed for the purpose. The waste heat itself drives a closed Brayton power cycle, which circulates cooled hydrogen from the radiators to the reactor and back so that no hydrogen is lost. This system reduces propellant waste by about 70% during each shutdown of the nuclear thermal rocket engine and saves approximately 100,000 lb (45,000 kg) of hydrogen on a crewed Mars mission.

The closed Brayton cycle has other uses when integrated with the propellant feed system (**Fig. 3**). If the reactor powers the closed Brayton cycle continually during space missions it never cools fully. Reactor life, which is controlled by physical fuel damage, is thereby extended. The closed Brayton cycle generates electricity throughout the mission and refrigerates hydrogen in the main propellant tank to prevent its boiling and subsequent loss into space. Thus, auxiliary electric power systems are eliminated from the spacecraft. Attitude control rocket engines are replaced by warm-hydrogen bleed valves with uncooled nozzles on the closed Brayton cycle. Reactor cool-down flow is circulated by electric pumps and stabilized throughout cooling channels of the nuclear thermal rocket engine by recuperated heat at all power levels that lie between turbopump and closed-Brayton-cycle capabilities.

Heterogeneous reactors. Nuclear fission power is produced by splitting uranium nuclei with neutrons. Daughter products from each fissioned nucleus total slightly less than the original mass, and this difference is liberated as energy, including more than one fast-moving neutron. If enough nuclei can be fissioned directly a self-sustaining, so-called fast neutron chain reaction occurs. If not, slower, more efficient thermalized neutrons are needed, and neutron moderator material must be included in the reactor core. Large, high-power reactors in nuclear thermal rocket engines favor thermal designs, because less (costly) fissionable material is needed.

Because neutrons have no electric charge, they are slowed by kinetic energy transfer through collisions with other light particles. Hydrogen has a single proton nucleus and is the best neutron moderator. Other good moderator materials are also hydrogenous, such as water, metallic hydrides, and plastics, but they all have low operating temperatures. Carbon (graphite) moderator was used in the NERVA reactor, because its properties enabled a high-temperature, homogeneous core design with fuel and moderator mixed within each fuel element. However, the carbon was rapidly attacked by hot hydrogen propellant during NERVA operation.

Heterogeneous reactors array unmoderated nuclear fuel assemblies within a field of unfueled, moderator material, recognizing that the best moderators are low-temperature materials and the highest-temperature fuels are poor moderators. Russian reactors designed for nuclear thermal rocket engines use thin fuel assemblies that contain uranium carbide fuel sticks in a field of thinner, zirconium-hydride moderator rods. Fuel and moderator forms enable compact, mutually supportive stacking and high-heat-flux cooling by hydrogen gas. Such thermal reactors with heterogeneous design provide higher specific impulse and longer life at reasonable cost, because they use less fissile material and superior materials throughout.

For background information SEE BRAYTON CYCLE; NUCLEAR REACTOR; ROCKET PROPULSION; SPECIFIC IMPULSE in the McGraw-Hill Encyclopedia of Science & Technology.

Donald W. Culver

Rogowski coils

Rogowski coils are used for measuring alternating current. They work by sensing the magnetic field caused by the current without the need to make an electrical contact with the conductor. For decades, these coils have been used in various forms for detecting and measuring electric currents, but only recently has their potential been realized on a commercial scale.

The coils operate on a simple principle. An air-core coil is placed around the conductor in a toroidal fashion so that the alternating magnetic field produced by the current induces a voltage in the coil. The coil is effectively a mutual inductance coupled to the conductor being measured, and the voltage output is proportional to the rate of change of current. To complete the transducer this voltage is integrated electronically (see **illus**.) to provide an output that reproduces the current waveform. This combination of coil and integrator provides a system which has a frequency-independent output and an accurate phase response, and can measure complex current waveforms. The output from the integrator can be used with any form of electronic indicating device such as a voltmeter, oscilloscope, protection system, or metering equipment. The coils are wound either on a flexible former that is subsequently wrapped around the conductor to be measured or on a rigid toroidal former.

Features. There are other devices that measure electric current without making electrical contact with the conductor. Many of these devices, including the conventional current transformer, use a ferromagnetic core and are subject to magnetic saturation effects that limit the range of measurable currents. However, a Rogowski coil is linear; that is, it does not saturate and the mutual inductance between the coil and the conductor is independent of the current.

Many of the useful features of Rogowski coil systems result from their linearity. They have a wide dynamic range in that the same coil can be used to measure currents ranging from a few milliamperes to several million amperes. Calibration is easier because the coil may be calibrated at any convenient current level and the calibration will be accurate for all currents including very large ones. They respond accurately to transient currents, which makes them an excellent choice for use in protection systems and for measuring current pulses. Finally, they are useful in situations where the approximate value of the current to be measured is not known beforehand.

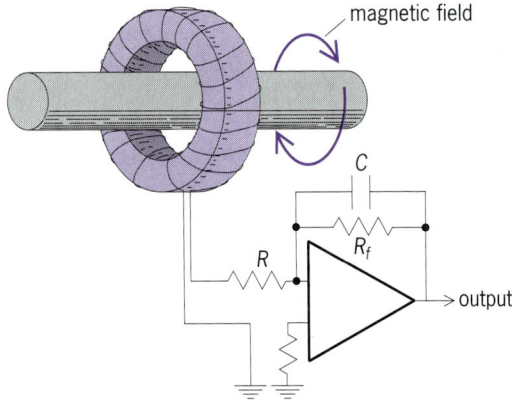

Arrangement of Rogowski coil and electronic integrator to give a complete transducer.

Coils wound on flexible formers have the additional unique feature that they can be wrapped around the conductor being measured. A long coil can be used as a compact portable device to measure the current in large conductors. Flexible coils can be manufactured with a cross section only a few millimeters (fraction of an inch) across and can be used where there is limited space around the conductor.

Development. In 1887, A. P. Chattock described the use of a long, flexible coil of wire wound on a length of india rubber as a magnetic potentiometer. The output of such a coil is proportional to the line integral of the magnetic field along its length, that is, proportional to the so-called magnetomotive force or the magnetic scalar potential between its ends. Chattock used the coil to measure the magnetic reluctance in iron circuits, but he calibrated it by bringing the ends together to encircle an electric current. This calibration method depended on Ampère's law, which states that the value of the line integral of magnetic field along a loop which completely encircles a current is equal to the current.

W. Rogowski and W. Steinhaus described the technique in 1912. They were also interested in measuring magnetic potentials. They described several ingenious experiments to test that their coil was providing reliable measurements, including using it to measure electric currents.

For accurate measurements using a Rogowski coil, it is essential that the winding be extremely uniform. According to Ampère's law, the output of a perfectly uniform coil encircling a current does not depend on the path the coil takes around the current or on the position of the conductor within the loop. It is only necessary that the ends of the coil be brought together accurately. Also, if the coil does not encircle a current, the output is zero even if the coil is positioned near a current-carrying conductor. These features are obviously highly desirable in an effective current-measuring transducer.

To achieve these ideal properties the coil must be wound with a constant number of turns per unit length on a former of uniform cross section. With a flexible coil the winding must remain uniform when the coil is bent. The more uniform the winding, the better the coil will approximate the ideal. Both Chattock and Rogowski had considerable difficulty achieving adequate coil geometry.

Practical systems. By using the right technique it is now possible to wind both flexible and solid coils with sufficient uniformity for them to be used in a wide range of applications, including those demanding precision measurements. The sensitivity of a complete system comprising a coil and integrator is the ratio between the voltage output, V_{out}, and the current being measured, i. This sensitivity is given by the equation below, where C and R represent the

$$\frac{V_{\text{out}}}{i} = \frac{M}{CR} \text{ (volts per ampere)}$$

capacitor and resistor in the integrator circuit (see illus.), and M is the mutual inductance between the coil and the conductor. For a given coil the sensitivity is adjustable over an enormous range by choosing suitable values of C and R. For example, with a typical flexible coil the sensitivity can be varied over a range greater than from 1 V/A to 1 μV/A. There is also ample scope for modifying the characteristics of the coils themselves by altering the turns density and cross-sectional area. The full range of permutations of coils and integrators provides an exceptionally versatile measuring system.

Although Rogowski coils are not suitable for measuring direct currents, by careful design systems can be built that measure at frequencies as low as 0.1 Hz. The high-frequency limit is determined by the self-resonance of the coil and depends on the coil design. High-frequency limits in the range from 20 kHz to 1 MHz are typical.

Very high frequency measurements can be made with a Rogowski coil by terminating the coil with a low impedance and by using the self-inductance of the coil to perform the integration. The output signal is then a current rather than a voltage. Coils operating on this principle have been used to measure currents at frequencies up to 100 MHz.

Applications. An area of applications where Rogowski coils have been particularly valuable is in the measurement of current transients. Conventional current transformers can become "confused" during the initial stages of a transient, especially if the transient contains an asymmetric component (sometimes referred to as a dc offset).

Transient measurements where Rogowski coils have been used include the following:

1. Monitoring the current in precision welding systems.

2. Measuring the plasma current in a controlled nuclear fusion experiment such as the Joint European Torus (JET) experiment in Culham, England.

3. Current measurement in arc melting furnaces. Arc furnaces use very large fluctuating currents, and they can be made more efficient by monitoring the current and appropriately regulating the arc.

4. Monitoring electric power systems for protection purposes. Rogowski coils give a more accurate measurement, particularly of the early stages of a fault current, and are suitable for interfacing with modern, all-electronic protection relays.

5. Measuring the current pulse in an electromagnetic launcher (rail gun). The current can be several million amperes and last a few milliseconds.

6. Testing the response of generators to sudden short circuits.

Rogowski coils have also been used to advantage for the measurement of steady currents. Energy-management systems that monitor the current consumption patterns of large buildings and industrial plants are becoming increasingly important. Some systems use Rogowski coils because of their versatility. They are useful for measurement of the harmonic components in electric currents because, being exceptionally linear, they faithfully reproduce the harmonic content. Rogowski coils are also used to measure currents with complex waveforms such as in thyristor circuits. They are used in the railway industry to monitor the signaling currents in railway lines. Flexible coils have been used to trace the currents induced in metal structures exposed to magnetic fields, for example, near a large transformer. The flexible coil has an educational value as an excellent practical demonstration of Ampère's law.

For background information SEE AMPÈRE'S LAW; CURRENT MEASUREMENT; INSTRUMENT TRANSFORMER; MAGNETIC FIELD in the McGraw-Hill Encyclopedia of Science & Technology.

David A. Ward

Bibliography. International Coil Winding Association Inc., *Proceedings of the 1992 Electrical Manufacturing and Coil Winding Conference,* Cincinnati, Ohio, September 21–24, 1992; D. A. Ward and J. LaT. Exon, Using Rogowski coils for transient current measurements, *IEEE Sci. J.*, pp. 105–113, June 1993.

Satellite navigation systems

Since the Global Positioning System (GPS) became operational in the early 1990s, its value to the military, commercial, scientific, and personal communities has exceeded all expectations. Some recent GPS applications are in personal navigation, vehicular tracking, aviation terminal guidance, and numerous scientific missions where application of GPS was unknown only a few years ago. Inexpensive GPS receivers are affordable by the public, and greatly improved accuracy has resulted from improved receiver designs and the rapidly growing use of differential GPS. SEE APPLICATIONS SATELLITES.

Global Positioning System. The GPS consists of 24 satellites, several Earth-based monitoring and control stations, and special radio receivers designed to receive information from the satellites and convert this information into the position and velocity of the receiver. The satellites circle Earth approximately every 12 h at an altitude of about 13,000 mi (21,000 km). Each satellite transmits its precise orbital position and velocity to the user's receiver, which is capable of accurately measuring the range to the satellite. A computer within the receiver combines the range measurements to several satellites with the known satellite position information to produce an accurate estimate of the receiver position.

A unique pseudorandom code, called the C/A code, is impressed upon the transmitted signal of each GPS satellite. This code uniquely identifies the satellite, permits the receiver to separate the signal from other GPS satellite signals, and is used in the measurement of range from the user's receiver to the satellite. The code also has the property of spreading the spectrum of the signal, thus greatly reducing susceptibility to interference.

GPS provides two navigational services: The Standard Positioning Service (SPS) available to all users, both civilian and military, uses the C/A code described above. The Precise Positioning Service (PPS) uses an additional code, the P code, which is encrypted and is therefore available only to users authorized by the military.

Personal navigation. Rapid improvement in GPS technology has spawned a variety of hand-held GPS receivers for personal use. Typical units can operate continuously for 10–20 h using AA-size batteries. A hiker, camper, or boater can locate his or her position day or night, anywhere in the world, with a typical accuracy of 150 ft (45 m) or better. (Even greater accuracy is possible by using differential GPS, described below.) In typical operation, the receiver automatically starts to search for satellite signals when it is turned on. When a signal is received, the ephemeris (precise orbital) data transmitted by the satellite are stored in the receiver. As soon as the ephemeris data from four satellites are available, the receiver automatically computes and displays the three-dimensional position (latitude, longitude, and altitude). Most receivers can display additional information, such as velocity, to within a few tenths of a mile (or kilometer) per hour, the direction of travel relative to true or magnetic north, the distance traveled, the direction to a destination, and very accurate time.

The receivers have many uses. For example, a hiker can locate the position of his or her vehicle before starting down the trail and store it in memory. When the hiker is ready to return, the GPS receiver can calculate the distance and direction back to the vehicle. During the return trip the GPS receiver continuously indicates the remaining distance to the vehicle and the direction to it. A more advanced mode of use involves the storage of positions, called way points, as the trail is traversed. Some hand-held receivers can even display the hiker's progress on a map which previously has been loaded into the receiver using small read-only memory (ROM) cartridges.

Vehicular applications. Navigational systems employing a GPS receiver combined with a map database are under development for use in automobiles and trucks. These systems can automatically locate the vehicle's position, display it on a map, and even issue voice-synthesized verbal instructions on how to reach a predetermined des-

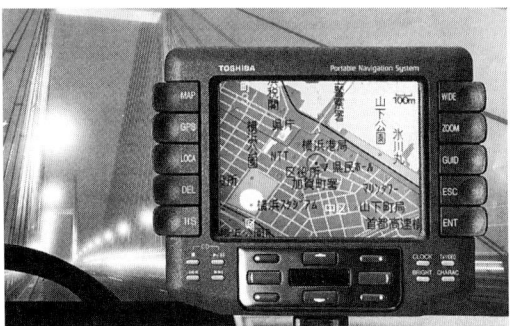

Fig. 1. GPS vehicular navigation system with map display of vehicle position. (*Magellan Systems*)

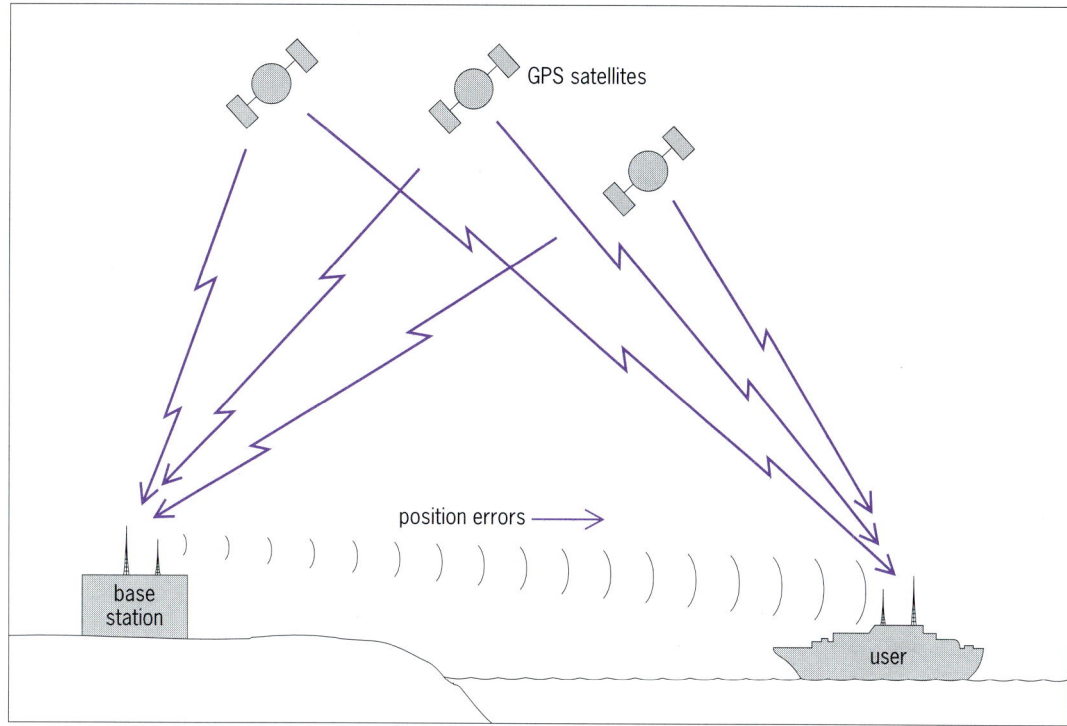

Fig. 2. Differential GPS positioning concept. The base station measures the position error of satellite signals, position errors are transmitted to users, and users correct their GPS positions with error data from the base station.

tination. The locations of the nearest restaurants, motels, and service stations, with hours of operation and types of service available, can also be continuously displayed (**Fig. 1**).

Businesses which use dispatched vehicles benefit from knowing the exact location of every vehicle in the fleet. In one application, ambulance service has reduced its emergency response time by using GPS-equipped ambulances which transmit their positions to the dispatch center at all times. Delivery and repair services are beginning to use a similar approach to improve operating efficiency. In one application, GPS permits the use of continuously roving tow trucks with an average emergency response time of 7 min, a dramatic reduction from the 40-min average wait when the towing vehicles were based at a fixed location.

Aviation guidance. After a long and controversial development period, it now appears that GPS can be used not only for aviation enroute navigation but also for guidance during landing. Although this application is still under development, tests have proved that category II landing approaches using differential GPS can be accomplished with better performance than approaches using the existing instrument landing system (ILS). [In a category II approach the pilot must have forward visibility of 0.25 mi (0.4 km) and must be able to manually complete the landing starting from an altitude of no less than 150 ft (45 m).] Major cost reductions accrue from the use of GPS, because it is no longer necessary to have a separate, relatively expensive ILS system at each airport.

Ultimately, category IIIa landings, in which the approach and touchdown are completely automatic, might be accomplished under computer control using only differential GPS to guide the aircraft. Positioning errors on the order of 3 ft (1 m) and velocity errors within a few tenths of a nautical mile (kilometer) per hour would be required. A major problem in automated landings is integrity monitoring, the ability of the system to reliably detect situations in which there is a malfunction that will cause excessive error in the aircraft position along the approach path, especially when the aircraft is near touchdown. The requirements for integrity monitoring are severe, because corrective action by the pilot often must be taken within seconds after a malfunction is detected.

To meet the high levels of accuracy required for aircraft approaches, a special wide-area differential GPS system, called the Wide Area Augmentation System (WAAS), has been proposed. In this system, the differential corrections from a network of 20 ground stations would be broadcast via INMARSAT satellites to all aircraft.

Differential GPS. Fearing that United States enemies might use the accuracy of the GPS to their advantage in delivering precision missile strikes, the federal government has intentionally degraded this accuracy to about 100–300 ft (30–90 m) by inserting random errors into the broadcast ephemeris data from each satellite and by so-called dithering of the satellite clocks to produce small timing errors, a procedure known as selective availability (S/A). This level of accuracy is adequate for many applications, such as most personal navigation, but is not sufficient for applications such as aviation terminal approaches or marine navigation in small or crowded harbor areas. Fortunately, a special way of using GPS, called differential GPS (DGPS), eliminates errors due to S/A and greatly reduces other errors, such as uncertainties in signal propagation time through the ionosphere.

In DGPS (**Fig. 2**), a base station receives signals from all GPS satellites in view. Because the base station's location is precisely known, the positioning error resulting from use of the received signals can be measured and broadcast via a radio link to all GPS users within a given distance of the base station. Since the user's position error is very nearly the same as that of the base station, the broadcast errors can be used to cancel out this error. Positioning errors of less than 15 ft (5 m) are common after differential corrections have been applied.

DGPS is effective only within a limited radius of the base station, because ionospheric errors are relatively constant only within this radius. For this reason, many base stations are needed to provide accurate positioning over a large area. Several services transmit differential corrections over FM broadcast stations. Additionally, the U.S. Coast Guard is installing equipment in its radiobeacon system to transmit differential corrections to marine users. In a promising approach called wide-area differential GPS (WADGPS), the base station broadcasts a model of ionospheric and tropospheric errors as a function of the user's approximate position. Thus, accurate differential corrections can be made over a much larger geographic area, and fewer base stations are needed.

Fig. 3. Use of GPS to map forest development. (*Magellan Systems*)

Scientific applications. GPS has been of inestimable value in the pursuit of many scientific missions. Some of these missions use very high-performance differential GPS receivers, capable of precise static positioning to within a few centimeters or even millimeters. With this accuracy, the movement of glaciers can be monitored over relatively short time periods. Crustal movements in earthquake-prone areas can also be accurately measured. Such measurements help geologists to understand earthquake formation, and may ultimately lead to accurate earthquake prediction. In the Arctic, special GPS-equipped buoys are used to study sea-ice deformations. Inexpensive GPS receivers are used to precisely track the location of weather balloons, including their altitude. Very small, low-power receivers are attached to wild animals to determine their foraging habits and migratory patterns. Highly accurate receivers mounted on the blades of bulldozers enable sculpting of land surfaces to precise specifications.

Numerous environmental studies are aided by GPS. By precise mapping, conservationists can develop accurate histories of forest growth and decay (**Fig. 3**). GPS-aided navigation can assist the precision aerial application of pesticides to guard against pollution of nearby water sources. In South America, large areas of impenetrable jungle can be efficiently mapped from the air to locate and monitor major areas of deforestation.

Attitude determination. In addition to providing accurate positioning and navigation, GPS can be used to determine the precise orientation of a platform, such as an aircraft or space vehicle, or an artillery unit. By measuring the difference in phase of the received GPS signal at several antennas located on the platform, the platform orientation along all three axes can be determined to within a small fraction of a degree, without the use of expensive gyrocompasses or relatively inaccurate magnetic compasses. Guns at battlefronts can be aimed far more accurately, and attitude sensor data for stabilization of orbiting satellites can be obtained cheaply and precisely.

For background information SEE AIR NAVIGATION; ELECTRONIC NAVIGATION SYSTEMS; INSTRUMENT LANDING SYSTEM (ILS); SATELLITE NAVIGATION SYSTEMS in the McGraw-Hill Encyclopedia of Science & Technology.

Lawrence R. Weill

Bibliography. *GPS World*, published monthly.

Shape memory alloys

The term shape memory alloys refers to all metal alloys that, after being deformed at a suitable temperature, fully or partially recover their original shape when subsequently heated to a high temperature.

Attributes. Shape memory alloys behave as if they remember their shape at a high temperature and prefer to attain it at that temperature. This attribute is called one-way (irreversible) shape memory, because cooling to the temperature at which the specimen was deformed does not restore the specimen to its deformed shape. When such an alloy is thermomechanically processed (thermally cycled while kept under a mechanical stress) or "trained," it remembers its shape at both the high and low temperatures, and is called a two-way shape memory alloy. After training, it bends on cooling and unbends on heating in a reversible manner when cycled between the two temperatures.

As the shape changes, the electrical resistivity, the magnetic susceptibility, the velocity of sound, the hardness, the mechanical modulus and damping capacity, the volume, and related thermodynamic functions all change irreversibly or reversibly. As **Fig. 1** illustrates, the austenite-martensite phase transformation occurs with change in physical properties at four characteristic temperatures M_s, M_f, A_s, and A_f, indicating the starting and end to the martensite (M) and austenite (A) phases. The separation of the heating and cooling curves, or hysteresis, allows either the austenite or martensite phase to be present at a certain temperature T_x, depending upon whether the sample was previously heated or cooled. The shape of the specimen may change, for example, from a bar to a coil and then reverse.

Recovery of the original shape of the alloy specimen may also occur at a high temperature when the stress or load that was used to deform it is removed. This phenomenon is called superelasticity. The full recovery of the shape here occurs with a substantial stress-strain (load-deformation) hysteresis. Some alloys also exhibit rubberlike flexibility after aging at a low temperature; when bars are bent, they spontaneously unbend upon release of stress.

For a material to have a shape memory it has to meet two conditions. First, it undergoes a structural transformation that causes a significant inhomogeneous strain. Second, the transformation is reversible and is subject to control, so that the inherent shape-strain caused by the transformation is converted to macroscopic deformation. In most shape memory alloys this transformation is martensitic, that is, it occurs by cooperative small displacement of atoms from their lattice sites, although rhombohedral-phase and bainitic (strain-producing) transformations can also occur.

Austenite-martensite transformation. Shape memory alloys deform by cooperative movement of lattice elements in a reversible manner. Thus, their behavior is distinguished from that of other alloys that deform by creation and motions of dislocations in an irreversible manner. The movement of lattice elements leads, on cooling, to formation of a new crystal structure referred to as the martensite phase (a substitutional or interstitial solid solution of a different crystal type) at the expense of the original parent phase. The parent phase is called the austenite phase, regardless of its crystal structure. The transformation between austenite and martensite

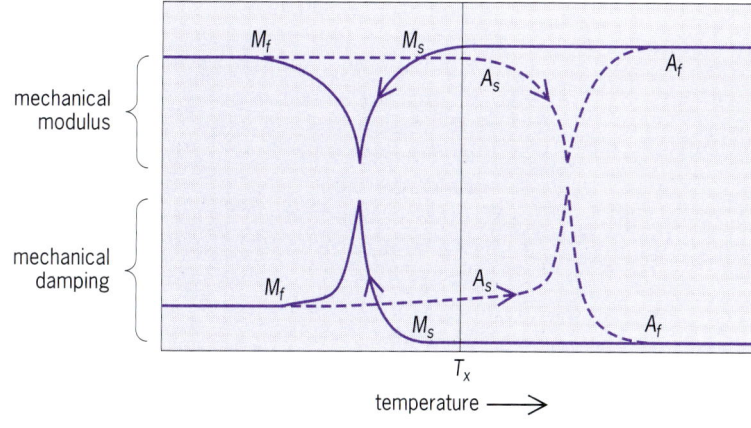

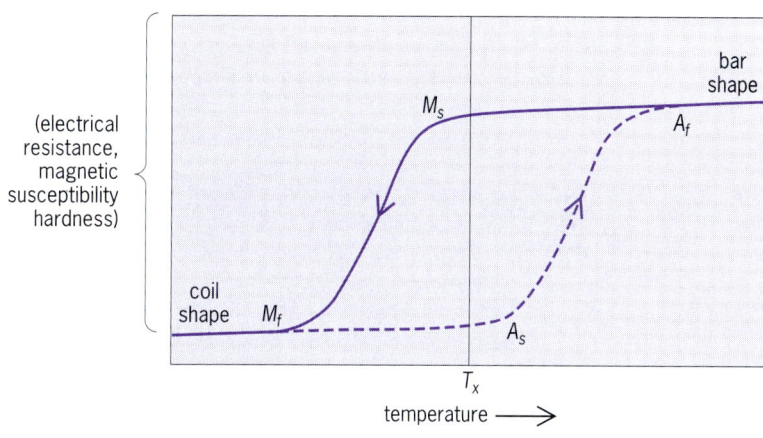

Fig. 1. Change in physical properties that occur during the austenite-martensite phase transformation dependent upon temperature T_x. (a) Mechanical clamping and mechanical modulus. (b) Electrical resistance and magnetic susceptibility hardness. M_s represents beginning of the martensite phase, and M_f, the end. A_s represents the beginning of the austenite phase, and A_f, the end.

phases is diffusionless (diffusion of atoms occurs over a distance less than a lattice spacing, and is therefore not long-range), and there is no change in the concentration of solute atoms from one phase to the other. Heat is evolved on the occurrence of the austenite-martensite transformation; volume usually contracts; Young's modulus goes sharply through a minimum value and mechanical damping through a maximum value; ductility continuously increases; and electrical resistance and magnetic susceptibility decrease (Fig. 1).

When the alloy specimen is cooled or stressed (loaded), crystal platelets of martensite phase form in the austenite matrix and grow continuously. The shape of the specimen changes as a result of coordinated shear displacement of microscopic sections of the specimen. When the alloy is heated or the stress removed, the martensite crystal platelets transform back to the austenite phase by an exact reverse path, and the original shape is regained. The reversibility of thermoelastic martensite is attributed mainly to elastic strains associated with the crystal structure change. The strains are low enough that the elastic limits of the martensitic and parent phases are not exceeded, and irreversible permanent deformation does not occur.

The growth of the martensite phase occurs along four directions (up and down, back and forth) of each of the six planes passing through the face diagonals of the cubic lattice of the austenite phase. Thus, the oriented growth can occur in 24 different ways known as martensite variants, and the small strains that build up on growth of martensite plates are effectively canceled out by forming groups of mutually accommodating plates. Ideally, a single crystal of the parent phase will ultimately form 24 orientations of martensite crystals on cooling. When this multiorientation configuration of martensite is deformed, a single orientation of martensite eventually results because of twinning and the movement of certain martensite interfaces. The same occurs when martensite-martensite interfaces move under an applied stress, and one martensite variant grows at the expense of the other. Eventually, the single remaining orientation of martensite is the variant whose shear or shape deformation will allow the greatest extension of the specimen in the direction of the applied stress.

The thermoelastic nature of the transformation, that is, the growth of martensite platelets on cooling or on application of stress and their disappearance by an exact reverse path on heating or on removal of the applied stress, originates from the necessity of a balance being maintained between two energy terms, thermal and elastic (hence the name thermoelastic), during the entire process of the transformation. The thermal or chemical free-energy difference between the parent and martensite phases favors the transformation. The elastic energy, which is developed as the microstructural units of the new martensite phase form in the parent phase matrix, tends to oppose the transformation. Thus, the transformation can be arrested if the elastic energy produced suddenly becomes too high or the thermal energy suddenly becomes too low.

Only when the temperature is below that of the start of the martensite phase (M_s) can the martensite crystallites form and grow (Fig. 1), and only on cooling below the end of the martensite phase (M_f) does this process finish. On heating the alloy above the start of the austenite phase (A_s), which is significantly higher than M_f, the martensite crystallites begin to disappear in favor of the austenite phase, and above the end of the austenite phase (A_f) all are converted to the austenite phase. Thus, a separation, or hysteresis loop between the cooling and heating curves, is produced. At any temperature within this loop different extents of both phases may be present, depending upon whether the alloy was previously heated or cooled. The properties of the alloy at a given temperature depend on the relative amounts of the two phases. For example, striking the wire of an alloy at temperature T_x (Fig. 1) will produce a ringing sound when the wire has been cooled from A_f and a thud when the wire has

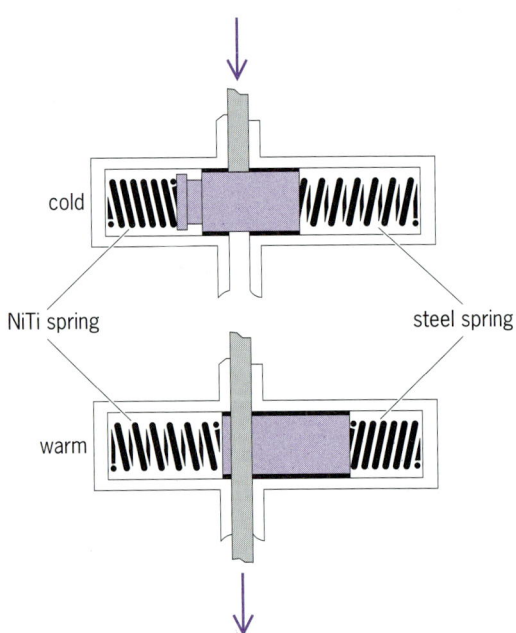

Fig. 2. Actuated governor valve for automatic transmissions fabricated from a nickel-titanium (NiTi) alloy. (*After D. Stoeckel and F. Tinschert, Temperature Compensation with Thermovariable Rate Spring in Automotive Transmission, SAE Tech. Pap. 910805, Society of Automotive Engineers, 1991*)

been heated from M_f. The temperature point M_s of alloys varies with the composition, as does the width of the hysteresis loop. Even for a single nickel-titanium alloy, NiTi, a 47–51 atomic % variation in Ni composition changes M_s from 75 to −160°C (167 to −256°F). Most alloys used in practice have a transformation range of ±150°C (±270°F) on either side of room temperature (around 25°C or 77°F) and a 20–100°C (36–180°F) hysteresis width.

Applications. Shape memory alloys have been used for a wide variety of applications, and worked into structures as so-called smart materials, where they are capable of sensing and responding to their environment in a predictable and desired manner. One interesting automotive application is the NiTi alloy control valve illustrated in **Fig. 2**, which regulates the flow of transmission fluid when the temperature changes and controls the shifting pressure in the automobile's automatic transmission, which in turn smooths shifting between gears. When the nickel-titanium (NiTi) spring is in the martensite phase at a low temperature, the steel spring easily compresses it. As the temperature increases, the NiTi spring transforms to the austenite phase, becomes extended and rigid as if remembering its extended shape and rigidity, and forces the steel spring to compress. Thus, the pathway for transmission fluid is opened through the valve. The alloy valve can go through millions of switching mechanisms (austenite-martensite transformation cycles) without fatigue. Another application is O. Deschamps's 0.6-m (2-ft) high dynamic sculpture

Le Skieur (*The Skier*), which crouches close to the ground when it is cold and straightens up when it is warm. In a cold environment the NiTi is in the martensite phase and the leg joints are bent at an angle. When it is warm, the NiTi alloy transforms to the austenite phase, and the leg joints become straight.

A wide variety of devices incorporate shape memory alloys to produce a shape change, to apply a stress, or as a combination of both. The most common use is in biomedical applications: orthodontic braces that apply a uniform stress when clamped, staples and bands to clamp broken bones together, and guide wires for arthroscopic surgery. Use of the alloys as temperature sensitive on-off switches in household devices, thermostatic radiator valves, automatic window openers for greenhouses, robotics, and eyeglass frames that may be bent and regain their original shape once the stress is removed are current applications.

For background information SEE ALLOY; CRYSTAL STRUCTURE; METAL, MECHANICAL PROPERTIES OF; SHAPE MEMORY ALLOYS; SOLID SOLUTION in the McGraw-Hill Encyclopedia of Science & Technology.

G. P. Johari

Bibliography. O. Deschamps, The application of shape memory alloys to sculpture, *J. Met.*, 43:64, March 1991; H. Funakubo (ed.), *Precision Machinery and Robotics*, vol. 1: *Shape Memory Alloys*, trans. by J. B. Kennedy, 1987; K. R. C. Gisser et al., Nickel-titanium memory metal: A smart material exhibiting solid-state phase change and superelasticity, *J. Chem. Educ.*, 71:334–340, April 1994; C. M. Wayman, Shape memory and related phenomena, *Prog. Mater. Sci.*, 36:203–224, 1992; C. M. Wayman, Some applications of shape memory alloys, *J. Met.*, 32:129–137, June 1980.

Shrimp

In many vertebrate and invertebrate animals, hardened or coagulated male substances (mating plugs) block or seal the genital openings of the female after mating. Mating plugs (also termed sperm or copulatory plugs) have evolved in males of many species to ensure successful insemination of the female. In the penaeoid shrimps, a group of marine crustaceans of considerable economic and ecological importance, there is considerable variation in the mechanics of insemination, with males of many species depositing various forms of mating plugs.

Penaeoid shrimps. The penaeoid shrimps are part of the crustacean order Decapoda (shrimps, crayfishes, lobsters, crabs). The suborder to which they belong, the Dendrobranchiata, consists of two subsuborders, the Sergestoidea, a planktonic group, and the Penaeoidea, with mainly benthic (bottom-dwelling) species. The Penaeoidea comprise the families of Solenoceridae, Aristeidae, Benthesicymidae, Penaeidae, and Sicyoniidae. Species in the family Penaeidae are of great importance both ecologically and commercially because of their large population sizes and high productivity. Most of the world's harvest of shrimps comes from either fishing or aquaculture of certain species of the genus *Penaeus*.

Reproductive morphology. The genital area of the female, located between the posterior 2–3 pairs of walking legs under the cephalothorax, is termed the thelycum. The thelycum consists of a variety of depressions, protuberances, plates, and other modifications of the exoskeleton which serve for either temporary attachment or prolonged storage of spermatophores (the sperm-bearing packages produced by the male) or as an area of genital contact with the male, leading to the openings of paired, internalized seminal receptacles (spermathecae) in which sperm are stored. An open thelycum, found in females of Aristeidae, Solenoceridae, and some species of Benthesicymidae and Penaeidae, allows the external attachment of spermatophores deposited by males. A closed thelycum, found in females of some Benthesicymidae, many Penaeidae, and all Sicyoniidae, either is a single, median external seminal receptacle (spermatheca) or consists of cuticular structures that guard the apertures to paired receptacles inside the female's body.

Spermatophores. The wide variation in form of the female thelycum is paralleled by that of the spermatophores. In penaeoids, the spermatophore may be defined as all the materials emitted from

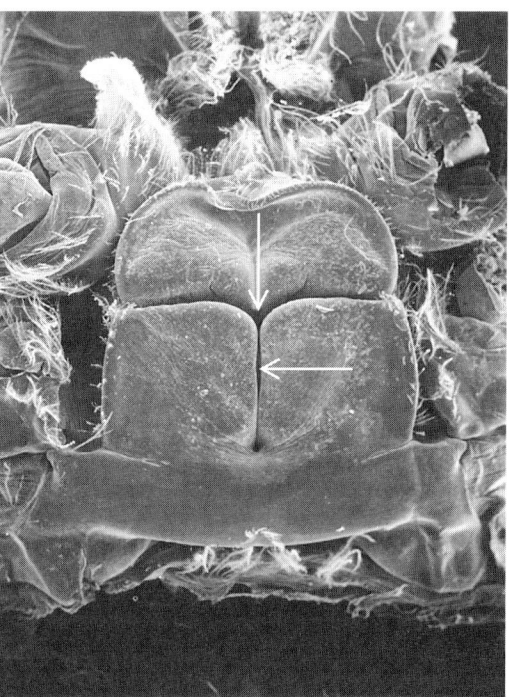

Fig. 1. External view of the thelycum of an uninseminated female of *Trachypenaeus similis*; slits or openings into the median pocket are indicated by arrows. (*After R. T. Bauer and L. J. Min, Spermatophores and plug substance of the marine shrimp* Trachypenaeus similis *(Crustacea: Decapoda: Penaeidae): Formation in the male reproductive tract and disposition in the inseminated female, Biol. Bull., 185:174–185, 1993*)

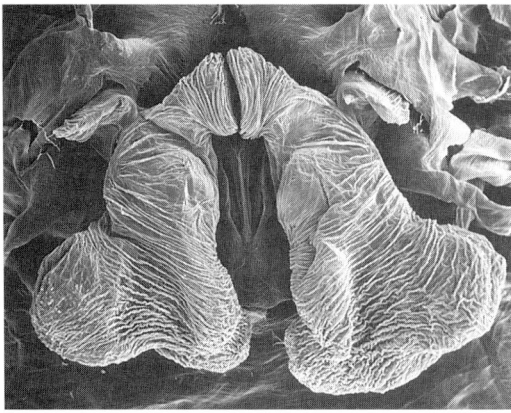

Fig. 2. Dorsal view of the paired seminal receptacles, located on the internal side of the thelycum, of a *Trachypenaeus similis* female. (*After R. T. Bauer and L. J. Min, Spermatophores and plug substance of the marine shrimp Trachypenaeus similis (Crustacea: Decapoda: Penaeidae): Formation in the male reproductive tract and disposition in the inseminated female, Biol. Bull., 185:174–185, 1993*)

the ejaculatory duct during copulation. In species in which females have open thelyca, such as *Penaeus* (subgenus *Litopenaeus*) *setiferus*, each of the two spermatophores is a sperm mass surrounded by several substances that attach the entire structure to the female. In some species with closed thelyca, such as *P.* (*Farfantepenaeus*) *aztecus*, the twin spermatophores are deposited within a median cuticular pocket enclosed by the lateral plates of the thelycum. The enclosed and protected spermatophores of *P. aztecus* are reduced in complexity (fewer accessory substances) compared to those of *P. setiferus*. In *Trachypenaeus similis*, the lateral plates of the thelycum also enclose a median pocket (**Fig. 1**), but sperm emitted by the male in small packets is stored within paired seminal receptacles inside the female's body (**Fig. 2**). The apertures into these receptacles are located within the median pocket, which after insemination is filled by a mating plug (**Fig. 3**), which is emitted from the male's gonopore just after the sperm is ejaculated. In sicyoniids, the male emits a fluid sperm mass from its gonopores, but no mating plugs are deposited on the female; the sperm are deposited directly into paired seminal receptacles through apertures which are completely exposed, that is, not within a median pocket protected by lateral plates.

Mating. The relationships among molting (periodic shedding of the exoskeleton), maturation of the ovary, and spawning are important in understanding when penaeoid females are receptive to mating. In open thelycum species, females are receptive when the ovary is mature (full of ripe eggs), and mating occurs during the intermolt period. In *P. setiferus* and *P. vannamei*, for example, spawning occurs within a few hours after the male deposits spermatophores on the female. In closed thelycum species, however, ovarian maturity and mating are not related, and females are receptive to courtship just after they molt. Several cycles of ovarian maturity and spawning may take place between molts. Sperm stored from the postmolt mating is used to fertilize the many thousands of eggs in these consecutive spawns. Females must mate again after a molt because any remaining sperm in the seminal receptacles are cast off with the molted exoskeleton.

Although the mating behavior of relatively few penaeoid species has been investigated in detail, it appears that the male mating system can be described as pure searching or scramble competition polygyny. Males actively search for females, mating when a receptive female is encountered. Unlike many other decapod species, penaeoid males do not appear to guard or defend females anytime prior to mating. Penaeoid males are usually smaller than females and have relatively small claws. They are generally unaggressive compared to many other decapod males, and they compete with each other for inseminations by scrambling to be the first to contact a receptive female.

The mating of penaeoids is a relatively brief affair. After contacting and trailing below a swimming or walking receptive female, the male often prods the female thelycum with its cephalic or head area. After following the female for a few seconds to minutes, the male flips upside down and transfers the spermatophores to the female, after which the pair quickly separate. Two or more males may push and jostle for position in an attempt to copu-

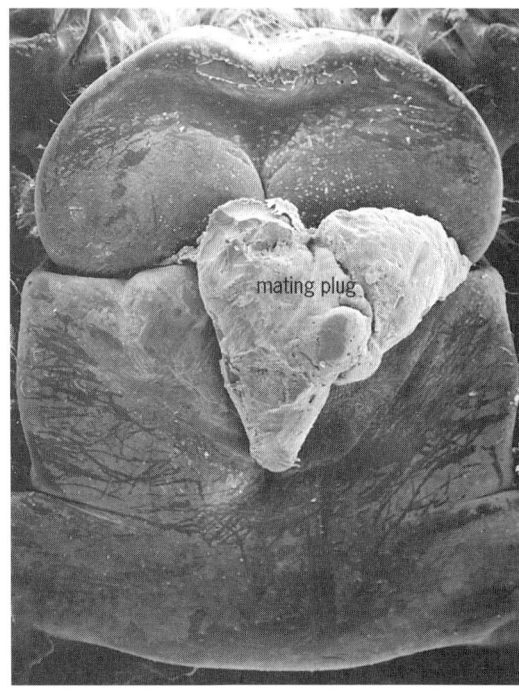

Fig. 3. External view of the thelycum of an inseminated female of *Trachypenaeus similis*, with the mating plug protruding from the median pocket. (*After R. T. Bauer and L. J. Min, Spermatophores and plug substance of the marine shrimp Trachypenaeus similis (Crustacea: Decapoda: Penaeidae): Formation in the male reproductive tract and disposition in the inseminated female, Biol. Bull., 185:174–185, 1993*)

late with the same female. Postcopulatory defense of the female by the mating male from other males has not been observed.

Insemination and sexual strategies. The term mating plug does not apply to open thelycum species of penaeoids since deposited spermatophores are attached to the external surface of the female and are fully exposed. Indeed, in species such as *P. setiferus* the spermatophores are easily dislodged from the thelycum. In contrast, in a closed thelycum species such as *P. aztecus*, the appendages of the spermatophores undergo a reaction with seawater, expanding and covering the aperture to the seminal receptacle, with subsequent hardening to form a seal or plug. In *T. similis*, first the small sperm packets are deposited into the median pocket of the thelycum, followed by the plug substance which apparently forces the sperm into the apertures of the paired, internalized seminal receptacles. The plug substance quickly hardens in the median pocket, sealing off the apertures of the receptacles from contact with the exterior of the female. Mating plugs may serve to stop sperm or spermatophoric materials from leaking back out to the exterior. They may also prevent deterioration of stored sperm by bacteria, fungi, and other microorganisms ubiquitous in the surrounding seawater. In addition, subsequent insemination of the female by another male is rendered impossible.

Penaeoid mating plugs thus function as paternity assurance devices. By ensuring that its sperm is stored and not mixed or replaced with that of other males, the inseminating male guarantees that the eggs of all spawnings by the female during the intermolt period will be fertilized by only its sperm. This arrangement may not be entirely adaptive for the female if the inseminating male does not supply enough sperm to completely fertilize all spawns or if the male is in some way genetically inferior to other possible mates.

In the sicyoniids *Sicyonia dorsalis* and *S. parri*, males do not produce a mating plug. A fluid sperm mass is directly deposited into the internalized seminal receptacles. Unlike penaeoid males, which produce mating plugs, a *Sicyonia* male can inseminate only one of the two receptacles with a single copulation. Furthermore, studies have shown that females may copulate with males without allowing insemination. Multiple copulations with one or more males may function as a courtship device by which females can choose a desirable or superior male. Since several males often crowd around a receptive female, it is quite likely that in nature, a female might choose to be inseminated by two different males (one filling each of the paired receptacles). Multiple male mating partners are possible since there is no mating plug to prevent sperm from being added to and mixing with a partially filled receptacle. Thus, it appears that sicyoniid females have some advantages over males regarding reproduction, the absence of a mating plug allowing them to mate repeatedly, thereby permitting a maximal filling of the receptacles and the possibility of insemination by more than one male. Multiple paternity of a female's spawnings has the advantage of increasing the genetic diversity of the numerous offspring—a benefit in changing, variable environments like the sea.

For background information SEE DECAPODA (CRUSTACEA) in the McGraw-Hill Encyclopedia of Science & Technology.

Raymond T. Bauer

Bibliography. R. T. Bauer, Repetitive copulation and variable success of insemination in the marine shrimp *Sicyonia dorsalis* (Decapoda: Penaeoidea), *J. Crust. Biol.*, 12:153–160, 1992; R. T. Bauer and J. W. Martin (eds.), *Crustacean Sexual Biology*, 1991; R. T. Bauer, and L. J. Min, Spermatophores and plug substance of the marine shrimp *Trachypenaeus similis* (Crustacea: Decapoda: Penaeidae): Formation in the male reproductive tract and disposition in the inseminated female, *Biol. Bull.*, 185:174–185, 1993.

Soil

A soil's ability to resist structural deterioration is one of its most important physical properties. An aggregate, a unit of soil structure, is a cluster or coherent association of primary particles (sand, silt, or clay) that has been cemented or bound together by organic or inorganic constituents. A quantitative measure of an aggregate's resistance to some applied disruptive force is its aggregate stability. Stability is usually measured by sieving aggregates in water, and is reported as the percent by weight of aggregates that remain clustered after sieving.

Aggregate stability. Aggregate stability affects the rate at which a soil erodes and the rate at which water infiltrates a soil's surface. A soil's susceptibility to erosion by either water or wind decreases as aggregate stability increases. Aggregate stability is also a quantitative measure of soil tilth. Soils with poor structure contain unstable aggregates. Such soils crust easily, compact readily, and may be poorly aerated. Crusting of the soil surface hinders or prevents seedlings from emerging, whereas sealing reduces infiltration and increases surface runoff; both conditions thus result in impaired crop production.

Deeper in the soil profile, unstable aggregates adjacent to existing soil pores can fracture when wetted, releasing primary particles and aggregate fragments that can constrict or obstruct pores. In contrast, soils with many large pores of diameters ≥ 1 mm (0.04 in.) contain relatively large stable aggregates. Large, continuous pores open to the soil surface increase the rate at which both water and air pass through the soil profile. A soil with a wide range of pore sizes is agriculturally productive because stable pores of all sizes are necessary for infiltration, drainage, water retention, root exploration, enhanced biotic activity, and gas exchange. Production practices that stimulate the formation, stabilization, and persistence of larger aggregates in

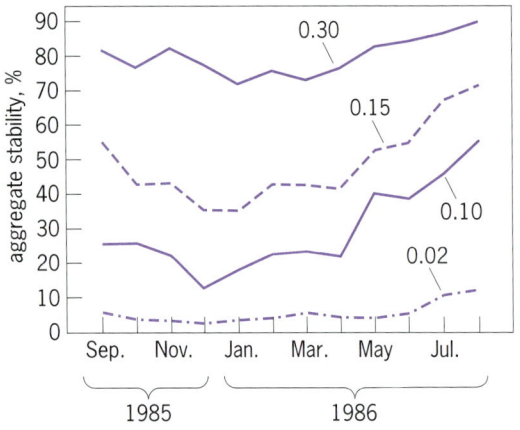

Aggregate stability as a function of time and presieved water content for a Portneuf silt loam. The values given represent kilograms of water per kilogram of dry soil. (*After M. S. Bullock, W. D. Kemper, and S. D. Nelson, Soil cohesion as affected by freezing, water content, time and tillage, Soil Sci. Soc. Amer. J., 52:770–776, 1988*)

soils result in greater productivity as well as resistance to crusting, erosion, and compaction.

Temporal variation. Aggregates are affected by many factors that cause stability to vary over time. Tillage, climatic processes (rainfall, freezing, and thawing), microbial activity, and crop residue management are all linked to temporal variation in aggregate stability. All must be managed or considered if soils are to support continued production of food and fiber.

Soil scientists have classified soils into different soil series based upon their kind and upon the arrangement of horizons; color, texture, and structure; and chemical and mineralogical properties. The **illustration** shows the annual variation in aggregate stability for one soil, a Portneuf silt loam, classified as a durixerollic calciorthid, from southcentral Idaho. In general, in early spring (April to May), aggregate stability increases rapidly from a winter low value. Through the remainder of the summer, aggregate stability continues to increase, but at a slower rate, to reach a maximum in August. In soils subject to freezing, aggregate stability in general decreases steadily from the fall to midwinter. The stability of Portneuf aggregates also varies within a season and from one year to the next. In short, changes in aggregate stability over time follow both short- and long-term trends, making characterization and prediction difficult.

Because of its correlation with erosion, aggregate stability has potential as a predictor of a soil's susceptibility to erosion. However, poorly characterized short- and long-term variation in the property has hampered attempts to employ it in that capacity. Better prediction of temporal variation in aggregate stability will greatly enhance its value for erosion prediction. Better understanding of the underlying processes affecting aggregate stabilization might lead to improvements of crop production practices or to irrigation techniques that would reduce erosion.

Physical and chemical factors. Many of the physical and chemical factors that cause aggregate stability to vary over time are related to climatic processes such as precipitation, evaporation, and freezing.

Wetting and drying. Wetting and drying affect the water content of an aggregate, which is important because water content is indirectly proportional to the aggregate's stability. Soils that have dried in the field are more stable, probably owing to inorganic bonding agents at particle-to-particle contact points. Drying also gathers and reorients clay particles and clay domains in a more orderly, parallel arrangement at the contact points between sand and silt particles. The water content of an aggregate also dictates the relative volume of air entrapped within the aggregate when it is rapidly wetted. Air trapped within an aggregate is first compressed, then forced to escape through the outer wetted and weakened layers, often breaking the aggregate. The illustration shows the dramatic effect that an aggregate's presieved water content (indirectly proportional to entrapped air volume) has upon its stability. Air entrapment during rapid wetting is obviously a primary cause of aggregate breakdown. Wetting and drying also cause swelling and shrinkage, respectively, thus generating stresses that can either break or form interparticle bonds.

Aggregates are wetted by irrigation, precipitation (rain or snow), or diurnal redistribution of soil water and dew. When aggregates at the soil surface are struck by drops of water, they can be weakened or broken, releasing primary particles or smaller aggregates that may seal the soil surface. A surface seal limits infiltration and can hasten surface ponding of water. When relatively dry aggregates are quickly hydrated, they are subjected to forces exerted by differential swelling, air entrapment, and localized heat of wetting. Aggregates in flowing runoff are subjected to the additional shear forces of the moving water and abrasion along the flow path. These forces develop, in the aggregates, planes of weakness along which they fracture when forces reach critical limits.

Diffusion and chemical precipitation. As aggregates dry, their stabilities often increase due to the diffusion and precipitation of cementing or bonding agents. As a soil dries, slightly soluble bonding agents, such as calcium carbonate, silica, or iron oxides, diffuse through the thin films of water that surround aggregates to the larger pockets of water that remain at the points of contact between primary particles or smaller aggregates. As drying continues, the concentration of the bonding agents in these water pockets increases. Ultimately, they precipitate, cementing the particles to one another. The steady increase in Portneuf aggregate stability from April through August (see illus.) results in part from this diffusion/precipitation process caused by repeated wetting and particularly drying over the summer.

Freezing and thawing. As a soil freezes, temperature differences cause liquid water to develop ice lenses through the unfrozen water films that surround soil particles. As this water flows, it carries some slightly soluble bonding agents to the contact points between soil particles. Once there, they precipitate, strengthening the aggregates. However, if a developing ice lens is near an aggregate, the enlarging lens can compress nearby constrained aggregates, either weakening or strengthening them depending on their water content, solute concentrations, and freezing history. In the portion of the soil profile supplying water to the ice lens, aggregates become drier and bonding agents precipitate at particle contact points. The fact that temperate soils often exhibit greater annual variation in aggregate stability than do tropical soils may, in part, be due to freezing and thawing. In addition, freeze-drying, a process in which ice sublimes from frozen aggregates on or near the soil surface, may affect the stability of aggregates from temperate soils in some areas.

Biological factors. Those factors that can cause variation include root growth, microbial activity, and residue management.

Root growth. Roots, as they elongate, move nearby aggregates and soil particles closer to one another. Contact points among them increase, thus providing more sites for potential stabilization. This process may account for some of the increase in aggregate stability often seen under no-till cultivation or in pastures. As roots and fungal hyphae grow, they quickly bind smaller aggregates into larger ones, which, if stabilized, improve soil structure and tilth.

As plants grow, roots withdraw large amounts of water from the soil to meet the plant's transpiration demand. This extraction of soil water by roots is yet another source of intensive and frequent soil drying. The precipitation process ensues, but in addition the water flowing to the roots concentrates the soluble bonding agents in the soil near the roots.

Microbial activity. Soil microbial activity intensifies in the spring and early summer as the soil warms and drains to field capacity. Microbes decompose plant residues, sloughed root cells, and root mucilage. In the process, they produce extracellular polysaccharides that, along with other organic compounds excreted by living roots, stabilize aggregates.

Residue management. Farmers can maintain high soil organic matter levels and increase aggregate stability by incorporating crop residues into the soil. The properties of the incorporated plant residue determine the extent of its effect on aggregate stability. If plant materials are easily decomposed, microbiological activity quickly produces extracellular polysaccharides that stabilize aggregates. As the growing season progresses, easily decomposable, carbonaceous material from plant residues may become limiting. Microbes then may use polysaccharides and other stabilizing materials as substrate, decreasing aggregate stability as the season continues. More recalcitrant plant residues can indirectly improve aggregate stability, but over a longer period of decomposition.

Cultural factors. The cultural factors responsible for temporal variation include tillage and irrigation.

Tillage. Disturbing soils by tilling them affects their structure according to the amount and kind of disturbance. As a tillage tool passes through a well-aggregated soil, it compresses some aggregates and shears others. These stresses either fracture aggregates or weaken them by creating potential failure planes. Aggregate stability often decreases immediately after tillage or cultivation. Tillage also influences aggregate stability by decreasing soil organic matter. The breaking of clods and mixing of soil by tillage dries the soil and increases its temperature, thus stimulating microbes to oxidize newly exposed organic matter. Tilling a wet soil is particularly damaging to the soil's structure and also increases its bulk density below the tillage depth. SEE AGRICULTURAL SOIL AND CROP PRACTICES.

Irrigation. Those aggregates on the soil surface are most likely to be affected by irrigation. Sprinkler irrigation affects aggregates in much the same manner as does natural rainfall. Raindrop or sprinkler-drop impact can disintegrate or weaken aggregates, leading to sealing of the soil surface. Surface irrigation, either furrow or flood, wets dry aggregates relatively quickly by immersing them in the water flowing over the soil surface. The escaping of entrapped air from within the aggregates often fractures them. Improper irrigation management or lack of drainage can also increase salt concentrations at the soil surface. Sodium accumulations disperse soil clays and destroy soil structure.

For background information SEE AGRICULTURAL SOIL AND CROP PRACTICES; EROSION; IRRIGATION (AGRICULTURE); SOIL; SOIL MICROBIOLOGY in the McGraw-Hill Encyclopedia of Science & Technology.

Gary A. Lehrsch

Bibliography. T. R. Ellsworth, C. E. Clapp, and G. R. Blake, Temporal variations in soil structural properties under corn and soybean cropping, *Soil Sci.*, 151:405–416, 1991; R. Horn and A. R. Dexter, Dynamics of soil aggregation in an irrigated desert loess, *Soil Till. Res.*, 13:253–266, 1989; G. A. Lehrsch and P. M. Jolley, Temporal changes in wet aggregate stability, *Trans. Amer. Soc. Agr. Eng.*, 35:493–498, 1992; J. M. Tisdall, Possible role of soil microorganisms in aggregation in soils, *Plant Soil*, 159: 115–121, 1994.

Solar system

The meteorites, the Earth, the Moon and probably all terrestrial planets are depleted in volatile elements (for example, potassium, rubidium, and lead) relative to solar or C1 chondritic abundances. Thus, the material forming the meteorites and planets has most likely been thermally processed at temperatures of about 1200–1400 K (1700–2100°F).

There are two possible mechanisms: either cold dust grains were heated by transient processes, in the solar nebula or during the accretion of the planets, and they lost volatile elements by evaporation, or the volatile elements never fully condensed from hot regions of the solar nebula onto the accreting dust grains. These two processes can be distinguished by heavy-isotope enrichment in residues of evaporation, but not during condensation, in the volatile elements. Such effects were searched for by high-precision measurements of the isotopic composition of potassium, an element that is significantly depleted in meteorites, the Earth, and the Moon relative to C1 chondritic abundances. The precision of measurement was sufficient to expose vaporization losses of 2% or more. No heavy isotopic enrichment of any solar system material has been found, with the exception of the lunar soils. Thus, vaporization of solids in the early solar system appears to have been an extremely restricted process, and most of the observable chemical effects in meteorites, the Moon, and the Earth were the result of condensation from a hot gas.

The formation of the solar system involved the collapse of a cold (about 10 K or –440°F) molecular cloud to produce a central star (the Sun) and an orbiting disk of gas and dust (the solar nebula), from which meteorites and planets formed. A plausible sequence of events includes the collapse of the cloud with resultant heating of the nebula; the destruction of interstellar dust grains by vaporization; the condensation of elements from the gas to produce new dust grains and possibly thermal processing of such grains; the accretion of grains to form small bodies (chondrite meteorites); and the accretion of small bodies to form planets (**Fig. 1**). Various theoretical models of the solar nebula yield thermal histories of matter ranging from 300 to 500 K (80 to 440°F) to temperatures sufficient to obliterate most preexisting grains (1400 K or 2100°F). Processes taking place on local scales in the nebula or during collisional aggregation of planets can significantly heat small quantities of dust to high temperatures, as well. The cumulative effects of thermal processes are recorded in the chemical and isotopic compositions of meteorites and planets, which are controlled by condensation of elements from the nebular gas or vaporization of solids. Condensation takes place only during grain growth from the nebular gas. Potential candidates for partially vaporized materials include interstellar dust that escapes complete destruction during nebular heating, condensates that have been melted by

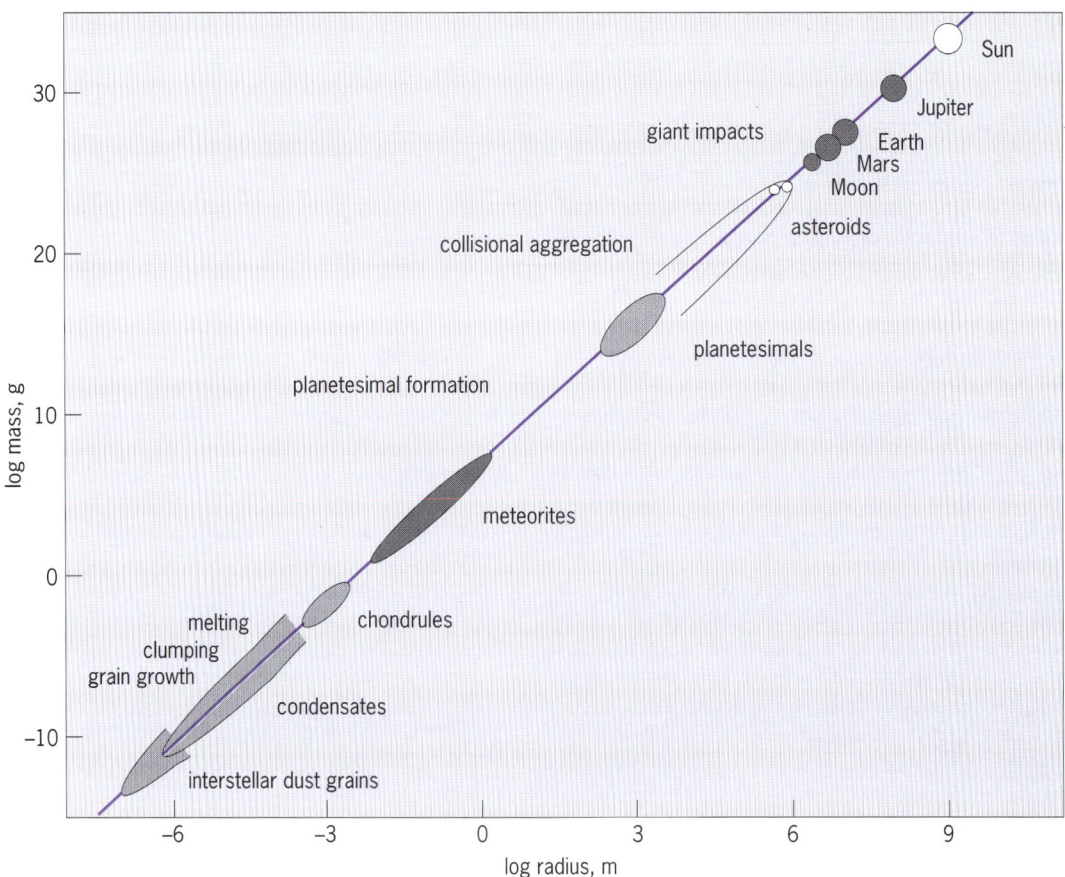

Fig. 1. Scale of solar system processes producing a range of materials from dust grains to planets. The materials are named on the right of the line and the processes responsible for their formation on the left. Meteorites are the fragments of asteroids (which in turn are surviving planetesimals), and are shown here to indicate the scale at which textural information is preserved and available for study.

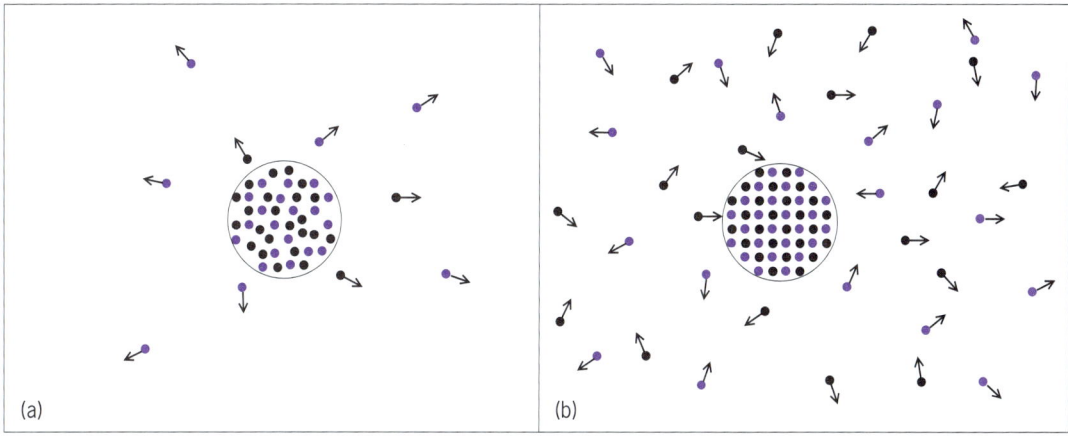

Fig. 2. Comparison of the isotopic separation taking place during (a) evaporation and (b) condensation. Isotopic molecules are shown as shaded spheres (black for heavy isotopes, colored for light isotopes).

brief, local heating events in the nebula, and impact molten planetesimals.

Processes of formation. The principal solids to condense from a hot gas of solar composition are magnesium silicates and iron-nickel metal. Elements that condense before these solids are termed refractory elements (for example, aluminum, calcium, titanium, strontium, zirconium, tantalum, rhenium, and uranium). Elements that condense after the magnesium silicates and iron-nickel metal are termed volatile elements and include manganese, sodium, potassium, silver, zinc, and sulfur. Some even more volatile elements are lead, mercury, carbon, nitrogen, and hydrogen, the last three condensing in the form of ices. The type 1 carbonaceous chondrites (C1) represent materials accreted from the solar nebula that contain all but the gaseous (hydrogen and noble gases) and icy (methane, ammonia, water) constituents in solar proportions. The condensation sequence is more or less reversed during the heating of chondritic (meteoritic) material to produce residues that are depleted in the more volatile elements.

A fundamental observation of the chemical compositions of the Earth, the Moon, and most classes of meteorites is that their compositions deviate from C1 abundances by being depleted in elements more volatile than magnesium silicates and iron-nickel metal. A first-order interpretation is that the materials constituting the planets and meteorites (exclusive of C1) have been thermally processed up to temperatures of 1200–1400 K (1700–2100°F). From chemical abundances alone, it is impossible to determine whether the various components that constituted chondrites (a stony meteorite containing chondrules) and planets never completely condensed a solar complement of volatile elements, or whether the volatile elements were partially lost by evaporation during high-temperature processes. It is possible to distinguish between these two types of processes by using isotopic criteria.

Condensation and evaporation. During condensation, a well-mixed gas containing molecules with two distinct isotopic species (light and heavy) is in equilibrium with a grain growing from the gas by addition of molecules to its surface. Since no physical process acts to separate molecules composed of the two isotopes, the grain inherits the isotopic composition (the proportion of light to heavy isotopes, shown in **Fig. 2** as equally abundant) of the gas. Now if this same grain were placed in a vacuum (a condition suitably approximated by a low-density gas), it would undergo thermal evaporation, that is, a net loss of molecules to space. Molecules at the same temperature have the same average kinetic energy; thus, molecules of the light isotope move faster than molecules of the heavy isotope. Molecules composed of the light isotopes, therefore, evaporate more readily from the grain surface than molecules composed of the heavy isotopes, and the residual grain becomes enriched in the heavier isotopes. The result is a kinetic isotope separation (Fig. 2).

Potassium isotopes. The relevant isotopic measurements have been performed for potassium (K), a volatile element condensing from 1200 to 1000 K (1700 to 1340°F), by secondary ion mass spectrometry. Analyses of representative samples from Earth's mantle and crust, and of seawater, indicate that there are no processes widely operating on Earth to fractionate potassium isotopes. A similar set of analyses found lunar rocks and various classes of meteorites to be isotopically identical within laboratory errors. The remarkable observation is that these sets are also identical to each other, providing powerful evidence against the operation of a cosmochemical process affecting the abundance of potassium (and other volatile elements) that is capable of fractionating the stable isotopes, ^{41}K and ^{39}K, to the extent of more than a few tenths of a per mill (‰).

The smallest detectable isotopic effect is calculated to arise from 2% evaporative loss of potassium. If chemical depletions in planets and meteorites are due to evaporation, isotope effects should have been produced in the range of several tens of per mill. Because no such effects are present, the process controlling volatile depletion is concluded to be incomplete condensation of volatile elements, presumably from a hot solar nebula. Three important implications emerge: (1) The process requires high temperatures (1200–1400 K or 1700–2100°F) in the inner solar system, which is contrary to viscous accretion disk models but required by proposed models that include compressional heating. (2) It requires that the chemical fractionations of parent-daughter systems such as uranium-lead and rubidium-strontium used in solar system chronology occurred in the nebula on time scales of a few million years and not during the longer period of accretional growth of planetesimals lasting about 100 million years. (3) It does not allow any known chondrite class to be the parental material of the Earth and other planets, since all known chondrite classes have higher volatile element abundances than the Earth, the Moon, and other planetary materials.

Lunar soils. The surface of the Moon is covered with a powdery soil, composed of finely ground rock and mineral fragments produced by micrometeorite bombardment. This pervasive bombardment fuses the fine portion of the soil to a glass that binds together tiny clumps of rock and mineral fragments termed agglutinates. There are also many submicrometer glass spheres produced by impact melting of the soil. The existence of large isotopic effects in many elements, including oxygen, silicon, sulfur, and potassium, in lunar soils was known early in the Apollo program. High-precision analyses of potassium by the new and more reliable technique have confirmed these isotopic effects in potassium, with heavy isotope enrichments of up to 13‰ (1.3%) found in bulk soils. These isotopic effects indicate the magnitude of heavy enrichment, because dilution by material never subjected to impact melting lowers the bulk soil isotopic compositions. The chemical depletions of potassium in the lunar soil are comparatively modest, at about 15% relative to lunar rocks. The lesson from the lunar soils is that, during vaporization processes, the magnitude of the heavy isotope enrichment is significantly more obvious than the chemical depletion. The production of potassium (and related sodium) vapor from the soils is a continuing process, and results in the presence of a tenuous atmosphere of sodium and potassium atoms around the Moon.

For background information SEE INTERPLANETARY MATTER; INTERSTELLAR MATTER; ISOTOPE; METEORITE; SOLAR SYSTEM in the McGraw-Hill Encyclopedia of Science & Technology.

Munir Humayun

Bibliography. A. P. Boss, Evolution of the solar nebula, II: Thermal structure during nebula formation, *Astrophys. J.*, 417:351–367, 1993; L. Grossman and J. W. Larimer, Early chemical history of the solar system, *Rev. Geophys. Space Phys.*, 12:71–101, 1974; M. Humayun and R. N. Clayton, Potassium isotope cosmochemistry: Genetic implications of volatile element depletion, *Geochim. Cosmochim. Acta*, vol. 59, 1995; S. R. Taylor, *Solar System Evolution: A New Perspective*, 1992.

Solution mining

Elemental sulfur and its product, sulfuric acid, are essential but largely unseen commodities in modern-day life. Sulfur can be derived from petroleum products, by roasting pyrite ore, or by Frasch production from native sulfur deposits.

Occurrence of sulfur. Sulfur is a very important industrial mineral. It has been said that its use is indicative of the status of a country's economy, a growing and vital economy requiring a great deal of sulfur as an additive to fertilizer and for production of sulfuric acid for use in producing fertilizer and in galvanizing metal. Deposits of elemental sulfur are found in solfataric deposits near volcanoes and hot springs and as native sulfur deposits replacing evaporite rocks of gypsum ($CaSO_4 \cdot 2H_2O$) and anhydrite ($CaSO_4$). The solfataric deposits have contributed only limited amounts of sulfur to the sulfur market. However, the evaporite-hosted deposits are a very important source of sulfur and occur in two settings: bedded evaporite deposits formed by drying up of ancient seas as represented by Permian age deposits (dating from 250 million years ago) in west Texas and New Mexico in the United States and in Poland, and Miocene age deposits (dating from 15 to 5 m.y.a.) in the Mediterranean area such as the Sicilian sulfur deposits. The other setting for evaporite-hosted sulfur deposits is in anhydrite cap rocks developed on salt domes in the Gulf Coast of Texas and Louisiana and in the crests of salt anticlines in the Isthmus of Tehuantepec of Mexico.

The genesis of evaporite-hosted sulfur takes place in the reaction below. Anaerobic bacteria, present in

$$CaSO_4 + CH_4 \xrightarrow{\text{bacteria}} H_2S + CaCO_3 + H_2O$$

ground water, attack anhydrite and gypsum and hydrocarbons [represented by methane (CH_4) in the reaction] to generate native sulfur and limestone ($CaCO_3$), thus replacing the evaporite units. The hydrogen sulfide (H_2S) produced is oxidized to native sulfur in the vugs and fractures of the limestone. This process requires a great deal of ground water and hydrocarbons, and the reaction takes place in sinkhole zones in the bedded evaporite deposits and in solution collapse zones in the salt dome caps.

Frasch process. Modern sulfur mining dates from the invention in the late nineteenth century of the Frasch process. In this process H. Frasch accomplished in-place sulfur mining by using superheated water. The process has come down to the present with only minor modifications, and it is currently being applied in sulfur deposits in west Texas, offshore Louisiana, Iraq, Poland, and Russia.

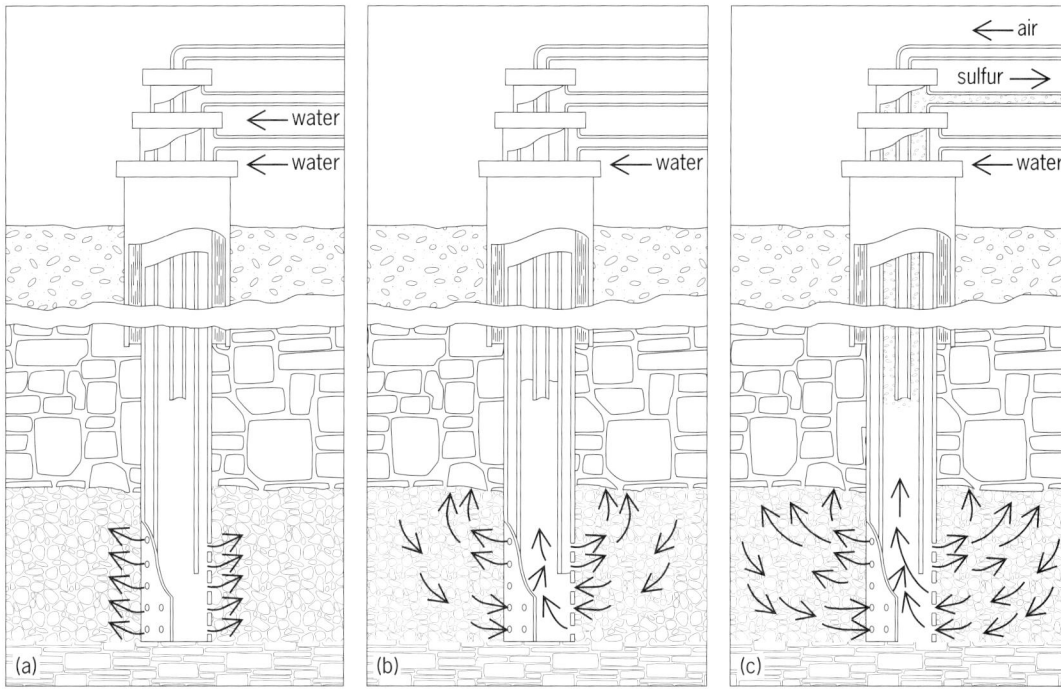

Sulfur production well in a Gulf Coast salt dome deposit. (*a*) Superheated water is pumped down the inner and outer casing strings. (*b*) Molten sulfur enters inner casing string. (*c*) Sulfur is jetted to the surface as a result of compressed air being pumped down the inner line.

Frasch mining begins by drilling wells into the sulfur deposit. Steel tubing (casing) is run into the drill hole to case off the barren overlying formation, and it is cemented in at the top of the sulfur deposit. Within this casing, three concentric strings of pipe are set within the sulfur deposit. The outermost string of pipe, 8 in. (20 cm) in diameter, is set to the bottom of the sulfur deposit. The lower 20 ft (6 m) of this pipe is perforated, providing access for drill-hole fluids to the sulfur deposit. Within this string of pipe is another, 4 in. (10 cm) in diameter, with its lower 10 ft (3 m) perforated. In the annulus between the two strings of pipe, a seal (or packer) is placed about 10 ft (3 m) from the bottom of the two casing strings. Thus, superheated water at 325°F (163°C) can be pumped down the annulus between the two strings of pipe to leave the casing string through the perforations above the packer, and to circulate in the sulfur deposit (**illus.** *a*). Native sulfur melts at 275°F (135°C) in a cone-shaped area of influence around the well, and because liquid sulfur is denser than water, it settles to the bottom of the well.

The pressure from the heated water forces the molten sulfur into the lower set of perforations and into the inner string of casing (illus. *b*). The molten sulfur rises to its hydrostatic head. This point is determined by practice, and a 0.75-in. (1.9-cm) titanium air line is set at this depth inside the 4-in. (10-cm) casing. Compressed air pumped through this inner tubing expands when it leaves the tubing, thus "jetting" the sulfur to the well collar, where it is collected in a tank (sulfur pan) [illus. *c*]. The illustration shows the procedure to initiate sulfur production by pumping hot water through the 8-in. (20-cm) and 4-in. (10-cm) casing, followed by changing to the production setup with sulfur and air traveling through the inner casing strings. The water used to mine the sulfur cools and is removed from the sulfur-bearing zone by bleed-water wells.

Main Pass Mine. A sulfur mine, then, consists of a power plant to produce the superheated water and compressed air, the field area over the deposit with the production wells, and liquid sulfur storage. Prior to the late 1970s, the sulfur was poured into forms, allowed to solidify, and then broken and shipped as solid sulfur. This procedure required considerable energy to remelt the sulfur at its point of consumption, and it also led to problems of fugitive dust. Shipping liquid results in fewer environmental problems and delivers the sulfur in a convenient form.

Freeport's Main Pass sulfur and oil deposit is located on the Main Pass Block 299 lease tract in the Gulf of Mexico, 30 mi (48 km) east of Venice, Louisiana. The mine facilities, consisting of 6 main platforms, 9 bridge supports, 13 bridges and 3 separate oil and gas platforms, are standing in 210 ft (130 m) of water. The discovery of the deposit was a significant geological triumph since it is the largest sulfur deposit discovered since the late 1960s. It is also a significant engineering triumph since the mine complex was designed for a 30-year mine life, to withstand winds of 180 mi/h (290 km/h) [category 5 hurricane], and to allow for subsidence associated with the sulfur extraction.

The Main Pass sulfur deposit consists of 6.7×10^7 long tons (6×10^7 metric tons) developed in a limestone cap rock that varies in thickness from 33 ft

(9.9 m) to more than 475 ft (143 m). The top of the limestone cap varies from 1234 ft (370 m) subsea to 1868 ft (560 m) subsea.

Frasch sulfur mining at Main Pass is being conducted from two production platforms. Each production platform has 76 slots, and each platform will support 15–20 continuously steaming wells with replacement wells drilled as needed. The power plant produces 10^{11} gal (4^{11} liters) of hot water per day. Daily sulfur production is targeted at 5500 long tons (5000 metric tons).

For background information SEE BORING AND DRILLING, GEOTECHNICAL; SALINE EVAPORITE; SALT DOME; SOLUTION MINING in the McGraw-Hill Encyclopedia of Science & Technology.

Alan F. Edwards

Bibliography. A. F. Edwards et al., *The Geology of the Main Pass Sulphur and Oil Deposit,* Amer. Inst. Mining Eng. Preprint 92-254, 1992; L. B. Green and J. R. Hooper, *Subsidence-Induced Sediment Motion: Implications for Production Facility Siting—Main Pass 299,* Offshore Technol. Conf. Paper 6664, 1991; J. C. Ruckmick, B. H. Wimberly, and A. F. Edwards, Classification and genesis of biogenic sulphur deposits, *Econ. Geol.,* 74 (2), 1979.

Sonoluminescence

When H. Becquerel exposed a photographic plate to a radioactive source in 1896, he attributed the darkening of the plate to rays that emanated from the metal compound. These radiations were later called radioactivity, and Becquerel was given credit for this important discovery. In a similar fashion, H. Frenzel and H. Shultes exposed a photographic plate to an intense sound field in 1934 and also observed a darkening of this plate, which they attributed to light emissions from the acoustic field. This production of light from sound came to be known as sonoluminescence. This phenomenon, which is associated with temperatures as hot as the surface of the Sun and pressures as high as those of the deepest ocean trench, can result in the production of exotic new materials and the destruction of harmful chemical compounds. Still not fully understood, it represents a rich but inexpensive new laboratory for the study of chemistry and physics.

Acoustic cavitation. A sound field propagating in a liquid is composed of periodic compressions and rarefactions that subject local regions to short intervals of negative pressure, or tensile stress, followed by similar-length intervals of positive pressure, or compression. Because these periodic fluctuations in the liquid pressure are not very large in an absolute sense (on the order of a few atmospheres of pressure or less), it is unreasonable to expect light emissions to result directly from them. Consequently, it has always been assumed that some other physical phenomenon, such as acoustic cavitation, is responsible for sonoluminescence.

Most liquids, particularly polar liquids such as water, possess large numbers of solid inhomogeneities that are not dissolved by the liquid. Embedded within these particles are small, microscopic gas pockets known as cavitation nuclei. When these pockets of gas are exposed to a pressure rarefaction that is larger than about 1 atm, the gas bubble grows very rapidly as the gas expands and the liquid interface evaporates. This bubble subsequently fills with vapor and grows explosively during the negative portion of the sound field cycle.

When the liquid pressure turns positive, the bubble is forced to smaller sizes. Because the pressure now becomes larger than the vapor pressure within the expanded cavity, this vapor condenses very rapidly, and the bubble begins to collapse. Only the small amount of gas that was contained within the original bubble prevents the bubble's implosion to zero size. The kinetic energy of the rapidly moving interface results in rapid heating of the internal contents of the bubble as the volume reaches its minimum value. Because the final size of the bubble is quite small, an increase in the concentration of energy by as much as 12 orders of magnitude results. Much of this mechanical energy is converted to thermal energy as the gaseous contents of the bubble are heated and compressed. Temperatures within collapsing cavitation bubbles have been indirectly measured to be on the order of 5000 K (9000°F) comparable to the surface of the Sun; furthermore, pressures of several kilobars (1 bar = 10^5 Pa $\cong$ 1 atm), larger than that under the deepest portion of the ocean, have also been indirectly measured. These high temperatures and pressures cause the internal contents of the bubble to luminesce and radiate light.

Sonochemistry. Such high temperatures and pressures within the liquid initiate chemical reactions that are otherwise not possible, producing localized regions of intense chemical activity. Sonoluminescence is thus associated with sonochemistry. A few developments in this area will be mentioned.

Acceleration of chemical reactions. The oxidation of potassium iodide into iodine, which might take hours in a conventional reactor, can be accelerated by ultrasound. The reaction time is reduced to a few minutes by the application of 20-kHz ultrasound, and to a few milliseconds by the combination of two different frequencies, 20 kHz and 1 MHz.

Destruction of harmful species. Chlorinated hydrocarbons, whose number and concentration in the atmosphere have increased by hundredfolds since the 1940s, are carcinogenic as well as damaging to the ozone layer. When aqueous solutions of these hydrocarbons are treated with ultrasound, significant breakdown to less hazardous chemicals results.

Production of exotic materials. When volatile metallic molecules are treated with ultrasound, the high temperatures and pressures within the cavitation bubble dissociate the individual elements and permit free metals to be formed. Because these high temperatures and pressures last for only a few nanoseconds, insufficient time is available for these materials to form crystalline solids. Consequently, amorphous metals such as iron are formed; such materials are extremely difficult to form in alternative procedures.

Multiple-bubble sonoluminescence. When an acoustic field is generated in the bulk of a liquid, many cavitation bubbles are generated that give rise to the high temperatures and pressures discussed above. Cavitation sustains itself through the slow diffusion of gas that is dissolved within the liquid into the expanding cavitation bubble. Consequently, when this bubble collapses, it typically shatters into several smaller bubbles, all containing small amounts of gas, these new nuclei initiate new cavitation bubbles, and so on. Eventually, the process reaches some steady state in which an equal amount of gas diffuses into the bubbles as dissolves back into the liquid; thus, the nuclei population remains relatively stable and the cavitation also reaches some steady-state behavior (see **illus.**).

It is suspected that the bubbles interact with each other and influence the collapse behavior. Furthermore, a collapsing interface is inherently unstable to the growth of perturbations on the surface, and thus it is suspected that the bubbles within a multiple-bubble cavitation region will not remain spherically symmetric during the final stages of collapse. These deviations from spherical collapse will probably result in the adiabatic heating of the principal contents of the bubble's interior. Furthermore, asymmetries in the collapse will inject liquid into the interior of the hot bubble, which will be heated to incandescent temperatures, thus allowing sonochemistry to occur within nonvolatile liquids. This behavior is very complicated and difficult to understand.

Single-bubble sonoluminescence. Recently, a form of sonoluminescence has been discovered that has provided new insight into this intriguing phenomenon. In 1988, it was discovered that, under certain specialized conditions, a single gas bubble could undergo a violent collapse each acoustic cycle without destroying itself in the process.

Some remarkable features of this bubble have been discovered. In particular, the duration of the sonoluminescence flash has an upper bound of only 50 picoseconds. This remarkably short time is difficult to explain on the basis of conventional understanding of acoustic cavitation. The best available theory expects a temperature elevation of a few thousand degrees for about 20 nanoseconds, some 400 times longer. Consequently, the conventional model of sonoluminescence, in which the interior of a bubble is heated by adiabatic compression, cannot be applicable.

It appears that the only plausible explanation for the short duration of the sonoluminescence flash is that an imploding shock wave is launched within the gaseous contents of the interior of the bubble sometime during bubble collapse. This imploding shock wave must stay spherically symmetric for a major portion of its collapse, and even more important, it must implode, rebound, and strike the advancing liquid interface to arrest its inward motion. It appears that only this imploding-shock-wave scenario will permit the bubble to oscillate repeatedly. This unusual scenario gives rise to some very interesting consequences.

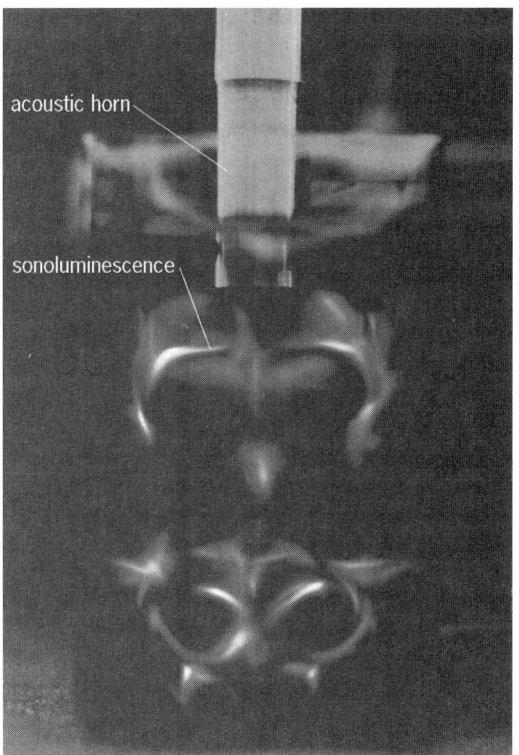

Sonoluminescence produced by ultrasonic acoustic horn driven at a frequency of 20 kHz. Because luminol was added to the water to produce more light in the visible region of the spectrum, these emissions are properly called sonochemiluminescence.

Picosecond synchronicity. If the time intervals between the sonoluminescence flashes are measured, they show a flash each acoustic cycle. Thus, if the acoustic frequency is exactly 20 kHz, the interval between flashes is exactly 50 microseconds. However, the variation in this interval is amazingly small, less than 50 ps, indicating a stability on the order of 1 part in 10^6. This stability is amazing, especially because the stability of a typical oscillator is normally not this precise. Somehow the sonoluminescing single bubble has a built-in self-stability feature.

Implosion thermodynamics. It is difficult to compute the thermodynamic behavior of the gas contained within the imploding shock wave. Computations suggest that the effective temperature should be on the order of 10^6 K, and the pressure on the order of several megabars. These quantities suggest the possibility of nuclear fusion under ideal conditions. Of course, it is difficult to obtain much reliable information concerning the thermodynamics of the bubble's interior.

Anomalous spectroscopy. The spectrum of multiple-bubble sonoluminescence in pure water displays emission bands characteristic of the hydroxyl (OH) free radical, a molecular species expected to be found within the heated gas-vapor mixture inside the collapsed cavitation bubble. Similarly, if a metallic salt, such as sodium chloride (NaCl), is dissolved in an aqueous liquid, the characteristic emission lines of sodium are seen. However, when the spectrum of single-bubble sonoluminescence is measured, only a broad-band spectrum similar to that of a blackbody is seen. No spectral bands characteristic of a particular species are observed. It appears that molecular species cannot exist within the imploding shock-wave core of single-bubble sonoluminescence.

Sensitivity to noble-gas doping. A pure nitrogen bubble emits relatively low levels of light in single-bubble sonoluminescence. However, if a trace amount of argon is inserted, for example, 0.1%, the luminosity significantly increases. If amounts corresponding to that of natural abundance are used, for example, on the order of 1%, the luminosity increases by more than an order of magnitude. This extreme sensitivity to noble-gas doping is difficult to explain.

Robust stability. Light-scattering measurements of the bubble radius have been obtained for several complete cycles of a bubble undergoing single-bubble sonoluminescence. These data demonstrate that the equilibrium radius of the bubble (the presumed value of the radius in the absence of a sound field) stays remarkably stable for many thousands of cycles. In fact, the system is so stable that once a bubble is made to undergo single-bubble sonoluminescence it will continue to emit light every cycle for as long as the principal variables such as acoustic pressure amplitude, driving frequency, liquid temperature, and dissolved gas content remain relatively fixed; stability periods of several hours are not uncommon.

For background information SEE CAVITATION; SHOCK WAVE; SONOCHEMISTRY in the McGraw-Hill Encyclopedia of Science & Technology.

Lawrence A. Crum

Bibliography. B. P. Barber and S. J. Putterman, Observation of synchronous picosecond sonoluminescence, *Nature,* 352:318–320, 1991; L. A. Crum, Sonoluminescence, *Phys. Today,* 47(9):22–29, September 1994; V. Kamath, A. Prosperetti, and F. N. Egolfopoulos, A theoretical study of sonoluminescence, *J. Acoust. Soc. Amer.,* 94:248–259, 1993; K. S. Suslick, Sonochemistry, *Science,* 247:1439–1445, 1990.

Space flight

In 1994, a number of significant events redefined the world's space programs. Dominating the news from the National Aeronautics and Space Administration (NASA) were the work of the newly repaired Hubble Space Telescope and the emerging International Space Station program's acquisition of Russia as a major partner. For both the United States and other countries involved in space flight, economic constraints reduced the number of new programs and made the need for international consolidation and cooperation more evident.

Despite program reductions, 1994 was one of the most active years in the United States quest to better understand the mechanisms that drive the climate and ecology of Earth and the effect of human activity on the environment since the inception of NASA's Mission to Planet Earth program in 1992. In astronomical exploration, important data were acquired by the Hubble Space Telescope and the *Compton Gamma-Ray Observatory* in Earth orbit, *Magellan* at Venus, *Ulysses* near the Sun, *Clementine* at the Moon, *Galileo* en route to Jupiter, and *Voyager 1* and *2* on their way into interstellar space.

Significant launches for 1994 are listed in **Table 1**, and the total number of launchings by the various countries and agencies are provided in **Table 2**.

United States Space Activity

The space shuttle continued to provide highly successful access to space for humans. At the same time, there were dramatic reshaping and redesign efforts in the space-station program.

Space shuttle. During the year, NASA launched seven shuttle flights, with a total flight time of more than 81 days. On these missions, which included the two longest ever flown in the program, the fleet of four orbiters lofted 42 astronauts into space, including crew members from Russia, Japan, and Europe. By the end of 1994, the total number of shuttle launches since *Columbia*'s first flight in 1981 had risen to 66.

STS 60. *Discovery,* from February 3 to 11, carried the first Russian cosmonaut aboard a United States spacecraft. The presence of Sergei Krikalev, who had lived 15 months in space on the *Mir* space station (*Soyuz TM-7* and *TM-12*), signaled the beginning of a three-phased cooperative effort between the United States and Russia in the development of the International Space Station. The crew attempted to deploy the first Wake Shield Facility (WSF), designed to create an extremely rarefied vacuum environment in its wake for growing thin semiconductor crystal films of gallium arsenide. The WSF could not be released from the robot arm, however, and another attempt will be made on a later mission.

STS 62. Carried into space by *Columbia* for 14 days from March 4 to 18 on its second flight, the U.S. Microgravity Payload (USMP 2) was a preview of the microgravity research work that will be conducted on the International Space Station. Its full complement of materials processing payloads included the Advanced Automated Directional Solidification Furnace for developing semiconductor materials, the Isothermal Dendritic Growth Experiment for the study of molten materials in the absence of gravity-driven fluid flows, and the French-supplied MEPHISTO furnace to process a bismuth-tin alloy in the study of temperature, velocity, and shape of solidifying materials in zero *g*.

STS 59. On *Endeavour,* the international Space Radar Laboratory (SRL) flew from April 9 to 20 on

Table 1. Some significant space launches in 1994

Mission designation	Launch date	Country	Main payload or mission
Soyuz TM 18	Jan. 8	Russia	Three-person Mir crew (fifteenth Mir visit)
Clementine	Jan. 25	United States	Lunar mapping probe, returning the first images from Moon since Apollo missions; failed flyby of asteroid Geographos
STS 60 (Discovery)	Feb. 3	United States	First Russian cosmonaut flying on a United States spacecraft
H-2	Feb. 4	Japan	First successful launch of the heavy-lift H-2 launcher, designed and built entirely in Japan
MILSTAR (DFS-1)	Feb. 7	United States	First launch of U.S. Air Force Titan IV/Centaur heavy-lifter, injecting heaviest payload ever in geosynchronous orbit, 10,000 lb (4500 kg)
Long March 3A (CZ)	Feb. 8	China	Sucessful maiden flight of China's heavy-lift CZ 3A to geosynchronous transfer, with large cryogenic third stage
STS 62 (Columbia)	Mar. 4	United States	Crew performed ground-breaking work for planned space station
Taurus	Mar. 13	United States	First launch of all-solid Taurus launcher developed by Orbital Sciences Corp.; two satellites, masses 1109 and 452 lb (503 and 205 kg)
STS 59 (Endeavour)	Apr. 9	United States	Six-person crew used advanced radar systems for environmental research as part of Mission to Planet Earth
GOES 8	Apr. 13	United States	The first United States geosynchronous meteorological satellite in 7 years, launched by Atlas-Centaur 73; mass of 4640 lb (2105 kg)
ASLV	May 4	India	First fully successful flight of India's ASLV launcher from Sriharikota; carried 250-lb (113-kg) SROSS C2 science satellite to orbit
Soyuz TM 19	July 1	Russia	Two cosmonauts, including one Kazakh, for a 126-day stay in Mir
STS 65 (Columbia)	July 8	United States	Crew of seven with first Japanese woman in space, conducted 82 microgravity experiments for more than 200 scientists worldwide
Long March 3A (CZ)	July 21	China	First commercial use of the Long March vehicle; launched communications satellite Apstar 1, owned by Chinese-controlled international consortium
STS 64 (Discovery)	Sept. 9	United States	Six-person crew used laser for atmosphere research, performed an untethered spacewalk for 6h 51min, the twenty-eighth of the shuttle program
STS 68 (Endeavour)	Sept. 30	United States	Second flight of international space radar laboratory for Mission to Planet Earth
Soyuz TM 20	Oct. 4	Russia	Crew of three to Mir, including European astronaut Ulf Merbold on his third space trip and the third Russian woman in the program
PSLV	Oct. 15	India	First flight of new Polar Satellite Launch Vehicle (PSLV), launching India's RS-P2 remote-sensing satellite
STS 66 (Atlantis)	Nov. 3	United States	Six-member crew, with a European astronaut, did atmospheric research and deployed-retrieved free-flying German satellite

the first of two flights, using two synthetic aperture (side-looking) radars and an atmospheric instrument in support of the Mission to Planet Earth. With multiple frequencies and polarizations of radar waves, images of land, water, snow, and ice surfaces were created. The data are being used in studies of Earth's water cycle, vegetation, volcanoes, and oceans. The flight also featured cell-biology experiments to help understand the effects of microgravity on growth of human bone and muscle cells in the weightlessness of space.

STS 65. The longest shuttle flight to date, the voyage of *Columbia,* lasted from July 8 to 23, carrying another major microgravity research project, the second International Microgravity Laboratory (IML 2). The seven-member crew of the Spacelab mission included the first Japanese woman in space. The crew conducted 82 experiments in areas such as materials science, fluid science, characterization of the zero-*g* environment, biology and bioprocessing, human physiology, and radiation biology.

STS 64. A new technique for remote sensing of Earth resources, used aboard *Discovery* from September 9 to 20, involved a powerful neodymium yttrium-aluminum-garnet (YAG) laser fired down through the atmosphere that measured the portion of laser energy reflected back up to the shuttle to observe clouds invisible to conventional weather satellites and to study the structure of a strong typhoon. The astronomy satellite *SPARTAN 201* was deployed and retrieved 40 h later. On an untethered 6-h 51-min spacewalk, the twenty-eighth of the shuttle program, two astronauts experimented with a crew rescue backpack called SAFER.

STS 68. The second Space Radar Laboratory flew on board *Endeavour* from September 30 to October 11 on a highly successful Mission to Planet Earth mission repeating many of the April investigations. This mission allowed observation of seasonal changes in different ecological settings. Measurements of carbon monoxide provided esti-

Table 2. Successful launches in 1994*

Country	Number of launches
Russia	49
United States (NASA, Department of Defense, commercial)	31
Europe (European Space Agency, Arianespace)	6
People's Republic of China	3
Japan	2
India	2
TOTAL	93

* Launches that achieved Earth orbit or beyond.

Fig. 1. New design of the International Space Station in its completed and fully operational state, with elements from the United States (front part of center core plus long truss and solar arrays), Europe (front left), Japan (front right), Canada (mobile servicer with robot arm, on truss), and Russia (rear part of core, smaller power truss and docked Soyuz-Progress vehicles). (*NASA*)

mates of the atmosphere's ability to cleanse itself of greenhouse gases.

STS 66. The mission of the shuttle *Atlantis* was the third in a series of Atmospheric Laboratory for Applications and Science (ATLAS) flights conducted in support of the Mission to Planet Earth program. During the flight from November 3 to 14, the lower concentrations of ozone and higher levels of ozone-depleting human-made chemicals in the Antarctic ozone hole were measured. The data clearly differentiated between artificially induced effects and those caused by natural atmospheric dynamics. The crew, which included a European, also deployed and later retrieved the German-built reusable atmosphere-research satellite *CRISTA-SPAS*.

International Space Station. In 1994 progress continued on the International Space Station program. A series of formal agreements brought Russia into the multinational partnership building the space station. A crucial review of the new space-station architecture was completed, the culmination of months of intensive work following President Clinton's order in February 1993 to substantially reduce the cost and time required to build the orbital laboratory.

February marked the end of the transition from the old Space Station *Freedom* program to a redesigned project (**Fig. 1**). In March, NASA managers, the international partners, and the contractor community conducted a systems design review involving a comprehensive look at the requirements, configuration, and maturity of the station's technical definition. In June, NASA and the Russian Space Agency RKA signed documents which put United States–Russian space cooperation on a firm basis and underpinned Russian participation in the program. Two packages of space-station hardware, each consisting of 45 solar energy panel modules, were shipped to Russia, and NASA took delivery of the spacecraft docking mechanism that will enable the space shuttle *Atlantis* to join up with the orbiting Russian *Mir* space station in 1995.

Astronomical observations. Several missions provided important findings for space sciences and astronomy.

Hubble Space Telescope. After NASA announced in January that the Hubble Space Telescope servicing mission aboard the space shuttle *Endeavour* in December 1993 had been a complete success, the orbiting observatory again turned its attention to the cosmos (**Fig. 2**). Results that touched on some of the most fundamental astronomical questions, such as the existence of black holes and the age of the universe, included compelling evidence for a massive black hole in the center of a giant elliptical galaxy 5×10^7 light-years away; observations of great pancake-shaped disks of dust (raw material for planet formation) swirling around at least half of the stars in the Orion nebula, evidence that the

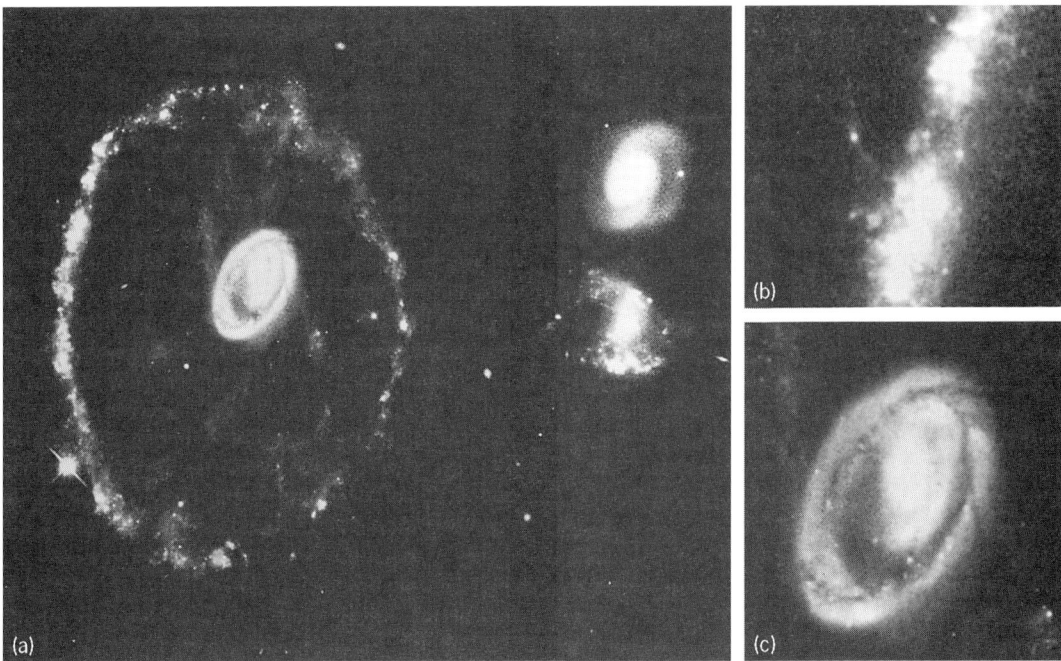

Fig. 2. Hubble Space Telescope's view of a ring world born in a head-on collision of two galaxies: the Cartweel Galaxy in the constellation Sculptor, 5×10^8 light-years away. (a) Galaxy and neighbors. The ring, containing several billion new stars not normally created in such a short time span and measuring 150,000 light-years across, resulted from a smaller galaxy (possibly one of the two objects on the right) careening through the core of the originally spiral-shaped host galaxy. (b) Details of knot-like structure of the ring. (c) Details of dust-rich core. (NASA)

process which may form planets is common in the universe; detection of primordial helium, confirming a critical prediction of the big-bang theory, that the chemical element helium was widespread in the early universe; discovery of a new quasar, not billions of light-years away like its known cousins, but a mere 6×10^8, hidden behind a dark band of dust around the nucleus of the galaxy Cygnus A; results in the determination of the age and size of the universe, showing it to be $8–12 \times 10^9$ years old, apparently conflicting with the age of some stars known to be much older; evidence ruling out a leading explanation for so-called dark matter, thought to make up over 90% of the mass of the universe and indicating that any invisible matter, required to explain the slowing expansion of the cosmos, probably consists of exotic subatomic particles or other unknown material; identification of primeval galaxies forming less than 10^9 years after the big bang as well as views of the earliest moments of the universe revealing a cosmic zoo of bizarre fragmentary objects in a remote cluster that are the likely ancestors of the Milky Way Galaxy; observations of the impacts of Comet P/Shoemaker–Levy 9 with Jupiter; and images of the planets Uranus and Saturn, revealing several of Uranus' 11 concentric rings, five of its inner moons, and bright clouds and a high-altitude haze above its south pole, as well as an Earth-size storm in the atmosphere of Saturn. SEE COMET.

Ulysses. The NASA–European Space Agency (ESA) spacecraft *Ulysses*, the first probe to explore the Sun's environment at high latitudes, completed the first phase of its primary mission of exploring the complex forces at work in the polar regions of the Sun when it overflew the Sun's southern pole on November 5, continuing a sweeping path that would pass over the north pole in June 1995. Surprising findings included the very high velocity of about 470 mi/s (750 km/s) of the solar wind flowing from the polar region, nearly double its speed at lower latitudes, the lack of clear evidence of a magnetic pole, and the unexpectedly low intensity of cosmic rays in the polar region.

Galileo. The deep-space probe *Galileo* continued to operate normally (though constrained to transmit its data via its low-gain antenna), remaining on schedule to reach Jupiter on December 7, 1995, when its probe will descend into the Jovian atmosphere. *Galileo* will then conduct 2 years of observations of Jupiter, its satellites, and its magnetosphere. Unlike all near-Earth observatories, *Galileo* had a direct view of the Shoemaker–Levy 9 impact sites on Jupiter, and its solid-state imaging system was able to take the only pictures of the impacts at the time they occurred. Data transmitted in March revealed that *Galileo* had discovered a natural satellite of the asteroid Ida during its flyby in August 1993. The tiny moon, about 1 mi (1.6 km) in diameter, was subsequently named Dactyl by the International Astronomical Union.

Magellan. NASA's *Magellan* spacecraft, which was launched in 1989 and had mapped 98% of the planet Venus with its synthetic aperture radar since September 1990, entered its last phase by conducting a unique experiment designed to return data about the upper atmosphere of Venus and the behavior of the spacecraft entering it. Its winglike solar arrays were turned in opposite directions, like windmill sails, to encounter pressure from molecules in the upper atmospheric regions; the measured torque needed to prevent the spacecraft from spinning on its axis provided information on aerodynamics and gas-surface interactions around Venus for future mission designs involving aerobraking maneuvers. On October 12, radio contact with *Magellan* was lost when the spacecraft started its spiraling descent, presumably burning up in the Venusian atmosphere within 2 days.

Compton. The *Compton Gamma-Ray Observatory,* launched April 5, 1991, continued its nearly flawless operations. One of its instruments discovered an unusually bright x-ray source, one of the three brightest in the sky, later named x-ray Nova Scorpii or GRO J1655-40, in the constellation Scorpio. The discovery led to further observations by radio telescopes that showed ejections of matter at velocities close to the speed of light. X-ray novae may be caused by matter spilling from a normal star onto and sucked up by a black hole. Throughout the year, *Compton* detected gamma-ray bursts at random locations, challenging current understanding, notably with regard to the isotropic distribution of these mysterious sources. SEE GAMMA-RAY BURSTS.

Clementine. Launched in January, the small spacecraft *Clementine*, built by the Naval Research Laboratory, provided the first images from the Moon since the Apollo missions. By using four cameras and a lidar laser transmitter, *Clementine* mapped the entire Moon in 11 spectral bands at spatial resolutions of 600–1000 ft (183–305 m), complete with altimeter data. *Clementine*'s second mission objective, to provide similar imaging during a flyby of the asteroid Geographos, was not accomplished because of premature mission termination due to an attitude control anomaly. SEE MOON.

Extreme Ultraviolet Explorer (EUVE). For the first time, an orbiting astrophysics satellite, the *EUVE* spacecraft, launched by NASA in 1992, was controlled by an artificial intelligence computer program. This expert system allows the *EUVE* science operations center to remain entirely unstaffed for extended periods. During these autonomous operations, the computer conducts maintenance tests on the *EUVE* science instrument. When Shoemaker–Levy 9 collided with Jupiter, the satellite detected neutral and ionized (charged) helium in its atmosphere, made visible by the impact energy.

Voyager 1 and 2. The two far-traveling probes continued their departure from the solar system into interstellar space, continuing to take data on fields and particles after 17 years of operation. By the end of 1994, *Voyager 1* had reached a distance of 5.6×10^9 mi (9×10^9 km) from Earth, and *Voyager 2* 4.4×10^9 mi (7.1×10^9 km). *Voyager 2* will pass *Pioneer 10* in early 1998 to become the most distant artificial object in the solar system. Both spacecraft are expected to transmit until at least 2015.

Wind. A small spacecraft called *Wind* was launched by NASA on November 1 into a highly elliptical Earth orbit for measuring the basic properties of the solar wind as it interacts with the Earth's magnetic field and atmosphere. The main scientific goal of the mission is to measure the mass, momentum, and energy of the solar wind that somehow is transferred into the space environment around the Earth.

Mission to Planet Earth. Besides the missions discussed above as part of the space-shuttle program, NASA, on behalf of the National Oceanic and Atmospheric Administration (NOAA), launched *GOES 8* on April 13, the first in a series of next-generation weather satellites. Capable of much longer and more precise atmospheric measurements than its predecessors, the NOAA spacecraft enables weather forecasters to more closely track severe storms over land and sea. Another environmental mission came to an end, when NASA's *Nimbus 7* satellite was retired in October after more than 15 years of operation. Its most visible success had been the Total Ozone Mapping Spectrometer (TOMS), which provided the first full view of the Antarctic ozone hole.

Department of Defense activities. The first of four MILSTAR communications satellites, designed for military communications with highy reliable jam-proof, low-data-rate capabilities during an all-out nuclear war, was launched in February on the first Titan IV/Centaur. The 10,000-lb (4500-kg) spacecraft represented the heaviest payload ever carried into geosynchronous orbit by a United States expendable launch vehicle.

The Department of Defense completed its NAVSTAR Global Positioning System (GPS) constellation with the launch of the twenty-fourth operational GPS satellite on March 9. Also in March, the U.S. Air Force launched the first in its series of Space Test Experiment Platform (STEP) satellites on the first flight of the medium-size launch vehicle *Taurus*, a four-stage solid-propellant rocket. For the second launch of the Department of Defense's *Miniature Sensor Technology Integration* (*MSTI*) satellite in May, NASA used the 118th and final vehicle of its highly successful Scout series of small launchers. (*MSTI* failed during orbital checkout.) SEE APPLICATIONS SATELLITES.

The *DC-X,* an innovative vertical launch-and-lander demonstrator built as an early experiment toward eventual reusable single-stage-to-orbit vehicles, completed its fourth successful flight in June, reaching an altitude of 2800 ft (850 m) in a flight lasting 136 s. The fifth flight, launched June 27, had to be terminated prematurely because of damage to the aeroshell caused by an explosion in the ground-support equipment. Further research efforts were turned over to NASA.

In December, the Air Force launched the *NOAA 14* weather observing satellite. The new spacecraft joined the *NOAA 11* and *12* satellites in polar orbit, where they have been providing data for weather prognosis and atmospheric research. It fills the void left by the failure of *NOAA 13* shortly after launch in mid-1993.

Commercial space activities. With the launch of the final operational GPS satellite on March 9, commercial use of this high-precision locator system took an upward swing with truckers, packaging companies, car manufacturers, and other users seeking new ways of applying space-based navigation technologies. SEE SATELLITE NAVIGATION SYSTEMS.

United States commercial companies launched eight expendable launch vehicles in 1994: 5 Atlas and 3 Delta II vehicles. On the international market, these two types are encountering three competitors: Europe's Ariane, Russia's Proton, and China's Long March. Even with five options, international demands for additional communications transponders will be difficult to satisfy quickly by the world's supply of launch vehicles.

Of particular commercial interest are satellite communications for mobile applications. By mid-1994, more than 20 major so-called mobile comsat (communications satellite) systems were proposed, some already in advanced planning. Iridium, a major mobile space-borne system, will consist of 66 satellites providing cellular phone service.

Russian Space Activities

Even with severe political and economic problems still pervading most of the former Soviet Union, Russia again succeeded in launching more spacecraft in 1994 than all other nations and entities combined: 49 versus 44. Thus, the space-station partnership with the United States, initiating cooperative operations in space, must be regarded as a significant accomplishment on both sides. Cosmonaut Krikalev's presence on board shuttle mission STS 60 opened the door to more joint activities in coming years.

Space station Mir. By the end of 1994, the *Mir* space station had been in orbit for 3237 days, commencing in February 1986. During that time, it circled Earth approximately 50,670 times, in an orbit inclined 51.65° to the Equator. Counting from its last brief period of nonoccupancy (September 1989) to the end of 1994, *Mir* has been inhabited continuously for 1943 days. Since its inception, it has been visited 17 times by two- to three-person crews. On January 10, *Soyuz TM 18* arrived with three cosmonauts, one of whom was still in space at the end of 1994. The other two returned to Earth after a six months' stay, shortly after the arrival, on July 1, of *Soyuz TM 19* with two new occupants. After *Soyuz TM 20* followed on October 4 with three more cosmonauts, including the German Ulf Merbold for ESA, the *TM 19* crew returned to Earth after 4 months on November 4, accompanied by Merbold (31 days). At the end of 1994, three occupants remained in *Mir,* one the third Russian woman in space (Elena Kondakova). In other space-flight areas, Russia continued its numerous launches of military, scientific, and telecommunications satellites, among them the new large data-relay satellite *Luch,* required for upcoming joint United States–Russian operations on *Mir,* on a Proton rocket on December 16. This heavy-lifter alone was used in a total of 13 launches in 1994, of which 10 carried geosynchronous payloads.

European Space Activities

Increasing economic constraints on the program of the ESA gave rise to growing concerns regarding Europe's future ability to maintain its commitment to the international space station with its *Columbus Orbital Facility* segment. Arianespace continued its French-based operation as the world's first commercial space transportation company, even though two of its eight launches in 1994 were failures (losing three satellites), while the payload of a third launch, *Telstar 402,* died shortly after reaching orbit on September 8. Successfully orbited payloads were an Intelsat VII satellite on June 17, *PAS 2* (*PanAmSat*) and *BS 3N* (Japan) on July 8, *Brasilsat B1* and *Turksat IB* on August 10, *Solidaridad 2* (Mexico) and *Thaicom 2* (Thailand) on October 7, and *Astra 1D* (Europe) on October 31. SEE COMMUNICATIONS SATELLITE.

Asian Space Activities

Japan, the People's Republic of China, and India continued their space activity in 1994.

Japan. Japan became a full participant in space flight in 1994 with two faultless launches of its powerful H-2 heavy-lift launch vehicle, the first on February 4, the second on August 28. The launches took place at the new Yoshinobu complex on Tanegashima Island. Unlike its predecessors, the N series and the larger H-1 vehicle (which were based on versions of the United States Delta rocket), the H-2 was designed and developed entirely by Japanese technology. Its first, or core, stage is powered by a cryogenic (liquid hydrogen-liquid oxygen) rocket engine quite similar to the United States space shuttle main engine, called the LE-7, and assisted by a pair of solid-propellant rocket boosters strapped to its sides. The upper stage, taken from the H-1, is also cryogenic, making the H-2 the world's only expendable launch vehicle with both liquid stages powered by liquid oxygen-liquid hydrogen.

Payloads on the maiden flight were an Orbital Reentry Experiment Vehicle (OREX) to test ceramic heat-shield tiles and GPS navigation for the planned HOPE spaceplane and a vehicle evaluation instrument package, totaling 7200 lb (3265 kg). Geosynchronous insertion of *ETS 6* (*Engineering Test Satellite*) on the second flight failed due to a problem with its kickstage. The H-2 is currently not economically competitive with the Atlas 2 or Ariane 4, but cost reduction will clearly be a priority in future space-flight activity.

China. The People's Republic of China joined the major space-transportation suppliers when its heavy-lift *Chang Zheng* (*Long March*) *3A* performed successfully on its maiden flight on February 8, using a CZ-2 with a stretched first stage and a new large cryogenic third stage and carrying a dummy mass and a Chinese scientific satellite. A second flight took place on July 3. The first commercial mission of a Long March (since December 1992 when a launch failure destroyed Australia's *Optus B2*) followed on July 21 with the launch of *Apstar 1,* owned by a Chinese-controlled international consortium. The replacement *Optus B3* was launched on August 27. A fifth flight took place on November 29 with the launch of the 4630-lb (2100-kg) *DFH-3* telecommunications satellite.

India. India launched its first fully successful medium-lift Augmented Space Launch Vehicle (ASLV) from Sriharikota in Anohra Pradesch province on May 4. It was already the fourth ASLV launch, since the first two (in 1987 and 1988) both failed and the third attempt in 1992 resulted in early mission termination because of insufficient fourth stage spin-up. The ASLV carried the 250-lb (113-kg) *SROSS C2* satellite (Stretched Rohini Satellite Series). Also successful was India's first launch, on October 15, of its new *Polar Satellite Launch Vehicle* (*PSLV*), designed to carry Sun-synchronous satellites into orbits over Earth's poles. The vehicle has a solid-propellant first stage with six strap-on boosters, a liquid-propelled second stage, and a solid third stage.

For background information SEE COMMUNICATIONS SATELLITE; METEOROLOGICAL SATELLITES; REMOTE SENSING; SATELLITE ASTRONOMY; SATELLITE NAVIGATION SYSTEMS; SPACE BIOLOGY; SPACE FLIGHT; SPACE PROBE; SPACE PROCESSING; SPACE SHUTTLE; SPACE STATION in the McGraw-Hill Encyclopedia of Science & Technology.

Jesco von Puttkamer

Bibliography. R. J. Cochetti, Mobile satellite services, *Via Satell.,* 9(11):26–36, November 1994; *Jane's Space Directory, 1994–1995;* Results from Hubble Space Telescope highlight 1994, *NASA News,* Release 94-216, December 20, 1994.

SQUID

The recent development of a viable technology to produce high-temperature superconducting (HTS) Josephson junctions (operating at 77 K or –321°F) has enhanced the opportunity to utilize SQUID magnetometers for biomedical applications and nondestructive evaluation. A SQUID (superconducting quantum interference device) has at its core a superconducting loop interrupted by either one or two weak-link Josephson junctions. A Josephson junction is formed when correlated pairs of superconducting electrons are able to tunnel through a thin insulating barrier, maintaining phase coherence in the process. This unique superconducting property gives rise to observations of macroscopic quantum tunneling, and somewhat more importantly has

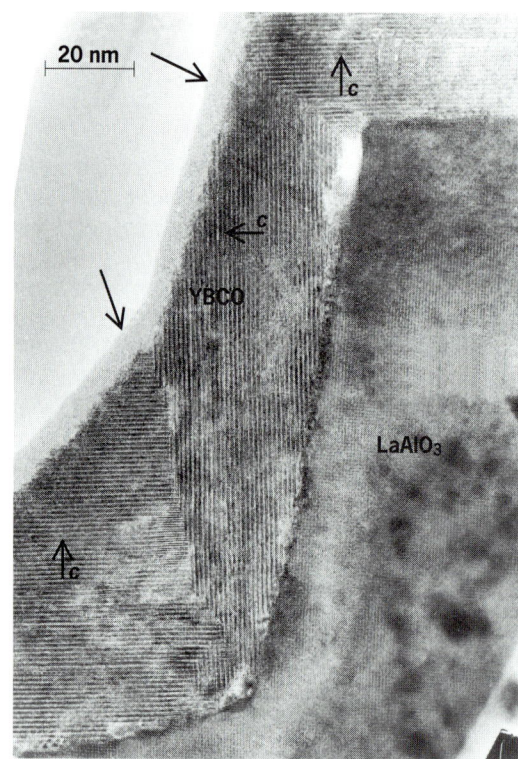

Fig. 1. Scanning electron micrograph of a grain-boundary, step-edge Josephson junction. Yttrium barium copper oxide (YBCO) film is deposited on a lanthanum aluminate (LaAlO$_3$) substrate with large-angle step. Directions of crystallographic *c* axis are indicated, showing changes in orientation near the bottom and top of the step. Arrows indicate discontinuities where this change occurs. (*After A. I. Braginski, KfA Juelich*)

led to the creation of SQUID-based magnetometers that are at least three orders of magnitude more sensitive than those based on any other physical phenomenon.

Fundamentally, a magnetic field produces a circulating current in a SQUID loop in which each Josephson junction is resistively shunted. When the voltage across the SQUID loop is suitably amplified by external (room-temperature) circuitry and the SQUID is operated in a feedback mode to maintain a constant local magnetic field environment, this voltage provides a precise measure of the ambient magnetic field. Low-temperature superconducting (LTS) SQUIDs (operating at 4 K or –452°F) have been fabricated which have a sensitivity to magnetic fields that is within a factor of three of the limit imposed by Heisenberg's uncertainty principle. In rugged commercial LTS SQUID systems, typical sensitivity (limited by thermal noise in the room-temperature part of the circuit) is about 1–10 femtotesla/Hz$^{1/2}$ over a broad frequency range of 1–10^5 Hz. By comparison, Earth's magnetic field is about 10^{-5} tesla, that is, 10^{10} femtotesla. This exquisite sensitivity has given rise to the relatively new field of neuromagnetometry, which recently has proven invaluable to neurosurgeons in their preparations for the removal of cancerous tumors, as well as of brain tissue in epileptics.

Practical considerations. If the SQUID were not properly shielded magnetically, it would be impractical to take advantage of its sensitivity. Normally, the SQUID loop itself is encased in a close-fitting superconducting enclosure which excludes all magnetic fields except that produced by a coil of superconducting wire introduced through a small opening in the enclosure. In a nearby unshielded region at the other end of this transformer coil are a few turns of wire, the sensing coils, placed as close as possible to the bottom of the (cryogenic) dewar vessel. The bottom of the dewar is placed as close as possible to the magnetic field source of interest, for example, a human head or a section of steel. Quite often these sources are much weaker than the Earth's field or that produced by other nearby magnetic field sources.

Although expensive magnetically shielded rooms are sometimes utilized for the most demanding applications, such as in neuromagnetometry, the standard method used to eliminate the effect of all unwanted background fields (in almost all cases) is to form the sensing coils into a configuration that measures only the gradient of the ambient field. To achieve this result the top and bottom turns of the sensing coils are wound in opposite directions. Thus, a field source that is uniform over the length of this gradiometer produces no net current in the coil circuit. Conversely, a field source that is close to the bottom of the gradiometer sensing coils induces a relatively large current in the bottom turns of wire and a relatively small current (moving in the opposite sense) in the upper turns of wire.

The key to keeping this gradiometer sensitivity high is to keep the standoff distance between the field source (under observation) and the gradiometer small compared to the gradiometer baseline, that is, the separation between the top and bottom turns. In a typical LTS SQUID system, the bottom turn of the gradiometer sensing coil is 0.5–1.0 cm (0.2–0.4 in.) from the exterior bottom surface of its dewar, and its baseline is 5–10 cm (2–4 in.) long. In cases for which the signal-to-noise ratio of the field source is not very high, the distance from the bottom of the sensing coil to the bottom of the dewar may be an inhibiting factor.

Development of HTS SQUIDs. After the discovery in 1987 of a superconducting phase of yttrium barium copper oxide, $YBa_2Cu_3O_7$ (YBCO), at a temperature above 90 K (−298°F), and hence above that of liquid nitrogen (77 K or −321°F), an intensive worldwide effort for 4–6 years ensued before a thin-film technology was developed to produce reliable Josephson junctions. At first it was difficult just to produce good-quality films with a high-enough superconducting transition temperature and a high-enough current-carrying capacity. It was soon discovered that the characteristic distance, the coherence length, over which a superconducting electron pair is correlated is on the order of only 1 nanometer, which is one to two orders of magnitude less than for LTS materials. Thus, for HTS films the weak-link insulating region in a Josephson junction must be made much narrower. Another complication is that YBCO and other HTS ceramic materials are highly anisotropic, with superconductivity most favored in the basal a-b copper-oxide plane of the orthorhombic structure. A number of ingenious methods have been developed to circumvent these daunting materials problems. Yet, no universally best way to make an HTS Josephson junction has been recognized, and some methods are not truly manufacturable. That is, the junctions can be made only on an individual basis, but not mass produced. One of the more effective techniques results in the step-edge junction (**Fig. 1**), in which YBCO is deposited on the steep slope and flat surfaces of a lanthanum aluminate ($LaAlO_3$) insulating substrate. The changes in orientation of the a-b planes (perpendicular to the c axis directions indicated) of the YBCO near the bottom and top of the step give rise to effective grain-boundary Josephson junctions.

Application of HTS SQUIDs. Manufacture of practical HTS SQUID instruments has just begun. However, since there is no suitable HTS wire technology available, fielding a usable instrument has been challenging. It has been possible to fabricate planar

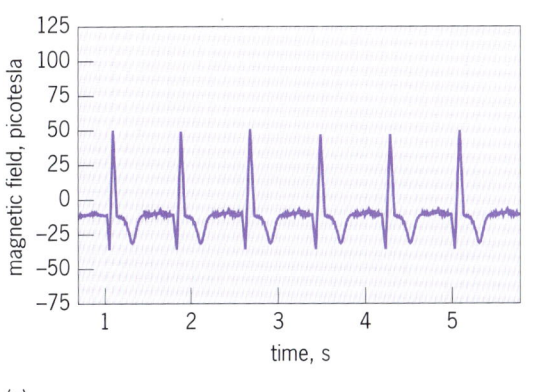
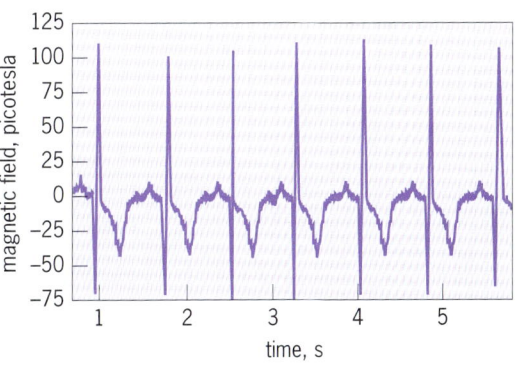

Fig. 2. Comparison of magnetocardiograms taken by (*a*) low-temperature superconducting (LTS) and (*b*) high-temperature superconducting (HTS) SQUIDs. (*After A. I. Braginski, KfA, Juelich*)

coils and planar gradiometers, with pressure contacts to bond a gradiometer-coil chip to a SQUID chip. Another method is to stack two or three planar HTS SQUIDs above each other, connected by copper wires, while using a signal-subtraction method to effect gradiometerlike results. An important feature of all these arrangements is that the SQUID systems have quite good magnetic-field sensitivity, often in the range of 20 fT/Hz$^{1/2}$, that is, within an order of magnitude of the best commercial LTS SQUIDs. More importantly, the sensing coils and SQUIDs can be placed closer to the chosen magnetic field source, resulting in higher output signals than for an LTS system with higher inherent sensitivity. The relative sensitivity of the HTS system is illustrated in **Fig. 2**, in which a magnetocardiogram has been taken on the same individual in the same environment. While the noise sensitivity of the LTS SQUID (25 fT/Hz$^{1/2}$) is almost half that of the HTS SQUID (40 fT/Hz$^{1/2}$), the response of the HTS system is comparable. Another important feature of HTS SQUIDs is that they can be hand-held and moved about at will and in any orientation for nondestructive inspections. The first commercial portable HTS SQUID system can be operated all day without refilling the small liquid-nitrogen dewar.

Such systems will soon find routine use in the monitoring of fetal heartbeats, in surveys of steel plates for signs of mechanical aging, and in the search for subsurface cracks and corrosion in the wings and bodies of aircraft. This last application will be extremely welcome because conventional eddy-current inspection techniques are limited to a lowest frequency of about 200 Hz, corresponding to a skin depth of less than 1 cm (0.4 in.) in aluminum, the major structural material used in aircraft manufacturing. SQUIDs may be used with no loss in sensitivity down to 1 Hz, corresponding to a 10-cm (4-in.) skin depth. With greater sensitivity and greater depth perception, HTS SQUIDs should soon be instrumental in ensuring the airworthiness of the aging aircraft of both the military and commercial fleets.

A promising application of SQUIDs in the medical world is the survey of magnetic field patterns associated with heart arrythmias, possibly replacing an invasive technique which places a patient at risk. Another potential application exists for exploration geophysicists. In the past, LTS SQUIDs had been used in geophysical prospecting for oil and minerals; they were abandoned primarily because of the difficulty and cost of transporting liquid helium to remote locations. This barrier is now removed with the advent of HTS SQUIDs in a smaller, technologically friendlier package. Either closed-cycle refrigeration may be used to refrigerate an HTS SQUID directly, or a small nitrogen liquefaction unit (pulling nitrogen from the air) may be operated locally.

For background information SEE BIOMAGNETISM; NONDESTRUCTIVE TESTING; SQUID; SUPERCONDUCTING DEVICES in the McGraw-Hill Encyclopedia of Science & Technology.

Harold Weinstock

Bibliography. W. G. Jenks, S. S. H. Sadeghi, and J. P. Wikswo, Jr., Use of SQUIDs in nondestructive evaluation, *J. Phys. D: Applied Physics*, 1996; H. Weinstock (ed.), *SQUID Sensors: Fundamentals, Fabrication and Applications*, Proceedings of NATO Advanced Study Institute, 1996; H. Weinstock and R. W. Ralston (eds.), *The New Superconducting Electronics*, 1993; J. P. Wikswo, Jr., SQUID magnetometers for biomagnetism and non-destructive testing: Important questions and initial answers, *IEEE Trans. Appl. Supercon.*, vol. 5, no. 2, June 1995.

Step pools

Steps and pools are the characteristic bedforms that dominate the channel morphology of steep mountain streams. The steps are generally composed of cobbles and boulders; they are separated by much finer materials forming the pools. Steps and pools alternate in the stream bed to produce a characteristic, repetitive sequence of bedforms, with a stepped longitudinal profile resembling a staircase (**Fig. 1**). Step pools serve a fundamental role in river systems because they provide an important measure of resistance in the channel. Stream energy is dissipated as water flows over the step elements and plunges into the pools below. In this respect, step pools counteract steep slopes, thereby preventing excessive erosion and channel degradation. Step-pool sequences and their associated hydraulics are also important in providing habitats for stream organisms.

Occurrence. Step-pool sequences occur because the size of bed materials, including vegetative debris, is large relative to the size of the channel. Therefore, step pools are commonly found in mountain streams and in headwaters of drainage basins where gradients are steep (generally more than 0.05) and materials are coarse. Step pools have been reported in a wide range of environments from humid streams to arid desert fluvial systems. In some cases, the boulders in step-pool streams are remnants of former glacial processes. Although steps composed of alluvial boulders are most common, step pools also form in bedrock streams where their occurrence may be controlled by lithology. A variation of the step-pool system occurs in heavily vegetated basins where channels incorporate large organic debris and provide the materials to form log steps.

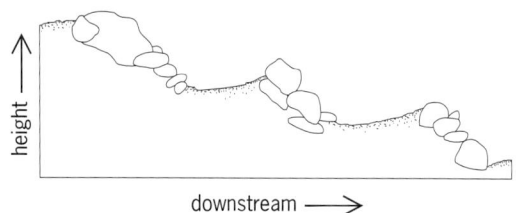

Fig. 1. Longitudinal profile of step-pool systems. (*After A. Chin, Step pools in stream channels, Prog. Phys. Geog., 13(3):104–120, 1989*)

Characteristics. The step-pool morphology is characterized by two well-defined dimensions: wavelength, to indicate the downstream dimension, and height, to reflect the vertical component (**Fig. 2**). Measurements conducted in the Santa Monica Mountains of southern California in the United States indicate that a wavelength (pool-to-pool) to height ratio of 11:1 typifies step-pool systems on slopes with gradients ranging from 0.04 to 0.12. Because scouring of the pools occurs at the bases of the steps, reverse-sloped sections are a characteristic component of these bedforms.

The step-pool structure is apparently controlled by channel slope. Step-pool wavelength is inversely related to channel slope, the wavelength increasing as the slope decreases. Therefore, steps appear more numerous at steep slopes, and they are spaced farther apart with decreasing slope. However, step height varies directly with slope. Higher steps are formed at steep slopes because larger rocks are generally found there; steps become smaller with decreasing slope along with a decrease in particle size. Because channel gradient generally decreases downstream, step pools can be expected to become smaller, less well developed, and less numerous in a downstream direction. Therefore, step pools are thought to represent a fluvial adjustment to channel slope.

Step-pool sequences exhibit a periodic spacing that tends toward regularity. The spacing of step pools, as with other similar fluvial features, is commonly expressed in units of channel widths because spacing is a function of channel size or flow discharge. In this context, step pools generally exhibit a spacing on the order of several channel widths. The most commonly reported values are within one to two channel widths, as typified by step pools in the Santa Monica Mountains with an average spacing of 1.6 channel widths, based on a sample of 464 sequences. The periodic spacing of step pools suggests that they are a fundamental characteristic of natural streams in a way similar to other rhythmic forms such as the meander and pool-riffle sequences found in lower gradient streams.

Functional importance. Step pools are functionally important in the fluvial system because they are energy-dissipating mechanisms. Their role is of special importance in small streams because narrow valleys prohibit lateral adjustments and energy dissipation by meandering and braiding. Without step pools to dissipate high energy in steep slopes, mountain streams would adjust vertically by erosion, and channel degradation would result. Energy dissipation also permits stable step-pool channel configurations to be maintained in the presence of sediment transport.

Energy dissipation by step pools is especially pronounced during low flows, where as much as 80–100% of the potential energy is reduced through vertical fall. As a result, little energy is available for bed and bank erosion and for sediment transport during these flows. However, step pools become less effective as energy dissipators at higher flows when

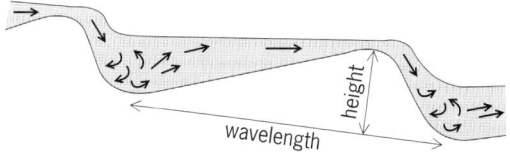

Fig. 2. Dimensions of the step-pool morphology (vertical exaggeration 2.5:1). The direction of water movement is shown by the arrows. (*After A. Chin, Step pools in stream channels, Prog. Phys. Geog., 13(3):104–120, 1989*)

pools are filled and steps become submerged. Because of diminished energy dissipation, step-pool streams can become very erosive at high flows. Thus, the role of step pools changes with changing flow conditions: they are more significant as energy dissipators at low flow, and they lose this capacity at high flows when they become submerged.

Processes of formation. Step pools are formed by high magnitude, low-frequency flood events. Well-developed steps in the steep headwaters are restructured by flows with frequency on the order of 50–100 years. Therefore, step pools are considered to be extremely stable structures that generally remain in place over the present time scale. However, steps composed of smaller particles can be mobilized by smaller flows that occur as frequently as every 1 or 2 years. Because steps generally decrease in size in the downstream direction, the less stable step pools are found downstream, and they can be observed to break down and reform over a period of several years.

The specific mechanism leading to the development of alternating step and pool sequences is not well understood. Because direct measurement of high-magnitude channel forming flows are difficult and opportunities are rare, much of the understanding of the step-pool origin has come from laboratory work where step pools are observed to form under a range of flow, sediment, and slope conditions. Recent flume experiments, in particular those conducted in Japan, have led to greater insights on the processes of step-pool development. At present, the prevailing theory to explain step-pool origin focuses on the formation of antidunes at high flow.

The antidune theory suggests that step-pool development is associated with a sequence of events that occurs with increasing flow. First, the stream (or flume) bed becomes active with particle movement until it is deformed in a wavelike manner. Second, antidunes form where the water surface is in phase with the flume bed. Third, the largest particles stop under the antidune crest, and they trap smaller debris to form a small dam or cluster of particles. Fourth, water spilling over these clusters creates a hydraulic jump in the antidune trough, scouring the bed and accentuating the pool. Fifth, a coarse layer of particles develops to armor and stabilize the bed. The resulting step-pool sequences are spaced to maximize resistance to flow.

The sequence described above requires two initial conditions. First, the stream bed must be composed of heterogeneous materials so that differential movement can lead to the resting of large particles and the trapping of smaller ones to create steps. Second, the magnitude of flow must be great enough to approach that necessary for the formation of antidunes. Therefore, the formation of step-pool sequences requires a combination of favorable basin or morphological parameters and the appropriate flow conditions. Given these conditions, the process is apparently not haphazard but effected by a similar mechanism that generates antidunes in sand-bed rivers.

Broader significance. In the broader context of river systems, step pools can be considered a type of gravel bedform that responds to the dominant controls of discharge and sediment load, independent variables that integrate the effects of climate, vegetation, soils, geology, and basin physiography. As such, they are most closely related to pool and riffle sequences found in lower gradient gravel-bed rivers. With similar periodic tendencies and variable grain-size distributions, step pools and riffle pools represent a continuum of gravel bedforms where step pools form the upper member and riffle pools the lower end. In this context, both represent fundamental adjustments: step pools to high-energy environments, and riffle pools to lower-energy conditions. Because pools and riffles are often considered to be a form of meandering in the vertical dimension, step pools may also be viewed as a type of meandering, the difference being the degree and the dimension in which the meandering operates.

For background information *SEE DEPOSITIONAL SYSTEMS AND ENVIRONMENTS; FLUVIAL SEDIMENTS; RIVER; STREAM TRANSPORT AND DEPOSITION* in the McGraw-Hill Encyclopedia of Science & Technology.

Anne Chin

Bibliography. A. Chin, Step pools in stream channels, *Prog. Phys. Geogr.*, 13(3):104–120, 1989; G. E. Grant, F. J. Swanson, and M. G. Wolman, Pattern and origin of stepped-bed morphology in high-gradient streams, Western Cascades, Oregon, *Geol. Soc. Amer. Bull.*, 102:340–352, 1990.

Storage rings

Storage rings dedicated to atomic and molecular physics have recently come into operation in Denmark, Germany, Sweden, and Japan. Advances in accelerator, beam cooling, and vacuum technology have made it possible to accelerate and store molecular and atomic ions in a wide range of charge states. Electron cooling of the stored ions provides monoenergetic beams with small size and divergence, which in turn allows for a wide range of precision experiments in atomic physics. An additional bonus is offered in the case of molecular ions: the vibrational excitations produced in the ions when they are created in an ion source relax by

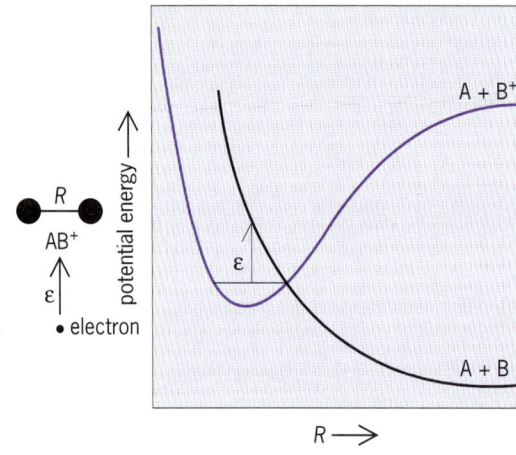

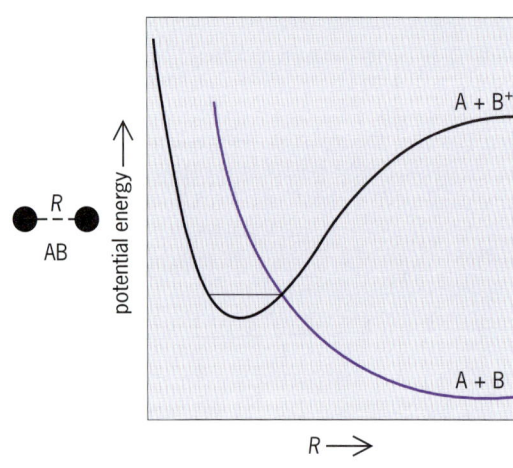

Fig. 1. Dissociative recombination of molecular ion AB$^+$ with an electron, illustrated with potential energy functions that depend on the internuclear distance R. The configuration of each system is shown to the left of its graph. (*a*) Molecular ion AB$^+$, and electron with kinetic energy ε, prior to collision. (*b*) System after collision. Electron has neutralized ion.

emission of infrared radiation when the ions are circulated in the ring. This effect facilitates detailed studies of low-energy collisions between cool molecular ions and electrons, collisions which usually lead to dissociative recombination.

Dissociative recombination. One of the most peculiar and complex elementary processes in molecular physics occurs when a molecular ion collides with an electron at low energy (that is, a collision energy much smaller than the binding energy of the molecular ion). This process can be written as AB$^+$ + e → A + B. The negatively charged electron is attracted by the positively charged molecule and attempts to attach to it. In order to do this, the free electron must slow down and deposit its kinetic energy somewhere. Lacking a third body to carry off the excess energy, the electron uses the molecular ion. The molecule converts the kinetic energy of the electron to motion of the nuclei through a rearrangement of the whole electronic cloud surrounding the molecule. The nuclei, originally vibrating in a bound state potential, are transferred to a free state that allows them to separate from each other.

Figure 1 illustrates the process with help of potential energy functions. The vibrational motion of the nuclei in the molecular ion AB^+ is determined by the bound potential energy function that dissociates to $A + B^+$ (Fig. 1a). The horizontal line at the bottom of the potential well shows the lowest vibrational level, and it is assumed that this is the only vibrational level which is populated. The collision of the ion with an electron results in the electron neutralizing the ion, and the motion of the nuclei is then determined by the repulsive potential energy function that dissociates to $A + B$ (Fig. 1b). This potential cannot support bound vibrational levels, and there is nothing to prevent the internuclear distance from becoming infinitely large, so the molecule AB dissociates very rapidly ($\sim 10^{-13}$ s) to the fragments A and B. The ability of the molecular ion to effectively transfer the kinetic energy of the free electron to the nuclei renders dissociative recombination a process with a large cross section.

Occurrence. Dissociative recombination assumes significance in a large variety of natural and laboratory-produced plasmas. Owing to its large cross section, dissociative recombination is the principal loss process for ionization in the Earth's thermosphere. Recombination of O_2^+ is a source of the atomic oxygen green (557.7-nanometer) and red (630.0-nm) lines in the atmospheres of Earth, Mars, and Venus, and in the Jovian atmosphere dissociative recombination of H_3^+ plays an important role. In interstellar space almost 100 different molecules have been observed. Molecular ions, in particular H_3^+, are very important for initiating chemical reactions that produce these molecules. Simultaneously, the ions are exposed to the risk of being destroyed by electrons. The uncertainty of the rate of destruction of H_3^+ by dissociative recombination has been a serious problem for models of interstellar molecular clouds.

Experimental difficulties. It is difficult to create and maintain ions at high density. An experiment aiming at the study of collision properties of ions must invariably deal with a target of low number density. When electron-ion collisions are being studied, the thin ion target is made to interact with a thin electron gas, and not many recombination events can occur in a given time. An additional problem is that the cross section for dissociative recombination is large only at very low collision energies (~ 1 meV). Thus, in order to study dissociative recombination it is necessary to exercise sufficient control of the ions and electrons in order that collisions at very low, well-defined energies can be achieved. A final difficulty is that state-selective molecular ions are difficult to obtain except in special cases and in small amounts, state-selective in this context meaning that only a single vibrational level (normally the lowest) of the ions is populated. The population of molecular ions extracted from ion sources for further use in various experiments is often distributed among a large number of vibrational levels.

Use of storage rings. The ions for a typical ion storage ring (**Fig. 2**) are produced in an external ion source, mass selected by a 90° magnet, accelerated by a radio-frequency quadrupole (RFQ) accelerator, and injected into the ring. The ion energy is further increased in the ring by radio-frequency acceleration. The maximum energy is determined by the properties of the bending magnets. Typically, the ring is filled with 10^7–10^8 ions.

Storage rings have considerable advantages over earlier-generation experiments using ion beams produced in standard accelerators. In earlier experiments, typically, the ions were passed once through the interaction region (single pass) and then collected in a Faraday cup. The time between ion creation and destruction was of the order of microseconds. In a storage ring, owing to an ultra-high vacuum system with a residual gas pressure of $\sim 10^{-11}$ torr (10^{-9} pascals), the ions can be circulated for tens of seconds. This time scale allows beam-cooling techniques to be used.

Electron cooling was demonstrated for protons as far back as 1974. In 1988 it was shown that beams of heavy atomic ions can also be electron cooled, and in 1992 molecular ions were electron cooled for the first time. In this process (Fig. 2b), the ions pass through a beam of cool electrons in one section of the storage ring, the electrons having the same average velocity as the ions. The low relative velocities between ions and electrons facilitate an effective energy transfer from the ion beam being cooled to the electron beam. Because the ion beam makes multiple passages through the electron beam (about 10^6 times per second), and new, cool electrons are constantly supplied, an ion beam of high density, narrow momentum spread, low divergence, and small beam size is obtained within a few seconds. Naturally, several ions are lost during the cooling phase because of recombination, but these losses are small compared with the total number of ions that are cooled.

Typical radiative lifetimes for excited vibrational levels in infrared-active molecular ions (that is, most ions except homonuclear diatomics) are of the order of tens to hundreds of milliseconds. Thus, if such ions are stored in a ring for time scales exceeding the vibrational lifetimes by two or three orders of magnitude, spontaneous emission of infrared radiation all around the ring will ensure that the ions become completely vibrationally relaxed.

After some tens of seconds excellent conditions for studies of dissociative recombination are at hand: a stored, monoenergetic, high-quality beam of cool molecular ions existing in a single vibrational level. The circulation of the beam provides a multiplicative enhancement of the beam current, and the high energy at which the ions are stored reduces disturbing background processes to essentially zero. The electron cooler is ready to serve its

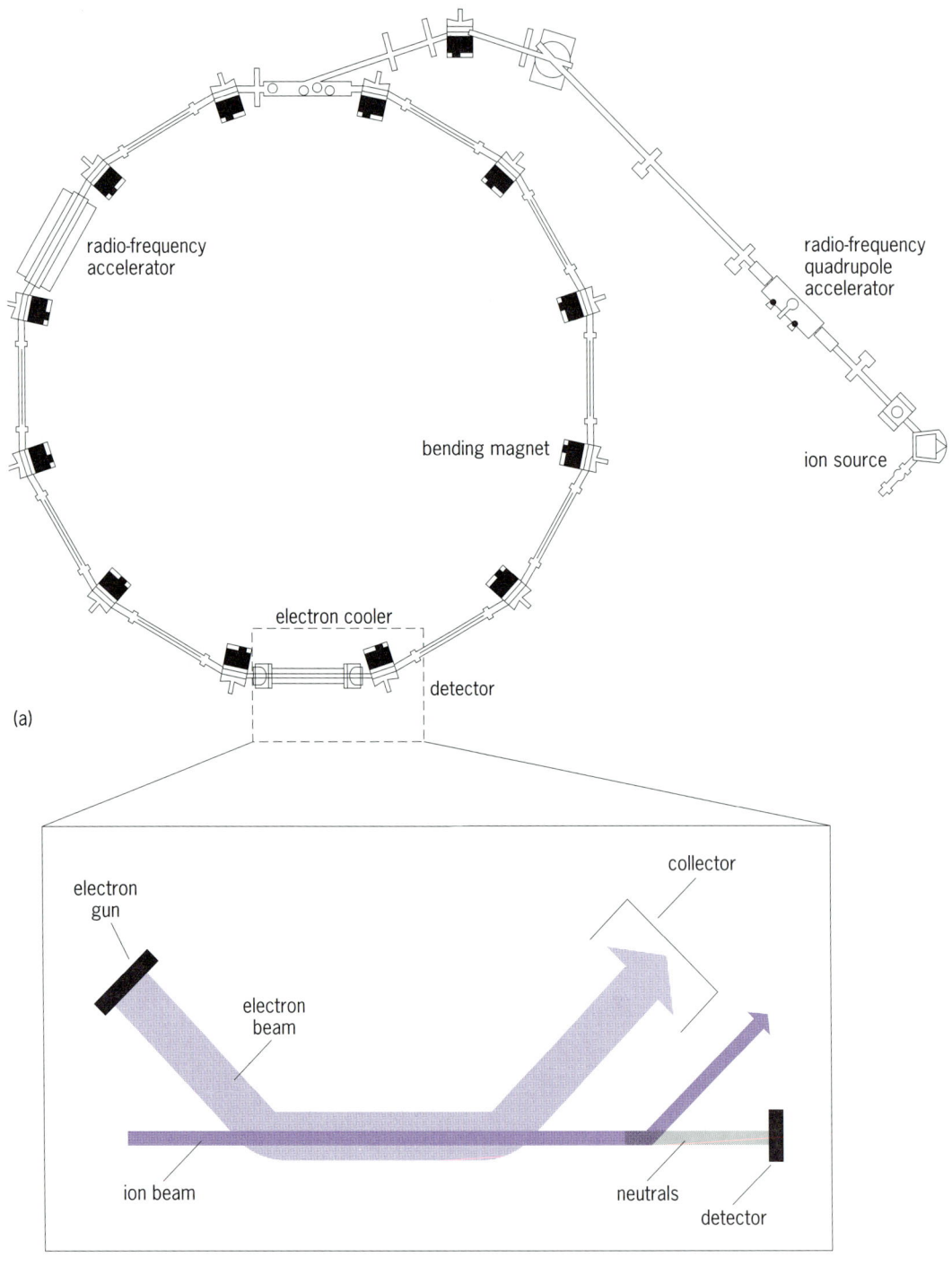

Fig. 2. Ion storage ring CRYRING in Stockholm, Sweden. (a) Diagram of ring. Ions are acelerated to 300 keV per atomic mass unit before injection into ring, and in the ring their energy is further increased to 96$(q/M)^2$ MeV/u, where q = charge (1 or 2 for molecules) and M = mass in atomic mass units (u). (b) Enlargement of electron cooler section.

second purpose as a target for low-energy electron-ion collisions.

Experiments. A powerful method of obtaining collisions at low energy is the merged-beam technique. Two beams of particles X and Y having almost the same mean velocities are collinearly merged so that the angle between the beams is 0°. In this way collisions between X and Y at very low energies in the particle's center-of-mass frame of reference can be induced.

In a storage ring the electron beam is merged with the ion beam in one straight section of the ring for cooling purposes. By switching the electron cooler acceleration voltage, the electrons are detuned away from the cooling energy, E_{cool}, to a value, E_{detune}, that corresponds to an energy in the

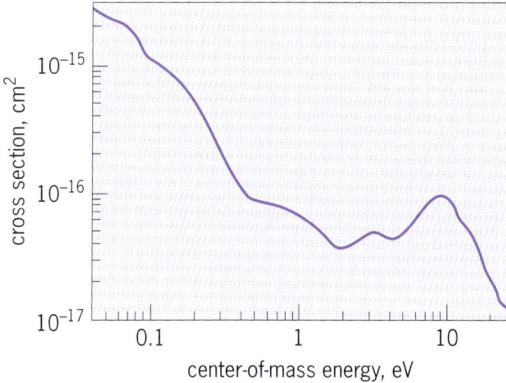

Fig. 3. Cross section for the process $H_3^+ + e \rightarrow H + H + H$ or $H + H_2$.

center-of-mass frame, $E_{\text{center of mass}}$, that satisfies the equation below. (In practice, the center-of-mass

$$E^{1/2}_{\text{detune}} = E^{1/2}_{\text{cool}} + E^{1/2}_{\text{center of mass}}$$

frame is identical to a frame of reference attached to the ions.) A simple numerical example illustrates the point. Ions stored at 10 MeV/u, where u = atomic mass unit, are cooled by 5.486-keV electrons. If the electron cooler is switched to 5.533 keV, collisions between the ions and electrons at $E_{\text{center of mass}} = 0.1$ eV will occur. Neutral particles produced in dissociative recombination events are separated from the ion beam in a bending magnet following the electron cooler section, and detected by a surface-barrier detector in the zero-degree direction of the electron cooler (Fig. 2b).

Figure 3 shows results for H_3^+ obtained at CRYRING. The peak at high energy (9.5 eV) occurs from a process similar to the one shown in Fig. 1, while the recombination at low energy (below 1 eV) is much more complex than Fig. 1 suggests and, in fact, not completely understood. From the cross section shown in Fig. 3, and from more detailed measurements at even lower energies (which involve some complications not discussed here), a rate coefficient of about 10^{-7} cm^3/s is deduced. This large value implies that dissociative recombination is an important loss mechanism for H_3^+ ions in interstellar space and other plasma environments.

Dissociative recombination is but one of many atomic and molecular physics processes of both applied and fundamental interest that can advantageously be studied in storage rings. Future advances in accelerator technology will be significant for the study of these processes.

For background information *SEE MOLECULAR STRUCTURE AND SPECTRA; PARTICLE ACCELERATOR; SCATTERING EXPERIMENTS (ATOMS AND MOLECULES)* in the McGraw-Hill Encyclopedia of Science & Technology.

Mats Larsson

Bibliography. M. R. Flannery, Electron-ion and ion-ion recombination processes, *Adv. Atom. Mol. Opt. Phys.*, 32:117–147, 1994; M. Larsson et al., Direct high-energy neutral-channel dissociative recombination of cold H_3^+ in an ion storage ring, *Phys. Rev. Lett.*, 70:430–433, 1993; B. R. Rowe, J. B. A. Mitchell, and A. Canosa (eds.), *Dissociative Recombination: Theory, Experiment and Applications*, NATO ASI Series, vol. 313, 1993; G. Sundström et al., Destruction rate of H_3^+ by low-energy electrons measured in a storage ring experiment, *Science*, 263:785–787, 1994.

Streptococcus

Streptococcus pyogenes, or group A streptococcus, is one of the most common and virulent of all bacterial pathogens. It is a member of the genus *Streptococcus*, which includes other species pathogenic for humans such as *S. agalactiae* (group B streptococcus), the leading cause of bacterial infection (sepsis) in newborns, and *S. pneumoniae*, the most common bacterial cause of pneumonia. The most frequently encountered group A streptococcal diseases are strep throat (pharyngitis) and a pustular infection of the exposed skin (impetigo), but for decades deeper and more widespread infections have also been known. Rheumatic fever, an important cause of heart disease, can also be attributed to infection with the group A streptococcus.

Although in recent decades the frequency of group A streptococcal pharyngitis and impetigo has remained relatively constant, rheumatic fever and other life-threatening acute bacterial infections have gradually declined in developed countries. This decrease can be attributed in part to antibiotics and improved hygienic practices, but to a great extent it is unexplained. Very recently, there have been resurgences of rheumatic fever in some locales and a worldwide increase in a severe form of invasive group A streptococcal infection known as streptococcal toxic shock–like syndrome (TSLS). Although the absolute numbers of both remain small, the prospect of further increases is a matter of great concern.

Streptococcal toxic shock–like syndrome. In 1987, reports of severe life-threatening group A streptococcal infections began to increase in the United States, Europe, and Australia. Cases often begin with seemingly innocuous minor skin trauma or soft tissue infection that progresses to widespread infection, shock, and multiorgan failure. In some instances, previously healthy adults die only a few days after symptoms begin. Loss of digits or limbs is related to the lay designation of the microorganisms as "flesh-eating bacteria." Other forms of infection that have been associated less often with toxic shock–like syndrome (TSLS) include pharyngitis and postpartum infections. In some cases, the group A streptococcus has been a secondary invader of chickenpox pustules.

Pathology. Similarities between group A streptococcal infections and toxic shock syndrome caused by *S. aureus* have caused this disease to be labeled toxic shock–like syndrome. However, in staphylococcal toxic shock syndrome, the systemic findings can be accounted for by the absorption of toxin produced at a local site (for example, the vagina or a wound); the local infection itself is often not prominent, and the organisms are rarely found beyond the primary site. Streptococcal TSLS has all the features of a highly invasive progressive bacterial infection because the bacteria are present in the bloodstream and multiple other organs in 60% of cases reported. The multiorgan involvement is due in part to the toxin circulating in the bloodstream, but the streptococci also spread to multiple sites where the invasive process begins again.

The pathogenic mechanisms involved in streptococcal TSLS are only partially understood. As compared to routine isolates, a much higher proportion of the TSLS strains produce one or more of a family of exotoxins called streptococcal pyrogenic exotoxin (SPE), which have actions and molecular structures similar to toxic shock syndrome toxin number 1 (TSST-1), the major staphylococcal pyrogenic exotoxin of toxic shock syndrome. The genes for the streptococcal pyrogenic exotoxins are encoded in bacteriophages that infect the streptococcus and integrate with the chromosome (lysogeny). Under the name erythrogenic toxin they have long been associated with scarlet fever, a toxic form of group A streptococcal pharyngitis.

Invasive potential. Although many of the clinical and pathogenic features of TSLS are explained by the action of the pyrogenic exotoxins, the enhanced invasive potential is not. Molecular epidemiologic studies of strains gathered worldwide have shown that TSLS-associated strains represent a relatively small number of clones. The upsurge of TSLS could represent bacteriophage-mediated transfer of the streptococcal pyrogenic endotoxin genes among recently emerged clones of group A streptococci that have an unexplained, enhanced invasive potential.

Although these changes could represent a recent shift in the virulence of the group A streptococcus, aggressive infections of this type have been recorded since the earliest observations of infectious diseases. For example, the childbed fever described by I. Semmelweis in nineteenth-century Vienna was a group A streptococcal infection. This aggressive form of infection took the lives of 10% of women delivering babies in the obstetrical unit of the Vienna General Hospital. Even the sequence of a minor abrasion progressing to death is not new. Although there is no question that such infections have increased recently, they are still too rare to justify any general recommendations for medical or public management of suspect group A streptococcal infections.

Resurgence of rheumatic fever. Group A streptococci are also responsible for rheumatic fever and one form of kidney inflammation called acute glomulonephritis. These poststreptococcal complications occur in only a small portion of group A streptococcal cases weeks after resolution of the acute infection. The disease mechanisms involve immunologic hyperreactivity to components of the organism that unfortunately also react against human tissues. Antibodies stimulated by a structural component of the organism (M protein) also bind to connective tissue membranes found in cardiac tissue.

Rheumatic fever is a recurrent disease that may be triggered by any subsequent group A streptococcal pharyngitis. It does not occur following streptococcal skin or soft tissue infections. Multiple episodes can lead to permanent damage, particularly to the heart (rheumatic heart disease). Rheumatic fever can be prevented by prompt antibiotic treatment of the pharyngitis.

The gradual decline in rheumatic fever accelerated dramatically in developed countries during the 1960s and 1970s, but in the mid-1980s, focal outbreaks occurred in the United States. The first clusters, reported from Salt Lake City, Utah, in 1985, were followed by clusters from the intermountain area and, by 1990, from areas all over the country. Most outbreaks involved fewer than 40 patients, but in Salt Lake City 200 cases were reported.

Some of these outbreaks have primarily affected children in suburban or rural locales. Although most cases occur in children 5 to 15 years old (which is usual for rheumatic fever), some cases have been reported in closed adult populations such as military training camps. There is no known link between the upsurge of rheumatic fever and that of TSLS.

The reasons behind these focal increases in rheumatic fever are not known. The strains involved have been of M protein types generally recognized as rheumatogenic on epidemiologic grounds. As with TSLS, the absolute number of cases is too small to generate alarm.

Treatment and prevention. To date, the development of resistance to antibiotics has not been a complicating feature for management of group A streptococcal disease. Penicillin remains the treatment of choice with a number of alternatives for individuals who are hypersensitive to penicillin. Treatment of group A streptococcal pharyngitis within 10 days of the acute episode prevents rheumatic fever; the threat of rheumatic fever remains the primary reason for physicians' culturing sore throats. Individuals with documented rheumatic fever are generally placed on continuous penicillin prophylaxis to prevent the immunologic reaction expected with any subsequent streptococcal infections. The development of a vaccine has been complicated because immunity is conferred by antibodies directed against M protein, the same molecule associated with the immune hyperreactivity of rheumatic fever. Progress promising for the development of a vaccine has been made by molecular and immunochemical dissections of regions of the protein responsible for its multiple biologic functions.

For background information *SEE ANTIBODY; PENICILLINS; RHEUMATIC FEVER; STAPHYLOCOCCUS; STREPTOCOCCUS; TOXIC SHOCK SYNDROME* in the McGraw-Hill Encyclopedia of Science & Technology.

Kenneth J. Ryan

Bibliography. C. W. Hoge et al., The changing epidemiology of invasive group A streptococcal infections and the emergence of streptococcal toxic shock–like syndrome, *J.A.M.A.*, 269:384–389, 1993; G. L. Mandel et al. (eds.), *Principles and Practice of Infectious Diseases*, 4th ed., 1994.

Superconductivity

The discovery of high-temperature superconductors has led to a great deal of excitement and activity in physics and engineering. Although no general consensus about the microscopic mechanism for superconductivity has been reached, there has been considerable progress in understanding the macroscopic behavior of these materials, in particular, the properties of vortex excitations in a magnetic field. The study of such behavior is largely independent of the specific microscopic mechanism of superconductivity. It is crucial for the application of these materials in high magnetic fields, including the operation of SQUID devices, high-resolution magnetic resonance imaging (MRI) systems, and levitated trains. *SEE SQUID.*

Meissner effect and mixed phase. The most striking property of a superconductor is perhaps the Meissner effect, where a small, externally applied magnetic field **H** is expelled from the bulk of a superconducting sample. In the Meissner phase, a sample has zero resistivity. In recently discovered cuprate materials such as $YBa_2Cu_3O_7$ (YBCO), this phase can exist up to a transition temperature (T_c) of about 90 K (–280°F). However, the Meissner phase of these high-T_c superconductors is destroyed in the presence of an external field exceeding a critical value $H_{c1}(T)$ of the order of several hundred gauss (several hundredths of a tesla). In that case a so-called mixed phase, in which magnetic fields penetrate the sample in the form of quantized flux lines, replaces the Meissner phase. A flux line consists of a vortex core of radius ξ within which the electrons are normal, surrounded by a larger flux tube of radius λ where magnetic field penetrates. Two flux lines repel each other if they are within a distance of order λ. Thus, flux lines stay as far away from each other as possible; consequently, normal and superconducting regions coexist in the mixed phase. As the external field is increased beyond H_{c1}, more and more flux lines penetrate into the system and are forced closer together. Naively, superconductivity might be expected to persist in the mixed phase until the external field was so high that the superconducting region in between flux lines was reduced to null. The field strength at which this effect occurs is called H_{c2}. For YBCO, $\xi \approx 1$ nanometer and $\lambda \approx 150$ nm, which leads to $H_{c2} \approx 100$ tesla. The following discussion will show that superconductivity actually breaks down at much lower fields and will cover various possible methods of preserving superconductivity at high fields.

Flux-line pinning. If the system is just above H_{c1} so that only one flux line penetrates, a small applied electric current **J** will generate a Lorentz force $\mathbf{J} \times \mathbf{H}$ which pushes the flux line laterally (**Fig. 1**). Since the core of the flux line consists of normal electrons, its motion causes dissipation and consequently destroys superconductivity. This motion can be preempted if the flux line is "pinned." Pinning can be accomplished by introducing microscopic crystalline defects, which damage superconductivity locally, that is, on the scale ξ (Fig. 1). Such a defect pins a flux line by reducing the vortex core energy.

In any high-field application of superconductors, many flux lines are involved. A small applied electric current then produces a Lorentz force acting on each flux line. In order to preserve the superconductivity of a sample deep in the mixed phase, that is, at a finite density of flux lines, it is important to find effective ways to pin the entire flux array. In 1957, A. A. Abrikosov presented a remarkable mean field theory of the mixed phase. In a clean sample, flux lines are found to form a regular lattice of parallel lines, like a tight bundle of uncooked spaghetti. The properties of the flux lattice can be described by classical elasticity theory in terms of a few elastic moduli, without regard to the quantum origin of the flux lines. If a few microscopic defects which pin a few flux lines are introduced, the entire flux array will be pinned due to the elasticity of the lattice. This method is analogous to holding a piece of wood fixed by putting a few nails through it.

Flux-lattice melting. In conventional (low-T_c) superconductors, the flux lattice is believed to exist

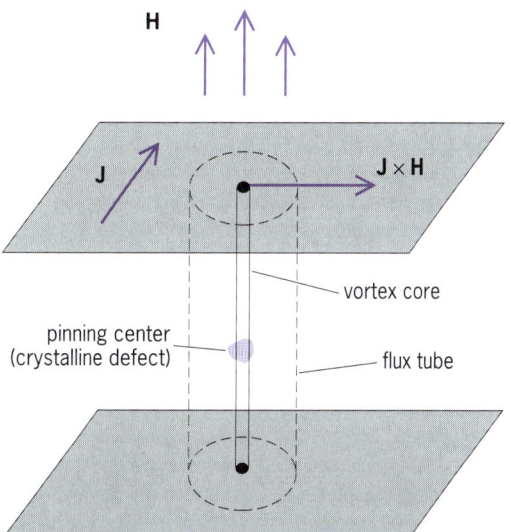

Fig. 1. Anatomy of a flux line. The flux line will move under a Lorentz force $\mathbf{J} \times \mathbf{H}$ unless it is effectively pinned by the defect.

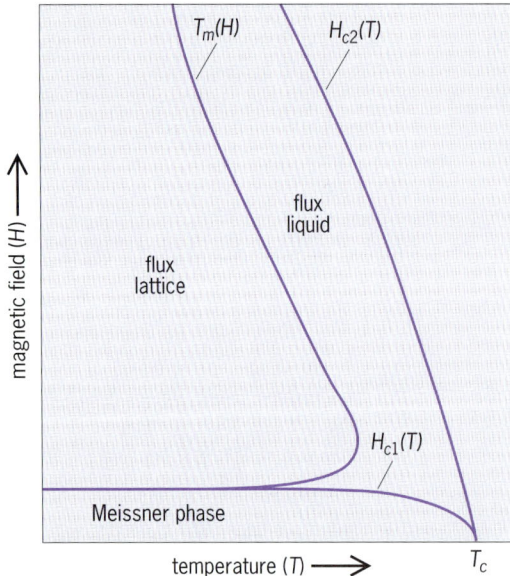

Fig. 2. Phase diagram of an ideal (that is, defect-free) superconductor, showing the Meissner phase, the flux-lattice phase, and the flux-liquid phase.

at essentially all temperatures up to $H_{c2}(T)$. Thus, the above pinning mechanism is effective in suppressing flux flow and preserving superconductivity. However, in the cuprate superconductors, because of a combination of stronger thermal fluctuations and weaker elastic moduli, the flux lattice can melt if either the temperature (T) or the magnetic field (H) is increased above a melting temperature $T_m(H)$ [**Fig. 2**]. It is no longer possible to pin the melted flux liquid by a few defects. As a result, flux lines can move freely in the liquid phase, causing dissipation and thereby destroying superconductivity. Fortunately, the melting temperature could be quite high (about 80 K or –296°F for YBCO in a field of 5 T). So the liquid phase does not dominate the bulk of the mixed phase.

Effect of weak disorder. Even the flux-lattice phase below the melting temperature is affected by the existence, in a realistic sample, of a finite density of unavoidable weak crystalline disorders, such as oxygen vacancies in the cuprates. This type of point disorder is not quite the pinning center described above, since the defects here are very weak and cannot effectively pin any flux line individually. However, the cumulative effect of such weak disorder becomes substantial over a large volume because the root mean square fluctuations of disorder increase rapidly as the square root of the volume. In the 1970s it was shown that arbitrarily weak point disorder always distorts the flux lattice enough to destroy the regular order of the lattice at large scales. Thus, the flux lattice cannot exist in a real sample even below the melting temperature.

It is crucial to understand what replaces the flux-lattice phase in the presence of weak disorder. If the disorder turns the lattice into a viscous flux liquid, it will not be possible to pin the flux array at all, and the entire mixed-phase regime will not be superconducting. Thus, of course, the application of these high-T_c materials will be limited. An intriguing alternative is that weak point disorders may collectively pin the flux array. If the flux lines were to take on specific configurations in order to take advantage of the (random) arrangement of the underlying disorder, it would be difficult for any part of the flux array to move, since a global rearrangement of the flux array would be required. Such sluggish dynamics are expected to suppress dissipation.

One useful characterization of the dynamics of the flux array is the current-voltage characteristic, where a current (of current density J) is applied through the superconductor and the voltage V developed across the sample is measured. As mentioned above, flux lines can flow freely under an applied Lorentz force in the liquid phase. Since the motion of magnetic field generates an electric potential difference V proportional to J, the liquid phase has a finite linear resistivity (it is not a superconductor). However, if the flux array is collectively pinned, the response to the applied current should be much reduced. On theoretical grounds, an exponentially weak dependence is expected, of the form given by the equation below, with the exponent μ

$$V \propto J e^{-(J_c/J)^\mu} \qquad \mu > 0$$

characterizing the strength of pinning, and J_c characterizing the current scale above which significant dissipation occurs. Characteristics of the above form have been observed in various YBCO samples below a certain critical temperature T_g, which is of the order 70–80 K (approximately –300°F) in a field of several teslas. Since the above equation implies a zero linear resistivity (that is, the ratio V/J approaches zero as J approaches zero), superconductivity is preserved over a significant portion of the magnetic field–temperature (H–T) phase dia-

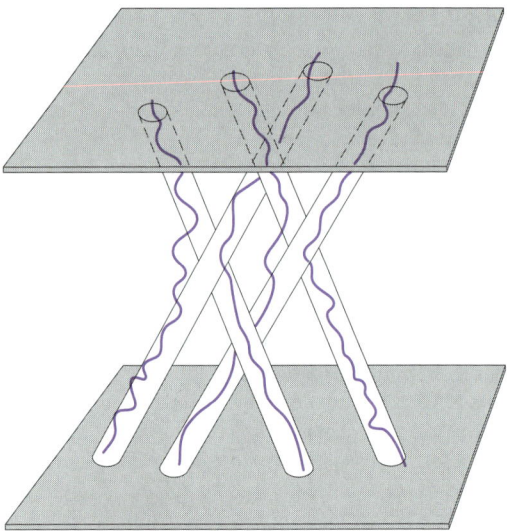

Fig. 3. Entangled flux-line configuration pinned to splayed columnar defects.

gram. The new phase of matter below T_g is called the vortex glass.

Flux pinning by columnar defects. The suppression of dissipation at small current is important in a number of applications. However, for power applications, it is desirable to suppress dissipation at high currents, or to increase J_c. Thus, experimentalists irradiated high-T_c materials with heavy ions, which cause parallel tracks of columnar defects, with radius ~ ξ and length up to 50 micrometers. Each one of these columnar defects is designed to trap one flux line. By matching the density of defects to the density of flux lines, it is possible to increase J_c by as much as 3 orders of magnitude. The theoretical and experimental aspects of flux pinning by columnar defects have been extensively investigated. It has been proposed that the pinning effect may be enhanced further by randomly splaying the columnar defects (**Fig. 3**). The exponent μ increases in this case. More importantly, the optimal configurations of the flux array obtained are highly entangled (Fig. 3). Since the flux lines repel each other, and therefore do not like to cut across each other, a few strongly pinned flux lines can impede the motion of many weakly pinned ones.

For background information SEE MEISSNER EFFECT; SUPERCONDUCTIVITY in the McGraw-Hill Encyclopedia of Science & Technology.

Terence Hwa

Bibliography. G. Blatter et al., Vortices in high temperature superconductors, *Rev. Mod. Phys.*, 66:1125, 1994; L. Civale et al., Vortex confinement by columnar defects in $YBa_2Cu_3O_7$ crystals: Enhanced pinning at high fields and temperatures, *Phys. Rev. Lett.*, 67:648–651, 1991; R. H. Koch et al., Experimental evidence for vortex glass superconductivity in YBaCuO, *Phys. Rev. Lett.*, 63:1511–1515, 1989; M. Tinkham, *Introduction to Superconductivity*, 1975, reprint 1980.

Surgery

Technology has propelled surgery into an age of minimally invasive (or minimal-access) operations. Surgeons in different specialties use this approach to operate in the abdominal cavity, chest, pelvis, and extremities by means of access ports placed through small incisions. In the abdomen, laparoscopy, a method of viewing the peritoneal cavity by means of an endoscope introduced through a small incision in the abdominal wall, has revolutionized the field of general surgery. Similarly, an operation guided by a telescope with fifteenfold magnification capabilities that is placed through a small incision in the chest is called thoracoscopy; in the pelvis, pelviscopy; and in the joints, arthroscopy. The first laparoscopic cholecystectomy (gallbladder removal) was performed in France in 1987, and the first in the United States in 1988. Since then, there has been an upsurge in popularity of laparoscopic surgery, and many other operations have been reported.

The major advantages of minimally invasive surgery include less pain, a shorter hospital stay, a more rapid return to work and full activity, and an improved cosmetic appearance (incisions are less than 0.47 in. or 12 mm in length). In addition, magnification of the operative field improves visualization of small structures and diseased tissue, and an enclosed working space may decrease intraoperative evaporative fluid losses and lessen the risk of infection.

Currently, a major drawback of laparoscopic surgery is that many such operations have not undergone long-term studies to validate their safety and efficacy. In addition, for surgeons who have not mastered the video-eye-hand coordination, basic skills such as suturing become insurmountable tasks. Furthermore, not all individuals are suitable for this approach: generally, an individual must be able to tolerate a traditional large incision in the event of a problem during the laparoscopic operation, must be at good risk for general anesthesia, and must have normal blood clotting factors. Most important, even though it is possible to perform an operation through a small incision, it may not be the best way to perform that operation.

Access. In the joints, orthopedic surgeons maintain a working space by distending the joint space with water after gaining access through a small incision. Constant irrigation not only opens a space but also washes away blood and debris from the operation. Unfortunately, water is useful for only a few operations done within small confined spaces such as the knee or wrist.

Traditional access to intraabdominal and intrathoracic organs requires large incisions. In open abdominal surgery, the view of the internal structures is afforded by retracting the abdominal wall laterally. In thoracic surgery, ribs are often broken or forced apart in order for the surgeon to see the operative field and introduce surgical instruments. In contrast, minimally invasive techniques create a working space within the abdominal cavity by injecting gas into the peritoneal cavity. The skin is punctured with a needle, and the abdomen is inflated with carbon dioxide. The gas lifts the abdominal wall away from the intestinal organs, permitting safe placement of the first access port. Ports consist of a removable trocar (a razor-sharp, pointed tip) within a hollow sheath. After the trocar punctures the abdominal wall, it is removed, and the sheath is secured. The ports are usually only 0.2–0.5 in. (5–12 mm) in diameter, and have valves that prevent gas leakage. A laparoscope is inserted through the initial port, and the abdomen is examined for abnormalities (see **illus.**). Additional ports are placed as needed to introduce surgical instruments.

Thoracic surgeons do not need to inflate the region with gas to create a working space in the chest cavity. By selectively placing a breathing tube in the opposite lung, the lung being operated on is temporarily collapsed, making a working space within the chest cavity. Ports without valves are

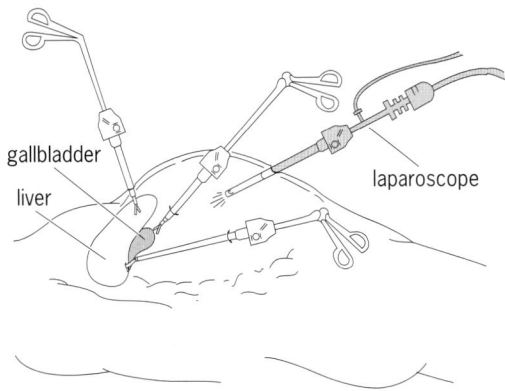

Port placement for laparoscopic cholecystectomy (removal of the gallbladder). The laparoscope (abdominal telescope) is inserted through a port at the umbilicus (navel). (*After N. J. Soper, L. M. Brunt, and K. Kerbl, Laparoscopic general surgery, N. Engl. J. Med., 330:409–419, 1994*)

used to gain access, and allow for the easy passage of instruments.

Access can be dangerous. Too much gas pressure may cause unwanted effects of hypercarbia (elevated carbon dioxide in the blood) and the potentially fatal effects of gas embolism (gas bubbles in the veins). Trocars can also inadvertently puncture major blood vessels, bowel, spleen, and liver. Such complications have caused several deaths.

The recent development of potentially safer external lift devices allows the surgeon to operate within the abdominal cavity, yet avoid using carbon dioxide. External lifts place a retractor under the skin, and the arm of the lift is raised mechanically to create a working space. Potential problems caused by these lift devices have not yet been investigated.

Visualization. Once a working space is established, a scope with an attached television camera is passed through the port. Video monitors and video cassette recorders (VCRs) allow the operating team to watch and record the operation. Fiberoptics illuminate the abdominal cavity for superior lighting. Scopes with fifteenfold magnification allow surgeons to detect abnormal pathology frequently missed by a radiologist using roentgenologic imaging. In the past, a surgeon might not know the extent of cancer spread without making a very large and painful incision. By using minimally invasive techniques, the surface of organs can be thoroughly visualized through a very small puncture wound. Surgeons can also see beyond organ surfaces by placing an ultrasound probe (sonar) directly over the critical organs.

In the future, examination of the abdomen may be facilitated with computers. Since many preoperative imaging modalities, such as computed axial tomography (CAT scan), magnetic resonance imaging (MRI), and angiogram, are digitized, information can be processed by a computer into color-enhanced three-dimensional representations. For example, a tumor in the brain might be colorized and then superimposed on the actual brain to facilitate removal of the tumor by the neurosurgeon. Likewise, this image could be projected outside the skull to permit the neurosurgeon to view and guide the operation around surrounding normal tissue. Currently, the U.S. military is developing robots that would operate through small incisions in a battlefield situation while the surgeons controlling the robots' movements would be safe behind the lines at computer terminals. In addition, virtual reality technology may soon allow surgeons to move around the abdomen just as architects today walk through computer-designed buildings.

Instrumentation. Before surgeons could operate through small ports, specialized instruments had to be developed. Today, many of the surgeon's traditional instruments have been adapted to fit through small ports, and many new innovations have been added. Probes which deliver electrocautery and laser energy coagulate and cut tissue. Ultrasonic energy resonates to divide tissue. Stapling devices allow surgeons to divide and reconnect tissues. Biological adhesives such as fibrin sealant glue tissues together with natural blood products. As instrumentation becomes more sophisticated, surgeons can accomplish more difficult tasks.

Operations. Many different types of operations are being performed using minimally invasive surgical techniques. Only a few of these operations have been practiced long enough to be rigorously evaluated with randomized outcome studies comparing the minimally invasive operation with the conventional open operation. More than 80% of all gallbladders are now removed through small incisions, and laparoscopic appendectomy (removal of an inflamed appendix) and laparoscopic inguinal hernia repairs have been shown to cause less pain and enable patients to return to normal activity more quickly. Other similar types of operations are gaining acceptance. Orthopedic surgeons, for example, can identify and remove torn cartilage in the knee joint; gynecologists can tie fallopian tubes to prevent future pregnancy; thoracic surgeons can do lung biopsies in order to diagnose different diseases; trauma surgeons can examine the abdomen to decide whether an operation is necessary; cancer surgeons can look into the abdomen to ascertain whether cancer has spread or if a patient has a fair chance for a surgical cure; urologists can remove a diseased kidney and fix a woman's bladder incontinence; and general surgeons can resect diseased colon, small intestine, adrenal gland, liver, stomach and pancreas, and cure problems such as reflux acid regurgitating from the stomach into the esophagus. Retrospective studies of these types of operations have identified many of their benefits and drawbacks, but they cannot determine whether the minimally invasive approach is safer and as effective as the traditional operation.

Other procedures are truly investigational and require advanced surgical skills. Laparoscopic techniques for vascular surgery (aortobifemoral bypass for aneurysms of the aorta), endocrine surgery (parathyroidectomy and thyroidectomy for removal of diseases in these glands), hepatobiliary surgery (pancreaticoduodenectomy and choledochojejunostomy for cancer of the pancreas), and fetal surgery (operations on the fetus while still in the uterus) are being developed. Moreover, investigators are working closely with industrial companies to develop three-dimensional laparoscopes, external lift devices, robotic camera holders, and laparoscopic training simulators.

The rapid evolution in surgery toward minimal invasion has been driven by the demand for more cosmetic and less painful incisions. Industry has responded to powerful market forces with new equipment and a substantial investment for research and development, and consequently, scientific and technological advances in minimally invasive surgery have been exponential. Established surgeons have had to learn new surgical techniques, and hospitals have been forced to redesign operating suites and retrain nurses to keep pace with the technological changes. Still, long-term follow-up and randomized outcome studies are needed to prove which minimally invasive operations are truly safer and more cost-effective.

For background information SEE ABDOMEN; SURGERY in the McGraw-Hill Encyclopedia of Science & Technology.

Daniel B. Jones; Nathaniel J. Soper

Bibliography. J. G. Hunter and J. M. Sackier (eds.), *Minimally Invasive Surgery*, 1993; N. J. Soper, L. M. Brunt, and K. Kerbl, Laparascopic general surgery, *N. Engl. J. Med.*, 330:409–419, 1994; N. J. Soper et al. (eds.), *Essentials of Laparoscopy*, 1994.

Tetrapoda

Humans, as primates and as mammals, belong, together with bats, whales, birds, crocodiles, dinosaurs, snakes, frogs and salamanders, to the group known as tetrapods. Members of this group can be traced to ancestors that had four legs and that held in common many unique features. Today, the nearest living relatives of the tetrapods consist of two groups of fish: lungfish, air-breathing fresh-water fish from the southern continents, and the coelacanth, a deep-water marine species from restricted parts of the Indian Ocean. These two groups and the tetrapods are the descendants of a once larger group called the sarcopterygians, or lobe-finned fishes. Until very recently, fossil evidence of the fish to tetrapod transition was sparse, but new finds have revolutionized scientific understanding of the animals involved.

Fish-to-tetrapod transition. Tetrapods are usually regarded as most closely related to the extinct lobe-finned fish called osteolepiforms, known first from the Middle Devonian period, about 385 million years ago (m.y.a.). The earliest fossil tetrapods are known from the Upper Devonian, only about 15 m.y. later. The first firm evidence of tetrapods consists of two parallel sets of footprints from the Upper Devonian of Australia, showing evidence of two very different animals. Whereas one set shows dragging feet and tail scrapes, the other shows the characteristic alternating pattern of fore and hind footprints and clear prints of toes of indeterminate number. It is not certain whether these tracks were made on land or in shallow water; however, it is clear that, by the time these tracks were made, limbs and digits had replaced lobed fins for this lineage of vertebrates.

Many other characters unique to tetrapods do not involve the legs but the structure of the snout and palate, ear region, and lower jaw; the development of a neck (fish have their heads fastened onto their shoulders by a set of bones lost in tetrapods); enlargement of the shoulder and hip girdles; substantial ribs; and loss of bony fin rays.

The order in which these characters were acquired, and the process by which they were acquired, is beginning to be understood. Although it is generally thought that legs evolved for walking, it seems more likely that they evolved first for use in water; only later did this adaption allow some creatures to exploit the much more difficult niche of terrestriality.

Ichthyostega. The first Devonian tetrapod to be discovered, *Ichthyostega,* was found in east Greenland during the 1930s. Until recently, *Ichthyostega* was known as the earliest "fish with legs." Early reconstructions show it with paddlelike hindlimbs

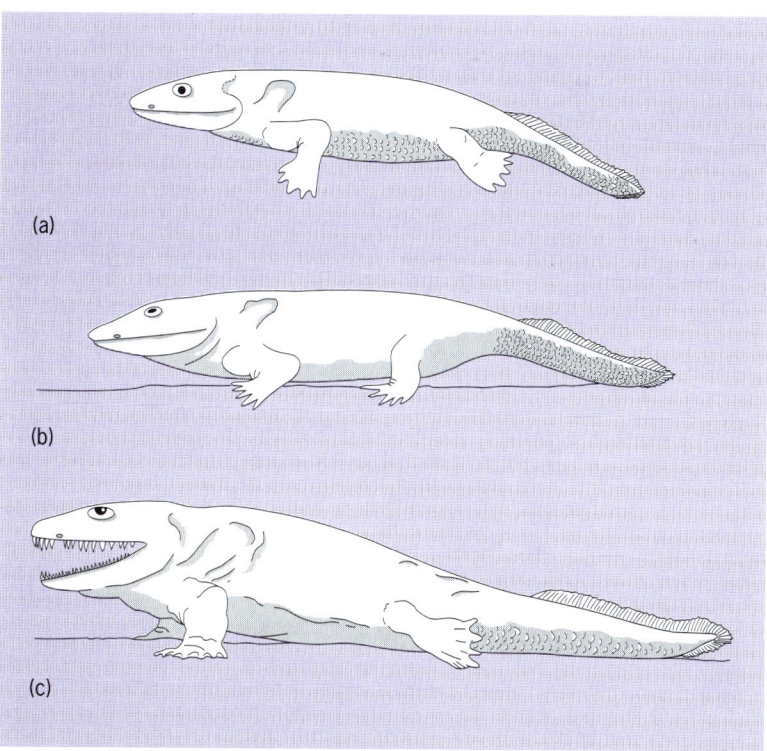

Fig. 1. Life reconstructions of *Ichthyostega*. (a) 1952 (after E. Jarvik, On the fish-like tail of the ichthyostegid stegocephalians, Meddelelserhom Grønland, vol. 114, no. 12, 1952). (b) 1980 (after E. Jarvik, Basic Structure and Evolution of Vertebrates, vols. 1 and 2, Academic Press, 1980). (c) 1994.

that are poorly adapted for walking. More recent findings have confirmed that the hindlimbs were indeed paddlelike, with knees and ankles almost unable to bend, and significantly smaller than the massive shoulders and forelimb (**Fig. 1**). Although the number of fingers is unknown, seven toes were discovered: four stout toes at the back and three small closely tied digits at the leading edge. The hindlimb of *Ichthyostega* resembles the front paddle of a modern dolphin and its proportions resemble those of a modern seal. Although it could have hauled itself out of the water by using its forequarters, its hindquarters would undoubtedly have dragged behind, being useful only when the animal swam. *Ichthyostega* is so unusual that it has shed little light on the evolution of other tetrapods.

Acanthostega. *Acanthostega*, an animal that was discovered from the same deposits as *Ichthyostega*, was first described in 1952 but remained known from only two partial skulls until the late 1980s, when more specimens were discovered. It measured about 3.3 ft (1 m) in length, including the tail. Analysis of the skeletal anatomy over the last 5 years has revealed *Acanthostega* to be far more primitive than *Ichthyostega* (**Fig. 2**). In many respects it is a transitional form, providing useful information on the fish to tetrapod transition. Although *Acanthostega* has about two-thirds of the characters used to define tetrapods, about one-third of its characters remain fishlike. Such fishlike characters include the structure of the nose and palate, some features of the ear region, the shoulder girdle, the neck region, and the

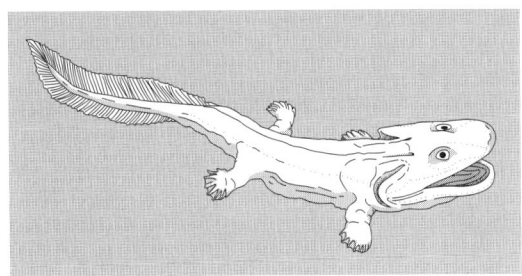

Fig. 2. Life reconstruction of *Acanthostega gunnari*. (*Courtesy of Richard Hammond*)

vertebral column. The most informative structures, however, are the gill skeleton, the tail, and the limbs.

Gill skeleton. *Acanthostega* retained heavily ossified, deeply grooved gill bars, unlike those of all other known tetrapods but similar to those of both modern and fossil fishes, especially lungfishes. The grooves accommodated the arteries for carrying blood toward the gills for oxygenation. Along the front edge of the shoulder girdle ran a flange, which forms the rear wall of an internal gill chamber in fishes. Both features suggest that *Acanthostega* retained functional internal gills and a fleshy rather than bony operculum, breathing like its fish ancestors. At the same time, it probably also had lungs which it used from time to time. A common assumption prior to this discovery was that all tetrapods had dispensed with internal gills.

The structure of the brain case and ear region in *Acanthostega* also was very fishlike. Its ears were

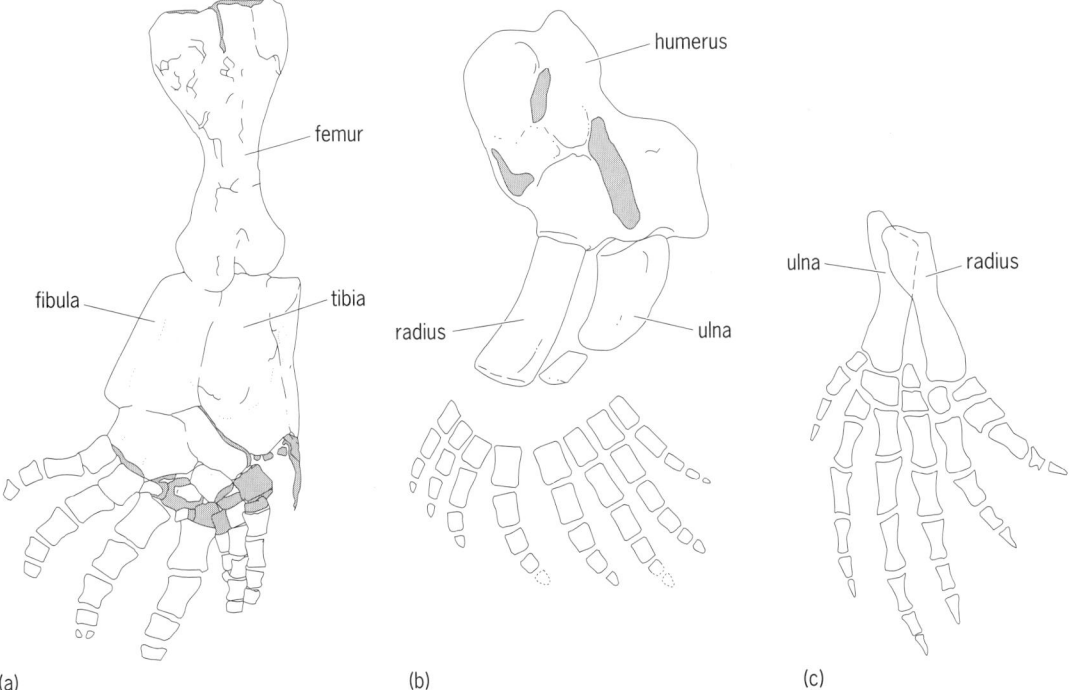

Fig. 3. Limbs of Devonian tetrapods. (*a*) Right hindlimb of *Ichthyostega*. (*b*) Left forelimb of *Acanthostega*. (*c*) Right forelimb of *Tulerpeton*. (Parts *a*, *b* after M. I. Coates and J. A. Clack, Polydactyly in the earliest known tetrapod limbs, Nature, 346:66–69, 1990; part *c* after O. Lebedev and M. I. Coates, The postcranial skeleton of the Devonian tetrapod Tulerpeton curtum, Zool. J. Linn. Soc. Lond., vol. 114, 1995)

used mainly for the sense of balance, and hearing would have been poor, limited to only low-frequency, water-borne sound. The equivalent of the human middle ear opened to the outside like a gill slit, and was used either in breathing or as a sensory organ, equivalent to the spiracle in some modern fishes.

Tail. The tail of *Acanthostega* bore a deep fin above and below, formed not just of flesh like a salamander's but of bony internal supports from the vertebral column and bony fin rays as in a fish, especially a lungfish. This type of tail is used in rapid acceleration by a lurking predator, not by an animal making frequent sorties onto the land, where the tail would dry out and become a nuisance. There are strong developmental reasons for believing that the presence of this tail was not a secondary adaptation to water but a primitive feature.

Limbs. Unlike other tetrapods, *Acanthostega* had a forearm with a spatulate radius that was significantly longer than its triangular ulna. The structure and proportions resembled those of the fossil osteolepiform fish *Eusthenopteron*, which are usually regarded as the fish most closely related to tetrapods. This observation supports the idea that *Acanthostega* is truly a primitive form. However, although conventional tetrapods have five digits, *Acanthostega* had eight. Its wrist was weak, and its limbs positioned so as to make them unlikely to have borne the animal's weight. Like the hindlimbs of *Ichthyostega*, these were paddles. Thus, they were probably used in water for pushing on vegetation or a soft muddy bottom of the river in which the animal lived.

Tulerpeton. *Tulerpeton* is a third genus of Devonian tetrapod, and was found in Russia. It is known only from skull fragments, girdles, and two complete, six-digited limbs. Thus, the remains of *Ichthyostega*, *Acanthostega*, and *Tulerpeton* indicate that the number of toes and foot elements was originally variable and stabilized only later (**Fig. 3**). Possibly the number five seen in conventional tetrapods was arrived at separately more than once.

Implications. Since the discovery of the *Acanthostega* fossils, more Devonian tetrapods have been discovered in Latvia, Scotland, and the United States. So far, most are known only from fragments, but when more complete specimens are found, conclusions about *Acanthostega* may be corroborated or refuted. Possibly, the connections between the skull and shoulder and the opercular bones were lost before gill breathing was abandoned. Digits may have evolved before wrists and ankles, and limbs, digits, and an enlarged pelvis may have evolved before walking, perhaps for some other, possibly genetic, reason. Only subsequently were limbs used on land; this development may be related to the establishment of the five-digit limb.

By the Upper Devonian fishlike tetrapods and tetrapodlike fish appear that are difficult to tell apart. Thus, it appears that tetrapods did not evolve much earlier than the lowest part of the Upper Devonian. Nevertheless, by the end of this period tetrapods were diverse and widely distributed. A coherent picture of their evolution is becoming possible as new discoveries of remains are made.

For background information SEE DEVONIAN; SKELETAL SYSTEM; TETRAPODA in the McGraw-Hill Encyclopedia of Science & Technology.

Jennifer A. Clack

Bibliography. P. E. Ahlberg and A. R. Milner, The origin and early diversification of tetrapods, *Nature*, 368:507–514, 1994; M. I. Coates and J. A. Clack, Fish-like gills and breathing in the earliest known tetrapod, *Nature*, 352:234–236, 1991; M. I. Coates and J. A. Clack, Polydactyly in the earliest known tetrapod limbs, *Nature*, 347:66–69, 1990.

Texaphyrins

Texaphyrins are porphyrin analogs that show great promise as potential diagnostic and therapeutic agents in oncology and possibly in cardiology. The texaphyrins, represented by the series of structures in **illus.** *a*, are a type of chemically enhanced porphyrin. Porphyrins, such as protoporphyrin IX (illus. *a*), are naturally occurring compounds whose molecules possess four pentagonlike rings (pyrrole rings) that are themselves tied up into an even bigger circular molecule, known as a macrocycle. In porphyrins the four pyrrole rings are joined at the sides so as to allow the nitrogen (N) atoms to point into the center of the macrocycle. Because these four nitrogen atoms can chelate, or hold, a metal atom, porphyrins play a very important role in the biology of humans (as well as of other living organisms). For instance, porphyrins hold the iron in human blood (as heme, illus. *b*) and give red blood cells their characteristic color. For reasons that are still not understood, porphyrins are also known to

Porphyrin structures. (*a*) Metallotexaphyrins; paramagnetic: M = gadolinium(III) and R = CH_2OH or $O(CH_2CH_2O)_2CH_3$; diamagnetic: M = lutetium(III) and R = CH_2OH or $O(CH_2CH_2O)_2CH_3$; the complexes are dicationic. (*b*) Protoporphyrin IX (M = 2H); H = hydrogen or heme [M = iron(II)]. (*After J. L. Sessler et al., Texaphyrins: Synthesis and applications, Acc. Chem. Res., 27:43–50, 1994*) localize selectively in tumors, leading researchers to suggest that this class of compounds could be used to detect and treat cancerous disorders.

Magnetic resonance imaging. One of the potentially most promising means of detecting cancerous lesions involves the use of magnetic resonance imaging (MRI). This type of medical imaging is based on the principles of magnetism and, in particular, uses an assessment of water-based proton spin relaxation rates to effect tissue differentiation. When placed in a magnetic field and "stretched" with a radiofrequency pulse, the nuclear spins of protons tend to "snap back" to a normal, so-called ground state in a certain characteristic time, defined by the corresponding relaxation rate. This snapping back, or relaxation, is actually not much different from what happens when the needle of a compass is spun off magnetic north: it tends to return quickly to its normal, north-pointing direction. In any case, the key idea is that by looking at the relaxation rates for the various water proton spins, it is possible, at least in principle, to tell the difference between a normal and a cancerous grouping of cells.

However, a major drawback in magnetic resonance imaging is that normal and cancerous cells look about the same. In other words, the spin relaxation rates in these two types of biological media are not so different from each other. Thus, it is hard to make the crucial differentiation needed for unambiguous diagnosis. To overcome this limitation, practitioners in the field are turning to the use of new drug substances known as contrast agents. These agents work by modulating the intrinsic relaxation rate of the proton spins that can change their effective "elasticity." When this modulation is carried out in a site-selective manner, it becomes possible to effect tissue-specific magnetic resonance imaging.

Rates of proton spin relaxation are adjusted in magnetic resonance imaging by adding a paramagnetic species (that is, one with unpaired electrons). Thus, at least as far as cancer detection is concerned, the site specificity problem has two aspects: (1) finding something, such as a porphyrin, that localizes to tumors, and (2) using it to deliver a paramagnetic species. However, a difficulty arises when binding the paramagnet to the porphyrin. The best atomic-scale paramagnets are metal cations and the best of these is the trivalent lanthanide cation, gadolinium(III). Unfortunately, this species is too big to fit into a porphyrin. Thus, for this approach to work a bigger porphyrinlike species is needed, and it is here that the texaphyrins have a role to play. The texaphyrins, in marked contrast to normal-sized porphyrins, contain 5 (not 4) nitrogen atoms (illus. *a*). They also possess a core size about 20% larger than that of the tetrapyrrolic porphyrins (illus. *b*). These two features, along with the starlike nature of the central core, combine to make the texaphyrins appear as true Texas-sized porphyrins. Of equal importance is the fact that these two properties also make this class of molecules ideal for the complexation of gadolinium(III) and other trivalent lanthanide cations.

Several gadolinium(III)-containing texaphyrins are undergoing preclinical testing as possible contrast agents for magnetic resonance imaging. One joint research program has identified two promising paramagnetic compounds (illus. *a*). These two drug candidates were found to be stable in the living organism and relatively nontoxic. Also, they were found to modulate significantly the rate of spin flipping, or relaxation, in the laboratory and to be highly effective contrast agents in living organisms. These systems were found to localize selectively to tumors and to effect the contrast enhancement in the magnetic resonance imaging. They were also found (data not shown) to target to atherosclerotic plaque. Thus, it is possible to conceive that these agents, or ones like them, will emerge as important tools for use in the diagnosis of oncologic and cardiovascular disease.

Photodynamic therapy. The second main way in which the science of porphyrins interacts with that of cancer treatment is in the area of photodynamic therapy; here too, it is anticipated that the texaphyrins will have an important role to play. Photodynamic therapy is an attractive new light-based means of treating tumors that was introduced by T. Dougherty of the Roswell Park Memorial Cancer Institute. It involves using a porphyrin or other photoactivable dye to produce so-called singlet oxygen, a species that is highly cytotoxic to both tumors and other tissues. In terms of basic physics, light is shined onto the porphyrin or dye, thus activating the molecule by ripping two paired electrons apart, so as to produce an excited molecular triplet state. This excited triplet reacts with normal atmospheric oxygen, which is also in the triplet state, to produce the singlet oxygen. This singlet oxygen is very reactive and effectively destroys whatever tissues it encounters. In terms of clinical practice, therefore, the goal would be to produce this singlet oxygen just at the tumor site and nowhere else. To do this, it is necessary that the dye makes contact with the tumor and that a wavelength of light specific for photoexciting the dye molecule in question is used.

Absorption problems. Porphyrins, especially a complex mixture known as hematoporphyrin derivative (a substance originally derived from heme; illus. *b*), show promise for photodynamic therapy. They suffer, however, from a serious drawback, namely, that they are the same color as human flesh. (The dominant pigments in the body, found in hemoglobin—red blood cells—and myoglobin—muscle tissue—are porphyrin-based.) An attempt to shine light deep into the body so as, for instance, to produce singlet oxygen at the site of a deeply embedded lung or colon tumor will lead to light absorption by the pigments in the body. As a consequence, little light will penetrate through to photoexcite the special porphyrin localized at the tumor. Thus, little if any singlet oxygen will be produced and little or no tumor eradication effected. Restated more succinctly, the porphyrins, with their inappropriate absorption characteristics, are simply not optimal for photodynamic therapy.

The texaphyrins offer a means of overcoming the absorption-related limitations because they contain a greater number of π electrons, making them act as if they are electronically bigger than the porphyrins and not just physically bigger than these simpler tetrapyrrolic systems. As a result, the texaphyrins absorb light further to the red end of the spectrum (at lower energy) than do the porphyrins and other naturally occurring body pigments (at around 750 nanometers versus ≤ 630 nm). Thus, it is possible to photoexcite texaphyrins deep within the body (that is, at a tumor site) by using red light. The reason is that red light is not absorbed extensively by the pigments of blood and tissue and thus it passes through body tissue relatively unattenuated until it reaches the texaphyrin dyes at the tumor locus; there, the light is absorbed by the texaphyrin, producing singlet oxygen.

Photosensitizing. For singlet oxygen to be produced by a texaphyrin or other big dye molecule, the molecule must have all its electrons paired (that is, it must be a diamagnetic system). This requirement is just the opposite of that required for effective function in magnetic resonance imaging. However, the texaphyrins are diverse in their binding characteristics. Thus, by replacing gadolinium(III) in the texaphyrin complex with the diamagnetic lanthanide cation lutetium(III), a tumor-localizing system that is photoactive is obtained. This compound [illus. b, where R = $O(CH_2CH_2O)_2CH_3$] represents an optimum photosensitizer. Indeed, it has proved possible to use this complex and red laser light (wavelength = 732 nm) to effect actual cancer cures in rodents. This result is exciting in its own right, but is made more interesting by the fact that it was obtained under excitation conditions where hematoporphyrin derivative is ineffective as a photosensitizer. Recent results suggest that this same general photodynamic approach allows for the light-based eradication of atherosclerotic plaque in specially conditioned rabbits and, by extrapolation, other animals, perhaps including humans.

Prospects. The great promise of the texaphyrins is that the same basic series of molecules could be used both to localize a tumor [possibly by using the gadolinium(III) system, where R = $O(CH_2CH_2O)_2CH_3$; illus. a] and cure it [by using the analogous lutetium(III) derivative], with the same holding true for certain types of cardiovascular disease. This, of course, would represent a new level of sophistication in the clinical control and treatment of these all-pervasive disorders. Thus, while difficulties remain as to the successful application of texaphrins to the treatment of disease, the promise inherent in this new class of porphyrin analogs makes them among the more interesting of all known porphyrin analogs.

For background information SEE LASER PHOTOBIOLOGY; MAGNETIC RELAXATION; MEDICAL IMAGING; PORPHYRIN; PYRROLE; TRIPLET STATE in the McGraw-Hill Encyclopedia of Science & Technology.

Jonathan L. Sessler

Bibliography. D. Dolphin (ed.), *The Porphyrins*, vols. 1–8, 1978; B. W. Henderson and T. J. Dougherty, How does photodynamic therapy work?, *Photochem. Photobiol.*, 55:145–157, 1992; J. L. Sessler et al., Texaphyrins: Synthesis and applications, *Acc. Chem. Res.*, 27:43–50, 1994; S. W. Young, *Magnetic Resonance Imaging: Basic Principles,* 1988.

Thyristor

Thyristors are the key building blocks of power electronics, the high-voltage, high-current counterparts of transistors used in ordinary electronic equipment. For power electronics applications, such as variable-speed drive motors or controls for utility transmission systems, thyristors provide much greater operating efficiency than can be achieved by transistors. An advanced type of thyristor with markedly improved performance characteristics, called the MOS-controlled thyristor, has recently become available as a commercial product.

Like other types of solid-state electronic devices, thyristors are built from layers of crystalline silicon that have been doped with small amounts of elements that either donate electrons to the crystal lattice or take them away. Current in an n-type layer is carried by electrons and in a p-type layer by positively charged holes formed when electrons are missing from the lattice. The junction between a p-type and an n-type layer will conduct current only when the electrons and holes are brought together from opposite sides of the junction by applying a positive voltage on the p side and a negative voltage on the n side.

Types. The simplest kind of thyristor, sometimes known as a silicon-controlled rectifier (SCR) (**Fig. 1**), consists of four layers of doped semiconductor material in a $pnpn$ configuration. When a positively charged electrode (anode) is connected to the p end of the device and a negatively charged electrode (anode) to the n end, the first pn junction acts as a diode rectifier, allowing current to flow in only one direction. To turn on this current, a positive voltage must be applied to the embedded p layer, which acts as a gate. A major disadvantage of silicon-controlled rectifiers is that they cannot be turned off until the voltage across the cathode and anode returns to zero.

A gate turnoff (GTO) thyristor (**Fig. 2**) overcomes this difficulty by having multiple gates and cathodes interspersed at one end of the device. Current flow can be interrupted by applying a negative voltage to the gates, which draws the current away from the cathodes. However, this configura-

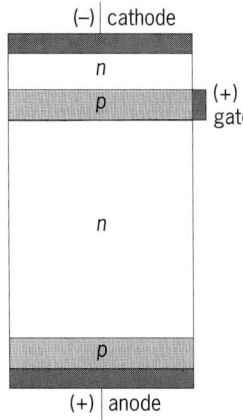

Fig. 1. Silicon-controlled rectifier (SCR).

tion is expensive to fabricate and relatively inefficient because of the need for a substantial gate current. Gate turnoff thyristors have therefore found only limited application.

The MOS-controlled thyristor (MCT) (**Fig. 3**) needs only a small gate current to turn the entire device off or on, thus significantly reducing energy loss and cutting the time required for turnoff to one-third that of gate turnoff thyristors. Device size can also be reduced considerably. An MOS-controlled thyristor achieves these advantages by placing a very thin metal oxide semiconductor (MOS) integrated circuit on the top surface of the high-power thyristor components. In this configuration, the gate is separated from the conducting layers of the thyristor by an insulating layer of oxide. Applying a positive or negative voltage to the gate turns the MOS-controlled thyristor off or on, respectively, by temporarily creating thin conduction channels in the p layer or n layer near the device surface.

Applications. The use of thyristors to replace mechanical control systems in applications ranging from hand-held tools to utility power delivery networks has been called the second electronics revolution. Already, about 40% of the electricity generated in the United States passes through some sort of power electronics device on its way to a load. Most of these functions currently depend on silicon-controlled rectifiers, but commercialization of the MOS-controlled thyristor is expected to further accelerate the power electronics revolution by opening new applications and reducing the cost of existing ones.

One of the most important uses of thyristors is to control the speed of electrical motors through power-conditioning equipment known as an adjustable-speed drive. Through the use of this equipment, the energy efficiency of industrial processes relying on blowers, compressors, and pumps can be improved by 50% or more over that of systems that use only fixed-speed motors. Inside an adjustable-speed drive, thyristors convert ordinary 60-Hz alternating current to an adjustable direct current and then back to alternating current with the desired combination of voltage and frequency to control motor speed.

Thyristors are also commonly used in devices designed to protect sensitive equipment against power interruptions or other power quality problems. Many companies, for example, use an uninterruptible power supply (UPS) to maintain computer operation in the event of a power failure. Each uninterruptible power supply contains batteries, which supply electrical energy in the absence of power from a utility line, and thyristors, which switch the load to battery power and convert direct current to alternating current. Other kinds of protection devices use thyristors to respond automatically to power disturbances, such as surges and spurious frequencies, and provide well-regulated voltage to equipment that might otherwise be damaged by the disturbances.

In utility applications, thyristors are coming into use to improve power-delivery networks at both the transmission and distribution system levels. Thyristor-controlled devices are being used to develop a flexible alternating current transmission system (FACTS), designed to maximize power transfer over existing power lines. On some lines, these power electronics controllers can increase available capacity by as much as 50%. Other thyristor-controlled devices are under development to enhance power delivery to customers through a program called custom power. The advantage of custom power is that it will reduce power disturbances at lower cost and with less energy waste than can be achieved by using an uninterruptible power supply.

MCT development. Silicon-controlled rectifiers have been in use for many years and currently control power up to levels of several megawatts. Gate-turnoff thyristors were first used at the megawatt level in 1995, to provide voltage support at a large utility substation, as part of the FACTS program. The first commercial MCT has a rating of about 120 kW, large enough for many applications in industry

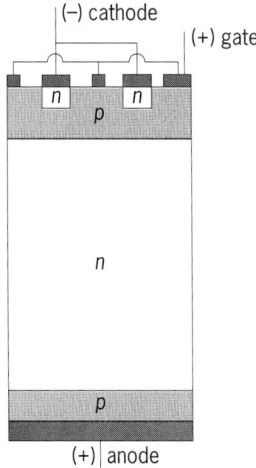

Fig. 2. Gate-turnoff (GTO) thyristor.

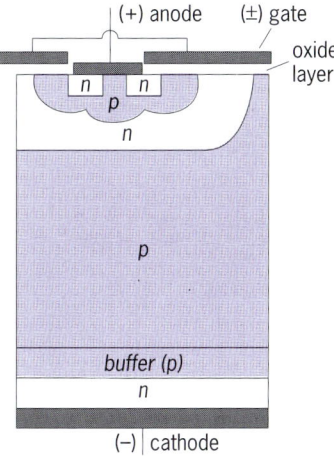

Fig. 3. MOS-controlled thyristor (MCT).

and electrical transportation, but still a long way from utility use.

By providing faster switching speeds with lower electrical losses, MOS-controlled thyristors are expected to revolutionize the use of power electronics in a variety of functions. One of the most promising is to replace some hydraulic systems in aircraft with thyristor-controlled electric motors, thus reducing their weight, improving reliability, and lowering maintenance costs. Such changes could simplify airframe design and extend flight capabilities. Thyristor controls are also expected to add significantly to the driving range of electric cars and to provide less expensive propulsion systems for ships.

To further increase the capacity of MOS-controlled thyristors, a National Power Semiconductor Interagency/Utility Consortium, formed in 1986, conducts a wide-ranging program of research and development in the United States. A primary technical target of the consortium is to develop very powerful MOS-controlled thyristors for utility applications. Specifically, development of a MOS-controlled thyristor that could switch a megawatt of power in less than a microsecond with less than a millijoule of gate energy would make FACTS and custom power very attractive for widespread utility use.

In addition to increasing power levels, a parallel research effort is aimed at reducing thyristor costs by improving their packaging. Currently, high-power thyristors are packed in modules that include a thick, heavy tungsten plate to disperse heat. These so-called hockey-puck modules can weigh 5 lb (2.25 kg) and represent a major sacrifice in cost and convenience. Because MOS-controlled thyristors produce less heat than other types of thyristors with similar capacity, they are potentially easier to package. The research consortium is thus sponsoring development of a hermetic packaging system for MOS-controlled thyristors, which would have only a small fraction of the weight and volume of current modules and could be manufactured less expensively.

As thyristor technology development progresses, silicon-controlled rectifiers are expected to continue to be used in applications where turnoff capability is not required. They are expected to remain less expensive than comparable MOS-controlled thyristors for some time. However, gate turnoff thyristors will probably be replaced by MOS-controlled thyristors because of the latter's capability to turn high currents off and on with very low energy loss.

For background information SEE CIRCUIT BREAKER; CONVERTER; ELECTRIC UNINTERRUPTIBLE POWER SYSTEM; MOTOR; SEMICONDUCTOR; SEMICONDUCTOR RECTIFIER; STATIC VAR COMPENSATOR in the McGraw-Hill Encyclopedia of Science & Technology.

Karl E. Stahlkopf

Bibliography. J. H. Douglas, Sealed in silicon: The power electronics revolution, *EPRI J.*, 11(9):5–15, December 1986; N. G. Hingorani and K. E. Stahlkopf, High power electronics, *Sci. Amer.*, 269(5):78–85, November 1993.

Tooth

Although phylogenetically derived from weakly anchored scales in the integument of jawless Paleozoic fishes, the teeth in most vertebrates have firm support from the underlying skeleton and serve to transfer force from the jaw muscles to the prey or food item. Teeth also concentrate stresses at points of transfer, thus placing mechanical demands on teeth that are not required of other skeletal structures. The result has been extensive evolutionary change in dental tissues, especially the enamel, in response to dietary changes.

Precise occlusion. Mammalian teeth have the greatest diversity of shape and the greatest individual complexity. The evolutionary event that is believed to underlie the complexity of mammalian teeth is the reduction of multiple tooth replacements (polyphyodonty), which characterize fishes, amphibians, and reptiles, to a single replacement dentition (diphyodonty). In polyphyodonty, teeth are replaced in a staggered order, and thus new teeth do not erupt in the precise positions of lost teeth. This prevents individual teeth from maintaining constant positional relationships with opposing teeth. In contrast, in mammalian diphyodonty positional stability can be so great that numerous wear facets produced by opposing teeth can be individually identified in homologous positions across the teeth of many different taxa. An advantage of this constancy of positional relationships is that evolutionary selection can modify the shapes of individual teeth to more precisely control the mechanical transfer of stress to food materials.

Mammalian dental requirements. By volume, enamel is 92% hydroxyapatite, making it the most highly mineralized of skeletal tissues in vertebrates. This high mineral density makes enamel the most resistant of all tissues to abrasion, and explains its widespread occurrence in vertebrates as the outer layer on teeth. Enamel allows the tooth to maintain its function as a stress-concentrating mechanism by conserving the sharp cusps and edges as they come in contact with resistant materials. Mammalian diphyodonty accentuates the need for such conservation because the single set of replacement teeth must last for the lifetime of the individual, and because the relatively high rate of energy and nutrient intake requires acceleration of the digestive process by fracturing food objects into small volumes.

This latter requirement is particularly acute in herbivores for two reasons: They have never been able to produce the enzyme (cellulase) that can digest cellulose and they have had to employ high stresses on the biting surfaces of their teeth in order to to fracture the tough cellulose of plant cells to reach the digestible cellular nutrients. Many taxa are able to digest some of the cellulose through the action of symbiotic bacteria that produce cellulase, but they must nevertheless first fracture the cellulose finely. The requirement to finely subdivide resistant plant fibers places a premium on the presence of numerous sharp enamel edges and the attainment of high stresses on biting surfaces of the teeth. High dental stresses are also attained in carnivores, either while consuming bones as part of their diet or when biting prey and striking resistant bone.

Enamel edges must be able to withstand the same stresses that are imparted to the food materials without fracturing or wearing excessively. However, the mineralization and hardness that make enamel resistant to abrasion also make it brittle. The fundamental weakness of enamel to fracture has led to its extensive evolutionary modification.

The difference in magnitude of the amount of work involved in the fracture of enamel under different loading directions results from a fundamental property of apatite: individual crystallites are most fracture-resistant to stress acting parallel to the crystal axis. Crystallites of mammalian enamel are organized in larger units called prisms; the average direction of crystallites in a prism is parallel to the long prism axis. Therefore, prismatic enamel is most easily fractured when tensile stresses act perpendicular to the long axes of the prisms. Also, enamel is least resistant to abrasion when the abrasive force acts perpendicular to the long axes of the prisms. This response of enamel to stress has provided mechanisms for the strengthening of mammalian enamel against both abrasion and fracturing through evolutionary changes in prism directions.

Abrasion resistance. It may not be possible for the structure of a given region of enamel to be optimized for both abrasion resistance and fracture resistance, because different types of reorganization are required for each. Optimum abrasion resistance is attained when the axes of the prisms are aligned parallel to the direction and magnitude of the abrasive force. Enamel prisms are deposited starting at the inner surface and ending at the outside surface, in parallel with the direction of movement of the enamel-depositing cells. Therefore, the prisms tend to be directed toward the outer surface of the enamel. However, the actual angle that prisms make with the outer surface has been found to be asymmetric, for example, on opposite sides of the tooth, and the difference correlates with local differences in the direction and magnitude of the chewing force with respect to the surface. Thus, selection appears to modify the directions of the prisms in order to improve the resistance of enamel to abrasion.

Fracture resistance. The crossing of prisms in parallel planes is the most frequently observed type of structural reorganization for improving fracture resistance of enamel. This type of structure, known as Hunter-Schreger bands, is characterized by prisms running in different directions in adjacent layers of enamel, the directions often differing by 90°. The boundaries of these layers form more or less parallel planes (called decussation planes). This structure is present in the teeth of many Primates (including humans), Perissodactyla, Carnivora, and Rodentia, as well as other mammalian orders. In cases where the directions of the principal chewing stresses within the enamel have been determined, the decussation planes tend to remain parallel to the directions in which the maximum tensile stresses act, even when the direction of the stress has radically changed in the course of evolution, as has occurred in the rhinoceroses.

Prism decussation has the effect of raising the level of stress required for crack propagation. Tensile stresses acting normal to the crack plane are responsible for driving cracks in brittle materials. Stresses acting in this direction are concentrated at the tip of the crack. For a given load, a longer crack or a smaller tip radius results in a higher stress at the crack tip acting to further extend the crack.

Because the orientation of prism decussation planes tends to be parallel to maximum stress directions, the prism plane is perpendicular to the crack path and therefore inhibits crack propagation in several ways. First, it forces an expanding crack that is initiated in the weakest direction (parallel to prism axes) to pass through prisms in an adjacent region in the most resisted direction (perpendicular to the prism axes). Second, as a crack propagates, smaller stresses acting perpendicular to the stresses that propagate the crack can separate prisms along their weak boundaries, creating small voids in regions where the prisms run at 90° (or at some oblique angle) to the crack plane. When the crack tip reaches such a void the radius of the tip enlarges, reducing the stress concentration at the tip and increasing the amount of tensile stress necessary to further extend the crack. Finally, the surfaces of cracks passing through enamel in planes perpendicular to the decussation plane are highly rugose, so that the surface area and the energy required for crack extension are greater than for crack surfaces running parallel to prism axes, thus making the crack more resistant to propagation.

In many herbivores the decussation planes are quite uniform in direction and both the chewing and loading directions are apparently constant. However, in many carnivores the decussation planes bend in a sinusoidal pattern. This pattern appears to be an adaptation to resist varying stress directions, because the greatest amount of stress that occurs in carnivore teeth results from striking bones when biting prey; the angles at which the tooth strikes bone tend to vary.

There are other ways in which enamel microstructure has been modified in response to stress. For example, in some taxa, including late Cenozoic horses, marsupials, and rodents, crystallites in the interprismatic region of the tooth enamel are at an angle of up to 90° from the direction of the prisms themselves, thus forming a finer type of decussation at the crystallite level. Many issues concerning how and why the different evolutionary strategies that strengthen enamel have developed remain.

For background information SEE TOOTH in the McGraw-Hill Encyclopedia of Science & Technology.

John M. Rensberger

Bibliography. H. O. Pfretzschner, Structural reinforcement and crack propagation in enamel, *Mem. Mus. Nat. Hist., Natur. Paris C*, 53:133–143, 1988; D. Stern, A. W. Crompton, and Z. Skobe, Enamel ultrastructure and masticatory function in molars of the American opossum, *Didelphis virginiana*, *Zool. J. Linn. Soc.*, 95:311–334, 1989; J. Thomason (ed.), *Functional Morphology in Vertebrate Paleontology*, 1995; W. von Koenigswald et al., Functional symmetries in the *schmelzmuster* and morphology of rootless rodent molars, *Zool. J. Linn. Soc.*, 110:141–179, 1994.

Transistor

Electronics has been based almost entirely on inorganic semiconductors, such as silicon or gallium arsenide; in fact, most active electronic devices, such as bipolar transistors and field-effect transistors (FETs), are fabricated by using silicon technology. However, it is possible to make an active device from other inorganic semiconductors, such as a gallium arsenide FET, or from superconductors, such as a Josephson junction gate. The range of materials employed in electronics may be considerably extended by two newly developed devices: thin-film transistors that are fabricated entirely from polymers (organic semiconductors) and the spin transistor, which is fabricated entirely from metals.

All-Polymer Thin-Film Transistors

Organic semiconductors have been considered chiefly as academic curiosities, but recently significant advances have opened a new area of flexible organic-based electronics with the potential of attractive new applications.

One of the key properties that define a semiconductor concerns the velocity at which electrical charges are able to move in this material under an applied field, which determines the response time of devices based on it. This velocity, expressed in terms of charge carrier mobility (the ratio of the charge carrier velocity in centimeters per second to the applied field in volts per centimeter), reaches values of the order of 10^3 $cm^2 V^{-1} s^{-1}$ in monocrystalline silicon, for instance, allowing operating frequencies in the gigahertz range, needed in fast computer electronics. In contrast, applications such as display are much less demanding, requiring switching times, in the microsecond range, which correspond to mobilities of the order of 1 $cm^2 V^{-1} s^{-1}$. Amorphous hydrogenated silicon (a-Si:H), which shows such mobility, is employed in the fabrication of thin-film transistors. These transistors are used in flat-panel display technology, where each of the million liquid crystal–based pixels, which compose the panel, are switched on and off by such a transistor.

Some organic materials also show semiconducting properties, which have been studied since the early 1950s for both their fundamental interest and their potential applications. Organic materials are inexpensive and can be easily processed, even on large surfaces. However, although extensive research brought about deepened understanding of the process of charge transport in organic materials, pessimistic conclusions were drawn at the end of the 1980s on their significance for practical applications. The measured mobilities were limited to some 10^{-5} $cm^2 V^{-1} s^{-1}$, which was thought to be an intrinsic limit in organic materials, excluding any application to electronics.

Steps toward organic devices. In 1990, however, two significant steps toward organic-based devices were announced. A large increase was achieved in the values of charge-transport properties of a new class of organic semiconductors, the oligothiophenes, owing to the tight structural control which can be exerted over the molecular organization in thin films made from these semiconductors. Oligothiophenes are short conjugated molecules which display much better stacking properties than their highly disordered counterparts, the polythiophenes. Thin-film transistors fabricated from sexithiophene thus showed mobilities as high as 5×10^{-3} $cm^2 V^{-1} s^{-1}$, better by a factor of almost 1000 than that of polythiophene. Other relevant work, carried out independently, showed that an organic semiconductor could be efficiently used as active layer in electroluminescent diodes. These two advances were the first evidence in support of the possibility of using organic materials as active layers in devices.

Subsequent work confirmed that the improvement in mobility was directly governed by the degree of molecular structural organization of oligothiophene thin films. Characterizations using x-ray diffraction, electron microscopy, and absorption spectroscopy showed that the values of electrical properties increased with the long-range order in these films. The long-range order could, in turn, be improved by modifying the experimental conditions used for film deposition (substrate temperature, rate of evaporation), which further increased the mobility to 2×10^{-2} cm^2 V^{-1} s^{-1}. This mobility was, however, still too low for practical applications. However, these results also pointed out the large dependence of the semiconducting characteristics on the conditions used for film deposition. They raised the question of whether it would be possible to (1) further increase the semiconducting properties of these oligomers, and (2) push the concept of an organic-based device to its ultimate bounds to realize a thin-film transistor involving only polymer materials.

Advantages of organic TFTs. Compared to their covalent inorganic counterparts, organic semiconductors, in which molecules are held together by weak van der Waals interactions, show a lower cohesion, which results in much easier processing under milder conditions. Inexpensive organic materials can thus be deposited over large surface areas, and, in the same way as polymers, their mechanical, optical, and electronic properties can be easily tuned through subtle chemical modifications. These advantages appear to meet demands which have emerged in the recent years for the development of so-called flexible electronics. Among the various envisaged applications are smart cards (used in banks, security applications, and so forth), on which more and more logic and memory functions are included. On a mechanical basis, the required flexibility of these cards is not compatible with the classical stiff chips actually fixed on them. Besides this application, manufacturers of automotive vehicles and aircraft have shown an interest in the development of flexible transparent displays which can be fixed on windshields. Finally, there has long been interest in a flexible display panel, which could be rolled up and down at will. These new technologies require the development of flexible, sometimes transparent devices compatible with large surfaces, and these requirements match the characteristics offered by organic materials.

Control of molecular organization. In order to try to further increase the semiconducting properties of these materials, a chemical approach has been used to increase the long-range structural order of molecules in a thin film. With this aim, alkyl groups, which are known to show aggregation properties based on lipohilic-hydrophobic interactions, have been covalently grafted at the end of the sexithiophene molecule previously studied. Thin films of dihexylsexithiophene, deposited at room temperature, showed an extremely well-ordered structure, in which molecules are perfectly stacked together, with a very long range order. This quasicrystalline organization has been obtained because of the intermolecular recognition properties brought by these alkyl groups, which induce self-assembly properties in the molecules. This result confirms the

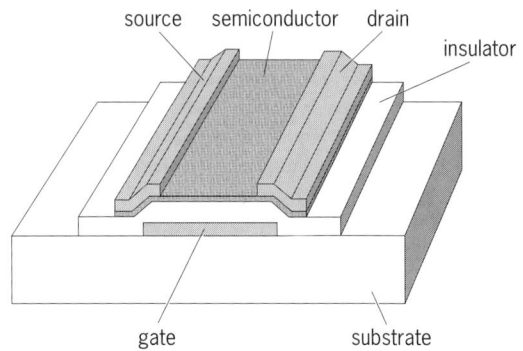

Fig. 1. Structure of all-polymer thin-film transistor.

power of a chemical approach to the structural control of molecular layers.

Thin-film transistors based on this semiconductor showed a mobility of 8×10^{-2} cm^2 V^{-1} s^{-1}, very close to that of amorphous hydrogenated silicon. An increase of mobility of organic semiconductors by a factor of almost 10,000 has thus been achieved. Logic gates based on these thin-film transistors have been also realized, confirming the potential of organic thin-film transistors.

All-polymer device. The second step toward construction of an all-polymer device was the substitution of the metallic source, drain, and gate electrodes by organic-based ones. Among the many possibilities, conducting graphite-loaded polymer inks were used. A thin-film transistor has thus been realized entirely from polymer materials, and fabricated under mild conditions comparable to those involved in printing technology (**Fig. 1**).

A commercially available thin polyester film forms the insulator on which the organic conducting pattern, gate, source, and drain electrodes are deposited through a mask. An adhesive polymer is then applied as substrate, providing stiffness. A very thin layer of organic semiconductor, dihexylsexi-

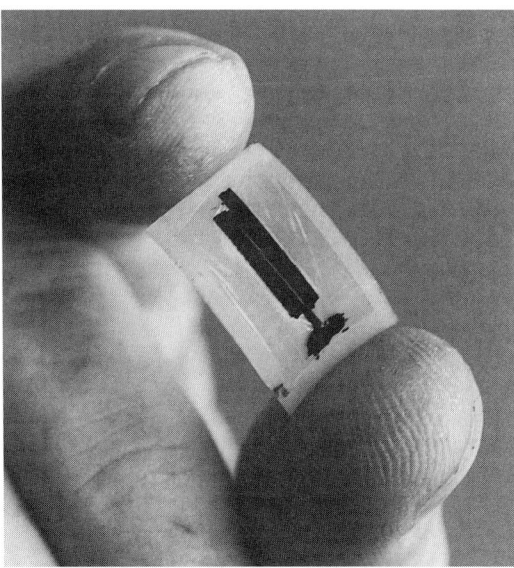

Fig. 2. Example of all-polymer thin-film transistor.

thiophene, is then deposited, either by evaporation at 320°C (608°F) or by sublimation at 200°C (392°F) under argon flow. The critical dimension in thin-film transistors is the channel length between the source and drain electrodes, typically some micrometers in classical technology. Because the spatial resolution of the mild printinglike techniques used here is limited to some 50 micrometers, this all-polymer device appears oversized (**Fig. 2**). However, some applications of large displays do not require micrometer-sized devices. Furthermore, micrometer-sized all-polymer thin-film transistors can also be fabricated by applying microlithographic techniques to conducting polymers, thus allowing microsized patterns to be realized for organic electrodes.

This all-polymer device has shown excellent amplification characteristics. The insensitivity of these characteristics to mechanical manipulations, such as bending and twisting of the device, highlights the device's flexibility. Other properties, such as high optical transparency, are achievable with these devices, owing to the ease with which the properties of organic materials can be controlled through molecular engineering.

Prospects. This first demonstration of an all-polymer device confirms the control which can be exerted on the molecular organization and on the charge-transport properties of organic semiconductors. An interesting potential for applications in flexible electronics also exists, although work is still needed before any real application can be envisioned. Improvements must still be made in the mobility of these organic semiconductors, which are actually lower by a factor of 10 than that of amorphous hydrogenated silicon, and other characteristics such as long-term stability must be confirmed. But the structure-property relationships which have been derived for these materials provide a reasonable hope that organic semiconducting materials equivalent to amorphous hydrogenated silicon will be achieved in the next few years, thus opening the field of flexible electronics. Thus, significant steps have recently been taken toward a new generation of organic-based devices, thin-film transistors and electroluminescent diodes, bringing a new dimension to organic materials.   *Francis Garnier*

All-Metal Spin Transistors

The spin transistor (or magnetic spin transistor) is a novel, active device, unique because it is fabricated entirely of metals. Still in an early stage of research and development, its basic embodiment as a switching device will lead to potential applications as a memory element or logic gate. Its characteristics favor miniaturization, and it may eventually be used wherever high packing densities are desired. A different category of applications, that is, as magnetic field sensors, will not be discussed.

Ferromagnetic and nonmagnetic metals. Ferromagnetic (F) metals differ from nonmagnetic (N) metals in the properties of their conduction electrons. An electron is a Fermi particle: it has a spin of ½, with two quantum states, up and down, and can be thought of as a magnetic dipole, with a moment equal to a Bohr magneton (β), pointing up or down. All metals have one or more bands in which valence electrons are available for conduction. In nonmagnetic metals the bands are spin-split symmetrically, so that at any energy, E, the number of spin-up conduction electrons per unit energy, $N_\uparrow(E)$, equals the number of corresponding spin-down electrons, $N_\downarrow(E)$. By contrast, in a band model of a transition-metal ferromagnet (such as nickel, iron, or cobalt), the two spin subbands of the $3d$ band are mutually offset by the exchange energy, U_s.

The exchange splitting of F has two consequences. First, the total number of electrons in one subband is greater than that in the other, resulting in a spontaneous magnetic moment, given by Eq. (1),

$$\mathbf{M} = \beta \int_0^{E_F} [N_\downarrow(E) - N_\uparrow(E)]dE \qquad (1)$$

where E_F is the Fermi energy, the energy of the highest filled state. Second, the density of electrons near the Fermi energy, $N(E_F)$, is different for the two spin subbands; that is, $N_\uparrow(E_F) \neq N_\downarrow(E_F)$. In a simplified model, the majority (down-spin) subband lies entirely below E_F so that $N_\downarrow(E_F) = 0$, but more generally the Fermi level, E_F, cuts across both subbands in an inequivalent way. Since electronic conduction involves only electrons with energies near E_F, more of the electric current in F is carried by one kind of spin than is carried by the other.

Spin-transistor operation relies on properties, recently discovered, of systems comprising contiguous ferromagnetic and nonferromagnetic metals. In 1973, tunneling conductance measurements showed that the electric current crossing the boundary of a ferromagnetic electrode had spin properties similar to those of the current internal to F. This surprising result was taken several steps further by the spin-injection experiment in 1985. It demonstrated the principle of charge-spin coupling, an effect predicted in 1980 by the theory of R. Silsbee.

Spin injection. Figure 3a depicts a thin ferromagnetic layer, F1, fabricated on a thicker, nonmagnetic metal, N. The battery, with voltage V, is drawn with a polarity such that electrons are driven clockwise, through F1, across N, and back to the battery. (Electric circuits are usually described, with unavoidable confusion, as though the charge carriers had positive charge. The diagrams in Fig. 3 use the opposite convention: current, voltage, and electric field are all defined with respect to electrons.) If the magnetization orientation $\hat{M}_1$ of F1 is down, the electrons carrying current in F1 have spin up. When the battery drives a circulation, the electrons crossing the F1-N interface have a net spin polarization which penetrates a relatively large distance in N. In the diffusive conduction process, each electron is constantly being scattered so that its mean free path l is quite short. The macroscopic current I results from a net drift of electrons when each electron nudges the

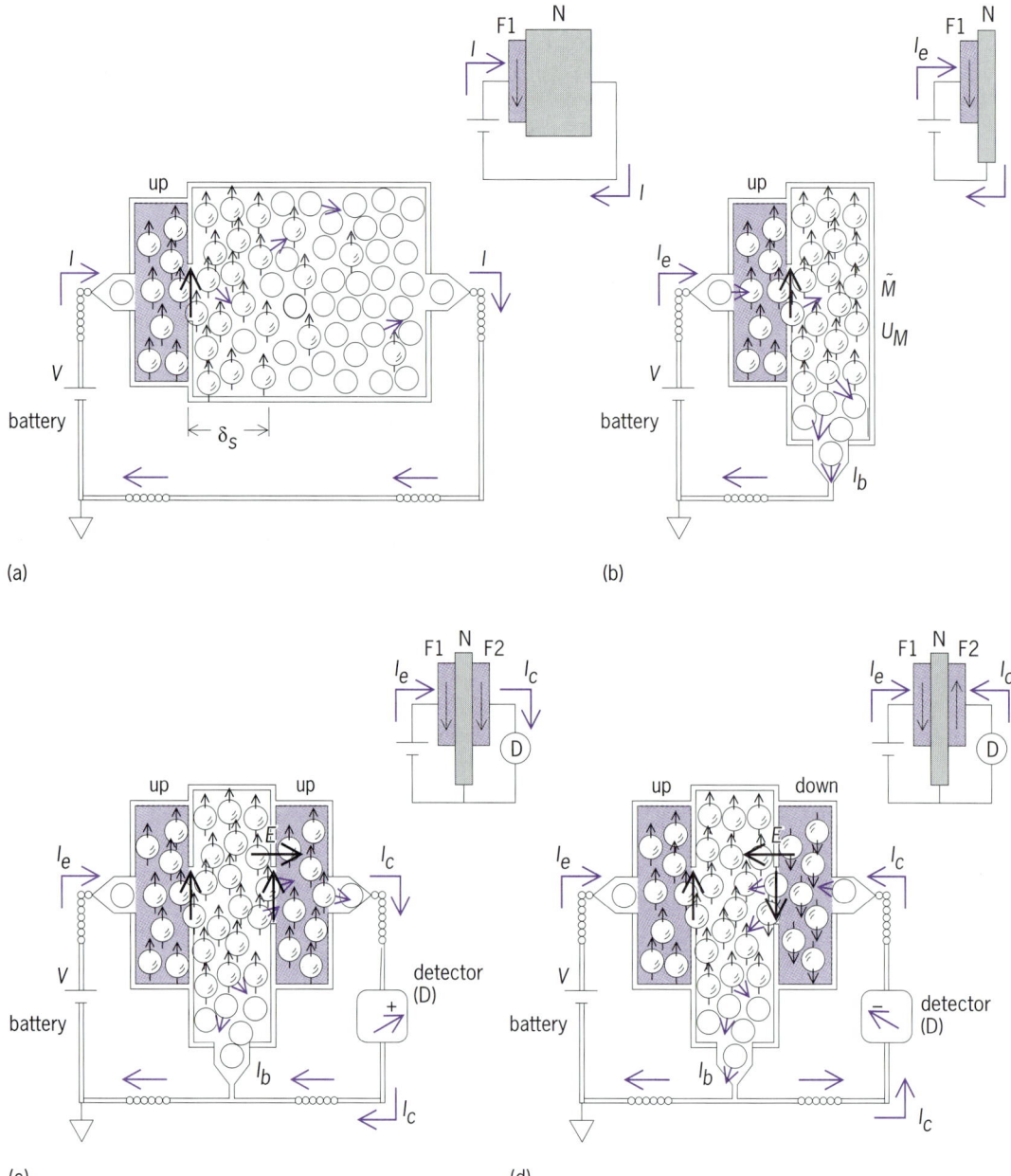

Fig. 3. Charge and spin transport in ferromagnetic-nonmagnetic (F-N) structures. A schematic circuit diagram is shown for each structure. Symbols are explained in text. (a) F-N bilayer in which thin ferromagnetic layer F1 is fabricated on thicker, nonmagnetic metal, N. (b) F-N bilayer in which the thickness of N is much smaller than the spin depth δ_s. $\tilde{M}$ = nonequilibrium magnetization and U_M = associated magnetic energy. (c) Three-terminal device made by adding F2. By analogy with a transistor, F1 is the emitter, N the base, and F2 the collector. Here, $\hat{M}_1$ and $\hat{M}_2$, the magnetizations of F1 and F2, are parallel. (d) Three-terminal device with $\hat{M}_1$ and $\hat{M}_2$ antiparallel.

one next to it. But only a small fraction, about 1 in 1000, of these scattering events alter the spin state of the electron. The spin state of an up-spin electron eventually becomes randomized, as an admixture of up-spin and down-spin states (depicted in Fig. 3 as plain circles), so that the spin polarization injected at the F1-N interface decays exponentially with a length scale called a spin depth, $\delta_s \approx 1000\ l$, of order 0.1–1 mm in a pure metal wire, and of order 1 μm in disordered metal films.

Transistor action. Figure 3b shows a F-N bilayer where the thickness of N is much smaller than δ_s. The electric current returned to the battery from the base of N, many spin depths away from the injecting interface, is labeled as the base current I_b and is unpolarized. But the current crossing the F1-N interface, the emitter current I_e, is spin-polarized; F1 provides to N a current of oriented dipoles at the rate $I_M = \beta I_e/e$, where I_e/e is the number current of carriers and e the electronic charge. Once in N, polarized dipoles lose their ordered orientation at a rate $1/T_1$, where T_1 is called the relaxation time. The resulting nonequilibrium distribution of spin-polarized electrons is determined by a balance between the rates of injection and relaxation. The resulting nonequilibrium magnetization $\tilde{M}$ is given by Eq. (2), where the volume is that occupied by the

$$\tilde{M} = \frac{I_M T_1}{\text{volume}} \qquad (2)$$

nonequilibrium spins. This distribution is described as nonequilibrium because in the absence of the bias current it would relax to the equilibrium condition.

The key to spin-transistor action is the nonequilibrium distribution. It is a highly ordered state, therefore low in entropy, and the low entropy is provided at a cost of energy, $U_M = \beta\tilde{M}/\chi$ (with χ the Pauli susceptibility), called an effective Zeeman energy. It represents the magnetic energy of each spin-polarized electron resulting from its interaction with the effective magnetic field ($\tilde{M}/\chi$) associated with all the other spin-polarized electrons. This energy generates current flow in the device.

A three-terminal device (Fig. 3c) can be made by fabricating a second ferromagnetic film, F2, on the other side of N, and connecting the back of F2 (the collector) to a detector (a current meter, a voltmeter, or an arbitrary load) and then to the ground of the battery. The resistance of the lower horizontal line can be ignored, and ground is identified with the base region N. When the orientation $\hat{M}_2$ is parallel with $\hat{M}_1$ (Fig. 3c), F2 acts as a reservoir of up-spin electrons. Each electron in the nonequilibrium spin distribution in N has excess energy U_M, whereas the spin population in F2 is an equilibrium distribution (held in equilibrium by the intrinsic ferromagnetism) with zero excess energy. A gradient of energy across the N-F2 interface pushes electrons into F2, thus lowering the total energy of the electrons in N. In slightly different words, the electrochemical potential of the up-spin subband in N is higher than that of F2. A gradient of electrochemical potential across the N-F2 interface is identical to an electric field, $\vec{E}$, forcing electrons from N into F2. A low-impedance ammeter measures a positive current flow (I_c) in the collector loop. If the detector is a high-impedance voltmeter, the electric field generates a voltage, $eV_s = U_M$, positive with respect to ground, at the back of F2. The phenomenon is called charge-spin coupling because the spin distribution of the particles results in forces that determine their motion, yet electronic transport is also affected because the same particle that carries spin (an electron) also carries charge.

If $\hat{M}_2$ is antiparallel to $\hat{M}_1$ (Fig. 3d), the energy of N can be lowered by pulling down-spin electrons from F2 into N, and a counterclockwise current is generated in the collector arm of the circuit. An ammeter will register a negative current, and a high impedance (voltmeter) in the collector arm will result in a voltage $-V_s$, negative with respect to ground, observed at the back of F2. Finally, it is noteworthy that U_M is inversely proportional to device volume. As the device shrinks in size the thermodynamic force (equivalently the electric field) gets larger, the efficiency of the switching action increases, and device performance improves.

Applications. The characteristics of spin transistors, such as switching speed and operating power, are comparable with those of semiconducting devices. Applications therefore relate to the unique attributes of all-metal construction.

Nonvolatile memories. Because of hysteresis, the orientation $\hat{M}_2$ (for example, in Fig. 3c) will remain unchanged until altered by application of an external magnetic field. This memory effect suggests a natural application of spin transistors as memory elements. For example, if F1 has a high coercivity (that is, a large external magnetic field is required to reverse its magnetization) so that $\hat{M}_1$ always remains down but F2 has a low coercivity, then a stored bit of information can be represented by the orientation $\hat{M}_2$, either parallel or antiparallel with $\hat{M}_1$. An array of such elements would use an array of write wires; sending small current pulses down the ith and jth write wires would uniquely address F2 of element (i,j), and the magnetic field from the current pulses would orient $\hat{M}_2$ either down or up. This orientation is stored indefinitely, without any further need of power. At any later time the stored bit is read by sending a small pulse down the ith read line, and sensing a voltage at the jth read line, positive for a "1" and negative for a "0."

This nonvolatile random access memory (NRAM) would have the advantages of RAM, such as fast access, read, and write times, along with the distinct advantage that the memory would use zero quiescent power. Dynamic RAM (DRAM), by contrast, must be refreshed every millisecond. NRAM could eventually replace magnetic disks and tapes, becoming the only kind of memory necessary in a computer.

Logic gates. Operated as a current switch, several spin transistors can be linked together to perform logic. When devices are fabricated to match 50 Ω transmission lines, the power per gate is estimated to be of the order of 10 microwatts, switching speeds are expected to be faster than a nanosecond, and thus the power dissipated per gate is comparable with, or smaller than, that of complementary metal-oxide-semiconductor (CMOS) and gallium arsenide (GaAs) devices. However, thermal conductivity in metals is much better than in semiconductors. Whereas the packing density of semiconductor devices is presently limited by heating effects, the metals used in spin transistors dissipate heat and packing densities can be increased. Furthermore, the charge-carrier population of semiconductors is thermally activated, so that device characteristics are changed if the temperature rises. The carrier population of metals is insensitive to temperature, allowing even greater packing densities.

Looking further into the future at the goal of fabricating nanoscale electronic devices, the high-charge carrier densities of metals are significant. The carrier density of highly doped semiconductors is of the order of 10^{18}–10^{19} carriers per cubic centimeter, so that a device with a typical dimension of 10 nanometers and an active volume of 10^{-18} cm^3 would have only 1–10 carriers available for operation, an inadequate number. Metals, however, have charge carrier densities of the order of 10^{22}–10^{23} carriers per cubic centimeter, so a device with the same active volume will have 10^4–10^5 carriers, insensitive to surface states or temperature. Although nanoscale devices are still far from realization, it is clear that metals have distinct fabrication advantages. The spin transistor is an active element fabricated solely from metals, with the additional advantage that device performance improves when the device is made smaller. SEE MAGNETIC SWITCH.

For background information SEE BAND THEORY OF SOLIDS; COMPUTER STORAGE TECHNOLOGY; ELECTROLUMINESCENCE; FERROMAGNETISM; FREE-ELECTRON THEORY OF METALS; POLYMER; SEMICONDUCTOR; TRANSISTOR in the McGraw-Hill Encyclopedia of Science & Technology.

Mark Johnson

Bibliography. J. H. Burroughes et al., Light-emitting diodes based on conjugated polymers, *Nature*, 347:539–541, 1990; F. Garnier et al., All-polymer field-effect transistor realized by printing techniques, *Science*, 265:1684–1686, 1994; F. Garnier et al., An all-organic thin film transistor with very high carrier mobility, *Adv. Mater.*, 2:592–594, 1990; F. Garnier et al., Molecular engineering of organic semiconductors: Design of self-assembly properties in conjugated thiophene oligomers, *J. Amer. Chem. Soc.*, 115:8716–8721, 1993; M. Johnson, The all-metal spin transistor, *IEEE Spectr.*, 31(5):47–51, May 1994; M. Johnson, The bipolar spin switch, *Science*, 260:320–323, 1993; M. Johnson and R. H. Silsbee, Interfacial charge-spin coupling: Injection and detection of spin magnetization in metals, *Phys. Rev. Lett.*, 55:1790–1793, 1985; P. M. Tedrow and R. Meservey, Spin polarization of electrons tunneling from films of Fe, Co, Ni, and Gd, *Phys. Rev. B*, 7:318–326, 1973.

Tree

The canopy is the combination of all leaves, twigs, and small branches in the forest, along with the collection of all the plant crowns. It is not simply the roof of the forest, but involves all the components, from top to bottom, often including the open spaces and atmosphere in the crowns. This definition is used because the covering layer is often difficult to designate and environmental gradients are not easily subdivided. The structure of the canopy is its organization in space and time, including the position, extent, quantity, type, and connectivity of the aboveground components of the forest.

The canopy is a unique climatic zone of the forest, a critical habitat for diverse specialized plants and animals and the primary site of important interactions between vegetation and the physical environment. An interface between the atmosphere and the biosphere, the canopy mediates enormous transfers of radiative and wind energy, as well as the exchange of water vapor (H_2O), carbon dioxide (CO_2), and pollutants.

The canopies of forests are distinct from those of other forms of vegetation because they are dense, extensive, elevated, and perennial. They have more biomass, a greater surface area, and a lower average leaf area density (the mean leaf area per unit volume). Their structural complexity makes them aerodynamically rougher than other forms of vegetation. Because of their stature, forest canopies harbor many specialized habitats. The extent and persistence of forest canopies indicate that they have a broad influence in climate modification.

Canopy structure affects both the atmospheric environment within the forest and its exchanges with the lower atmosphere in two general ways. First, canopy surfaces act as passive drag elements and exchange surfaces for the absorption of wind energy, the dissipation of turbulence, and the exchange of radiation. Second, canopy surfaces actively participate in exchanges of biologically important compounds, such as CO_2 and water vapor.

Components. The proximate units of canopy structure are the crowns of trees; the ultimate components are leaves and twigs. The foliage may be clumped, or arranged in clusters, branch systems, and crownlets. Even crowns are sometimes grouped. Canopy structure can be characterized by the mean height of the dominant trees, the ratio of the total one-sided leaf area to the projected ground area (the leaf area index), and leaf area density or LAD. Another common descriptor of canopy structure is the height distribution of leaf area. In young forests and plantations, the foliage is concentrated near the top of the canopy; in mixed-species or multicohort stands, many other types of organization are possible.

Canopy surface area is an important feature of canopy structure, not only for atmospheric interactions but also for habitat quality. Although the biomass of stem tissues far exceeds that of leaves, leaf area dominates the total aboveground surface area at all canopy levels. Forests generally have a higher leaf area index (from 4 to 8 m^2/m^3 in broadleaved stands to 12 m^2/m^3 or more under conifers) and lower leaf area density (from 0.2 to 0.4 m^2/m^3) than other types of vegetation. The dry weight of canopy leaves can range from 5.4 to 41 short tons/acre (2 to 15 metric tons/hectare).

Vertical organization. Canopy elements are often differentiated by their height, which is caused by species differences in growth form and shade tolerance. In some stands, tree heights, species, or leaf biomass are organized into distinct strata, often with little overlap. Such stratification is most easily recognized in young, monospecific forests. Stratification was earlier thought to be a conspicuous characteristic of tropical rain forests and older forests. However, vertical layering is not always found in such stands.

Several categories of canopy levels (stories) are often recognized. Crowns that are fully illuminated (dominant) or largely illuminated from above (codominant) compose the overstory. The understory (or subcanopy) includes stems in the shady lower layers. The midcanopy is a transitional region between understory and overstory and has crowns that are partly illuminated (intermediate) or overtopped (suppressed). The ground layer includes the seedlings of woody plants and other herbaceous vegetation just above the forest floor. An irregular zone of extremely tall crowns (emergents) may rise above the main canopy in some forests. The outer canopy is the zone immediately adjacent to the atmosphere; its undulating surface may be steeply sloped and even intersect the ground. The complex topography and immense surface area of this region is difficult to envision from the ground.

Height is the most convenient axis for representing canopy structure. But distance from the ground is only a crude proxy for environmental characteristics in forests. Proximity to the outer canopy is a more important aspect of location. Many characteristics of canopy components vary with height: leaf size, thickness, shape, orientation, and tissue chemistry differ among canopy levels.

Forests are heterogeneous at various scales, with foliage-free voids of many sizes. Light gaps are holes in the canopy that extend to the forest floor and which permit some direct radiation to penetrate to the understory. They originate from the death of whole trees or branches; large gaps are less common than small ones. In some forests, individual subcrowns and crowns are clearly separated by foliage-free spaces; this pattern is called crown shyness.

Canopy dynamics. Canopy structure changes seasonally in all forests, but is most dramatic in completely deciduous stands. Slower changes occur during succession. Although the total amount of leaf area stabilizes early in stand development, its vertical and horizontal distribution alter gradually. The stand is said to become closed when growing crowns come into contact with each other. As the stand ages, crowns elongate and species differentiate in height. The appearance of understory and shade-tolerant species initiates another layer of leaves. As overstory trees begin to die, the stand leaf area may decline slightly. In older stands, crowns may be distributed at every canopy level.

Following external disturbances (such as wind storms) leaf area recovers rapidly, usually in less than 5 years (1–2 years in many tropical forests). However, the recovery of the original species composition can be very slow, especially if there was much mortality or reduction of sprouting capacity. The recovery of a forest's vertical and horizontal structure can be extremely slow, depending on the severity of the disturbance.

Microclimate. Above and within the canopy are several distinct aerodynamic regimes of the surface boundary layer. Above the stand, the wind generally flows horizontally, with low levels of turbulence. Close to and within the forest, profiles of environmental variables are complex, and the wind field is rather complicated.

Most incident short-wavelength radiation is absorbed in the outer canopy. This layer reradiates heat more readily than do lower levels and cools rapidly at night. The cooling at the top can establish a stable inversion, which effectively divides the canopy into two thermal zones. The zonation ceases when heating resumes in the morning.

The environment just above and within the canopy is a transition region between the free atmosphere and conditions near the forest floor. The average values of many quantities change continuously (although often in a complicated way) with changes in height. Furthermore, the range of environmental variation also differs between canopy layers. The top layer has a marked diurnal fluctuation in almost every characteristic; variation is progressively reduced with depth in the canopy. Relative to the outer canopy, the understory is reliably moist, dark, and still. Additional microclimatic features of forests are not well treated as averages by level because these features depend on local conditions.

Precipitation. Rain, snow, sleet, hail, fog moisture, and cloud water are intercepted, retained, and redistributed by the canopy. Water ultimately evaporates from the canopy (interception), drips through (throughfall), or runs down the stems (stemflow) to the forest floor. In cloud forests, additional water may be captured from clouds by the canopy and transferred to the forest floor in throughfall and stemflow. In general, between 10 and 30% of incident precipitation is intercepted and evaporated from the canopy. From 0.04 to 0.12 in. (1 to 3 mm) of water can be retained in the canopy at a given time (interception capacity), with greater amounts held in coniferous than in broadleaved stands. At least three-quarters of the water eventually reaching the forest floor arrives as throughfall.

The chemical composition of precipitation is also altered during passage through the canopy. Most solutes have higher concentrations in forest throughfall and stemflow than in incident precipitation. This loss of material from the canopy derives both from the cycling of nutrients within the forest and from the wash off of the material that is deposited on the canopy between precipitation events.

Radiation. Because of their surface roughness, forest canopies are efficient absorbers of short-wave radiation. The reflection coefficient (albedo) of temperate forests is 0.08–0.13 in conifers, 0.10–0.12 in growing season hardwoods, and approximately 0.13 for tropical rainforests. Albedo also decreases with stand height.

Forest leaves absorb most ($\geq 80\%$) of the incident short-wave radiation (<700 nanometers), transmitting and reflecting the remainder. More than half the radiation of the longer wavelengths

(>700 nm) penetrates canopy leaves; the remainder is largely reflected. The absorption and transmission spectra of leaves depend on species, age, and the angle of light.

Light transmission decreases markedly with each leaf contact (typically 40–60% depletion in irradiance) and continues to decline rapidly below the level of canopy closure. Average light levels decline with depth thereafter, but there is substantial spatial variability, particularly in the overstory. At the forest floor, illumination levels are typically 1–3% for hardwood trees in leaf (30–40% when leafless) and >5% for most conifers. Rainforest understories can be very dark (<1% illumination).

Wind speed. Wind speed rapidly decelerates in the layer just above the forest; average wind speeds decline evenly with depth. The pattern of wind-speed attenuation depends on aerodynamic characteristics of the stand, which are determined by the amount and organization of canopy material. At deeper canopy layers, the wind field can be very complex. At forest margins and in some stands of simple structure, wind speeds accelerate in the lowermost canopy levels.

The canopy acts to filter wind turbulence. High-frequency gusts are arrested, but large eddies may penetrate to various depths. Much of the total transport of energy and material in forests and the ventilation of lower layers occurs during the short-term incursion of such large eddies.

Temperature and humidity. Temperature and humidity gradients tend to be weak in forest canopies. The absolute humidity may change little at a forest level over a period between the passage of weather systems. The drying power of the air also changes throughout the forest, from highly unsaturated during the day to near saturation at some canopy levels at night, stimulating dew formation.

Gases and particles. Mean concentrations of gases, vapors, and particles are typically lower in the understory than in the overstory. For CO_2, however, the active layer of the canopy is an enormous sink, and daytime CO_2 concentrations are often slightly depressed in the overstory. The layer near the ground is a source of CO_2, arising from root, litter, and soil respiration. Pronounced CO_2 concentrations often develop early in the morning understory, especially under stable conditions.

For background information SEE BIOMASS; CARBON DIOXIDE; FOREST ECOSYSTEM; FRONT; TREE in the McGraw-Hill Encyclopedia of Science & Technology.

Geoffrey G. Parker

Bibliography. B. A. Hutchison and B. B. Hicks (eds.), *The Forest-Atmosphere Interaction*, 1985; R. Lee, *Forest Microclimatology*, 1983; M. Lowman and N. Nadkarni, *Forest Canopies: A Review of Research on a Biological Frontier*, 1995; M. W. Moffet, *The High Frontier: Exploring the Tropical Rainforest Canopy*, 1994; G. Russel, B. Marshall, and P. Jarvis (eds.), *Plant Canopies: Their Growth, Form and Function*, 1989.

Tumor

The field of immunology has seen advances in the elucidation of the mechanisms of natural resistance to tumors and in the development of antitumor vaccines.

Immune system. One of the major challenges facing medical research is that of controlling, treating, and possibly vaccinating individuals against cancer. Unlike acquired immune deficiency syndrome (AIDS), a relatively new disease that has directed much attention to the importance of the immune system, the immunologic implications of cancer are not as obvious.

It was clear from the discovery of AIDS that the immune system is centrally involved and critically affected by the virus that induces this syndrome. CD4+ T cells, the population of lymphocytes that is essential for providing enhancing helper signals for effector cell function of the immune system, are destroyed in progression to AIDS. However, although there are indications of a loss of immune function in experimental animals and individuals with certain cancers, it is not known whether loss of immune function precedes the appearance of cancer or is somehow impaired by the presence of the tumor. Resolving this question could significantly affect the direction of cancer research, therapy, and the possibility of immunization or of boosting an ineffective immune system against tumors. Significantly, loss of immunity prior to the appearance of tumors would suggest that the immune system plays a surveillance role against newly arising tumors.

Immune system components. The immune system can be broadly divided into two major components, cellular and humoral immunity. Cellular immunity involves the destruction of virus-infected cells and tumor cells by leukocytes or white blood cells, many of which possess the ability to kill their target cells via molecules that damage cell membranes. Natural killer (NK) cells are able to destroy tumor target cells without having been previously immunized to the tumor. Lectin-activated killer (LAK) cells can be activated with strong nonspecific stimuli to destroy tumor cells. Cytotoxic T lymphocytes (CTLs) require immunization against specific proteins or antigens for maturation and expansion; they are specific for particular antigens, and can be selectively expanded by immunization against the antigens for which they are specific. The humoral arm of the immune system is affected by protein molecules known as antibodies, which are produced by B lymphocytes. Antibodies specific for tumor antigens have been isolated; these antibodies can destroy certain tumors under experimental conditions.

Immune regulation. The cellular and humoral components of the immune system are modulated by molecules known as cytokines. Immunoregulatory cytokines are produced by a variety of different cell types, including T and B lymphocytes, natural killer

Table 1. Nonspecific host defense adjuvants

Category	Example
Protein	Purified protein derivative
	Muramyl dipeptide
	Muramyl derivatives
Viral	Viral oncolysates
Bacterial	Bacillus Calmette-Guérin
	Corynebacterium parvum
	Cellular components of bacillus Calmette-Guérin
	Nocardia cell wall skeleton
Glucan	*Pachymaran* (fungus)
	Glucan (yeast)
Chemical	Polymers (pyan) MVE-2
	Ribi reagents
	Inosinic-cytidylic acid

cells, and monocyte/macrophages. They can be divided into two broad categories, types 1 and 2. Type 1 cytokines include interleukin 2 (IL-2), interleukin 12 (IL-12), and interferon γ, and they mainly, but not exclusively, enhance T effector functions such as cytotoxic T lymphocytes, natural killer, and lectin-activated killer activity. Type 2 cytokines include interleukin 4 (IL-4), interleukin 5 (IL-5), interleukin 6 (IL-6), and interleukin 10 (IL-10), and they mainly augment B-cell function and facilitate antibody production. Certain type 1 and 2 cytokines have cross-regulatory properties in that they not only enhance one component of the immune system but down-regulate the other component. Thus, although researchers might hope to optimally enhance both arms of the immune system simultaneously, this possibility may be difficult to achieve.

Immune protection. It now appears that the cellular arm is the more protective against human immunodeficiency virus (HIV) infection and progression to AIDS. Neutralizing antibodies against parts of the AIDS virus have little or no effect against naturally occurring strains of the virus. It is possible that a dominant type 1 cytokine profile and a strong cellular immune response to HIV, and not a dominant type 2 cytokine profile and strong humoral immunity, are protection against HIV infection and against progression to AIDS. This pattern of differential responses is due to the cytokine regulation noted above, and is the foundation of the type 1 and 2 model of resistance, and of susceptibility to HIV infection.

It also appears that the cellular arm of the immune system is deficient or dysregulated in certain cancers. In fact, the pattern of immune defects recently reported for Hodgkin's lymphoma is remarkably similar to that seen in HIV infection prior to progression to AIDS. Other human cancers in which loss of cellular immune function may occur include melanomas, prostate cancer, and possibly breast cancer. It is possible that immune dysregulation rather than immune suppression occurs such that an increase in type 2 cytokines contributes to the loss of cellular immunity. If this suggestion is correct, it may be possible to treat tumors with type 1 cytokines or antibodies against type 2 cytokines. It may be significant that interleukin 12 (IL-12) has been found to be defective in leukocytes from HIV-infected individuals, and has also been found to be effective in treating several tumors in mice.   *Gene M. Shearer; Mario Clerici*

Antitumor vaccines. The concept and evolution of antitumor vaccines followed the overwhelming success of vaccines directed against bacterial and viral organisms. As with vaccines against infectious diseases, cancer vaccines stimulate the patient's immune system by introducing attenuated organisms, cells, or cellular components. However, cancer vaccines are administered after the advent of disease, rather than prophylactically before the disease develops. Although the field is still in its infancy, rapid advances in biotechnology are allowing researchers to identify and target antigenic components of tumor cells for killing by the patient's immune system.

Evolution of antitumor vaccines. The work of W. B. Coley, initiated more than 100 years ago, forms the basis for nonspecific cancer immunotherapy. Coley and previous workers had observed cancer regressions in individuals undergoing acute bacterial infections, and this association led Coley to develop mixed bacterial vaccines (Coley's toxins) that could reproduce aspects of an acute infection without introducing live bacteria.

Specific cancer immunotherapy was introduced at the turn of the century with tumor cell vaccines. Although these early vaccines were nothing more than rudimentary suspensions of either viable attenuated same host (autologous) tumor cells, different host (allogeneic) tumor cells, or tumor cell extracts, several reports linked their use to tumor regression.

The subsequent addition of immune-stimulating adjuvants to vaccine regimens produced a dramatic improvement in the rate of response. Nonspecific protein, viral, bacterial, and chemical agents often can amplify the effect of a vaccine (**Table 1**). Some adjuvants, such as bacillus Calmette-Guérin (a modified form of the tubercle bacillus) and levamisole (a synthetic agent used to kill intestinal worms) have also been used alone in specific types of cancers. Most vaccine protocols use only one immunomodulator. However, use of multiple adjuvants might improve the response, depending on the site of secondary growth and the antigen profile of the individual's tumor.

Vaccines were then formulated to consider tumor-associated antigens (TAA) that were identified and characterized through study of metastatic tumor clones. It was recognized that vaccine preparations that included many different immunogenic tumor-associated antigens were more likely to stimulate clinically effective increases in anti-TAA antibody production and cytotoxic T cells.

In general, results of specific vaccine immunotherapy have been inconclusive because of the incidence of immunosuppression produced by tumor-derived factors, differences in study populations, and a failure to select vaccine antigens eliciting adequate cell-mediated and humoral responses. Although tumor regression, impaired tumor doubling time, and eradication of residual tumor cells have been reported, these findings did not necessarily translate into an increase in disease-free or overall survival.

Melanoma. Although the effects of antitumor vaccines have been studied in leukemia and sarcoma and in brain, colorectal, genitourinary, gynecologic, and pulmonary tumors, most investigations have focused on melanoma because this tumor is linked to host immunity (**Table 2**). For example, the incidence of spontaneous regression is higher in melanoma than in other tumors.

More than 25 years ago, the possibility of specific humoral and cellular immune reactions against melanoma encouraged investigations leading to the discovery of tumor-associated antigens found on the tumor cells of most individuals with melanoma.

Table 2. Antitumor vaccines used in clinical trials

Neoplasm	Type of vaccine
Melanoma	Ganglioside
	Anti-idiotype (mirror image)
	Melanoma cell lysate with DETOX (a bacterial adjuvant)
	Plasma membrane of melanoma cells
	Irradiated whole melanoma cells
	Vaccinia viral lysates of melanoma cells
Brain	Tumor cell vaccine
Colorectal	Autologous tumor cell vaccine with bacillus Calmette-Guérin
Gynecologic	Allogeneic tumor cell vaccine with bacillus Calmette-Guérin
Genitourinary	Autologous tumor cell vaccine with *Corynebacterium parvum*
Leukemia	Allogeneic tumor cell vaccine with bacillus Calmette-Guérin
Lung	Tumor cell vaccine
Sarcoma	Allogeneic tumor cell vaccine with bacillus Calmette-Guérin

Recently, the intense interest in active specific immunotherapy of human melanoma has produced a number of vaccines based on melanoma tumor-associated antigens. The most practical approach seems to be an allogeneic cell-based vaccine derived from genetically different but species-related cells that contain a broad spectrum of known and possibly unknown melanoma tumor-associated antigens.

Of particular interest is a whole-cell preparation that was recently tested. A polyvalent, allogeneic melanoma cell vaccine (MCV) was administered to individuals with melanoma metastatic to regional lymph nodes, subcutaneous sites, or distant sites. Humoral responses and development of skin sensitivity were both significantly correlated with survival. Melanoma cell vaccine therapy caused a threefold increase in median survival time for individuals with stage IV melanoma (7.5 months for individuals who did not receive the melanoma cell vaccine versus 23.1 months for those who did). Stage IIIA individuals receiving melanoma cell vaccine also had a better median survival (43.9 months) than those not receiving melanoma cell vaccine (28.7 months). In addition, when stage IV individuals were categorized by site of distant secondary growth, those receiving melanoma cell vaccine survived significantly longer than those receiving non-MCV therapies. Respective median survivals were 21.5 versus 8 months for lung metastases, 38 versus 11.5 months for soft-tissue metastases, and 16 versus 4.4 months for liver and brain metastases. These promising results await validation in a randomized trial comparing the vaccine with chemotherapy for stage IV melanoma.

Problems and variables of vaccine therapy. Although the current staging systems stratify individuals according to disease progression and prognosis, comparisons within the same stage may be misleading. For example, the status of a patient whose stage IV melanoma has been completely cut out is not the same as that of a patient with extensive tumor.

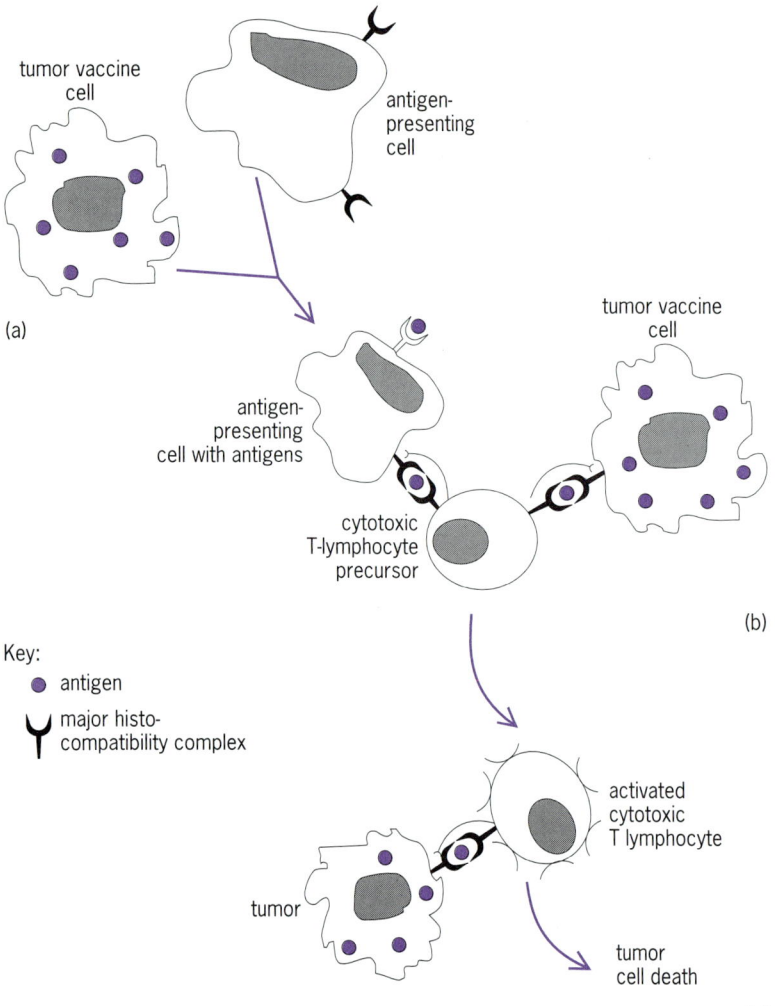

Possible mechanisms for T-cell response to the vaccine's tumor-associated antigens. (*a*) Tumor vaccine cell interacts with an antigen-presenting cell, which presents antigen to the cytotoxic T-lymphocyte precursor. (*b*) Tumor vaccine cell interacts directly with the cytotoxic T-lymphocyte precursor. (*c*) Induction in *a* and *b* of an activated cytotoxic T-lymphocyte that destroys an in-place tumor. (*After A. M. Barth, R. F. Irie, and D. L. Morton, Update on immunotherapy for advanced melanoma, Contemp. Oncol., 4(10):52–60, 1994*)

In fact, a large tumor significantly impairs the immune response to active specific immunotherapy. In these cases, it is not yet clear whether aggressive cytoreductive surgery or chemotherapy can improve the response to vaccine therapy by increasing overall survival.

Another variable is the tumor's rate of growth, which varies even among individuals who have the same stage and site of disease. A tumor with a rapid doubling time may require chemotherapy or surgery to slow its progression prior to immunotherapy.

A third variable is the orientation and presentation of the vaccine's tumor-associated antigens. Melanoma vaccines have been shown to induce both antibody responses and T-cell responses to common melanoma tumor-associated antigens. T-cell induction may occur through recognition of tumor-associated antigen peptides associated with major histocompatibility complex (MHC class I) molecules on the surface of intact melanoma cells, or tumor-associated antigens may be processed by antigen-presenting cells prior to their recognition as foreign proteins (see **illus**.). Recent developments suggest that certain humoral responses can impair tumor-associated antigen recognition and the response to vaccine immunotherapy, possibly by antibody blockade of antigens or by configurational changes that seclude antigens from interaction with the cellular components of the immune system.

Future approaches. Although current vaccines for some solid tumors have demonstrated survival advantages, the ideal vaccine has yet to be developed. The perfect adjuvant or combination of adjuvants for specific tumors and tumor sites also requires further evaluation.

Enriching a vaccine's tumor-associated antigen pool may improve its immunogenicity. Isolation of tumor-specific antigens presented with MHC class I molecules that are responsible for rejection of foreign grafts may also improve vaccine efficacy. Tumor-specific antigens associated with MHC class I molecules increase the potential for inducing the cytotoxic T lymphocytes that lead to tumor regression. An example is the recently discovered *MAGE* gene family that codes for tumor-specific antigens; exploitation of *MAGE* genes might enhance immunogenic responses to biologic therapy.

Combining antitumor vaccines with cytokines and other effectors, either by concurrent administration or by cell transfection, may alter the local environment of the tumor cells, thus enhancing the presentation of antigens to the immune system or inducing specific cytotoxic T lymphocyte clones. Transfection of tumor cells with foreign cell-surface molecules such as B7 may increase the host's ability to reject tumors. Locally acting transfected interleukin 2, tumor necrosis factor, or interferons may also improve the response to active specific immunotherapy, inducing more cytotoxic T lymphocytes or downregulating suppressor T lymphocytes.

Even as whole-cell and cell lysate or cell extract cancer vaccines enter advanced-stage clinical trials, technological advances are allowing cancer scientists to explore newer vaccines based on mirror-image antibodies and recombinant antigens produced by genetic engineering techniques.

For background information SEE ACQUIRED IMMUNE DEFICIENCY SYNDROME (AIDS); ANTIBODY; ANTIGEN; CANCER (MEDICINE); CYTOKINE; IMMUNOLOGY; HISTOCOMPATIBILITY; TUMOR in the McGraw-Hill Encyclopedia of Science & Technology.

Ralph C. Jones; Rishab K. Gupta; Donald L. Morton

Bibliography. A. M. Barth, R. F. Irie, and D. L. Morton, Update on immunotherapy for advanced melanoma, *Contemp. Oncol.,* 4(10):52–60, 1994; M. J. Brinda, Interleukin 12, *J. Leukocyte Biol.,* 15:280–288, 1994; J. F. Holland et al. (eds.), *Cancer Medicine,* 3d ed., 1993; D. L. Morton et al., Prolongation of survival in metastatic melanoma after active specific immunotherapy with new polyvalent melanoma vaccine, *Ann. Surg.,* 216(4):463–482, 1992; D. M. Pardoll, A new look for the 1990s, *Nature,* 369:357–358, 1994; S. A. Rosenberg, The immunotherapy and gene therapy of cancer, *J. Clin. Oncol.,* 10:180–199, 1992.

Turfgrass

Recently, turfs such as lawns and golf courses have been viewed negatively because of the potential detrimental effects of fertilizer and pesticides on the environment. However, research has demonstrated that turfgrasses provide more than aesthetics to the environment, for example, by effectively dissipating heat and decreasing noise levels. Well-maintained turfgrasses and landscapes contribute to increased property values, and serve as safety buffer zones on highway right-of-ways, airfields, and correctional facility surrounds.

Psychological and physical well-being. Studies have demonstrated that individuals working in highly stressful environments show a marked reduction in their stress levels when exposed to landscapes with turfgrasses, ornamentals, and trees. Individual productivity has been found to decrease and anxiety levels to increase in situations devoid of greenspaces.

Sports and recreational activities such as golf, softball, baseball, football, rugby, soccer, lawn bowling, croquet, polo, field hockey, track and field, and horse racing use turfgrasses for their functional surfaces. In the late 1960s and through the 1970s, many sports turfgrass facilities were converted from natural to artificial turfs. More recently, this trend has been reversed because of growing concerns over injuries associated with artificial surfaces. Thus, a number of professional and college sports facilities have replaced their artificial surfaces with natural turfgrass. Natural turfgrass provides a resilient surface which enhances safety and reduces the potential for injury, particularly for contact sports such as football. A 3-year study of 26 teams in the National Football League found that the mean number of major and minor injuries per game was 2 on natural turfgrass and 2.7 on artificial turfgrass. This differences is statistically significant ($P < 0.01$), indicating a lower injury rate on natural turfs. In addition, natural turfgrass surfaces provide cooler playing conditions during high temperature periods.

Environmental benefits. Turfgrasses provide many benefits to our environment. They effectively dissipate heat, abate noise, enhance air quality, filter and entrap potential environmental contaminants, function in soil quality restoration and carbon storage, stabilize soils, and minimize soil erosion.

Temperatures in urban environments are generally 3–7° warmer than nearby rural surrounds. The lower rural temperatures result from the cooling effect of vegetation and enhanced air movement associated with these rural areas. Turfgrasses and other allied plant materials can moderate urban temperatures through the process of evapotranspiration. For example, at midday in Lincoln, Nebraska, the air temperature at 59 in. (150 cm) was 95°F (35°C). At the same time, temperatures on a paved street surface and an adjacent turfgrass surface were 131°F (55°C) and 104°F (40°C), respectively. The difference in temperature between the two surfaces marks a considerable difference in heat load. A temperature probe placed in the asphalt paving indicated that its internal temperature differed only slightly from the surface temperature. However, the turfgrass canopy surface temperature differed by more than 10° from its internal temperature of 84°F (29°C), and the soil temperature at the 1 in. (2.5 cm) depth was even lower at 73°F (23°C). The cooling process of evapotranspiration associated with the turfgrass ecosystem serves to dissipate the high heat loads associated with urban environments. The cooling effects of turfgrasses and allied landscape plant materials may result in reduced energy inputs and costs required for associated mechanical cooling.

Turfgrass stands have rough surfaces that function in noise abatement and glare reduction. Studies have demonstrated that a grass strip, 70 ft (21 m) wide, along a roadside reduces traffic noise by as much as 40%. Similarly, the rough surface of a grassed area results in multidirectional light reflection that reduces glare.

An actively growing turf enhances air quality through its contribution of oxygen to the atmosphere as a result of respiration, removal of carbon dioxide via photosynthesis, and assimilation of gaseous pollutants. The dense canopies of turfgrasses also entrap dust and other particulate matter that may foul air quality. Turfs are maintained around airport runways to help protect sensitive jet engines from dust and other particulate matter that might affect the engines' performance and longevity. Regularly mowed and maintained lawns reduce allergy-related pollen problems by removing pollen sources and by entrapping airborne pollen grains in the rough canopy surface. Turfgrasses serve as an inexpensive means of stabilizing soils and minimizing wind and water erosion. They virtually eliminate dust and mud problems in and around homes, schools, businesses, and industrial sites.

The extensive fibrous root system associated with perennial turfgrasses contributes substantially to the restoration of soil quality by increasing soil organic matter content. The organic matter produced by turfgrass improves soil structure on sites where the structure has deteriorated during the process of home building or construction of business and industrial facilities. Improved soil structure is very important because it enhances water infiltration and percolation and reduces water runoff, thus minimizing soil erosion and surface water contamination.

Turfgrasses greatly contribute to carbon storage in soils as a result of their fixing carbon from atmospheric carbon dioxide during the process of photosynthesis. The enhancement of carbon storage contributes to soil quality through increased microbial activity as well as through the associated physical improvements. A large diverse population of microorganisms and macroorganisms are an integral part of the turfgrass ecosystem, which supports these organisms through the contribution of carbon and other nutrient sources from the decay of roots, leaves, stems, and other vegetation. These organisms supply one of the most active and effective biological systems for the degradation of pesticides and other organic chemicals that might be trapped in the turfgrass canopy.

Not only do turfgrasses enhance soil infiltration and percolation rates as mentioned above, but their dense irregular canopies and allied thatch accumulation help trap water and reduce runoff. Many impervious surfaces found in urban environments result in a large amount of water runoff that carries many pollutants. Turfgrasses offer one of the most inexpensive and effective ecosystems for detainment of urban runoff, and for the entrapment and dissipation of associated pollutants. Documented studies indicate that runoff volumes from properly managed turfgrass ecosystems are minimal. This

Annual flow-weighted nitrate-nitrogen concentration from various land uses

		NO_3-N, µg/liter	
Land use	Fertilizer source	1987	1988
Corn			
Cover crop	Urea	15.3	8.1
No cover crop	Urea	14.9	15.6
Corn	Manure	4.2	17.5
Lawn	Urea:ureaform-aldehyde	1.6	0.3
Lawn	None	0.2	0.2

SOURCE: A. J. Gold et al., Nitrate-nitrogen losses to groundwater from rural and suburban land uses, *J. Soil Water Cons.*, 45:305–310, 1990.

attribute is important in enhancing ground-water recharge, improving surface water quality, and reducing the need for costly flood control structures. Furthermore, pesticide and nutrient fate studies show that turfgrass ecosystems are highly conducive to the assimilation and degradation of organic contaminants. Nitrate (NO_3^-) leaching studies have demonstrated that lawns, whether fertilized or unfertilized, have very low nitrate-nitrogen (N) levels (<1.7 micrograms per liter) in ground water (see **table**).

Safety role. Turfgrasses also serve a safety role. They reduce fire hazards by providing succulent vegetation which serves as a fire break. Turfgrasses along roadsides not only reduce soil erosion on banks and inclines but contribute to safe vehicle operation and human safety by providing an important stabilized zone for emergency stoppage of vehicles and improved line-of-sight visibility and view of road signs and potential hazards. Well-maintained turfs reduce the potential for such problems as ticks, chiggers, snakes, rodents, and allergic reactions to weed pollen.

Robert C. Shearman

Bibliography. J. B. Beard and R. L. Green, The role of turfgrasses in environmental protection and their benefits to humans, *J. Environ. Qual.*, 23: 452–460, 1994; R. C. Shearman, Turfgrasses, *Ency. Agri. Sci.*, 4:413–420, 1994; A. J. Turgeon, *Turfgrass Management*, 3d ed., 1991.

Underground mining

In mining terminology, that portion of mill waste (tailings) or other rock waste materials that can be reintroduced into subsurface excavations to meet specific mine engineering needs is defined as backfill. A variety of types of backfill exist, including processed mill tailings, recovered surface aggregates (for example, sands and gravels), and even waste rock produced in mining or quarrying operations.

For various reasons, all or some of the tailings (generally the fines fractions, or fine particles) that can be produced at mine sites may not be acceptable for reuse as backfill. Typically, some fraction may be too contaminated by chemical reagents used in the ore extraction process or may develop adverse physical characteristics, necessitating its storage on the surface. Optimum tailings use occurs when all materials can be placed underground to fill mine voids, with none being discarded on the surface.

Types of backfill. Backfill plays an integral part in the exploitation of underground mines. In principal mining districts of North America, for example, up to 70% of underground mines utilize backfill to satisfy a variety of engineering purposes. Two major categories of fill materials currently exist: hydraulically placed aggregate materials and coarse aggregate materials (rock fill).

Hydraulically placed aggregates may consist of mill tailings, alluvial sands, or mixtures of both that are prepared on the surface and are transported through pipelines for underground placement. Aggregates are normally mixed with water at 65–75% by weight dry solids and are pumped underground in the form of a slurry or liquid mixture of solids and water. In order to effect transport and to prevent coarse particle settlement within pipelines, a minimum slurry flow velocity approximating 5 ft/s (1.6 m/s) is required. Typically, these mixtures must also exhibit reduced fines (particles smaller than average in a mixture of particles varying in size) contents to aid water drainage from placed backfill within mine excavations. Rock fill materials, alternately, consist of waste rock fragments, produced from either surface quarries or underground, which exhibit much larger average particle sizes than tailings fill. Rock fill requires transport either mechanically or by gravity action through vertical rock passages (drop fill raises) into excavations which are to be filled.

Noncemented hydraulic backfill accounts for 14% of all backfill products placed in mines and is generally composed of angular particles grading in size from 0.033 in. (840 micrometers) to 0.0002–0.0004 in. (5–10 μm) in diameter. The primary problem associated with use of noncemented hydraulic backfill is the high water content necessary for promotion of slurry transport. Following slurry placement, water is permitted to drain naturally through the backfill volume and is collected for return to surface. Where significant quantities of liquids are retained within the fill mass, the possibility of backfill liquefaction or poor support strength capability exists. Noncemented waste rock fill, accounting for approximately 12.5% of all mine backfills, consists of angular rock fragments, at sizes up to 24 in. (60 cm) in diameter, which normally exhibit much higher porosity and strength conditions than hydraulically placed fill materials.

Both backfill types can be strengthened (consolidated) through the addition of cementing agents to the backfill aggregate mixtures. Typical cementing agents used by mines include normal portland cement or other pozzolanic agents such as fly ash or ground smelter slags, which are mixed at proportions generally ranging between 3 and 9% of the total dry backfill mix weight. Cemented hydraulic backfill, representing approximately 46% of all backfill placed, is the most common form of mix used. Cemented rock fills, alternately, account for approximately 13% of total backfill products.

Functions of backfill. A common factor associated with backfill use is the necessity for mines to dispose of surface tailings in an environmentally responsible way. This necessity establishes a requirement for underground void filling, thereby reducing surface environmental hazards and minimizing the need to design and provide for costly surface tailings storage facilities.

The primary engineering functions that backfill must satisfy include provision of excavation wall support, creation of bearing surfaces during ore recovery operations, and stabilization of exposed backfill faces during pillar ore extraction.

Backfill technological innovations. Numerous detrimental manufacturing processes and backfill physical conditions exist to restrict efficient utilization of backfill in underground mining operations. The inclusion of high fines contents in total tailings backfills has been a principal deterrent to effective fill use in mines. Innovative techniques, which concentrate primarily upon development of alternate backfill manufacturing technologies to permit acceptable use of total tailings products, are presently being developed. Additional research has also been conducted concerning the utilization of cheaper and more environmentally responsible binder-agent materials as replacements for costly and nonsustainable cementing agents.

High-density fills. The use of high-density fill has gained considerable attention from the mining industry. High-density slurry backfills consist of mixtures of classified tailings, with the <0.0008 in. (20 µm) fines fractions removed, and water at high pulp densities. High-density backfills, depending upon the constituent solids materials, have been successfully pumped at pulp densities approximating 80–82% by weight solids over vertical heights exceeding 3650 ft (1100 m) and horizontal travel paths up to 3000 ft (915 m). Such high-density slurries must, however, be kept in motion at all times in order to prevent particle settlement, water separation, and pipeline plugging. When placed in underground excavations, water will separate from the backfill solids and will require removal by pumping. However, the volumes of water produced by high-density backfills are considerably reduced compared to normal slurry backfill materials.

Paste backfills. Paste backfills are composed of high fines content materials that are capable of retaining water at levels above their saturation limits when transported through pipelines at low velocity without suffering particle settling and pipeline plugging problems. The presence of fines material of <0.0008 in. (20 µm) provides paste fill with the capability to retain matrix water when it is placed within stopes (horizontal workings in the shape of a flight of stairs) and when such fills experience stopped flow in pipelines. In experiments, paste fill materials have been shown to remobilize full slurry flow conditions even after stoppages of up to several hours within delivery pipelines. Paste fills exhibit higher flow resistance than high-density slurry fill materials, although such resistance has not been observed to exceed delivery pipeline operating pressure limits of approximately 1000 lb/in.^{2} (6.9 megapascals) or to cause excessive pipeline wear.

Typical paste backfill pulp densities range between approximately 75 and 88% by weight solids. Such backfill materials are characteristically less porous than high-density slurry fill materials (averaging 30% for paste fill versus 35% for high-density slurry fill), and yield significantly reduced water runoff when placed underground in stopes. The decreased backfill porosity and water loss achieved by paste fill materials yields higher in-place densities and consequent material strength. Where cement additions are contemplated, less cement loss is also attributed to the effect of reduced water decantation.

Flocculated total tailings. Research effort is presently being directed toward the use of flocculating agents with total tailings slurries to agglomerate fine aggregate fractions. Where fine and coarse tailings particles may be chemically bonded, losses of fines in the stream of decanted water may be reduced. Investigations to date have shown limited success because of the time-dependent nature of the flocculation process. Flocculants must be added to the slurry stream prior to the point of final deposition so that sufficient time is available to permit complete reaction between aggregate particles and the flocculating agents within the slurry stream. With late mixing, inadequate flocculation of solids occurs before the decantation of water from the stopes takes place. In such cases, fine particle losses result. With early mixing, flocculation occurs too soon within the slurry stream, and the consequent formation of coarse, agglomerated particles may severely hinder efficient turbulent flow of backfill slurries.

Water gelling additives. The use of rapid-setting water gelling agents is also being reviewed as a means of physically binding backfill solids within the liquid fraction of slurries. Where effective, the liquid portion of a slurry, commonly representing 30–35% by weight of a typical hydraulic backfill mixture, may be retained within stopes without the need to rehandle this volume for transport back to the surface. Thus, problems resulting from decantation of fines or cement agent materials within the decant water will be avoided.

The short reaction times of most gelling agents, generally in seconds, requires that slurry mixing occur at or close to the point of backfill discharge into stopes. Gelled backfill materials also exhibit poorer strength response relative to similar backfill materials mixed without chemical additives.

Pelletized total tailings. Mining operators, typically in the iron ore industry, have utilized material handling processes such as pelletizing to fabricate ore products exhibiting uniform size and chemical composition. Backfill research has been directed toward subjecting total unclassified tailings to similar pelletizing processes as a means of preparing a backfill product that can be easily stored, transported, and placed underground. By physically bonding total tailing aggregates into pellet form, it may become possible to transport such materials underground more cost effectively (by using pneumatic or gravity transport techniques) and to store them more easily at the surface prior to use.

Manufacturing trials have successfully demonstrated the feasibility of pelletizing unclassified and classified tailings materials. However, no field trials have yet been attempted to review placement techniques or mill capabilities to manufacture large volumes of pelletized backfill agents.

Alternate binder agent materials. Research has also focused on the replacement of standard backfill cement binder agents, such as normal portland cement, with ground glass materials. When ground to the fineness of cement, waste glass, retrieved from municipal landfills, has been demonstrated to induce cementing reactions at a significant fraction of that attributed to portland cement materials. Since large volumes of glass are discarded annually to municipal landfill sites, it may be possible to establish a larger and more economical source of effective cementing agent material for mine backfill. Additionally, by utilizing waste glass binder additions to backfill, the mining industry may be capable of establishing positive environmental stewardship for a presently nonrenewable resource.

For background information SEE FLOCCULATION; FLUID-FLOW PRINCIPLES; PIPELINE; PORTLAND CEMENT; UNDERGROUND MINING in the McGraw-Hill Encyclopedia of Science & Technology.

James F. Archibald

Bibliography. J. F. Archibald, P. Lausch, and H. Zhe-Xiang, Quality control problems associated with backfill use in mines, *Canad. Inst. Mining Metall. Bull.*, 86(972):53–57, July 1993; W. Bawden and J. F. Archibald (eds.), *Innovative Mine Design for the 21st century*, 1993; P. Campbell, D. Ames, and C. Graham, *Backfill Practices and Trends in Ontario Mines*, Annu. Gen. Meet. Canad. Instit. Min. Metall., May 1987.

Urban heat storage

Heat storage occurs as the result of transport of energy into and out of a material. The heat transfer and the thermal climate of a solid medium is governed by four interrelated thermal properties: heat capacity, the ability of a substance to store heat and express the temperature change produced as a result of gain or loss of heat; thermal conductivity, a measure of the ability to conduct heat; thermal diffusivity, the ability to diffuse thermal influences, which controls the speed at which temperature waves move through the medium and the depth of thermal influence; and thermal admittance, the ability of a surface to accept or release heat. Typical values for each of these properties are presented in standard reference works. In the simplest case, for a single homogeneous medium the vertical transfer of heat is proportional to the temperature gradient and the thermal conductivity of the medium.

When related to the surface of the Earth, the conduction of energy into the ground is often called the soil heat flux (Q_G). In general, the daily pattern of the soil heat flux is tied to the diurnal pattern of radiative forcing (specifically the net all-wave radiation, Q^*): energy is transferred into the soil by day and back out at night. Annually, a similar pattern occurs, with a net gain in the warm season and loss in the cool season.

In a natural environment, most surfaces do not consist of a single homogeneous medium but rather a complex mix of substrate materials and moisture contents. Consequently, an accurate assessment of soil heat flux is dependent on how well the substrate thermal properties can be determined. Soil moisture is often the key factor in estimating the temporal changes in soil heat flux, because of the high heat capacity of water and the spatial variability (horizontally and vertically).

Urban areas. Urbanization significantly changes the materials and morphology of the surface. The results are a new environment with very different aerodynamic, moisture, radiative, and thermal properties. These factors lead to the formation of so-called urban climates, characterized by increased surface roughness, decreased surface moisture, and enhanced temperatures (often referred to as the urban heat island). A significant consequence of urbanization is an increase in heat stored by urban materials compared to that of typical rural terrain.

Urban areas consist of a myriad of surface types, both anthropogenic (such as asphalt or concrete) and natural (such as grass, trees, or open water). The combination of varied surface materials and their three-dimensional geometry complicates the determination of heat storage. The three-dimensional nature of the urban surface means heat energy is stored in vertical as well as horizontal surfaces. Incoming solar radiation (controlled by the solar zenith angle) tends to be spatially consistent over simple, open horizontal surfaces. However, this is not the case for vertical surfaces, where height and aspect are important. Shadows cast by nearby structures are variable throughout a day, depending on latitude, time of day, and day of the year.

When the heat storage of buildings or neighborhoods is of interest rather than that of individual components (such as a section of road), it is more appropriate to view heat storage in the context of a surface volume (see **illus.**) rather than as a surface plane. The upper boundary of the volume, above the roof and tree level, is the urban canopy layer, which can be viewed as the active surface, that is, the site of energy transfer into and out of the system. The resulting volume consists of the subsurface, all structures, and the overlying air. The net storage heat-flux density ΔQ_s is defined as the net uptake or release of heat energy including latent-heat (heat required to change a liquid into a vapor without a change in temperature) and sensible-heat (heat that manifests itself as a change in temperature) changes in the air, buildings, vegetation, and ground extending from above roof level to a depth in the ground, where the net heat exchange over the period of study is negligible.

Methods for determination. Several approaches are used to obtain estimates of the net storage heat flux in urban areas, including direct measurement, calculation as a residual of the surface energy bal-

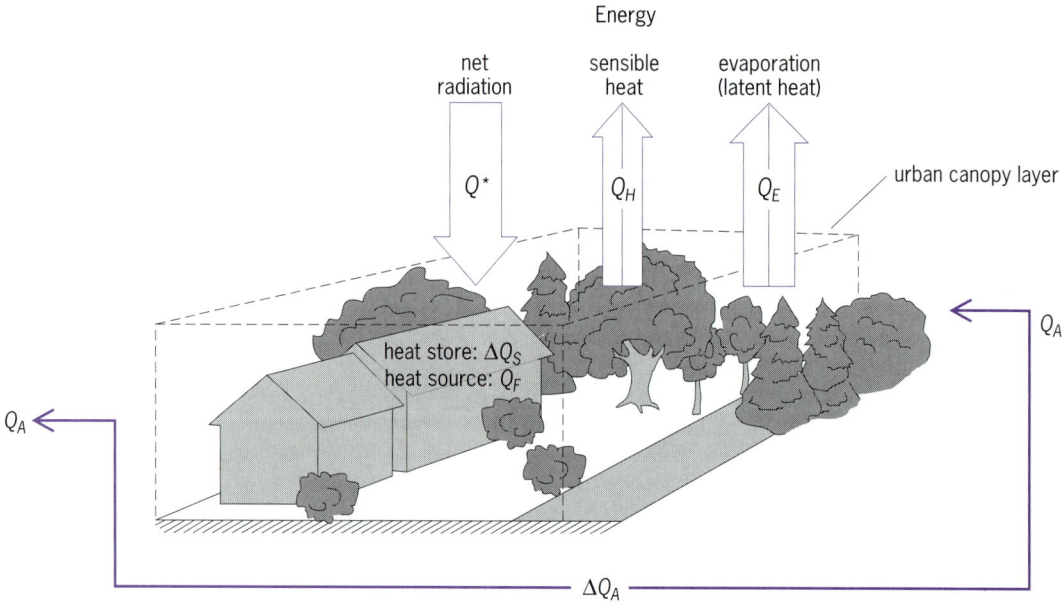

Heat storage of an urban surface volume.

ance, and empirical or physically based modeling. Each approach has its own limitations.

Direct. Direct measurements can be obtained for homogeneous surfaces by measuring temperature gradients and thermal conductivities or through the use of soil heat-flux plates. Many environments, especially urban ones, are not flat and contain a complex mixture of surface types and geometries (see illus.). Thus, if determination of heat storage in neighborhoods, cities, or regions of several land-use types is of interest, use of soil heat-flux plates present major problems of sampling/integration. Normally, it is not feasible to measure a sufficient sample of all the surface types and orientations composing the urban fabric to obtain direct measurements at this scale.

Residual. Currently, the most common approach in urban areas is to estimate the net storage heat-flux density (ΔQ_S) as a residual of the surface energy balance. This approach is based on the principle of the conservation of energy, and it can be calculated by using the equation below where Q^* is the

$$\Delta Q_S = (Q^* + Q_F) - (Q_H + Q_E + \Delta Q_A)$$

net all-wave radiation flux density, Q_F is the anthropogenic heat-flux density, Q_H and Q_E are the turbulent flux densities of sensible and latent heat, respectively, and ΔQ_A is the net advective (horizontal) heat-flux density (see illus.). However, this approach incorporates all the instrumental measurement errors of the other fluxes into the value for ΔQ_S. If all the errors are compounded in the same direction, the probable error in the residual is about 50% by day and 25% at night, although typical errors would be less. Much of the previous research on ΔQ_S has assumed that Q_F and ΔQ_A are negligible. The contributions of Q_F to the energy balance are typically small, peaking during the winter as a result of heat loss by poorly insulated buildings or during the summer air-conditioning season. The term ΔQ_A is dependent upon the scale of interest. If there is an extensive area of similar land use it may be considered to be insignificant. However, this assumption may not be valid for all cases, which may have significant implications for the value of ΔQ_S when determined as a residual.

Numerical. Given the problems of direct measurements and the residual technique, numerical models are often used. Several empirical and physically based approaches have been developed for the urban environment. Many of these approaches draw on work conducted in forested and agricultural environments and involve determination of a driving force (for example, temperature, net all-wave radiation, and soil heat flux) and the characteristics and areal fractions of the surface materials in the area of interest. Each model characterizes these variables differently. Such models range from linear parameterizations used to determine daily values, to hourly models that incorporate differential timing (hysteresis) in the Q^* and ΔQ_S fluxes and the three-dimensional surface characteristics.

The more physically based models use temperature gradients in conjunction with thermal properties of the materials. The most sophisticated models simulate the temperature gradients, by using sky-view geometry and regional radiative forcing.

Modeling of heat storage in urban areas, either empirical or physically based, tends to be scale limited. Because of the complex nature and distribution of surface materials in an entire urban region, heat storage is modeled for subareas and extrapolated for larger urban areas.

Current results suggest increased heat storage in cities compared with their rural environments. However, to date there have been limited attempts to validate these various modeling approaches against real data.

Remote-sensing techniques offer a large-scale spatial view of surface characteristics. As ground and spectral resolution improves, information satellites should provide valuable data on heat storage. However, many questions still exist when using satellite data in such a context. For example, what is the actual surface that is being sensed and how does it vary with changes in the view angle of the sensor from directly overhead (nadir) to an oblique angle?

Importance. By the year 2000, it is estimated that 47% of the world's population will live in urban areas. In 1990, 69 cities had a population over 3×10^6, and this number is expected to increase to 85 cities by the year 2000. Many of the world's major cities are faced with problems of deteriorating air quality, fresh-water shortages, and increased energy demand.

In many cities, the World Health Organization guidelines for acceptable air quality are regularly exceeded, in some cases to a great extent. The primary source of air pollution is the combustion of fossil fuels by automobiles and through industrial processes. The dispersion and deposition of airborne pollutants are controlled by the meteorological and topographic conditions. Like mountains, buildings can act as barriers, trapping air pollutants downwind of source areas, particularly when the regional air flow is perpendicular to the street canyon geometry. Air quality is often at its worst in the early morning hours, when inversions are shallow and automobile use is at a peak as people drive to work. During the afternoon hours the mixing depth increases as the boundary layer grows, because of the addition of sensible heat from the surface. A greater mixing depth allows for the dilution of pollutants, thus improving air quality. The amount of sensible heat available to allow growth of the boundary layer is inversely related to how much heat energy is stored at the surface.

The high concentration of people in urban areas places an additional demand on the availability of fresh water, particularly in the arid and semiarid regions of the world, where the supply of fresh water is already meteorologically limited. Water is often transported into urban centers from surrounding regions. Much of this water is lost to the atmosphere through evapotranspiration from open water surfaces (such as rivers, lakes, or canals) or from irrigated surfaces (such as agricultural crops, parks, or lawns). There is an inverse relationship between the magnitude of evapotranspiration and the amount of heat stored in the surface.

The thermal inertia provided by storage is often considered as key to the formation of the urban heat island. Accurate estimates of heat storage are essential for predicting energy consumption rates related to the urban heat island. Warmer urban temperatures have their greatest economic impact in the summer season, when the demand for cooling peaks.

With expected global population increases, urban areas will continue to expand at the expense of natural surfaces. A major result of urbanization is an increase of heat stored by urban materials compared to that of typical rural terrain. This storage of heat has important implications for the growth of the boundary layer, the magnitude of evapotranspiration, the formation of the urban heat island, and thus the energy necessary to heat or cool commercial/residential establishments.

For background information SEE AIR POLLUTION; CONDUCTION (HEAT); HEAT BALANCE; HEAT CAPACITY; RADIATIVE TRANSFER in the McGraw-Hill Encyclopedia of Science & Technology.

Mark D. Hubble; C. Susan B. Grimmond

Bibliography. American Society of Heating, Refrigerating, and Air-Conditioning Engineers, *ASHRAE Handbook: Fundamentals,* 1994; C. S. B. Grimmond, H. A. Cleugh, and T. R. Oke, An objective urban heat storage model and its comparison with other schemes, *Atmos. Environ.,* 25B:311–326, 1991; T. R. Oke, *Boundary Layer Climates,* 1987; World Health Organization, *Urban Air Pollution in Megacities of the World,* 1992.

Visual agnosia

Information received by the eyes is transmitted to a number of different subcortical and cortical brain structures, and normal visual perception involves the orchestrated activity of these structures. Damage to any part of this complex system results in impaired visual perception. In rare cases, when the regions of the brain responsible for visual object recognition are damaged, a form of visual impairment known as visual agnosia results.

Object recognition. Individuals with visual agnosia have adequate acuity, depth perception, and color perception for perceiving objects but cannot recognize many objects by vision alone. The problem appears to be specific to visual object recognition, as these individuals have no difficulty identifying objects that they can hear or touch. The discrepancy between their preserved visual capabilities and their object-recognition ability can be quite striking. For example, visual agnosics are often able to see a drawing clearly enough to produce a good copy of it, yet do not know what they have copied.

Historically, visual agnosia has been a controversial topic, its very existence even being discounted. It was suggested that the appearance of disproportionate difficulty with visual object recognition could be explained by a combination of mild perceptual impairments and mild general intellectual impairments. The rarity of visual object agnosia and the almost invariable presence of some degree of concommitant visual or intellectual impairment prevented a rapid resolution to this debate. Eventually, a number of carefully conducted case studies demonstrated convincingly that agnosic patients are no more impaired in their elementary visual capabilities and their general intellectual functioning than many patients who are not agnosic, and that agnosia must therefore result from damage to neural mechanisms specialized for object recognition. Therefore, most current research focuses on a new set of questions: What type of visual processing is impaired in visual agnosia? Are there different types of agnosia, affecting recognition of different types of object? What brain regions are critically involved in visual object recognition?

Visual processing. The nature of the underlying visual impairment resulting from visual agnosia varies from individual to individual, but certain generalizations are nevertheless possible. Two broad classes of agnosia have been distinguished on the basis of an individual's ability to copy or match simple shapes and drawings: apperceptive visual agnosia and associative visual agnosia with the subclass prosopagnosia. Different types of visual impairment appear to underlie each.

Those individuals who are able to copy and match shapes have traditionally been said to have associative visual agnosia. This form of impairment occurs at the highest levels of visual processing, where perception makes contact with an individual's ability to remember shapes and appearances of known objects.

Individuals with apperceptive visual agnosia have trouble copying even simple shapes, although their visual acuity and ability to perceive color and texture may be normal. The **illustration** shows the attempts of an apperceptive agnosic to copy some simple shapes. The underlying impairment in apperceptive agnosia appears to lie in the processes that group local elements of the visual field into whole regions and objects.

Prosopagnosia. The scope of the associative agnosic deficit also varies from individual to individual. Some individuals with relatively good perception encounter difficulty mainly with face recognition. These individuals have prosopagnosia. They may be so profoundly impaired with regard to face recognition that they fail to recognize their own spouses and children, and even fail to pick themselves out of a group photograph. Yet their recognition of common objects may be only minimally impaired or in some cases normal, and they may be able to read without difficulty.

Prosopagnosia has been interpreted in two ways with respect to theories of normal human vision. The most straightforward interpretation is that face recognition requires some specialized recognition mechanism, which is not needed for the recognition of objects and printed words, and that this mechanism has been damaged in individuals with prosopagnosia. The alternative interpretation is that faces are generally more difficult to recognize than other objects or words, and that prosopagnosia results from slight damage to a general-purpose visual recognition mechanism. Direct tests of these hypotheses have supported the first interpretation. Further support for the existance of face-specific processing mechanisms comes from brain-damaged individuals with the opposite pattern of visual abilities: good face recognition with poor object recognition; this pattern is clearly inconsistent with the idea of a single recognition mechanism for which faces are harder to recognize than objects.

Affected regions of the brain. If visual agnosia is regarded as a single undifferentiated category, it is impossible to make generalizations about the regions of the brain responsible for visual object recognition. Although the damage generally affects the posterior regions of the brain, which are known to be involved in visual peception, different specific regions are affected in different cases. Separating the apperceptive from the associative forms of agnosia reveals some degree of localization of function. Whereas apperceptive agnosics have generally suffered diffuse damage to the posterior parts of the brain, including the occipital cortex, associative agnosics tend to have more focal lesions affecting the boundary zones between the occipital and temporal cortexes on the inferior surface of the brain. Distinguishing those individuals with cases of associative agnosia who do and do not have difficulty with faces allows even further specificity for the localization of function. When face recognition is impaired (prosopagnosia alone or with object agnosia), the lesions are generally bilateral, whereas when face recognition is spared (object

Drawings made by an individual with apperceptive visual agnosia.

agnosia alone), the lesions are confined to the left or dominant hemisphere.

For background information SEE AGNOSIA; BRAIN; VISUAL IMPAIRMENT in the McGraw-Hill Encyclopedia of Science & Technology.

Martha Farah

Bibliography. H. D. Ellis et al. (eds.), *Aspects of Face Processing*, 1986; M. J. Farah, *Visual Agnosia: Disorders of Object Recognition and What They Tell Us About Normal Vision*, 1990; R. A. McCarthy and E. K. Warrington, Visual associative agnosia: A clinico-anatomical study of a single case, *J. Neurol., Neurosurg., Psychiat.*, 49:1233–1240, 1986.

Water

Important issues in water purification are the health implications of microbiological and chemical contaminants and methods to identify and reduce potentially hazardous chemical disinfection by-products.

The importance of water purification has been recognized for thousands of years, as documented in Sanskrit medical writings, ancient Egyptian inscriptions, and Hippocrates' discourses on public health. Although methods for removing turbidity from water supplies improved over time, water-borne diseases such as cholera and typhoid continued to cause extensive illness and death in the industrialized world until the turn of the twentieth century. The first documented evidence of an association between a contaminated water supply and disease in a human population appeared in 1855, when J. Snow published an epidemiological study linking the spread of cholera in a district of London to the consumption of polluted water. Beginning in the early 1900s, the disinfection of water supplies with chlorine became a standard practice, leading to a dramatic decline in the incidence of waterborne diseases in the United States and other industrialized nations.

Treatment. Other important steps in the treatment process include aeration to remove odors and gases, coagulation and flocculation followed by sedimentation to remove settleable particles, and filtration to remove particles and microorganisms. Water treatment plants may also use adsorption to remove organic contaminants and color, softening to remove hardness-causing chemicals, stabilization to prevent scaling and corrosion, and fluoridation to supply protection for the tooth enamel of the consumer. Chlorine continues to be the most commonly used disinfectant for treatment of drinking water because of its biocidal effectiveness, low cost, ease and relative safety of application, and ability to maintain a residual in the distribution system. A growing number of water utilities now use alternative disinfectants, particularly ozone, chlorine dioxide, or chloramine.

When properly maintained and operated in compliance with federal regulations, modern water treatment systems are considered to be highly effective in producing water that is safe for public consumption. Nevertheless, the continued occurrence of outbreaks of water-borne disease has demonstrated that drinking water supplies are vulnerable to contamination with microbial pathogens that can cause serious illness or even death. Additionally, public health concerns have been raised concerning chemical contaminants in the drinking water supply. Surface- and ground-water sources may be contaminated with many different natural and artificial toxic substances, and the water treatment process itself leads to the formation of a large number of organic by-products. The toxic effects of many of these contaminants are unknown or poorly understood. To ensure the safety of drinking water in the United States, the federal government is required by the 1986 Amendments to the Safe Drinking Water Act to establish standards that limit public exposure to specific contaminants in drinking water. Water suppliers must meet these standards lest they face enforcement actions by their respective state governments.

Microbiological contaminants. Contamination of source waters with pathogenic bacteria, viruses, and parasites may pose a serious public health threat if treatment is inadequate (see **table**). During 1991–1992, a total of 34 outbreaks of disease associated with drinking water were reported in the United States. It is highly likely that many other outbreaks are either unrecognized or unreported. Most of the known outbreaks have been attributed to a lack of treatment or to problems with the treatment process. In the majority of outbreaks, the causative agent has not been identified. The largest documented outbreak of disease associated with contamination of a public water supply in the United States occurred in 1993, when an estimated 400,000 people in Milwaukee, Wisconsin, experienced acute gastrointestinal illness caused by the protozoan parasite, *Cryptosporidium*. This pathogen is of special concern because of its high infectivity at low doses, its environmental persistence, its resistance to disinfection, and its ability to cause life-threatening disease in immunocompromised individuals such as acquired immunodeficiency syndrome (AIDS) patients.

Effective drinking water treatment processes, protection of watersheds, active surveillance of waterborne disease, and continued research are necessary to ensure that the risk to the public from microbial pathogens in drinking water is minimized.

Chemical contaminants. Drinking water is a complex mixture of many chemicals. Although most natural contaminants (for example, salts and humic acids) pose little public health risk, other naturally occurring substances (such as arsenic and sulfate) may cause adverse health effects if present in sufficiently high concentrations. Human activities may lead to the contamination of surface- or ground-water sources with a variety of toxic substances, including nitrates and pesticides from agricultural runoff and organic solvents from industrial activi-

Water-borne diseases, microbial agents, and general symptoms

Disease	Microbial agent	General symptoms
Typhoid fever	Bacterium (*Salmonella typhi*)	Fever, headache, constipation, nausea, vomiting, rash
Campylobacteriosis	Bacterium (*Campylobacter jejuni*)	Fever, abdominal pain, diarrhea
Cholera	Bacterium (*Vibrio cholerae*)	Watery diarrhea, vomiting, muscle cramps, dehydration
Shigellosis	Bacterium (*Shigella* species)	Fever, diarrhea, bloody stool
Cryptosporidiosis	Protozoon (*Cryptosporidium parvum*)	Diarrhea, abdominal discomfort
Giardiasis	Protozoon (*Giardia lamblia*)	Diarrhea, abdominal discomfort
Hepatitis	Virus (hepatitis A)	Fever, chills, abdominal discomfort, jaundice
Viral gastroenteritis	Viruses (Norwalk, rotavirus, and others)	Fever, headache, gastrointestinal discomfort, vomiting, diarrhea

ties. Lead is a particularly toxic contaminant that may enter the drinking water because of corrosion of lead-bearing materials in the water distribution system and in household plumbing. The nature and magnitude of the effects of these substances depend upon the toxicity of the specific contaminant, the exposure concentration, the duration of exposure, and the ability of the exposed individual to detoxify and eliminate the agent.

The discovery in 1974 that chlorine reacts with natural organic substances in the raw water during the treatment process to form chloroform and other potentially harmful by-products raised immediate concerns about the safety of drinking water treatment practices. It is now known that chlorination produces many chemical by-products, such as trihalomethanes, haloacetonitriles, and halogenated acids. Recent research indicates that different profiles of by-products are formed during the treatment process when alternative disinfectants such as ozone or chlorine dioxide are used.

Public health implications. The public health implications of the presence of low levels of disinfection by-products in drinking water are uncertain. Many disinfection by-products have been shown to cause cancer and other toxic effects under experimental conditions involving high-dose exposures of laboratory animals, whereas the effects of other by-products are very poorly characterized. A large number of epidemiologic studies have attempted to evaluate the association between cancer and the long-term consumption of chlorinated and unchlorinated water from various sources. Although some studies have reported elevated rates of cancer of the bladder and possibly other organs, the interpretation of these studies is difficult. Most experts consider current scientific evidence to be inadequate to conclude that the chlorination of water poses a significant health threat. However, it is clear that the resolution of this issue will require continued research of health effects, water treatment, and analytical methods for many years to come.

Fred S. Hauchman

Alternative treatments. Chlorine is still by far the most popular disinfectant in the United States, but it produces potentially harmful trihalomethanes and other chemical by-products, such as aldehydes and carboxylic acids, and inorganic by-product ions, such as bromate, which have been linked to cancer in laboratory animals. As a result, the U.S. Environmental Protection Agency (EPA) established a maximum contaminant level (MCL) for total trihalomethanes at 0.10 mg/liter, and new regulations are being proposed for other potentially harmful by-products. Because ozone and chlorine dioxide do not produce the high levels of trihalomethanes that chlorine does, some drinking water treatment facilities have changed their treatment process from chlorine to ozone or chlorine dioxide in order to more easily meet the maximum contaminant level. However, until recently, little was known about the disinfection by-products from these alternative disinfectants.

Ozone. Ozone is a powerful oxidant, and is more effective than chlorine for killing harmful microorganisms in drinking water. Ozone, which exists as an unstable gas, has been used for more than 80 years to disinfect drinking water and to control taste and odor problems. Because ozone does not persist in water, microorganisms can regrow in the distribution system; therefore, a residual disinfectant, such as chlorine or chloramine, must also be added before the water is sent through the distribution system to the consumer.

The major disinfection by-products identified from ozonation of natural water are aldehydes and carboxylic acids. Aldo and keto acids also have been observed. More recently, laboratory studies have been carried out to determine the by-products of ozone in the presence of bromide. Bromide is present in the raw water of coastal cities, and it is also present in ground water that has a high mineral content. When bromide is present, the structures of the ozonation by-products change dramatically, with most becoming brominated. Two of the most common brominated by-products are bromoform and bromate, both of which have raised health concerns. Bromoform is currently regulated under the maximum contaminant levels for total trihalomethanes, and bromate has recently been proposed for regulation.

Other brominated disinfection by-products that have been identified include a series of brominated alcohols, called bromohydrins. Identification of the bromohydrins by conventional gas chromatography/mass spectrometry (GC/MS) is not possible, however, because their spectra are not located in the databases commonly used to identify environmental contaminants. As a result, more complicated analyses are required, including gas chromatography coupled with chemical ionization mass spectrometry (GC/CI-MS), high-resolution mass spectrometry, and Fourier-transform infrared spectroscopy (GC/FT-IR). Chemical ionization mass spectrometry allows molecular weights to be determined for unknown compounds when that information (the molecular ion) is missing from a typical GC/MS spectrum. High-resolution mass spectrometry is used to determine an exact empirical formula for the unknown compound when many possibilities are available from the low-resolution GC/MS data. Finally, GC/FT-IR provides functional group information. In the case of the bromohydrins, it revealed that the functional group was an alcohol group adjacent to the bromine atom.

An interesting feature of this study was the identification of several diastereomers and enantiomers. Diastereomers look similar when their structures are drawn on paper, but their structures are not superimposable. Enantiomers are the mirror images of these diastereomers (analogous to an individual's right and left hands). Because diastereomers typically have different boiling points, they are separated easily by gas chromatography. However, a chiral GC column is necessary to see the enantiomers. When the chiral column is used, each diastereomer peak turns into two enantiomer peaks. Sometimes knowing which enantiomer is present can be important, because receptors in the human body are often specific for one enantiomer over another. Thus, one enantiomer can be toxic and the other nontoxic. Because the bromohydrins are so unusual, it is not known at this time whether they are a health concern. Also, because the bromohydrins studied here were generated in laboratory studies, concentrations in actual drinking water treatment plants remain uncertain.

Chlorine dioxide. Like ozone, chlorine dioxide is a powerful oxidant, even more effective in killing harmful microorganisms than chlorine. Chlorine dioxide, which exists as a yellow-green gas, is currently used in approximately 400 drinking water plants in the United States and in several thousand in Europe. A residual disinfectant, such as chlorine or chloramine, must be added before the treated water is sent through the distribution system to the consumer.

Until recently, only limited laboratory studies have been performed to determine potential disinfection by-products of chlorine dioxide. However, the presence of two inorganic by-product ions, chlorite and chlorate, has been known for some time, and some health concerns have been raised about them. Chlorate can be controlled through proper operation of the chlorine dioxide generators. Chlorite is proposed for regulation by the EPA.

A recent study assessed the organic semivolatile disinfection by-products produced at a pilot drinking water treatment plant that uses chlorine dioxide. Because chlorine is also likely to be used as a residual disinfectant at some chlorine dioxide treatment plants, its conjunctive use with chlorine dioxide was also investigated. In total, more than 40 disinfection by-products were identified. Because many were difficult identifications to make, the combination of mass spectral and infrared techniques mentioned above was used.

The most prevalent by-products identified were a series of carboxylic acids, the same as those commonly observed with ordinary chlorination of drinking water. Carboxylic acids are believed to arise from natural plant material present in the raw water. The acids identified do not have any known health concerns associated with them. Other compounds observed included a series of halopropanones and a series of maleic anhydrides, which were believed to have been formed from maleic acids during the extraction and concentration process. Many other halogenated compounds, including halomethanes, also were observed, but only when chlorine was used as a residual disinfectant in conjunction with the chlorine dioxide. As a result, it may be possible to totally eliminate the halomethanes by using a different residual disinfectant than chlorine, for example, monochloramine, which does not produce significant levels of halomethanes.

Although more than 40 chlorine dioxide disinfection by-products were identified, they were of extremely low concentrations (1–10 parts per trillion), much lower levels than those commonly found in chlorinated drinking water.

For background information SEE WATER-BORNE DISEASE; WATER PURIFICATION; WATER SUPPLY ENGINEERING; WATER TREATMENT in the McGraw-Hill Encyclopedia of Science & Technology.

Susan D. Richardson

Bibliography. R. J. Bull and F. C. Kopfler, *The Health Effects of Disinfectants and Disinfection By-Products,* 1991; G. F. Craun (ed.), *Safety of Water Disinfection: Balancing Chemical and Microbial Risks,* 1993; W. H. Glaze and H. S. Weinberg, *Identification and Occurrence of Oxonation By-products in Drinking Water,* Amer. Water Works Ass. Res. Found., 1993; S. D. Richardson et al., Multispectral identification of chlorine dioxide disinfection byproducts in drinking water, *Environ. Sci. Technol.,* 28:592–599, 1994.

Water-jet cutting

Continuously dripping water can erode and penetrate rocks. The ancient Chinese and Romans used sand-mixed water for cleaning and tunneling. In the early 1930s, a low-pressure and high-flow-rate water jet was introduced for coal mining in the former Soviet Union. In a technology introduced in the 1970s, the dripping water is replaced by high-speed water droplets and a mixture of water and abrasive. Water accelerated up to twice the speed of sound can penetrate and cut rocks in a few seconds.

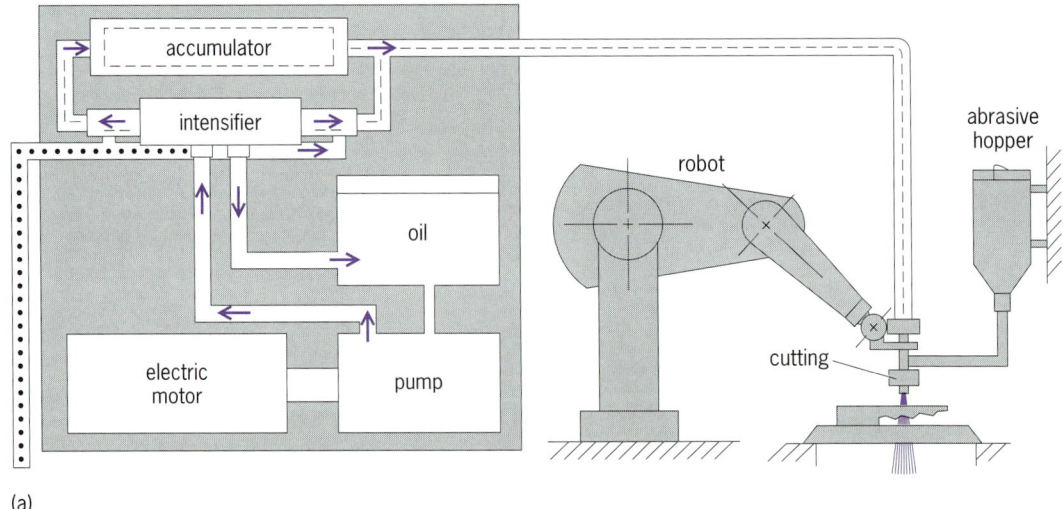

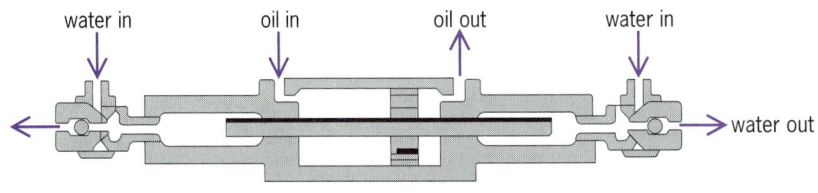

Fig. 1. Water-jet cutting system. (*a*) Water-jet pump and motion control system. (*b*) Double-acting intensifier pump.

Operation. A typical ultrahigh-pressure water-jet cutting system consists of two stages of pumping equipment, a high-pressure water-jet delivery system, a mechanical manipulator for motion control, and a discharge catcher system.

Pump. The first stage of pumping is carried out by a radial displacement pump, which can pressurize the hydraulic oil up to 3000 lb/in.^{2} (20.7 megapascals). The hydraulic oil then drives an intensifier pump (**Fig. 1**). The intensification ratio of this piston type of pump is 1:20, which is the area ratio of two ends of the piston. The intensifier pump can pressurize water up to 60,000 lb/in.^{2} (414 MPa). The tap water supplied to the intensifier pump passes through several stages of filters (10–0.5-micrometer range) at a boosted water-feed pressure of 80 lb/in.^{2} (0.555 MPa). The high-pressure water from the intensifier pump is channeled through an accumulator to level off pressure fluctuations created by the plunger motion of the pump. The water is then carried to a cutting station by means of stainless-steel high-pressure tubing and swivel joints or by a flexible high-pressure hose which can withstand pressures up to 100,000 lb/in.^{2} (690 MPa).

Jet delivery system. The high-pressure water delivered to a nozzle assembly is first converted into a high-speed jet through an orifice assembly which houses an on-off control valve and orifice. The jet created through the orifice can be manipulated by motion-control equipment such as a robot or *x-y* table to cut a desired shape of workpiece. For cutting hard and dense materials such as steel, stone, and ceramics, the water jet is mixed with abrasive particles in a mixing chamber (**Fig. 2**), before being discharged through the nozzle. The orifice is made of sapphire or diamond, and lasts up to 200 h and 2000 h, respectively. The nozzle (often called the mixing tube) is made of tungsten carbide, which has a life of 5 h under a normal operating conditions. A new nozzle material, a composite carbide, can extend the life of the nozzle up to 50 h. The coherent water jet emerging from the tip of the nozzle can attain a speed of 2000 ft/s (610 m/s) at a normal operating pressure of 45,000 lb/in.^{2} (310 MPa).

Process control parameters. The precision and efficiency of water-jet cutting can be controlled by 18 different system parameters. The process parameters include the water pressure; abrasive type, size, and flow rate; orifice and nozzle size; stand-off distance; cutting angle; traverse rate; and target material strength (Fig. 2). The velocity of the jet stream as well as the velocity of the entrained abrasive (usually 20% less than the water-droplet velocity) is primarily governed by the water pressure. The jet velocity can be calculated from the equation below,

$$v = K\sqrt{\frac{2P}{\rho}}$$

where P is the water pressure, ρ is the water density, and K is a dimensionless system constant which includes the effects of water compressibility, orifice

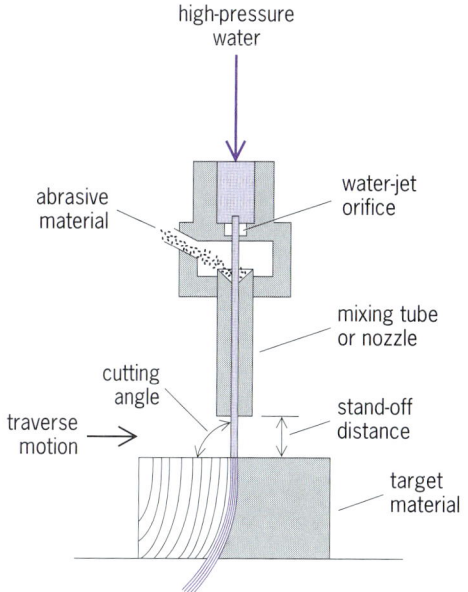

Fig. 2. Typical abrasive water-jet cutting head.

efficiency, and the abrasive and water mass flow rates. The water flow rate is controlled by the water pressure and orifice configuration. The cutting efficiency generally increases as the water flow rate increases. Since the water flow rate is limited by the pump capacity, the orifice size for typical commercially available pumps (20–100 hp, 15–75 kW) ranges between 0.004 and 0.022 in. (0.1 and 0.55 mm). The corresponding range of water flow rate is from 0.5 to 2 gal/min (1.89 to 7.57 liters). The optimized pairing in size selection for orifice and nozzle is also an important factor in increasing the efficiency of the cutting operation. The optimum pairing ratio (orifice diameter/nozzle diameter) is between 0.3 and 0.4.

Abrasives. The abrasives are used for cutting hard materials. Garnet is the most commonly used abrasive in industrial cutting operations. Other abrasives such as aluminum oxide, silicon carbide, silica sand, glass bead, and steel grit are also used in special-purpose cutting and cleaning operations. The rate of workpiece erosion, and therefore the cutting efficiency, is very much dependent on the material used. Aluminum oxide is an ideal abrasive for cutting brittle materials such as advanced ceramics and stones, yielding a twofold increase in cutting power as compared to garnet. However, aluminum oxide is not much more effective than garnet for cutting ductile materials. The abrasive size and flow rate also greatly affects the quality of the cut surface as well as the cutting efficiency. The range of mesh sizes commonly used for industrial applications is 50–120, and the range of abrasive flow rates is 0.5–1.5 lb/min (3.8–11.4 g/s).

Standoff and traverse speed. The standoff distance of the nozzle tip from the workpiece ranges from 0 to 0.2 in. (0 to 5 mm). A standoff in this range has a minimal influence on the cutting efficiency. The traverse speed of the cutting head is the most critical parameter. The depth of cutting as well as the surface finish is primarily governed by the traverse speed of the nozzle tip, which is controlled by mechanical manipulation. Therefore, the traverse speed is used as a single control parameter in most nonprecision cutting operations.

Cutting principle. The water-jet pump and its delivery system are designed to produce a high-velocity jet stream within a relatively short trajectory distance since the kinetic energy of the water and abrasive particles is directly proportional to the square of the jet velocity. In abrasive jet cutting applications, the abrasives entrained in the jet stream usually attain approximately 80% of the water-droplet velocity at the tip of the nozzle. The impacting jet cuts the target material by a rapid erosion process when its force exceeds the compressive strength of the material. Erosion mechanics is a highly dynamic and material-dependent phenomenon involving shear and tensile failure due to localized stress fields. Since the area eroded by the impacting abrasive is also swept by the water stream, the heat generated during the cutting is dissipated immediately, resulting in a small rise in temperature (less than 90°F or 50°C) in the workpiece. Therefore, no thermal distortion or work hardening is associated with water-jet cutting. The cutting by rapid erosion also significantly reduces the actual force exerted on the target material, enabling the water jet to cut fragile or deformable materials such as glass and honeycomb structures. Unlike traditional cutting and machining such as turning and drilling, where the cutter is fed by a continuous and constant level of energy during the entire cutting operation, the abrasive water-jet stream loses energy along its cutting path. The cutting power of the jet stream decreases from the top of target material to the bottom, leaving a tapered kerf and striation marks on the lower portion of the cut surface. This phenomenon is typical of high-energy beam-cutting applications such as laser and electron-beam cutting.

Applications. Among the methods of cutting metal and nonmetallic materials, pure and abrasive water-jet cutting have a distinct advantage because of their versatility and speed (see **table**). They can cut all materials including hard-to-machine materi-

Cutting speeds of abrasive water jets		
Material	Thickness of material, in. (mm)	Cutting speed, in./min (mm/min)
Aluminum	0.25 (6.25)	9–40 (225–1000)
	1.00 (25)	1–10 (25–250)
Steel	0.25 (6.25)	5–20 (125–500)
	1.00 (25)	0.5–5 (12.5–125)
Titanium	0.25 (6.25)	5–30 (125–750)
	1.00 (25)	1–7 (25–175)
Superalloy	0.25 (6.25)	8–15 (200–375)
	1.00 (25)	0.1–0.5 (2.5–12.5)
Kevlar	1.00 (25)	2–10 (50–250)
Graphite composite	1.00 (25)	3–10 (75–250)

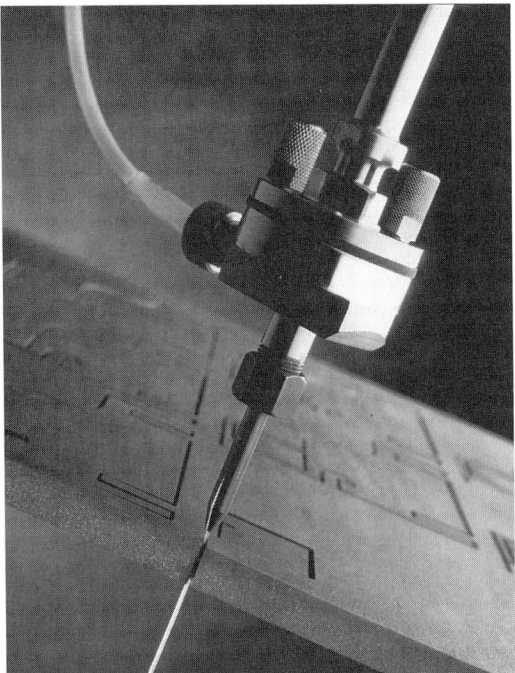

Fig. 3. Abrasive water-jet cutting of 0.5-in.-thick (12.5-mm) titanium at pressure of 45,000 lb/in.² (310 MPa). (*Flow International Corp.*)

als such as superalloy, Kevlar, and boron carbide. They can also easily cut aerospace materials such as graphite composite and titanium, and brittle materials such as granite, marble, and glass (**Fig. 3**). The pure water jet is used by the food industry to cut candy and chocolate bars, meats, vegetables, and fruits. It is also being tested for orthopedic surgery applications in bone cutting and scaling the flesh from bones. The abrasive water-jet cutter is also used in turning and milling brittle materials such as advanced ceramics, marble, and glass. The advantages of pure and abrasive water-jet cutting are (1) absence of thermal distortion and work hardening; (2) noncontact during cutting, thus eliminating tool wear and contact force; and (3) omnidirectional cutting, allowing the cutting of complex shapes and contours. Although the use of the water-jet system is rapidly growing, the technique has some drawbacks and limitations. Water jet technology has not yet developed fully for high tolerance and precision machining. The initial capital investment for the system, including the motion-control equipment and operating costs, is relatively high. The noise level (80 adjusted decibels) is also high, but the system can be specially designed to isolate the noise source.

For background information SEE ABRASIVE; JET FLOW; MACHINABILITY OF METALS; MACHINE TOOLS in the McGraw-Hill Encyclopedia of Science & Technology.

Thomas J. Kim

Bibliography. M Hashish (ed.), *Proceedings of the 7th American Water Jet Conference*, Water Jet Technology Association, 1993; R. A. Tikhomirov, *High Pressure Water Jet Cutting*, 1992.

Water resources

In the 1970s and 1980s, investments in municipal sewage treatment plants and industrial waste treatment systems greatly reduced pollutant loading into surface and ground waters from point sources. Pollutants from point sources typically enter surface waters at specific points where pipes carrying wastes from municipalities and industries empty into streams, rivers, or lakes. With reductions in point source pollutant loading, it soon became evident that another major pollutant transport pathway and set of pollutant sources existed. The pathway is that of water as it moves through the hydrological cycle, and the pollutant sources, termed nonpoint sources, are those associated with various land-use activities.

Land-use activities greatly influence the kinds and concentrations of both the dissolved and suspended materials that move with water. For many important types of pollutants (such as sediments, nutrients, and pesticides) nonpoint sources are more significant than point sources. Of the various land-use activities that contribute to nonpoint pollutant loading, agriculture is the most important. However, agriculture's impacts on water resources extend well beyond those of being simply a source of nonpoint pollutants. In many locations, agriculture's physical effects on aquatic habitats have greater impacts on aquatic biota than agriculturally derived chemical pollutants.

Erosion and sedimentation. Over geological time, the erosion of land surfaces by water and wind shapes much of the land surface. Conversion of land from native vegetation to agricultural land uses often greatly accelerates erosion. Erosion not only moves soil particles downslope or downwind but also changes the particle size composition of soils. Typically, fine-grained materials are suspended and removed from soil by erosion, leaving coarse-textured materials behind. Since the fine-grained materials have greater surface areas per unit weight and more binding sites for nutrients, erosion often decreases the fertility of remaining soils.

More recently, it has become evident that off-site damages from cropland erosion exceed damages associated with the loss of cropland fertility. Many of these damages are associated with deposition of the eroded sediments in drainage ditches, streams, reservoirs, lakes, and harbors. Sedimentation reduces the channel capacity of drainage ditches and streams, thereby aggravating flooding problems; it also decreases the water storage capacity of reservoirs and lakes, reducing their utility for flood prevention and water supplies. Navigation channels and harbors often become filled with sediment, requiring frequent and expensive dredging.

Sedimentation within streams and rivers has adverse impacts on habitats for aquatic organisms. Silty and clayey sediments often accumulate in and on the stream bottom, burying gravel substrates. Since many aquatic organisms require gravel substrates for reproduction, burying these substrates with fine-grained sediments has major impacts on the composition of aquatic communities.

Other habitat impacts. Sedimentation is not the only adverse impact of agriculture on aquatic habitats. Many of the wooded corridors along streams and rivers have been removed, resulting in higher water temperatures and decreased movements of organic matter from the stream edges into aquatic food chains. The flow regimes of streams have also changed. Conversion of forests, prairies, and wetlands to cropland increases both the amounts of surface runoff and the speed with which it reaches stream systems. Consequently, peak flood discharges have increased. With drainage of wetlands and less water infiltration, ground-water levels have often decreased, reducing water discharge during periods of low stream flow. This altered stream flow adversely impacts many biological communities of streams and rivers.

Nutrients as water pollutants. To maintain high crop yields from cropland, it is often necessary to increase the nutrient levels in soils and to replace the nutrients that are removed from cropland by harvested crops. Fertilizers and manure are added to cropland to build up and maintain soil nutrient levels. These nutrients are among the chemicals that move from cropland into surface and ground waters, often adversely impacting water resources. Two nutrient elements, phosphorus (P) and nitrogen (N), can be particularly troublesome in aquatic environments because they can contribute to water-quality problems associated with excessive growth of algae or rooted aquatic plants (eutrophication). Nitrate, one of the common forms of nitrogen, can also pose human health threats in drinking water supplies. *See* WATER.

Phosphorus is an essential element for plant growth, both on land and in water. In most freshwater environments, phosphorus is a limiting nutrient for plant growth. Thus, increases in phosphorus concentrations in water generally result in increases in the growth of algae and rooted aquatic plants. Although algae and aquatic plants are important as a basis of the food chain in aquatic ecosystems, their rapid growth often has adverse effects on streams and lakes. An excessive growth of algae can cause taste and odor problems in surface waters that are used for drinking water. Bacterial decay following the death of large populations of algae and rooted aquatic plants can deplete the water of oxygen, particularly in shallow lakes. In addition, the lack of oxygen can result in the death of fish and in basic changes in the types of organisms which inhabit the bottoms of streams and lakes. Most of the phosphorus that moves from cropland into surface water is attached to sediment particles. Consequently, the problems of sedimentation in lakes and reservoirs are often combined with the problems of overenrichment by phosphorus.

In marine and estuarine environments, nitrogen-containing compounds, including nitrate, ammonia, and organic nitrogen, often limit plant growth. Excessive loading of these compounds from cropland into marine estuaries and bays can stimulate excessive algal growth in these habitats. The Chesapeake Bay is an example of an estuary where nitrogen loading from agricultural sources adversely impacts the surrounding water resources.

Nitrate can pose direct human health problems in drinking water supplies, particularly for infants, when its concentration exceeds its safe drinking water standard of 10 mg/liter as nitrate-nitrogen. In contrast with phosphorus, which is predominantly attached to sediment, nitrate is highly soluble. Thus, nitrate can be a problem in both ground and surface water. Shallow ground-water-bearing rock (aquifers) in areas of irrigated cropland with highly permeable soils are particularly vulnerable to nitrate contamination, as are aquifers where sinkholes have developed in shallow limestone bedrock. Within the Midwest, aquifers underlying less than 2% of the land area contain nitrate at levels exceeding drinking water standards. Rivers draining cropland with extensive tile drainage, such as those in northwestern Ohio, may also periodically contain nitrate in excess of drinking water standards.

Pesticides. Of the various types of agricultural pollutants, the public is often most concerned about the use of pesticides. The current generation of pesticides used by farmers in the United States differs in important ways from earlier-generation pesticides, such as dichlorodiphenyltrichloroethane (DDT), whose use was banned in the United States in 1973. They are much less persistent in the environment than earlier-generation pesticides, and show a lesser tendency to bioaccumulate along food chains. Consequently, their use appears to pose smaller risks to ecosystems and to human health. In 1989, herbicides, which are used for weed control, made up about 60% of the total amount of pesticides used in the United States, whereas insecticides composed 20%, and fungicides, another 8%.

Because of concerns regarding possible adverse effects on human health and ecosystems, pesticides now undergo extensive toxicity testing as part of their registration and subsequent review by the U.S. Environmental Protection Agency (EPA). Stringent safety factors are incorporated between pesticide doses that have no observed adverse effects on animals and doses that are deemed safe for human drinking water supplies (see **illus.**). Since pesticide standards include a margin of safety, the EPA does not believe that drinking water for short durations at concentrations somewhat above lifetime standards poses significant risks to the public.

Pesticide monitoring programs indicate that current-generation pesticides are also among the chemicals that move from cropland into surface and ground waters. In general, pesticide concentrations in surface waters are considerably higher than those in ground waters. However, even in areas of intensive pesticide use, less than 1% of the population consumes drinking water contain-

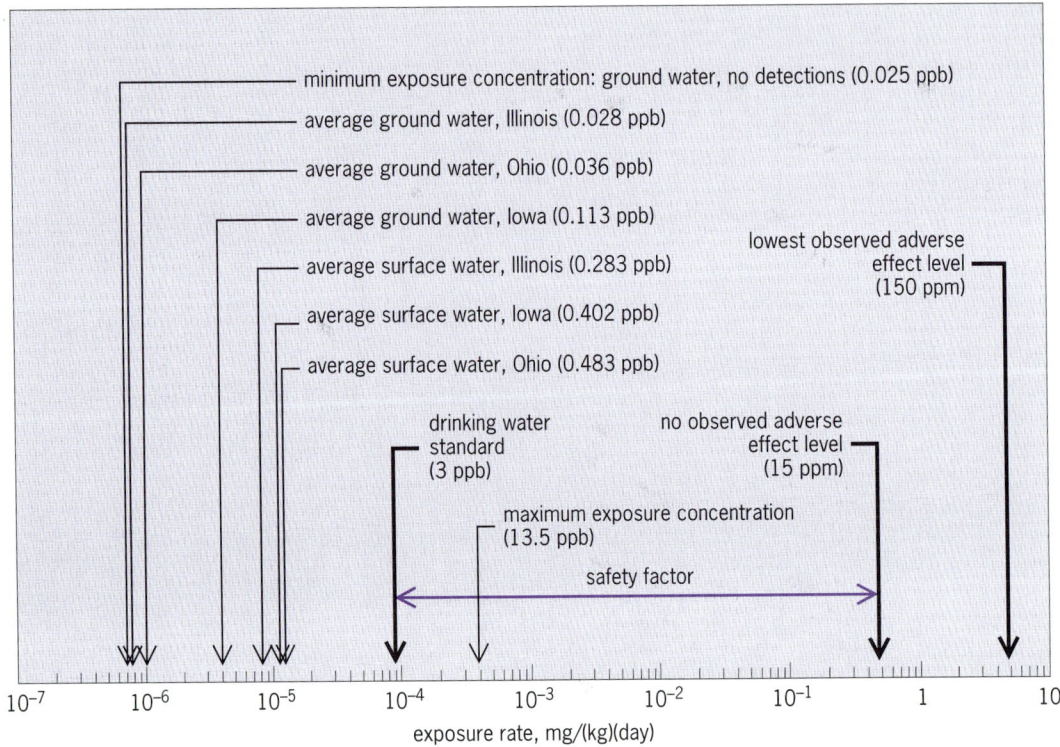

Relationship between the drinking water standard for the herbicide atrazine, the no-observed adverse effect level for atrazine in animal toxicity testing, and the average concentrations of atrazine in surface and ground waters of Ohio, Illinois, and Iowa.

ing pesticides in excess of safe drinking water standards. In addition, where drinking water standards are exceeded in public water supplies, federal regulations require that steps be taken to lower the concentrations.

Pollution abatement programs. The general approach to reducing the adverse impacts of agriculture on water resources is to foster the voluntary adoption of best management practices (BMPs). Best management practices represent sets of site-specific farming practices that reduce the movement of pollutants from cropland into surface and ground waters, while generally maintaining or increasing farm profitability. To reduce erosion, farmers apply various types of reduced tillage and no-till systems. These tillage practices retain crop residue on the soil surface, consequently reducing erosion and the subsequent movement of sediment and sediment-associated pollutants into surface waters. Other best management practices involve more careful management of fertilizers and pesticides and the establishment of buffer strips between cropland and waterways. SEE AGRICULTURAL SOIL AND CROP PRACTICES.

For background information SEE AGROECOSYSTEM; FRESH-WATER ECOSYSTEM; HERBICIDE; PESTICIDE; WATER CONSERVATION; WATER POLLUTION in the McGraw-Hill Encyclopedia of Science & Technology.

David B. Baker

Bibliography. F. J. Humenik, *Water Quality: Agriculture's Role*, Council Agr. Sci. Tech. Rep. 120, 1992; R. P. Richards et al., Atrazine exposures through drinking water: Exposure assessments for Ohio, Illinois, and Iowa, *Environ. Sci. Tech.*, 29:406–412, 1995; A. N. Sharpley et al., Managing agricultural phosphorus for protection of surface water, *J. Environ. Qual.*, 23:437–451, 1994; R. F. Spalding and M. E. Exner, Occurrence of nitrate in groundwater: A review, *J. Environ. Qual.*, 22:392–401, 1993.

Water supply engineering

As the demand for water continues to increase in the United States, water conservation and the reuse of municipal wastewater for beneficial purposes will play important roles in the planning, development, and management of water resources. Even in water-rich regions, local water shortages may occur due to urbanization, industry, and occasional drought conditions. Water shortages of varying degrees, resulting from droughts or inadequate development of water supplies, have occurred in almost all areas of the country, and it has been estimated that by the year 2000 more than 20% of the United States will occasionally experience serious water supply shortages. Water conservation and water reuse reduce the demand on fresh-water supplies, delaying or eliminating the need to develop new water supplies.

Conservation. The major use categories for fresh water in the United States are agricultural irrigation, industrial and commercial, public and domestic, and thermoelectric (see **illus.**).

Agricultural irrigation. This use accounts for 40% of the total offstream water withdrawal of approximately 4×10^{11} gal (1.5×10^9 m^3) per day and represents a prime area in which to conserve water. Conserving water in irrigated agriculture is accomplished by proper farm management (soil moisture monitoring and irrigation system evaluations), capital improvements (land leveling, proper soil drainage, and reduction of seepage and evaporation from conveyance systems), irrigation system conversions (optimization of irrigation, for example, by using drip irrigation, water use can be reduced by as much as 60%), computerized irrigation scheduling, and advances in biotechnology (development and use of drought-resistant crops).

Industrial and commercial. Cooling water needs account for more than half of all industrial water use. Replacement of once-through cooling processes with recirculating systems has been shown to provide some large water savings and rapid investment payback. Often, cooling water is uncontaminated and can be reused for landscape irrigation, cleanup, and other purposes. Other common conservation measures implemented by industry include water-use monitoring, recycling, reuse, equipment modification, improved landscape irrigation, and employee education.

Public and domestic. The average total water usage in an urban potable water system is approximately 180 gal per capita per day (gpcd) [680 liters per capita per day (lpcd)], of which 120 gpcd (450 lpcd) is for combined residential and public uses. Residential water demand can be further categorized as indoor use, which includes toilet flushing, cooking, laundry, bathing, dishwashing, and drinking, or outdoor use, which consists primarily of landscape irrigation. Outdoor use accounts for approximately 30% of the residential demand, whereas indoor use represents approximately 70%.

The U.S. Energy Policy Act, adopted in 1992, set national water efficiency requirements for toilets, urinals, showerheads, and faucets manufactured after January 1994. When the act is fully implemented, the estimated current water use of 46 gpcd (174 lpcd) for toilets, showerheads, and faucets is expected to be reduced by at least 50%.

Water for irrigating grass and other plants accounts for as much as half of the total urban demand, up to 45 gpcd (170 lpcd) in some areas. Summer lawn sprinkling and other landscape irrigation cause demand peaks requiring as much as twice the amount of water needed for typical winter use. Improved irrigation practices and technologies have helped to increase the efficiency of water applied to turf and landscaping, usually reducing the demand 30–60%. Environmental benefits and economic savings are also realized as a result of reduced use of fertilizer, fuel, herbicide, and labor. Xeriscape, which is defined as water conservation through creative landscaping, is practiced in many

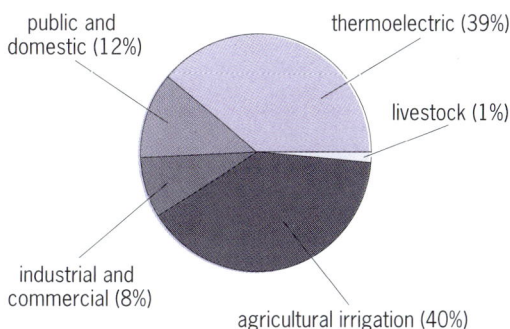

Categories of fresh-water use in the United States and their demands. (*After W. B. Solley, C. F. Mesk, and R. R. Porce, Estimated water use in the United States in 1985, U. S. Geol. Surv. Circ. 1004, Denver, Colorado, 1988*)

areas of the country. Xeriscape landscaping includes water-conserving design, use of drought-tolerant plants, reduction in turf areas, efficient irrigation methods, soil improvements and use of mulches, and proper maintenance methods.

Thermoelectric. Some methods used by utilities to conserve water include construction of new reservoirs to develop maximum dependable yield of rivers and streams, protection of watersheds from development practices that can cause degradation of quality and dependable yield, reduction of leakage and system losses, universal metering to reduce waste, installation of devices to reduce water consumption, rate adjustments, water-use restrictions, and implementation of public education programs.

Rates and pricing. Conservation-oriented water utility rate structures and pricing mechanisms have been adopted or are being considered by a growing number of water supply agencies. Rate and pricing strategies that encourage conservation include, inclining rates, seasonal quantity (peak) charges, quantity surcharges, and various incentives. For example, one water utility company in California charges a $2.00 per month conservation incentive fee for single-family homes that have not installed low-volume toilets and showerheads and $1.30 per month for each nonretrofitted apartment unit. The incentive fee acts as a surcharge on the water bill until the changes have been made.

Advantages and disadvantages. In addition to reducing water demand, conservation has other benefits, such as energy savings from reduced pumping and heating of water, reduction in wastewater flows, protection of the environment by increasing stream flows and reducing drawdown of ground-water levels, and reduction of costs associated with treating and pumping both water and wastewater. Potential drawbacks that should be considered when planning a water conservation program include the possible reduction of water-utility revenues, growth inducement, and increase in drought vulnerability.

Water reuse. With increasing water demands in areas with limited supplies of fresh water, the reuse of treated municipal wastewater may be the only feasible way of supplementing water resources.

Using reclaimed water can significantly reduce the fresh-water demand and is often the most cost-effective option for increasing a community's water resources. Water reclamation and reuse may be less costly than treatment for disposal of effluent, and it often offers an expedient approach to pollution abatement. Under such conditions, reuse not only provides an additional water supply but reduces overall costs to the community and eliminates a source of contamination in surface waters. In water-short regions of the country, it is becoming increasingly more common for state, regional, and local regulatory agencies to mandate water reclamation and reuse.

Historically, reclaimed water was primarily used where a high-quality effluent was not required, for example, pasture irrigation; reclaimed water use was often perceived as a method of wastewater disposal. In recent years, the trend has shifted toward higher-level uses such as urban irrigation, toilet flushing, industrial uses, and ground-water recharge, where the greater benefit of reuse became the value of the reclaimed water. Many types of water reuse are currently practiced in the United States.

Although commercial and domestic water use constitutes only about 10% of the total water demand, water reuse is more likely to be cost-effective in or near urban areas, where reclaimed water is used to conserve or replace existing potable water for various applications:

1. Landscape irrigation: parks, cemeteries, golf courses, roadway rights-of-way, school grounds, greenbelts, and residential lawns.
2. Agricultural irrigation: food crops; fodder, fiber, and seed crops; nurseries; sod farms; silviculture; and frost protection.
3. Industrial: cooling, boiler feed, stack scrubbing, and processing water.
4. Ground-water recharge: recharge aquifers and salt-water intrusion control.
5. Potable water supply augmentation: ground-water recharge and surface-water augmentation.
6. Nonpotable urban (other than irrigation): toilet and urinal flushing, fire protection, air conditioner cooling, vehicle washing, street cleaning, and decorative fountains.
7. Impoundments: ornamental and recreational.
8. Environmental: stream augmentation, marshes, wetlands, and fisheries.
9. Miscellaneous: aquaculture, snow making, soil compaction, dust control, equipment washdown, and livestock watering.

Although up-to-date statistical information on water reuse in the United States has not yet been compiled, it is likely that the use of reclaimed water exceeded 1.5×10^9 gal (5.7×10^6 m^3) per day in 1994. California and Florida are the leading states in both number of water reuse projects and quantity of water reused. An average of 3×10^8 gal (1.1×10^6 m^3) per day of municipal wastewater was reclaimed in California in 1993, representing about 13% of the total wastewater produced in the state. Of the reclaimed water, 15% was used for ground-water recharge, 60% for agricultural irrigation, and 15% for landscape irrigation. In Florida, 2.9×10^8 gal (1.1×10^6 m^3) per day, or about 30% of the state's municipal wastewater, was reused in 1992; 38% of the reclaimed water was used for landscape irrigation, 30% for agricultural irrigation, and 14% for ground-water recharge.

The impacts or constraints on reuse from physical parameters (such as pH, color, temperature, and particulate matter), chemical constituents (such as chlorides, sodium, heavy metals, and some trace organic compounds), and microbiological organisms (such as bacteria, viruses, and parasites) are well known; recommended limits have been established for many of the parameters and constituents. Making reclaimed water safe and suitable for reuse applications is achieved by eliminating or reducing the concentrations of microorganisms and chemical constituents of concern through wastewater treatment and by limiting exposure of the public or personnel to the water by means of design or operational controls.

Although it may be technically possible to produce reclaimed water of almost any desired quality, some health authorities and others have been reluctant to allow or support indirect potable water reuse via augmentation of ground water or surface water with reclaimed water; they generally subscribe to the concept of using waters derived from the most protected sources as raw water supplies. Public health issues notwithstanding, indirect potable reuse is receiving increasing attention. For example, considerable research efforts are being directed toward the use of membrane processes to remove health-significant constituents, alternatives to the use of chlorine for disinfection to prevent the possible formation of carcinogenic chlorinated by-products upon disinfection, and fail-safe treatment reliability.

There are no federal regulations governing water reclamation and reuse in the United States; hence, the regulatory burden rests with the individual states. No states have promulgated regulations that cover all potential uses of reclaimed water, but several states have extensive regulations or guidelines that prescribe requirements for a wide range of uses. Other states have regulations or guidelines that focus on land treatment of wastewater effluent, emphasizing additional treatment or effluent disposal rather than beneficial reuse, even though the effluent may be used for irrigation of agricultural sites, golf courses, or public access lands. In 1992, 18 states had some form of water reuse regulations, 18 had guidelines, and 14 had neither regulations nor guidelines. The U.S. Environmental Protection Agency published water reuse guidelines in 1992 that provide guidance to states that have not developed their own criteria or guidelines.

For background information SEE IRRIGATION (AGRICULTURE); SEWAGE DISPOSAL; WATER CONSERVATION; WATER SUPPLY ENGINEERING in the McGraw-Hill Encyclopedia of Science & Technology.

James Crook

Bibliography. J. Crook, D. A. Okun, and A. B. Pincince, *Water Reuse Assessment,* 1994; W. O. Maddaus, *Water Conservation,* American Water Works Association, 1987; U.S. Environmental Protection Agency, *Guidelines for Water Reuse,* EPA/625/R-92/004, 1992.

Wave phenomena

Wave phenomena and liquid surface motion in general occur in many interesting applications in the universe, from the collapse of stars to the fall of raindrops. Other specific applications include the interaction of ships with surface waves and the motion of liquids in containers such as oil tankers, railroad tank cars, and fuel tanks of various vehicles including spacecraft and satellites. The nature of free surface motion is strongly influenced by the magnitude of gravitational forces. Consequently, significantly different behavior is observed for liquid surfaces in the greatly reduced gravitational field in many space applications than is encountered in applications on Earth. Such phenomena as the collision and merging of bubbles, and liquid-vapor interfacial motion in reduced and fluctuating gravities are of great interest in space applications.

It is desirable for design purposes to develop the capability to predict the behavior of fluid motion under the various conditions that are expected to occur in applications. Determination for many types of fluid flows uses computer simulations based on numerically solving the equations believed to govern the phenomena under study. Despite advances in computing power, a numerical simulation for a three-dimensional free surface flow such as the sloshing motion of a liquid in a partially filled container is still a formidable challenge. The complexity arises from several sources. Motions of most interest are usually variable with regard to time and are three-dimensional, thus straining existing computer resources. Numerical solutions of such flows involve setting up a system of grid points within the domain of interest, discretizing the governing equations at each grid point or computational cell, and solving them at each time step in the unsteady problem. In some applications, the container holding the liquid is itself moving with respect to a reference frame fixed to the Earth, thus adding to the complexity of the numerical formulation. The spatial grid and time steps must be kept small to maintain accuracy, consequently requiring a substantial amount of computer memory and central processing unit time.

In addition, the free-surface position is usually not known beforehand and must be determined as part of the solution. Here, the free surface is defined as the interface between a liquid and a second fluid, which is often a gas. The free-surface motion can also be influenced by interfacial phenomena such as surface tension. The exact physical mechanisms which

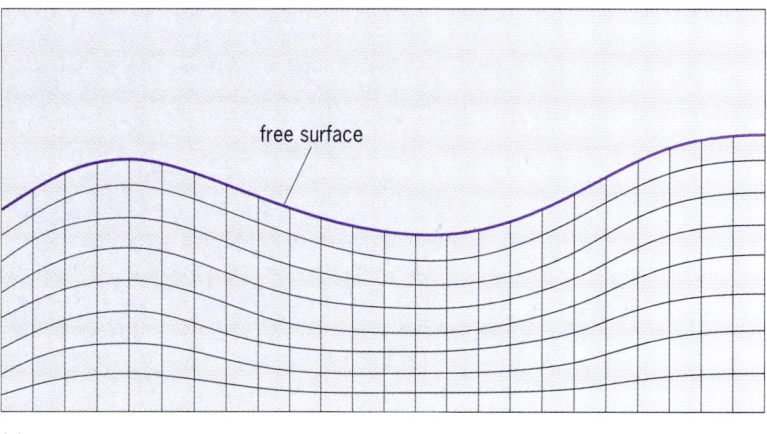

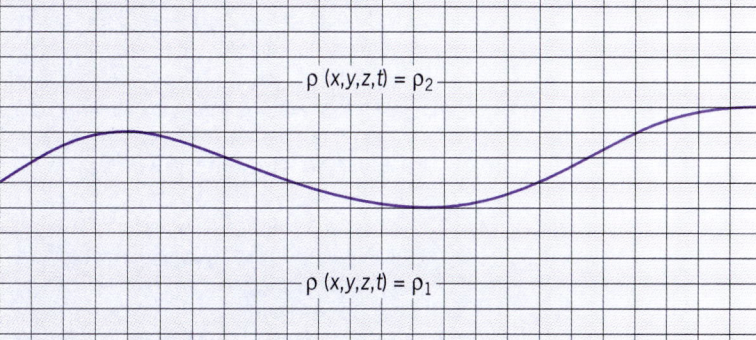

Fig. 1. Methods for handling of free surfaces numerically. (*a*) Surface fitting. (*b*) Surface tracking. (*c*) Surface capturing. (*After F. J. Kelecy, Numerical simulation of two and three-dimensional viscous free surface flows in partially-filled containers using a surface capturing approach, doctoral dissertation, Iowa State University, 1993*)

determine the detailed interactions between the free surface and the container wall are themselves not well established in a form that can be resolved in a numerical simulation along with resolving the bulk motion of the fluid on computers currently available. In brief, this interaction involves the apparent contradiction that the fluid particles must adhere to the wall of the container (no-slip condition), but yet the free surface appears to move along the container wall. To satisfy these conditions, the implication is that a thin layer of fluid must remain on the wall as the fluid slides down the wall of the container in a typical sloshing motion. The net effects of this behavior are approximated in the simulations but not resolved in detail because of incomplete understanding of the phenomenon and the lack of computer resources. Often, the free-surface position at the container wall is obtained by extrapolation of the free-surface position determined adjacent to the walls. The expectation is that the influence of such an approximations on the overall flow is small.

Locating the free surface. Numerical approaches for handling flows with free surfaces can be grouped into three broad categories: surface fitting methods, surface tracking methods, and surface capturing methods (**Fig. 1**).

Surface fitting. With surface fitting, (Fig. 1a) the calculations are carried out in a transformed coordinate system that conforms to the shape of the free surface. In other words, calculations are carried out within only the liquid region and the free surface becomes a boundary of the computational domain. Appropriate boundary conditions are applied at the free surface, and the surface location changes with time as dictated by the velocities computed at grid points falling on the free surface. To locate the new free-surface position, it is assumed that particles on the free surface remain there. Generally, the grid points are moved after each computational time step as a new surface-conforming grid is constructed. The new grid adapts to the changed geometry of the computational domain. However, the task of constructing the new grid at each time step adds to the computational effort required for the simulation.

Surface tracking. Typically, surface-tracking methods use a grid that remains fixed throughout the duration of the calculation. Thus, the grid must be established over a sufficiently large domain initially so that it will be adequate for all liquid configurations that may evolve in the problem. Thus, for liquid motion in a container, it is normally required that the grid cover the entire container. The computational cells (Fig. 1b) are marked according to whether they contain all liquid (L), all gas (G), or free surface (F). The governing equations are solved in liquid and free-surface cells only. The movement of the free surface must be tracked by applying the

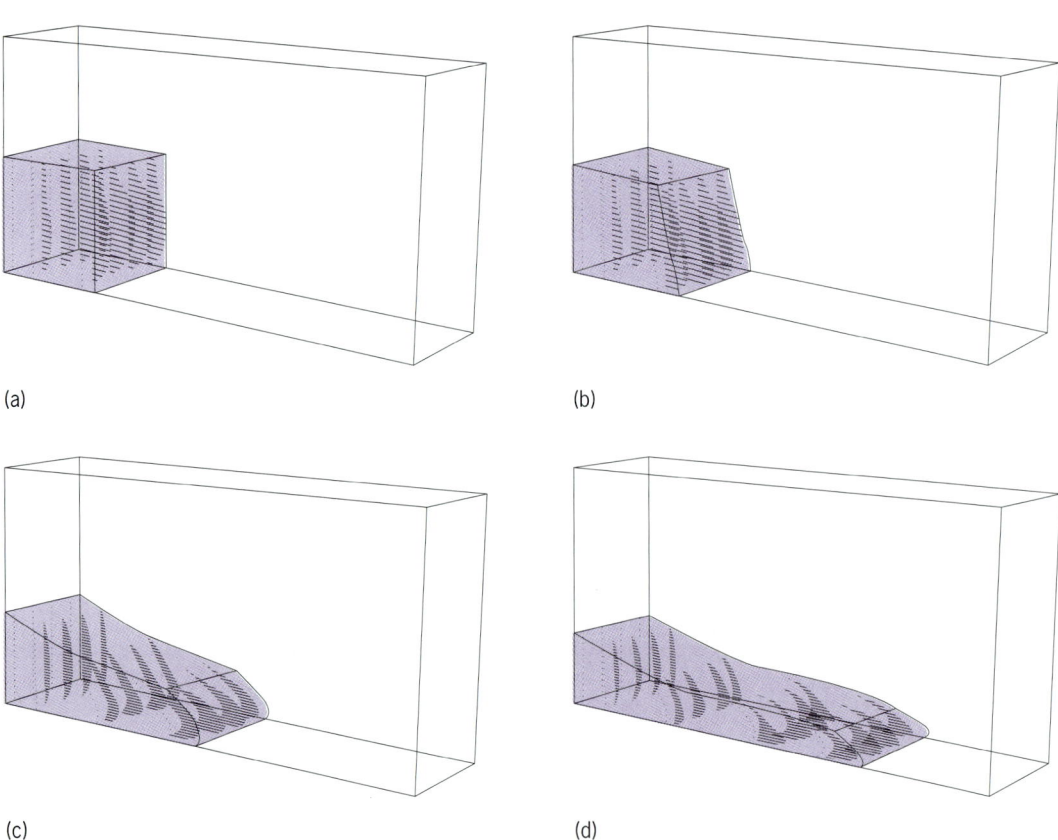

Fig. 2. Collapse of a water column. Computed free-surface shape and velocity vectors are shown at the nondimensional times (*a*) 0.1, (*b*) 0.8, (*c*) 1.6, and (*d*) 2.4. (*After S. Babu, Simulation of three-dimensional incompressible flows with free surfaces including fluid-structure interaction and microgravity flows, doctoral dissertation, Iowa State University, 1994*)

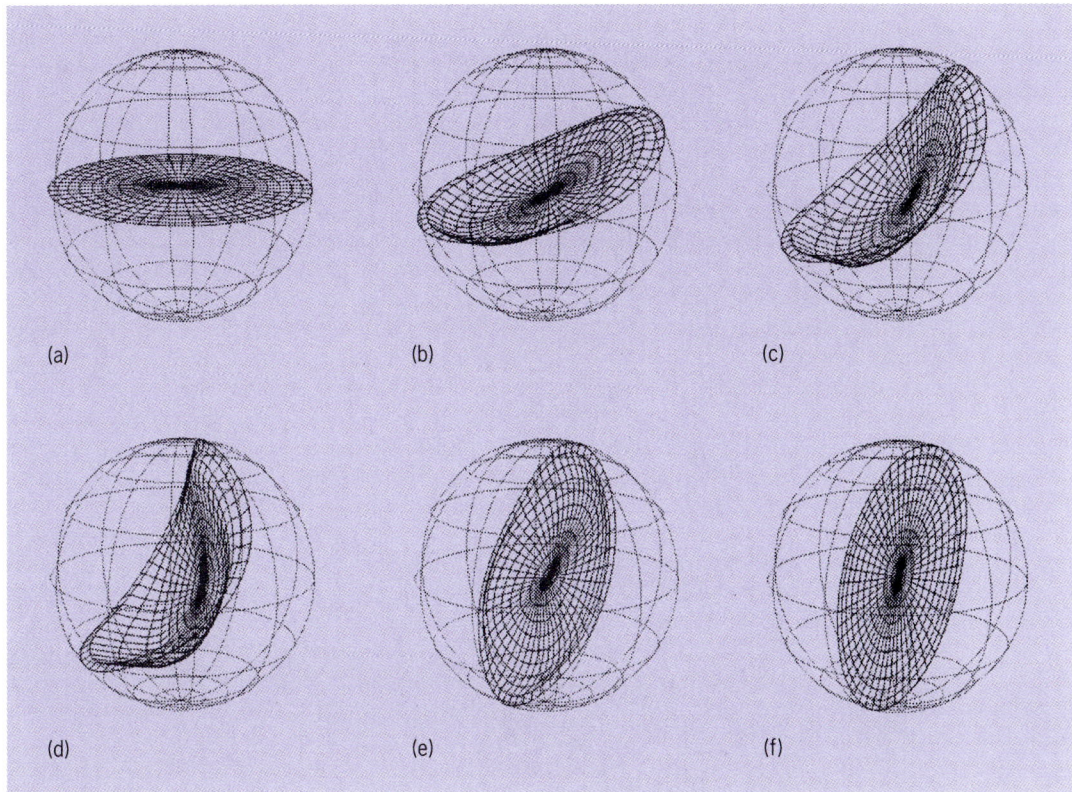

Fig. 3. Free-surface profiles during the orbital motion of an initially capped half-filled spherical tank at the nondimensional times (a) 0.0, (b) 3.0, (c) 4.5, (d) 6.0, (e) 15.0, (f) 40.00. (*After S. Babu, Simulation of three-dimensional incompressible flows with free surfaces including fluid-structure interaction and microgravity flows, doctoral dissertation, Iowa State University, 1994*)

physical principle that particles on the surface remain on the surface although they move. A self-consistent scheme must also be devised to apply the governing equations to cells containing the free surface, which should be considered as only partially filled with liquid.

Surface capturing. Surface-capturing methods also use a fixed grid that generally fills all the space that might become occupied by the liquid. However, in contrast to the tracking method, in the surface-capturing method the regions of a container not occupied by the liquid are considered to be filled with a second fluid, usually a gas in most applications, and the governing equations are solved throughout the fixed grid. This result becomes possible because the conservation principles for the problem are augmented by a conservation-of-mass principle in a form which allows for a variable density. In most applications of the method, both fluids have been treated as incompressible, which in principle implies a discontinuity in the density function (ρ) at the interface between the two fluids (Fig. 1c). This discontinuity is said to be captured in the calculation much as a shock wave is captured in the numerical simulation of high-speed compressible flows.

Computational results. A number of investigators have reported simulation results for two-dimensional free-surface flows. These results include flows such as the the reflection of a solitary wave impinging upon a wall, the sloshing of a liquid in a rectangular tank, and the mixing motion that results when a relatively dense fluid is placed over a less dense fluid (the Rayleigh-Taylor instability of two immiscible fluids). Three-dimensional results are less common.

Broken-dam problem. This problem involves the flow resulting when a restrained column of liquid is allowed to collapse by the removal of a restraining wall or partition (dam). In 1952, this problem was studied experimentally in a channel 0.05715 m (2.25 in.) wide and 0.28575 m (11.25 in.) long. A column of water in the shape of a cube 0.05715 m (2.25 in.) on a side was constrained in the channel by a so-called paper dam (**Fig. 2**). As the experiment began, the dam was removed, allowing the column of water to collapse and flow along the the channel floor. Measurements were made of the advance of the surge front along the channel floor and the column height as functions of time.

The collapse of the liquid column has recently been computed by solving the three-dimensional Navier-Stokes equations. The computed advance of the surge front along the channel floor was in good agreement with the experimental measurements. The computed free-surface shape and velocity vectors are shown in Fig. 2 for several nondimensional times, which are real times that are made nondimensional by multiplying them by a reference velocity and dividing them by a reference length. For the physical dimensions given above, each nondimensional time unit corresponds to 0.0764 s. In the numerical simulation, using 10,571 grid points, 310 min of Cray Y-MP supercomputer time was required to advance the solution to a point corresponding to 0.1833 s after the dam broke in the experiment.

Liquid sloshing in a spherical tank. The stability of spin-stabilized satellites can be influenced by the sloshing motion of any liquids carried on board in partially filled tanks. Such liquids might be the fuel required by the propulsion devices used to make small adjustments in the position of the satellite. Motivated by such concerns, experimental and computational studies have been carried out recently to study the effects of sloshing liquids on rotating systems. In one such study, the motion of a liquid (glycerin) in a half-filled spherical tank has been simulated for the start-up transient that occurs when the liquid is initially constrained by a membrane to remain level until the tank is brought up to steady orbital motion and then the membrane is removed. The movement of the free surface under such conditions is illustrated in **Fig. 3**. The radius of the tank in the simulation was 7.62 cm (3 in.), the distance from the axis of rotation to the center of the tank was 73.91 cm (29 in.), and the gravitational field was that of Earth. The tank was in orbital motion at 60 revolutions per minute. Each unit of nondimensional time in Fig. 3 corresponds to 0.0164 s. A steady-state condition (Fig. 3*f*) was reached in approximately 0.656 s which corresponds to the time required for the tank to move about 65% of the way around the orbital path. About 50 min on a Cray Y-MP was required for the simulation, which employed 6171 grid points within the liquid.

For background information SEE OCEAN WAVES; OPEN CHANNEL; SPACECRAFT PROPULSION; WAVE MOTION IN LIQUIDS in the McGraw-Hill Encyclopedia of Science & Technology.

Richard H. Pletcher

Bibliography. K.-H. Chen, F. J. Kelecy, and R. H. Pletcher, Numerical and experimental study of three-dimensional liquid sloshing flows, *J. Thermophys. Heat Trans.*, 8:507–513, 1994; J. M. Floryan and H. Rasmussen, Numerical methods for viscous flows with moving boundaries, *Appl. Mech. Rev.*, 42:323–340, 1989; K. N. Ghia, U. Ghia, and D. Goldstein (eds.), *Advances in Computational Methods in Fluid Dynamics*, 1994; C. L. Mader, *Numerical Modeling of Water Waves*, 1988; J. C. Martin and W. J. Moyce, An experimental study of the collapse of liquid columns on a rigid horizontal plane, *Phil. Trans. Roy. Soc. London A*, 244:312–324, 1952.

Wavelets

An elaborate waveform, such as might be generated by a microphone responding to a passage of music, can be fully represented by its Fourier series, harmonic analysis, or spectrum. Thus, any subtlety in the original passage, even the briefest pizzicato, is captured in the spectrum. But it would be hard to locate a spectral feature that corresponded to plucking the string. If such a feature could be found, it would not be obvious, from just looking at the spectrum, at what moment the violin string was plucked; the Fourier series coefficients are constants, independent of time. If only the power spectrum were available, it would be impossible to say when the string was plucked because the mean power at any particular frequency is independent of choice of the origin of time and does not change if some harmonics are translated in time with respect to others. The moment of plucking is encoded in the phase of the complex coefficients distributed around the natural frequency of the string (196 Hz if the G string is plucked). Thus, something that is obvious to the ear need not be obvious from the Fourier analysis. Conversely, a faint hum may be hard to see in a recorded waveform of the highest fidelity. A purpose of analysis into wavelets is to convey to the eye what the ear can hear.

Figure 1 shows examples of wavelets. They may be of any frequency, amplitude, phase, or frequency drift rate, and indeed of wave shapes other than gaussian. Wavelet analysis aims to split a given complicated waveform into such components.

The ear is equipped for effectively analyzing sound into a bank of frequency bins by the use of several thousand flexible hair cells of graded sizes excited by a tapered fluid-dynamic waveguide, the cochlea. But the ear does not merely perform a Fourier analysis; the purpose of the operations performed behind the ear is to extract information while the music is in progress. So the frequency-analysis filter bank continually furnishes changing individual outputs from which the listener constructs an impression of the actions that originally created the sound. Musical notation suggests very aptly the sensation of pitch and duration experienced by the listener. The notes of high pitch are high on the staff, while time flows from left to right. The same convention of frequency increasing upward and time increasing to the right is the basis for the dynamic spectrum. SEE PSYCHOACOUSTICS.

Dynamic spectrum. As soon as developments in electronics permitted, complicated waveforms, especially of human speech, bird calls, and naturally occurring hertzian waves, were analyzed for visual presentation in the spirit of musical notation. **Figure 2** is an example of a dynamic spectrum of 3-s duration and frequency band from 0 to 6 kHz. If the signal were presented to the ear as sound, the listener would hear three things. First, a deep continuing rumbling, broken into surges of irregular duration would be heard, represented by the dark band at the bottom. Then, at irregular intervals clicks would occur, sometimes isolated and sometimes grouped, represented by the straight vertical stripes. Finally, and most conspicuously, a loud whooshing whistle descending steeply in pitch would be followed by a number of fainter ones. The visual presentation clearly registers the train of audible events and provides a record for later study

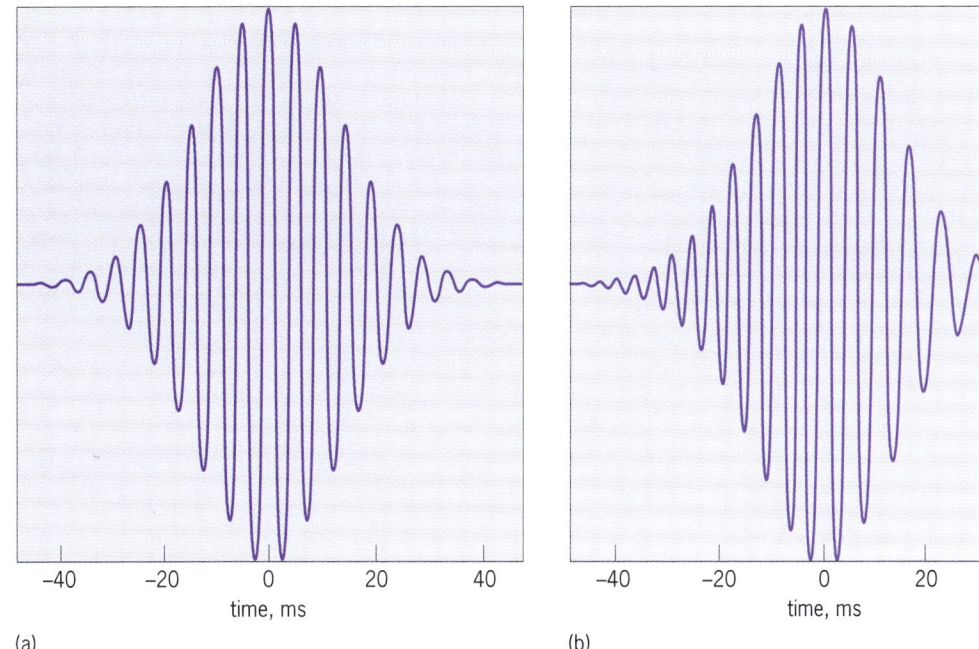

Fig. 1. Wavelets. (*a*) Gabor elementary signal, with frequency of 200 Hz, bandwidth of 50 Hz, and equivalent duration of 10 ms. (*b*) New elementary signal, the chirplet, with a frequency drift rate of −4500 Hz/s.

of details that would not be apparent to the ear. Neither the waveform nor the Fourier series would be as readily interpretable as the dynamic spectrum, in which moment-to-moment changes in the spectrum of brief excerpts can be examined.

The record in Fig. 2 was obtained with a radio receiver working in the very low frequency band. The background rumble is due to noise generated in an amplifier, the clicks are static interference due to distant lightning, and the whistles are due to lightning flashes in the Southern Hemisphere. Some of the hertzian wave energy generated by the electric discharge has penetrated the ionosphere, traveled well beyond Earth's atmosphere and, guided by Earth's magnetic field, has returned to Earth at the receiver. The few electrons along the path of the waves contribute to the guiding and delay the lower frequencies so that what was originally a momentary impulse is dispersed into a long drawn out whistle. The value of the dynamic spectrum as a tool for analyzing waveforms is apparent from this example. Examples could also be given for other natural phe-

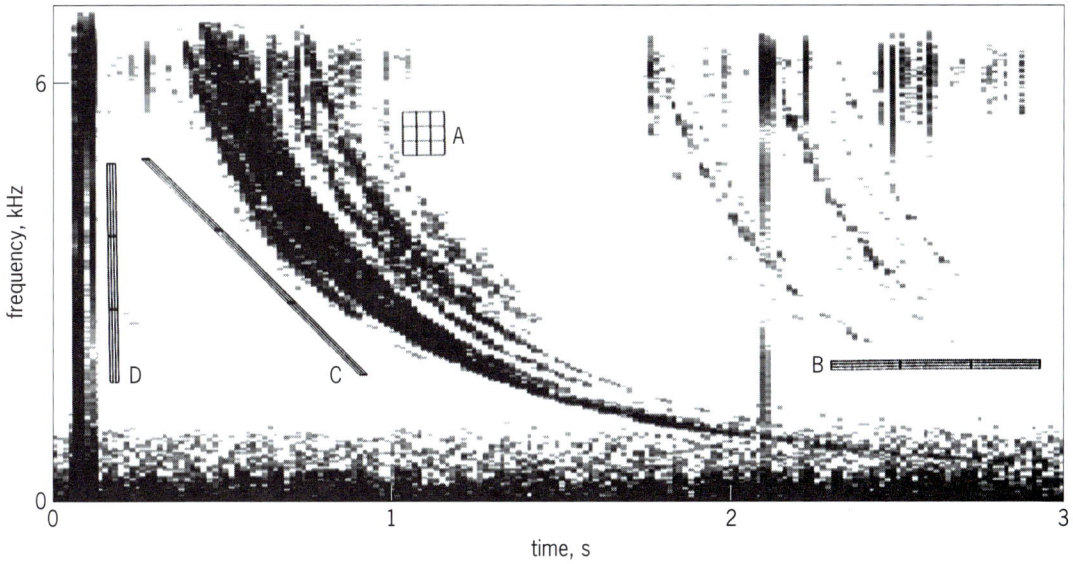

Fig. 2. Dynamic spectrum from 0 to 6 kHz showing impulses (vertical stripes) due to lightning, and whistlers (curved features) resulting from the dispersion of hertzian wave emission by lightning. The nine cells at A show the square cells (somewhat magnified) used for the presentation; the other cell sets (B, C, and D) have the same area of 0.5 but have aspect ratios, and in the case of C a slope, appropriate to improved resolution of fine structure occurring in their vicinity. *(Courtesy of Jerry Yarbrough)*

nomena such as seismic waves, ocean waves, underwater sound generated by sea mammals, and hertzian waves from the Sun. Hertzian waves is the term for those electromagnetic waves first generated by H. Hertz, including millimeter waves, microwaves, waves used for radio communication and television, and waves in the very low frequency and ultralow-frequency bands (the bands in which whistlers and magnetic disturbances occur).

Dynamic spectrograph. Commercial instruments for dynamic spectral analysis appeared in the 1950s. One design took an incoming electrical waveform in the audible range of frequencies, and split it into several hundred copies to be passed through narrow-band filters (bandwidth about 50 Hz) tuned to frequencies evenly spaced across the frequency range. The 88 taut strings of a piano constitute an analogous bank of narrow-band acoustic filters, except that the frequencies are spaced by one semitone (a frequency ratio of approximately 16:15) rather than by a fixed increment. Each filter output was detected and used to blacken a strip of paper or film to produce the dynamic spectrogram with a time resolution of about 10 ms. The product of the bandwidth and resolution time was about 0.5. Figure 2 shows at A, enlarged about a dozen times, a pattern of nine cells illustrating the size and shape of the domains occupied by the wavelets.

Wavelet theory. D. Gabor in 1946 related the operation of a dynamic spectrograph to Heisenberg's uncertainty principle in quantum mechanics. A signal can be analyzed into narrower 10-Hz frequency bins but only at the sacrifice of time resolution; the cells then stretch to 50 ms, as at B in Fig. 2. Conversely, to achieve 1-ms time resolution, it is necessary to accept a frequency resolution as poor as 500 Hz, as at D. A gaussian wave packet (the quantum mechanical term) occupying a domain of area 0.5 is known as an elementary signal, which is Gabor's term, or as a wavelet in recent terminology. To construct a gray-level diagram, as in Fig. 2, a numerical value is attached to each cell as follows: An elementary signal is formed, as in Fig. 1a, whose frequency and time origin are given by the coordinates of the cell center and whose bandwidth and duration are given by the base and height of the rectangular shape chosen for that cell. The required value is obtained by integrating the product of the incoming signal waveform with this elementary signal.

The original elementary signals of Gabor have been supplemented by another elementary signal, the chirplet, whose instantaneous frequency drifts at a fixed rate throughout its duration (Fig. 1b). Chirplets, occupying oblique domains as at C in Fig. 2, are suitable for resolving trains of signals of time-varying pitch that cannot be resolved by standard wavelets. It is now possible to alter the wavelet domain shape spatially so as to adapt to local structure. Improved frequency resolution in the neighborhood of B, improved time resolution at D, and the use of suitable chirplet analysis around C, would all reveal previously obscure structure. To attach a numerical value to an oblique cell, as at C in Fig. 2, an elementary signal is formed as before but a frequency drift rate is introduced, equal to the slope of the oblique cell that is chosen (Fig. 1b).

Synthesis. Owing to a mathematical property of nonorthogonality pointed out by Gabor, gaussian wavelets, although useful for signal analysis (as they have been for several decades), are not adapted to strict synthesis of signals, and much attention has been drawn to the theoretical development of orthogonal wavelets. In addition, frequency spacing in fractional steps, as on a piano, has been advocated as a way of improving time resolution at the high-frequency end of the spectrum (while sacrificing time resolution at low frequencies). In effect this proposal dissects the time-frequency plane of Fig. 2 into elementary signals whose bandwidth-duration product is held at the theoretical minimum of 0.5 but whose bandwidth is forced to increase in proportion to frequency; the term wavelet analysis is often taken to imply this restriction of bandwidth to a fixed fraction of the midfrequency. The ultimate outcome will be to adapt the bandwidth as best suited to each frequency, as well as each time. Finally, the drift rate also can advantageously be chosen to suit the local texture in the time-frequency-plane by making use of the extra degree of freedom permitted by chirplet theory.

For background information SEE ELECTROMAGNETIC WAVE TRANSMISSION; FOURIER SERIES AND INTEGRALS; HARMONIC ANALYZER; MUSICAL ACOUSTICS; RADIO-WAVE PROPAGATION; SPEECH PERCEPTION in the McGraw-Hill Encyclopedia of Science & Technology.

Ronald N. Bracewell

Bibliography. L. Cohen, Time-frequency distributions: A review, *Proc. IEE*, 77:941–981, 1989; I. Daubechies, Orthonormal bases of compactly supported wavelets, *Comm. Pure Appl. Math.*, 41:909–996, 1988; D. Gabor, Theory of communication, *J. Inst. Elec. Eng. London*, 93(III):429–441, 1946; Y. Meyer, *Wavelets: Algorithms and Applications*, 1993; Y. Meyer, *Wavelets and Operators*, 1993; D. Mihovilovic and R. N. Bracewell, Whistler analysis in the time-frequency plane using chirplets, *J. Geophys. Res. Space Phys.*, 97:17,199–17,204, 1992.

Whey

Cheese production in the United States exceeds 2.5×10^6 tons (2.3×10^6 metric tons) per year. Each ton of cheese requires the use of about 10 tons of milk and produces about 9 tons of whey as a by-product. The result is a total annual production of more than 5.3×10^9 gal (2×10^{10} liters) of whey. Whey comprises the water, milk proteins, milk sugars, and nutrients remaining after the butterfat and proteins are removed during cheese manufacturing.

Production. Some cream and cottage cheeses are made by coagulating milk with phosphoric acid. The rest, and all hard cheeses are made by coagulating the milk proteins and butterfat with various bacterial cultures. The residues from the two processes are referred to, respectively, as acid and sweet whey.

Whey is primarily used directly as livestock feed, is concentrated or dehydrated for use in human and animal food manufacture, or is disposed of by land application. Large cheese processing plants are able to concentrate or dehydrate sweet wheys into marketable by-products economically. Smaller plants and plants that use the phosphoric acid process find the high energy costs of removing water from the whey prohibitive.

Temporary conditions may arise in the larger plants when animal feeding, dehydrating, or concentrating whey is not possible as a disposal choice, such as during periods of process malfunction. In such instances land application is usually the method of whey disposal.

The chemical composition of whey is such that whey not only is a nutritious human and animal food supplement (see **table**), but, when properly applied, an enhancer of soil fertility and physical properties. Its disadvantages for use on soils are that it must be removed from the cheese plants on a daily basis and that it possesses a high water-to-nutrient ratio. Like any fertilizer or soil amendment, improper or excessive whey application has the potential of producing environmental problems.

Use as fertilizer. Land-applied whey at a rate of about 53,000 gal per acre (500,000 liters per hectare), which is equivalent to a 2-in. (5-cm) application depth, will supply nitrogen at a rate of 53–107 lb per acre (60–120 kg/ha), phosphorus at a rate of 27–53 lb per acre for sweet whey (30–60 kg/ha) or 160–220 lb per acre (180–250 kg/ha) for acid whey, and potassium at a rate of 80–135 lb per acre (85–150 kg/ha). Whey can be applied up to about 3 times this rate if it is applied over the growing season and the site is properly managed.

The nitrogen is primarily in the amine form as part of the milk protein, and consequently it will become available to plants as the soil organisms decompose the whey. The phosphorus in whey also occurs primarily in organic compounds in the sweet whey. In this form, it is more mobile than in the orthophosphate form, and can move deeper into the soil. Most of the phosphorous in acid whey is present as orthophosphate from the phosphoric acid. Most phosphorus fertilizer is applied in the orthophosphate form. The potassium is primarily in the mineral form, and it is readily available to plants. Whey also contains the essential micronutrients for plant and animal growth in organic forms, and these are readily available to plants as the whey decomposes in the soil.

Sodic soil reclamation. A parameter known as the saturation extract electrical conductivity is used to assess soil salinity; it is obtained by measuring the electrical conductivity of a solution made from a soil sample. Soils with saturation extract electrical conductivities less than 4 decisiemens/m and sodium absorption ratios greater than 13 are classified as sodic. In sodic soils, high pH (>8.3) and high sodium concentrations cause the soils to become dispersed, which leads to reduced air and water entry, poor tilth, and excess surface crusting. To reclaim this soil, it is necessary to lower the pH, temporarily increase the soluble salts, and replace the exchangeable sodium with other cations. Low-sodium wheys, particularly the acid wheys, contain sufficient calcium, magnesium, and potassium along with the low pH (due to the phosphoric acid) to flocculate the dispersed clay in the sodic soils. Thus, soil aggregates form and infiltration increases, allowing leaching of the unwanted sodium from the soil. Sodic soils not affected by a high ground-water level can be reclaimed with 2–5 in. (5.0–12.5 cm) of acid whey followed by leaching with good quality irrigation water or 20–30 in. (50–75 cm) of rain. For best results, the whey should be applied in 2-in. (5.0-cm) increments with irrigation or rain following each treatment. The main obstacle to this sodic soil reclamation method is the transportation cost if the whey must be hauled an appreciable distance.

Erosion control. Under furrow irrigation of row crops such as corn, beans, potatoes, and sugar beets, furrow erosion is often a problem. Up to 98% of this soil loss can be eliminated by treating the furrows with whey after each cultivation. The simplest method is to add just enough whey to each furrow to coat the sides. A second method is to add whey to furrows that have had straw scattered in the row at a rate of about 700 lb/acre (780 kg/ha). A more sophisticated and effective method that uses less whey involves mechanically applying the straw to the furrow and then spraying the straw with whey. The first two methods require about 6000 gal per acre (56,000 liters/ha). The last uses only about 2000 gal per acre (19,000 liters/ha). All three methods essentially eliminate soil erosion from the irrigated fields and at the same time increase water infiltration and soil tilth. The beneficial action is initially due to the stickiness of the whey and subsequently to the decomposition of the milk proteins and sugars by soil organisms. The increased biolog-

Typical whey composition and properties		
Property	Sweet whey	Acid whey
Water	92%	92%
Milk solids	8%	8%
Chemical oxygen demand (COD)	50,000 ppm	50,000 ppm
pH	3.8–4.6	3.3–3.8
Electrical conductivity	7–12 dS/m	7–8 dS/m
Total nitrogen	0.10–0.20%	0.10–0.20%
Total phosphorus	0.05–0.10%	0.30–0.40%
Total potassium	0.15–0.25%	0.15–0.25%
Calcium	840 ppm	840 ppm
Magnesium	100 ppm	100 ppm
Sodium*	Highly variable	600 ppm
Sodium adsorption ratio*	Varies	3–4
Micronutrients	†	†

* The sodium concentration and the sodium adsorption ratio vary with the amount of salt used in the various cheese manufacturing processes and the fraction that ends up in the whey.
† Concentrations are about the same as in milk.

ical activity produces polysaccharides that bond or cement soil particles together.

Environmental concerns. There are several environmental concern involved in land application of cheese whey. First, as with any industrial or food processing waste, whey should not be allowed to enter any surface or subsurface water without proper pretreatment.

Second, whey contains up to 60,000 parts per million by weight of chemical oxygen demand (COD); COD is a measure of the amount of oxygen a material consumes during decomposition in soil or water. The acceptable COD loading rate for a site is affected by soil temperature, moisture, hydraulic conductivity and porosity, application frequency, and the depth to the water table. Exceeding the COD loading rate reduces free oxygen levels and may cause foul odors associated with the free ammonia or hydrogen sulfide. Exceeding the COD loading rate may lead to plant death. Also, iron and manganese may be reduced, solubilized, and leached into ground waters.

Third, whey contains appreciable quantities of soluble salts (40,000–80,000 ppm), and if the salts are applied faster than they are leached from the soil by precipitation or irrigation, the salts accumulate in the soil surface and limit plant growth. Highsodium whey also presents additional problems if not properly managed.

Fourth, the phosphorus in whey that comes from the milk is primarily in organic form, and it is much more mobile in the soil than is orthophosphate. Excessive whey application presents the potential of leaching organic phosphorus below the root zone, before it is changed to orthophosphate by soil organisms, thus allowing it to leach into the ground water.

If no more than an 8-in. (20-cm) depth of whey (216,000 gal/acre or 2×10^6 liters/ha) is applied annually in 1-in. (2.5-cm) increments to soils with a dry surface that is not immediately wetted by rain or irrigation, these four factors will not likely present a problem. This application rate is less than the rates set in the United States by most state regulatory agencies. When greater application rates are used, soil and ground-water monitoring should be included as part of the management system.

Nitrogen use by crops and the high carbon-to-nitrogen ratio in the decomposing whey will usually cause sufficient nitrogen immobilization to keep nitrate leaching from being a hazard. Whey is of food-grade quality and will not present a hazard with respect to toxic metals or radioactive materials.

For background information SEE AGRICULTURAL SOIL AND CROP PRACTICES; CHEESE; EROSION; MILK; SOIL; SOIL CHEMISTRY; SOIL FERTILITY in the McGraw-Hill Encyclopedia of Science & Technology.

Charles W. Robbins

Bibliography. M. J. Brown and C. W. Robbins, Combining cottage cheese whey and straw reduces erosion while increasing infiltration in furrow irrigation, *J. Soil Water Conser.*, 1995; S. B. Jones, C. W. Robbins, and C. L. Hansen, Sodic soil reclamation using cottage cheese (acid) whey, *Arid Soil Res. Rehabil.*, 7:51–61, 1993; G. A. Lehrsch, C. W. Robbins, and C. L. Hansen, Cottage cheese (acid) whey effects on sodic soil aggregate stability, *Arid Soil Res. Rehabil.*, 8:19–31, 1994.

Wind

The word katabatic means a movement down a slope. Such a movement occurs in the atmosphere when the surface radiation balance becomes negative and the air near the surface cools. A so-called temperature inversion develops. The surface air, being cooler and hence heavier than the air at the same level but at some horizontal distance away from the slope, starts to move downslope in the absence of a synoptic pressure gradient. The katabatic force is dependent mainly on the inversion strength and the inclination of the slope. Such gravity drainage is quite common in mountainous terrain, for example, it occurs as down-valley flow for most clear nights. High wind speeds occur when the temperature contrast becomes large, for example, in winter between the cold air over the Dinarich Alps and the warm air above the Mediterranean Sea, where the resulting katabatic wind is called the bora. Such a wind is characterized by a high gust and a high directional constancy.

Gravity flow in Antarctica. Antarctica is a continent covered with snow and ice. The polar location together with the high albedo (reflection factor) of snow, which reflects most of the solar radiation back to space, results in a negative surface radiation balance for most of the continent most of the time. Exceptions are experienced for only some limited ice-free coastal areas during midsummer. Less than 3% of Antarctica is not permanently ice covered. Hence, a semipermanent surface temperature inversion develops to a degree unknown on any other continent. Furthermore, since interior altitudes can be in excess of 13,000 ft (4000 m), and the surface roughness is low because of the absence of vegetation, conditions for the development of the katabatic wind are ideal. Nowhere else does a single meteorological parameter have such a domineering influence on the surface climate of an entire continent. All Antarctica, with the exception of the inland ice domes, is under the influence of this gravity flow, which results in very high wind speeds in coastal areas as the fetch (area over which the wind blows) exceeds more than 600 mi (1000 km). Funneling of wind through mountains can further enhance its speed, and in such areas winds are the strongest found anywhere near sea level on Earth. D. Mawson, wintering at Cape Denison, eastern Antarctica, between 1911 and 1913, was the first to give a popular account of these winds. His observations have been confirmed by automatic weather stations that have been in place since the mid-1980s.

Wind speed and direction. Katabatic winds have a very high directional constancy, blowing steadily not directly from the south but, because of the Coriolis force, from a southeasterly direction. The constancy is defined as the mean vector wind divided by mean wind speed. Monthly values in excess of 0.95 are being found. This constancy is higher than observed for the trade winds, so well known for steadiness in direction. Along the King George V and the Adélie coasts, where Mawson made his observations, mean annual wind speeds of around 45 mi/h (20 m/s) have been observed. Normally, wind speed declines to a minimum during the summer months, as the temperature contrast between the interior and the coastal areas is less severe. However, the maximum wind speed is frequently found not in midwinter but in autumn, when the interior has cooled down but the coastal areas are still relatively warm because of proximity to open water. Later in the winter, sea ice hinders the heat transfer from the relatively warm ocean to the atmosphere, and the temperature difference decreases. At Port Martin, an abandoned French station some 36 mi (60 km) from the place where Mawson originally landed, a monthly wind speed of 58 mi/h (26 m/s) was recorded for a fall month. Storms with wind speeds exceeding 100 mi/h (44 m/s) can last days and even weeks. Absolute maxima are, of course, much higher and wind speeds of up to 215 mi/h (96 m/s) have been recorded; gusts were estimated to exceed 250 mi/h (112 m/s).

Modeling and observations. The phenomenon of the katabatic wind has attracted the attention of many investigators. A recent team of United States and French scientists carried out a detailed field program, including simultaneous measurements at different locations. The measurements were not limited to the surface but were also obtained with balloons, kites, and drones as carriers that telemetered the data to a ground station. Thus the entire boundary layer could be probed.

Observations confirmed model calculations that the wind maximum is normally found not at the coast but some distance inland. Normally, the slope angle close to the edge of the continent becomes quite steep, and the gustiness and depth of the katabatic layer increases. The increased thickness of the layer somewhat reduces the surface wind speed at the coast. There is normally a well-developed jet of wind speed in the boundary layer. In summer, this maximum in wind speed is well developed during the night, with a typical height of 230–400 ft (70–120 m) above the surface. During the day, with the weakening of the surface inversion, the maximum decreases in intensity but increases in height.

A unique event sometimes observed in the coastal area is the so-called hydraulic jump. The katabatic wind decouples from the steep surface slope, and a wall of blowing snow extending into the sky can be observed. Although the winds are extremely violent upslope from this event, downslope they are benign and can, on occasion, reverse their direction.

Air-ocean interactions. Katabatic winds can be strong enough to drive the sea ice from shore, even in midwinter, a phenomenon that could be studied in detail with synthetic-aperture radar satellites, which have a high resolution (66 ft or 20 m) and the ability to penetrate clouds and darkness. However, as the winds decrease rapidly in speed once they reach the flat sea ice, these coastal polynyas (ice clearings) are typically only about 6 mi (10 km) wide. In these areas, the energy transfer between the warm ocean and the cold atmosphere is immense. Furthermore, the rate of formation of new sea ice, which drifts steadily away from the coast, is large.

Eolian processes. Strong winds transport snow. If this process is limited to the surface, this is known as drifting snow, which starts at about 22 mi/h (10 m/s). At higher wind speeds, blowing snow extends to higher and higher levels, reducing visibility to several feet at a wind speed of 45 mi/h (20 m/s). At very strong wind speeds, the visibility can be reduced even further. Such winds create an extremely dangerous situation.

Large amounts of snow can be transported. This eolian transport plays a major role in the accumulation of snow. An array of measuring stations from the coast to a height of more than 10,000 ft (3000 m) over a distance of 600 mi (1000 km) showed negative accumulation (ablation) at different points for several years, primarily because of wind transport. Areas in which the winds are frequently very high might be free of snow for at least part of the year. Such winds are normally found in the coastal areas. However, snow transport is only one of the parameters that determine the snow mass budget at a specific location, the other two being ablation (from melting and sublimation) and precipitation. In addition, in coastal regions with strong wind speeds blue ice areas, for example, inland from Port Martin, form. All the snow that has fallen during the year might blow away, leaving the blue glacier ice below visible. Hence, after a storm of several days' duration, the intensity of the blowing snow might decrease, even though the winds are unabated in strength, because the amount of blowing snow is a function not only of the wind speed but also of the available snow.

Originally, the amount of blowing snow was measured with mechanical snow traps; recent measurements are made electronically. The strong winds break up the snowflakes, and the typical size of the particles of blowing snow is 0.004–0.008 in. (0.1–0.2 mm). Every second a few hundred to a few thousand particles are transported per square centimeter perpendicular to the wind direction. With increasing wind speed not only the number of particles but also the size distribution increases; thus, very strong winds support not only more but also larger particles.

Estimates of the export of snow from Antarctica in the windiest region are of the order 4×10^8 lb per year per foot of coastline [6×10^8 kg/(yr)(m)]. This large amount has a major impact on the mass balance on a regional basis, for example, on the establishment of blue ice areas. However, as the areas with extreme wind are not extensive, this amount of erosion is of limited importance for the total mass budget of the Antarctic ice sheet.

For background information SEE ANTARCTICA; POLAR METEOROLOGY; SEA ICE; WIND in the McGraw-Hill Encyclopedia of Science & Technology.

Gerd Wendler

Bibliography. T. R. Parish and D. H. Bromwich, The surface winds over the Antarctic ice sheets, *Nature*, 328:51–54, 1987.

X-ray optics

X-ray optics is one of the most exciting and rapidly growing fields of modern optics. Within the past few years, significant advances have been made in optical figuring, fabrication, and mounting technologies. These developments have made it possible to produce x-ray optical components capable of reflecting, focusing, and imaging x-rays and extreme ultraviolet (EUV) radiation (0.01–40 nanometers). Recent advances in optical polishing methods have yielded the ultrasmooth mirror substrates needed for low-scatter grazing-incidence x-ray mirrors. Sophisticated coating methods used in conjunction with these ultrasmooth mirror substrates have made possible the fabrication of multilayer coatings capable of efficiently reflecting soft x-ray and extreme ultraviolet radiation (3–40 nm) at normal angles of incidence (~90°). Multilayer coatings can be applied to grazing-incidence mirrors to produce x-ray optics capable of reflecting hard x-rays with wavelengths as short as 0.01 nm (corresponding to a photon energy of 120 keV).

Kumakhov polycapillary optics. Over the past few years, there have also been important advances in the development of Kumakhov-lens polycapillary optics. The Kumakhov polycapillary optic is composed of bundles of hollow glass fibers that behave as waveguides to propagate x-rays by multiple internal reflections. The fibers can be bent or tapered so as to collect and concentrate the radiation. Although polycapillary optics are not ideal for widefield, high-resolution imaging applications, they offer a broadband response and can be used efficiently for concentrating, collimating, or focusing x-rays and neutrons. The Kumakhov lens systems have great potential for x-ray spectroscopy and high-intensity x-ray sources, as well as for a number of medical and industrial x-ray applications.

Grazing-incidence optics. Since the spatial-resolution limitations imposed upon an optical system by Fraunhofer diffraction are directly proportional to the wavelength, imaging optical systems capable of operating at x-ray wavelengths offer the potential for spatial resolution that is far better than can be achieved with microscopes and telescopes which operate at optical wavelengths (400–700 nm). High spatial resolution is also possible with extremely small x-ray telescopes. For example, an x-ray telescope with a 2-in. (5-cm) diameter and operating at 6.3 nm is theoretically capable of diffraction-limited spatial resolution comparable to that of the 200-in. (5-m) Palomar telescope operating at 630 nm (red light).

Grazing-incidence and multilayer x-ray mirrors are the primary types of reflecting elements used in x-ray optical systems designed for high-resolution imaging applications. Grazing-incidence x-ray optics have been in use since the early 1950s. Grazing-incidence mirrors operate on the principle that at x-ray wavelengths, the index of refraction of air or vacuum is exactly 1, whereas that of materials such as gold or glass is slightly less than 1. Hence, x-rays which strike the mirror surface are traveling from a regime of higher index to one of lower index of refraction. In a manner analogous to the total internal reflection of visible light in a prism, the x-rays of a given wavelength which strike the mirror surface at an angle less than the so-called critical angle undergo total internal reflection. The critical angle ranges from arc-minutes to a few degrees, depending upon the mirror material and the x-ray wavelength. Since the x-ray must strike the surface at such a small (that is, grazing) angle of incidence to be efficiently reflected, mirrors that operate on this principle are called grazing-incidence x-ray optics. The mirror material and the grazing angle at which the mirror operates determine the so-called short-wavelength cutoff of the mirror. Below this wavelength the mirror does not efficiently reflect the radiation. Since grazing-incidence mirrors are capable of reflecting a broad range of x-ray wavelengths above the cutoff, they are particularly desirable for synchrotron beam-line optics and broadband x-ray telescopes for astrophysical research. SEE X-RAYS.

Grazing-incidence x-ray telescopes were flown for the solar research on the *Skylab* space station and used for astronomical research by the *Einstein Observatory*. Large grazing-incidence x-ray mirrors are now being developed for the *Advanced X-Ray Astrophysical Facility* (*AXAF*). The optical systems used in imaging x-ray telescopes are usually based upon the configuration devised by H. Wolter in 1952. The most commonly used x-ray telescope is the Wolter I design, which employs two mirrors which are conic surfaces of revolution. The x-rays are first reflected by an internally reflecting paraboloidal mirror and then reflected to the prime focus of the telescope by a coaxial and confocal, internally reflecting hyperboloidal mirror. Since Wolter I mirrors operate at grazing incidence, the active region of the mirror is only a very thin annulus. A small collecting area is obtained even for a mirror of large diameter. This problem can be addressed somewhat by nesting several Wolter I mirrors of progressively smaller diameters (one inside the other). Nested mirrors must have very thin walls and are difficult to fabricate without large figure errors that seriously degrade the spatial resolution.

Sophisticated mounting technologies enable the surface figure to be maintained under varying gravitational effects. This flexibility is necessary for nested grazing-incidence telescopes, such as that being developed for the *AXAF*, to approach diffraction-limited performance. The off-axis resolution of grazing-incidence Wolter I x-ray mirrors also suffers severely from inherent optical aberrations. Grazing-incidence mirrors are also very sensitive to x-ray scattering effects.

Advanced polishing methods, such as bowl-feed, ion, and advanced flow-polishing techniques have been used to construct ultrasmooth grazing-incidence x-ray mirrors with very low scatter. However, they are still strongly affected by contamination and must be rigorously maintained in clean-room conditions to minimize scattering degradation. When these polishing methods are applied to materials such as hemlite-grade sapphire and fused silica it has been possible to produce flat, concave, or convex mirror substrates with subangstrom root-mean-square surface smoothness. The application of advanced coatings to these ultrasmooth mirror substrates (using radio-frequency-diode sputtering, electron-beam evaporation, and atomic-layer or molecular-beam epitaxy methods) has made possible an entirely new class of x-ray optical components, multilayer optics.

Multilayer optics. Multilayer x-ray optics are the result of the pioneering discoveries made in the mid-1980s by E. Spiller and T. W. Barbee, Jr. Working independently, they demonstrated that multilayer x-ray mirrors could be fabricated by precisely depositing (onto an ultrasmooth substrate) a coating consisting of a stack of many alternating layers of high-atomic-number (high-Z) diffractor material separated by layers of a low-atomic-number (low-Z) spacer material. Since the multilayer coating constitutes a synthetic Bragg crystal, the layers must be very uniform and of precisely repeatable thicknesses d_1 and d_2, respectively. These coatings can be deposited on flat mirrors or gratings or on concave or convex figured surfaces. X-ray reflection from the multilayer coating is obtained by the process of Bragg diffraction. When the slight refraction effects are ignored, the wavelength at which the peak of the reflectivity occurs is given by the Bragg relation, shown below, where $D = d_1 + d_2$,

$$n(\lambda) = 2D \sin(\theta)$$

n is the order of diffraction, and θ is the angle at which the radiation strikes the mirror, as measured from the mirror surface. For mirrors operating at normal incidence ($\theta = 90°$), the equation becomes $\lambda = 2D$, where λ is the wavelength of peak reflectivity of the first-order Bragg diffracted light. For example, a multilayer coating designed to reflect 4.4-nm x-rays could be produced as a stack of alternating 1.1-nm thick layers of a high-Z material (such as tungsten carbide) separated by 1.1-nm thick layers of carbon (serving as the low-Z spacer material).

Each of these layers is only a few atomic dimensions in thickness and must follow the contours of the substrate upon which it is deposited. Hence, it is obvious that there are very stringent requirements upon the substrate smoothness as well as the uniformity of all the layer pairs in the multilayer coating. These coatings make it possible for the x-ray mirror to operate at normal incidence rather than grazing incidence. Since the entire mirror area is used, rather than a thin annulus, normal-incidence x-ray optics provide a much larger collecting area than a grazing-incidence mirror of the same diameter. Normal-incidence multilayer x-ray mirrors are also extremely insensitive to scattered light resulting from particulate contamination.

Since the multilayer mirrors operate by Bragg diffraction, they reflect only the radiation in a very narrow waveband. Choice of the mirror materials and the coating parameters allows the mirrors to be tuned to isolate a narrow wavelength regime centered upon a single spectral line or multiplet of lines. This property of multilayer x-ray mirrors is of great value for the development of imaging x-ray microscopes designed to operate in the biologically significant water window, which lies between the K-absorption edges of oxygen (2.33 nm) and carbon (4.37 nm). Within this narrow wavelength regime, water is relatively transparent and carbon is relatively opaque. Consequently, a narrow-band multilayer x-ray microscope operating between these wavelengths offers the potential of producing high-resolution, high-contrast images of carbon-based structures within the aqueous environment of living cells.

The narrow band pass afforded by multilayer mirrors is also of great value for solar research, where telescopes can be designed to isolate emission lines characteristic of different temperature regimes within the solar chromosphere, transition region, or corona. Thus, a probe for temperature diagnostics of the solar atmosphere is made possible. The Stanford/MSFC Rocket X-Ray Spectroheliograph, launched on October 23, 1987, produced the first high-resolution, full-disk solar image (see **illus**.) ever obtained by a normal-incidence multilayer x-ray telescope. This 17.1–17.5-nm image was produced by a telescope using a concave spherical primary mirror in conjunction with a convex spherical secondary. The x-ray mirrors were coated with molybdenum-silicon layer pairs of 3.68 nm and 5.52 nm, respectively. This image clearly revealed faint polar coronal plumes and established that multilayer mirrors could produce high-resolution, high-contrast images with very low x-ray scatter. The mirrors in this telescope had a measured reflectivity of 35%. With coating technology now available, it is possible to achieve normal-incidence reflectivities approaching 70% at selected wavelengths. Since the radiation is reflected by Bragg diffraction, multilayer optics provide the potential for good spectral resolution coupled with superb spa-

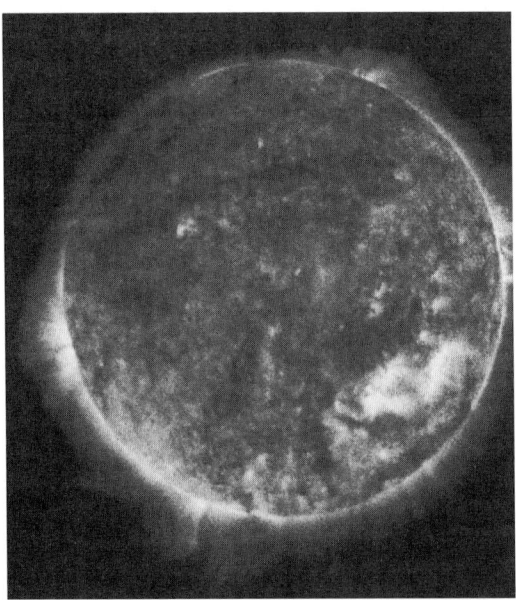

Solar corona at 10 K photographed by multilayer Cassegrain x-ray telescope. Image at 17.1–17.5 nanometers is dominated by Fe IX and Fe X emission lines. (*Courtesy of A. B. C. Walker, Jr., R. B. Hoover, T. W. Barbee, Jr., and J. F. Lindblom*)

tial resolution and mirrors that are far less sensitive to the effects of x-ray scattering than grazing-incidence systems. These combined properties of high efficiency, low x-ray scatter, and inherently good spectral resolution coupled with the potential for ultrahigh spatial resolution are of tremendous value to instrumentation now being developed for astronomy, microscopy, polarimetry, and projection x-ray lithography.

For background information *SEE SUN; X-RAY ASTRONOMY; X-RAY DIFFRACTION; X-RAY MICROSCOPE; X-RAY OPTICS; X-RAY TELESCOPE* in the McGraw-Hill Encyclopedia of Science & Technology.

Richard B. Hoover

Bibliography. T. W. Barbee, Jr., Advances in multilayer x-ray/EUV optics: Synthesis, performance, and instrumentation, *Opt. Eng.*, 29(7):711–720, July 1990. A. B. C. Walker, Jr., et al., Soft x-ray images of the solar corona with a normal-incidence Cassegrain multilayer telescope, *Science*, 241:1781–1787, 1988.

X-rays

Since their discovery by W. Roentgen in 1895, x-rays have provided a unique tool for examining the structure, composition, chemical bonding, and other properties of matter ranging from semiconductor chips to proteins and similar biological macromolecules. The brightest x-ray sources readily usable for this type of research are electron accelerators known as storage rings, which emit beams of synchrotron radiation far more intense than the radiation produced by the x-ray tubes found in laboratories and hospitals. Beginning in 1993, the first of a new generation of facilities, so-called third-generation synchrotron-radiation sources, came into operation and began producing vastly brighter beams of x-rays.

Brightness (which, along with flux, is one of the two basic parameters that measure the performance of a source) is the key feature of third-generation synchrotron sources. Flux is the number of photons generated per second. Two sources may have the same flux but vary vastly in brightness: a small, laserlike source emitting a narrow cone of radiation is much brighter than a large source, such as a light bulb, emitting in all directions with the same flux. The advantages of brightness are well illustrated by the revolutionary impact the needle-thin beams of lasers have had on visible and infrared spectroscopy since the early 1960s.

Third-generation synchrotron sources are in many ways like x-ray lasers. More than 20 such facilities worldwide are being planned, are under construction, or are already operating. Two of these are in the United States: the Advanced Light Source (ALS) at Lawrence Berkeley Laboratory in California, completed in 1993 (**Fig. 1**); and the Advanced Photon Source (APS) at Argonne National Laboratory in Illinois.

Synchrotron radiation sources. Synchrotron radiation is generated by electrically charged particles following a curved trajectory. The wavelength range of the spectrum produced depends on the particle energy, its rest mass, and the radius of curvature of its path. High energy, low mass, and small radius of curvature all extend the spectrum to shorter wavelengths.

Several features of synchrotron radiation motivated the initial interest in it as a source of light for scientific research. First, for relativistic electrons traveling at nearly the speed of light, the synchrotron radiation spectrum is broad, continuous, and most intense at ultraviolet and x-ray wavelengths. Second, relativistic effects compress the radiation into a narrow forward cone tangential to the electron path. Finally, the pulsed nature of the beam in high-energy accelerators also means that synchrotron radiation is pulsed. This combination of a continuous spectrum, high intensity, collimated beam, and pulsed time structure is not available from conventional ultraviolet and x-ray sources.

First-generation synchrotron sources are electron accelerators designed and operated for high-energy physics, which provide synchrotron radiation as a by-product. The second generation includes dedicated synchrotron sources built around electron storage rings. Second-generation storage-ring designers increased the brightness by decreasing the emittance of the electron beam. The emittance is a composite parameter that measures the electron beam size and divergence, quantifying how far away the electrons in the beam are from the positions and directions of travel they would have in an ideal orbit. Since the electron beam is the radiation source and the radiation is emitted tangential to the electron path, reducing the emittance increases the brightness in two ways: bringing electrons' positions closer to the ideal orbit reduces the source size of the beam, and bringing electrons' angles of travel closer to the ideal reduces the opening angle of the emitted radiation cone. The emittance is determined by the number, strength, and arrangement of magnets (dipoles, quadrupoles, and sextupoles) that guide and focus the electron beam in the curved arcs of the storage ring.

Undulators. A further increase in brightness can be obtained by using special magnetic structures called undulators that fit in the otherwise empty straight sections of a storage ring, between the curved arcs. The use of undulators and similar devices, along with still lower emittance than second-generation sources (by a factor of 10 or so), characterize the third generation of synchrotron radiation sources, although second-generation sources have also been retrofitted with undulators. Third-generation sources have long straight sections for undulators, since brightness scales with undulator length. The most common undulator (**Fig. 2**) bends the electrons into a roughly sinusoidal trajectory in the horizontal plane of the electron-beam orbit. Each magnetic pole in the undulator acts as a source of synchrotron radiation that adds coherently with its neighbors to increase the brightness dramatically. In the ALS, for example, the brightness of undulator radiation is about 10,000 times that of bend-magnet radiation.

The brightness of undulator radiation arises from three factors. First, the magnetic structure includes a large number of poles, each of which adds to the brightness. Second, the broad spectrum of bend-magnet radiation is compressed into a highly peaked spectrum comprising a fundamental and several harmonics, so the flux in a narrow wavelength range can be much higher. Third, the radiation cone is narrower than that of bend-magnet radiation.

The wavelengths of the fundamental and its harmonics can be tuned by varying the undulator magnetic field. This result is accomplished by mechanically adjusting the gap between the top and bottom portions of the undulator. However, the tuning range is narrower than the bend-magnet spectrum, and it is difficult for a single synchrotron radiation facility to cover effectively the entire wavelength range from the ultraviolet through the x-ray with undulators. Consequently, the third-generation facilities specialize in producing either short-wavelength x-rays (also known as hard x-rays) or long-wavelength (soft) x-rays and ultraviolet radiation. The undulator spectrum shifts to shorter wavelengths as the electron energy increases. Thus, the storage rings required for hard x-ray facilities tend to be considerably larger and more expensive than those specializing in soft x-rays and ultraviolet.

Applications. Benefits from this heavy investment in new facilities are expected to cover a broad range from quite basic research in atomic physics to such industrial applications as analyzing semiconductor wafers for trace impurities. In fact, industry is expected to be a prime beneficiary of third-generation sources.

Fig. 1. Advanced Light Source (ALS) at Lawrence Berkeley Laboratory, the first third-generation synchrotron radiation facility in the United States. *(Lawrence Berkeley Laboratory)*

Most current drugs were discovered by trial and error. To make drug development more efficient, structure-based drug design begins with detailed information about the positions of the atoms in a molecule known to play a role in a disease. The structure is determined by protein crystallography, which requires intense beams of highly collimated x-rays.

To further shrink the size of circuit features on computer chips and the spaces between them, the use of extreme-ultraviolet (EUV) projection lithography is being investigated, in which EUV light shines through a stencillike mask and lenses to start the process of etching circuits onto silicon wafers. EUV light from undulators will be used to develop mirrors and lenses needed for this technology.

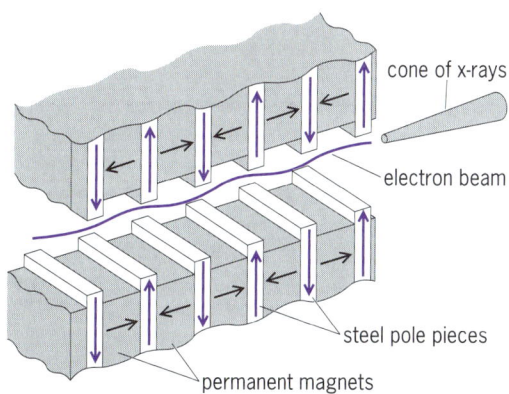

Fig. 2. Undulator, based on high-field permanent magnets that produce alternating magnetic dipoles along the device. *(After A. L. Robinson, Third-generation synchrotron light sources, Beam Line, 24(1):17–27, Spring 1994)*

Microelectromechanical machines (MEMS) include motors that can pass through the eye of a needle and tweezers that can grasp a single-celled organism. Many of these devices will be manufactured by a new technique, called deep-etch x-ray lithography, that uses synchrotron radiation.

Spatially resolved spectroscopy. A very important advance made practical by high brightness is the addition of spatial resolution to the various spectroscopic techniques already in existence. Collectively, these techniques are called spatially resolved spectroscopy or spectromicroscopy. In traditional experiments, the x-ray beam illuminates a comparatively broad area, 50 micrometers or more in diameter at the ALS, of the sample surface, which is often inhomogeneous in structure, composition, chemical bonding, and other properties on a much smaller distance scale. The measured signal therefore contains a mixture of information about the various regions illuminated.

To obtain spatial resolution, either scanning or imaging methods can be used. In the first case, x-ray lenses focus the beam to a small spot, illuminating the area of interest; the spot can then be scanned across the surface to build up an image. In the second, a large area is illuminated, but optical or electromagnetic lenses preserve the spatial relationship of the photons or electrons that constitute the signal until they reach a position-sensitive detector. The third-generation sources based on undulators will turn these spatially resolving techniques from cumbersome demonstration experiments into accessible tools. Depending on the technique used and the information desired, the spatial resolution may be as small as 5 nanometers. SEE X-RAY OPTICS.

Surface areas of the order of 1–10 nm^2 mark the frontier for many technologies, including microelectronic circuitry, magnetic storage devices, and catalysts for chemical reactions. Many of the catalysts used in industry consist of small metal particles, ranging in size from 1 to 100 nm, supported on a microporous oxide with a very high surface area. To improve catalysts further, it is necessary to understand their structure and behavior at the individual particle level, a task well matched to spectromicroscopy with third-generation synchrotron sources.

Other spectroscopic techniques. Other techniques that will emerge at the forefront of applications at third-generation synchrotron sources include soft x-ray fluorescence spectroscopy and x-ray magnetic circular dichroism spectroscopy. Soft x-ray fluorescence, the emission of a long-wavelength x-ray following x-ray absorption, is an important method of studying structures located well below the sample surface, such as the interface between a coating and the substrate it covers. So-called buried interfaces are widespread in technologically important materials. Circularly polarized x-rays are essential for studying the structure of magnetic materials, such as those used for computer disks, as well as molecules whose physical structure has a handedness, such as the helical biomolecules.

For background information SEE INTEGRATED CIRCUITS; SYNCHROTRON RADIATION; X-RAY CRYSTALLOGRAPHY; X-RAY FLUORESCENCE ANALYSIS; X-RAY SPECTROMETRY; X-RAYS in the McGraw-Hill Encyclopedia of Science & Technology.

Arthur L. Robinson; Alfred S. Schlachter

Bibliography. A. Jackson, The challenges of third-generation synchrotron light sources, *Synchro. Radiat. News,* 3(3):13–20, May–June 1990; E. E. Koch (ed.), *Handbook on Synchrotron Radiation,* vol. 1, 1983; A. L. Robinson, Third-generation synchrotron light sources, *Beam Line,* 24(1):17–27, Spring 1994; H. Winick, Synchrotron radiation, *Sci. Amer.,* 257(5):88–99, November 1987.

Yeast infection

The interest in yeast infections, which are mainly caused by *Candida albicans,* has grown considerably since the mid-1950s. Research on the microorganisms, on the pathology of the diseases which they cause, and on the search for new antifungal agents has been stimulated worldwide because of an increase in morbidity and mortality. The exact figures on morbidity are not known because yeast infections are not required to be reported to health agencies; however, they are a common problem in medical practice. For example, vaginal candidosis in the United States has been estimated to have doubled from 1980 to 1990 and to have affected about 1.3×10^7 individuals in 1990. The prominence of yeast infections in acquired immune deficiency syndrome (AIDS) patients as an early manifestation of the disease has further stimulated yeast research.

Yeast organisms. Yeasts are ubiquitous eucaryotic organisms found in the natural environment, for example, in animals, plants, and soil, sometimes in very defined niches. Approximately 200 species within 25 genera at one time or another have been isolated from humans, mainly from sputum, skin, or feces. Most yeasts are transitory, but some species, such as *C. albicans* and *C. glabrata,* occur frequently (20–40 and 5–20%, respectively) in the indigenous microflora. *C. albicans* is isolated in a small percentage from the hands of hospital physicians and nurses, and various species, from food products and beverages.

Approximately 80 different yeast species within several genera have been isolated from yeast infections, often in single cases. Species belonging to the genus *Candida* are the most frequent causes of yeast infections; altogether, about 20 *Candida* species have been isolated. *C. albicans* is the most prominent and is the etiological agent in about 70–80% of all cases. *Candida glabrata, C. tropicalis, C. krusei,* and *C. guillermondii* are examples of other species that cause yeast infections. Most yeast infections are considered endogenous, but 7–10%

of cases in hospitals are reported to be acquired during the hospital stay.

Cryptococcus neoformans, which formerly rarely caused infections, has gained in importance as a complication in AIDS patients. The source of infection is always external. Its serotype A-D has often been found in pigeon droppings, and type B has been found in connection with eucalyptus trees of Australian origin. An increase has also been observed in infections from *Trichosporon beigelii,* and the genus *Malassezia* has recently attracted the attention of investigators.

The ability of yeast to adhere to skin and mucosal surfaces, tissue penetration, enzyme production (proteinases, phospholipases), morphogenesis, and genetic change are factors in pathogenesis and decisive virulence. *Candida albicans* is the most virulent species, followed by *C. tropicalis* and *C. parapsilosis.*

Predisposing factors. Yeasts are of low virulence. Therefore, a prerequisite for yeast infections is a decrease of microbial defense mechanisms; such decrease compromises the individual. Local disruption of tissues (trauma, burns, and maceration) also facilitates infections. Microbial infections, various hormonal disorders (for example, diabetes mellitus), and diseases where the immune apparatus is affected predispose individuals for yeast infections. The humoral immune system, cell-mediated immunity, and the phagocytic cell system (neutrophils, monocytes, macrophages, and eosinophils) are all parts of the defenses against yeast infections, although the humoral system may play a lesser role, especially in mucocutaneous infections. Therefore, any deficiencies in these systems confer susceptibility for infections as is seen in individuals with AIDS, leukemias, lymphomas, certain categories of cancer, chronic granulomatous disease, and chronic mucocutaneous candidosis. Physiological states such as pregnancy, infancy, and old age present a higher frequency of yeast infections.

In modern medicine, factors caused inadvertently by a physician or by medical treatment play an essential role in yeast pathology. The yeast flora increases in individuals in the hospital for treatment, but this increase may be due to the lower resistance of hospitalized individuals. If an individual receives antibacterial drugs, the bacterial flora decreases, thus promoting the growth of drug-resistant yeasts. However, a high count of yeast flora does not necessarily lead to yeast infections, which may be due to the low virulence of some strains or to the lack of predisposing factors. Therapy with corticosteroids is immune suppressive and adds to the risk of infections. The treatment of malignant diseases (cancer, leukemias) with neutropenia-inducing drugs (drugs that promote a deficiency in neutrophils, a specific type of white blood cell) suppresses the immune apparatus, and some of the drugs cause disruption of intestinal mucosa, as in the case of cytosine arabinoside where penetration of low pathogenetic bacteria and of yeasts is facilitated.

Surgery (and general anesthesia), especially advanced surgery such as open heart surgery and transplantations, are also risk factors. Foreign bodies such as intravenous catheters and dentures, to which *Candida albicans* can adhere, may become foci of fungal growth.

Infections and symptoms. Yeast can infect any organ and become disseminated. Skin lesions may affect smaller or larger areas of hairless skin, anal regions and the diaper area. Infections are typically characterized by blisters surrounded by reddened, scaly skin, but lesions can have other appearances. Nails with painful reddened swelling along the paronychial edge and discolored and striated nails can be due to infection by yeasts; however, the presence of other organisms may result in similar lesions. Mucosa of the oral cavity covered by a white-gray pseudomembrane clearly indicates a *Candida albicans* infection (thrush). Chronic mucocutaneous candidosis involves persistent infections of skin, nails, and mucous membranes, but the clinical picture can vary considerably. Infections in the genitals (vaginitis, balanopostitis), urinary tract, bronchi and lungs, central nervous system, eyes, bones, heart and endocardium, intestines and stomach, and liver and spleen present symptoms that are not specific for yeast. The medical history taken from the individual together with underlying disease or states will lead the clinician to suspect a yeast infection. The difficulty in diagnosing disseminated forms of acute and chronic candidosis constitutes a serious problem. One or several different organs may be involved, often the liver and spleen; symptoms are often uncharacteristic with protracted fever as the only sign.

Diagnosis. In order to make a diagnosis, a sample must be cultured in order to determine the species and, where such methods have been developed and if hospital-related infections are suspected, the type of fungus. A positive culture from sites that are normally sterile is diagnostic. When cultured from areas with a normal microflora, assessment of the finding can be difficult and must be related to clinical symptoms.

In suspected disseminated infections involving the visceral organs, a diagnosis can be difficult to obtain. Diagnosis must be made quickly, especially in individuals with leukopenia, because persistent infections have a poor prognosis. Unfortunately, blood cultures are often negative in some of these cases, and ordinary blood culture techniques are slow and not very sensitive. A recent improvement is the lysis-centrifugation method, a faster and more sensitive technique that involves the destruction of blood cells to release the attached microorganisms, which are then concentrated by centrifugation and cultured. Organ imaging by radiology, computed tomography, magnetic resonance, or ultrasound may disclose infected foci from which biopsy material can be obtained for histological examination and culture.

Nonculture diagnostic measures include serology. Antibodies to *Candida albicans* can be measured by using various *Candida* antigen preparations, but because a certain level of antibodies is always found in normal individuals, only a significant rise in antibody level during illness is useful in supporting the diagnosis. The presence of fungal products, such as fungal wall components, metabolites, or deoxyribonucleic acid (DNA), in the blood characterizes individuals with fungemia. Attempts have been made using various techniques to measure these components in the blood of individuals with disseminated *Candida* infections and to evaluate the findings as a diagnostic aid. Promising results have been obtained, but the methods are still experimental and are not available for routine use. The detection of cryptococcal capsule antigen in body fluids in cryptococcosis has been a useful diagnostic tool for many years.

Treatment. Many antifungal agents are used for treatment of yeast infections. Some are suitable only for topical application; others are used for treatment of deep-seated and disseminated candidoses and other yeast infections. The most important agents belong to three groups: polyenes (amphotericin B, liposomal amphotericin B), pyrimidinanaloges (flucytosine), and azoles. Ketoconazole is an imidazole; fluconazole and itroconazole are triazoles. These agents have a wide antifungal spectrum and affect most pathogenic yeasts. Apart from the side effects, many problems connected with antifungal therapy remain, although the great therapeutic value of the agents is undisputed. Because it is difficult to measure whether a strain is sensitive or resistant, especially with the azoles, susceptibility testing with these agents does not provide a good guide to the clinician as to choice of drug. Measuring the drug level in the blood and at the infection site in order to prescribe the proper drug dosage is not useful.

The relative sensitivity of strains can be determined in the laboratory, indicating if a strain has gained in resistance. In order to observe whether a strain has gained in resistance in individuals during treatment, the yeast should ideally be tested before treatment begins and cultured and tested again during treatment. It has been determined that resistance seldom develops to amphotericin B but often to flucytosine. Resistance to ketoconazole is known to develop during long-term treatment of *Candida albicans* and other yeasts. Simultaneously, cross resistance develops to other azoles. Less-sensitive strains have been noted during treatment with fluconazole, in particular, and with itraconazole, and also with cross resistance to other azoles. Although azole resistance may develop and cross resistance may be a general phenomenon, the extensive use of these drugs in the future may be jeopardized because the resistant strains will spread within the community.

Antifungal agents may be used prophylactically in individuals with a high risk of developing yeast infection in order to minimize yeast in the normal flora, and they have been used with partial success in specific categories of individuals (for example, those with leukemia). Flucytosine is not useful because resistance can be developed; this condition does not exist with amphotericin B, which can be given by mouth without toxic side effects.

New approaches in fungal therapy are promising for individuals seriously ill with neutropenia, such as types of cancer and malignant blood disorders, who have a great risk of acquiring disseminated yeast infections. Use of antifungal agents is not sufficient (unless the granulocyte counts start to rise as a sign of remission of the underlying disease), but administration of cytokines such as granulocyte colony-stimulating factor (G-CSF) and granulocyte macrophage colony-stimulating factor (GM-CSF) interleukin 3, and stem-cell factor can reduce the duration of neutropenia and decrease the risk of fungal infection. The treatment can be combined with antifungal agents. Cytokines not only affect stem cells but also enhance the ability of mature monocytes and macrophages to kill fungal organisms. Clinical experience with cytokine treatment is still limited, but the results merit further investigation.

For background information SEE ACQUIRED IMMUNE DEFICIENCY SYNDROME (AIDS); CLINICAL PATHOLOGY; MEDICAL MYCOLOGY; YEAST in the McGraw-Hill Encyclopedia of Science & Technology.

Aksel Stenderup

Bibliography. G. P. Bodey, *Candidiasis: Pathogenesis, Diagnosis, and Treatment*, 1992; F. C. Odds, *Candida and Candidosis: A Review and Bibliography*, 1988; E. Reiss and C. J. Morrison, Nonculture methods for diagnosis of disseminated candidiasis, *Clin. Microbiol. Rev.*, 6:311–323, 1993; T. J. Walsh, J. Hiemenz, and P. A. Pizzo, Evolving risk factors for invasive fungal infections: All neutropenic patients are not the same, *Clin. Infect. Dis.*, 18:793–798, 1994.

Contributors

Contributors

The affiliation of each Yearbook contributor is given, followed by the title of his or her article. An article title with the notation "in part" indicates that the author independently prepared a section of an article; "coauthored" indicates that two or more authors jointly prepared an article or section.

A

Ahl, Dr. Jonna S. B. *Department of Molecular and Cell Biology, The University of Connecticut, Storrs.* CRUSTACEAN—coauthored.

Akerberg, Peeter M. *Department of Electrical Engineering, University of Houston, Texas.* NEURAL NETWORK—coauthored.

Aldridge, Dr. Richard J. *Department of Geology, University of Leicester, United Kingdom.* CONODONTS.

Ames, Dr. Doreen E. *Geological Survey of Canada, Ottawa, Canada.* MARINE GEOLOGY—in part.

Archibald, Prof. James F. *Department of Mining Engineering, Queens University, Kingston, Canada.* UNDERGROUND MINING.

Awschalom, Prof. David D. *Department of Physics, University of California, Santa Barbara.* MAGNETIC SWITCH; QUANTUM WELLS—both coauthored.

B

Baker, Dr. David B. *Director, Water Quality Laboratory, Heidelberg College, Tiffin, Ohio.* WATER RESOURCES.

Bauer, Prof. Raymond T. *Department of Biology, The University of Southwestern Louisiana, Lafayette.* SHRIMP.

Birmingham, Dr. Jeannette M. *Senior Biologist Specialist, Mycology and Botany Department, American Type Culture Collection, Rockville, Maryland.* MICROBIAL COMPOUND.

Block, Prof. Gene D. *Director, NSF Science and Technology Center for Biological Timing, University of Virginia, Charlottesville.* BIOLOGICAL CLOCKS.

Blundell, Dr. B. G. *Department of Electrical and Electronic Engineering, University of Canterbury, Christchurch, New Zealand.* ELECTRONIC DISPLAYS—in part.

Bracewell, Ronald N. *Professor Emeritus, Electrical Engineering, Stanford University, California.* WAVELETS.

Bramble, Prof. Dennis M. *Department of Biology, The University of Utah, Salt Lake City.* RESPIRATORY SYSTEM.

Brearley, Dr. Adrian J. *Department of Geology, Institute of Meteoritics, The University of New Mexico, Albuquerque.* MINERALOGY—in part.

Brick, Prof. Mark A. *Department of Soil and Crop Sciences, Colorado State University, Fort Collins.* PLANT MOVEMENTS.

Briggs, Prof. Derek E. G. *Department of Geology, University of Bristol, Queens Road, United Kingdom.* ANOMALOCARIS.

Brooke, Dr. Stefan. *Department of Neurology and Neurological Sciences, Beckman Center for Molecular and Genetic Medicine, Stanford University Medical Center, California.* AUTOIMMUNITY—in part.

Bryant, Dr. J. Daniel. *Department of Vertebrate Paleontology, American Museum of Natural History, New York, New York.* PALEOCLIMATOLOGY and MAMMALIA.

Businger, Prof. Steven. *School of Ocean and Earth Science and Technology, Department of Meteorology, University of Hawaii at Manoa.* APPLICATIONS SATELLITES.

C

Caldwell, Prof. Allen C. *Department of Physics, Columbia University, Irvington, New York.* ELEMENTARY PARTICLE.

Carson, Dr. John W. *President, Jenike & Johanson, Inc., Westford, Massachusetts.* FLOW OF SOLIDS.

Castleman, Prof. A. Welford, Jr. *Department of Chemistry, College of Science, The Pennsylvania State University, University Park.* MOLECULAR MATERIALS.

Chin, Dr. Anne. *Department of Geography, University of Southern California, Los Angeles.* STEP POOLS.

Chinsamy, Dr. Anusuya. *Department of Animal Biology, The School of Veterinary Medicine, University of Pennsylvania, Philadelphia.* DINOSAUR—in part.

Chomel, Dr. Bruno B. *Department of Population Health and Reproduction, School of Veterinary Medicine, University of California, Davis.* CAT SCRATCH DISEASE.

Choptuik, Dr. Matthew W. *Department of Physics, Center for Relativity, University of Texas, Austin.* GRAVITATIONAL COLLAPSE.

Chuang, Dr. Catherine C. *Global Climate Research Division, Lawrence Livermore National Laboratory, Livermore, California.* ATMOSPHERIC AEROSOLS.

Clack, Dr. Jennifer A. *University Museum of Zoology, Cambridge, United Kingdom.* TETRAPODA.

Clare, Dr. Anthony S. *The Marine Biological Association of the United Kingdom, The Laboratory, Citadel Hill, Plymouth, United Kingdom.* BARNACLE.

Clark, Dr. Edward A. *Department of Microbiology, Regional Primate Research Center, University of Washington, Seattle.* CELLULAR IMMUNOLOGY—coauthored.

Clerici, Dr. Mario. *Càttedra di Immunologia, Università degli Studi, Milano, Italy.* TUMOR—in part.

Conway, Dr. Clay M. *U.S. Department of the Interior, Geological Survey, Branch of Western Mineral Resources, Flagstaff, Arizona.* PROSPECTING.

Cooperman, Richard S. *Flitecom, Silver Spring, Maryland.* RADIO COMMUNICATIONS.

Crook, Dr. James. *Director, Water Reuse Department, Black and Veatch, Cambridge, Massachusetts.* WATER SUPPLY ENGINEERING.

Crum, Dr. Lawrence A. *Department of Acoustics and Electromagnetics, Applied Physics Laboratory, University of Washington, Seattle.* SONOLUMINESCENCE.

Culver, Dr. Donald W. *Science Applications International Corp., McLean, Virginia.* ROCKET PROPULSION.

Cummings, Dr. John F. *Department of Anatomy, College of Veterinary Medicine, Cornell University, Ithaca, New York.* HORSE—coauthored.

Curtin, Dr. Denis. *Research Scientist, Research Centre, Agriculture Canada, Swift Current, Saskatchewan, Canada.* PLANTS, SALINE ENVIRONMENTS OF—in part.

D

Dalziel, Prof. Ian W. D. *Senior Research Scientist, Institute for Geophysics, University of Texas, Austin.* PALEOGEOGRAPHY.

Dawson, Dr. David M. *West Roxbury Veterans Affairs Medical Center, Massachusetts.* CARPAL TUNNEL SYNDROME.

Deckett, Martin. *Vice President, International Development, Orbcomm, Dulles, Virginia.* DATA COMMUNICATIONS.

de Jongh, Prof. L. Jos. *Faculty of Mathematics and Natural Sciences, Department of Astronomy and Physics, Leiden University, Leiden, The Netherlands.* METAL CLUSTER COMPOUND.

de Lahunta, Dr. Alexander. *Department of Anatomy, College of Veterinary Medicine, Cornell University, Ithaca, New York.* HORSE—coauthored.

DeMeis, Richard. *Associate/Technical Editor, Penwell Publishing Company, Nashua, New Hampshire.* AIRCRAFT NOISE.

DeSimone, Dr. Joseph. *Department of Chemistry, Venable Laboratories, University of North Carolina at Chapel Hill.* PLASTICS—coauthored.

Desurvire, Dr. Emmanuel. *Alcatel Alsthom Recherche, Route de Nozay, Marcoussis, France.* OPTICAL FIBER AMPLIFIERS.

Diaz, Prof. Robert J. *Virginia Institute of Marine Science, The College of William & Mary, Gloucester Point, Virginia.* POLYCHAETA—coauthored.

Dorn, Prof. Ronald I. *Department of Geography, Arizona State University, Tempe.* RADIOCARBON DATING.

E

Edwards, Alan F. *Freeport-McMoRan Inc., New Orleans, Louisiana.* SOLUTION MINING.

Edwards, Dr. Clive A. *Department of Entomology, The Ohio State University, Columbus.* EARTHWORM.

Eesley, Dr. Gary L. *Physics Department, General Motors Research and Development Center, Warren, Michigan.* LASER.

Eidson, Dr. Millicent. *American College of Veterinary Preventive Medicine.* PLAGUE.

F

Farah, Prof. Martha. *Department of Psychology, University of Pennsylvania, Philadelphia.* VISUAL AGNOSIA.

Fenoglio-Preiser, Dr. Cecilia M. *Department of Pathology and Laboratory Medicine, University of Cincinnati College of Medicine, Ohio.* CLINICAL PATHOLOGY—coauthored.

Finch, Prof. Robert D. *Department of Mechanical Engineering, University of Houston, Texas.* NEURAL NETWORK.

Fishman, Dr. Gerald J. *Chief, Gamma Ray Astronomy Branch, National Aeronautics and Space Administration, George C. Marshall Space Flight Center, Alabama.* GAMMA-RAY BURSTS.

Forbes, Dr. Andrew M. G. *Scripps Institution of Oceanography, University of California, San Diego.* ACOUSTIC THERMOMETRY.

Frederick, Dr. Christin A. *Laboratory of X-Ray Crystallography, Dana-Farber Cancer Institute, Boston, Massachusetts.* ENZYME—coauthored.

Fredericksen, Rick S. *Echo Bay Alaska, Inc., Juneau.* BORING AND DRILLING (PROSPECTING)

G

Gallier, W. Thomas. *Manager, Wastewater Management Division, City of Fresno, California.* BIOSOLIDS.

Garnier, Prof. Francis. *Laboratoire des Matériaux Moléculaires, Centre National de la Recherche Scientifique, Thiais, France.* TRANSISTOR—in part.

Garrett, Dr. J. D. *Scientific Director, Holifield Radioactive Ion Beam Facility, Oak Ridge National Laboratory, Tennessee.* RADIOACTIVE BEAMS.

Gauld, Dr. Suellen C. *Research Associate, Department of Anthropology, University of California, Los Angeles.* NEANDERTALS.

Gedney, Dr. Richard T. *Manager, ACTS Project, National Aeronautics and Space Administration, Lewis Research Center, Cleveland, Ohio.* COMMUNICATIONS SATELLITE—in part.

Geissman, Prof. John W. *Department of Earth and Planetary Sciences, The University of New Mexico, Albuquerque.* PALEOMAGNETISM.

Gelman, Dr. Andrew. *Biomaterials Bioreactivity Characterization Laboratory, Department of Pathology and Laboratory Medicine, University of California, Los Angeles Medical Center.* MOLECULAR TRANSPORT—coauthored.

Gider, Dr. Savas. *Department of Physics, University of California, Santa Barbara.* MAGNETIC SWITCH—coauthored.

Gorzelnik, Mr. Eugene F. *North American Electric Reliability Council, Princeton Forrestal Village, Princeton, New Jersey.* ELECTRIC UTILITY INDUSTRY.

Gould, Dr. Harvey A. *Lawrence Berkeley Laboratory, Berkeley, California.* FUNDAMENTAL INTERACTIONS.

Gozani, Dr. Tsahi. *Science Applications International Corp., Santa Clara, California.* NONDESTRUCTIVE TESTING.

Green, Prof. David M. *Department of Psychology, University of Florida, Gainesville.* PSYCHOACOUSTICS.

Grimmond, Prof. C. Susan B. *Department of Geography, Indiana University, Bloomington.* URBAN HEAT STORAGE—coauthored.

Guidry, Prof. Michael W. *Department of Physics and Astronomy, University of Tennessee, Knoxville.* ATOMIC NUCLEUS.

Gupta, Dr. Rishab K. *John Wayne Cancer Institute, Santa Monica, California.* TUMOR—in part.

H

Hall, Dr. Brian K. *Research Professor, Department of Biology, Dalhousie University, Halifax, Nova Scotia, Canada.* EVOLUTIONARY DEVELOPMENTAL BIOLOGY and EMBRYOGENESIS.

Harris, Dr. W. Lamar. *National Program Leader, Engineering and Energy, U.S. Department of Agriculture, Agricultural Research Service, Beltsville, Maryland.* BIOMASS.

Harry, Dr. John E. *Department of Electronic and Electrical Engineering, Loughborough University of Technology, Leicestershire, United Kingdom.* HAZARDOUS WASTE.

Hauchman, Dr. Fred S. *Associate Director, United States Environmental Protection Agency, Health Effects Research Laboratory, Research Triangle Park, North Carolina.* WATER—in part.

Havener, Dr. George. *Principal Engineer, Micro Craft Technology, Arnold Air Force Base, Tennessee.* FLOW VISUALIZATION.

Hawthorne, Prof. Frank C. *Department of Geological Sciences, The University of Manitoba, Winnipeg, Manitoba, Canada.* MINERALOGY—in part.

Hebbar, Dr. K. Prakash. *Biocontrol of Plant Diseases Laboratory, U.S. Department of Agriculture, Agricultural Research Service, Beltsville, Maryland.* PLANT PATHOGENS—coauthored.

Heymann, Dr. Dieter. *Department of Geology and Geophysics, Rice University, Houston, Texas.* FULLERENE.

Hoffert, Dr. Marvin J. *John R. Graham Headache Centre, Faulkner Hospital, Boston, Massachusetts.* HEADACHE.

Hoffman, Dr. Eric P. *Department of Molecular Genetics and Pediatrics, University of Pittsburgh School of Medicine.* HORSE—coauthored.

Hoover, Dr. Dallas G. *Department of Food Science, University of Delaware, Newark.* FOOD ENGINEERING—in part.

Hoover, Dr. Richard B. *Space Science Laboratory, National Aeronautics and Space Administration, Marshall Space Flight Center, Huntsville, Alabama.* X-RAY OPTICS.

Hsiao, Dr. Yu-Ling. *Department of Chemistry, Venable Laboratories, University of North Carolina at Chapel Hill.* PLASTICS—coauthored.

Hubble, Dr. Mark D. *Department of Geography, Indiana University, Bloomington.* URBAN HEAT STORAGE—coauthored.

Huber, Dr. Michael J. *College of Veterinary Medicine, Oregon State University, Corvallis.* ENDOMETRIAL CUPS.

Hughes, Dr. Walter T. *Department of Infectious Diseases, St. Jude Children's Research Hospital, Memphis, Tennessee.* PNEUMOCYSTIS.

Humayun, Dr. Munir. *Department of Terrestrial Magnetism, Carnegie Institution of Washington, Washington, D.C.* SOLAR SYSTEM.

Hwa, Prof. Terence. *Department of Physics, State University of New York at Stony Brook.* SUPERCONDUCTIVITY.

I

Imaino, Dr. Wayne. *IBM Research Division, Almaden Research Center, San Jose, California.* COMPACT DISK—coauthored.

J

Jaggard, Prof. Dwight L. *Associate Dean, School of Engineering and Applied Science, University of Pennsylvania, Philadelphia.* FRACTALS.

Jansen, Ben H. *Department of Electrical Engineering, University of Houston, Texas.* NEURAL NETWORK—coauthored.

Jeanloz, Prof. Raymond. *Department of Geology and Geophysics, University of California, Berkeley.* EARTH.

Jessen, Dr. Poul. *Optical Sciences Center, The University of Arizona, Tucson.* ATOMIC CRYSTAL.

Johari, Prof. G. P. *Department of Materials Science and Engineering, McMaster University, Hamilton, Ontario, Canada.* SHAPE MEMORY ALLOYS.

Johnson, Dr. Mark. *Department of the Navy, Naval Research Laboratory, Washington, D.C.* TRANSISTOR—in part.

Jones, Dr. Daniel B. *Research Fellow in Surgery, Washington University Institute for Minimally Invasive Surgery, St. Louis, Missouri.* SURGERY—coauthored.

Jones, Dr. Ralph C. *Marine Corps., U.S. Navy, John Wayne Cancer Institute, Santa Monica, California.* TUMOR—in part.

Jurica, Dr. Paul. *Department of Psychology, University of California, Berkeley.* MEMORY.

K

Kahn, Prof. Olivier. *Laboratoire de Chimie Inorganique, Université de Paris-Sud, Orsay, France.* MOLECULAR MAGNET—coauthored.

Kazmierczak, Dr. James J. *Bureau of Public Health, Madison, Wisconsin.* LYME DISEASE.

Kellman, Dr. Philip J. *Department of Psychology, University of California, Los Angeles.* COGNITION.

Khabbaz, Dr. Rima F. *Associate Director for Medical Science, Division of Viral and Rickettsial Diseases, National Center for Infectious Diseases, Department of Health and Human Services, Atlanta, Georgia.* HANTAVIRUS.

Kim, Prof. Thomas J. *Director, Waterjet Laboratory, Department of Mechanical Engineering and Applied Mechanics, University of Rhode Island, Kingston.* WATER-JET CUTTING.

Klaus, Dr. Stephen J. *Department of Microbiology, Regional Primate Research Center, University of Washington, Seattle.* CELLULAR IMMUNOLOGY—coauthored.

Kobayashi, Prof. George S. *Department of Internal Medicine, Washington School of Medicine, St. Louis, Missouri.* MEDICAL MYCOLOGY.

Kossovsky, Dr. Nir. *Director, Autopsy Service, Deputy Coroner, Los Angeles County, Biomaterials Bioreactivity Characterizations Laboratory, Department of Pathology and Laboratory Medicine, University of California, Los Angeles Medical Center.* MOLECULAR TRANSPORT—coauthored.

Kumar, Dr. K. Sharvan. *Division of Engineering, Brown University, Providence, Rhode Island.* INTERMETALLIC MATERIALS—coauthored.

Kwitowski, Dr. August J. *Supervisory Physical Scientist, U.S. Department of the Interior, Bureau of Mines, Pittsburgh Research Center, Pennsylvania.* COAL MINING.

L

Labandeira, Dr. Conrad. *Department of Paleobiology, Smithsonian Institution, Washington, D.C.* INSECTS.

Larsson, Dr. Mats. *Physics Department, The Royal Institute of Technology, Stockholm, Sweden.* STORAGE RINGS.

Laufer, Dr. Hans. *Department of Molecular and Cell Biology, The University of Connecticut, Storrs.* CRUSTACEAN—coauthored.

Lehrsch, Dr. Gary A. *Soil Scientist, U.S. Department of Agriculture, Northwest Irrigation and Soils Research Laboratory, Kimberly, Idaho.* SOIL.

Lentz, Dr. Rodrick D. *Soil Scientist, Kimberly Research & Extension Center, University of Idaho, Department of Agriculture Engineering, Kimberly.* IRRIGATION (AGRICULTURE); PRECIPITATION (METEOROLOGY).

Levine, Dr. R. D. *The Fritz Haber Research Center for Molecular Dynamics, The Hebrew University of Jerusalem, Israel.* PHOTOSELECTIVE CHEMISTRY—coauthored.

Lieber, Prof. Charles M. *Department of Chemistry and Division of Applied Sciences, Harvard University, Cambridge, Massachusetts.* CARBON NITRIDE.

Lienhard, Prof. John H., V. *Department of Mechanical Engineering, Massachusetts Institute of Technology, Cambridge.* EXTREMELY HIGH HEAT FLUX.

Lowenstein, Dr. Charles J. *Division of Cardiology, Department of Medicine, Johns Hopkins University, Baltimore, Maryland.* NITRIC OXIDE.

Lucas, Dr. Spencer G. *Curator of Paleontology, New Mexico Museum of Natural History, Albuquerque.* DINOSAUR—in part.

Lumsden, Dr. R. D. *Biocontrol of Plant Diseases Laboratory, U.S. Department of Agriculture, Agricultural Research Service, Beltsville, Maryland.* PLANT PATHOGENS—coauthored.

Luu, Dr. Jane. *Solar and Stellar Physics Division, Harvard-Smithsonian Center for Astrophysics, Cambridge, Massachusetts.* COMET.

M

Maas, Dr. Eugene V. *Plant Physiologist, U.S. Department of Agriculture, Agricultural Research Service, Riverside, California.* PLANTS, SALINE ENVIRONMENTS OF—in part.

McCarthy, Prof. Michael J. *Department of Food Engineering, University of California, Davis.* FOOD ENGINEERING—in part.

McShea, Prof. Daniel W. *Museum of Paleontology, The University of Michigan, Ann Arbor.* EVOLUTION.

Madon, Dr. Pierre J. *Vice President, Engineering and Research, Intelsat, Washington, D.C.* COMMUNICATIONS SATELLITE—in part.

Marsh, Prof. Richard F. *Animal Health and Biomedical Sciences, University of Wisconsin, Madison.* MAD COW DISEASE.

Mascarenhas, Dr. E. J. P. *Thorn Security Ltd., Product Management Department, Sunbury-on-Thames, Middlesex, United Kingdom.* LIGHTING SYSTEMS.

Mascarenhas, Prof. Joseph P. *Chairperson, Department of Biological Sciences, University at Albany, State University of New York.* REPRODUCTION (PLANT).

Meade, Dr. Dale M. *Deputy Director, Plasma Physics Laboratory, Princeton University, New Jersey.* NUCLEAR FUSION.

Meese, Prof. Jon M. *Department of Electrical and Computer Engineering, University of Missouri, Columbia.* OPTICAL COMPUTING.

Mohammed, Dr. Hussni. *Department of Anatomy, College of Veterinary Medicine, Cornell University, Ithaca, New York.* HORSE—coauthored.

Morton, Dr. Donald L. *John Wayne Cancer Institute, Santa Monica, California.* TUMOR—in part.

Mountz, Dr. John D. *Department of Medicine, Division of Clinical Immunology and Rheumatology, The University of Alabama, Birmingham.* AUTOIMMUNITY—in part.

N

Noffsinger, Dr. Amy E. *Department of Pathology and Laboratory Medicine, University of Cincinnati College of Medicine, Ohio.* CLINICAL PATHOLOGY—coauthored.

Nyman, Dr. Matthew W. *Senior Research Associate, Department of Earth and Planetary Sciences, The University of New Mexico.* METAMORPHISM.

P

Parker, Dr. Geoffrey G. *Smithsonian Environmental Research Center, Edgewater, Maryland.* TREE.

Pei, Dr. Yu. *Laboratoire de Chimie Inorganique, Université de Paris-Sud, Orsay, France.* MOLECULAR MAGNET—coauthored.

Pennaz, Mr. Thomas J. *Deluxe Corporation, St. Paul, Minnesota.* INK.

Petrie, Dr. Steven E. *Unocal Agriproducts, Boise, Idaho.* FERTILIZER—in part.

Phadke, Prof. Arun G. *Director, Center for Power Engineering, The Bradley Department of Electrical Engineering, Virginia Polytechnic Institute and State University.* ELECTRIC POWER SYSTEMS.

Pious, Dr. Donald. *Department of Pediatrics, University of Washington School of Medicine, Seattle.* ANTIGEN.

Pletcher, Prof. Richard H. *Department of Mechanical Engineering, Iowa State University of Science and Technology, Ames.* WAVE PHENOMENA.

Pohl, Dr. D. W. *IBM Research Division, Zurich Research Laboratory, Ruschlikon, Switzerland.* MICROSCOPE—in part.

Poinar, Dr. George O., Jr. *Department of Environmental Science, Policy, and Management, Division of Entomology and Plant and Soil Microbiology, University of California, Berkeley.* BACTERIA.

R

Remacle, Dr. F. *Departement de Chimie, Liège University, Belgium.* PHOTOSELECTIVE CHEMISTRY—coauthored.

Rensberger, Dr. John M. *Burke Memorial Washington State Museum, University of Washington.* TOOTH.

Richardson, Dr. Susan D. *U.S. Environmental Protection Agency, Environmental Research Laboratory, Athens, Georgia.* WATER—in part.

Robbins, Dr. Charles W. *Soil Scientist, U.S. Department of Agriculture, Agricultural Research Service, Kimberly, Idaho.* SOIL POLLUTION.

Robinson, Dr. Arthur L. *Lawrence Berkeley Laboratory, University of California, Berkeley.* X-RAYS—coauthored.

Rosen, Dr. Hal. *IBM Research Division, Almaden Research Center, San Jose, California.* COMPACT DISK—coauthored.

Rosenberg, Dr. Michael. *Parallel Program, University of Delaware, Wilmington.* PREHISTORIC DOMESTICATION OF ANIMALS.

Rosenberg, Rutger. *Güteborg University, Fiskebäckskil, Sweden.* POLYCHAETA—coauthored.

Rosenzweig, Dr. Amy C. *Laboratory of X-ray Crystallography, Dana-Farber Cancer Institute, Boston, Massachusetts.* ENZYME—coauthored.

Ross, Dr. Malcolm. *U.S. Department of the Interior, Geological Survey, Reston, Virginia.* ENVIRONMENTAL POLLUTION.

Rubin, Dr. Kurt. *IBM Research Division, Almaden Research Center, San Jose, California.* COMPACT DISK—coauthored.

Ryan, Dr. Kenneth J. *Professor and Associate Head, Department of Pathology, University of Arizona, Arizona Health Sciences Center, Tucson.* STREPTOCOCCUS.

S

Sachdev, Prof. M. S. *Department of Electrical Engineering, University of Saskatchewan, Saskatoon, Canada.* ELECTRIC PROTECTIVE DEVICES.

Sage, Prof. Andrew P. *First American Bank Professor and Dean, School of Information Technology and Engineering, George Mason University, Fairfax, Virginia.* REENGINEERING.

Sandwell, Dr. David T. *Institute of Geophysics and Planetary Physics, Scripps Institution of Oceanography, University of California at San Diego, La Jolla.* MARINE GEOLOGY—in part.

Sarid, Dr. Dror. *Optical Sciences Center, University of Arizona, Tucson.* MICROSCOPE—in part.

Sastry, Prof. Sudhir K. *Department of Agricultural Engineering, The Ohio State University, Columbus.* FOOD ENGINEERING—in part.

Schlachter, Dr. Alfred S. *Lawrence Berkeley Laboratory, University of California, Berkeley.* X-RAYS—coauthored.

Schwartz, Dr. Mel. *Materials Science Consultant, Madison, Connecticut.* MATERIALS PROCESSING.

Schwarz, Mr. A. J. *Department of Electrical and Electronic Engineering, University of Canterbury, Christchurch, New Zealand.* ELECTRONIC DISPLAYS—in part.

Sessler, Prof. Jonathan L. *Department of Chemistry and Biochemistry, University of Texas, Austin.* TEXAPHYRINS.

Shearer, Dr. Gene M. *Division of Cancer Biology, Diagnosis and Centers, Experimental Immunology Branch, National Cancer Institute, Bethesda, Maryland.* TUMOR—in part.

Shearer, Dr. Peter M. *Institute of Geophysics and Planetary Physics, Scripps Institution of Oceanography, La Jolla, California.* EARTH INTERIOR.

Shearman, Prof. Robert C. *Executive Director, National Turfgrass Evaluation Program, Department of Horticulture, University of Nebraska, Lincoln.* TURFGRASS.

Sherwood, Prof. John L. *Department of Plant Pathology, Division of Agricultural Sciences and Natural Resources, Oklahoma State University, Stillwater.* PLANT VIRUS.

Shochet, Prof. Melvyn J. *The Enrico Fermi Institute, The University of Chicago, Illinois.* QUARKS.

Singh, Dr. Ashbindu. *Regional Coordinator, Environment Assessment Programme-North America, EROS Data Center, Sioux Falls, South Dakota.* FOREST.

Singh, Dr. Prithipal. *CEO and Chairman of the Board, ChemTrak, Sunnyvale, California.* MEDICAL TESTING.

Sojka, Dr. R. E. *U.S. Department of Agriculture, Agricultural Research Service, Kimberly, Idaho.* AGRICULTURAL SOIL AND CROP PRACTICES.

Soper, Dr. Nathaniel J. *Associate Professor of Surgery, Washington University School of Medicine, St. Louis, Missouri.* SURGERY—coauthored.

Specter, Dr. Herschel. *RBR Consultants, Inc., White Plains, New York.* RISK ANALYSIS.

Spier, Dr. Sharon J. *Department of Medicine, School of Veterinary Medicine, University of California, Davis.* HORSE—coauthored.

Sponsler, Dr. Edward. *Biomaterials Bioreactivity Characterization Laboratory, Department of Pathology and Laboratory Medicine, University of California, Los Angeles Medical Center.* MOLECULAR TRANSPORT—coauthored.

Stahlkopf, Dr. Karl E. *Vice President, Power Delivery Group, Electric Power Research Institute, Palo Alto, California.* THYRISTOR.

Steinman, Dr. Lawrence. *Department of Neurology and Neurological Sciences, Beckman Center for Molecular and Genetic Medicine, Stanford University Medical Center, California.* AUTOIMMUNITY—in part.

Stenderup, Prof. Aksel. *Institute of Medical Microbiology, University of Aarhus, Denmark.* YEAST INFECTION.

Steppuhn, Dr. Harold. *Research Centre, Agriculture Canada, Swift Current, Saskatchewan.* PLANTS, SALINE ENVIRONMENTS OF—in part.

Steyn, Dr. Julian J. *Energy Resources International, Inc., Washington, D.C.* NUCLEAR FUELS.

Surguy, Dr. Paul W. H. *Principal Engineer, Central Research Laboratories Ltd., Thorn EMI Ltd., Hayes, Middlesex, United Kingdom.* ELECTRONIC DISPLAYS—in part.

T

Thomson, Dr. James. *Jason Consultants International, Inc., Washington, D.C.* PIPELINE.

Tonomura, Dr. Akira. *Advanced Research Laboratory, Hitachi, Ltd. Hatoyama, Saitama, Japan.* ELECTRON HOLOGRAPHY.

Trotter, Dr. John A. *Department of Anatomy, The University of New Mexico, School of Medicine, Albuquerque.* MUSCULAR SYSTEM.

Tulchinsky, David. *Department of Physics, University of California, Santa Barbara.* QUANTUM WELLS—coauthored.

Turro, Prof. Nicholas J. *Department of Chemistry, Columbia University, New York.* DEOXYRIBONUCLEIC ACID (DNA).

U

Usherwood, Dr. Noble R. *Vice President, Foundation for Agronomic Research, Norcross, Georgia.* FERTILIZER—in part.

V

von Itzstein, Dr. Mark. *Victorian College of Pharmacy, Monash University, Parkville, Victoria, Australia.* DRUG DESIGN.

van Keuren, Dr. David K. *History Office, Naval Research Laboratory, Washington, D.C.* MOON.

von Puttkamer, Dr. Jesco. *National Aeronautics and Space Administration, Washington, D.C.* SPACE FLIGHT.

W

Waldman, Dr. Jeffrey. *Head, Metallics and Ceramics Branch, Naval Air Warfare Center, Aircraft Division Warminster, Pennsylvania.* INTERMETALLIC MATERIALS—coauthored.

Ward, Dr. David A. *Rocoil Limited, Harrogate, North Yorkshire, United Kingdom.* ROGOWSKI COILS.

Weill, Prof. Lawrence R. *Department of Mathematics, California State University, Fullerton.* SATELLITE NAVIGATION SYSTEMS.

Weinstock, Dr. Harold. *Air Force Office of Scientific Research, Physics and Electronics, Bolling Air Force Base, Washington, D.C.* SQUID.

Wendler, Prof. Gerd. *Geophysical Institute, University of Alaska, Fairbanks.* WIND.

Whitaker, Prof. Michael. *Department of Physiological Sciences, The Medical School, University of Newcastle, Newcastle upon Tyne, United Kingdom.* FERTILIZATION.

Wilczek, Prof. Frank. *Institute for Advanced Study, Princeton, New Jersey.* ANYONS.

Index

Index

Asterisks indicate page references to article titles.

A

Abalone 116
Abrikosov, A. A. 335
Acanthamoeba 24
Acanthostega 340–341
Accelerator mass spectrometry: radiocarbon dating 288–290*
Acid rain 90
Acoustic cavitation 318
Acoustic monitoring 224–226
Acoustic thermometry 1–3*
   long-range measurement 1
   technical approach 3
   ten-year program 1–2
   two-year demonstration phase 2–3
Acoustic Thermometry of Ocean Climate program 1–3
Acoustics: aircraft noise 6–8*
   psychoacoustics 279–280*
   wavelets 378–380*
Acquired immune deficiency syndrome: drug design 75–76
   medical mycology 185–187*
   Pneumocystis 268–269*
   tumor 354–357*
   vaccine 215
   yeast infection 388–390*
Actinomycetes 27
ACTS *see* Advanced Communications Technology Satellite program
Advanced Automated Directional Solidification Furnace 320
Advanced Communications Technology Satellite program 59–61
   future field trials 60–61
   new systems 61
   user program 59–60
Advanced Light Source 386–387
Advanced Photon Source 386
Advanced X-Ray Astrophysical Facility 384–385
Aerial photography: forest monitoring 132–133
Aerosol, atmospheric 17–19

Afrovenator 71
Agnosia, visual 363–365
Agricultural soil and crop practices 4–6*
   advantages and disadvantages 5
   domestication of plants and animals 274–275
   efficacy assessment 4
   energy from biomass 32–35
   fertilizer 118–121*
   irrigation 162–165*
   plant pathogens 261–262*
   plants in saline environments 263–266*
   soil 311–313*
   special considerations 5–6
   subsoiling methods 4–5
   subsoiling purpose 4
   water resources 370–372*
   whey 380–382*
Agrobacterium 261
AIDS *see* Acquired immune deficiency syndrome
Air pollution: urban area 363
Aircraft: navigation systems 305
   office-in-the-sky 60–61
   satellite-based radio broadcasts 285–286
Aircraft noise 6–8*
   in cabin 7–8
   engine design 6–7
   engine location 7
   levels 6
Albedo: atmospheric aerosols 17–19*
Algae: presence in amber 27–28
Alkaloids 199
All-metal spin transistor 349–352
Allosaurus 73
Alloy: intermetallic materials 158–162*
   shape memory 306–309
Almaz 1 (satellite) 133
Alternating current: Rogowski coils 302–304*
Altimetry 178–180
Aluminides 159–161
Alxasaurus 71

Amanita muscaria 198
Amber: algae in 27–28
   bacteria preserved in 27–29
   insects preserved in 156–158
Ambiortus 71
Ammonium nitrate 120–121
Ammonium polyphosphate 120
Ammonium sulfate 120
Ammonoids, sutures in 110
Amnesia 189–190
Amosite 103
Amphibole 207
Amphotericin B 186–187, 390
Amyotrophic lateral sclerosis 151–152
Angiosperm 156, 293–294
Animal feed 173–174, 380–381
Animal husbandry 274–275
Anomalocaris 8–10*
   distribution 8
   mode of life 9
   morphology 8
   new Chinese examples 8–9
   relationships 10
Anomalocaris canadensis 8
Anoxia: tolerance in Polychaeta 270–271
Antarctica 278–279, 382–384
Antenna (electromagnetism): radio communications 285–286*
Antibody: antigen 10–12*
   autoimmunity 24–27*
   cellular immunology 45–47*
   against plant viruses 262–263
Antidunes 329–330
Antifungal agents 186–187, 390
Antigen 10–12*
   autoimmunity 24–27*
   cellular immunology 45–47*
   DM genes 11–12
   hantavirus 145–147*
   linkage and function 12
   MHC genes and gene products 10–11
   MHC peptide complexes 11
   tumor 354–357*
Antiquark 97–98

Antitumor vaccines 355–357
  evolution 355–356
  future approaches 357
  melanoma 356
  problems and variables of therapy 356–357
Antiviral drugs 74–75
Anyons 12–13*
  bosons and fermions 12
  generalized exclusion principle 13
  properties of matter 12–13
  statistics 13
Aphrodite aculeata 271
Aplasia 32
Aponeuroses 219
Apoptosis 25–27
Appert, N. 129
Applications satellites 13–17*
  atmospheric aerosols 17–19*
  atmospheric water vapor 13–14
  electric power systems 83–85*
  field experiment 15–16
  meteorological signals 14–15
  prospects 16–17
  satellite altimetry 178–180
  satellite navigation systems 304–306*
Aqua ammonia 120
Arc discharge: hazardous waste destruction 149
  met-car formation 212
Archaeopteryx 71
Arthritis 24–25
Arthropodin 30
Arthropods: fossils 289
  insects 155–158*
Asbestos: environmental pollution 102–105*
Asia: space activities 325–326
Astra 1D (satellite) 325
Astrophysics: gamma-ray bursts 142–143*
  studies using radioactive beam 287–288
Atlantis (space shuttle) 322
Atmosphere: applications satellites 13–17*
  comet impact on Jupiter 55–56
  water vapor 13–17
Atmospheric aerosols 17–19*
  assessment of effects 19
  composition 17–18
  gaseous sulfur emissions 18
  link to clouds 17
  various types of emissions 18
Atmospheric Laboratory for Applications and Science 322
ATOC *see* Acoustic Thermometry of Ocean Climate program
Atom cluster: molecular materials 211–213*
Atomic crystal 19–21*
  atom-atom interactions 21
  optical lattice 19
  quantum motion 20–21
  spectroscopy 19–20
Atomic force microscopy 206

Atomic nucleus 21–24*
  fermion dynamical symmetries 22–23
  global and dynamical symmetries 22
  interacting boson model 22
  nuclear shell model 21–22
  unified picture of nuclear structure 23–24
Atomic structure: fundamental interactions 139–142*
Auditory filter 279–280
Augmented Space Launch Vehicle 326
Austenite-martensite transformation 307–308
Australia 278–279
Australopithecus robustus 177
Autoimmunity 24–27*
  activation and apoptosis signaling 27
  defective apoptosis 25–26
  genetic basis of autoimmunity in mice 26–27
  infection and autoimmunity 24
  possible mechanisms 24–25
Automobile: automatic transmission 308
  emergency reporting from passenger cars 68
  navigation systems 304–305
  tire disposal 90, 149
AXAF *see* Advanced X-Ray Astrophysical Facility

B

B cells 354–355
  autoimmunity 24–27*
  cellular immunology 45–47*
Bacillus 28–29, 128
Backfill 359–361
Bacteria 27–29*
  DNA studies 28–29
  flesh-eating 333–334
  presence in amber 27–28
Balanus amphitrite 30
Ballasts: lighting systems 169–171
Barbee, Jr., T. W. 385
Barnacle 29–31*
  cyprid gregarious settlement 30
  cypris larvae 29–30
  ecological implications 30–31
  pheromone recognition 30
  settlement pheromones 30
Bartonella henselae 43–44
Baryonyx 73
Becquerel, H. 318
Bee: *Bacillus* DNA 28–29
Békésy, G. von 279
Big LEOs (satellites) 66
Binnig, G. 202
Biodiesel 34–35
Bioelectronics: molecular magnet 208–211*
Bioindicator: earthworm 83

Biological clocks 31–32*
  cellular and molecular mechanisms 32
  environmental signals to set clocks 31
  photoreceptors and pacemakers 31–32
Biological complexity: evolution 108–110*
Biological control: plant pathogens 261–262
Biomass 32–35*
  biodiesel 34–35
  electricity 35
  ethanol from 34
  ethanol from grain 33–34
Biosensor, DNA-based 70
Biosolids 35–37*
  anaerobic digestion 35–36
  current challenges 36
  rangeland management 37
  recent innovations 36–37
  silviculture 37
Bipolar spin switch *see* Magnetic switch
Bird: dinosaurian ancestry 71, 74
Black hole 143–145
Black smoker 181
Blastomyces dermatitidis 185–187
Blood substitute 214
Bolide impact 72
Bone: dinosaur 72–74
  paleoclimate reconstruction 241–243
  paleodietary analyses 176–177
Borehole 38–40
Boring and drilling (prospecting) 38–40*
  advanced techniques 38–39
  directional drilling in minerals exploration 39–40
  standard technology 38
Borrelia 24, 171–173
Bose-Einstein statistics: anyons 12–13*
Boson 12–13, 22, 98, 141
Bottom quark 141, 283
Boundary-layer flow: extremely high heat flux 112–115*
Brain: headache 150–151*
  memory 189–190*
  Neandertal 221–222
Branchiostoma 64
Brasilsat B1 (satellite) 325
Brayton cycle 301–302
Breast cancer 48
Breathing 295–297
Bridge: acoustic monitoring 224–226
Brockhouse, Bertram N. 228
Bromohydrins 366–367
Brucella 24
Bubonic plague 257
Building: acoustic monitoring 224–226
Bulk-handling machines: flow of solids 121–124*
Bulla 32

Burgess Shale 8–10
Burkholdia cepacia 261
Burnet, F. M. 25
Burst and Transient Source Experiment 142–143

C

Calcium: plant nutrition 265
   wave in fertilization 116–118
Calcium ammonium nitrate 120
Campylobacter 24–25
Cancer (medicine): clinical pathology 47–49*
   detection of minimal residual disease 48–49
   diagnosis 48
   environmental pollution 102–105*
   immunostimulant therapy 197–198
   magnetic resonance imaging 342
   photodynamic therapy 342–343
   prognosis of malignant tumor 48
   tumor 354–357*
Candida 185–187, 388–390
Canine Lyme borreliosis 171–173
Canning technology 129–130
Canopy (forest) 352–354
   dynamics 353
   microclimate 353–354
   vertical organization 352–353
Cantor bar 135
Carbocations 228
Carbon: fullerene 138–139*
   met-cars 211–213
   radiocarbon dating 288–290*
Carbon dioxide: ancient 242–243
   atmospheric 32–33
   forest canopy 354
   supercritical 266–267
Carbon nitride 40–42*
   composition 41
   key properties 42
   preparation of materials 40–41
   structure 41
Carboxylic acids 367
Carcinogens 102–105
Carnivore: teeth 346
Carpal tunnel syndrome 42–43*
   assessment 42
   pathophysiology 43
   prevalence 42–43
   treatment 43
Cassubia 8
Casting: intermetallic materials 161
Cat, plague-infected 257–258
Cat scratch disease 43–44*
   clinical signs 44
   diagnosis 44
   epidemiology 44
   etiology 43–44
   treatment 44
Cathayornis 71
Cattle: domestication 274–275
   mad cow disease 173–174*
Cave bear 177

Cavitation: sonoluminescence 318–320*
CD-ROM *see* Compact disk
Cell membrane: fertilization 115–118*
Cell polarity: embryogenesis 100–101*
Cellular immunology 45–47*, 354
   antigen 10–12*, 45
   B- and T-cell costimulation and crosstalk 46–47
   CD40 ligand 45–46
Cereals 274–275
CFI *see* Computational flow imaging
Chang Zheng 3A (satellite) 326
Chaos: fractal applications 137–138
Charge-spin coupling 349
Chattock, A. P. 303
Cheese: whey 380–382*
Chemical bonding: carbon nitride 40–42*
Chemical furnace 184
Chemotaxis 115–116
Chengliang fauna 8–9
Chicxulub impact structure 72, 138–139
Childbed fever 334
China: space activities 325–326
Chirplet 380
Chitin 289
Chlorination: water treatment 365–366
Chlorine dioxide: water treatment 367
Chlorite 367
Cholesterol: medical testing 188–189
Chrysotile 103
Circadian rhythm: biological clocks 31–32*
Claviceps 199
Clay minerals 207
Clean Air Act (1990) 90
Cleavage (embryology) 100–101
Clementine (satellite) 216–218, 324
Climate: forest canopy 353–354
   magnetism and climate records 251
   modeling with acoustic thermometry 1–3
   paleoclimatology 240–243*
   radiocarbon dating 289–290
   urban heat storage 361–363*
Clinical pathology 47–49*
   cancer 48–49
   cat scratch disease 43–44*
   diagnosis of infectious disease 49
   hyperkalemic periodic paralysis 153
   molecular diagnostic techniques 47–48
   prenatal diagnosis of hereditary disease 49
   yeast infection 388–390*
Clostridium 128–129
Cloud physics 17–19
Coal mining 49–52*
   advanced technology 51–52

Coal mining—*cont.*
   ergonomics 51
   highwall mining 49–50
   highwall mining system extraction 50–51
   prospects 52
   sensors 51
   teleoperation 50
   teleoperator control 51
Coccidioides immitis 185–187
Coelophysis 71
Coercivity: molecular magnet 210–211
Cognition 52–53*
   infant perception 52–53
   methodological advances 52
   physical and social knowledge 53
Coley, W. B. 355
Collagen 219–220
Columbia (space shuttle) 320–321
Columbus Orbital Facility 325
Comet 53–56*
   impact phenomena 55–56
   Jupiter 54–55
   Shoemaker-Levy 9 54–56, 323–324
   unresolved questions 56
Communications satellite 56–61*
   ACTS program 59–61
   ACTS system 59
   data communications 66–68*
   future field trials 60–61
   INTELSAT VII and VIII spacecraft 57
   new systems 61
   radio communications 285–286*
   space flight 320–326*
   user program 59–60
Compact disk 61–63*
   optical storage 61–62
   read-only disks 62
   writable optical disk 63
Complement: inhibitors 198
Composites, intermetallic matrix 162
Compton Gamma-Ray Observatory 142, 324
Computational flow imaging: flow visualization 124–127
Computer: compact disk 61–63*
   drug design 74–76*
   ferroelectric liquid-crystal displays 94–95
   nonvolatile memory 351
   optical computing 236–238*
Conduction (heat): extremely high heat flow 112–115*
   urban heat storage 361–363*
Conodonts 63–64*
   architecture and function 64
   biological affinities 64
   morphology 63–64
Continental drift: paleogeography 243–247*
   prospecting 275–279*
Continuous-wave laser excitation 204
Contrast agents 342
Control systems: electric power systems 83–85*

Convection (heat): extremely high
    heat flow 112–115*
Cooling system: industrial water use
    373
  liquid-jet-impingement cooling
    112–115
Coordination chemistry: metal cluster compound 190–193*
Coriolus versicolor 197
Cosmic censorship hypothesis 145
Crenothrix 27
Cretaceous: fullerene 138–139*
CRISTASPAS (satellite) 322
Cristobalite 104
Crocidolite 103
Crops *see* Agricultural soil and crop practices
Crosstalk: optical storage 62
Crustacean 65–66*
  juvenile hormone 65
  male morphotypes in spider crabs
    65
  reproductive behavior of male
    morphotypes 65–66
  shrimp 309–311*
Cryolophosaurus 71
Cryptococcus neoformans 185–187,
    389
Cryptosporidium 365
Crystal structure: atomic crystal
    19–21*
  intermetallic materials 158–162*
  metal cluster compound 190–193*
  shape memory alloys 306–309*
Cyanobacteria 27
Cypris larvae 29–30
Cyrtolophosis 28
Cytochrome P450 105
Cytokine: autoimmunity 25
  cellular immunology 45–47*
  nitric oxide 226–228*
  treatment of yeast infections 390
  tumor 354–357*

D

Darwin, Charles 81, 111, 258
Data communications 66–68*
  applications 68
  communications satellite 56–61*
  networks 67
  ORBCOMM 66
  position determination 68
  satellites 66–67
  signal routing 67–68
DC-X (launch-and-land demonstrator) 324
Decussation planes 346–347
Deep-etch x-ray lithography 388
Deforestation 133–134
Dehydration: food engineering
    132
Delay line: architecture 237–238
  circuits 237–238
14-Deoxymyriocin 198
Deoxyribonucleic acid (DNA)
    68–70*
  biological relevance 70

Deoxyribonucleic acid (DNA)—*cont.*
  fossil vs. extant microbe studies
    28–29
  hybridization techniques 47
  insect amber fossils 156–158
  molecular light switch 69–70
  molecular transport 215–216
  as molecular wire 68
  photoinduced electron transfer
    68–69
  polymerase chain reaction
    47–48
Department of Defense: space
    activities 324–325
Depth perception 53
Deschamps, O. 308–309
Desertification 18
Deuterium-tritium experiments:
    nuclear fusion 234–235
DFH-3 (satellite) 326
Diabetes 187–188
Diamond 76, 115
  cell 77
Diet: ancient mammals 176–177
  Neandertal 222–223
Differential Global Positioning System 305–306
Digital optical computer 236–238
Dihydroergotamine 150–151
Dimethyl sulfide 18
Dinosaur 70–74*
  bird origins 71
  bone 72–74
  distribution 71
  earliest dinosaurs 70–71
  evolutionary relationships 70
  extinction 72
  footprints 71–72
  metabolism 71
  nests and hatchlings 71
Diphyodonty 345–346
Directional couplers 236–237
Discovery (space shuttle) 320–321
DNA *see* Deoxyribonucleic acid
Dog: locomotor-respiratory integration 296
  Lyme borreliosis 171–173
  plague-infected 257–258
Doppler-tuned spectrometer 140
Double helix 68–70
Dougherty, T. 342
Drilling: boring and drilling
    (prospecting) 38–40*
Drosophila 32, 100
Drug design 74–76*, 387
  anti-influenza drugs 75
  computer-assisted design of
    antiviral drugs 74–75
  future developments 75–76
  migraine therapy 150–151
Dryosaurus 73–74

E

Earth 76–79*
  crustal and mantle minerals 76–80
  experimental mineral transformations 76–78

Earth—*cont.*
  extrapolation and interpretation
    78–79
  mineral reactions 206–208
  perovskite mineral structures 78
Earth crust: metamorphism
    193–196*
  prospecting 275–279*
Earth interior 79–81*
  anomalies 81
  mantle models 80–81
  observation methods 80
  velocity discontinuities 79–80
Earth Observing System 134
Earthworm 81–83*
  bioindicators of environmental
    contamination 83
  role in organic waste management 81–82
  soil amelioration and land reclamation 82–83
Ectoedemia 156
Eggs: activation 116–118
  dinosaur 71
  elongate 100
  embryogenesis 100–101*
  fertilization 115–118*
  size 112
  spherical 100–101
EGRET experiment 142–143
Einstein, A. 143
  field equations 144
Einstein Observatory 384
Eisenia fetida 81–83
Electric field:
  interference microscopy 92–93
  transmission lines 90
Electric power systems 83–85*
  applications 84–85
  communications to central site
    84
  electric protective devices 85–86*
  electric utility industry 86–91*
  electricity from biomass 35
  positive-sequence measurements
    83
  role of measurements 83
  synchronized phasor measurements 83–84
Electric protective devices 85–86*
  adaptive relaying 85–86
  open-system relaying 86
  relay testing 86
Electric utility industry 86–91*
  alternate energy applications 90
  comparability 87–88
  division by plant and utility type
    88
  electric and magnetic fields 90
  integration of IPPs 88
  mergers 89–90
  nitrous oxide regulations 90
  nuclear cancellations 90–91
  power marketers 88
  retail wheeling 86–87
  rotating blackouts 89
  RTG formation 87
  transmission case settled 88–89

Electrical engineering: fractals 134–138
Electron beam fluorescence 124
Electron capture from pair production 141
Electron charge density: metal cluster compound 193
Electron cooling 331
Electron holography 91–93*
  experimental method 91
  high-resolution microscopy 91
  interference microscopy 91–93
Electron interferometry 91–93
Electron spin: magnetic switch 174–176*
  molecular magnet 208–211*
Electron transfer: through DNA 68–70
  photoinduced 68–69
Electronic displays 93–97*
  applications 95
  color and shading 94–95
  ferroelectric liquid-crystal displays 94–95
  nonuniformities 97
  operating principle 94
  static volume systems 96
  status 97
  swept volume systems 96
  volumetric three-dimensional displays 95–97
Electronics: aircraft 7
Elementary particle 97–100*
  diffraction 97
  experimental data 99–100
  fundamental interactions 139–142*
  hadron structure 97–98
  high-energy scattering 98–99
  implications for the strong force 100
  quarks 283–285*
Elephant: ivory source identification 177
Eleutherodactylus 111
Embryogenesis 100–101*
  elongate eggs 100
  evolutionary developmental biology 110–112*
  somitomeres 101
  spherical eggs 100–101
Emu Bay Shale 9
Emulsions: food engineering 131–132
Enamel (tooth) 345–347
Endeavour (space shuttle) 320–322
Endocrine system: crustacean 65–66*
Endometrial cups 101–102*
  function 102
  pregnancy loss 102
  research 102
  role of eCG 102
  structure 102
Endomysium 219–220
Endoplasmic reticulum 117
Energy source: biomass 32–35*
  electric utility industry 86–91*

Engine, aircraft 6–8
Engineering: food 127–132
  reengineering 290–293*
  water supply 372–375
Engineering Test Satellite 325
Environmental management: earthworm 81–83*
  turfgrass 357–359*
Environmental pollution 102–105*
  asbestos minerals and disease 103–104
  crystalline silica and disease 104
  implications 104
  school building exposures 104
  standardized test 83
Enzyme 105–108*
  chemical analysis of the diiron center 106
  implications of the hydroxylase structure 106–107
  methane monooxygenase system 105
  pressure-induced denaturation 128
  related iron-containing proteins 105
  structure of the hydroxylase 106
Eolian transport: snow 383–384
Eoraptor 71
EOS *see* Earth Observing System
Epidemiology: equine motor neuron disease 152
  hantavirus 146
  mad cow disease 174
  plague 257
  Pneumocystis 268–269
Epimysium 218
Epitope 263
Epstein-Barr virus 24–25
Equine chorionic gonadotropin 102
Equine motor neuron disease 151–152
  causative factors 152
  clinical signs 151
  epidemiological findings 152
  pathology 151–152
Erbium-doped fiber amplifier 239–240
Ergonomics: remote operator's control room 51
Ergotamine 150
Erosion: soil 162–165, 311–313*, 370–371, 381–382
  step pools 328–330*
ERS 1 (satellite) 133
Estrogen 102
Estrous cycle 102
Ethanol: from biomass 34
  from grain 33–34
Europe: space activities 325
Eusthenopteron 341
EUVE *see* Extreme Ultraviolet Explorer
Evolution 108–110*
  Anomalocaris 8–10*
  complexity and evolutionary development 108
  definitions of complexity 108

Evolution—*cont.*
  dinosaur 70–74*
  empirical studies 109–110
  evolutionary developmental biology 110–112*
  evolutionary trends 108–109
  implications 110
  insects 155–158*
  Mammalia 176–177*
  Neandertals 221–224*
  Tetrapoda 339–341*
  see also Fossil
Evolutionary developmental biology 110–112*
  evolution 111
  history 110–111
  indirect to direct development 111–112
  phyla 111
Exclusion principle 13
Extreme Ultraviolet Explorer 324
Extreme-ultraviolet projection lithography 387
Extremely high heat flux 112–115*
  heat fluxes 112
  limitations on heat flux 112–114
  liquid-jet impingement 114
  prospects 115

F

Face recognition 364
Faraday rotation 281–282
Farkas, D. 128
Fault zone 196
Federal Energy Regulatory Commission 86–90
Feldspar 207
FERC *see* Federal Energy Regulatory Commission
Fermion 12–13
  atomic nucleus 21–24*
Ferroelectric liquid-crystal displays 94–95
  applications 95
  color and shading 94–95
  comparison of types 94
  operating principles 94
Ferromagnetism 349
  magnetic switch 174–176*
Fertigation 121
Fertilization (reproduction) 115–118*
  calcium wave 116–118
  chemotaxis 115–116
  fusion and activation 116
  gametes 115
  species-specific recognition 116
Fertilizer 118–121*
  biosolids 36–37
  nitrogen 120–121
  nutrients as water pollutants 371
  phosphorus 119
  potassium 118–119
  production 120–121
  types 120
  whey 381–382
  worldwide requirements 119–120

Finite-differencing technique 144
Fish: evolution 339–341
　oxygen isotopes in otoliths 241
Flea: control 258
　plague vector 257–258*
Flow of solids 121–124*
　bin inserts 122–123
　feeders 122
　flow problem 121–122
　impact of flow patterns 122
　measures of flowability 122
　purge and conditioning vessels 123
　quality-control testers 123–124
　tumble blending 123
Flow visualization 124–127*
　benefits 125–127
　development of CFI 124–125
　new formats 127
　practical issues in CFI 124
　visualization methods 124
Flower 293–294
Fluconazole 187, 390
Flucytosine 187, 390
Fluorescence: flow visualization 124–127*
Fluorescence spectroscopy: atomic crystal 20
Fluorescent lamps 168–171
Fomes amnosus 261
Food engineering 127–132*
　dehydration 132
　design 130
　emulsions 131–132
　high hydrostatic pressure 128
　history 128–129
　magnetic resonance imaging 130–132
　mass and energy transport 131
　ohmic heating 129–130
　practical considerations 128–129
　principles 128, 130–131
　processing at ultra-high pressure 127–129
　products 130
　technologies for sterilization 129–130
　temperature profiles in solid foods 132
Food web: Mammalia 176–177*
Forest 132–134*
　data analysis and interpretation 133
　forest assessment 133
　future trends 133–134
　limitations 133
　microwave sensing systems 133
　remote sensing 132–133
　satellite sensing 133
　tree 352–354*
Fossil: *Anomalocaris* 8–10*
　arthropods 289
　bacteria in amber 27–29*
　conodonts 63–64*
　dinosaur 70–74*
　insects 155–158*
　Mammalia 176–177*
　Neandertals 221–224*

Fossil—*cont.*
　paleodietary analyses 176–177
　Tetrapoda 339–341*
　see also Evolution
Fourier series and integrals: wavelets 378–380*
Fractals 134–138*
　application areas 137–138
　construction and dimension 134–136
　electrical engineering examples 136–137
　prospects 138
Frasch process 316–317
Free-radical polymerization: plastics 266–267
Freezing/thawing cycle: soil 312–313
Frenzel, H. 318
Friction: flow of solids 121–124*
Frog: development 111–112
Fullerene 138–139*
　K-T boundary sites 138–139
　occurrences 139
Fundamental interactions 139–142*
　new types of atomic collisions 140–141
　prospects 141–142
　quantum-electrodynamic effects 140
Fungi: medical mycology 185–187*
　microbial compound 196–199*
　plant pathogens 261–262*
　yeast infection 388–390*
Fusarium 199, 260–261

G

G proteins 228–229
Gabor, D. 91, 380
Gadolinium(III) 342
Gaeumannomyces graminis 260
Galileo (space probe) 323
Gamma-ray bursts 142–143*
　Compton Gamma-Ray Observatory 324
　counterparts 143
　distribution on the sky 143
　observation 142
　spectral characteristics 142–143
　time profiles 142
Ganoderma lucidum 198–199
Gasification technology 35
Gate-turnoff thyristor 343–345
Gauge boson 98
Gene expression: during pollen development 293–294
　MHC genes 10–12
Generalized exclusion principle 13
Geoid height 179–180
Geomagnetism 247–251
Geosat (satellite) 179
Germ layers 101
Gigathermy 71
Gill skeleton: *Acanthostega* 340–341
Gilman, Alfred A. 228–229
Ginocchio model 23
Glacial chronology 289–290

Global Positioning System 13–17, 179, 304–306, 324–325
　differential 305–306
　measurement of atmospheric water vapor 15–16
　meteorological signals 14–15
　power-system monitoring 83–85
Glomerulonephritis 334
Glucans 197–198
Glucose: medical testing 187–188
Gluon 97–99
Goats: domestication 274–275
GOES 8 (satellite) 324
Gold: mining 39–40
Gondwana 249–250
GPS *see* Global Positioning System
Graphite 76
　preparation of carbon nitride materials 40–41
Grass: C3 and C4 242
　turfgrass 357–359*
Gravitational collapse 143–145*
　computational techniques 144
　dynamics of general relativity 144
　results of simulations 144–145
Gravity: Moon 217
Green manure 261
Greenhouse effect: acoustic thermometry 1–3*
　applications satellites 13–17*
　atmospheric aerosols 17–19*
GSTAR III (satellite) 285
Guanylate cyclase 226–227

#

H-2 heavy-lift launch vehicle 325
Hadron 97–99
Hall effect: anyons 12–13*
Halophyte 265
Hamiltonian symmetries 22–23
Hammer, M. 291
Hantavirus 145–147*
　characteristics 145–146
　clinical illness 146–147
　epidemiology 146
　preventive measures 147
　pulmonary syndrome 146–147
Harmonic oscillator 20
Hartmanella 24
Hazardous waste 147–149*
　arc discharge 149
　electric discharges 147–148
　existing incineration methods 147
　glow or corona discharge 148–149
Headache 150–151*
　causes 150
　symptoms 150
　treatment 150–151
Heard Island Feasibility Test 1–3
Hearing: psychoacoustics 279–280*
　wavelets 378–380*
Heat balance: acoustic thermometry 1–3*
　urban heat storage 361–363*
Heat-transfer coefficient 113–115
Heinrich event 290
Helmholtz, H. 279

Hemorrhagic fever with renal syndrome 146
Herbicide 371–372
Herbivore: teeth 346
Herding 274–275
Herpes simplex virus 24–25
Herrerasaurus 70–71
Hertz, H. 380
Heteromastus filiformis 270
Higgs boson 141
Highwall mining 49–52
Histocompatibility *see* Major histocompatibility complex
Hite, B. 128
Holography, electron 91–93*
Homeotic mutation 158
Hominid: evolution 177, 223–224
Homo erectus 223–224
Horse 151–153*
   endometrial cups 101–102*
   equine motor neuron disease 151–152
   fossil 242
   hyperkalemic periodic paralysis 152–153
Hospital waste 149
Hubble Space Telescope 56, 322–323
Human chorionic gonadotropin 189
Human-factors engineering: teleoperated coal mining 51
Humoral immunity 354
Hunter-gatherers 222, 274
Huxley, Thomas Henry 110
Hybridization techniques: DNA 47, 49
Hydrocarbon 211–213, 228
Hydrogen: isotopes in bones and teeth 241–242
   isotopes in precipitation 241
Hydroides dianthus 30
Hydrothermal deposit *see* Ocean-floor hydrothermal deposits
20-Hydroxyecdysone 65–66
Hyperkalemic periodic paralysis 152–153
Hypersurface 144
Hypoxia: tolerance in Polychaeta 270–271

I

Iberomesornis 71
Iceberg armada 290
Ichthyostega 339–340
Identical band problem 24
Image compression 137
Image enhancement 125
Image processing: fractals 134–138*
Immunity: autoimmunity 24–27*
   immune system 354–355
   nitric oxide in immune killing 227–228
Immunodiagnostics 189
Immunoglobulin: antibodies against plant viruses 262–263
Immunological deficiency: cat scratch disease 43–44*
   Pneumocystis 268–269*

Immunology: antigen 10–12*
   tumor 354–357*
Immunomodulator 197
Immunostimulants 197–198
   high molecular weight polysaccharides 197–198
   low molecular weight substances 198
Immunosuppression: medical mycology 185–187*
   suppressants 198–199
   yeast infections 388–390*
Immunotherapy 199, 355–356
Impact crater 138
   lunar craters 218
Incineration: hazardous waste 147
India: space activities 326
Industrial health and safety: asbestos-related disease 103–104
   carpal tunnel syndrome 42–43*
   ink 153–155*
   silica-related disease 104
Infant: cognition 52–53*
Infection and autoimmunity 24
Infectious disease: diagnosis 49
   drug design 74–76*
   medical mycology 185–187*
Influenza virus 75
Information processing (psychology): cognition 52–53*
Ingot processing: intermetallic materials 162
Ink 153–155*
   benefits 155
   lithography process 153–154
   solubility conversion 154–155
   water-reducible resins 155
Insects 155–158*
   fossil evidence 155–156
   lepidopteran fossils 156
   modern evidence 158
   phylogenetic interpretation 156
   preservation in amber 156–158
Instrument landing system 305
Instrumentation: coal mining 49–52*
   medical testing 187–189*
   surgery 338
   ultrashort laser technology 168
Insulin 187–188
Integrated circuits: electric protective devices 85–86*
   magnetic switch 174–176*
Integrated services digital network services 59
INTELSAT spacecraft 56–59, 325
Interference microscopy 91–93
Interferon γ 11, 25
Interleukin 25–26, 46, 355
Intermetallic materials 158–162*
   intermetallic matrix composites 162
   ordered structures 158–161
   production 161–162
International Global Aerosol Program 19
International Global Positioning System 13

International Microgravity Laboratory 321
International Space Station 320, 322
International Thermonuclear Experimental Reactor 235–236
Interplanetary matter 313–316
Interstellar matter 313–316
Iridium 138
Iridium system (global communications) 61
Iron aluminides 160
Irrigation (agriculture) 162–165*
   application strategies 164–165
   environmental safety 163
   furrow irrigation 164
   plants in saline environments 263–266
   soil 311–313*
   soil erosion 162–163, 381–382
   soil interactions 163–164
   water conservation 373–374
   water-soluble polyacrylamide 163
Isaria sinclairii 198
ISDN *see* Integrated services digital network services
Isothermal Dendritic Growth Experiment 320
Isotope: nuclear fuels 231–233*
   paleoclimatology 240–243*
   paleodietary analyses 176–177
   solar system 313–316*
Isotopic age 251
Isotopic spin 23
ISP-I (immunosuppressant) 198
ITER *see* International Thermonuclear Experimental Reactor
Itraconazole 187, 390
Ivory: identification of elephant source 177
Ixodes 171–172

J

Japan: space activities 325
JERS 1 (satellite) 133
Jet flow: water-jet cutting 367–370*
Jet noise 6–8
Josephson junction 326
Jupiter: impact of Comet Shoemaker-Levy 9 53–56
Juvenile hormone 65

K

Katabatic wind 382–384
Ketoconazole 390
Kochia scoparia 265
Kondakova, Elena 325
Korean hemorrhagic fever 145–146
Krikalev, Sergei 320
Kumakhov polycapillary optics 384

L

Lactarius flavidulus 199
Land Mass satellites 58–59

Landscape: irrigation 373–374
  precipitation (meteorology) 271–273*
Landsat satellites 133
Laparoscopy 337–339
Large Hadron Collider 141
Larva, cypris 29–30
Laser 165–168*
  applications 167–168
  carbon nitride production 40–42
  high-speed scanning tunneling microscopy 203
  optical fiber amplifiers 238–240*
  photoselective chemistry 251–254*
  time-resolved measurements 166
  ultrashort light pulses 165–166
  ultrashort pulsed lasers 166–167
Laser cooling: atomic crystal 19–21*
Leaf: plant movements 258–260*
Lehman, R. 87
Lentinula edodes 197
Lepidoptera 156
Lepton 283
Leukemia, chronic myelogenous 48
Levy, David 54
Libinia emarginata 65
Ligand shell: metal cluster compound 192
Lighting systems 168–171*
  compact fluorescent lamps 169
  electromagnetic compatibility 168
  electronic ballasts 169
  electronic transformer 168–169
  fluorescent ballasts 169–170
  high-intensity discharge lamps 170
  improving ballast performance 171
  induction lamps 170
Limbs (legs): evolution 110, 341
Ling Zhi-8 (protein) 198
Liquid crystal displays: active matrix 94
  ferroelectric 94–95
Liquids: wave phenomena 375–378*
Listeria monocytogenes 25
Lithography 153–155, 387–388
Little LEOs (satellites) 66–68
Locomotion: *Anomalocaris* 9–10
  dinosaurs 71–72
  locomotor-respiratory interactions 295–297
Logic gate 351–352
Loimia medusa 271
Luch (satellite) 325
Lucianosaurus 70
Lungs: cancer 103–104
  Pneumocystis 268–269*
  respiratory system 295–297*
Luteinizing hormone 189
Lutetium(III) 343
Lyme disease 171–173*
  clinical manifestations 172
  diagnosis 172
  distribution 171
  etiologic agent 171
  prevention 172–173

Lyme disease—*cont.*
  treatment 172
  vectors 171–172

M

Machining: intermetallic materials 162
  water-jet cutting 367–370*
Macroevolution 111
Mad cow disease 173–174*
  clinical signs 173
  diagnosis 173
  epidemiologic intrarelationships 174
  etiology 173
  implications 174
  species susceptibility 174
  transmissible encephalopathies 173–174
Magellan (spacecraft) 323–324
Magic-number cluster 191–192
Magnetic field: interference microscopy 92–93
  Rogowski coils 302–304*
  transmission lines 90
Magnetic materials:
  metal cluster compound 192
  molecular magnet 208–211*
Magnetic resonance imaging: emulsions 131–132
  food engineering 130–132
  mass and energy transport 131
  principles 130–131
  texaphyrins 342
Magnetic susceptibility: metal cluster compound 190–193*
Magnetic switch 174–176*
  bipolar spin device 175
  miniaturization and stability 175
  prospects 176
  spin transistor 351–352
Magnetostratigraphy 251
Maiasaura 71
Main Pass sulfur deposit 317–318
Major histocompatibility complex: cellular immunity 45–47*
  genes and gene products 10–12
  peptide complexes 11
  tumor-specific antigens 357
Malacoceros fuliginosus 271
Malassezia 185, 389
Mammalia 176–177*
  ancient hominids 177
  cave bear 177
  dental requirements 345–347
  elephant ivory sources 177
  future research 177
  locomotor-respiratory integration 295–297
  trace element and isotope systematics 176
  trace elements and isotopes in paleodietary analysis 176–177
Marginocephalian 70
Marine geology 178–182*
  alteration of oceanic crust 181–182

Marine geology—*cont.*
  applications 180
  chimney and mound structures 180–181
  hydrothermal plumes 181
  hydrothermal sites and mineral precipitation 180
  measurements 179–180
  ocean charting 178
  Ocean Drilling Program 181
  ocean-floor hydrothermal deposits 180–181
  ocean surface bumps 178–179
  research gains and future directions 182
  satellite altimetry 178–180
Martensite 307–308
Masking of sound 279–280
Mass flow 122–123
Mass spectrometry 205
Massospondylus 73
Mastotermes 158
Materials processing 182–185*
  combustion synthesis 182
  control of the reaction 183–184
  densification 183
  reaction hot pressing 184
  reaction temperature 183
  reactive sintering 184
  shock compression 184–185
  synthesis of refractory materials 182–183
Materials science: studies using radioactive beams 288
  ultrashort laser technology 167–168
Mating: plug 309–311
  shrimp 310–311
Mawson, D. 382–383
Median nerve: carpal tunnel syndrome 42–43*
Medical imaging: fractal techniques 137
  texaphyrins 341–343*
Medical mycology 185–187*
  diagnosis and treatment 186–187
  microbial compound 196–199*
  opportunistic fungal infections 186
  primary systemic pathogens 186
  yeast infection 388–390*
Medical testing 187–189*
  cholesterol levels 188–189
  glucose measurement 187–188
  immunodiagnostics 189
Megaevolution 111
Meissner effect 335
Melanoma 356
Memory 189–190*
  future classification 190
  procedural and declarative memories 189–190
  semantic and episodic memory 190
Memory element: spin transistor 351
MEPHISTO furnace 320
Meson 141

Mesothelioma 103
Met-cars see Metallocarbohedrenes
Metabolism: dinosaur 71
   metabolic switching in polychaetes 271
Metal cluster compound 190–193*
   evolution to metallic properties 192–193
   giant magic-number clusters 191–192
   large clusters 191
   monodisperse particles 191
Metallocarbohedrenes 211–213
   binary metal met-cars 212–213
   composition and structure 211–212
   mechanisms of formation 212
   status and prospects 213
Metallogeny 275–277
Metamorphism 193–196*
   interrelationship with deformation 194
   mechanistic relationships 195–196
   spatial and temporal relationships 194–195
Metamorphosis: crustacean 65–66*
Meteorite: fullerene 138–139*
   solar system 313–316*
Meteorology: applications satellites 14–15
   space flight 320–326*
Methane: oxidation to methanol 105–108
Methane monooxygenase: enzyme 105–108*
Methanol: methane oxidation to 105–108
Methanotroph 105–108
Methyl farnesoate 65–66
Methylococcus capsulatus 105–106
MHC see Major histocompatibility complex
Microbial compound 196–199*
   immunostimulants 197–198
   immunosuppressants 198–199
   immunotherapy 199
Microevolution 111
Microgravity Payload 320
Microorganisms: in amber 27–29
   autoimmunity 24–25
   bacteria 27–29*
   biosolids 35–37*
   inactivation in food processing 128
   soil 313
   water-borne disease 365–367
Microprobe sampling methods 205–206
Microscope 199–204*
   design 200–201
   examples of imaging 201
   high-speed scanning tunneling microscopy 202–204
   near-field optical microscopy 199–202
   near-field optics 201–202
   optical modulation of samples 203–204

Microscope—*cont.*
   optical switching of electrodes 204
   principles 199–200
   principles of STM 202–203
   use of lasers 203
Microwave sensing system: forest monitoring 133
Mid-Oceanic Ridge 180–181
Migraine headache 150–151
Milankovitch, M. 289
Milk: whey 380–382*
MILSTAR communications satellites 324
Mineralogy 204–208*
   atomic arrangements 206
   chemical composition 205–206
   mineral surfaces 206
   reactions in the Earth 206–208
Minerals: directional drilling in exploration 39–40
   Earth 76–79*
   hydrothermal deposits 180–182
   metamorphism 193–196*
   mineral dust aerosols 18
   prospecting 275–279*
Miniature Sensor Technology Integration 324
Miniaturization 175, 347–352
Mining: coal 49–52
   solution 316–318
   underground 359–361
Mir space station 325
Mission to Plant Earth 322, 324
Molecular beam epitaxy 281
Molecular light switch 69–70
Molecular magnet 208–211*
   design strategy 208
   ferromagnetic ordering of one-dimensional ferrimagnets 208
   interlocking of two-dimensional networks 208–210
   large coercivity 210–211
Molecular materials 211–213*
   binary metal met-cars 212–213
   mechanisms of formation 212
   met-car composition and structure 211–212
   status and prospects 213
   structures 212
Molecular mimicry: autoimmunity 24–25
Molecular transport 213–216*
   materials 214
   present and future applications 214–216
   self-assembly 213–214
   stabilizing film 214
   surface interactions 214
Molting, crustacean 65–66
Molybdenum disilicide 161
Monoammonium phosphate 120
Monoclonal antibodies: against plant viruses 262–263
Monoculture 260–261
Monolophosaurus 71
Mononykus 71
Moon 216–218*
   satellite 216

Moon—*cont.*
   shape and internal structure 216–217
   soils 316
   surface features 217–218
Mountain meteorology 271–273
Mouse: genetic basis of autoimmunity 26–27
   moth-eaten 26
   T-cell receptor transgenic mouse 25–26
MRI see Magnetic resonance imaging
Mulch 261
Multilayer optical storage: compact disk 61–63*
Multilayer x-ray optics 385–386
Multiple-bubble sonoluminescence 319
Multiple sclerosis 24–25
Murine hepatitis virus 24
Muscular system 218–220*
   aponeuroses 219
   eccentric contractions 219
   geometric models 219–220
   hyperkalemic periodic paralysis 152–153
   nitric oxide 226–228*
   shear properties 220
   strain energy 219
Musical acoustics: wavelets 378–380*
Mutagens 102–105
Mycobacterium tuberculosis 25
Myelin basic protein 24–25

Nacelle 6
Nassula 28
NAVFAC array 2–3
Neandertals 221–224*
   anatomical characteristics 221–222
   artifact assemblages 223
   geographic and geological associations 221
   origin 223
   relationship to *Homo sapiens* 223–224
   sites 222
   social behavior 223
   subsistence 222–223
Near East: domestication of animals 274–275
Near-field optical microscopy 199–202
   examples of imaging 201
   microscope design 200–201
   near-field optics 201–202
   principles 199–200
Nephtys 271
NERVA program 300–302
Nervous system (vertebrate):
   equine motor neuron disease 151–152
   nitric oxide 226–228*
Nest: dinosaur 71

Neural network 224–226*
    artificial neural network 224–225
        data collection 225
        fault detection 225–226
        prospects 226
Neuroblastoma 48
Neuropeptide Y 150
Neurospora 32
Neurotransmitter 227
Neutron-capture reactions 287–288
Neutron diffraction 228
Neutron halo 287
Nickel aluminides 159–160
Nickel-titanium alloy 308–309
Nitrate 226, 371
Nitric oxide 226–228*
    nitric oxide synthetase 226
    physiological actions 226–228
Nitrogen: fertilizer 120–121, 371
Nitrogen oxides 17, 90, 148
Nival system 289
NOAA 14 (satellite) 324–325
Nobel prize 228–229*
    chemistry 228
    physics 228
    physiology or medicine 228–229
Noble-gas doping: sensitivity of sonoluminescence 320
Noise, aircraft 6–8
Nondestructive testing 229–231*
    fast neutron interrogation 229
    PFNA operation 231
    PFNA system 229–231
    SQUID 326–328*
Nonrelativistic quantum theory:
    atomic crystal 19–21*
    atomic nucleus 21–24*
North America: paleogeography 246–247
Nuclear force 97–100
Nuclear fuels 231–233*
    HEU conversion to LEU 232–233
    weapons-grade uranium 232
Nuclear fusion 233–236*
    deuterium-tritium experiments 234–235
    fusion process 233
    prospects 235–236
    tokamak development 234
    tokamak system 233–234
Nuclear magnetic resonance: food engineering 130–132
    mineralogy applications 206
Nuclear power industry 90–91
    risk analysis 297–299*
Nuclear reactor 232–233
    heterogeneous 302
    rocket propulsion 299–302*
Nuclear Regulatory Commission 297–298
Nuclear shell model: atomic nucleus 21–22
Nucleation: atmospheric aerosols 17–19*
Nucleic acid *see* Deoxyribonucleic acid
Nucleon 287
Nyctinastic leaf movements 258

O

Object recognition 363–364
Occlusion (teeth) 345
Ocean: acoustic thermometry 1–3*
    marine geology 178–182*
Ocean Drilling Program 181–182
Ocean-floor hydrothermal deposits 180–182
    alteration of oceanic crust 181–182
    chimney and mound structures 180–181
    hydrothermal plumes 181
    hydrothermal sites and mineral precipitation 180
    Ocean Drilling Program 181
    research gains and future directions 182
Ohm, G. 279
Ohmic heating 129–130
    history 129
    products 130
    technologies for sterilization 129–130
Olah, George A. 228
Oligochaeta: earthworm 81–83*
Oligothiophenes 347–349
Oncogene 47–48
Opabinia 10
Opportunistic infection: medical mycology 185–187*
    Pneumocystis 268–269*
    yeast infection 388–390*
Optical communications: optical fiber amplifiers 238–240*
    quantum wells 281–283*
Optical computing 236–238*
    analog optical computers 236
    delay-line architecture 237–238
    delay-line circuits 238
    demonstration computer 238
    directional couplers 236–237
Optical fiber amplifiers 238–240*
    development 239–240
    impact of EDFAs 240
    types 239
Optical isolator 281–282
Optical lattice 19–21
Optical microscope 199–202
Optical pulse: ultrashort light pulses 165–166
Optical storage: compact disk 61–63*
Optics, x-ray 384–386
ORBCOMM (satellite system) 66–68
Orbital Reentry Experiment Vehicle 325
Organ transplantation: immunosuppressants 198–199
    Pneumocystis 268–269*
Organic farming 261
Organic semiconductors 347–349
Organic waste: breakdown by earthworms 81–82
Ornithopod 70
Orodromeus 71

Orogeny: metamorphism 193–196*
Orthodontic braces 309
Otolith: paleoclimate reconstruction 241
Oxygen isotopes: in bones and teeth 241–242
    in fish otoliths 241
    in precipitation 241
Ozone: water treatment 366–367

P

Pacemaker: biological clocks 31–32
Pachyrhinosaurus 73
Pacific Ocean: acoustic thermometry 1–3*
Pain: headache 150–151*
Paleoclimatology 240–243*
    ancient carbon dioxide 242–243
    isotopes in continental precipitation 241
    oxygen and hydrogen isotopes in bones and teeth 241–242
    oxygen isotopes in fish otoliths 241
Paleogeography 243–247*
    implications 247
    North America 246–247
    paleomagnetism 247–251*
    Pangea 243
    Rodinia 243–246
Paleomagnetism 247–251*
    basic principles 247–248
    growth and deformation of continents 250
    magnetism and climate records 251
    magnetostratigraphy and isotopic age 251
    paleogeography 243–247*
    paleomagnetic reconstructions 248–250
    records in rocks 248
    remagnetization 250–251
Pangea 243
Paracoccidioides brasiliensis 185–187
Paraconodonts 64
Paramecium 28
Paraprionospio pinnata 270–271
Parinaud's oculoglandular syndrome 44
Particle accelerator: elementary particle 99–100
    fundamental interactions 139–142*
    quarks 283–285*
    radioactive beams 286–288*
    storage rings 330–333*
Particle image velocimetry 124–125
Particle trap: atomic crystal 19–21*
Parton 98–100
Pas 2 (satellite) 325
Patagopteryx 74
Pathology, clinical pathology 47–49
PCB *see* Polychlorinated biphenyls
Penaeoid shrimp 309
Penaeus 310–311

Penicillin 334
Penicillium 199
Penrose, R. 145
Peptide transporter 11
Perimysium 218–219
Perovskite 78–79
Pesticide 371–372
Petrography: metamorphism 193–196*
Peytoia 8–9
Peziza vesiculosa 197
PFNA *see* Pulsed fast neutron analysis
Phanerozoic ore deposits 277–278
Pharyngitis 333–334
Pheromone: barnacle 30
Phialophora 261
Phosphate rock 119
Phosphorus: fertilizer 118–120, 371
Photochemistry 167–168
Photodynamic therapy 342–343
Photoinduced electron transfer: deoxyribonucleic acid (DNA) 68–70*
Photoperiodism: biological clocks 31–32*
Photoselective chemistry 251–254*
   excited electronic states 252–253
   laser chemistry 252
   perspectives 254
   selective excitation 251–252
   specificity and intramolecular vibrational energy redistribution 253–254
Photovoltaic plant 90
Phyla 111
Phylogenetic relationships: *Anomalocaris* 10
   dinosaur 70–74*
   insects 156
Phytophthora parasitica 198
Pigs: domestication 274–275
Pineal gland 32
Pipeline 255–256*
   off-line replacement 256
   on-line replacement 255–256
PIV *see* Particle image velocimetry
PIXE *see* Proton-induced x-ray emission
Plague 257–258*
   clinical symptoms 257
   diagnosis 257–258
   epidemiology 257
   infectious agent 257
   pathogenesis 257
   prevention and control 258
Planar laser-induced fluorescence 124–127
Plant mineral nutrition: fertilizer 118–121*
   plants in saline environments 263–266
Plant movements 258–260*
   genotype variability 259–260
   heliotropic movements 258–259
   pulvinus 259
   relation to productivity 260

Plant movements—*cont.*
   response to environmental factors 259
Plant pathogens 260–262*
   biological agents 261
   implications 261
   induction of suppression 261
   mechanisms of disease suppression 260–261
   plant virus 262–263*
   suppressive soils 260
Plant virus 262–263*
   monoclonal and polyclonal antibodies 262
   uses 262–263
Plants, saline environments of 263–266*
   salt tolerance 264
   sulfate and chloride tolerance 264–266
Plasma physics: nuclear fusion 233–236*
Plasma torch 149
Plastics 266–268*
   heterogeneous free-radical polymerizations 267
   homogeneous free-radical polymerizations 266–267
   prospects 268
   supercritical carbon dioxide 266
   various polymerizations 267–268
Plate tectonics: Earth interior 79–81*
   paleogeography 243–247*
   prospecting 275–279*
   satellite altimetry 180
PLIF *see* Planar laser-induced fluorescence
Pneumocystis 268–269*
   epidemiology 268–269
   the organism 268
   pathology 269
   prevention 269
   treatment 269
Pneumonia: *Pneumocystis* 268–269*
Pneumonic plague 257
Podarkeopsis levifuscina 271
Polar Satellite Launch Vehicle 326
Polaron 282
Pollen 293–294
Pollution: air 363
   environmental 102–105
   water resources 370–372*
Polyacrylamide: soil treatment 163–165
Polychaeta 270–271*
   adaptations 270–271
   population dynamics 271
   tolerance to hypoxia and anoxia 270
Polychlorinated biphenyls 147–149
Polymer: plastics 266–268*
   polyacrylamide 162–165
   thin-film transistors 347–349
Polymerase chain reaction 47–49
Polyphyodonty 345
Polyphysia crassa 270–271

Polysaccharides: immunostimulants 197–198
Pomeron 97–100
Porphyrin: texaphyrins 341–343*
Porphyroblasts 194–195
Potassium: fertilizer 118–120
   isotopes in solar system 315–316
   mining 118–119
   plant nutrition 265
Powder metallurgy: intermetallic materials 162
Power systems: electric power systems 83–85
   electric protective devices 85–86*
   electric utility industry 86–91*
Praseodymium-doped fiber amplifier 240
Precipitation (meteorology) 271–273*
   forest canopy 353
   isotopes in 241
   large-scale impacts 272
   microscale impacts 273
   small-scale effects 272–273
   wind-terrain interactions 272
Precipitation polymerizations 267
Precise Positioning Service 304
Predation: *Anomalocaris* 8–10*
Pregnancy: horse 101–102
   test 189
Prehistoric domestication of animals 274–275*
   Near Eastern domesticates 274–275
Pressure-sensitive paint 124–127
Printing: ink 153–155*
Prion 173–174
Priroda satellites 133
Prism decussation 346–347
Process reengineering 291
Process to significantly reduce pathogens 36–37
Product reengineering 291–292
Profile analysis: psychoacoustics 279–280
Progesterone 102
Projectile fragmentation technique: production of radioactive beams 286–287
Prosopagnosia 364
Prospecting 275–279*
   boring and drilling 38–40*
   controls on ore deposits 277–278
   metallogeny 275–277
   proposed Australia-Antarctic connections 278–279
Protein: reactions in food processing 128
Proteosome 11
Proterozoic provinces (southwestern United States): metallogeny 275–279
Protoavis 71
Protoceratops 73
Proton-capture reactions 287–288
Proton-induced x-ray emission: mineralogy applications 205
Protorthoptera 155–156

Protozoa: presence in amber 28
Pseudeurythoe paucibranchiata 271
Pseudomonas fluorescens 261
Psychoacoustics 279–280*
   complex sounds 279–280
   simple sounds 279
   wavelets 378–380*
Public health: drinking water 366
   environmental pollution 102–105*
Pulsed fast neutron analysis: nondestructive testing 229–231*
Pulsed laser excitation 203–204
Pulvinus 259
Pump: water-jet cutting 368
Purge vessel 123
Pyroxene 207

Q

Quality control: flow of solids 123–124
Quantum chromodynamics 97–100
Quantum confinement 211
Quantum dots 282–283
Quantum electrodynamics: fundamental interactions 139–142*
Quantum mechanics: atomic crystal 19–21*
Quantum-size effects 190–191
Quantum statistics: anyons 12–13*
   atomic nucleus 21–24*
Quantum wells 281–283*
   diluted magnetic semiconductors 281
   magnetic wells 281–282
   quantum dots 282–283
   well structure 281
Quarks 97–99, 141–142, 283–285*
   experimental apparatus 283–284
   producing the top quark 283
   prospects 285
   top-quark search 284–285
Quartz 76, 104
Quasiparticle: anyons 12–13*

R

RADARSAT (satellite) 133
Radio communications 285–286*
   aircraft equipment 285–286
   system operation 285
   system reliability 286
Radioactive beams 286–288*
   materials science studies 288
   nuclear astrophysics studies 287–288
   nuclear structure studies 287
   production 286–287
Radiocarbon dating 288–290*
   climatic change 289–290
   fossil arthropods 289
   glacial chronologies 289–290
Rain shadow 271–273
Rainfall *see* Precipitation (meteorology)
Raman spectroscopy 20
Rangeland conservation 37

Rare-earth doped fiber amplifiers 239–240
Rat: plague 257–258*
Rathole (flow problem) 121
Rayleigh scattering 124
Receiver array: acoustic thermometry 3
Reengineering 290–293*
   process 291
   product 291–292
   systems-management 291
Refractory materials: synthesis 182–183
Regge theory 99
Relativistic Heavy Ion Collider 141
Relativity: gravitational collapse 143–145*
Relay 84–86
Remagnetization 250–251
Remote sensing: forest monitoring 132–133
   space flight 320–326*
Reproduction (animal):
   horse 101–102
   shrimp 309–311
   spider crabs 65–66
Reproduction (plant) 293–294*
   gene expression during pollen development 293–294
   interactions in self-incompatibility 294
   morphological events of pollen development 293
   pollen tube structure and growth 294
Reptile: locomotor-respiratory conflict 295
Resact peptide II 116
Resin: lithographic ink 155
Respiratory system 295–297*
   hantavirus pulmonary syndrome 146–147
   implications 297
   locomotor-respiratory conflict in reptiles 295
   locomotor-respiratory coupling 295–297
Reverse engineering 292
Revueltosaurus 70
Rhabdodon 73
Rheumatic fever 334
Rheumatoid arthritis 24
RHIC *see* Relativistic Heavy Ion Collider
Rhizoctonia solani 261
Riffle pools 330
Rioaribasaurus 71
Risk analysis 297–299*
   application to surveillance testing 298
   deterministic regulation 297
   distribution of resources 298
   probabilistic safety assessment 297–298
   risk-based regulation 297
   win-win situations 298–299
River: step pools 328–330*
Rock fill 359

Rock magnetism 247–251
Rock phosphate 119
Rocket propulsion 299–302*
   heterogeneous reactors 302
   nuclear thermal rocket engines 299–302
   space-flight cost 299
   specific impulse 299
Rocks: evidence for Rodinia 244–246
Rodbell, Martin 228–229
Rodent: *Borrelia burgdorferi* reservoir 171
   hantavirus reservoir 145–147
   plague 257–258*
Rodinia 243–246
Roentgen, W. 386
Rogowski, W. 303
Rogowski coils 302–304*
   applications 303–304
   development 303
   features 302–303
   practical systems 303
Rohrer, H. 202
Root: growth 313
Russia: space activities 325
Rutherford, E. 97–98

S

Saccharomyces cerevisiae 197
Saline environment, plants in 263–266
Satellite: applications satellites 13–17*
   communications satellite 56–61*
   forest monitoring 133–134
   radio system 285–286
Satellite altimetry 178–180
   applications 180
   measurements 179–180
   ocean charting 178
   ocean surface bumps 178–179
Satellite navigation systems 304–306*
   attitude determination 306
   aviation guidance 305
   differential GPS 305–306
   Global Positioning System 304
   personal navigation 304
   scientific applications 306
   space flight 320–326*
   vehicular applications 304–305
Sauropodomorph 70
SCADA *see* Supervisory control and data acquisition
Scanning tunneling microscopy 176, 202–204, 206
Scattering from fractal surfaces 136
Schizophyllum commune 197
School building: asbestos 104
Sclerosis, amyotrophic lateral 151–152
Scrapie 174
Scytonema 27
Sea-floor spreading 243
Sea ice 383
Sea urchin 111–112

Secondary ion mass spectrometry 205
Sedentism 274–275
Sedimentation: eroded soil 370–371
Segmentation: vertebrate body 101
Segregation (flow problem) 121
Seismic energy: frequency content 81
Seismology: Earth interior 79–81*
Seismonastic leaf movements 258
Self-assembly: molecular transport assembly 213–214
Self-incompatibility: plants 294
Self-phase modulation 167
Semantic memory 190
Semibalanus balanoides 30
Semiconductor: quantum wells 281–283*
   transistor 347–352*
Semmelweis, I. 334
Sensor: teleoperated coal mining 51
Sepsis: nitric oxide in 227
Septicemic plague 257
Serological assays: plant viruses 263
Serotonin agonist 150–151
Sessile animal: barnacle 29–31*
Settlement behavior: barnacle 29–31*
Sewage: biosolids 35–37*
   earthworm treatment 81–83
   pipeline 255–256*
Shape memory alloys 306–309*
   applications 308–309
   attributes 306–307
   austenite-martensite transformation 307–308
Shear zone 196
Sheep: domestication 274–275
Sheet silicate 207
Shipping container: nondestructive inspection 229–231
Shock compression 184–185
Shock wave: sonoluminescence 318–320*
Shoemaker, Carolyn 54
Shoemaker, Eugene 54
Shrimp 309–311*
   insemination and sexual strategies 311
   mating 310–311
   penaeoid 309
   reproductive morphology 309
   spermatophore 309–310
Shull, Clifford G. 228
Shultes, H. 318
Shuvosaurus 71
Sialidase 75
Sicyonia 311
Sierpinski gasket 135–136
Signal processing: fractal techniques 137
Silica dust: environmental pollution 102–105*
Silicate minerals 76–79
Silicides 161
Silicon-controlled rectifier 343–345
Silsbee, R. 349

Silt tillage 5
Silviculture 37
SIMS *see* Secondary ion mass spectrometry
Simulation: flow visualization 124–127*
Singularity 144–145
Sinornis 71
Sintering 184
Skeleton: conodonts 63–64
   evolution 339–341
Skull: Neandertal 222
Skylab (space station) 384
Slime mold: presence in amber 28
Slow neutron spectroscopy 228
Smart cards 348
Smart materials 308–309
Smuggling: nondestructive testing 229–231*
Sneak mating 66
Snow: eolian transport 383–384
   snowfields 289
Snow, J. 365
Social behavior: Neandertals 223
Sodic soil 381
Sodium channel 153
Soft x-ray fluorescence spectroscopy 388
Soil 311–313*
   aggregate stability 311–312
   biological factors 313
   cultural factors 313
   earthworm 81–83*
   erosion 162–165, 370–371
   erosion control 381–382
   heat flux 361–362
   lunar soils 316
   physical and chemical factors 312–313
   plant pathogens 260–262*
   plants in saline environments 263–266*
   sodic soil reclamation 381
   subsoiling 4–6
   suppressive soils 260
   temporal variation 312
   turfgrass 357–359*
   urban heat storage 361–363*
   see also Agricultural soil and crop practices
Solar energy 90
Solar nebula 314
Solar radiation: forest canopy 353–354
   plant movements 258–260*
Solar system 313–316*
   lunar soils 316
   processes of formation 315–316
Solid-state chemistry: carbon nitride 40–42*
Solidaridad 2 (satellite) 325
Solids, flow of 121–124
Soliton 240
Solubility: ink 154–155
Solution mining 316–318*
   Frasch process 316–317
   Main Pass mine 317–318

Solution mining—*cont.*
   occurrence of sulfur 316
Solvent: lithographic industry 153–154
   plastics production 266–268
Somites 101
Somitomeres 101
Sonoluminescence 318–320*
   acoustic cavitation 318
   multiple-bubble sonoluminescence 319
   single-bubble sonoluminescence 319–320
   sonochemistry 318–319
Soot-rich clays 138–139
Southern blotting 47
Space, infant perception of 53
Space flight 320–326*
   Asian space activities 325–326
   astronomical observations 322–324
   commercial space activities 325
   cost 299
   Department of Defense activities 324–325
   European space activities 325
   International Space Station 322
   Russian space activities 325
   space shuttle 320–322
   space station *Mir* 325
   United States space activity 320–325
Space probe 320–326
Space Radar Laboratory 320–321
Space shuttle 320–322
Space station 322, 325
Space Test Experiment Platform 324
Spaceway (satellite) 61
Spaerotilus 27
SPARTAN 201 (satellite) 321
Spatial dither technique 95
Specific impulse 299
Spectromicroscopy 388
Spectroscopy: atomic crystals 19–20
   carbon nitride 41
   single-bubble sonoluminescence 319–320
Speech perception: wavelets 378–380*
Sperm: fertilization 115–118*
Spermatophore 309–311
Sphingomyelinase 27
Spider crabs: male morphotypes 65–66
Spiller, E. 385
Spin transistor 349–352
   applications 351–352
   ferromagnetic and nonmagnetic metals 349
   spin injection 349–350
   transistor action 350–351
Sporophytic self-incompatibility 294
Sports facility: turfgrass 357–358
SPOT satellites 133
Squeezed quantum state: atomic crystal 19–21*

SQUID 175, 326–328*
   application of HTS SQUID 327–328
   development of HTS SQUID 327–328
   practical considerations 327
   superconductivity 335–337*
SROSS C2 (satellite) 326
Stachybotrys complementi 198
Standard Positioning Service 304
Stanford/MSFC Rocket X-Ray Spectroheliograph 385
Staphylococcus aureus 25
Steinhaus, W. 303
Stemonites 28
Step pools 328–330*
   broader significance 330
   characteristics 329
   functional importance 329
   occurrence 328
   processes of formation 329–330
Sterilization: foods 129–130
Steyla 101
Stone tools 223
Storage rings 330–333*
   dissociative recombination 330–331
   experiments 332–333
   use of storage rings 331–332
   x-rays 386–388*
Strain energy: muscular system 219
Stream: agricultural impacts 371
   step pools 328–330*
Streblospio benedicti 270–271
Strepsiptera 158
Streptococcus 333–335*
   resurgence of rheumatic fever 334
   streptococcal toxic shock–like syndrome 333–334
   treatment and prevention 334
Streptomyces scabies 261
Stroke 227
Subsoiling *see* Agricultural soil and crop practices
Substance P 150
Sudbury impact crater 139
Sulfate: plant tolerance 264–266
Sulfur: aerosols 17–18
   mining 316–318
   occurrence 316
Sulfur dioxide 148
Sumatriptan 150–151
Sun compass 31
Superantigen 25
Superconducting quantum interference device *see* SQUID
Superconductivity 167–168, 335–337*
   effect of weak disorder 336–337
   flux-lattice melting 335–336
   flux-line pinning 335
   flux pinning by columnar defects 337
   Meissner effect and mixed phase 335
   SQUID 326–328*
   vortices in superconductors 93
Supercontinent 243–246

Supercritical fluids: polymerizations involving 266–268
Supertwist liquid-crystal displays 94
Supervisory control and data acquisition 68
Suprachiasmatic nucleus 31–32
Surface capturing 377
Surface fitting 375
Surface physics: microscope 199–204*
Surface tracking 376–377
Surgery 337–339*
   access 337–338
   instrumentation 338
   operations 338–339
   visualization 338
Surveillance testing: nuclear power plants 298
Swimming: *Anomalocaris* 9–10
SXRF *see* Synchrotron x-ray fluorescence microprobe
Symmetry laws (physics): atomic nucleus 21–24*
Synchrotron radiation: x-rays 386–388*
Synchrotron x-ray fluorescence microprobe 205
Syntarsus 73
Synthetic aperture radar 133
Systemic lupus erythematosus 26
Systems engineering: reengineering 290–293*

T

T-cell receptor 24–27, 45–46
T cells 354–355
   autoimmunity 24–27*
   cellular immunology 45–47*
Tadpole 111
Tail: *Acanthostega* 341
TAP genes 11
Taxonomy: dinosaur 70–74*
Technosaurus 70
Tecovasaurus 70
Teeth *see* Tooth
Teleoperation: coal mining 49–52*
Telstar 402 (satellite) 325
Temporal dither technique 95
Tendon 219
Terrestrial radiation: atmospheric aerosols 17–19*
Terrestriality 339–341
Tertiary: fullerene 138–139*
Tetrapoda 339–341*
   Acanthostega 340–341
   fish-to-tetrapod transition 339
   Ichthyostega 339–340
   implications 341
   Tulerpeton 341
Texaphyrins 341–343*
   magnetic resonance imaging 342
   photodynamic therapy 342–343
   prospects 343
TFTR *see* Tokamak Fusion Test Reactor
Thaicom 2 (satellite) 325

Theiler's murine encephalitis virus 24
Thermal explosion 183
Thermite reactions 182
Thermometry, acoustic 1–3
Theropod 70–71, 73
Thin-film transistor 347–349
   advantages of organic TFTs 348
   all-polymer device 348–349
   control of molecular organization 348
   prospects 349
   steps toward organic devices 347–348
Thoracoscopy 337–339
Three-dimensional displays 95–97
Thyreophoran 70
Thyristor 343–345*
   applications 344
   MCT development 344–345
   types 343–344
Tick: Lyme disease 171–173*
   repellent 172
Tillage 4–6, 313
Time-delay neural network 225
Titanium aluminides 160–161
Tokamak: Fusion Test Reactor 234–235
   Physics Experiment 235–236
   system 233–235
Tolpocladium inflatum 198
Tooth 345–347*
   conodonts 63–64*
   mammalian dental requirements 345–347
   paleoclimate reconstruction 241–243
   paleodietary analyses 176–177
   precise occlusion 345
Top quark 283–285
Topography: Moon 216–217
Total Ozone Mapping Spectrometer 324
Toxic shock–like syndrome 333–334
Trachypenaeus similis 310–311
Transistor 347–352*
   advantages of organic TFTs 348
   all-metal spin transistors 349–352
   all-polymer device 348–349
   all-polymer thin-film transistors 347–349
   applications 351–352
   control of molecular organization 348
   ferromagnetic and nonmagnetic metals 349
   prospects 349
   spin injection 349–350
   steps toward organic devices 347–348
   transistor action 350–351
Transition metal: met-cars 211–213
Transition metal-based polymerization 268
Transmission electron microscopy 207
Treadwell mine 39–40

Tree 352–354*
 canopy dynamics 353
 components 352
 forest 132–134*
 microclimate 353–354
 vertical organization 352–353
Trentepohlia 28
Triassamoeba alpha 28
Trichoderma 261
Trichosporon beigelii 389
Trichothecenes 199
Tridymite 104
Trihalomethanes 366
Trilobita: *Anomalocaris* 8–10*
Trimethoprim-sulfamethoxazole 269
Troodon 73
Truck: nondestructive inspection 229–231
 tracking and monitoring traffic 68, 305
Trypanosoma 24
Tulerpeton 341
Tumor 354–357*
 antitumor vaccines 355–357
 immune system 354–355
Tumor-associated antigens 355–357
Tumor necrosis factor α 25
Tumor-specific antigens 357
Turbofan: aircraft noise 6–8*
Turfgrass 357–359*
 environmental benefits 358–359
 psychological and physical well-being 357–358
 safety role 359
Turksat IB (satellite) 325
Tyrannosaurus 73

U

Ultrafast molecular processes: laser 165–168*
Ultrashort pulsed laser 165–168
Ulysses (spacecraft) 323
Underground mining 359–361*
 backfill technological innovations 360–361
 functions of backfill 359–360
 types of backfill 359
Underwater sound 1–3
United States: space activity 320–325
Uranium: nuclear fuels 231–233*
Urban heat storage 361–363*
 importance 363
 methods for determination 361–362
 urban areas 361
Urea: fertilizer 120–121

V

Vaccine: antitumor vaccines 355–357
 molecular transport assembly 215
Vapor lamp: lighting systems 168–171*
Vasodilator: nitric oxide 227
VEGETATION instrument 133

Vertebrata: conodonts 63–64*
 evolution 110
Very large scale integrated circuits technology 85
Vibrio parahaemolyticus 129
Video disk recording: compact disk 61–63*
Virus: antiviral drug design 74–76
 hantavirus 145–147*
 plant virus 262–263*
 Visna 24
Visual agnosia 363–365*
 affected regions of the brain 364–365
 object recognition 363–364
 prosopagnosia 364
 visual processing 364
Vitamin E 152
Vitrification 149
VLSI *see* Very large scale integrated circuits technology
Volcano: aerosols from eruptions 18
 undersea 180–182
Volumetric three-dimensional displays 95–97
von Békésy, G. 279
Voyager (space probe) 324

W

Wake Shield Facility 320
Waste management: biosolids 35–37*
 hazardous waste 147–149*
 using earthworms 81–83
Water 365–367*
 alternative treatments 366
 conservation 372–375
 gelling additives 360
 pollution 370–372
 reuse 373–375
 treatment 365–367
 water resources 370–372*
Water-borne disease 365–367
Water-jet cutting 367–370*
 applications 369–370
 operation 368–369
 principle 369
Water resources 370–372*
 erosion and sedimentation 370–371
 habitat impacts 371
 nutrients as water pollutants 371
 pesticides 371–372
 pollution abatement programs 372
Water supply engineering 372–375*
 conservation 372–373
 water reuse 373–375
Water utility: rates and pricing 373
Water vapor: atmosphere 13–17
Wave phenomena 375–378*
 computational results 377–378
 locating the free surface 376–377
Wavelets 378–380*
 dynamic spectrograph 380
 dynamic spectrum 378–380
 fractals 137
 synthesis 380

Wavelets—*cont.*
 wavelet theory 380
Weather forecasting 13–17
Weathering 207
Welding and cutting of metals: intermetallic materials 162
Wet milling: making ethanol from grain 33–34
Wetting: soil 312
Whey 380–382*
 environmental concerns 382
 erosion control 381–382
 production 380–381
 sodic soil reclamation 381
 use as fertilizer 381
White smoker 181
Wide Area Augmentation System 305
Wind 90, 272–273, 382–384*
 air-ocean interactions 383
 eolian processes 383–384
 forest canopy 354
 gravity flow in Antarctica 382
 modeling and observations 383
 wind speed and direction 383
Wind (spacecraft) 324
Wolter, H. 384
Wrist: carpal tunnel syndrome 42–43*

X

X-ray absorption near-edge structure 206
X-ray fluorescence: mineralogy applications 205
X-ray magnetic circular dichroism spectroscopy 388
X-ray optics 384–386*
 grazing-incidence optics 384–385
 Kumakhov polycapillary optics 384
 multilayer optics 385–386
X-ray spectrometry 387–388
X-ray telescope 384–385
X-rays 386–388*
 applications 387–388
 synchrotron radiation source 386–387
 undulators 387
XANES *see* X-ray absorption near-edge structure
Xeriscape landscaping 373

Y

Yeast infection 388–390*
 diagnosis 389–390
 infections and symptoms 389
 medical mycology 185–187*
 predisposing factors 389
 treatment 390
 yeast organisms 388–389
Yersinia pestis 257–258

Z

Zeitgeber 31
ZEUS detector 99